McGraw-Hill Yearbook of Science and Technology 1971 REVIEW

1972 PREVIEW

McGraw-Hill Yearbook of

McGRAW-HILL BOOK COMPANY

NEW YORK	ST. LOUIS
SAN FRANCISCO	
DUSSELDORF	NEW DELHI
JOHANNESBURG	PANAMA
KUALA LUMPUR	RIO DE JANEIRO
LONDON	SINGAPORE
MEXICO	SYDNEY
MONTREAL	TORONTO

REFERENCE

65621

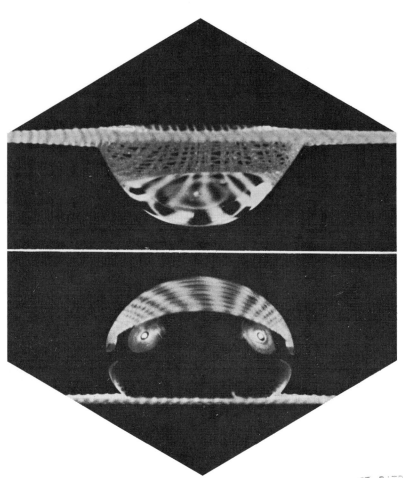

ST. PATRICK'S SEMINARY LIBRARY

Science and Technology

COMPREHENSIVE COVERAGE OF
THE IMPORTANT EVENTS OF THE YEAR
AS COMPILED BY THE STAFF OF THE
McGRAW-HILL ENCYCLOPEDIA OF SCIENCE AND TECHNOLOGY

1971

On preceding pages:

Left. General Electric Research and Development Center photo of sulfur-ion tracks from an Apollo test helmet. Replicas of the tracks were used to calibrate the helmets for cosmic ray detection. (E. Lifshin, Science, vol. 172, no. 3979, Apr. 9, 1971)

Right. Upper photo, untreated 100-mesh nylon fabric lets water droplet soak through. Lower photo, fabric coated with Silanox 101, a water-repellent powder made by Cabot Corp., causes water to form beads. (Chemical Week, 109(10):39, Sept. 8, 1971)

McGRAW-HILL YEARBOOK OF SCIENCE AND TECHNOLOGY Copyright © 1972, by McGraw-Hill, Inc. All Rights Reserved. No part of this publication may be reproduced, stored in a retrieval system, or transmitted, in any form or by any means, electronic, mechanical, photocopying, recording, or otherwise, without the prior written permission of the publishers. Philippines Copyright, 1972, by McGraw-Hill, Inc.

Library of Congress Catalog Card Number: 62-12028

International Standard Book Number: 07-045250-4

Table of Contents

PREVIEW OF 1972   *1–78*

urban fires   Stanley B. Martin 1–9

science in art   W. T. Chase 10–23

solid-waste management   Richard D. Vaughan 24–33

risk evaluation in engineering   Leon E. Borgman 34–43

energy sources in galaxies and quasars   William H. McCrea 44–55

surface physics   Homer D. Hagstrum 56–67

pathology of heavy metals   N. Karle Mottet 68–78

photographic highlights   79–96

REVIEW OF 1971   97–424

list of contributors   425–430

index   431

McGRAW-HILL YEARBOOK OF SCIENCE AND TECHNOLOGY

Editorial Advisory Boards

1960—1971

Dr. Sidney D. Kirkpatrick
Formerly Editorial Director,
Chemical Engineering and *Chemical Week*
McGraw-Hill Publishing Company

Dr. Joseph W. Barker
Consulting Engineer
President and Chairman of the Board (retired)
Research Corporation, New York

Dr. William W. Rubey
Department of Geology and
Institute of Geophysics
University of California, Los Angeles

Dr. George R. Harrison
Dean Emeritus, School of Science
Massachusetts Institute of Technology

Dr. Detlev Bronk
President (retired), The Rockefeller University

1968—1971

Dr. Neil Bartlett
Professor of Chemistry
University of California, Berkeley

Dr. Alfred E. Ringwood
Department of Geophysics and Geochemistry
Australian National University

Dr. Richard H. Dalitz
Department of Theoretical Physics
University of Oxford

Dr. Koichi Shimoda
Department of Physics
University of Tokyo

Dr. Freeman J. Dyson
The Institute for Advanced Study
Princeton, N.J.

Dr. A. E. Siegman
Professor of Electrical Engineering
Stanford University

Dr. Leon Knopoff
Professor of Physics and Geophysics
University of California, Los Angeles

Dr. Hugo Theorell
The Nobel Institute
Stockholm

Dr. H. C. Longuet-Higgins
Department of Machine Intelligence and Perception
University of Edinburgh

Lord Todd of Trumpington
University Chemical Laboratory
Cambridge University

Dr. E. O. Wilson
Professor of Zoology
Harvard University

Editorial Staff

Daniel N. Lapedes, *Editor in Chief*

David I. Eggenberger, *Executive Editor,*
Professional and Reference Books Division

Gerard G. Mayer, *Director of Design and Production*

Marvin Yelles, *Senior editor*

Joe Faulk, *Copy manager*
George Ryan, *Copy editor*
Patricia Walsh, *Editing assistant*
Kathleen Skultety, *Editing assistant*

Edward J. Fox, *Art Director*
Richard A. Roth, *Art editor*
Ann D. Bonardi, *Art coordinator*
Donna Zloty, *Art/traffic*

Consulting Editors

Gustave E. Archie. *Assistant to Vice-President of Exploration and Production Research, Shell Development Company.* PETROLEUM ENGINEERING.

Dr. Roger Batten. *American Museum of Natural History.* PALEOBOTANY AND PALEONTOLOGY.

Prof. Theodore Baumeister. *Consulting Engineer; Stevens Professor of Mechanical Engineering, Emeritus, Columbia University; Editor in chief, "Standard Handbook for Mechanical Engineers."* MECHANICAL POWER ENGINEERING.

Prof. Jesse W. Beams. *Professor of Physics, University of Virginia.* ELECTRICITY AND ELECTROMAGNETISM; LOW-TEMPERATURE PHYSICS.

Prof. Raymond L. Bisplinghoff. *Deputy Director, National Science Foundation, Washington, D.C.* AERONAUTICAL ENGINEERING.

Dr. Salomon Bochner. *Chairman, Department of Mathematics, Rice University.* MATHEMATICS.

Dr. Walter Bock. *Professor of Zoology, Columbia University.* ANIMAL ANATOMY; ANIMAL SYSTEMATICS; VERTEBRATE ZOOLOGY.

Waldo G. Bowman. *Black and Veatch, Consulting Engineers, New York, N.Y.* CIVIL ENGINEERING.

Prof. Ludwig Braun. *Professor of Electrical Engineering, Polytechnic Institute, Brooklyn, N.Y.* CONTROL SYSTEMS.

Dr. John M. Carroll, *Associate Professor of Computer Science, Department of Computer Science, University of Western Ontario.* RADIO COMMUNICATIONS.

Dr. Jule G. Charney, *Department of Meteorology, Massachusetts Institute of Technology.* METEOROLOGY AND CLIMATOLOGY.

Prof. Glen U. Cleeton. *Dean Emeritus, Graphic Arts Management, Carnegie-Mellon University.* GRAPHIC ARTS.

Cyril Collins. *Technical consultant.* TELECOMMUNICATIONS.

William R. Corliss. *Technical Consultant.* PROPULSION; SPACE TECHNOLOGY.

Richard B. Couch. *Naval Architecture Research Office, The University of Michigan.* NAVAL ARCHITECTURE.

Dr. Arthur Cronquist. *New York Botanical Gardens.* PLANT TAXONOMY.

Prof. Kenneth P. Davis, *David T. Mason Professor of Forest Land Use, School of Forestry, Yale University.* FORESTRY.

Dr. C. J. Eide. *Department of Plant Pathology, Institute of Agriculture, University of Minnesota.* PLANT PATHOLOGY.

Dr. Francis C. Evans. *Department of Zoology, University of Michigan.* ANIMAL ECOLOGY.

J. K. Galt. *Bell Telephone Laboratories, Murray Hill, N. J.* PHYSICAL ELECTRONICS.

Prof. Newell S. Gingrich. *Professor of Physics, University of Missouri.* CLASSICAL MECHANICS AND HEAT.

Dr. Edward D. Goldberg. *Scripps Institution of Oceanography, La Jolla, California.* OCEANOGRAPHY.

Prof. Julian R. Goldsmith. *Professor of Geochemistry and Associate Dean, Division of the Physical Sciences, University of Chicago.* GEOLOGY (MINERALOGY AND PETROLOGY).

Prof. Roland H. Good, Jr. *Department of Physics, Iowa State University of Science and Technology.* THEORETICAL PHYSICS.

Prof. David L. Grunes. *United States Department of Agriculture.* SOILS.

Dr. H. S. Gutowsky. *School of Chemical Sciences, University of Illinois.* PHYSICAL CHEMISTRY.

Prof. Howard L. Hamilton. *Department of Biology, University of Virginia.* GROWTH AND MORPHOGENESIS.

Dr. J. Allen Hynek. *Chairman, Department of Astronomy, Northwestern University.* ASTRONOMY.

Prof. H. S. Isbin. *Professor of Chemical Engineering, University of Minnesota.* NUCLEAR ENGINEERING.

William F. Jaep. *Consultant, Engineering Department, Du Pont Corporation.* THERMODYNAMICS.

Dr. R. E. Kallio. *Director, School of Life Sciences, University of Illinois.* MICROBIOLOGY.

Dr. Donald R. Kaplan. *Department of Botany, University of California, Berkeley.* PLANT ANATOMY.

Dr. Joseph J. Katz, *Senior Chemist, Argonne National Laboratory.* INORGANIC CHEMISTRY.

Dr. Charles E. Lapple. *Senior Scientist, Stanford Research Institute.* FLUID MECHANICS.

Frank K. Lawler. *Editor in chief, "Food Engineering."* FOOD ENGINEERING.

Prof. Harry A. MacDonald. *Professor of Agronomy, Cornell University.* AGRICULTURE.

Consulting Editors (continued)

Prof. Robert W. Mann. *Department of Mechanical Engineering, Massachusetts Institute of Technology.* DESIGN ENGINEERING.

Dr. Edward A. Martell. *National Center for Atmospheric Research, Boulder, Colorado.* GEOCHEMISTRY.

Dr. Harold B. Maynard. *President, Maynard Research Council, Inc.* INDUSTRIAL AND PRODUCTION ENGINEERING.

Dr. Howard Mel. *Donner Laboratories, University of California.* BIOPHYSICS.

Dr. Bernard S. Meyer. *Professor and Chairman, Department of Botany and Plant Pathology, The Ohio State University.* PLANT PHYSIOLOGY.

Dr. Jacob Millman. *Department of Electrical Engineering, Columbia University.* ELECTRONIC CIRCUITS.

Dr. William Mosher. *Chairman, Department of Chemistry, University of Delaware.* ORGANIC CHEMISTRY.

Dr. N. Karle Mottet. *Professor of Pathology and Director of Hospital Pathology, University of Washington.* ANIMAL PATHOLOGY.

Dr. Royce W. Murray. *Assistant Professor of Chemistry, University of North Carolina.* ANALYTICAL CHEMISTRY.

Dr. Robert H. Noble. *Optical Science Center, University of Arizona.* ELECTROMAGNETIC RADIATION AND OPTICS.

Dr. Harry F. Olson. *Staff Vice President, Acoustical and Electromechanical Laboratory, RCA Laboratories.* ACOUSTICS.

Dr. Jerry Olson. *Oak Ridge National Laboratory.* CONSERVATION; PLANT ECOLOGY.

Dr. Guido Pontecorvo. *Imperial Cancer Research Fund, London.* GENETICS AND EVOLUTION.

Prof. K. R. Porter. *Chairman, Department of Molecular, Cellular and Developmental Biology, University of Colorado.* CYTOLOGY.

Prof. C. Ladd Prosser. *Head, Department of Physiology, University of Illinois.* COMPARATIVE PHYSIOLOGY.

Brig. Gen. Peter C. Sandretto. *Director, Engineering Management, International Telephone and Telegraph Corporation.* NAVIGATION.

W. C. Schall. *Goldwater, Valente, Fitzpatrick & Schall, Members of the New York Stock Exchange.* COMPUTERS.

Dr. Bradley T. Scheer. *Head, Department of Biology, University of Oregon.* GENERAL PHYSIOLOGY.

Prof. Frederick Seitz. *President, The Rockefeller University.* SOLID-STATE PHYSICS.

Dr. Raymond Siever. *Department of Geological Sciences, Harvard University.* GEOLOGY (SURFICIAL AND HISTORICAL); PHYSICAL GEOGRAPHY.

C. Dewitt Smith. *Mining consultant, Dewitt Smith and Co., Inc., Salt Lake City, Utah.* MINING ENGINEERING.

Dr. Mott Souders. *Formerly Director of Oil Development, Shell Development Company.* PETROLEUM CHEMISTRY.

Prof. William D. Stevenson, Jr. *Department of Electrical Engineering, North Carolina State of the University of North Carolina at Raleigh.* ELECTRICAL POWER ENGINEERING.

Dr. Horace W. Stunkard. *Research Associate, Invertebrate Zoology, American Museum of Natural History.* INVERTEBRATE ZOOLOGY.

Dr. E. L. Tatum. *The Rockefeller University.* BIOCHEMISTRY.

Dr. Aaron J. Teller. *Teller Environmental Systems, New York, N.Y.* CHEMICAL ENGINEERING.

Dr. Garth Thomas. *Center for Brain Research, University of Rochester.* PHYSIOLOGICAL AND EXPERIMENTAL PSYCHOLOGY.

C. N. Touart. *Research Physicist, Air Force Cambridge Research Laboratory.* GEOPHYSICS.

Dr. Henry P. Treffers. *Professor of Microbiology, Yale University School of Medicine.* MEDICAL MICROBIOLOGY.

Prof. H. H. Uhlig. *Department of Metallurgy, Massachusetts Institute of Technology.* METALLURGICAL ENGINEERING.

Prof. William W. Watson. *Department of Physics, Yale University.* ATOMIC, MOLECULAR, AND NUCLEAR PHYSICS.

Contributors

A list of contributors, their affiliations, and the articles they wrote will be found on page 427.

Preface

As the 1972 *McGraw-Hill Yearbook of Science and Technology* goes to press we look back at the previous ten Yearbooks and see the forward sweep that has taken place. The reports of research and development included this year are indicators of the continuing ferment in science and technology. Some of these reports are concerned with progress in the basic sciences of ecology, genetics, cosmology, and biophysics and in the technologies of lasers, liquid crystals, and electric power generation and transmission.

This Yearbook, like its predecessors, continues to analyze and report significant scientific and technical achievements of the previous year. The Yearbook performs two functions. As an individual book it is a ready reference source to what has taken place in science and technology. As a supplement it updates and enriches the basic material in the *McGraw-Hill Encyclopedia of Science and Technology*, third edition (1971), and previous editions. Following the organization established in previous Yearbooks, the 1972 edition begins with major articles on seven subjects selected for their broad interest and growing significance. The second part, a selective pictorial section, features a number of the outstanding scientific photographs of the past 12 months. The third part consists of alphabetically arranged articles on advances, discoveries, and developments in science and technology during the past year.

The choice of the subject matter in another year of accelerating scientific achievement was the work of 67 consulting editors and the editorial staff of the *McGraw-Hill Encyclopedia of Science and Technology*. But most of the credit should go to the 158 eminent specialists who contributed to the present volume. Their interest, knowledge, and writing ability make them the real creators of the 1972 Yearbook.

DANIEL N. LAPEDES
Editor in Chief

Urban Fires

Fire has been an unrivaled benefactor of mankind and a catalyst of human civilization. And paradoxically, fire has become a pervasive, ever-present threat to urbanized man—the product of the very civilization fire has made possible. As the population grows and becomes more concentrated in metropolitan areas, the tendency toward destructive fires and the difficulty of protecting against them grow too. The main contributing factors in the urban environment are the proximity of buildings, traffic congestion, hazardous practices, and expanding use of flammable materials.

Stanley B. Martin is manager of the fire research program at Stanford Research Institute, Menlo Park, Calif. He has contributed to the knowledge of incendiary effects through fundamental studies of ignition and combustion and has pioneered in using systems analysis techniques for fire defense.

2 URBAN FIRES

Fig. 1. The great fire of London of 1666. Fanned by a strong wind, the fire burned over an area of 430 acres (1.7 km²). (*National Fire Protection Association*)

Fig. 2. The New York fire of 1835. Gale-driven flames destroyed 700 warehouses. (*National Fire Protection Association*)

Fig. 3. The great Chicago fire of 1871 is ranked as one of the greatest disasters in American history.

Man's defense against destructive fire has been largely improvisational, and although this approach has been reasonably effective in the past, it has not kept pace with urbanized and industrialized life. Quite recently, however, an impressive array of scientific and engineering disciplines has been brought to bear on the complex practical problems of fire defense.

History abounds with examples of great urban conflagrations that destroyed large parts or the entirety of cities. Rome, London, Moscow, New York, Chicago, San Francisco, and Tokyo have all seen many square miles of property consumed and hundreds—even thousands—of lives lost in fire catastrophes (Figs. 1 to 3). But major urban conflagrations occur relatively infrequently, and it is property lost to fire each year in noncatastrophic fires that amounts to such a substantial economic drain. For example, the total annual cost of fire in the United States has been estimated at $5,000,000,000 or more. This estimate includes loss from actual physical damage plus the cost of fire department operations and other direct forms of fire protection. However, it does not include resources wasted in constructing buildings to replace those lost by fire, nor does it include the losses resulting from the disruption of productivity and from human suffering. If these costs were included, the estimate would double. Of far greater concern, however, is the annual loss of some 12,000 lives in fire, or about 1000 lives per month.

For several decades in the United States, urban fires and other fires in structures have accounted for 80–85% of all fire losses, with the remaining 15–20% split between forest and transportation fires. Figure 4 is a sample breakdown of United States fire losses by type of occupancy; it shows that residential occupancies represent over 25% of the total fire losses each year, and the combination of residential and industrial occupancies amounts to almost 50%. The industrial category includes the most expensive fires of all. For example, in recent years the average industrial fire damage was about $5000 per fire, in contrast to an average of only about $800 per fire for all occupancies. In about 1 in every 500 industrial fires, the fire loss may reach $250,000 or more in a single fire. In the last decade, such disasters have accounted for approximately 33% of all industrial fire losses.

Another way of looking at the urban fire problem is in terms of the causes of fires. Figure 5 indicates the distribution of causes of fires and reveals the interesting fact that over 33% of all United States building fires have unknown or undetermined causes. It may reasonably be suspected that many of these fires are the result of undetected arson.

It is easy to conclude from these statistics that fire is an exceptionally serious threat to mankind in an urban environment. In fact, no other destructive element in the history of man has been responsible for as much devastation as fire.

Further interpretation of the statistics reveals that the greatest losses in life occur in the areas where people live and congregate rather than where they work; that is, in residences, theaters, hotels, hospitals, homes for children and the aged, schools, and other institutions. These losses in life, as well as the large losses in property, occur main-

URBAN FIRES 3

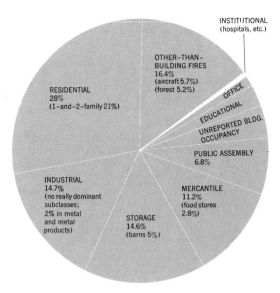

Fig. 4. Fire losses in the United States according to type of occupancy. Percentages are on the basis of total dollar value. Dominant subclasses appear in parentheses. (*National Academy of Sciences*)

Fig. 5. Causes of building fires in the United States in 1965. Percentages are in terms of total dollar value. (*National Academy of Sciences*)

DEVELOPMENT OF FIRE PROTECTION PRACTICES

Although fire-protection practices throughout the world have many points in common, this discussion concentrates on those of the United States.

From colonial times there has been a particularly serious fire problem in America. In the 17th century, Peter Stuyvesant attempted to limit fire damage in New Amsterdam through the enactment of strict laws of housekeeping and chimney maintenance. The record indicates that these laws, the first example of fire ordinances among the Colonies, were moderately successful. But in spite of such efforts, the early history of the United States was marked by a succession of disastrous fires.

Sometime during the 17th centry, in Boston, the first paid fire department was established to protect the city against fires and incendiarism. At about the same time, citizens of other towns in colonial New England banded together to form societies for mutual fire protection. These societies were the forerunners of our present-day volunteer fire departments.

ly at night, in spite of the fact that only about 30% of the fires occur at night. This suggests that there is a very great need for early detection and warning of fire so that occupants will have time to evacuate a burning building and so that fire-suppression facilities and fire-fighting services can be brought to bear on the fire while it is still small.

Fire losses per person in the United States are the highest in the world (Fig. 6). These proportionately higher losses are due to a combination of factors, which include the highly urbanized and industrialized character of the United States today, the extensive use of combustible building materials combined with excessive amounts of flammable contents, and a generally apathetic attitude among the citizens.

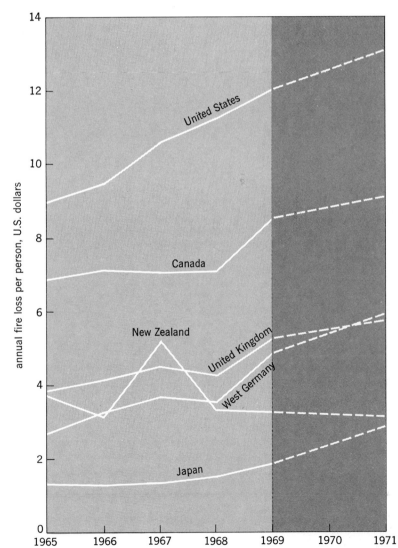

Fig. 6. Annual fire loss per person for some major countries. Broken lines indicate estimated values. (*National Fire Protection Association*)

Fig. 7. Steam powered pump "Long John" was used in the great Chicago fire.

Methods of fire fighting have evolved gradually since the days of the colonial fire-fighting forces, except for a few well-marked occasions when a new technique made a major advance possible. One such advance was the adoption of steam power to pump water, replacing manual pumping. About 1840 the first steam-operated pumper was built in the United States, based on an English design, and was put into service in New York City. It could throw a stream of water to a very great height, but unfortunately this version was so heavy and awkward to use that it never gained acceptance by the fire department. However, about 10 years later, a lighter American-designed steam fire engine, "Old Joe Ross," was completed and tested in Cincinnati, and soon other versions were in use in other cities (Fig. 7).

The added weight of the steam-powered pumper forced the fire departments to use horses in place of men to pull the fire engines, and by the time of the Civil War virtually all of the fire engines in major municipalities were horse-drawn. The first telegraph fire alarm was invented and installed in Boston in the 1850s. In the 1870s the aerial ladder was developed (Fig. 8), and it was quickly followed by the aerial hose (Fig. 9).

Another major technological advance was the development of a separate system of mains and hydrants used only for fighting fires. The first such system was installed in Rochester, N.Y., in 1874. In 1904, Philadelphia installed a similar system in its downtown area, and in 1908, New York City installed a high-pressure system in lower Manhattan. Finally the internal-combustion, gasoline engine replaced the horse as a means of moving fire-fighting vehicles, making possible quick response to fire alarms.

Technological advances of lesser importance include more powerful pumps, improved hoses, more versatile nozzles, radio communications, and electrically powered tools.

MODERN FIRE-FIGHTING METHODS

The fundamental techniques used by firemen today consist, as in the past, primarily of putting water on a fire. Water serves to cool a burning material down to a point where it does not produce gases that burn. Until recently, the water was applied principally by hose nozzles delivering solid streams of water.

Fog or spray nozzles. We know now that water cools more efficiently if it is broken up into tiny droplets which expose more water surface to heat; hence the increasing popularity of the so-called fog or water-spray nozzles. Wetting agents and other surfactant materials are also being employed to make water more efficient.

With the introduction of fog and spray nozzles, there is a tendency to use smaller hose lines, principally the 1½-in. (38.1-mm) booster lines, and smaller quantities of water to fight fires in dwellings, apartments, and other smaller occupancies. Fog nozzles have been used successfully on nearly all types of fires, even oil, tar, grease, and electrical fires where water had previously been considered ineffectual or dangerous. It also has been found that water spray or fog absorbs some of the toxic gases of the fire and tends to clear away the smoke. As a result, the sprays are widely used now in fighting smoky interior fires.

Firemen generally agree that the way to fight a fire is to get inside the building, find the source, and get water directly on it. However, when alarms are delayed and the fire is fully developed before the arrival of the firemen, hose streams may be required to reduce the fire to the point where it can be approached.

Ventilation. The fire-fighting tactic least understood by laymen is the one of building ventilation. Observers often criticize the firemen's practice of chopping holes in roofs and breaking windows in portions of the building which do not appear to be involved in the fire. In fact, however, the firemen are trained to ventilate the building and clear it of heavy smoke and heated gases so that men can enter the building and reach the fire. Ventilation reduces hazards to life and damage due to smoke and heat. But ventilation should only be attempted by experienced fire officers, since improper ventilation can spread the fire and increase the damage and can even introduce the risk of explosion. In recent times, fire departments have begun to use electrically driven fans to help clear buildings of smoke.

FIRE PREVENTION AND LOSS REDUCTION

Although the practices of fire prevention are less dramatic and less advertised than the activities of fire fighting, their contribution to fire-loss reduction is potentially much greater. Fire-prevention and loss-reduction measures take many forms, including fire-safe building codes, fire-detection and automatic fire-suppression systems in industrial and public buildings, the substitution of flame-retardant materials for their more flammable counterparts, insurance to offset the financial loss of fire, and the investigation of fires of suspicious origin, serving to deter the fraudulent and illegal use of fire.

The revision of building codes, the establishment of meaningful fire-resistive tests, and many other activities of fire-protection engineering constitute a field in themselves. Similarly the subject of fire insurance is highly specialized and complex.

As Fig. 5 suggests, there is much interest in improving methods for investigating fires of unknown origin. Fire insurance companies employ or retain investigators to determine the cause of a large number of fires of suspicious origin each year. In addition, many firemen are now trained to recognize the clues pointing to arson.

Fig. 8. Aerial ladder, invented in the 1870s, for fighting fires in heavily urbanized areas.

A century ago, a large percentage of fires could be traced directly to the work of arsonists. Today, only a small proportion, less than 1%, of the fires of known origin is traceable to arson. Nevertheless the very large portion of the fires with "unknown" or "undetermined" causes may include a substantial incidence of arson. Although some fire protectionists ascribe the reduced number of known cases of arson to effective methods of investigation, critics counter that the arsonist, especially the one who burns his own property to defraud the insurance company, is just much more cunning today.

NATURE OF TODAY'S FIRE PROBLEMS

Fire involves the chemical reaction between oxygen (air) and a fuel that is heated to its ignition temperature. This definition may be depicted by the fire triangle (Fig. 10), which shows the three ingredients required for a fire. The reduction or removal of a sufficient amount of any one of these will stop the fire. This concept formed the rudimentary beginnings of the technology of fire control.

Classification of fires. Fires are now divided into four classes according to the type of combustible involved.

Class A. Fires involving solids such as wood, paper, and most plastics are termed class A fires and were almost the only type that concerned mankind until quite recently. Water was, and still is, the most effective, the cheapest, and the most easily applied extinguishing agent for class A fires.

Class B. These are fires of organic liquids such as gasoline, fuel oil, benzene, and acetone. Such fires became important with the advent of the petroleum industry and other modern chemical processing industries. These fires can be controlled by removing the air with a blanketing agent, such as carbon dioxide or a water-based foam. Water in the form of high-pressure fog is an effective extinguisher, but conventional water streams are unsuitable because they spread the fire.

Class C. A class C fire is any fire in which energized electrical equipment is involved. Because of the hazard of electric shock, any extinguishing agent may be used that does not form electrically conductive paths, including high-pressure water fogs. Aqueous foams are not suitable.

Class D. Fires fueled by sodium, magnesium, titanium, and other reactive metals or their reactive compounds, such as sodium hydride, compose this class and are latecomers to the fire scene. These fires are best controlled by removal of air by a blanket of unreactive powder, such as sodium chloride or graphite. Carbon dioxide, sand, and water cannot be used either to blanket or cool reactive-metal fires because these substances are a source of oxygen for these fuels.

New fire problems in cities. Many practices in cities tend to modify and substantially increase the problems of fire protection. For example, there is an increasing use of such highly flammable materials as gasoline, liquefied fuel gases, and organic solvents; of high-wattage electric appliances, with the attendant risks of fire from electrical short-circuiting or electrical overheating; and of certain synthetic polymers that decompose in a fire to produce toxic gases, adding to the already severe life hazards.

A recurring problem is the use of natural-wood roof coverings. This material has been responsible for many of the major destructive fires that have occurred in North America since very early times. Wood shingles and wood shakes that have not been treated to give them fire resistance are easily ignited by sparks, embers, burning brands, and other burning materials, and they in turn are carried by the wind as firebrands. It is a remarkable fact that such materials are still permitted in building practice in the United States.

A growing problem is the increased numbers of high-rise buildings. These buildings often cannot be readily evacuated in the event of a fire, nor can the firemen get to fires in upper stories in time to achieve early control. In fact, merely finding the fire is frequently a problem. Since many of the modern skyscrapers are totally air-conditioned by recirculating air systems, the smoke is carried throughout the building, thus compounding the fire-locating problem.

Also, increasing traffic problems impede the fire service. The public often delays the arrival of firefighting equipment and then sometimes even interferes with the work the firemen are there to do.

Future hazards. Fire has been discussed so far in terms of the fuel involved without regard to the oxidant because it has been assumed that the oxidant is supplied by the air. Fires that involve oxi-

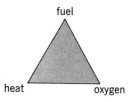

Fig. 10. Triangle of essential ingredients of a fire. A sufficient reduction of any ingredient stops the fire.

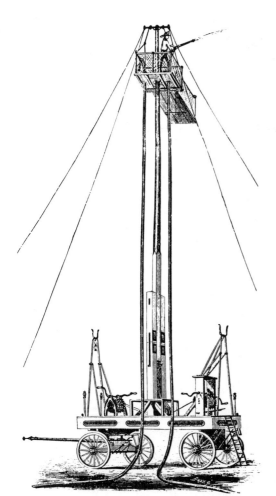

Fig. 9. Aerial hose tower was invented in the 1870s.

dizing agents other than the oxygen in air will become more frequent. Oxidizers such as oxygen, fluorine, and chlorine as liquefied gases and concentrated solutions of hydrogen peroxide are finding increased industrial use and will be transported by highway truck or railroad car in growing quantities, finding their way into the urban scene with increasing regularity.

These materials are hazardous because they oxidize, or burn, many common materials of construction, such as wood, paints, plastics, and rubber, and because the temperature required to ignite such a mixture may be lower than that required for a fire in air alone. Finally, increased use of concentrated oxygen atmospheres and high-pressure, or hyperbaric, atmospheres constitutes a new danger.

TRENDS IN FIRE RESEARCH

In the last analysis, effective fire protection depends on (1) the use of fire-retardant materials and practices, (2) early detection of fires that cannot be prevented, (3) fast response to these fires with countermeasures tailored to the fire-suppression job, (4) careful planning for the occasional fire that resists early control, and (5) an efficient fire service trained and equipped to meet new challenges. Fire research is now beginning to make contributions in all these areas.

Fire-retardant materials. Modern fire prevention has two principal objectives: to prevent fire from breaking out at all, and if this fails, to confine the resulting fire. To prevent fire is to rule out all possibility of its initiation; this means building with fireproof or slow-burning materials, using intrinsically safe installations and equipment, observing regulations when using flammable materials, maintaining strict supervision on the premises, and checking faults as soon as they are discovered.

Confinement of a fire is mainly ensured by constructional measures, such as dividing a building into fire compartments and using fireproof or fire-retarding materials.

Fire research reveals many new fire-retardant materials for urban use. Perhaps the most noteworthy progress has been made by the National Aeronautics and Space Administration (NASA), spurred by the tragic Project Apollo fire in 1967. Since then, NASA scientists have screened more than 5000 materials potentially capable of resisting fire, or even suppressing fire, in the oxygen-rich environment of the Apollo space capsule. Only three materials have been found to satisfy all the criteria imposed by the Apollo program; however, a large number of materials were found to possess a degree of fire-protection capability that could be entirely advantageous in the normal atmosphere of the urban environment. Many of these materials are special polymeric formulations designed to release retardant chemical species into the air-surface boundary on heating, thus inhibiting fire development. Some of the materials, however, are essentially standard materials treated by some simple, inexpensive procedure to endow them with properties of nonflammability.

In a similar vein, the Office of Civil Defense has actively supported, for several years, research seeking a fuller understanding of the fundamental chemical mechanisms accompanying ignition and combustion and of the changes produced in them by fire-retardant additives. This research has shown that: (1) Many inorganic substances reduce the flammability of cellulosics, even in quite low add-on concentrations, by profoundly altering the chemistry of cellulose pyrolysis. (2) Acidic additives have a very different chemical effect than basic and neutral substances. (3) High-molecular-weight inorganic additives, especially those containing large numbers of oxygen atoms per molecule, strongly enhance the production of char during cellulose pyrolysis, thereby limiting the production of flammable, carbon-rich volatiles.

Such research, while not intended to lead directly to the development of practical fire retardants, promises the insights necessary before major breakthroughs can be made.

Automated fire protection. Once a fire does start, the essential problem is one of detection and quick response. The earlier the fire is detected, the more effective the fire-suppressing system or fire-fighting activities will be.

The automatic fire alarm system provides a continuous and untiring watch. Today there are a number of extremely effective devices for detecting fire at an early stage. The oldest (and slowest) of these is the fusible metal seal for the sprinkler systems that protect many industrial and public buildings. At the other extreme are optical detectors that can react in milliseconds after the initiation of the fire. Such detection systems, when combined with an automatic suppression system, become an automated fire-suppression system. Such systems are now becoming standard equipment in certain kinds of high-risk occupancies. In fact, today only cost stands in the way of automated fire protection for homes, stores, and other small buildings. Nevertheless, impelled by the upsurge of competition, prices are coming down, so that one can look for much greater acceptance within the next few decades.

The kind of fire that can be expected largely determines the choice of the detector, but other factors such as ceiling height and the presence of any interfering influences in the room must be considered. In every fire, combustion gases are the first evidence. In smoldering fires, they may be, for a considerable period of time, the only clues. Electronic combustion-gas detectors therefore form the basis of any fast-response fire alarm system. They are supplemented when necessary by other types of detectors.

The five principal types of fire detectors in common use today are as follows.

Electronic combustion-gas detector. This device detects the combustion gases at an early stage of a fire, often long before the smoke would be noticed. Such a detector can usually monitor in excess of 500 ft^2 (50 m^2 or more) of floor area.

Radiant-emission detector. Such a detector responds to infrared or ultraviolet radiation. In some cases, this detector is designed to respond to rapid fluctuations in the radiation, the rate of fluctuations being chosen to correspond approximately to the flicker rate of fires but to exclude the background radiation from other sources. Some versions have a response delay to further screen out background signals. Such a detector can monitor as much as 1000 m² of floor area if its field of view is unobstructed.

Fixed-temperature detector. This detector triggers an alarm upon reaching a preset temperature. Typical versions respond at a temperature level of 70°C. They are also available in temperatures ranging from as low as 50°C to as high as 130°C. The fixed-temperature detector ordinarily will protect an area of about 10 m² or somewhat more.

Rate-of-temperature-rise detector. Such a type responds to a quick rise in room temperature. A typical version responds when the rate of rise exceeds about 0.2°C/sec; first, however, the temperature must exceed a certain fixed value. A rate-of-rise detector usually monitors about 20 m² of floor area.

Light-scattering (or refracting) detector. This device appears in various versions, each employing a different principle. One commonly used detector of this type is based on the scattering of light by smoke particles. A new device uses the refraction of a multiply reflected laser beam to detect convection cells produced by a fire. The smoke detectors are relatively slow and they sample only small volumes of room air, but the laser-beam approach appears to suffer neither disadvantage.

Fire-suppresion agents. In recent years, a new general approach to fire control has developed, involving use of flame inhibitors. Unlike older fire-extinguishing materials such as water and carbon dioxide, these agents operate indirectly in that they interfere with those reactions within a flame that lead to a sustained release of heat. As a result, the temperature of the system falls below the ignition temperature.

Chemical flame inhibitors offer one of the greatest hopes for better control of fires of all sizes. These agents have achieved such importance that they have, in effect, added a new dimension to the fire triangle, making it a tetrahedron, with one additional mechanism for interfering with the combustion process (Fig. 11).

Chemical extinguishers are of two types: liquid (or liquefied gas) and dry powder.

Liquid chemical extinguishers. The most effective liquids are the halogenated hydrocarbons such as chlorobromomethane (CB) and bromotrifluoromethane (better known as Halon 1301).

Colorless, odorless, and electrically nonconductive, Halon 1301 is remarkably effective. Automatic Halon 1301 extinguishing systems are ideally suited to situations in which damage of equipment or furnishings by water or dry chemicals would cause irreparable harm. Insurance companies, banks, universities, and other organizations with computer facilities and valuable and irreplaceable records on data-processing cards or tapes are installing automated fire-protection systems that use Halon 1301 to protect equipment, subfloor electrical wiring, and record libraries.

Tests with Halon 1301 have been successful in hyperbaric (pressurized) atmospheres as well. Like the Apollo capsule and other space vehicles and deep-underwater exploration vehicles, the atmospheres in hyperbaric chambers for medical purposes are oxygen-rich if not pure oxygen. Materials that are fire-resistant in normal air burn enthusiastically in such oxygen-rich environments, but can be extinguished by Halon 1301 if they can be detected in time.

Dry-powder chemical extinguishers. Of the dry-powder chemical extinguishers, ammonium dihydrogen phosphate is most useful and is rated for class A, B, and C fires. Other dry-powder inhibitors are salts of the alkali metals (which include lithium, sodium, potassium, rubidium, and cesium). For example, the dry powders based on sodium bicarbonate and the more effective potassium bicarbonate are rated as extinguishers for class B and C fires. The rubidium and cesium salts have not been studied in detail, and lithium salts have no appreciable activity. Dry chemicals can be effective even when they are applied as a fog in water solutions. Their mechanism of action is not completely understood but is chemical in nature and is directly related to particle size, with smaller particles being more effective than larger ones.

Water. Notwithstanding the success of chemical inhibitors, water is still the prime suppressant because of its great cooling power, general availability, and low cost. However, application in a solid stream is now regarded as wasteful of water and damaging to the components and contents of structures. Water spray and fog have been highly effective, as has been mentioned earlier, even in quenching gasoline fires, for which water is normally unsuitable. Moreover, spray and fog minimize water damage.

Foams. Foams are now widely used. Protein-type, low-expansion foams are particularly useful in quenching burning volatile petroleum products and are used in crash-rescue operations. High-expansion foams are available for fire suppression in enclosed areas. Some foams of this type are generated at a rate of 15,000 ft³/min (424.8 m³/min). They can contain sufficient air to allow a man to breathe inside of them.

Light water. A film-forming solution of a specific fluorocarbon surfactant in water, known as light water, was developed by the U.S. Navy for use with dry chemicals to fight aircraft crash fires. It may be used either as a liquid or a low-expansion foam to interfere with the release of flammable vapors from the burning fuel. Light water is also useful in extinguishing petroleum storage tank fires and may find application to urban fires once the cost is no longer prohibitive.

FUTURE OF FIRE RESEARCH

Additional research is needed to develop improved, more versatile chemical agents for controlling fire. Research yet to be undertaken should clarify the role of chemical elements and compounds in extinguishing fire. This understanding will lead to the formulation of suppressants many times more effective than those of today. Such suppressants promise to be one of the biggest contributions to fire protection that fire research will make.

Water is likely to remain the most widely used suppressant for urban fires, but it will be increasingly used in the form of fog or fine spray and with additives. Water and other extinguishing agents, including foams and powders, will be used separately or in combinations according to the nature

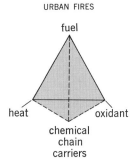

Fig. 11. Chemical flame inhibitors add another dimension to the fire triangle, forming the fire tetrahedron.

Fig. 12. Progress of a fire at an urban renewal site in full-scale testing.

cal model is often in itself a sufficient accomplishment of the research. Once this understanding has been achieved, it may not be necessary to use the physical model at all.

Operations research. Aside from its potential to provide improvements in fire-fighting materials and hardware, research can contribute to fire-fighting strategy, tactics, and resource allocation.

Operations research on fire is concerned with these problems. For example, it is often desirable to have some quantitative understanding of the potential return in lives and property saved with a given expenditure and various alternative ways of spending the same money. Increasingly more fire chiefs and government officials are asking: "What are the potential savings in lives, property, and community resources of getting the firemen to the fire a minute earlier?"—and in the same vein: "What are the trade offs between response time and response strength?" Such questions are hard enough to answer, but a host of them must be

of the hazard and the damage potential. The exploration of this concept is a task for fire research.

Although more effective detectors with earlier response are needed, there is also a need to investigate possible trade offs between the speed and reliability of sophisticated detectors and the increased coverage that simpler, low-cost detectors might provide. When and where is investment in many local detector alarms better than installation of a smaller number of more elaborate and reliable systems?

Modeling. To design more fire-resistant buildings and to test building materials for fire resistance meaningfully, it is important to study how fire spreads within structures. It is a much-sought-after long-term goal of fire research to predict, from a knowledge of the behavior of the component parts, the overall behavior of fire in a building without actually burning the building. However, there is a very large amount of uncharted territory to be explored before this goal can be realized. As a result, one of the potentially most rewarding forms of fire research is the full-scale testing of structural fires (Fig. 12), accompanied by attempts at physically modeling such fires on a reduced scale (Fig. 13). Frequently, urban renewal projects provide sites for full-scale tests, in the form of buildings slated for demolition.

Modeling is not as simple as it appears to the layman. Valid modeling is achieved in the research sense only when the ratios of certain of the forces involved, such as those of buoyancy, turbulence, and viscous sheer, and ratios of certain characteristics of the heat and gas-flow processes are the same in both the model and its prototype. It is hardly surprising that in a situation as complex as a burning building the number of such similarity ratios that must be considered is very large. Ordinarily it is not possible, or even practical, to match such a large number of ratios, and therefore the modeling research consists, to a large extent, of identifying which ratios are important to the simulation and which may be disregarded without introducing serious error in the final simulation. It is the nature of this sort of research that the level of understanding needed to properly design a physi-

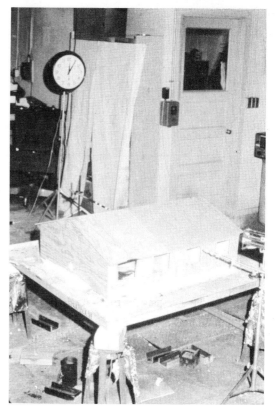

Fig. 13. Reduced-scale modeling of a fire.

studied, through the methods of operations research, before it will be possible to assess with assurance such broad matters as the relative values of additional men and equipment, on the one hand, and the adoption of innovative methods, on the other.

The methods of operation research do not provide new concepts for fire-fighting techniques and procedures. They do, however, allow objective selection of the best one of several alternative concepts that may be under consideration for application to a specific task. Operations research is also a valuable tool to aid with such other decisions as when and how an industry should self-insure.

The operations research method involves the construction of a model, usually mathematical in form, to simulate the functional dependencies of the interacting elements of the system under study. In general, there are two different approaches to the construction of such a model. The first of these, the one most often employed, uses statistical data from past experience to evaluate the characteristics of the system. The second method, a much more ambitious one, models the time course of a fire under various conditions of fire-suppression activity. This model must be able to describe the fire in physically significant terms. The empirical information required to build the second type of model is much more difficult to obtain than the statistical information required for the first, but the second model is much more useful and any conclusions derived from it would be much more reliable, assuming the empirical descriptions had been suitably developed.

As such research continues through the next few decades, it will make possible dramatic changes in fire prevention and fire protection, in methods as well as in materials. These changes, of course, are imperative as increasing industrialization and urbanization escalate fire hazards. But change does not depend on the creative innovations of the scientific community alone; before change can be implemented, much effort will be demanded of the public, the fire service, and the various fire-protection specialists.

[STANLEY B. MARTIN]

Science in Art

Before science, man created art. The earliest bone sculptured "Venuses" or the paintings in the Lascaux Cave still compel us with their magical power, a strength which evokes our deep response, as does any great work of art.

Art long stood at the forefront of technology. With the Renaissance, science arose as a serious study; for the first time, research employed the experimental method, and science gradually assumed its modern place of prominence. In some respects art changed, became more precious and contrived, as photography took over as the image-maker and

W. T. Chase is head conservator of the technical laboratory of the Freer Gallery of Art. He trained at the Conservation Center of the Institute of Fine Arts of New York University. He has participated in various projects involving examination of art objects, including the manufacturing process of Egyptian blue, Chinese bronzes and weapons, Sasanian silver, and paint pigments. Currently he is working on a complete study of the Chinese belt-hooks in the Freer.

Fig. 1. The David Litter Laboratories, which are used in investigating the stability and working properties of new materials for artists. (*The Artists Technical Research Institute, Inc.*)

as mass production and the corporation (rather than the craftsman) took over as maker and creator of things. Yet even through our modern culture runs a strong tradition of handcraftsmanship and devotion to the object. We turn to the arts of the past for their insights into the men of the past, and to the arts of the present for their insights into us and our culture. While the scientific, technological, and artistic realms seem on the surface to be divided into separate, distinct islands, between them flows the sea of human experience.

The adjective which arises most easily when one attempts to talk about science in art is "interdisciplinary." In another context someone warned that the term "interdisciplinary studies" was usually an introduction to some muddled thinking! Here we will try to avoid this pitfall and talk concretely about the roles science and modern technology

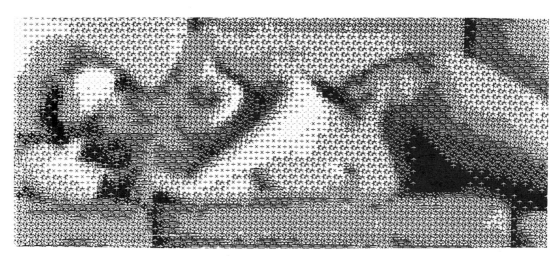

Fig. 2. *Studies in Perception, 1*, a computer-processed photographic print by Leon D. Harmon and Kenneth C. Knowlton, was exhibited at the Museum of Modern Art, New York City, in 1968. (*Courtesy of L. D. Harmon and K. C. Knowlton, Bell Laboratories*)

have played and will play in the making and interpretation of works of the visual arts, concentrating most heavily on the technical examination of art objects.

MODERN MATERIALS IN TRADITIONAL ART

In recent years advances have been made in materials available to artists, the most important being the introduction of the so-called "polymer" colors. In these paints a pigment is dispersed in an acrylic emulsion; various additives are used to promote stability and to improve working qualities. Polymer colors are flexible enough to be used on canvas, dry quickly, and can be brushed on thickly, but they are no substitute for oil colors. Ralph Mayer, in *The Artist's Handbook of Materials and Techniques*, said that "artists whose styles require the special manipulative properties of oil colors, including finesse and delicacy in handling, smoothly blended or gradated tones or control in the play of opacities and transparencies, find that these possibilities are the exclusive properties of oil colors."

The durability of the new materials is not yet proven, but the record of the past decade has shown no great problems in their lasting qualities (Fig. 1). As R. J. Gettens has pointed out, oil paints have so many poor qualities that the polymer colors may well be better, and probably will not be worse.

In addition to water-based emulsions, paints based on plastics dissolved in organic solvents are available. Their properties are closer to those of oils. Other new materials for painting include high-durability synthetic pigments and luminescent pigments, including the Day-glo colors.

NEW ESTHETIC EXPERIENCES

In recent years, artists and technologists have become interested in wedding art and modern technology to yield new esthetic experiences. In 1968 a group of artists, scientists, and businessmen formed Experiment in Art and Technology (EAT) in New York City to develop an effective collaboration between engineer and artist. The purpose of the association is the creation of works which are not the preconception of either the engineer or the artist, but which are the result of the exploration of the human interaction between them. EAT held a competition and exhibited some works in "The Machine as Seen at the End of the Mechanical Age" at the Museum of Modern Art in New York (Fig. 2) and held its own show in 1970 at the National Collection of Fine Arts, Washington, D.C. Many interesting ideas and creations have arisen from this group's activities.

A similar but more ambitious show, "Art and Technology," was held recently at the Los Angeles County Museum of Art, with 19 works by 16 artists. At that show Rockne Krebs used the crisscrossing lines produced by red and green lasers, and Newton Harrison employed five Plexiglas columns, 12 ft high, filled with gases lit by programmed electrical discharges (Fig. 3). Other technological devices used in the show included large parabolic mirrors to produce optical illusions, stroboscopic lights, corrugated cardboard, and plastics. Leroy Aarons, in an article in *The Washington Post*, June 1, 1971, concluded that "with the help of the catalog, we are able to see the ways in which the process has become as important as the product. From conception, design, construction, modification to finished work, the interaction of artist and his collaborators is emerging as a value to be reckoned with in the future direction of the arts."

The works exhibited in the "Art and Technology" show bring up the old question "what is art?" The dividing lines between art, nonart, and antiart continue to shift. Andy Warhol's mechanically reproduced paintings, the poster revolution (heavily relying on Day-glo colors), and the "earthworks" school all try to shift the boundaries of art. The whole idea of "minimal" art, along with environment as art (elegantly displayed in the recent EAT show with ultraviolet and sodium-vapor light rooms), shifts these boundaries further. There is a constant emphasis on process rather than product. Some of these art works are not objects at all, but scripts for performances or drawings for proposed projects.

In answering the question "what is art?" we must fall back on the tautology that art is what artists do, and try to be satisfied.

"SCIENTIFIC" ART

A blurring of boundaries between art and science can occur in other ways. The etchings of

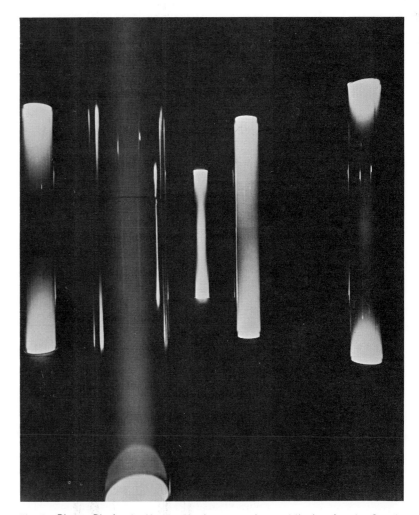

Fig. 3. *Plasma Display*, by Newton Harrison, was shown at the Los Angeles County Museum in 1971. (*Los Angeles County Museum of Art*)

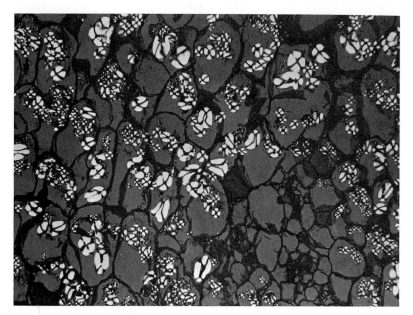

Fig. 4. Photomicrograph of potato scab. (*From Lewis R. Wolberg, Micro-Art, Harry N. Abrams, Inc., 1970*)

Table 1. Some characteristics, or "dimensions," of the Ling Fang-i (FGA 30.54)

Number	Dimension	Quantity
1	Height	35.6 cm
2	Width	24.7 cm
3	Length	22.5 cm
4	Dynastic period	Early Chou (ca. 1000 B.C.)
5	Weight	9.92 kg
6	Type of object	Metal
7	Copper content	78%
8	Tin content	22%
9	Lead content	1%
10–32	(Trace elements)	(Not listed here)
33	Major part of patina	Tin oxide
34	Cuprite present?	Yes
35	Paratacamite present?	Yes
36	Incipient bronze disease?	Yes
37	Azurite on inside?	Yes
38	Soil accretions inside foot?	Yes
39	Chaplets?	Yes
40	Symmetrically placed chaplets?	Yes
41	Rectangular chaplets?	Yes
42	Knob on top of lid?	Yes
43	Is knob clay-cored?	Yes
44	Mold marks on flanges?	Yes
45	Inscription?	Yes
46	Inscription in intaglio?	Yes
47	Inscription in both lid and bottom?	Yes
48	Number of characters—top	187
49	Number of characters—bottom	188
50	Ancient repairs	Patch in bottom inscription
51	Low-relief bands in foot?	Yes
52	Central ridges on bands?	Yes
53	Brackets in juncture of foot?	Yes

Maurits C. Escher have given new insights into symmetry and repeating patterns in crystallography and mathematics.

A recent book by Lewis R. Wolberg explores the artistic qualities of images revealed by photomicrography. Wolberg's craftsmanship and careful choice reveal the hidden beauty of the microworld (Fig. 4), a beauty akin to that of abstract painting.

The famous pioneer micrographer Robert Hooke's *Micrographia*, published in 1665, today seems more like an art book full of beautiful engravings than a scientific treatise.

SCIENTIFIC INVESTIGATION OF ART OBJECTS

The fact that art fakes and forgeries generally have had a life of only about one generation before they are easily detected could be attributed to the ignorance, on the part of the forger and the art public, of one or more dimensions of the work which later scholars stress. These dimensions are lacking, but their absence goes unnoticed at first in the fake. The *Ling Fang-i*, a bronze ceremonial vessel at the Smithsonian Institution's Freer Gallery of Art, can be used as an example of the multidimensional nature of a genuine art object (Table 1 and Fig. 5). The complete table lists more than 50 dimensions and yet does not list the decor of the bronze, the content of the inscriptions, the men who made it, the collections it has been in, and so forth. The list could be almost endless. The table therefore demonstrates two points: (1) No object can be described (or even perceived) completely; usually we are shown only what the investigator is interested in. (2) Comparisons of small numbers of features ignore a rich and diverse field of data contained in the object; the more dimensions considered, the better and more definitive the comparisons will be.

This multiplicity of dimensions and characteristics raises one more point: the polythetic nature of groups of art objects. A polythetic arrangement places together objects that have the greatest number of shared features; no single feature is essential to group membership or sufficient to make an object a member of the group (Table 2). In the world of modern art, examples abound, such as "paintings" composed of blank framed canvases, "sculptures" from found objects, and "prints" made by embossing without ink. The idea of polythetic classification also applies to earlier epochs of art. No single feature or dimension could be found to define Sasanian silver; some Chinese "bronzes" are made from lead; some Japanese paintings are done primarily with reproductive stamps. As a general rule, no one characteristic can be taken to define or differentiate groups of artifacts or clearly distinguish between genuine objects and forgeries. A whole constellation of features must be taken into account.

The investigator, however, can test only one hypothesis at a time. Let us survey some of the modern scientific techniques of investigation and then return to the larger view of classes of objects to see what these techniques have told us and may tell us in the future. We will further break down the techniques into chemical, physical, and structural dating studies. It is impossible here to give a complete list of scientific techniques as applied to art and archeological objects, but we can suggest some of the diverse realms of interest.

Table 2. Characters composing a polythetic group (individuals 1–4) and a monothetic group (individuals 5–6)

Individuals	Characters						
1	A	B	C				
2		B	C	D			
3	A	B		D			
4	A		C	D			
5					F	G	H
6					F	G	H

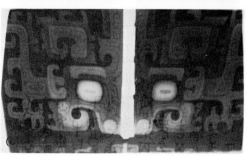

Fig. 5. Chinese bronze ceremonial vessel *Ling Fang-i*. (a) Side of vessel showing design, flanges, and corrosion. (b) Detail of knob on top of vessel showing mold mark at right. (c) Detail of radiograph of top made at 250 kV showing decor and inscription on inside of vessel. (*Freer Gallery of Art, Smithsonian Institution*)

Chemical techniques. Chemical techniques, as the term is used here, are procedures that give information about the elemental composition of an object. Apart from the chemical techniques discussed here, polarography, colorimetry, atomic absorption, and mass spectrometry have been used to examine works of art.

Gravimetric analysis. Classical "wet" chemistry still plays a large part in the analysis of art objects, particularly ancient bronzes. The accuracy and reproducibility of classical gravimetric analysis, combined with the unique nature of archeological samples, ensure a place for this field of endeavor in the art laboratory.

Although the gravimetric analyses of the 121 Chinese ceremonial vessels in the collection of the Freer Gallery of Art have not given any clear chronological or geographical clues, we hope that with the gradual accumulation of data, the picture will become clearer.

Optical emission spectrometry. This technique continues to be employed for the analysis of art objects. About 12,000 analyses by emission spectrography have been done by the Arbeitsgemeinschaft für Metallurgie des Altertums at the Römisch-Germanisches Zentralmuseum, Mainz, Germany, with the idea of delineating the Bronze Age metallurgy of Europe. Some minor doubts have been cast on this enterprise because of the use of only one sample per object and the "univariate" approach to grouping (using a dichotomous key of element concentrations). Nevertheless the approach has yielded some archeologically interesting conclusions.

Laser microprobe. A useful accessory to the spectroscope is the laser microprobe. This apparatus directs a laser beam down through the objective lens of a microscope to impinge on a sample or an object (Fig. 6). The laser pulse raises a plume of vaporized material which passes between two electrodes of the spark gap of the spectrograph and is analyzed in the conventional manner. The laser removes material from a craterlike area, about 60 μm deep. Both conductive and nonconductive material can be analyzed, and the instrument bridges the gap between conventional sampling analysis and the electron microprobe.

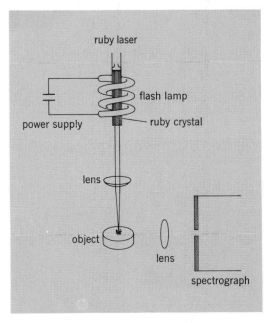

Fig. 6. Diagram of laser microprobe. (*Adapted from F. Brech and W. J. Young, in Application of Science in Examination of Works of Art, Museum of Fine Arts, Boston, 1967*)

Neutron activation. In neutron activation the sample is exposed to the neutron flux of an atomic reactor. It is a very sensitive technique, capable of measuring a large number of elements down to the parts-per-million range. The analytical technique has become more practical with the introduction in the past few years of new types of detector crystals, as has the newer energy-dispersive x-ray spectrometry described below. To determine a large number of elements to the required accuracy, computers must be employed to reduce the data.

Two realms for intensive application of neutron activation analysis have been the analysis and grouping of ancient pottery and the analysis of Sasanian silver. Here we will consider pottery. I. Perlman and F. Asaro have concentrated on the method of analysis, the method of data reduction, and criteria for using the data to divide pottery into groups. They have analyzed more than 1000 shards from many different sites and have found that, in all cases, two analyses drawn from the same pot agree better than two analyses from separate pots in the same archeological group. They infer that (1) the potter's lump of clay was well homogenized; (2) they have not yet encountered two pots made from the same lump of clay (a common practice of potters in Japan and elsewhere); (3) local clay sources have internal variations in composition; and (4), most important, the method of analysis is sensitive enough to make these distinctions. In some cases of remarkably close agreement, the two pieces analyzed proved to be from the same pot!

The statistics of creating and matching groups of pottery, even if handled by computer, is extremely complex and beyond our scope here. The following two examples illustrate neutron activation as applied to more specific problems. The hypothesis tested by E. V. Sayre, L. Chan, and J.

Fig. 7. Two energy-dispersive x-ray fluorescence analyzer systems in use. Unit on left is used to analyze glass, ceramics, bronzes, and so on and is capable of quantitative determination of elements above potassium in the periodic system. Unit on right is employed for routine analysis of silver objects for silver, copper, lead, gold, zinc, and mercury. (*The Henry Francis du Pont Winterthur Museum*)

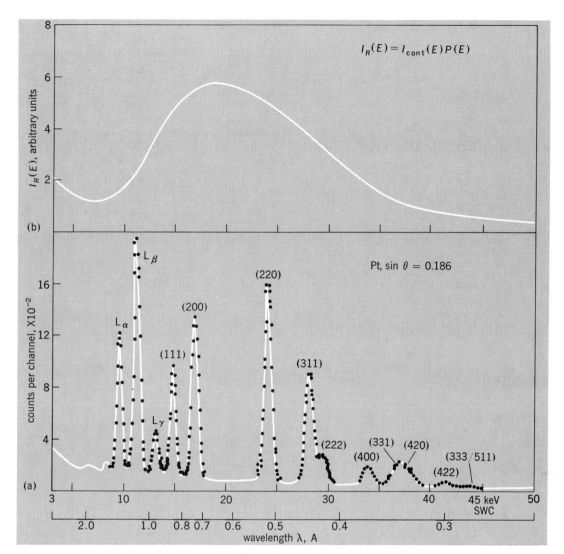

Fig. 8. Chart of the fluorescent and diffracted spectrum of platinum sheet (below) and the response of the detector to the x-ray beam (above). (B. C. Giessen and G. E. Gordon, X-ray diffraction: New high-speed technique based on x-ray spectrography, Science, 159:973–975, Mar. 1, 1968)

A. Sabloff had to do with the place of manufacture of Mayan Fine Orange Pottery. They concluded from their analyses that Mayan Fine Orange ware was made in one place for several centuries and exported throughout Mexico and Yucatan. They also found that the Mayan Fine Gray ware has the same place of origin as the Mayan Fine Orange, and that all of the local coarse wares analyzed were somewhat different. These conclusions interested American archeologists greatly.

Jacqueline S. Olin and Sayre investigated whether two shards from Drake's Bay in California were made of English Devonshire pottery or not. If they were, greater credence could be attached to reports of Sir Francis Drake's landing there. In the course of the investigation a number of shards of English and American Colonial pottery were examined. Although the Drake's Bay shards are clearly not North Devon ware, the English and American groups could be differentiated easily. Thus, even if the primary conclusions are negative, the data such an investigation produces will be useful in later comparisons.

X-ray fluorescence spectrometry. In examination of works of art, x-ray fluorescence has followed two separate directions. In one approach, conventional, wavelength-dispersive x-ray optics are used; samples are usually taken from the objects; and accuracy, precision, and reduction of sample size are pushed to the limit. The work of Maurice Salmon in the Conservation-Analytical Laboratory of the Smithsonian Institution's United States National Museum illustrates this clearly. With a 10-mg sample, Salmon can equal or surpass accuracy and precision of classical gravimetric wet chemistry on bronzes. His technique involves dissolving the sample, adsorbing it onto cellulose powder, making a pellet from the powder, and analyzing the pellet.

The second approach involves the use of energy-dispersive x-ray analyzers. In this approach, the entire x-ray spectrum from the object (excited by either a radioactive source or by an x-ray beam) is analyzed and results are displayed at once (Fig. 7). The systems use high-resolution crystals and can analyze a large number of elements simultaneously. The method is applied to simple questions which require a quick qualitative answer, often without removing a sample from the object. For answering such questions as to whether the white

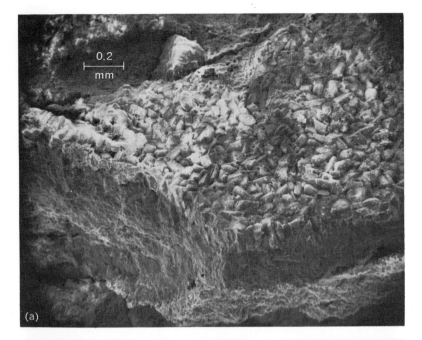

Fig. 9. Scanning electron micrographs of crystalline corrosion, probably atacamite, on the surface of a bronze from Nimrud. (*Freer Gallery of Art, Smithsonian Institution; photo by Sperry-Rand*)

tem. To do this, one employs a conventional x-ray beam source, a lithium-drifted silicon or germanium detector, a pulse-height analyzer, and a computer. The detector is usually set on a circular mount and moved with respect to the x-ray beam. In two different positions the diffracted lines move, but the fluorescent (elemental) lines stay at the same place (retain the same energy) on the pulse-height analyzer (Fig. 8). Although the system has, as yet, been little used with art objects, its potential is vast, for a great deal of information could be obtained from an object without sampling. Pigments could be identified unambiguously, and metal objects could be analyzed for content and fabrication technique. If this system lives up to its potential, it will be one of the most important developments in the next few years.

Physical techniques. "Physical" techniques are those that tell us about the structure, rather than the composition, of objects. Under this heading a great many investigative techniques should be covered: infrared study of the structure of organic materials (linseed oil, oriental lacquer, and so forth); x-ray diffraction studies; micrography of wood; studies of micro- and macrostructure of ivory, bone, or horn; metallography; electron micrography; and so on. The structure of a work of art determines its shape and properties, enables it to persist, and carries the record of the manufacturing process and the alterations caused by time. Here we will discuss only a few recent structural studies.

Scanning electron microscope. A promising new technique for studying works of art is the scanning electron microscope. Although most scanning electron microscopes have small chambers and a piece therefore must be taken from the object, instruments with larger chambers are being developed. The ability of the scanning electron microscope to look at the surface of objects and fractures at great magnification and in depth can give a whole new view on the mechanisms of corrosion or the structure of the surface (Fig. 9). For greater magnification, especially of particulates such as pigments, transmission electron microscopy is an invaluable tool.

Patina. The nature of corrosion and corrosion processes over a long term is beginning to be investigated by the techniques of physical chemistry and x-ray diffraction. The latter is particularly helpful in determining the authenticity of patina on metals. A thick layer of cuprite (cuprous oxide, Cu_2O) right next to the metal, intergranular corrosion, and a layered structure seem to point to the genuineness of a patina, but studies into the fundamental mechanisms of patina formation are necessary. Seymour Z. Lewin and J. B. Sharkey have studied the mechanism of formation of atacamite and paratacamite, and conclude that if the bulk of the patina on an object contains atacamite (with or without paratacamite) and if the object is not deeply corroded, the patina is probably false. Although other investigators take exception to this conclusion, the reasoning behind it rests on fundamental understanding of corrosion mechanisms, and it may well be a valid criterion under certain circumstances. In any case, this sort of study will continue and intensify in the next few years, and we should see some real advances in understand-

paint on an object is lead white or titanium white or whether a particular alloy is brass or bronze, this approach is ideal. It may be that for strict quantitative use, the sampling approach has advantages. Of course, prepared pellets or solutions could be analyzed by energy-dispersive means. So far the most successful applications have been in sorting objects and in qualitative analyses of surfaces.

Spanning the gap between "chemical" and "physical" techniques is another application of energy-dispersive x-ray analysis: the simultaneous determination of elemental and structural information of simultaneously viewing fluorescent and diffracted energies with an energy-dispersive sys-

SCIENCE IN ART 19

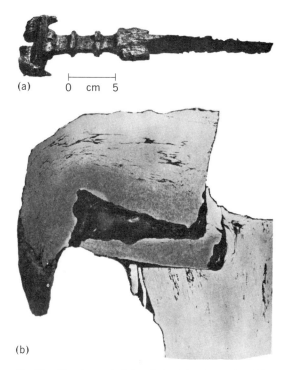

Fig. 10. Short sword of classic Luristan type, about the 8th century B.C., after electrolytic cleaning. (a) Overall view. (b) Metallographic section through end of hilt, pommel, and attached heads. Both the handle and the decorative head show slag lines terminating abruptly at the end with almost no deformation at the surface; they were probably ground down rather than cut with chisel. (*C. S. Smith, Science and Archaeology, MIT Press, 1971*)

the tools for insights into structure. Cyril S. Smith's recent research on ironworking and steelworking in early Iran as seen through the objects shows the power of this approach. The Luristan smith used some unique technical processes, but did not select and use harder metal for edges or where strength was needed, nor did he use softer metal for decorative parts. He was adept at forging, cutting, and interlocking parts, but knew nothing of welding, nor of quench hardening. This peculiar combination of knowledge and ignorance suggests that these smiths were working in a transitional period of iron production. More interesting than the general conclusions, however, is the specific light that these structural studies throw on how an object was made (Fig. 10). Casting or raising from sheet, heat treatments, amount of working, types of tools used, surface treatment and corrosion can all be inferred from studies of the structure of the metal.

Structural studies can thus contribute to our understanding of any object. By studying microsections of oil paintings one can see the artist's technique, detect repaint, and sometimes discern the changes the artist made as he worked. Microsections of pottery enable us to ascertain the materials and manufacturing techniques employed.

Gilding. Heather Lechtman made another recent application of metallography in examining the gilding techniques of Near Eastern and Peruvian silver. By microscopic examination and the use of the electron microbeam probe she has established beyond any doubt the use of mercury leaf and mercury amalgam gilding on the Near Eastern objects, and a depletion gilding method, entirely different in technique but superficially similar in the end result, on the Peruvian objects. The clear evidence that she has produced has begun to clarify one of the puzzles of ancient metallurgy, and has defined the tools necessary to study these gilding processes (Fig. 11).

ing the information contained in the corrosion crusts on objects.

Mechanical structure. Metallography and microscopy, coupled with modern instruments such as the electron microbeam probe, are the most important, purest, and probably most intriguing of

 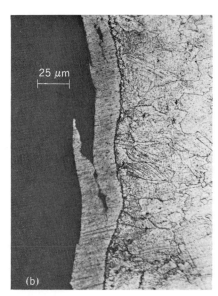 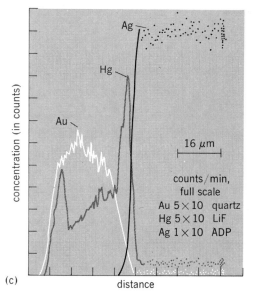

Fig. 11. Gilded Sasanian silver rhyton with representation of the goddess Anahita, early 4th century A.D. (a) Overall view (*The Cleveland Museum of Art, gift of Katharine Holden Thayer*). (b) Cross section of gilding; (c) electron microbeam probe trace across surface shown in b (*from R. H. Brill, Science and Archaeology, The M.I.T. Press, 1971*).

Fig. 12. Thermoluminescence data apparatus. (*University Museum, Philadelphia*)

Microprobe analysis. Along with the techniques of classical wet chemistry and x-ray diffraction, the electron microbeam probe was also used to establish the meteoritic origin of the iron blades of two Chinese bronze weapons. The microprobe showed that the minute uncorroded areas in the iron blades correspond in composition with kamacite and taenite, two iron-nickel alloys usually found in octahedrite meteorites but very rarely in appreciable amounts in terrestrial samples. These objects were very probably made in about 1000 B.C., and are the earliest extant objects made from meteoritic iron.

DATING AND CHRONOLOGICAL STUDIES

One of the most frequent questions asked of the museum analyst is "how old is it?" Often the question is asked to determine the authenticity of an object. The museum analyst has a very vital interest in dating techniques, but a brief general discussion of the meaning of the question is in order before we turn to techniques themselves.

At first glance, it appears that the age question has a definite answer. In the case of some objects it does; we know the date on which Pablo Picasso painted *Guernica*, or the date on which certain cuneiform tablets were inscribed (although converting this date into our system may be difficult). Many objects were completed on a definitive date, and this may be determinable. Other objects have gone through a constant process of "manufacture" or change even up to the present. Architectural structures are a case in point; a sample of timber or mortar taken to date a building must be chosen with great care.

Similarly, care must be taken to distinguish between the age of an object and the age of the material from which it was made. Ivory statuettes, for instance, may be carved from mammoth ivory. When the dating laboratory returns a date of 30,000 B.C. for a Chinese Ming dynasty statuette, the curator may be somewhat shocked, until he is told that the ivory is indeed that old but the carving is not.

The difference between absolute and relative dating must also be stressed. Relative dating defines "before-and-after" relationships; archeological stratigraphy is an example of this method. Enough samples and a close view of the fine structure in the stratigraphy can enable very accurate relative chronologies to be drawn up. But these must still be tied to absolute chronologies by the finding of dated pieces or by some of the absolute dating methods. These methods (such as carbon-14, thermoluminescence, dendrochronology, lead isotopic decay, fission-track dating, and obsidian hydration dating) depend either on regular deposition of something (including radiation) in the object or on regular radioactive decay, and can be said to date a specific event.

In the case of thermoluminescence, the firing of the clay is the event dated. In carbon-14 dating, the death of the organism starts the clock. Dendrochronology matches the patterns in the growth rings of trees and attempts to extend this pattern back from modern times into the past. A list has been constructed for oaks of the Main and Rhine valleys back to A.D. 800, and one exists for the bristlecone pines of the Sierra Nevada which extends back to 5142 B.C. For dating wood objects, including polychrome sculpture and easel paintings, dendrochronology can be extremely useful.

Carbon-14 dating methods are constantly being refined, and can be cross-checked with methods such as dendrochronology. It remains one of the archeologist's standard methods for dating.

Thermoluminescence dating has been in existence for a number of years but recently has been refined to the point where its results are becoming very useful to the archeologist and art historian. Since thermoluminescence primarily dates pottery and since pottery is the universal archeological material, it is a method of great general usefulness. To obtain a thermoluminescence date, a piece of pottery or ceramic material is ground up and then heated at a controlled rate; the light it gives off as it is heated is carefully measured (Fig. 12). The amount of light is proportional to the age of the object, and comes from electrons which have been bumped into higher energy states by radiation, and which drop back to the ground state on the addition of heat, emitting light. The radioactivity of both the fired clay comprising the object and the soil surrounding the burial, and the radioactive susceptibility of the clay must also be known. If they are known accurately enough, the date will be within 10% of the actual age of the piece.

This method is currently being used for authenticity and dating studies on Chinese pottery, the fired clay cores of Chinese bronzes, and other objects; its wider application and refinement will continue in the next few years. Figure 12 shows a linearly programmed heating system and glow-curve-measuring apparatus.

The method that probably has the greatest general applicability is the relative method of comparison or correlation, which also includes seriation. An example of this approach is the analyst's response to the question of how old an object is. The analyst first, instinctively and without hesitation, asks himself what the object resembles that he knows something about. Then he samples it, analyzes it, x-rays it, and so on in order to determine the differences or similarities.

Now, this approach, although it seems self-evidently simple, combines a large number of problems. The analyst must have an adequate information retrieval system to find the data on similar objects. He should know what confidence he can put in the published data. And most important, the analyst should draw a conclusion based on as many variables, or "dimensions," of the object as possible.

Unfortunately the necessary information often lies buried in someone else's files or has never been derived from the objects; or the necessary comparisons within the body of material have never been made. Often, however, definitive conclusions can be drawn, and with more intensive studies and more sophisticated information retrieving and correlating systems, even more successes should be forthcoming. The Museum Computer Network, now in its initial stages, should eventually make it much easier to find data on objects of any particular type.

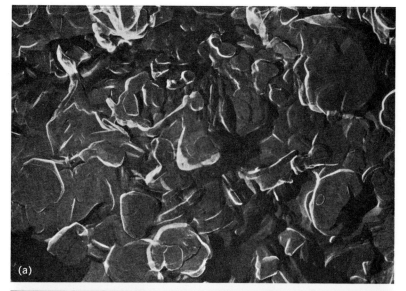

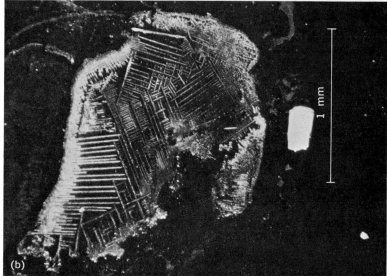

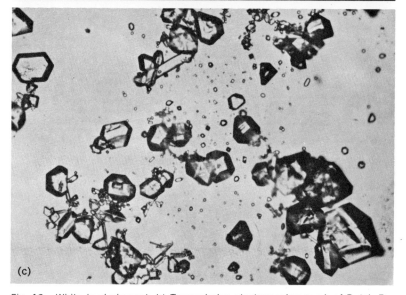

Fig. 13. White lead pigment. (a) Transmission electron micrograph of Dutch Boy modern white lead paint (*Smithsonian Institution Department of Paleobiology; photo by Ken Towe*). (b) Particle of white lead paint and lead nitrate residue from James McNeil Whistler's painting *The Music Room* under reflected light. (c) Lead nitrate crystals from b after recrystallization under transmitted light.

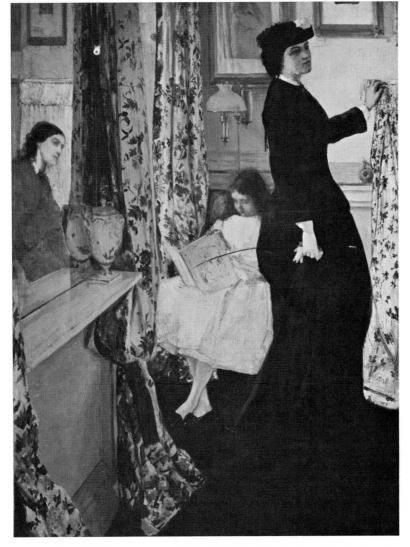

Fig. 14. *The Music Room*, by James McNeil Whistler. (*Freer Gallery of Art, Smithsonian Institution*)

INVESTIGATIONS OF CLASSES OF OBJECTS

The three investigations mentioned in this section have been chosen from the hundreds possible simply because I have been closely associated with them. Like scientists in general, museum analysts and art conservators each have their special loves, and while they may work on a wide range of materials, they tend to concentrate effort on subjects of particular interest.

Compilation of data. For the reasons of comparison and correlation, the publication of reliable data on specific objects has a high priority in art-technical literature. To further this purpose, R. J. Gettens set out in 1961 to publish a handbook on analysis of materials of paintings, now tentatively entitled *Handbook for the Identification of Materials of Paintings*. The handbook has a double emphasis: to publish reliable identification methods for painting materials and to list the notable occurrences of these materials in paintings. To date, six pigments have been covered in serial publication. Figure 13 shows some representative photographs from the entry on white lead, and Fig. 14 shows a 19th century painting from which samples were taken.

The first compiled volume of Gettens's handbook is expected to be ready in 1973. It is certain to be an invaluable reference for all conservators and museum scientists.

Sasanian silver. A topic of real interest to collectors and museums is Sasanian silver. The Sasanians ruled Iran from A.D. 227 to 651 and produced some beautiful and powerful works in silver. Only one Sasanian silver object has been found to this date in a controlled excavation in Iran. On the other hand, Sasanian silver objects have been pouring into the United States (and other countries) in increasing numbers since the 1950s. No satisfactory chronology of Sasanian silver has yet been made. There is increasing suspicion that some of the objects now in our museums may be fakes or later (Islamic) productions.

Three small conferences of interested parties have been held in the past few years, and many worthwhile ideas have arisen from these meetings. We now have a fairly good idea of the methods of production shown in this body of material, the gilding methods (already described), the metal contents, and some interesting data on gold-silver ratios in the objects. Other methods of investigation are under development, for example, neutron activation analysis for trace elements such as iridium, microsampling for metallography, and local polishing of an area for metallography without removing a sample.

Unfortunately, we have found no panacea and do not seem to be any closer to constructing a reliable chronology or even being able to say that a particular Sasanian silver object is genuine or not. Many questions remain to be answered; we hope that with more data and the correlation of the technical data with stylistic and art-historical variables, logical groupings will arise. The present state of Sasanian silver studies is a good example of a complicated problem which may be solved within the next few years.

Chinese bronzes. The technical investigation of Chinese bronzes has been going on for a good deal longer than that of Sasanian silver, and much more is known about the subject. In China many archeological finds have been made and published; much analytical data are in the literature. In fact, the Freer Gallery has recently published a two-volume catalog on its Chinese bronze ceremonial vessels. Like most investigations, this one has raised more questions than it has answered.

Just after publication of the Freer catalog, the technique of thermoluminescence dating began to be applied to Chinese bronzes, with very interesting results. Now there is a possibility of absolutely dating about two-thirds of the bronze vessels in the Freer collection, which may make it possible to correlate date, composition and corrosion.

Other studies on Chinese bronzes now in progress include a comprehensive study of Chinese belt-hooks at the Freer, a study of Chinese bronze weapons at the Royal Ontario Museum, and a study of the casting methods used in the bronzes from An-yang (of the Shang period, about 1100 B.C.) at the Academia Sinica, Taiwan. In the near future, comprehensive laboratory studies are expected to begin on the thousands of ceremonial vessels in various collections on Taiwan, including the very important imperial collection in the National Palace Museum and the excavated, fully-attested material in the Academia Sinica. These studies will be of immense value in defining ancient Chinese bronze foundry practice, and the results will aid in evaluating other Chinese bronzes.

CONSERVATION OF ART

Science also interacts with art in its conservation and preservation. An object can be seen as constituting a process. It is made, lives, changes as it ages, corrodes, and finally disintegrates. Modern conservation practice aims to preserve the object unchanged as long as possible. To do this, the environment (light, temperature, relative humidity, gaseous and particulate pollution) must be controlled and the object must be made structurally sound and handled correctly. New materials from science, especially synthetic resins, and new techniques of determining condition and composition of artifacts all aid the conservator in these tasks. Possibly the most important idea that the conservator can glean from the scientist is that the objects he is dealing with are immensely complex and structured at an infinite number of levels, and the conservator should try to change them as little as possible. Evidence is easily destroyed by almost any treatment, and great care should be taken not to change the object any more than the minimum necessary for its health.

CONCLUSION

The interaction of science and art has been for some time one area of mixed cultures. The "interfacial" nature of this interactive field makes it an exciting place to work, on any level. The adaptation of new materials to old uses in art and to new processes in art conservation can be beautiful and useful. Watching artists create with and comment on the new technology can be exciting. And working with artifacts from the past can impart a sense of man's whole history seen all at once and a sense of time and our existence in it.

This last aspect of science in art is perhaps the most significant in human terms. The realization of time as an actual dimension of an artifact and the transcendence of time by the maker and the viewer of an artifact is best summed up by another art, that of poetry:

    Only by the form, the pattern
Can words or music reach
The stillness, as a Chinese jar still
Moves perpetually in its stillness.
Not the stillness of the violin, while the note lasts,
Not that only, but the co-existence. . . .
                 (From "Burnt Norton")

Men's curiosity searches past and future
And clings to that dimension. But to apprehend
The point of intersection of the timeless
With time, is an occupation for the saint —
No occupation either, but something given
And taken, in a lifetime's death in love,
Ardour and selflessness and self-surrender. . . .

Here the impossible union
Of spheres of existence is actual.
Here the past and future
Are conquered, and reconciled. . . .
            (From "The Dry Salvages")
            T. S. Eliot, *Four Quartets*,
        Harcourt Brace Jovanovich, Inc.
                           [W. T. CHASE]

Solid-Waste Management

Until the relatively recent past, Americans have not seemed greatly concerned with the problems of environmental pollution, and least of all with the pollution resulting from inadequate solid-waste management. The most convenient disposal method—usually an open dump—would suffice. The land was large, the population pressure was small, and natural resources were seemingly without limit. However, the 20th century, and particularly the period since World War II, has witnessed dramatic changes in the United States—most importantly its transformation from a

Richard D. Vaughan is deputy assistant administrator for solid-waste management programs of the U.S. Environmental Protection Agency. Apart from the management of solid wastes, he has had much experience in water-pollution control.

predominately rural population of approximately 76×10^6 in 1900 to a population now over 200×10^6. Of particular significance is the fact that over 70% of this population was counted as urban in the 1970 census. As striking as the increase in the population and its shift from rural to urban character has been the growth in the productivity of American agriculture and industry.

TYPES AND VOLUME OF WASTES

All these population changes have had profound effects upon the types and volume of solid wastes generated in the United States and encourage their accumulation in urban areas.

Municipal wastes. The rise of certain industries and marketing techniques, such as the packaging industry and the wide acceptance of packaged and canned foods (primarily for convenience), has contributed to a very great increase in per capita generation of solid wastes. For example, in the United States in 1920 on a per capita basis, an average of 1.25 kg of wastes were collected per day; today this figure has grown to 2.4 kg, and by 1980 it is estimated that this will rise to 3.6 kg. In a typical year Americans will discard over 27.2×10^6 metric tons of paper, 3.6×10^6 metric tons of plastics, 48×10^9 cans, and 26×10^9 bottles. The numerous disposable products on the market and an economy which frequently makes it cheaper to replace worn items than to repair them, contribute to the high per capita waste production.

In addition to increasing waste volumes, these technological changes have also had an important impact upon the characteristics of refuse, and in some cases the point at which the refuse is generated. For example, the trend toward consumption of processed foods (canned and frozen) is largely responsible for a relatively low percentage of putrescibles and high percentage of paper, plastics, and metals in the waste stream. Fruit and vegetable trimmings now accumulate in enormous quantities at canneries and food-processing plants, whereas they formerly accumulated in the kitchens of individual dwellings.

Although there is some geographical variation in characteristics of municipal solid wastes, social, political, and economic factors cause the major variation in physical and chemical composition. However, with standardization of data, reporting methods, and so forth, it has been possible to develop data meaningful from a national standpoint. In gross terms, the following list is descriptive of the physical composition of United States municipal solid waste, by weight: 50% paper; 10% metal; 10% glass; 20% food wastes; 3% yard waste (grass clippings, tree trimmings); 1% wood; 1% plastic; 1% cloth and rubber; and 4% inert material.

Agricultural wastes. The great rise in productivity of American agriculture is largely due to newer and more efficient methods of stock raising and cultivation of the land. These same methods, however, have tended to aggravate the problem of solid-waste disposal in rural areas. Herds of cattle and other animals, once left to graze over large open meadows, are now often confined to feedlots, where they fatten more rapidly for the market. However, on feedlots they produce enormous and concentrated quantities of manures that cannot be readily and safely assimilated into the soil, as under conventional open grazing practices (Fig. 1). Animal manures from concentrated livestock production are associated with such undesirable effects as fish-kills, eutrophication of lakes, off-flavors in surface waters, escessive nitrate contamination of aquifers, odors, dusts, and the wholesale production of flies and other noxious insects. Some agricultural wastes, such as those from concentrated poultry and egg production enterprises, may contain elements which, if allowed to leach into ground or surface waters, may constitute a localized threat to public health. Examples of these are the arsenic, manganese, and zinc substances normally found in poultry manure. The problem of agricultural-waste disposal is compounded by the fact that the superior convenience and nutrient value of chemical fertilizers have resulted in a lessened demand for animal manures to be used as soil fertilizers.

Mineral and fossil-fuel wastes. During the past 30 years, more than 18.1×10^9 metric tons of mineral solid wastes have been generated in the United States by the mineral and fossil fuel mining, milling, and processing industries. Because of transportation costs, solid-waste problems in the minerals industries are largely local in nature. Slag heaps, culm piles, and mill tailings generally accumulate in proximity to the extraction or processing operation. Prior to 1965 an estimated 18,000 km², or approximately 2×10^6 hectares, were covered with unsightly mineral and solid fossil fuel mining and processing wastes or otherwise devastated as a result of mineral and fuel production.

Although some 80 mineral industries generate quantities of solid waste, 8 major industries are responsible for 80% of the total. These are copper, which contributes the largest waste tonnage, fol-

Fig. 1. Changes in agricultural practices have served to increase production but have also aggravated agricultural waste problems. For example, raising livestock on feedlots, rather than in open pasture, has resulted in large accumulations of manure which cannot be readily assimilated by the soil. (*U.S. Environmental Protection Agency*)

Fig. 2. Open dumps provide food and harborage for rats, flies, and other pests. Nearly half of all open dumps contribute to water pollution, and three-fourths to local air pollution. (*U.S. Environmental Protection Agency*)

lowed by iron and steel, bituminous coal, phosphate rock, lead, zinc, alumina, and anthracite industries.

Total solid wastes. In summary, the total solid-waste load generated from municipal and industrial sources in the United States amounts to more than 326×10^6 metric tons annually; this figure includes 227×10^6 metric tons of household, commercial, and municipal wastes and 99×10^6 metric tons resulting from industrial activities. The annual total of agricultural wastes, including animal manures and crop wastes, is estimated to be over 1.8×10^9 metric tons. The present annual rate of mineral solid-waste generation is an estimated 10^9 metric tons, with an anticipated rise to 1.8×10^9 metric tons by 1980. Even this projection for mineral wastes may prove low if ocean and oil shale mining become large-scale commercial enterprises. Altogether, over 3.2×10^9 metric tons of solid wastes are generated in the United States every year.

PUBLIC HEALTH, NATURAL RESOURCES, AND ECONOMY

Nature has demonstrated its capacity to disperse, degrade, absorb, or otherwise dispose of unwanted residues in the natural sinks of the atmosphere, inland waterways, oceans, and the soil. But the major concern are those residues that may poison, damage, or otherwise affect one or more species in the biosphere, with resultant changes in the ecological balance.

Public health. The relationship between public health and improper disposal of solid wastes has long been recognized. Rats, flies, and other disease vectors breed in open dumps and in residential areas or other places where food and harborage are available (Fig. 2). An extensive search of the medical literature has indicated an association between solid wastes and 22 human diseases. Perhaps the most obvious and direct relationship between solid wastes and human health can be observed for that segment of the population which experiences occupational exposure — namely, refuse collectors and processing/disposal plant operators. A study performed jointly by the American Public Works Association and the National Safety Council revealed the startling fact that the work injury (frequency) rate among solid-waste employees is nearly nine times greater than the average for all United States industries.

Implications for public health and other problems associated with water and air pollution have been linked to mismanagement of solid wastes. Leachate from open refuse dumps and poorly engineered landfills has contaminated surface and groundwaters. Contamination of water from mineral tailings may be especially hazardous if the leachate contains such toxic elements as copper, arsenic, and radium. Open burning of solid wastes or incineration in inadequate facilities frequently results in gross air pollution. Many residues resulting from mismanagement of solid wastes are not readily eliminated or degraded. Some are hazardous to human health; others adversely affect desirable plants and animals. Although attempts to set standards for air and water quality have been made and are continuing, in many cases harmful levels of many individual contaminants and of their combined or synergistic effects are not yet known.

Natural resources. The Earth's mineral and other resources are not unlimited. Man extracts metals from ores and transforms other materials from a natural state to man-made products, but when these products have fulfilled their usefulness and are classified as solid wastes, the valuable and nonrenewable materials are often lost. Iron, for example, is a nonrenewable resource concentrated in ores over periods involving millions of years of geological processes. This element is extracted, processed, and widely dispersed over the Earth and serves many useful purposes, from cans to automobiles. When discarded, these objects rust away and that amount of iron may never be available again for use by future generations. Shortages in resources for some of the less common elements have been created already by many indiscriminate waste disposal practices.

Economic aspects. Implicit in public attitudes toward waste materials is the problem of economics. National concern must transcend both the concept of what the public can "afford" to pay and the question of why the expenditure of about 4.5×10^9 each year for solid-waste collection and disposal has not staved off the present mounting problems with solid wastes. The truth is that the national standard of living depends to a significant degree upon processes leading to waste generation. It is reasonable to presume that a departure from traditional wastefulness of resources might reduce the volume of wastes to be managed. Thus there is a need for finding ways in which wastes themselves may be salvaged, reworked, and recycled back as a part of resources that are being processed.

A system for managing solid wastes must be economically as well as technologically feasible. In waste management, as in many other fields, a particular facility or item of equipment may be technically suitable but prohibitively expensive. Many communities are unable or unwilling to pay the price necessary to take advantage of the best presently available solid-wastes disposal systems or devices.

Location and quality of operation of solid-wastes disposal facilities may have an important impact upon the economics of an area, affecting land-use patterns and the value of surrounding property. Although incinerators, sanitary landfills, and compost plants generally do not enhance the value of surrounding property, there are isolated examples in southern California and other areas in which expensive residential neighborhoods are found adjacent to sanitary landfill sites, planned for ultimate conversion to parks, golf links, and other recreational uses. Three-quarters of the 4.5×10^9 yearly cost of United States solid-waste disposal is attributed to the collection and transport of solid wastes. Yet this expenditure has purchased only hit-or-miss collection systems, with little satisfaction for the majority of American communities. The one-quarter of expenditures currently allo-

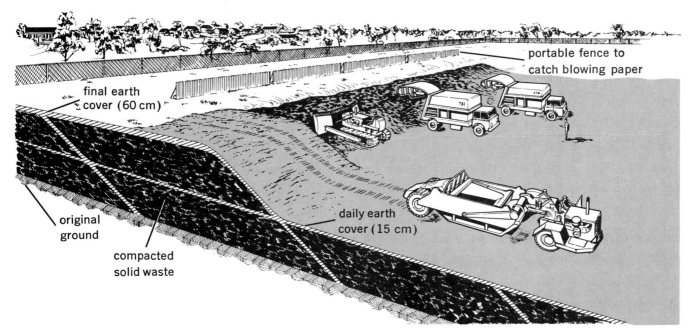

Fig. 3. Area method of sanitary landfill. The bulldozer spreads and compacts solid wastes. The scraper (foreground) is used to haul the cover material at the end of the day's operations. A portable fence catches any blowing debris. This is used with any landfill method. (*U.S. Environmental Protection Agency*)

SOLID-WASTE MANAGEMENT

Fig. 4. Trench method of sanitary landfill. The waste collection truck deposits its load into the trench, where the bulldozer spreads and compacts it. At the end of the day the dragline excavates soil from the future trench; this soil is used as the daily cover material. Trenches can also be excavated with a front-end loader, bulldozer, or scraper. (*U.S. Environmental Protection Agency*)

cated for processing and disposal of wastes has, despite obvious public objections, resulted in the use of the open dump as the predominant disposal method.

MANAGEMENT TECHNIQUES

Solid-waste management differs in important respects from air- and water-pollution control. The difference derives from the fact that there are two pollutant transport systems: natural (air and water) and artificial (vehicular transport). In general, air and flowing water carry pollutants across political boundaries in response to natural laws that are not subject to legislative repeal. In contrast, solid wastes must be left where they are generated or transported by mechanical means.

The bulk of solid wastes are deposited on land, and disposal tends to be a local problem. Thus the principal solutions to solid-waste management lie in providing operational systems that employ physical procedures rather than in regulation. Such handling, along with reclamation and reuse as the solid-waste management goal, offers the ultimate solution.

Although open dumping—with all of its attendant problems—is the predominant means of solid-waste disposal in the United States, several acceptable alternatives to open dumping and open burning are presently available.

Sanitary landfill. The sanitary landfill is defined by the American Society of Civil Engineers as a method of disposing of refuse on land without creating nuisances or hazards to public health or safety by utilizing the principles of engineering to confine the refuse to the smallest practical area, to reduce it to the smallest practical volume, and to cover it with a layer of earth at the conclusion of each day's operation or at such more frequent intervals as may be necessary.

Such a landfill is a well-controlled and truly sanitary method of disposal of solid wastes upon land. It consists of four basic operations: (1) The solid wastes are deposited in a controlled manner in a prepared portion of the site. (2) The solid wastes are spread and compacted in thin layers. (3) The solid wastes are covered daily or more frequently, if necessary, with a layer of earth. (4) The cover material is compacted daily.

Two general methods of landfilling have evolved: the area method and the trench method (Fig. 3 and 4). Some schools of thought also include a third, the slope, or ramp, method (Fig. 5). In some operations, a slope or ramp is used in combination with the area or trench methods.

In an area sanitary landfill the solid wastes are placed on the land; a bulldozer or similar equipment spreads and compacts the wastes; then the wastes are covered with a layer of earth; and finally the earth cover is compacted. The area method is best suited for flat areas or gently sloping land, and is also used in quarries, ravines, valleys, or where other suitable land depressions exist. Normally the earth cover material is hauled in or obtained from adjacent areas.

In a trench sanitary landfill a trench is cut in the ground and the solid wastes are placed in it. The solid wastes are then spread in thin layers, compacted, and covered with earth excavated from the trench. The trench method is best suited for flat land where the water table is not near the ground surface. Normally the material excavated from the trench can be used for cover with a minimum of hauling. A disadvantage is that more than one piece of equipment may be necessary.

In the ramp or slope method (a variation of the area and trench landfills) the solid wastes are dumped on the side of an existing slope. After spreading the material in thin layers on the slope, the bulldozing equipment compacts it. The cover material, usually obtained just ahead of the working face, is spread on the ramp and compacted. As a method of landfilling, this variation is generally suited to all areas. The advantage of utilizing only one piece of equipment to perform all operations

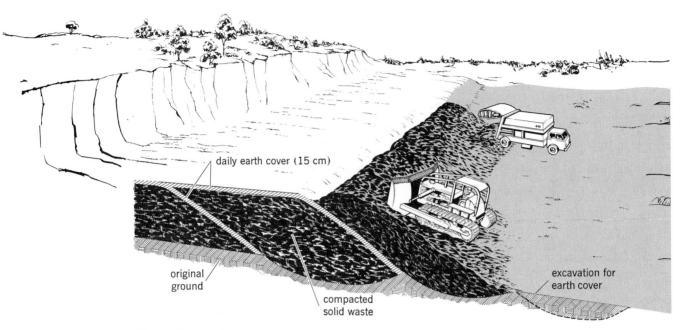

Fig. 5. Ramp variation of sanitary landfill. Solid wastes are spread and compacted on a slope. The daily cell may be covered with earth scraped from the base of the ramp. This variation is often used with either the area or trench method. (*U.S. Environmental Protection Agency*)

Fig. 6. Incineration of solid wastes in a modern plant can effectively reduce the volume of wastes to be disposed. However, this process is comparatively expensive and is therefore used primarily by large cities. (*U.S. Environmental Protection Agency*)

makes the ramp or slope method particularly applicable to smaller operations. The slope or ramp is commonly used with either area or trench sanitary landfill.

Completed landfills have been used for recreational purposes, for example, in parks, playgrounds, or golf courses. Parking and storage areas or botanical gardens are other final uses. Because of settling and gas problems, construction of buildings on completed landfills generally has been avoided; in several locations, however, one-story rambling-type buildings and airport runways for light aircraft have been constructed directly on sanitary landfills. In such cases it is important for the designer to avoid concentrated foundation loading, which can result in uneven settlement and cracking of the structure.

Incineration. Incineration is a controlled combustion process for burning solid, liquid, or gaseous combustible wastes to gases and to a residue containing little or no combustible material. In this regard, incineration is a disposal process because incinerated materials are converted to water and gases that are released to the atmosphere. The end products of municipal incineration, however, must be disposed of. These end products include the particulate matter carried by the gas stream, incinerator residue, siftings, and process water. Incinerator residue consists of noncombustible materials, such as metal and glass, as well as combustible materials not completely consumed in the burning process.

The advantages of incineration are numerous, especially where land within economic haul distance is unavailable for disposal of solid waste by the sanitary landfill method. A well-designed and carefully operated incinerator may be centrally located and has been found acceptable in industrial areas so that haul time and distance can be shortened (Fig. 6). The solid waste is reduced in weight and volume, and the residue produced can be nuisance-free and satisfactorily used as fill material. In a properly designed incinerator, the operation can be adjusted to handle solid waste of varying quantity and character.

An incinerator requires a large capital investment, and operating costs are higher than for sanitary landfill. Skilled labor is required to operate, maintain, and repair the facility. Thus capital and operational costs must be compared with the costs of alternate disposal methods, and full consideration must be given to the effects of the methods on the community and its neighbors.

Oversized or bulky burnable wastes (logs, tree stumps, mattresses, large furniture, tires, large signs, demolition lumber, and so on) usually are not processed in a municipal incinerator since they are either too large to charge, burn too slowly, or contain frame steel of dimension and shape that could foul grate operation or the residue removal systems. A few incinerators include grinding or shredding equipment for reducing incinerable bulky items to sizes suitable for charging. In recent years, special incinerators have been designed and constructed to handle portions of bulky, combustible solid wastes without pretreatment. Unless these materials can be incinerated, their bulk and abundance will add greatly to the amount of land necessary for final disposal. Other discarded large items, such as washing machines, refrigerators, water heater tanks, stoves, and large auto parts, cannot be handled by incineration and they too add considerable volume to a fill.

Stringent air-pollution restrictions will require the installation and maintenance of air-pollution control devices on municipal incinerators. The best means of preventing air pollution is to provide efficient combustion; that is, adequate grate surface and combustion air, proper temperature, and constant operation. Since even the well-designed, well-operated incinerators may occasionally pro-

Fig. 7. Although composting is a minor method of solid-wastes disposal, it has the advantage of permitting use of organic wastes as a soil conditioner. Here a mechanical device is turning a compost heap to aerate it and thereby promote the process of biological degradation of the organic wastes. (*U.S. Environmental Protection Agency*)

duce particulates or odor because of uncontrollable variations in the refuse, most incinerators will include fly ash (airborne particles) removal systems. The larger particles are removed by means of screens or baffles, and slow the velocity of the gases so that large airborne particles fall out. To reduce the particulate load even further, water sprays or water-wetted impingement baffles may be installed. Cyclones (devices which remove particles through centifugal force) may also be used for removal of fine particulates. Electrostatic precipitators are used in some incinerator plants.

Composting. Composting is a method of handling and processing solid-waste material which produces a sanitary, nuisance-free humuslike substance, which may be used as a soil conditioner (Fig. 7). Technically composting is a biological degradation process, employing microorganisms to decompose the raw organic materials under controlled conditions of ventilation, temperature, and moisture.

Although many different methods of composting have been developed, there are certain common operations in all:

1. Sorting and separating. This operation is carried out manually or mechanically or both to remove bulky wastes, metals, glass, cloth, cardboard, and other materials that will not compost or that have salvage value.

2. Reduction of particle size. Reduction is achieved by grinding, shredding, rasping, or tumbling, sometimes accompanied by preliminary screening to develop a near homogeneous mass with uniform particle size, which will degrade more rapidly.

3. Composting (or biological decomposition) in the presence of air to prevent anaerobic conditions. Composting is accomplished in either enclosed chambers containing mechanical means of

turning the ground mass and introducing air, or in open rows (windrows) which are turned at intervals to maintain aerobic conditions.

4. Screening and further grinding of the composted material to prepare it for use as a soil conditioner. The larger undesired particles are either rejected or returned to the process.

A basic problem area in composting is development of a market for the product. Compost is not a fertilizer but a soil conditioner which has been marketed primarily to some segment of agriculture. If sold in small bagged lots to the nursery industry or to the homeowner, municipal compost must compete with animal manures, peat moss, and other similar products. Furthermore the quality of the compost has to be more stringently controlled if it is to compete with these organic garden products.

As a method for treatment of solid wastes, composting is not widely used in the United States, and only a very small percentage of the total waste load is subjected to this process.

RECYCLING AND REUSE OF SOLID WASTES

A traditional view of the solid-waste problem focuses on difficulties inherent in collection and disposal operations. However, solid wastes must also be regarded as a problem in the proper management of resources. This difference in point of view could well hold revolutionary implications for those involved in solid-waste management. Increasingly, solid wastes are being viewed as a "resource out of place," to be recovered and reused whenever possible.

If the meaning of this concept is widely applied, it will result not only in conserving natural resources but in reducing to a minimum the amount of waste material that must be ultimately disposed. In the long run we may have no choice but to accept and apply this principle. As waste tonnages accumulate and even greater demands are placed upon the Earth's resources, every indication is that we will run short of both disposal sites and certain mineral and forest products.

The primary barrier to reusing valuable elements in municipal wastes lies in the expense and difficulty of hand separation. Two possibilities exist for overcoming this barrier: separation and segregation of waste materials at the source, and development and improvement of mechanical separation techniques. The first solution would require cooperation on the part of the entire community to place household wastes into separate containers reserved for paper, glass, tin cans, aluminum cans, putrescibles, or other categories. While this method is simple and can be made to work, the suggestion sometimes meets with opposition from disposal service patrons. This approach also increases collection costs, due to the necessity for keeping various classes of wastes separate.

The alternate approach—efficient mechanical separation—is presently the object of considerable research and development effort.

Whether or not an increased fraction of the solid-waste load will be salvaged and reused depends upon the existence of secondary materials markets. Improved technology in the area of separation will have no practical effect unless salvaged materials can be sold and utilized. It is likely that in many areas of the country a potential, but as yet unexploited, market already exists for certain materials.

In some cases it may be necessary to develop hitherto untried uses for solid-waste materials and to stimulate new markets for them. For example, it is possible by a combination of various chemical, physical, and biochemical processes to convert cellulosic wastes, such as paper or bagasse (sugarcane waste), into protein. This source of protein could be a valuable supplement for animal feeds or perhaps even foods for human consumption.

Experimental use of crushed glass as a road paving material is another example of current efforts being made to develop new markets for a common waste material.

While important technological and economic barriers serve to limit the amounts of solid waste that are now recycled and reused, it seems certain that these limits can be greatly extended. However, new technology and imaginative ideas in developing markets for waste materials are needed.

FUTURE

The Federal government, under the Solid Waste Disposal Act of 1965 (P.L. 89-272) as amended, has provided impetus for better solid-waste disposal

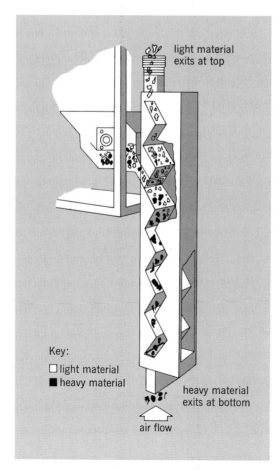

Fig. 8. Schematic diagram of a device for classifying solid wastes in a high-velocity airstream. Light materials, such as paper and metal foil, are carried up through the zigzag column and are thus separated from falling heavy materials. (*Stanford Research Institute*)

planning and has sponsored an extensive research and development program to improve technology in all aspects of solid-waste management.

Fluidized-bed incinerator. One of the most promising research and development projects in the solid-waste field is the CPU-400 now being developed under contract to the U.S. Environmental Protection Agency (EPA) by the Combustion Power Co. of Palo Alto, Calif. Basically the CPU-400 is a fluidized-bed incinerator that burns solid waste at high pressure to produce hot gases to power a turbine, which in turn drives an electrical generator. Municipal solid wastes constitute a better fuel than generally imagined—having a heating value of 2268 Btu/kg, or approximately one-half that of a good grade coal. As designed, the CPU-400 should produce approximately 15,000 kW of electric power while burning 363 metric tons of municipal refuse daily. The generator unit should supply 10% of the electric power requirements of the community providing the refuse, thus offsetting part of the cost of waste disposal. Under the concept envisaged with the CPU-400, refuse haul distances could be greatly reduced by locating the units at strategic points in the urban area. For example, five such units would be required to process all of the refuse from the city of San Francisco, while 40 would be required for New York City. In addition to power generation, up to three-fourths of the heat in the gas-turbine cycle would be available for such auxiliary functions as steam production, drying sewage sludge, or saline water conversion. Developmental work on the CPU-400 also contemplates use of the vacuum produced by the gas turbine. This vacuum should be useful in conveying refuse to the incinerator from collection points in the city through pipes buried in the streets.

Classification of wastes. In an effort to automate the separation of mixed wastes into homogeneous batches for salvage and reuse, EPA has contracted with the Stanford Research Institute, Menlo Park, Calif., for development of an "air classification" process. Waste materials are separated by a high-velocity airstream passing upward through a vertical zigzag column; particles in a waste mixture are separated as a function of density, size, and aerodynamic properties (Fig. 8).

Paper fiber recovery. One of the most advanced systems for recovery of municipal wastes is being demonstrated by the city of Franklin, Ohio, with EPA grant support. The basic technology applied in the system was developed for use in producing paper from pulp. Mixed refuse is fed by conveyor belt to a large tank of water with rotating blades at the bottom. Large and heavy materials are removed from the bottom of the tank and passed under an electromagnet, which separates the ferrous metals. The water slurry which contains the smaller and lighter materials is passed through a battery of screens and centrifuges, which extract cellulose fiber—for use in making paper—and a separate mixture of glass, aluminum, and other nonferrous metals.

An additional step will involve extraction of glass from this gritty mixture, with separation into various colors by an optical sorting device. The remaining mixture has a relatively high percentage of aluminum, which has potential for reclamation by the aluminum industry.

Conclusion. It is unlikely that a research "breakthrough," affording some universally applicable solution to the solid-waste problem, will ever be achieved. The varying nature and quantities of wastes generated in different locations and the economic differences prevailing in different parts of the country will require varying approaches to solution. Ideally, solid-waste systems should allow for maximum salvage and reuse of waste materials, with hygienic and pollution-free collection and disposal for the remaining fraction of the waste load.   [RICHARD D. VAUGHAN]

Risk Evaluation in Engineering

Risk is unavoidable in most engineering operations. There is almost always a small probability that the forces of the environment will combine, perhaps in unusual ways, to damage a structure or impede its operation. Very few residential houses, for example, are designed to withstand the direct onslaught of a tornado. Nor are offshore oil drilling platforms intended to resist the battering force of a barge hurled against the platform by a hurricane. Even when the engineering loads are not so extreme, it may not be economically feasible to reduce the risk to negligible proportions.

Leon E. Borgman is a professor of geology and statistics at the University of Wyoming and a consultant on storm risk in coastal engineering activities. Currently he is conducting research on the maximum wave heights in storms with time-varying intensity and on various problems in mathematical geology and mining operations research.

Evaluating such risks will never be an exact science. Nevertheless, mathematical techniques are available for evaluation, and when they are used skillfully, they can give a workable analysis of the trade-offs among risk, economics, and a variety of other factors. Risk evaluation is therefore becoming an integral part of engineering design in such critically important services as communication and transportation, as well as in extraction of natural resources.

In order to cope with the thorny questions implicit in predicting risk and establishing acceptable risk levels, the engineer must develop many skills, including perhaps some ability in extrasensory perception. After estimating the probability of occurrence of various events hazardous to the engineering design, the engineer must use the rules of probability to construct a mathematical model for the probability of failure. From this he can predict the financial or human loss that might result from the various design decisions from which he can choose. Finally he must convince upper management that the risk levels he has incorporated into his planning are reasonable and acceptable. Management naturally prefers to take no risk. If risk is unavoidable, managers correctly and quite properly wish to examine very carefully the alternatives available, with their associated cost and risk.

A variety of topics will be outlined as an introduction to the techniques of risk evaluation. These will include a test to measure your ability to estimate probabilities, a brief introduction to probability concepts, an outline of some elementary risk models and some of the typical measures of risk such as return period and encounter probability, the basic elements of the statistics of extreme values, structural design, and gaming techniques, including a game illustrating some of the risk decision situations in offshore oil production.

APTITUDE TO GUESS PROBABILITIES

Although problems in risk evaluation in engineering are based on carefully scrutinized data and are treated mathematically, the psychological makeup of the estimator can influence the results. The subjective judgment of probability values is a remarkably complex psychological process and is influenced by the age, cautiousness, sex, and background of the person making the judgment. Since it is instructive and somewhat amusing to learn what your own bias is in risk estimation, a general test of probability guessing follows.

There are four answers to each of the 10 multiple-choice questions, and hence, on the average, you might expect to answer one-fourth of the questions correctly by guessing. If you get substantially more than 2 correct, you are exhibiting good judgment. If you get more than 5 correct you are doing very well, and 8 or more correct would indicate fantastic ability. A score substantially lower than 2 might be taken as an indication of an overly conservative or liberal bias.

The test should not be taken too seriously, of course, since there is a high element of chance involved in many of the questions. In any event, the results of your efforts may provoke some thoughts regarding the difficulty of subjectively estimating probabilities.

Answers are given at the end of this section.

1. Assuming all locations in the United States are equally likely, what is the probability of an airplane crashing within 50 ft of a given location (your home, for example) during a 20-year interval? (a) 0.0004. (b) 0.000 0004. (c) 0.000 000 0004. (d) 0.000 000 000 04.

2. Consider a person who received his bachelor's degree in the 1967–1968 academic year. What is the probability that his degree was in an engineering field? (a) 0.056. (b) 0.012. (c) 0.13. (d) 0.25.

3. What is the probability that a United States citizen is a California resident? (a) 0.30. (b) 0.088. (c) 0.053. (d) 0.22.

4. What is the probability that a United States citizen is 21 years old or older? (a) 0.75. (b) 0.42. (c) 0.60. (d) 0.18.

5. Assuming that the national risk average holds, what is the probability that a person traveling 10,000 mi a year in an automobile will be killed in an auto accident during that year? (c) 0.011. (b) 0.0073. (c) 0.0012. (d) 0.00024.

6. What is the probability of a 15-year-old boy selected at random in the United States being married? (a) 0.005. (b) 0.001. (c) 0.0005. (d) 0.0001.

7. If a person flies 100,000 mi in 1971 on a commercial scheduled airline in the United States, what is the probability of his being killed in an airline crash during that year? (a) 0.0031. (b) 0.0009. (c) 0.0003. (d) 0.00005.

8. A person is selected at random from the world's population. What is the probability that he speaks English? (a) 0.44. (b) 0.092. (c) 0.055. (d) 0.009.

9. If your location in the United States is randomly selected, what is the probability that you will be bitten by a venomous reptile in 1972 and require medical attention? (a) 0.003. (b) 0.0003. (c) 0.00003. (d) 0.000 003.

10. What is the probability that a person in the United States, picked at random, will be killed by a tornado in 1972? (a) 0.000 064. (b) 0.000 00064. (c) 0.000 000 064. (d) 0.000 000 0064.

Answers: 1b, 2a, 3b, 4c, 5d, 6b, 7c, 8b, 9c, 10b.

ESSENTIAL CONCEPTS OF PROBABILITY

Probability is an idealization of the relative frequency with which an event occurs. If enough data are taken, probabilities can be estimated with fair accuracy. However, only an infinite amount of data will suffice for an exact determination.

Because it is impossible to make an infinity of observations, mathematicians have shied away from the definition of probability as relative frequency. Instead, they define it as a set of quantities in a probability model which have the properties of relative frequency and which must be estimated in any particular application. The properties required are that (1) the probabilities are positive, (2) the probability of an event which is sure to happen is 1.0, and (3) if two events can never logically happen together (like the roll of a pair of dice yielding 7 and 6 simultaneously), then the probability of one or the other occurring is equal to the sum of the probabilities of the two events. These three postulates together with the techniques of set theory lead to many additional properties, such as (4) probabilities are always numbers between zero and one, and (5) the probability of an impossible event is zero.

In most applications, attention is centered on numbers rather than events; that is, some quantity is measured and it is different in each observation. Such randomly varying numbers are called random variables. The probability that a random variable X is less than or equal to some specified number x is called the distribution function for X and is denoted by $F_X(x)$. The subscript indicates the random variable, and the argument, x, is the specified number. The distribution function is usually plotted on graph paper as a function of x (Fig. 1).

From its definition, it can be shown that $F_X(x)$ is a nondecreasing function of x as x increases, that $F_X(-\infty) = 0$, and that $F_X(\infty) = 1$.

Probability can be visualized as the load on a horizontal beam in engineering mechanics. The distribution function gives the load at or left of the point x on the beam. The load on a beam can be point loading or continuous loading. In probability, point loading corresponds to a random variable that assumes only discrete values, such as the total number of spots turning up in the roll of two dice (in which case the possible values are the integers from 2 to 12). The usual way to specify the probabilities for a discrete random variable is to list the set of numbers the random variable may assume, together with the probabilities of each number happening. Sometimes the probabilities can be represented by a formula rather than a list. A good example is the Poisson probability function, Eq. (1), where $x = 0, 1, 2, 3, \ldots$, $p_X(3x)$ is the probability the random variable X is equal to x, t denotes time, and λ is constant.

$$p_X(x) = e^{-\lambda t}(\lambda t)^x / x! \qquad (1)$$

Continuous loading on a beam is usually treated by working with the load density (load per foot of length). Similarly, for continuous random variables, the distribution function can be differentiated with respect to x to obtain a probability density, $f_X(x)$. If $\rho(x)$ is the load density on the beam, then $\rho(x)dx$ is the load in a small interval of length dx. Similarly $f_X(x)dx$ is the probability that an observation of X will assume a value somewhere within the dx interval. A good example of probability density is the exponential density in Eq. (2).

$$f_X(x) = \begin{cases} \lambda e^{-\lambda x} & \text{for } x \geq 0 \\ 0 & \text{for } x < 0 \end{cases} \qquad (2)$$

Mean and variance. In engineering mechanics the concepts of the first moment (the center of mass) and the second moment about the center of mass (the moment of inertia) are extremely helpful. In probability and statistics, similar concepts are no less valuable. Here the first moment is called the mean, while the second moment about the mean is called the variance. The square root of the variance is the standard deviation. These concepts are expressed in symbols in Eq. (3) for continuous random variables, and in Eq. (4) for discrete random variables.

$$\text{Mean} = \mu = \int_{-\infty}^{\infty} x f_X(x)\, dx$$

$$\text{Variance} = \sigma^2 = \int_{-\infty}^{\infty} (x - \mu)^2 f_X(x)\, dx \qquad (3)$$

$$\text{Standard deviation} = \sqrt{\sigma^2}$$

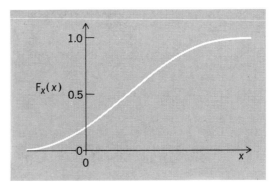

Fig. 1. Distribution function.

$$\mu = \sum_x x P_X(x)$$
$$\sigma^2 = \sum_x (x - \mu)^2 p_X(x) \qquad (4)$$

With these relations, it can be shown that the mean and variance of the Poisson formula are both λt, while the mean and variance of the exponential density are $1/\lambda$ and $(1/\lambda)^2$, respectively.

The sample moments corresponding to these theoretical quantities are given in Eq. (5), where $x_1, x_2, x_3, \ldots, x_n$ denote the observations in the sample.

$$\bar{x} = \frac{1}{n} \sum_{i=1}^{n} x_i$$
$$s^2 = \frac{1}{n-1} \sum_{i=1}^{n} (x_i - \bar{x})^2 \qquad (5)$$

Joint distribution function. Two events are said to be statistically independent of each other if the probability that both occur is the product of their separate probabilities. Two random variables X and Y are said to be independent if Eq. (6) holds, where $P[\ ]$ denotes probability of the event described within the brackets. The left side of Eq. (6) is called the joint distribution function of X and Y. Hence independence means that the joint distribution function factors into the product of the two separate distribution functions.

$$F_{X,Y}(x,y) = P[X \leq x \text{ and } Y \leq y]$$
$$= P[X \leq x]\, P[Y \leq y] = F_X(x) F_Y(y) \qquad (6)$$

ELEMENTARY RISK MODEL

The Poisson formula, Eq. (1), is often used to represent the probability that x events occur in the time interval t. From the discussion following Eq. (4), the theoretical mean of X is λt. Hence λ is the average number of events per unit time, and λt is the average number in the time interval $(0, t)$.

Suppose that events hazardous to a given engineering operation occur on the average once every r years. Then $\lambda = 1/r$ would be the average number per year. If the operation is planned to continue for a life of L years, the probability of x hazardous events during that time would be as shown in Eq. (7), where $x = 0, 1, 2, 3, \ldots$

$$p_X(x) = e^{-L/r}(L/r)^x / x! \qquad (7)$$

The probability of no events during the life would be as shown in Eq. (8), and the probability of at least one hazardous event would be as shown in Eq. (9), where E is the encounter probability.

$$p_X(0) = e^{-L/r} \quad (8)$$
$$E = 1 - e^{-L/r} \quad (9)$$

What would be the distribution function for the time interval T between the occurrence of hazardous events? The probability that T is less than or equal to t is exactly the probability that at least one hazardous event will occur in the interval $(0,t)$. This is precisely the encounter probability with L replaced by t, given in Eq. (10).

$$F_T(t) = \begin{cases} 1 - e^{-t/r} & \text{for } t \geq 0 \\ 0 & \text{for } t < 0 \end{cases} \quad (10)$$

The probability density for T would be the derivative of $F_T(t)$ with respect to t, Eq. (11), which is the exponential density with $\lambda = 1/r$. Again from the discussion after Eq. (4), the mean and variance of T for this case would be r and r^2, respectively. This conclusion makes good sense intuitively, since it shows that on the average a hazardous event "returns" every r years but the interval between events may vary randomly.

$$f_T(t) = \begin{cases} (1/r)e^{-(t/r)} & \text{for } t \geq 0 \\ 0 & \text{for } t < 0 \end{cases} \quad (11)$$

The quantity r is called the return period for the hazard. Equation (9) relates return period to engineering life, while Eq. (10) relates return period to waiting time. Both relations are illustrated in Fig. 2. The encounter probability for the situation where the life of operation is equal to return period of the hazard is 0.63. Alternatively one can reverse this reasoning and work backward to the value of L/r that gives a preselected encounter probability. For example, for an encounter probability of 0.05, L/r would be 0.051 and the operation therefore should be designed so that it can withstand events with return periods less than $L/0.051$.

At this point in the risk evaluation, the engineer would introduce the distribution function for the random cost associated with the occurrence of a hazardous event and would then develop the probability law for the total cost during the operational life.

STATISTICS OF EXTREMES

Risk considerations almost always involve questions related to the largest or the smallest value of some environmental or engineering variables. Thus the statistical theory of extremes is highly relevant to risk evaluation.

Suppose that each threatening event has a random intensity associated with it. Suppose further that x_m is the most severe intensity measured in the mth year of data at a given location, where $m = 1, 2, 3, \ldots, n$. This gives a string of data, $x_1, x_2, x_3 \ldots, x_n$, with number being the largest intensity for its year. What probability law would be appropriate for the largest intensity in the n-year interval?

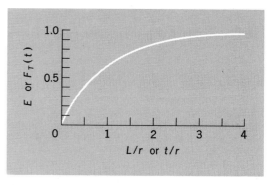

Fig. 2. Relations between return period r, operational life L, encounter probability E, and the waiting time distribution function $F_T(t)$.

If $F_X(x)$ is the distribution function for the largest intensity in a single year and if the maximum intensities for successive years are statistically independent of each other, the question has an easy answer. The maximum n-year intensity M is less than or equal to x if and only if the intensities are less than or equal to x for every one of the n years. Thus Eqs. (12) and (13) hold if it is assumed that the distribution function of X_i does not change from year to year.

Geometric probability. Suppose that a threatening event with an intensity greater than x_0 will be hazardous to the engineering operation. What will be the average return period between years containing hazardous events?

The answer to this question leads to the geometric probability law. Let $p_0 = 1 - F_X(x_0)$ be the probability that any given year will contain a hazardous event. To say that the probability that $(n-1)$ years pass without a hazardous event occurring but that a hazardous event occurs on the nth year would be the same as saying that the waiting time T to the first hazardous event is n years. The probability function for T is therefore as in Eq. (14).

The theoretical mean of T would be, from Eq. (4), that shown in Eq. (15) after some algebraic manipulation involving the geometric infinite series. The symbol $r(x_0)$ has already been used to denote the return period or theoretical average time between years containing hazardous events. Hence, if p_0 is defined in terms of the distribution function $F_X(x)$, Eq. (16) holds.

$$r(x_0) = \sum_{n=1}^{\infty} n(1-p_0)^{n-1} p_0 = (1/p_0) \quad (15)$$
$$r(x_0) = 1/[1 - F_X(x_0)] \quad (16)$$

If $F_X(x)$ is known, the return period for an intensity exceeding any given level can be computed from Eq. (16). Similarly the exact distribution function for the largest intensity in any n-year period can be calculated from Eq. (13). Unfortunately Eq. (13) requires that $F_X(x)$ be known quite accurately

$$P[M \leq x] = P[X_1 \leq x \text{ and } X_2 \leq x \text{ and } X_3 \leq x \text{ and} \ldots \text{ and } X_n \leq x]$$
$$= P[X_1 \leq x] P[X_2 \leq x] P[X_3 \leq x] \ldots [PX_n \leq x] \quad (12)$$
$$F_M(x) = [F_X(x)]^n \quad (13)$$
$$P[T=n] = P[X_1 \leq x_0] P[X_2 \leq x_0] P[X_3 \leq x_0] \ldots P[X_{n-1} \leq x_0] P[X_n > x_0] = (1-p_0)^{n-1} p_0 \quad (14)$$

for large x, and this is precisely where there is usually very little data.

Several techniques have been developed to assist in making predictions for such extremes. All suffer from the basic principle that statistics cannot manufacture information but can only distill relationships from data and make them more obvious and useable. However, an engineer is often in the position of having to use the available data to make a decision immediately. Under such circumstances, the following techniques can be useful.

Approximate methods. Suppose that constants $a_n > 0$ and b_n, for $n = 1, 2, 3, 4, 5, \ldots$ can be found that satisfy Eq. (17).

$$\lim_{n \to \infty} [F_X(a_n y + b_n)]^n = G(y) \qquad (17)$$

Then for large n, Eq. (18) holds.

$$[F_X(a_n y + b_n)]^n \approx G(y) \qquad (18)$$

As a consequence, Eq. (19) holds, so that $G(y)$ can be used as an approximating formula for the distribution of the largest annual intensity.

$$F_M(x) = [F_X(x)]^n \approx G\left(\frac{x - b_n}{a_n}\right) \qquad (19)$$

It has been shown mathematically that there are only three possible functional forms for $G(y)$ that do not degenerate into a step function. These are given in Eq. (20).

$$G_1(y) = e^{-e^{-y}} \text{ for } -\infty < y < \infty$$

$$G_2(y) = \begin{cases} e^{-y^{-a}} & \text{for } y \geq 0 \\ 0 & \text{for } y < 0 \end{cases} \qquad (20)$$

$$G_3(y) = \begin{cases} e^{-(-y)^a} & \text{for } y \leq 0 \\ 1 & \text{for } g > 0 \end{cases}$$

These formulas therefore are natural first guesses with which to approximate the distribution of the largest of a large number of independent random quantities all drawn from a common population.

These asymptotic distributions should be used with caution. Although they are quite useful in many applications, there are many other situations where their premises are not satisfied. It has been shown that lack of uniformity in the parent populations from which the observations arise cause significant deviations from the asymptotic forms. The maximum of a set of observations taken from a time-varying population may converge to other formulas. Needless to say, much remains to be done in this area of research.

Another approach that often leads to excellent asymptotic forms in specific cases has been suggested by the mathematician H. Cramer. First, $W_n(x)$ is defined as in Eq. (21), where $F_X(x)$ and n are given in Eq. (13). Then Eq. (22) holds, and hence Eq. (23) as $n \to \infty$. If the approximate probability function $F_X(x)$ is known, the specific form of Eq. (23) may suggest ways to plot data against observations so that the parameters in the distribution function may be estimated.

$$W_n(x) = n[1 - F_X(x)] \qquad (21)$$

$$F_X(x) = \left[1 - \frac{W_n(x)}{n}\right] \qquad (22)$$

$$F_M(x) = \left[1 - \frac{W_n(x)}{n}\right]^n \approx e^{-W_n(x)} \qquad (23)$$

Straight-line plotting of data against cumulative frequency is one of the more commonly used techniques for extrapolating from the region where there are much data up to larger values of the variate where data are scarce.

RISK ESTIMATION IN STRUCTURAL DESIGN

So far this article has dealt with risk evaluation of a finished design for environmental extremes such as floods or hurricanes. However, probabilistic considerations can be introduced into the design procedure itself. The two basic random variables involved in such a probabilistic design are the resisting ability R of the structural member and the load S imposed by the environment on the structural member. If both R and S are deterministic—if they are known in advance—then the ratio R/S is the safety factor for a structural element with resistance R when the maximum load the element will experience is S. Presumably if R/S is less than one, the element will fail.

But if R and S are random quantities, additional complexities arise. The probabilistic behavior for R/S can be stated mathematically in terms of the probability laws for R and S and derived for various assumed probability distributions. The probability of failure will then be $P[(R/S) < 1.0]$. Alfredo H.-S. Ang and Mohammad Amin have suggested that a better and more conservative criterion for failure should be $P[(R/S) < \gamma]$, where γ is a number greater than or equal to one. The quantity γ takes the place of the conventional safety factor.

The whole area of probabilistic structural design is currently in a state of rapid flux. Probabilistic design codes are being considered carefully, and useful procedures no doubt will arise from all the discussions.

GAMING TECHNIQUES

To be realistic, the models for risk evaluation must be tailored to treat all the features of the engineering operation. Such factors as economics, structural constraints, and maintenance must be taken into account.

In order to compute risk from such realistic but highly complicated models, a "Monte Carlo," or game, approach is sometimes used. In this procedure the engineering situation is set up as a game incorporating the essential details and decisions of the real design operation. Then the game is played again and again following specified rules of operation. The randomness of various factors in the real operation is simulated with random devices, random number tables, or computerized random number generators.

An illustration of the game approach is the Offshore Oilman's Game (devised by Paul Aagaard and presented here with the permission of the Chevron Oil Field Research Co., La Habra, Calif.). This game simulates some of the decision and risk factors involved in offshore oil operations, but it is not intended to be a really comprehensive or realistic management game. The costs, damages, and other figures in the game are fictitious and are only intended to indicate in a general way the various

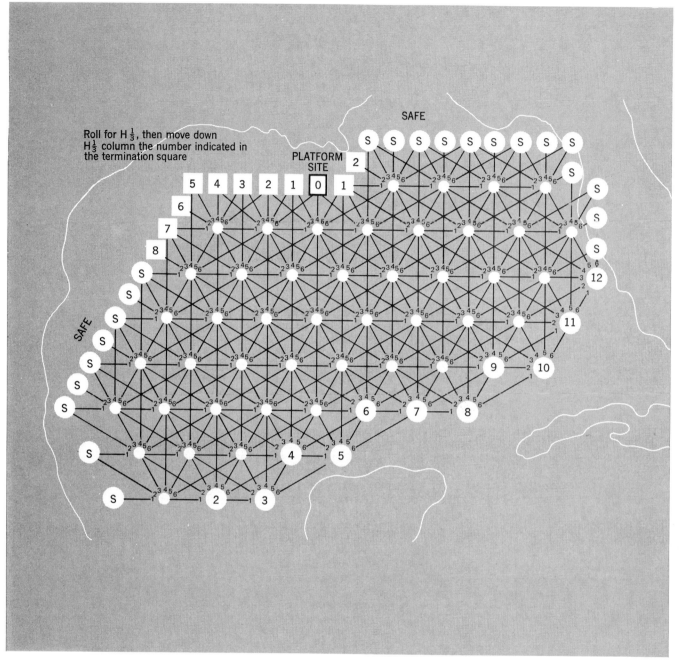

Fig. 3. Hurricane movement chart for the Offshore Oilman's Game. *(Copyright © Chevron Research Co.)*

risk penalties and rewards. The game should not be taken with great seriousness, but should be played for fun with the idea that it illustrates in an elementary form the features of more elaborate computerized gaming procedures.

The objective of the game is to maximize profit for a 20-year period during which three hurricanes each year enter the Gulf of Mexico and threaten the player's 20-well, oil-producing platform, located southwest of New Orleans in the Gulf of Mexico. The player borrows money at a specified interest rate to construct his platform and drill the 20 wells from it by directional drilling. He may build a stronger platform which will withstand more intense hurricanes by paying more money or he can put less money into construction and take greater risk of hurricane damage. His income each year is determined from the roll of a die and the use of an income chart. He can insure any portion of his investment by paying a premium at the beginning of the year.

The hurricane action each year is determined by rolling dice. For the first storm, the player rolls a pair of dice to determine the initial hurricane position in the circles on the map of the Gulf of Mexico in Fig. 3. For example, if he rolls a seven, the hurricane starts in circle 7. He rolls the dice again to determine the initial wind-speed position in the right hand column of the chart shown in Fig. 4. For example, if he rolls a nine, the initial position is the

RISK EVALUATION IN ENGINEERING 41

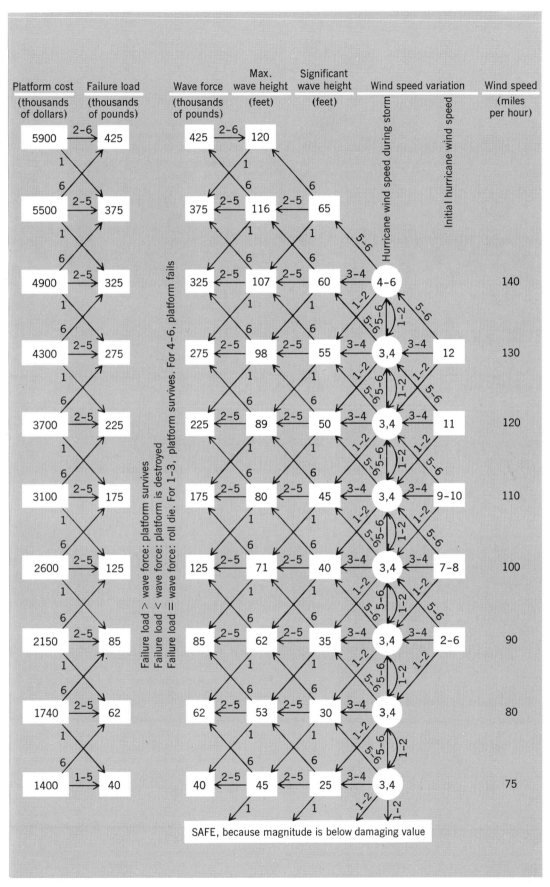

Fig. 4. Structural strength and hurricane wave force chart for the Offshore Oilman's Game. This chart is used in combination with the chart shown in Fig. 3. (*Copyright © Chevron Research Co.*)

42 RISK EVALUATION IN ENGINEERING

Table 1. Income chart for the Offshore Oilman's Game

Year	Die outcome		
	1–2	3–4	5–6
1	1900	2150	2400
2	1750	1900	2150
3	1550	1750	1900
4	1400	1550	1750
5	1250	1400	1550
6	1130	1250	1400
7	1000	1130	1250
8	910	1000	1130
9	820	910	1000
10	740	820	910
11	660	740	820
12	590	660	740
13	530	590	660
14	470	530	590
15	430	470	530
16	380	430	470
17	340	380	430
18	300	340	380
19	270	300	340
20	240	270	300

SOURCE: Copyright © 1971 Chevron Research Co.

Table 2. Cost charts for the Offshore Oilman's Game

SALVAGE COST
If platform is destroyed, roll die for cost of clearing out the wreckage

Die outcome	Salvage cost, dollars
1	800,000
2–5	1,500,000
6	3,000,000

WELL COST
(a) Initial cost: $2,400,000 for 20 wells
(b) If platform is destroyed, roll die for number of wells lost and redrilling cost

Die outcome	Number wells lost	Redrilling cost
1–2	None	
3–5	10 wells	Redrill for $1,200,000 or cut income 50% for wells lost
6	20 wells	Redrill for $2,400,000

REMOVAL COST
After 20 years of production, the cost for removal of the platform to eliminate a navigational hazard is $500,000

INSURANCE COST
Yearly policy is available; the rate is 5% of the insured amount

INTEREST RATES
Interest on money borrowed for platform construction, well drilling, and other purposes

Loan amount, thousands of dollars	Annual interest, thousands of dollars
0–999	60
1000–1999	120
2000–2999	180
3000–3999	240
4000–4999	300
5000–5999	360
6000–6999	420
7000–7999	480
8000–8999	540
9000–10,000	600

SOURCE: Copyright © 1971 Chevron Research Co.

9–10 square, which corresponds to a wind speed of 110 mph.

The player now rolls a single die to determine the circle to which the hurricane next moves on Fig. 3. The hurricane moves along the numbered line corresponding to the number on the die. With each step of the hurricane position on Fig. 3, the player also rolls a die to determine the time-varying wind speed. The wind-speed value moves up and down the second column from the right (the circles) in Fig. 4 according to the outcome of the die roll.

This procedure is repeated until eventually the hurricane either drops to the safe-magnitude square at the bottom of Fig. 4, reaches a safe-location circle (labeled S in Fig. 3), or reaches a threatening location (the numbered square in Fig. 3). If a safe outcome is attained, the procedure is repeated for the second and third hurricanes of the year.

However, if a threatening square is reached, the player must take additional steps to see if his platform is damaged. To do this, he rolls a die to determine the significant wave height, or $H_{\frac{1}{3}}$, which is defined as the average of the highest one-third of the waves in the storm. According to the roll, the player moves from the column of circles to the $H_{\frac{1}{3}}$ column in Fig. 4. The resulting value is the most severe $H_{\frac{1}{3}}$ value for the hurricane.

To correct this value for the distance from the storm center to the platform, the player moves down the $H_{\frac{1}{3}}$ column by the number of squares specified in the terminal location box for the storm on Fig. 3. After determining the $H_{\frac{1}{3}}$ value at the platform site, the player rolls again and follows the line indicated by the number on the die to discover the maximum wave height at the platform site. Finally he rolls a die and follows the indicated line to get the wave force against the platform produced by the waves.

The failure load which will destroy the platform is determined mainly by the platform cost initially selected by the player. However, the actual strength varies somewhat from the design values. Hence the player rolls a die and moves from the cost column at the left of Fig. 4 to the appropriate square in the second column from the left to obtain the failure load. If the wave force exerted by the hurricane is greater than the failure load, the platform is destroyed. If the wave force is less than the failure load, the platform survives. If the wave force equals the failure load, the die is rolled again. If the outcome is one, two, or three, the platform survives. If the die yields four, five, or six, the platform is destroyed.

If the platform survives, the next hurricane is initiated. If the platform survives all three hurricanes, the annual income from the oil production is determined by the year number and a roll of a die according to the income chart in Table 1. The player pays interest on his indebtedness and pays his insurance premiums if he wishes insurance, and the next year's experience commences.

If any of the hurricanes destroy the platform, the salvage cost for removing the wreckage is computed by rolling a die and consulting the cost charts in Table 2. Similarly the number of wells lost and the cost of redrilling is determined from the roll of a die and the well cost section of Table 2. If the player is insured, he can collect the amount he is insured for, but not more than the amount of damages experienced. The player can rebuild his destroyed platform to withstand greater or smaller loads if he wishes by paying the appropriate construction costs.

At the end of the 20 years, the oilfield has been pumped dry and the player must pay $500,000 to remove the well from the shipping lanes so that it will not be an unnecessary navigational hazard. At this point, the player may be quite wealthy (on paper, at least) or he may be deep in the red. In any case, it is hoped that he will have developed a greater understanding of the risk complexities in a large engineering operation.

[LEON E. BORGMAN]

Energy Sources in Galaxies and Quasars

A normal galaxy is composed mainly of stars, and its radiation is almost entirely the summation of optical radiation from these stars. In recent times, however, astronomers have discovered that certain galaxies transmit other radiation in similar or even greater amounts: Some galactic nuclei emit much optical radiation that is not ordinary starlight; radio galaxies emit much energy in radio frequencies (often from regions devoid of visible matter); some galaxies radiate principally in the infrared; and some radiate surprisingly large amounts of energy in ultraviolet and x-ray frequencies. Quasi-stellar objects (also called quasars or QSOs) apparently emit still greater amounts in some or all of the same frequency ranges.

William H. McCrea is research professor of theoretical astronomy at the University of Sussex, England. He has been a visiting staff member at various astronomical centers in the United States and has worked in many branches of theoretical physics and astrophysics.

Table 1. Examples of emission rate and energy output*

Type	Example	Age or lifetime, years	Mass, g	Predominant emission	Emission rate, J/sec	Energy output, J
Large spiral galaxy	Milky Way	1.5×10^{10}	2.2×10^{44}	Mainly optical	4×10^{36}	2×10^{54}
Large elliptical galaxy	Virgo A (M 87)	1.5×10^{10} (as optical galaxy)		Optical Radio X-ray	18×10^{36} 0.05×10^{36} 0.5×10^{36}	8×10^{54}
Strong radio galaxy	Cygnus A	10^6 to 10^9		Radio	50×10^{36}	1.5×10^{51} to 1.5×10^{54}
X-ray "galaxy"	Sources in Coma and Perseus associations			X-ray	10×10^{36} to 20×10^{36}	
Seyfert galaxy	NGC 1068	10^9		Infrared	10×10^{36}	0.3×10^{54}
Quasar	3C273	10^6		Optical Radio X-ray Infrared	1000×10^{36} 2500×10^{36} 300×10^{36} $20,000 \times 10^{36}$	10^{54} at observed emission rate; 0.01×10^{54} if mean rate is 1% of observed rate; 0.1×10^{54} if mean rate is 10% of observed rate

*Approximate estimates expressed in round numbers using solar mass = 2×10^{33} g and solar luminosity = 4×10^{26} J/sec; 1 year ≈ 3×10^7 sec.

Fig. 1. Spiral galaxy M 64. The dark region is produced by dust such as may cause infrared radiation from some galaxies. (*Hale Observatories*)

In this article I shall discuss possible sources of large amounts of such "nonstellar" radiation from galaxies and of radiation in general from quasars, the sources almost certainly being similar in both of these categories. In assessing the energy, I shall consider observations in all frequency ranges. However, in discussing possible sources, I shall focus attention on the possible mechanisms for producing the required total energy. When energy is liberated in any of the suggested ways, it may be assumed that it will sooner or later escape from the system mostly as electromagnetic radiation; however, we will not here delve into details of the particular processes by which it does so.

EMPIRICAL EVIDENCE

The first step toward understanding nonstellar sources is to acquire a knowledge of the kinds of radiation they emit. Fortunately, observations of such galaxies have yielded much information on which to base quantitative estimates of the rates of energy emission, as summarized in Table 1. Qualitatively, such information enables us to characterize the various types of galaxies as follows.

Optical galaxy. Although normal galaxies radiate mostly in optical frequencies, measurements made from NASA's *Orbiting Astronomical Observatory 2* show that some emit with unexpected strength in the ultraviolet, and measurements from other instruments carried to great heights show that some galaxies emit strongly in the infrared. In many such cases, astronomers believe, interstellar material merely redistributes some of the stellar radiation through the spectrum; the total energy is still closely equal to the sum of the stellar contributions. A photograph of spiral galaxy M 64 illustrates this possibility (Fig. 1).

In addition, various components of an ordinary galaxy emit in radio and x-ray frequencies and

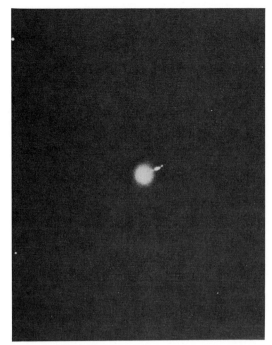

Fig. 2. Left photo shows giant elliptical galaxy M 87, which is identified with the radio source Virgo A (*Hale Observatories*). Right photo is a shorter exposure of M 87 showing the nuclear jet (*Royal Astronomical Society*).

produce gamma rays and cosmic rays, but without making significant contributions to the total energy output. In other sorts of galaxies, or other states of a galaxy, some of these emissions may predominate.

Radio galaxy. Many discrete radio sources are associated with objects similar to elliptical or irregular galaxies, but in a disturbed state. Such an object is called a radio galaxy. Its radio emission may come from compact regions in its central parts, or from diffuse regions outside its visible parts. Figure 2 shows two views of a radio galaxy, the giant elliptical galaxy M 87. In the lower view, a shorter exposure than the top view, a nuclear jet is evident. This jet apparently is the main source of the radio emission. Figure 3 shows the radio galaxy Cygnus A.

Typically the energy of radio emission from a radio galaxy is comparable with the optical emission. The radio spectrum is usually such that, within most of the frequency range concerned, the intensity varies as some negative power of the frequency. It is interpreted as due mainly to synchrotron emission, in which relativistic particles (those moving at nearly the speed of light) interact with a magnetic field, giving off radiation.

Infrared galaxy. There appear to be some galaxies for which about 80% of the emission is in the far infrared. It is not known whether these galaxies are extreme cases of the sort of infrared emission from some normal galaxies, or some different type of object producing synchrotron radiation or inverse Compton radiation (produced when relativistic particles give up some of their energy to long-wavelength radiation, converting it to a shorter wavelength) predominantly in the infrared part of the spectrum.

X-ray galaxy. Astronomers working with the NASA *Uhuru* satellite recently have discovered

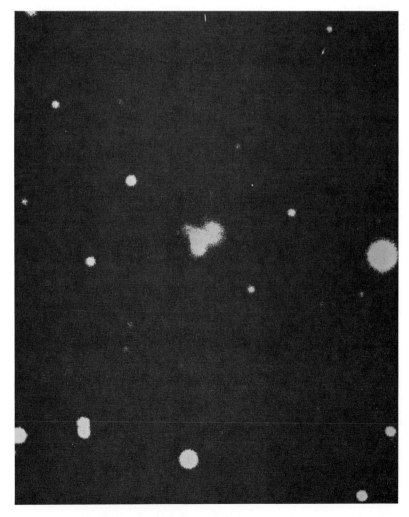

Fig. 3. Cygnus A radio galaxy. (*Hale Observatories*)

Fig. 4. Seyfert galaxy NGC 4151. (*Hale Observatories*)

sources of x-radiation emitting at "galactic" rates. These sources were correlated with associations of galaxies; it is not yet known what particular objects are responsible.

Seyfert galaxy. A Seyfert galaxy is a spiral galaxy having a compact nucleus that is many times more luminous than the central part of an otherwise similar spiral (Fig. 4); the nucleus emits about as much energy as the whole of a galaxy like the Milky Way. It is estimated that 1 or 2% of all spirals, which they resemble except for the characteristic nucleus, are Seyferts. It is tentatively thought that a Seyfert galaxy is a temporary state of a certain type of spiral, a state which endures for about 10^9 years.

The continuous optical spectrum of a Seyfert nucleus is different from that of a normal galaxy; it is much flatter, and could be largely optical synchrotron radiation. Some Seyferts emit most strongly in the infrared. Some, but not all, give fairly strong radio emission.

The astonishing discovery about Seyfert nuclei is that the continuous emission is variable, at least in some cases. For example, NGC 4151 has been found to fluctuate by about a factor 4 in a time of about 1.5 years. Thus the amplitude of the change in this short time is of the order of the luminosity of the entire Milky Way galaxy.

Quasar. It is estimated that there are about 10^7 quasars within range of existing optical telescopes; a few hundred have been observed individually in some detail (Fig. 5).

A quasar possesses all or most of the following properties, the combination of which distinguishes it from other astronomical objects.

1. It produces a starlike image on a photographic plate, with a trace of nebulosity or jet in a very few cases.

2. It has a very flat optical continuous spectrum, which appears to be fairly standard for a number of cases. (Because of different redshifts—which will be discussed—this spectrum is known into the far ultraviolet.)

3. It exhibits strong radio emission in about one percent of the cases. Most such sources are quite compact. (Optically similar quasars may differ greatly in radio power.)

4. It produces a radio spectrum generally similar to that of a radio galaxy.

5. It emits strongly (sometimes most strongly) in the infrared in some cases.

6. It emits strong x-rays in one case and presumably in others.

7. Its optical emission may fluctuate with a characteristic time ranging from several years down to a few days, and with amplitude ranging from tenths of a magnitude up to over a magnitude. Also there may be some secular (long-term) changes.

8. Its radio emission may fluctuate with a characteristic time of months, along with apparent secular changes.

9. Its optical and radio variability are uncorrelated in any detail.

10. Its optical line spectrum exhibits many characteristic features in emission and absorption.

From such properties, an "anatomy" of a typical quasar can be inferred. Most of the continuous optical emission comes from a single source, or a small number of sources, each no more than the order of a light-month in linear extent. The emission line spectrum comes from tenuous filaments in the vicinity of the sources, but of much greater extent. The absorption spectrum is probably produced by shells or jets moving inward or outward from the sources. The radio emission must come from a relatively extended plasma (with localized concentrations) carrying a magnetic field. Relativistic electrons injected into this magnetic field lose energy in the form of synchrotron radiation and through inverse Compton interaction with such radiation. The x-ray emission probably comes mainly from repeated Compton scattering. The infrared emission also may be synchrotron or inverse Compton radiation, or else the system may contain large dust clouds that reemit in the infrared.

Most studies suggest that a quasar, as such, has a lifetime on the order of 10^6 years. This means that a system rises from relative obscurity until it emits energy at a greater rate than any other sort of object, and returns to obscurity after about 10^6 years of splendor. It is then an economy of hypothesis to suppose that the life history is about the same for every quasar and that the great variety of intrinsic properties results from seeing quasars at all stages in their careers. This hypothesis leads to the inference that the luminosity rises to a very sharp peak, the mean over the lifetime being only a few percent of the peak value.

REDSHIFT

It is well known that the line spectra of the various kinds of celestial objects show shifts of various amounts toward the red. In the case of optical galaxies these redshifts lead to Hubble's law (redshift is correlated with apparent magnitude in such a way that when redshift is translated into recession speed and apparent magnitude into distance, the recession speed is found to be nearly proportional to the distance) and the notion of the expansion of the universe. The interpretation of redshift in accordance with Hubble's law is called cosmological.

In the case of radio galaxies there is not much systematic information, but there is no reason to doubt the cosmological interpretation. In the case of quasars, the matter is more controversial, but no satisfactory alternative to the cosmological interpretation is forthcoming. In this article the cosmological interpretation is therefore used as a working hypothesis. The properties quoted for quasars are derived accordingly by using, when required, the value $H_0 = 75$ km s^{-1}Mpc^{-1} for Hubble's constant.

One of the main aims of studying the energy emission of quasars is to understand their intrinsic properties well enough ultimately to obtain distance estimates that are essentially independent of the redshift.

COMPARATIVE PROPERTIES

Values such as those listed in Table 1 show that the sequence

optical galaxies — radio galaxies — Seyfert nuclei — quasars

is a sequence of increasing rate of energy output, although with considerable overlap from one category to the next. When objects are ordered in this way, the variation along the sequence of any particular intrinsic property is found to be fairly continuous. This continuity provokes the following three remarks.

1. If the cosmological interpretation of the redshifts were mistaken, this continuity would be lost.

2. The sequence strongly indicates that systems in the various categories are all manifestations of the same set of phenomena, although the relative importance of each phenomenon varies considerably among the categories. The sequence is not a simple evolutionary one, but it does not exclude the possibility that a particular object might be classed in more than one category at different stages in its career.

3. If this comparison of properties is significant, quasars must belong to the general class of "galaxies." The occurrence of quasars should then be correlated with the clustering of galaxies, but the evidence of this remains inconclusive.

TOTAL ENERGY OUTPUT

The only way to estimate total energy requirements of a source is to estimate the times for

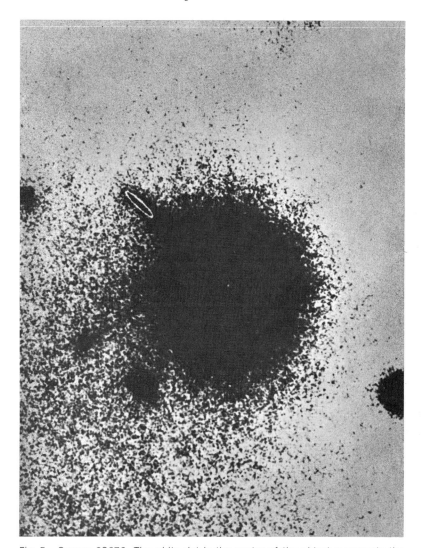

Fig. 5. Quasar 3C273. The white dot in the center of the object represents the quasar component B; white oval over the jet represents component A. The object is overexposed in order that the jet may be seen. (*Royal Astronomical Society*)

which the observed emission rates must endure. (Something may be inferred about the energy content at any epoch that permits the observed energy escape, but this information helps only a little.) Such estimates are given in Table 1.

Astrophysics shows that an optical galaxy must have maintained approximately its observed energy output for 10^{10} to 2×10^{10} years. If there were a large energy release in the process of forming the galaxy, during that phase it would be classed in some other category. In the case of a radio galaxy, the life can be estimated only from theories of the processes involved. The size attained by the system may give a way of checking on the estimate to some extent. The estimated mean life of Seyfert nuclei, as we have seen, is 10^9 years.

Quasars present the most uncertain case. The mean life could scarcely be less than 10^5 years; otherwise there would be an embarrassing number of extinct quasars. Most theories give a mean life of about 10^6 years, which is plausible but is not entirely independent of the estimate of energy requirements. If the most powerful quasars maintain their output for about 10^6 years, the energy re-

Table 2. Nuclear energy output of a galaxy according to various hypotheses*

Hypothesis	Fraction of total mass converted to energy	Energy output, J
1. Usual estimate for nuclear processes in stars in past life of galaxy	2×10^{-4}	$4 \times 10^{54} M/M_G$
2. All existing helium, etc., synthesized in galaxy	2.8×10^{-3}	$6 \times 10^{55} M/M_G$
3. Maximum possible from nuclear fusion	10^{-2}	$2 \times 10^{56} M/M_G$
4. Total annihilation	1	$2 \times 10^{58} M/M_G$

*M = mass of galaxy, M_G = mass of Milky Way galaxy $\approx 2.2 \times 10^{44}$ g.

quirement is very great. However, to allow for variability and possibly anisotropic emission, a more significant estimate is obtained by averaging over the emission rates of all quasars of relatively small redshift (so as to include the intrinsically faint objects) and multiplying by 10^6 years.

ENERGY SOURCES

Known nuclear processes could supply most of the energy requirements for galaxies. Although gravitational collapse in one form or another could draw upon an adequate energy store, it is doubtful whether the required initial conditions would ever be realized. More novel speculations offer little help. We will now consider each of these possible sources in greater detail.

Nuclear energy. In the conversion of protons into alpha particles, about 0.7% of the mass goes into energy; the greatest amount of energy that can be obtained by fusion processes from a given mass of matter (protons) is the energy equivalent of just under 1%. The energy equivalent of 1 g $\approx 9 \times 10^{13}$ joules (1 erg = 10^{-7} J).

Table 2 gives energy estimates for several hypotheses in regard to nuclear processes. It is virtually certain that a galaxy like the Milky Way has produced the amount of energy given for hypothesis 1 in Table 2. If effectively all the helium and heavier elements were synthesized in the Galaxy (and were not already in its raw material), the object would have produced the amount for hypothesis 2; but most of this would have been liberated in an early stage at a rate much higher than that for a normal galaxy. It would then be probable that there are some cases in which there was an even greater production of helium and heavier elements, so that the amount for hypothesis 3 would be approached; no nuclear process in ordinary matter is known that could give more. Standard physics cannot admit any process that gives more than that for hypothesis 4.

An object in a given category could not suffer the annihilation of a large part of its mass, even if a process were available that could do this, because its category would be altered. For such an object, then, the gap between the best that nuclear fusion could achieve and the utmost that is compatible with standard physics is no more than a factor of about 50. Since uncertainties in distances, masses, and other factors are so considerable, we therefore cannot assert simply that some particular system indubitably demands more energy than can be produced by known nuclear processes. Moreover, the estimates in Tables 1 and 2 show that the total energy requirement of any object may perhaps not exceed that energy known to have been liberated already by nuclear reactions in any large galaxy.

Gravitational energy. Given mass M of some standard material forming a body having characteristic linear dimension R, according to classical physics, the ratio of the gravitational energy liberated in attaining this state to the nuclear energy that could be extracted from the body is a constant multiple of M/R. Two examples of this principle are given in Table 3. Here W/E is the ratio of lost gravitational energy to maximum possible energy from nuclear fusion; a solar mass (1 $M_\odot$) $\approx 2 \times 10^{33}$ g; and a solar radius (1 $R_\odot$) $\approx 7 \times 10^{10}$ cm. If masses 1 and 10^7 $M_\odot$ form uniform spheres of equal density 1.4, then in the first case (the Sun) the lost gravitational energy is minute compared with the available nuclear energy, while in the second case the gravitational is much greater than the nuclear energy. Provided that R does not vary too much, gravitational energy can always be made to exceed nuclear energy simply by taking a large enough mass of material. Relativity theory ensures that the gravitational energy shall never exceed the energy liberated by complete annihilation.

About a century ago it was suggested that the energy of solar radiation is supplied by the Sun's gravitational contraction. But it was then realized that this could not sustain the luminosity for even 1% of the Sun's past life. A more prolific source had to be discovered and thermonuclear energy generation met the need. However, the situation may be reversed for a sufficiently large mass. In fact, three main mechanisms have been proposed for gravitational sources of radiation: collapse, "black hole," and cluster processes.

Collapse. If a body of about solar mass or less is under no influence other than its own gravitation (and if nuclear energy generation is ignored), the body will contract. It will contract slowly through a sequence of states of near equilibrium. About half the lost gravitational energy is radiated away, with the rest going to heat up the material. In any such state, if a slight squeeze were applied over the boundary of the body, it would, of course, resist and simply adopt a slightly modified configuration. If the mass exceeds about 10^7 $M_\odot$, the body at first contracts as before until it reaches a stage when the internal pressure is mainly radiative. According to classical theory, if a slight squeeze were then applied, the body would contract indefinitely without offering appreciable resistance. According to relativity theory, however, the more the body is squeezed, the less able it becomes to offer resistance. The body therefore would go into a state of collapse, instead of contraction through quasi-equilibrium states. The body loses energy, and its

Table 3. Comparison of gravitational and nuclear energy

Mass	Radius	Density	W/E
1 $M_\odot$	1 $R_\odot \approx 7 \times 10^{10}$ cm	1.4 g cm^{-3}	1.3×10^{-4}
10^7 $M_\odot$	1 astronomical unit $\approx 1.5 \times 10^{13}$ cm	1.4 g cm^{-3}	6

mass equivalent, at an enhanced rate. Ultimately the mass remaining at any epoch will be almost within its Schwarzschild boundary (the sphere of smallest radius permitted by relativity theory for the mass concerned). In course of time, any positive amount of the initial proper mass presumably could be radiated away. However, little is known about the rate of this process.

There are serious difficulties with the collapse mechanism. It is implausible to suppose that in nature a body of the required mass could have such small angular momentum as to permit indefinite contraction without rotational breakup. In fact, it appears to be the destiny of any such large mass of material to resolve itself into bodies of stellar mass—not to form a single body.

Black hole. Because of the difficulties with the collapse mechanism, it is natural to consider the gravitational properties of a system composed of more than one body, in fact consisting of material falling on to a central gravitating body of mass M.

Given M, relativity theory imposes a minimum radius R_0, such that an infalling particle would strike the surface with the speed of light; the surface is precisely the Schwarzschild boundary mentioned above. There seems to be no objection, in principle, to assuming that the body has a radius, perhaps not precisely R_0, but only slightly larger. If there is diffuse material in the vicinity, this will tend to fall into the central body and, if the energy of infall can be converted into radiation, nearly all the proper mass of the material would reappear as radiant energy, with little addition of mass to the central body.

A more feasible mechanism, however, takes into account the fact that portions of the diffuse material tend to collide with each other, thus losing energy and becoming more closely bound gravitationally to the central mass. According to general relativity, there is a circular orbit of radius $3R_0$ which is the orbit of lowest energy. For a particle of mass m it gives a binding energy of nearly $0.06mc^2$, where c is the speed of light. This orbit is unstable; if the particle is disturbed, it will spiral into the central body. It need not lose further energy, in which case about $0.94m$ is added to the central mass. The energy $0.06mc^2$ would be imparted to the surrounding material and radiated away. This is the mechanism of the black hole. As a means of converting mass into energy, it would be about six times more effective than nuclear fusion.

In order to get, say, 10^{53} J from this 6% conversion of mass into energy, the black hole would have to swallow about $10^7\ M_\odot$. A quasar with a life of 10^6 years would require an average infall of 10 $M_\odot$ per year, and this would demand a dense surrounding medium, by astronomical standards. Also, with this infall, the central mass would be at least about $10^7\ M_\odot$ on the average.

The difficulty is to explain how a black hole could come into existence in the first place and how it could find itself in cosmic surroundings from which it could swallow material of some $10^7\ M_\odot$ or more. Some scientists therefore suggest, by implication, a combination of the two mechanisms: The collapse of a large mass as considered above could produce a "large" black hole, perhaps near the center of a galaxy, and this black hole could swallow a further large mass from its surroundings, both stages causing the release of large quantities of energy. This model has attractive features, such as compactness of energy source, but it still faces the objections mentioned. (The possible collapse of individual stars will be mentioned later in this article; it is not likely to be significant in regard to the present mechanism.)

Cluster processes. Several astronomers have asked whether the benefit of gravitational energy as a source of radiation can be had from a mass that is already fragmented, presumably a star cluster.

If a cluster behaves solely under the gravitational attractions of its member stars, in the course of time some members will escape, leaving the rest more tightly bound and in possession of greater kinetic energy. Such evolution is very slow for known clusters (furthermore the tidal effect of other systems is more significant). But if a cluster were much more massive and condensed than these—say $10^9\ M_\odot$ inside a radius of a few parsecs—then in astronomical intervals of time the cluster could lose much of its mass, leaving surviving members moving at high speed inside a small volume. A possibility (to which we shall return) is the occurrence at this stage of "soft" collisions in which colliding members coalesce into a much smaller number of more massive stars.

Another possibility, one resulting from slightly different initial conditions, is that the members attain even greater speeds (approaching the speed of light) before collisions become frequent. Then "hard" collisions occur in which enormous amounts of energy are converted into radiation in a very short time. Since the mechanism is proposed as one for liberating mainly gravitational energy, the final product must again be matter in some collapsed state.

If this hard-collision mechanism occurs in actuality, the only known possible site would be the nucleus of a galaxy. Some workers have suggested the occurrence as the death throes of such a nucleus; they offer this as an explanation of the quasar phenomenon. The merit of the mechanism is that it releases a great deal of energy in a short time, and since it relies upon randomly occurring collisions, it gives a fluctuating output as required.

Among several serious disadvantages of the model is its requirement of exceedingly stringent initial conditions for the "cluster."

Gravitation and magnetism. Magnetism appears to play an essential part in the radiative operation of the sources considered here. It could also make some of the proposed mechanisms more effective, by facilitating the transfer of angular momentum, by assisting the capture of material by a black hole, and by making a larger collapsed mass into a magnetic rotator, which appears to be an efficient means of accelerating electrons to relativistic energies. However, magnetism provides no fresh energy source.

Other sources and processes. Among speculations regarding other conceivable energy sources are: matter-antimatter annihilation; delayed "bangs" after the first in big-bang cosmology; processes involving quarks or other hypothetical particles; and variability of the constants of physics, particularly the gravitation constant. Another suggestion is that the quasar phenomenon may

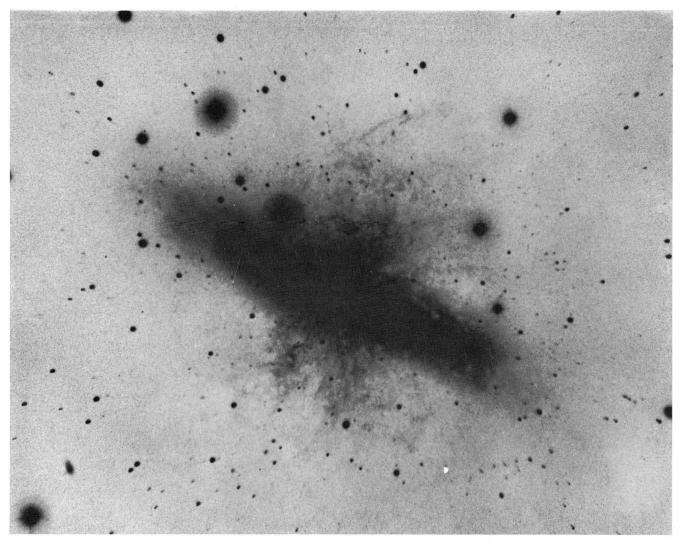

Fig. 6. View of irregular galaxy M 82 showing "streamers," an example of a violent event. (*Hale Observatories*)

denote the birth of a galaxy, or cluster of galaxies, from matter in some hitherto unknown form; but the observed liberation of energy could still come from orthodox physical processes. However, none of these speculations has yet led to a well-defined model.

THEORETICAL WORKING MODELS

A somewhat more "operational" approach will now be followed by asking how known astronomical phenomena — rather than physical processes — may help to solve the problems. The energy sources of such phenomena may then be noted.

Certain galaxies show evidence of relatively recent outbursts of tremendous violence (Figs. 2 and 6), and we have already noted the occurrence of enormous fluctuations. Such happenings have produced a new outlook in astronomy by suggesting that big systems may evolve through big and rapid changes instead of gradual ones.

Supernova. Until recent discoveries, the most violent phenomenon known was the supernova outburst. At maximum, a supernova can outshine the whole galaxy in which it occurs (Fig. 7), and so it does produce a fluctuation of a relevant amplitude. It is natural to investigate whether the same sort of astronomical phenomenon may be responsible for some of the manifestations that we have been discussing.

Several types of supernovae are recognized. One type definitely appears to be the explosion of a star of mass well in excess of $10\ M_\odot$. This massive type is what we shall refer to as a supernova; estimates of its properties are given in Table 4. Here

Table 4. Properties of massive supernova or temporary star

Mass	$50\ M_\odot \approx 10^{35}$ g
Steady radiation before outburst	
Life t	10^6 years
Luminosity l	$> 10^5\ L_\odot = 4 \times 10^{31}$ J/sec
Energy emission lt	1.5×10^{45} J
Outburst	
Time of fading	1 year
Peak luminosity (optical) l_p	$5 \times 10^9\ L_\odot \approx 2 \times 10^{36}$ J/sec
Energy liberated in outburst	10^{45} J
Whole career	
Total energy liberated e	2.5×10^{45} J

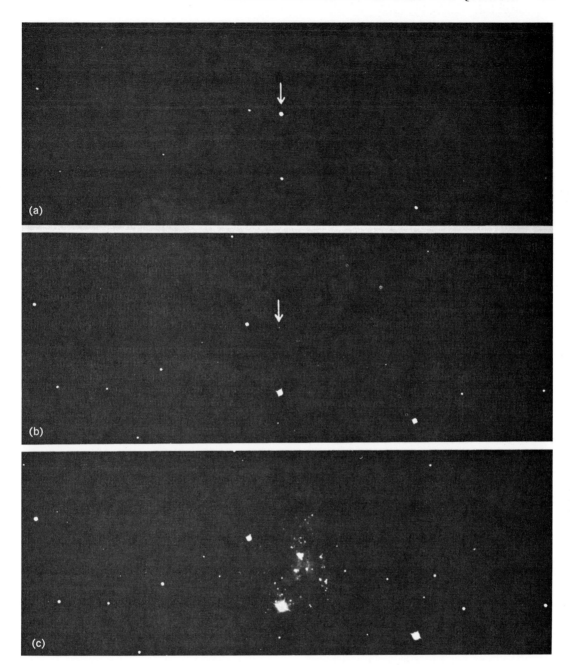

Fig. 7. Supernova in galaxy IC 4182. (a) View taken in 1937. The supernova (indicated by the arrow) outshines the galaxy. (b) View taken in 1938. The supernova is faint. (c) By 1942 the supernova is too faint to observe, even though this exposure is considerably longer than in the other views. (Hale Observatories)

all values are approximate; the mass is assumed to be 50 times the solar mass; 1 $L_\odot$ = solar luminosity $\approx 4 \times 10^{26}$ J/sec. The peak luminosity lasting 1 year yields 6×10^{43} J, which agrees well with some estimates of the observed radiation from a supernova, but is less than the estimated outburst energy of about 10^{45} J. In the case of an isolated supernova, the energy excess presumably does not go into observed radiation. The value given for energy emitted in the outburst, incidentally, is supported by the latest estimates of the amount of energy released in forming a neutron star and that imparted to ejected material.

Astrophysicists agree that a star more massive than about 60 $M_\odot$ is susceptible to pulsational instability. For a star of more than about 100 $M_\odot$, it appears that the instability would take effect so quickly that a "star" would scarcely form. In the case of a star between about 60 and 100 $M_\odot$ the instability sets in very slowly. A typical time for the outset seems to be some 10^6 years, during which the luminosity is 10^5 $L_\odot$ or more. At the end of this interval, the instability causes the star to break up in some way. Whether this breakup has the character of a supernova outburst is uncertain. But if it be asked, conversely, what a massive supernova was like before outburst, it has to be concluded that it was a short-lived, highly luminous star generally similar to that described. These properties are therefore included in Table 4.

Table 5. Typical values (approximate) for Seyfert nucleus

Life T	10^9 years
Emission rate in all frequencies L	10^{37} J/sec
Total energy output $E\ (=LT)$	3×10^{53} J
Radius of nucleus	10^{20} cm
Mass of nucleus M	$5 \times 10^9\ M_\odot \approx 10^{43}$ g

Seyfert nucleus. The next question is whether the behavior of a Seyfert nucleus summarized in Table 5 could be reproduced by a "battery" of supernovae of the sort described in Table 4. The answer is yes, as detailed in Table 6. Here we assume that effectively all the outburst energy of the supernovae ultimately escapes from the system as a whole in the form of radiation. In this table, the only empirical properties for a Seyfert nucleus that have been employed are the total energy output E and the lifetime T. From the parameters adopted for a supernova, values have been inferred for the mass, steady and mean luminosities, and the frequency and amplitude of the peaks for a Seyfert nucleus. All these are in satisfactory agreement with the further empirical properties in Table 5.

The model is also plausible on general grounds. In a spiral galaxy like the Milky Way, a supernova occurs about once a century, or more often. In order to produce the appearance of a Seyfert nucleus, this rate need be increased by a factor of no more than about 5 (admittedly with most of this activity concentrated in the nucleus) for only about 1/20 of its past life.

Quasar. In the model of Table 6, there are two independent characteristic times: t the lifetime of a supernova before outburst, and T the time during which supernovae continue to be produced. Whichever happens to be the longer gives the lifetime of the model. If T is sufficiently large compared with t, the model is reasonable for a Seyfert nucleus. If the inequality is reversed, it is natural to ask if it yields a quasar.

The first consequence is that the lifetime of every quasar would be about 10^6 years. If there is no other intrinsic difference, from the model in Table 6, then the consequences for the model quasar as a battery of supernovae are as shown in Table 7. This numerical example appears to match a powerful quasar; it indicates that for an average quasar on this model the values of total energy output and mass of nucleus and of number of supernovae required (from Table 6) should be divided by about 100. The observable luminosity of the model would fluctuate because of the randomness in the overlapping of outbursts. Actually some workers have recently been able to reproduce fluctuations like those of 3C273 with an average rate of 15 ± 5 "pulses" per year, with an average pulse length of 3.2 ± 1 years, giving an average of 48 ± 15 overlapping outbursts. This is quite near enough to show that the model can give the required kind of fluctuation. The very rapid fluctuations of some quasars would be due to some different cause. But if there are a large number (about 10^8) of outbursts of supernovae born in an interval small compared with the lifetime of 10^6 years—say 1/10 of the lifetime—then the outbursts themselves will tend to crowd toward a peak large compared with the mean—say 10 times greater. Thus the model does satisfactorily produce the general characteristics of a quasar.

THEORY

To progress from the working model to a theory, it is necessary to account for the occurrence of massive stars.

Several astronomers contemplate the possibility that these are produced by coalescence of ordinary stars in the final stages of evolution of a galactic nucleus. This process was mentioned in our discussion of clusters.

Alternatively, an early stage in the genesis of a nucleus may be involved. If a normal galaxy possesses a nucleus, it consists of some 10^7 to 10^9 stars having about the same age, which appears to be less than the age of the galaxy. Therefore the raw material must have accumulated in the appropriate region and then must have resolved itself into stars. This process of star formation would be tentative, since mechanisms of fragmentation and condensation take no account of the requirements for the resulting bodies to form working stars. If, for example, the first attempt produces many fragments of more than 100 $M_\odot$, they would be short-lived. The system must keep trying until it gets down to bodies not exceeding 50 to 100 $M_\odot$. Were there a stage when the fragments are largely in this range, according to the preceding section, a quasar would result. While the quasar behavior is in progress, it would be appropriate to describe the galaxy as experiencing a violent event. This would presumably eject diffuse material to considerable distances, and so it might produce a radio galaxy. To produce a Seyfert nucleus, the process of star formation would have to be more protracted.

Table 6. Seyfert nucleus model as battery of supernovae*

Number of supernovae required $N\ (=E/e)$	1.2×10^8
Average rate of appearance N/T	1 supernovae in 8 years
Average number of preoutburst supernovae radiating at any epoch $n\ (=Nt/T)$	1.2×10^5
Steady luminosity $L\ (=nl)$	5×10^{36} J/sec
Mean luminosity $\bar{L}\ (=Ne/T)$	10^{37} J/sec
Change in luminosity produced by each outburst (l_p) during 1 year	Increase by 2×10^{36} J/sec (40% increase on steady luminosity L; 20% increase on mean luminosity $\bar{L}$)
Total mass of supernovae $M\ (=50NM_\odot)$	$6 \times 10^9\ M_\odot$

*Values of t, l, l_p, e are from Table 4, and values of T, E are from Table 5; values of L, M in Table 6 are calculated from these and are to be compared with those in Table 5; all values are approximate.

Table 7. Quasar model as battery of supernovae*

Life T	10^6 years
Number of supernovae required $N\ (=E/e)$	1.2×10^8
Average rate of appearance N/T	120 supernovae per year
Average number of preoutburst supernovae radiating at any epoch $n\ (\approx N)$	1.2×10^8
Mean emission rate $\bar{L}\ (=Ne/T)$	10^{40} J/sec
Mean number of overlapping outbursts at any epoch	120

*Using same value of E as in Table 6, but different value of T.

The attraction of such a theory is that astronomers know that the stars must have been formed out of diffuse raw material; they know that the process must have been tentative and that if the system went through a stage of forming many "temporary" stars, they can be fairly sure that it would look like a quasar or a Seyfert nucleus.

If this is correct, the energy output in this phase comes from the "steady" radiation of the temporary stars, which must be thermonuclear, and the energy of their outbursts, which is partly nuclear but may also include a considerable gravitational contribution if the remnants are highly condensed like neutron stars. [WILLIAM H. MC CREA]

Surface Physics

The surface of a solid, in its most general sense, is the termination of the solid at its interface with vacuum, a gas, a liquid, or another solid. The surface region is clearly one of the remaining frontiers in our understanding of solid-state physics and chemistry. Indeed, it is only within the last decade or so that experimental techniques and theoretical studies have developed to the point where a sophisticated and all-encompassing science of surface physics may be said to be emerging. In this respect, surface physics is perhaps 20 years behind the physics of bulk solids.

Homer D. Hagstrum is head of the surface physics research department at Bell Telephone Laboratories, Murray Hill, N.J. His research in surface phenomena has dealt principally with electron ejection by slowly moving ions. In recent years, he has applied a spectroscopy (which he developed during the course of his work on surfaces) to the study of electronic energy levels in adsorbed complexes on surfaces.

58 SURFACE PHYSICS

Fig. 1. Apparatus used in surface physics experimentation has modular construction in which peripheral parts can be used for different surface physics experiments. By means of a rotating mechanism, the solid surface to be studied can be presented in turn to each of the four experimental ports.

The gap can be expected to close rapidly, however, under the impetus of the strong current interest in surface physics. It is now recognized that many interactions—electronic, atomic, and chemical—of very great scientific and technological importance occur at the surfaces of solids. Examples are electrical transport in electronic devices, crystal growth, atomic abrasion, heterogeneous catalysis, and corrosion.

Because the field of surface phenomena is so broad, it is perhaps not possible to produce a definition of surface physics which distinguishes it from surface chemistry, for example, and which is acceptable to all. Surface physics is probably best defined in terms of its goals, which will be discussed presently.

The surface physicist, like all physicists, strives for detailed electronic and atomistic understanding of matter and its interactions, based on the ideas of quantum mechanics. He is also the developer of new experimental tools which are rooted in physical phenomena.

Although surface physics is in many respects in an early stage of development, it is not really new. Work which might meaningfully be classified as surface physics goes back at least to that of Irving Langmuir in the 1920s. It has included theoretical and experimental work on the emission of electrons from solids and on how this phenomenon is affected by the adsorption of foreign atoms on the solid's surface. Studies of electron diffraction at surfaces, too, go back to the demonstration in the late 1920s of the wave nature of the electron by C. J. Davisson, L. H. Germer, and G. P. Thomson and have yielded information on surface crystallography. Nevertheless, understanding has evolved slowly until recent years.

GOALS OF SURFACE PHYSICS

Surface physics has two principal goals. The first is to characterize the ideal, atomically clean surface of a crystalline solid, and surfaces derived from it, by known chemical or structural modifications. The second is to understand the fundamental interactions of these surfaces with the elementary constituents of matter and radiation. These two goals are inextricably intertwined, since the fundamental interactions provide the basic experimental understanding upon which the characterization of the surface is based.

Surface characterization. The surface region of a crystalline solid can be thought of as a fourth state of matter, clearly distinguishable from a liquid, a gas, or the interior of a bulk solid. However, under various conditions of chemical composition, temperature, or electric field, a surface may have characteristics which are solidlike, liquidlike, or which resemble local atoms or molecules interacting with each other and with the underlying solid.

In characterizing the surface of a solid, the physical scientist wants to know the chemical identity of the constituent atoms at the surface, their geometrical or crystallographic arrangement in space, and the structure of energy levels which electrons can occupy in the surface region. He also wants to know about surface energetics—about such quantities as atomic binding energies, electronic work functions, and electronic orbital energies—and about the mechanics of motion of surface atoms, the electric fields that naturally exist at surfaces, and many-electron or plasma phenomena.

Fundamental surface interactions. The entities with which a surface may interact include electrons, atoms and molecules, ions, excited atoms or molecules, photons, and electric or magnetic fields. These entities are, in general, most readily provided in a vacuum or in a gas for interaction with the surface from the outside; therefore it is not an accident that surface physics has concentrated on the vacuum-solid or gas-solid interface. A notable exception is electrochemistry, in which surface interactions in a liquid electrolyte are studied. However, surface physics is not restricted—in fact, must not be restricted—to interactions with entities solely from outside the solid. Electrons, atoms, and ions can arrive at a surface from the interior of the solid, and atoms can diffuse over surfaces.

SURFACE EXPERIMENTS

To conduct surface experiments properly we must meet certain conditions. The experiments must be done in a controlled vacuum environment on adequately prepared surfaces with experimental techniques that produce and control electron, ion, or neutral atom beams and provide for the detection and energy analysis of ions or electrons formed in the experiments.

Evacuated apparatus. Modern surface physics experiments dealing with the gas-solid or vacuum-solid interface must be conducted in ultrahigh vacuum. This is necessary not only to give us a sufficiently long mean free path between collisions with background gas, but also to reduce the molecular arrival rate at surfaces within the apparatus. A sufficiently long mean free path for directed beams of particles is readily achievable; at a pressure of 10^{-4} torr (1.33×10^{-2} N/m² or 10^{-7} atmospheres) the mean free path is 50 cm. Reduction of molecular arrival rate at surfaces to an acceptable level requires a background pressure about a mil-

SURFACE PHYSICS 59

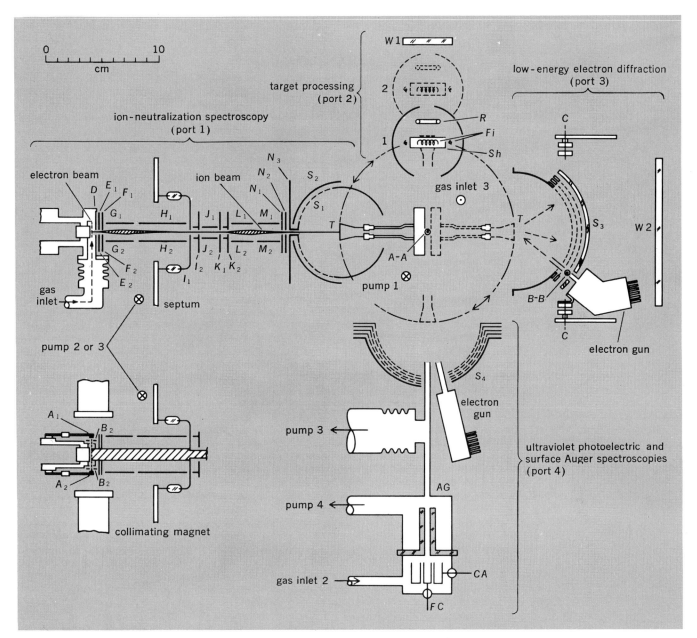

Fig. 2. Schematic diagram of the experimental arrangements included in the apparatus of Fig. 1. Port 1 provides a focused low-energy ion beam and electron-energy analyzer for the study of ion-neutralization phenomena. Port 2 provides the means of cleaning the surface by ion bombardment, or sputtering, of the surface. Port 3 provides a low-energy electron diffraction (LEED) apparatus for determining surface crystal symmetries. Port 4 provides two experiments: photoemission through use of a windowless resonance lamp, and surface Auger spectroscopy for surface atom identification.

lion times less than this. If every gas molecule striking the surface were to stick, a surface at a pressure of 10^{-10} torr (1.33×10^{-12} N/m^2) would be completely covered in about 30 hr, a condition that would provide the experimentalist with a reasonable period, perhaps a fraction of an hour, in which a specially prepared surface could be studied before it became so covered with adsorbed gas as to require reconstitution.

The last 10 to 15 years have witnessed a remarkable improvement in vacuum pumps, vacuum gages, vacuum chambers, and operating procedures. One versatile apparatus for surface physics studies, for example, permits preparing the surface, examining its crystallography and chemical purity, and performing two surface-characterizing experiments, all on the same crystal surface in the same vacuum ambient (Figs. 1 and 2).

The vacuum pumps used in surface experimental apparatus are of several types. Among the most important are mercury diffusion pumps pumping through traps cooled with liquid nitrogen; sputter-ion pumps which adsorb or ionize and then bury their intake; and titanium sublimation pumps which adsorb certain gases onto freshly produced titanium surfaces deposited (in many cases) on cooled substrates. Pressure-measuring devices have been greatly improved and diversified, some being capable of reliable measurement of pressures below 10^{-12} torr (1.33×10^{-14} N/m^2). Valves,

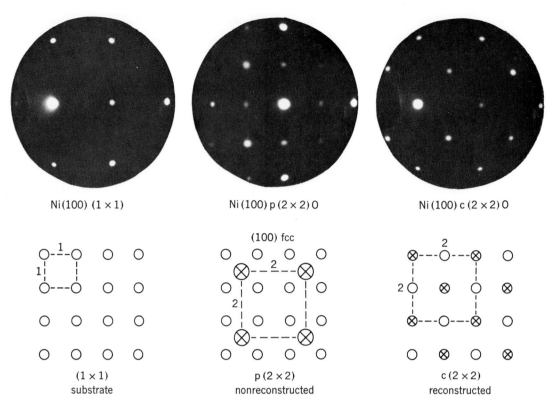

Fig. 3. LEED patterns and their interpretation. (*Top*) Photographs showing LEED patterns for three different nickel surfaces in the (100) orientation. The surface in the left-hand photograph is clean and has a (1×1) atomic arrangement. The patterns shown in photos at center and right are for Ni(100) with adsorbed oxygen in, respectively, the p(2×2)O and c(2×2)O ordered atomic arrays. (*Bottom*) Diagrams showing surface atomic positions corresponding to the LEED patterns directly above them. Foreign atoms are indicated by the encircled crosses. (*Photos by H. D. Hagstrum and G. E. Becker*)

connectors, electrical feed-throughs, and mechanical motion transmitters now available are compatible with pressures below 10^{-10} torr (1.33×10^{-12} N/m²).

Surface preparation. As important as the provision of the proper vacuum ambient in surface physics experiments is the preparation of the surface itself. Ideally the surface should be atomically clean, atomically plane, and defect-free over a large area, perhaps 1 cm². This surface could then be modified in known ways. Needless to say, this ideal has not been achieved. The closest approach to it is reached when certain nonmetallic crystals are mechanically cleaved in vacuum along specific, so-called cleavage planes, but even these surfaces have steps on them and include dislocations, vacancies, and impurity atoms.

Another method of surface preparation involves mechanical and electrolytic polishing outside the apparatus, followed by ion bombardment and annealing in the apparatus after a good vacuum has been achieved. This procedure produces surfaces that approach the ideal over an area large enough for some of the important surface interactants. A properly chosen ion, for example, will undergo electronic transitions a few angstrom units (10 Å = 1 nm) away from a surface with only a few atoms in its immediate neighborhood. On the other hand, a meaningful electron diffraction experiment requires that the surface be made up largely of well-ordered patches, each of which contains upward of hundreds of atoms. Field desorption, which tears atoms from a surface by an electric field applied at a sharp tip, can produce very well-ordered, clean, and plane surfaces of many crystallographic orientations, but the largest such surfaces are limited to a single patch containing in total less than 100 atoms. Finally, chemical interactions are used to prepare clean surfaces by the removal of chemically bonded impurities.

Other experimental techniques. Surface physics, like other branches of physics, may require in its experiments means for producing and controlling beams of electrons, ions, or neutral atoms and for detecting and analyzing the electrons or ions formed in the experiments. Thus the surface physicist works with electron and ion guns, electron and ion optics, low- and high-resolution energy analyzers, and sensitive current-measuring devices and data-gathering systems. Those experimental components which are incorporated inside the vacuum apparatus must be so designed and operated that the hard-won cleanness of the vacuum environment is not degraded by their presence and use.

BASIC SURFACE PHENOMENA

The understanding of a basic surface phenomenon can provide a way of fundamentally characterizing a surface. Thus a method of study which utilizes the phenomenon is a tool for studying a variety of conditions. Some representative surface studies—many of which involve the development of new tools—will be discussed.

Electron diffraction. The phenomenon of electron diffraction can be observed at surfaces for slow electrons (10–200 eV) at near-perpendicular incidence, and for fast electrons (10–50 eV) at grazing incidence. These are known as low-energy electron diffraction (LEED) and high-energy electron diffraction (HEED), respectively. Diffraction occurs because of the wave nature of the electron and was first observed in the Nobel-prize-winning experiments of Davisson and Germer (low energy) and Thomson (high energy) in the late 1920s.

The wave field of the incident electron beam may be thought of as causing surface atoms of the crystal to reradiate coherent waves of elastically reflected electrons, which constructively interfere to produce diffracted beams only in certain directions. The directions in which these diffracted beams appear, their relative intensities, and the variation of these parameters with incident electron velocity or wavelength are the basic data obtained from the diffraction experiments.

LEED. Very soon after the discovery of electron diffraction, Germer used LEED in a study of gas adsorption. H. E. Farnsworth and others have used the technique to study surfaces over the intervening years.

The last decade has seen a remarkable growth of experimental work with LEED, made possible by the development of apparatus which, in an arrangement of successive spherical grids, separates the elastically reflected, back-diffracted electrons from those which have lost energy, and then accelerates the elastically reflected electrons to make their termination points visible on a spherical fluorescent screen. The apparatus is shown diagrammatically at port 3 in Fig. 2.

Photographs of representative LEED patterns obtained from the fluorescent screen of this apparatus are reproduced in Fig. 3, along with drawings of the corresponding surface atomic arrangements, derived from the (100) face of a face-centered cubic (fcc) crystal. The atom arrangement and its LEED pattern for the clean, so-called (1 × 1) substrate associated with this face are shown at the left in the figure. Here the atoms are in a square array, the minimum square being called the (1 × 1) unit mesh of the surface. The center view shows the same surface with a superstructure of atoms in a so-called primitive (2 × 2), or p(2 × 2), arrangement adsorbed upon it. Because of the reciprocal relation that exists between atom spacing on the crystal surface and diffracted beam separation on the LEED pattern, new spots appear on the p(2 × 2) photo in Fig. 3 at all points halfway between the spots of the (1 × 1) LEED pattern. This forms an unequivocal signature of the p(2 × 2) structure of the surface. The structure contains one-quarter of a monolayer of adsorbed atoms, where one monolayer is defined as the number of surface metal atoms in the topmost layer of the (1 × 1) substrate structure.

At the right in Fig. 3 is shown a centered (2 × 2), or c(2 × 2), structure of adsorbed atoms containing one-half of a monolayer of foreign atoms. This particular surface is termed reconstructed because the foreign atoms actually replace surface metal atoms. The LEED pattern for the c(2 × 2) structure is simpler than for the p(2 × 2), showing only an additional spot in the center of the square array of spots for the (1 × 1) substrate. The LEED patterns of reconstructed and nonreconstructed

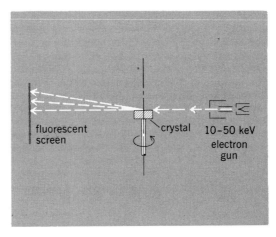

Fig. 4. Geometrical arrangement for grazing-incidence, high-energy electron diffraction (HEED). The crystal can be rotated in azimuth (as indicated by curved arrow), and the pattern on the fluorescent screen can be photographed.

structures are the same. Thus the LEED pattern alone reveals the symmetry of the superstructure (referred to as the surface crystal repeat pattern) of adsorbed atoms, but does not reveal the exact positions of the adsorbed atoms with respect to the substrate atoms. The idea of adsorptive reconstruction of surfaces has important consequences for many surface phenomena, but as yet LEED cannot unequivocally decide between these two possible interpretations—whether primitive or centered—of a given surface symmetry.

HEED. High-energy electron diffraction (HEED) for surface studies employs the geometrical arrangement shown in Fig. 4, in which high-energy electrons skim the surface and the diffracted beams impinge on a fluorescent screen. HEED can determine the surface crystal repeat pattern, but for this requires diffraction patterns at more than one azimuth; the crystal therefore must be rotated as shown in the figure. HEED presents important surface information that is difficult, if not impossible, to obtain with LEED. HEED can "see" small, randomly oriented crystalline imperfections if they are present at the surface, and it is more sensitive than LEED to surface contaminants in the form of agglomerated islands on the surface. A most important characteristic of HEED is its ability to observe surface smoothness. The photographs of a HEED fluorescent screen in Fig. 5 show the use of HEED in the study of surface topology. The separation of the bright vertical lines is a measure of the distance between atoms of the solid in one particular azimuth. The lumpiness of these lines in the photo at the left in Fig. 5 indicates that the grazing electron beam is passing through the solid protrusions of a rough surface. The uniformity in the right-hand photo shows that the surface is now much smoother. The weaker spot structure in the right-hand photo derives from the specifically surface atoms and is used in the analysis of their crystalline order.

Importance of LEED and HEED. The study by LEED or HEED of the surface structures in which various adsorbate-substrate combinations arrange themselves is an important part of surface physics

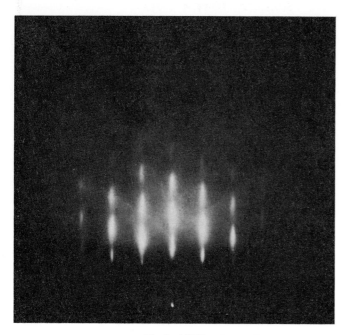
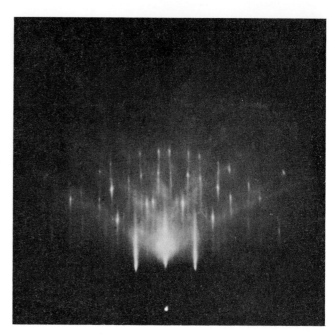

Fig. 5. HEED patterns. A rough surface is represented at left and a smooth one at right. (*Courtesy of A. Y. Cho*)

and chemistry. These techniques have also been used to study the surface phase transformation from one surface crystalline arrangement to another and the surface atomic vibrations as evidenced by the component of diffuse electron scattering. But LEED and HEED also provide the information on surface crystallography and topology necessary to many other surface experiments. Thus LEED has been an important adjunct in the study of the energy spectrum of electrons in chemisorption bonds. And the special characteristics of HEED are used to study crystal growth at surfaces by the epitaxial deposition from atomic beams of the species composing the crystal.

Information concerning local atomic positions at surfaces—as opposed to the symmetry of the surface structural arrangement—should be derivable from LEED data, but practical application must await further development of theory. Suggestions as to local bonding positions come from the interrelation of atomic structure to energy-level structure. It also appears possible to obtain information about surface atom positions from the magnitude of backscattering of fast ions incident on the surface.

Surface Auger spectroscopy. It was shown in 1954 by J. J. Lander that the secondary electron emission from a solid caused by primary electrons in the thousands-of-electron-volts range has a very weak component arising from so-called two-electron or Auger processes. These processes result from the ionization by the primary beam of inner-shell electrons in the atomic core. The three principal types of Auger process are shown in Fig. 6. In each case a vacancy or "hole" (indicated by an open circle) is created in an inner shell at energy $-\epsilon_c$ by core-electron ionization. Following this, an Auger-type electronic rearrangement occurs; one outer electron (indicated by a solid circle) falls into the inner vacancy, releasing an amount of energy which, when absorbed by a second electron (upper solid circle), is sufficient to eject it from the solid. The electrons may come from the valence band, which lies between the energies $-\phi$ and $-\epsilon_0$, or from one of the higher lying core levels. The measured kinetic energy E_k of the ejected electron can

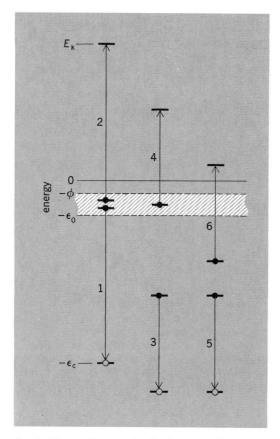

Fig. 6. Energy diagram showing by arrows the energy changes experienced by electrons as they undergo paired Auger transitions in the surface atoms of a solid. The two arrows in each pair—(1,2), (3,4), and (5,6)—are of equal length, indicating process is radiationless.

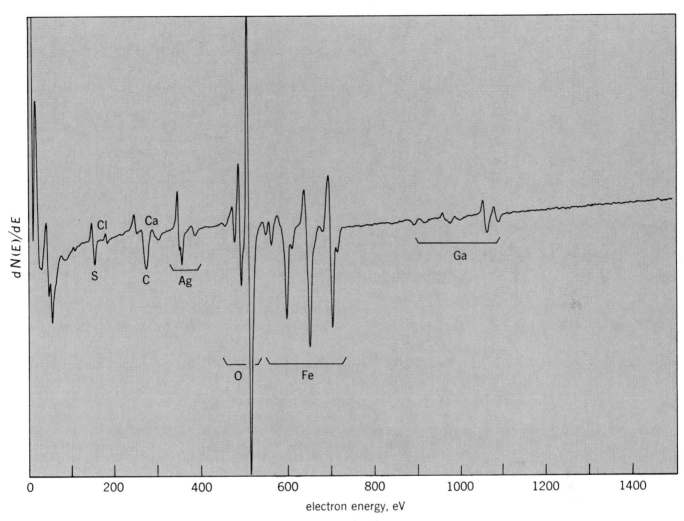

Fig. 7. Plot of the derivative, $dN(E)/dE$, of the energy spectrum, $N(E)$, of Auger electrons ejected from atoms in an impure iron film on gallium arsenide. The paired up-down peaks in this spectrum appear at the energies E_k of Fig. 6 and lead to the chemical identifications indicated. (*Courtesy of J. C. Tracy, Jr.*)

be used to determine the energies of the levels in which the initial vacancy occurs and from which the electrons making the Auger transitions are drawn.

This identification of some of the core energy-level structure is, in most cases, sufficient to identify the atom in which the process has occurred. It is this fact that has made possible the development by a number of investigators of surface Auger spectroscopy as a powerful and versatile identifier of atoms present in the surface layers of a solid. A derivative Auger spectrum is shown in Fig. 7. The paired up-down peaks in this spectrum appear at the energies E_k of Fig. 6 and lead to the chemical identifications indicated.

The required energy analysis can be carried out in the multigrid system at port 4 in Fig. 2 or in deflection-type analyzers. Surface Auger spectroscopy has already been used in many interesting experiments concerned with the identify of impurities at specific surfaces, the segregation of impurities at surfaces, and the effect on other surface experiments of the presence of known impurities.

Other means of chemical identification of surface atoms include thermal or field desorption into a mass spectrometer, observation of x-ray fluorescence from core-level deexcitation, and detection of electrons photoejected directly from core levels. Recently it has been shown that the measurement of the energy loss of fast, back-reflected ions can also be used to identify the surface atom with which the incident ion collided.

Field emission. A sufficiently large field (4×10^7 V/cm) applied at the surface of a solid reduces and thins the normal barrier to electron escape to such an extent that electrons from inside the solid can tunnel directly into vacuum. A convenient way to produce such a large field with manageable voltages and interelectrode distances is used in the field-emission microscope devised by E. W. Mueller in 1936 (Fig. 8). Here the voltage is applied to the crystalline tip of radius approximately 10^{-5} cm. Electrons are ejected from the tip, traveling in straight lines to the fluorescent screen, where they produce an image, magnified a million times, of the emission probability as it varies over an area of the tip (which contains a number of crystal planes). Variations of this emission pattern from one crystal face to another can be related to variations in the minimum energy (the work function) required to remove an electron from a solid through a specific crystal face. Studies of the effect on work

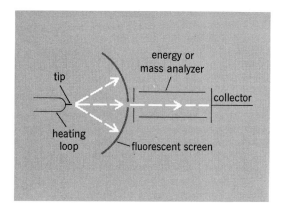

Fig. 8. Schematic diagram of a field-electron or field-ion microscope. A voltage of several kilovolts and appropriate sign is applied between tip and screen. The emission pattern may be viewed on the screen, or particles may pass into an analyzer through a hole in the screen. For electrons, energy is analyzed. For ions, mass is analyzed for chemical identification.

function of adsorbed gases have been carried out with this apparatus. In recent work, electrons from a specific face are allowed to pass through a hole in the fluorescent screen into an energy analyzer. Careful measurement of the variation of electron current with energy reveals a structure which can be attributed to the enhanced tunneling of electrons at the energy levels of atoms resident on the surface of the tip.

In a later version of the field-emission microscope known as the field-ion microscope, Mueller reversed the electric field and admitted a gas such as helium. Helium atoms approaching the tip are preferentially ionized at the positions of protruding atoms on the surface where the field is enhanced. The positive ions thus formed are accelerated in the same field toward the fluorescent screen,

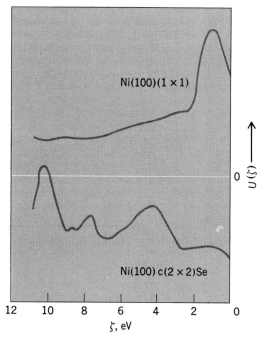

Fig. 9. Energy (level spectral) functions determined by ion-neutralization spectroscopy. (H. D. Hagstrum and G. E. Becker)

where they produce images with sufficient resolution to show the positions of individual atoms.

This microscope is the most powerful yet devised. It can be used to study surface imperfections and migration or diffusion of atoms over surface planes and to observe the phenomenon of field desorption, in which atoms of the solid are pulled off when the field is high enough. Removal of many such field-desorbed layers produces surfaces with a high degree of atomic order. In a recent development, field-desorbed atoms are allowed to pass through the fluorescent screen into a time-of-flight mass analyzer, where their chemical identity is determined. This analyzer of surface chemical species has the ultimate sensitivity, since it can detect individual atoms.

Ion neutralization. When a slowly moving ion of sufficiently large neutralization energy (ionization energy of the parent atom) is neutralized close to a solid surface by an electron from the solid, energy is released which can eject a second electron. The electronic transitions involved are of an Auger type resembling those of the left-hand set in Fig. 6. However, in this case the initially vacant level is not a core level of a surface atom but is the vacant ground state of the incoming ion.

This ion neutralization phenomenon has been used in an electron spectroscopy which generates information about the energy levels of electrons at the surfaces of clean solids or in the chemisorption bonds that hold foreign atoms to the surface. The experiment is performed in the apparatus of port 1 shown in Fig. 2 and can thus be correlated with the results of the experiments performed at the other three ports.

The spectrum of electron-level energies, such as that for the $c(2 \times 2)$ selenium surface shown in Fig. 9, is related both to the local atomic structure and to electrical charge shifts in the surface molecule formed by the adsorbate atom and one or more metal atoms presented by the substrate. Thus the functions plotted in Fig. 9 are closely related to the local density of electronic states in and just outside the topmost layer of the solid. The energy is measured from zero at the Fermi level positively downward into the filled electron band. The upper curve is for atomically clean Ni(100) and shows the surface band of d electrons lying between 0 and 2 eV. The lower curve shows a much modified function, obtained when a $c(2 \times 2)$ layer of selenium atoms is present on the Ni(100) surface. The peaks on this curve at 4, 8, and 10 eV energy indicate the energy levels of electrons in the local surface molecule formed of nickel atoms and the selenium adsorbate. Specification of the energy-level structure for such surface compounds determined by this or other methods is clearly a fundamental part of an adequate characterization of the surface region of a solid. This is an interesting extension of our knowledge of moleculelike structures because the surface molecules formed in chemisorption can differ from any known to exist in free space since the solid constrains the positions of the metal atoms incorporated into the molecule.

Photoelectric emission. The photoelectric effect, whereby electrons are ejected from solids or molecules by the absorption of electromagnetic radiation, has a venerable history. As interpreted by Albert Einstein, it figured prominently in establishing experimentally that light consists of individual quanta of energy, $h\nu$, called photons (h is Planck's constant; ν is the frequency of the radiation). Electron transitions induced by photon absorption in a semiconductor are shown in Fig. 10. Here the unshaded energy range (from $-\epsilon_c$ to $-\epsilon_v$) is the so-called forbidden gap between conduction and valence bands in which bulk electrons cannot reside. The left-hand transition indicates the energy gained by a valence-band electron on absorbing a photon. The right-hand transition indicates direct photon absorption by an electron in a surface state.

The measurement of the kinetic energies of photoelectrons provides another kind of electron spectroscopy, known as photoelectron spectroscopy, which has been applied with spectacular success to both bulk solids and molecules.

As far as specifically surface phenomena are concerned, photoelectric emission has been used in many ways. For metals, photoemission has been used to determine work functions by finding the threshold light frequency at which emission is first observed. For semiconductors, the interpretation of photothreshold measurements is more complicated. These measurements, combined with independent work function measurements, give the energies of the electron band edges at the surface, as indicated in Fig. 10. Then comparison with band-edge energies in the bulk of the solid provides information on the energy and density of electrons in surface states.

Direct photoemission from surface states has also been observed. Also, it has recently been shown for metals that photoemission, particularly in the range of photon energies 20 to 40 eV, is sufficiently specific to surface conditions to detect electrons in atoms adsorbed at the surface. Photoejected electrons are generated somewhat deeper in the solid (several atomic layers) than are the electrons released by ion neutralization outside the surface. Thus the two methods complement one another in an interesting fashion.

Semiconductor surfaces. Because a semiconductor contains far fewer electrical carriers than does a metal, the conditions prevailing at its surface can control the electrical characteristics deep into the bulk. This fact underlies both the development of surface electronic devices and the strong role the surface plays in the functioning of bulk devices that employ internal interfaces between regions of differing impurity concentrations (*pn* junctions).

Intensive study of semiconductor surfaces followed John Bardeen's hypothesis in 1947 concerning the existence of surface states and their role in controlling many surface-sensitive phenomena. (A surface state at a semiconductor surface is the state of an electron in which it is spatially limited to the surface region and may have an energy in the range forbidden to electrons in the bulk solid.) Much sophisticated understanding of semiconductor electronics resulted from surface studies during the 1950s and later. The nature of surface photoeffects, characteristics of surface electronic traps, and the role of minority carriers in surface-sensitive behavior were all studied with great effect.

However, little of the physical characterization of the surface discussed in this article was attempted in this early work. Surfaces were covered

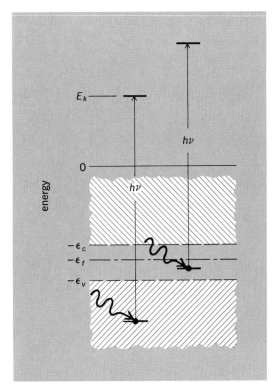

Fig. 10. Energy diagram of electron transitions induced by photoabsorption in a semiconductor.

with many layers of essentially unknown composition and crystallographic arrangement. Electron diffraction and photoemission studies constituted the first attempts to bring into semiconductor surface research the kind of control and characterization developed for metal surfaces. Now scientists are on the threshold of a new and important synthesis of previously disparate approaches to the study of semiconductor surfaces.

Flash desorption. If a metal filament is cleaned by heating to high temperature and then cooled and exposed to a gas, it will adsorb, in different binding states, gas molecules, or atoms derived from them. When at a later time the filament is heated again, preferably with a rapid linear rise in temperature, atoms and molecules reappear in the gas phase by the process of thermal or flash desorption. The temperature at which desorption occurs is related to the energy with which the atom or molecule is bound. This phenomenon has served as the basis for a surface experimentation tool and much work has been done with it.

A flash desorption spectrum for nitrogen on tungsten is shown in Fig. 11. The three rises in gas density (γ, α, and β) as temperature is increased indicate that nitrogen, in differing amounts, is held by the surface in three binding states of differing binding energy. Another possible interpretation, currently under investigation, would attribute some of the multiplicity of such desorption peaks to the desorption process itself.

The rate at which gas accumulates on a surface during the cold interval can also be measured by flash desorption, providing a way to determine the probability that an incident molecule will remain on the surface when it strikes. Work of this type is being extended to single crystal faces, to the use of

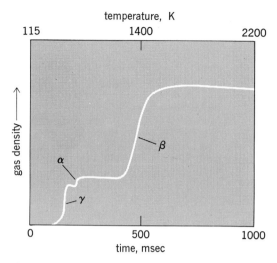

Fig. 11. Plot of nitrogen gas density or pressure above a polycrystalline tungsten surface as it is heated from 115 to 2200K. (*Courtesy of G. Ehrlich*)

molecular beams to dose the surface, and to studies of the kinetics of adsorption and desorption for various gas-solid and vapor-solid systems.

Atomic vibrations at surfaces. It was stated earlier in this article that the so-called thermal diffuse scattering component in low-energy electron diffraction is the result of surface atomic vibrations. Another way to study surface vibrational modes involves the absorption of infrared radiation. This radiation is in the proper frequency range, and the absorption of such energy into vibrational modes has been observed. Recent work with single crystals shows that it will be possible to get away from the earlier use of powderlike samples and to work with surfaces which can be controlled and characterized properly.

Measurement of the energy loss experienced by a well-monochromated electron beam scattered by a surface has also been used to detect vibrational modes of surface atoms. The application of this technique to well-controlled surfaces is in its infancy. It is important work, nonetheless, because vibrational modes are another indicator of surface bonding structure.

Surface theory. An article on surface physics cannot be concluded without at least some mention of surface theory. Like other physical theory, surface theory can be of a macroscopic nature (statistical mechanics or thermodynamics) or of a microscopic nature (detailed electronic and atomic models). Monumental efforts have been made in macroscopic theory that illuminate many aspects of surface phenomena, but microscopic theory is essential to the further development of surface physics. The electronic theory of surfaces is an effort to understand how electrons behave at surfaces, or, in the terms of quantum mechanics, to determine the electronic wave function at surfaces. These efforts are embodied in the theory of surface electronic states, of low-energy electron diffraction, of electron emission phenomena, and of the escape of electrons across and through surface barriers. The theory of elementary excitations at surfaces deals with many-electron effects such as surface plasmons and atomic vibrations in the form of surface phonons, as well as electronic states. The theory of adsorption of foreign atoms on surfaces is now beginning to utilize sophisticated approaches devised to solve problems encountered in the theory of the bulk solid. The nature of the surface electric potential barrier has also been the subject of much theoretical effort in which there is renewed interest.

The theory of surfaces is more difficult than that for the infinitely extended bulk solid because the symmetry at the surface is much reduced from that in the bulk. Thus, in terms of difficulty, it more nearly resembles molecular theory. We can expect a rich variety of surface phenomena demanding theoretical understanding, because each bulk solid can have many surfaces differing significantly in crystallographic form and in electronic character and, further, because it appears possible to dope surfaces with a wider range of foreign atoms than can be dissolved into the bulk solid.

FUTURE

The best way to assess the future of surface physics is to look carefully at present trends. It has already been pointed out that, in many instances, the availability of vacuum processing and measuring equipment has made possible better surface control and characterization. This availability will continue to spread to the study of other surface phenomena not yet studied for well-characterized surfaces. We shall see an increase in "multiple" surface experiments, in which the same surface is observed and studied by several techniques in the same vacuum ambient. Detectors of surface crystallography and chemical identity will increasingly become essential components of every basic surface experiment.

Strong efforts, both theoretical and experimental, will continue to be made to enable electron diffraction to yield unequivocal structural information like that provided for bulk solids by x-ray diffraction. Much effort will be spent in the immediate future to measure and understand the character of the electronic energy spectra of surfaces and its relation to structure both for clean surfaces and for surfaces with adsorbed atoms resident in or upon them. Clearly more study of semiconductor surface states is needed for surfaces which are well characterized both as to chemistry and structure.

Solid-state electronics should certainly feel the impact of surface physics research in the form of new crystal growth procedures, such as molecular-beam deposition; these methods will make possible new or improved device structures. We may expect surface theorists to attack in a fundamental way the electronic structure not only of the clean surface but of this surface with an ordered array of a known foreign atom adsorbed upon it.

In addition to specific contributions of the types mentioned above, the ideas and interpretations of surface physics should influence in a less direct way those fields involving surfaces and their interactions for which much of the technological information is empirical—the field of catalytic chemistry, for example. A radical breakthrough in such fields may be a great deal to expect, but there is little doubt that the conceptual structure of surface physics, based as it is on fundamental experiments, will be important to technologists in these fields. Perhaps it is not too much to hope that the ideas, if not the techniques, of surface physics will be helpful in biology also, where surfaces and interfaces are of paramount importance. The pursuit of surface physics is not particularly easy either experimentally or theoretically, but the great importance of surface phenomena to mankind justifies the effort to understand them in a fundamental way.

[HOMER D. HAGSTRUM]

Pathology of Heavy Metals

Man has evolved in an environment
containing metals and has developed
protective means of defense against
natural concentrations. Indeed,
some metals have become essential for
life. However, industry is pouring
metals, particularly heavy metals, into
man's environment at an unprecedented
and constantly increasing rate.
As a result, technological society is
exposing the population to some
metals in unnaturally high concentrations,
in unusual physical or chemical forms,
and through unusual portals of entry.

This trend is certainly likely
to continue. It is vital, therefore,

*N. Karle Mottet, a graduate of Yale
University, is director of hospital
pathology and professor of pathology at
the University of Washington School of
Medicine. His current research
centers on cell development.*

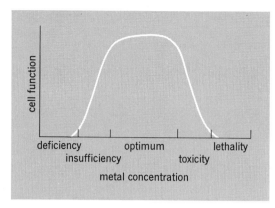

Fig. 1. Effects of metals on cells. Metals have an optimum range of concentrations within which a cell is healthy. The actual values and range vary from metal to metal.

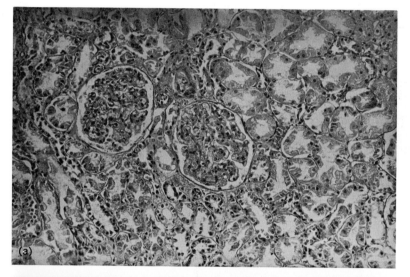

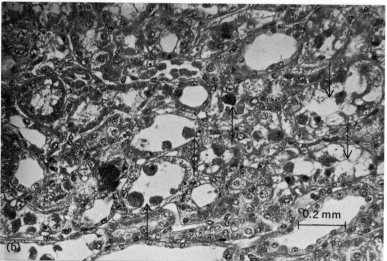

Fig. 2. Kidney cells. (a) Normal kidney cells. (b) Methyl mercury–poisoned cells. Arrows indicate dead cells.

that we understand the effects of metals on living cells as a first step toward controlling the problems. In this article, I will summarize our present understanding. I will not attempt to catalog the pathologic processes produced by all metals. Rather, my aim is to selectively describe the pathology of some environmentally important heavy-metal poisonings that illustrate the diversity of pathologic lesions.

NATURE OF METAL TOXICITY

Heavy metals, the principal subject of this article, are arbitrarily defined as those metals having a density at least five times greater than that of water. Although metals have many physical properties in common, their chemical reactivity is quite diverse, and the toxic effects of metals on biological systems is even more diverse.

A metal can be regarded as toxic if it injures the growth or metabolism of cells when it is present above a given concentration. Almost all metals are toxic at high concentrations and some are severe poisons even at very low concentrations. Copper, for example, is a micronutrient, a necessary constituent of all organisms, but if the copper intake is increased above the proper level, it becomes highly toxic. Like copper, each metal has an optimum range of concentration, in excess of which the element is toxic (Fig. 1). When the optimum range for a particular metal is narrow, the risk of toxicity increases; thus even a minor environmental increase can be serious.

Metals exert their toxicity on cells by interfering in any of several ways with cell metabolism. The most important of these affects enzyme systems. The more strongly electronegative metals (such as copper, mercury, and silver) bind with amino, imino, and sulfhydryl groups of enzymes, thus blocking enzyme activity. Another mechanism of action of some heavy metals (gold, cadmium, copper, mercury, lead) is their combination with the cell membranes, altering the permeability of the membranes. Others displace elements that are important structurally or electrochemically to cells which then can no longer perform their biologic functions.

Conditions for toxicity. The toxicity of a metal depends on its route of administration and the chemical compound with which it is bound. For example, mercury is highly toxic when injected intravenously or inhaled as a vapor, but is less toxic when taken by mouth. Cadmium, chromium, and lead are also highly toxic when they are injected intravenously, whereas a large number of metals (gold, cobalt, manganese, tin, thallium, nickel, zinc, and others) require greater intravenous doses to be toxic. When taken by mouth, gold and iron are only slightly toxic; however, copper, mercury, lead, and vanadium are much more toxic. Tantalum, on the other hand, is nontoxic no matter how it is administered.

The combining of a metal with an organic compound may either increase or decrease its toxic effects on cells. For example, the combination of mercury with a methyl organic radical makes the element more toxic, whereas the combination of the cupric ion with an organic radical, such as salicylaidoxine, makes the metal less toxic. Generally the combination of a metal with a sulfur to form a sulfide results in a less toxic compound than the corresponding hydroxide or oxide, because the sulfide is less soluble in body fluids than the oxide.

Many other heavy metals besides copper are essential to life, even though they occur only in trace amounts in the body tissues. They are taken up by the living cell in the form of cations and are admitted into an organism's internal environment in carefully regulated amounts, under normal circumstances, thus avoiding the effects of toxic levels. Toxicity results (1) when an excessive concentration is presented to an organism over a prolonged period of time, (2) when the metal is presented in an unusual biochemical form, or (3) when the metal is presented to an organism by way of an unusual route of intake.

Specificity. Metals have a remarkably specific effect; seldom can an excess of one essential metal prevent the damage caused by the deficiency of another. In fact, such an excess often increases the injurious effect of the deficiency. The heavy metals that appear to be essential for living cells are copper, iron, manganese, molybdenum, cobalt, and zinc. Some lighter metals, such as calcium, magnesium, sodium, and potassium, are also essential. There is some evidence that other metals, such as aluminum, cadmium, chromium, and vanadium, are essential. Many other metals found in the human body tissues have no apparent function.

Cellular response. On a molecular level of organization, metals produce diverse responses in cellular behavior. If the toxic action of a metal on a cell is interference with an essential part of cell metabolism, the cell will, of course, die. Figure 2, for example, shows two microscopic views of kidney cells. In Fig. 2a two glomeruli are surrounded by cross sections of normal kidney tubules, and Fig. 2b shows kidney tubule cell degeneration and necrosis (death) 72 hr after exposure to methyl mercury. The degenerating cells are swollen and vacuolated. The necrotic cells (indicated by arrows) are small, round, and dense and have small nuclei. Cell death provokes, in turn, an inflammatory response by the body and an attempt to repair the damage. Indeed, some metals (beryllium, for instance) provoke an excessive reparative process consisting of extensive proliferation of connective-tissue cells (histiocytes) that produce nodules of inflammatory tissue (granulomata) and extensive scarification. The microscopic views of lung tissue in Fig. 3 illustrate this effect. Figure 3a shows normal lung cells, with characteristically thin alveolar septa surrounding the air sacs. Figure 3b shows cells after exposure to beryllium. Nodular proliferations of cells and thickened septa (arrows) are evident.

Lesser degrees of injury to cells may alter their structure and function and may be associated with degenerative changes such as accumulation of fat or may limit the cells' principal biologic activity, which in turn may alter the performance of the organism as a whole. For example, degeneration of kidney tubule cells by mercury degrades the excretory function of the kidneys and leads to severe fluid and electrolytic imbalance in an individual, and possibly to death.

Carcinogens and teratogens. Less well understood but perhaps of equal significance are the carcinogenic properties of some metals. Nickel, cobalt, and cadmium have clearly been shown to produce cancers of the muscle cell type (rhabdomyosarcoma) when injected subcutaneously. However, many other metals similarly injected do not produce aberrant new cell growth (neo-

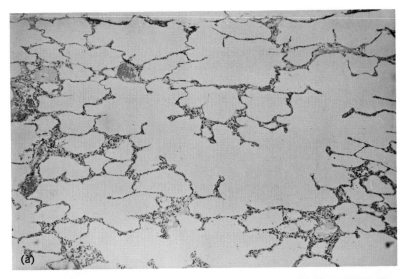

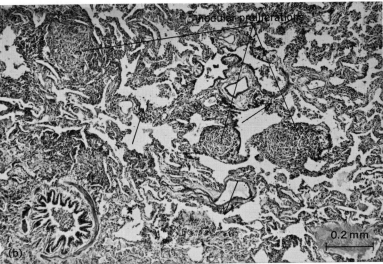

Fig. 3. Lung cells. (a) Normal cells. (b) Cells exposed to beryllium. Nodular proliferations of cells are indicated. Thickened septa are shown by arrows.

plasms). Nickel compounds also produce cancers of the respiratory tract. Lead has been shown to produce kidney cancers in experimental animals. The carcinogenic potential of many other metals is suspected but not proven.

Birth defects (teratogenesis) as a consequence of excessive metal intake by pregnant women do occur, but their precise nature and frequency are poorly documented. Chick embryos, when exposed to low doses of thallium or chromium, subsequently hatch with a high incidence of defective long-bone formation. Lead and cobalt have been shown to produce defective brain development in chick embryos. Many diverse metals produce growth inhibition in developing chick embryos. Manganese is a powerful mutation-producing agent for the bacterium *Escherichia coli* and the T4 bacteriophage; it is known to alter the nuclear enzyme DNA polymerase in lower forms of life, but whether it similarly affects mammals is not known. Mercury produces brain damage and mental retardation in children born to mothers exposed to excessive amounts of that metal. Lead, too, has teratogenic effects; specific congenital skeletal

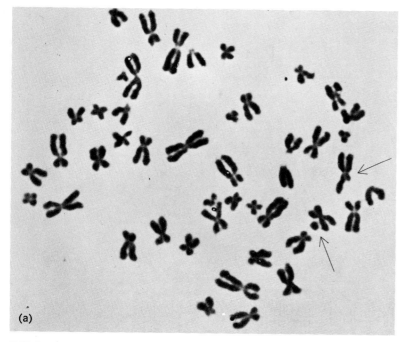
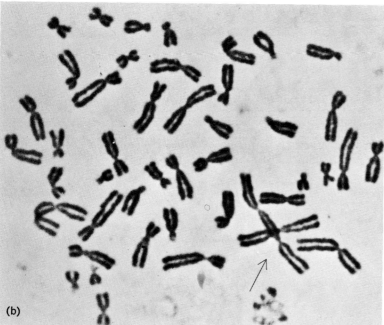

Fig. 4. Chromosomes damaged by mercury poisoning. (a) Arrows show a broken chromosome and an extra fragment. (b) Arrow shows chromatid fragments without centromeres. (From S. Skerfving, Chromosome breakage in humans exposed to methyl mercury through fish contamination, Arch. Env. Health, 21:133–139, 1970)

malformations have been induced in hamster embryos by treating the pregnant mother with various salts of lead.

Chromosome damage. In addition to producing injury to the developing embryo, some metals produce damage to chromosomes. Both mercury and lead compounds have been shown to produce breaks in the chromosomes of somatic cells. Figure 4 shows chromosome damage from mercury. The lymphocytes have a broken chromosome and extra fragment (Fig. 4a), and three sister-chromatid fragments that lack centromeres (Fig. 4b). Steffan Skerfving, of the National Institute of Public Health, Stockholm, Sweden, cultured the lymphocytes from the blood of nine patients who were known to have excessively high tissue levels of methyl mercury from eating contaminated fish. He found a statistically significant rank correlation with chromosome breaks and mercury concentration. The full significance of these findings remains to be determined.

Many questions remain to be answered regarding the effects of metals on chromosomes. Many agents produce chromosome breaks in the test tube, but do they have the same effect in living cells? Are chromosome breaks produced in germ cells as well as somatic cells? Does the genetic damage lead to cell death, carcinogenesis, or teratogenesis? Whereas the effects of the deficiency of many trace elements on the developing embryo have been extensively studied, little is known about the teratogenic effects of an excess of metals. These questions will undoubtedly be answered as research proceeds.

CADMIUM DISEASE

Cadmium occurs in trace quantities—less than 1 part per million (ppm)—throughout the Earth's crust. There are no known deposits rich enough to justify separate mining of the metal, but it is frequently found in zinc and lead ores. Industrially cadmium has important and diverse uses in electroplating and in alloys, solders, batteries, paints (heat-resistant pigments), and metal bearings. It is also used as a neutron absorber in nuclear reactors and as an insecticide for fruit trees.

Cadmium also occurs naturally in trace quantities in many plant and animal tissues (Table 1). It therefore forms part of our normal diet, although the exact amount normally ingested is not known. Cadmium is very slowly absorbed from the gastrointestinal tract and is eliminated in both urine and feces. In healthy human adults the cadmium levels are low (Table 2), except for the kidneys,

Table 1. Metals in our environment in parts per million

Metal	Rock	Coal	Sea water	Plants (dry wt.)	Animals (dry wt.)
Cadmium	0.2	0.25	0.0001	0.1–6.4	0.15–3.0
Chromium	100.0	60.0	0.00005	0.8–4.0	0.02–1.3
Cobalt	25.0	15.0	0.00027	0.2–5.0	0.3–4.0
Lead	12.5	5.0	0.00003	1.8–50.0	0.3–35.0
Mercury	0.08	—	0.00003	0.02–0.03	0.05–1.0
Nickel	75.0	35.0	0.0045	1.5–36.0	0.4–26.0
Silver	0.07	0.1	0.0003	0.07–0.25	0.006–5.0
Thallium	0.45	0.05–10.0	0.00001	1.0–80.0	0.2–160.0
Vanadium	135.0	40.0	0.002	0.13–5.0	0.14–2.3
Gold	0.004	0.125	0.00001	0–0.012	0.007–0.03

Table 2. Distribution of some metals in mammalian tissues in parts per million dry weight

Metal	Skin	Lung	Liver	Kidney	Brain
Cadmium	1.0	0.08	6.7	130.0	3.0
Chromium	0.29	0.62	0.026	0.05	0.12
Cobalt	0.03	0.06	0.23	0.05	0.005
Lead	0.78	2.3	4.8	4.5	0.24
Mercury	–	0.03	0.022	0.25	–
Nickel	0.8	0.2	0.2	0.2	0.3
Silver	0.022	0.005	0.03	0.005	0.04
Thallium	0.2	0.3	0.4	0.4	0.5
Vanadium	0.02	0.05	0.04	0.05	0.3
Gold	0.2	0.3	0.0001	0.5	0.5

which are uniquely high in cadmium. Much lower concentrations occur in the tissues of the human fetus and newborn. With increased age, the tissue cadmium level progressively increases.

Ingestion of 15 ppm of cadmium in food, or about 15 times the normal amount, can produce mild symptoms of poisoning. Food poisonings from contamination of food and drink by cadmium-plated containers have been recorded frequently. For example, several outbreaks have occurred as the result of the action of the acids in fruit juice acting on cadmium-plated ice trays. Cadmium is soluble in weak acids such as acetic acid (vinegar), citric acid, and organic acids commonly found in food. The principal human risk is from industrial exposure, namely, inhalation of cadmium fumes. Cadmium is primarily a respiratory poison and has higher lethal potential than most other metals with a mortality rate of 15% in poisoning cases.

Recent occurrence. A remarkable case occurred in 1969 when a 28-year-old Japanese girl, Takako Nakamura, who had been working as a lathe operator in a zinc fabricating factory, hurled herself before a speeding train. Her death was listed as suicide and nothing more was thought of it for a time. Recently her body was exhumed and an autopsy performed. After correlating the levels of cadmium within the tissues with the anatomic lesions observed, the pathologists reported that she was a victim of cadmium poisoning.

Symptoms. Mild cadmium intoxication causes smarting of the eyes, dryness and irritation of the throat, and tightness in the chest and headache. With increased exposure, the respiratory distress increases in severity with uncontrollable cough and gastrointestinal pain, nausea, retching, vomiting, and diarrhea. Takako, the suicide, recorded in her diary the early symptoms: "The doctors cannot diagnose my disease; I am afraid it is cadmium poisoning. It is running through my whole body, pain eats away at me; I feel I want to tear out my stomach; tear out all my insides and cast them away."

Tissue reaction to the inhalation of cadmium fumes results in the initial deposition of cadmium in the lungs. From there the metal is widely distributed throughout the body and accumulates in the liver, kidneys, and spleen. The high concentrations may remain in the tissues for many years after the cessation of exposure. Excretion in the urine is extremely low.

Cadmium acting on the alveoli of the lungs produces a severe proliferation of alveolar septal cells which, in some cases, may completely fill the air spaces. Chronic exposure to cadmium produces emphysema, an irreversible lesion of the lungs characterized by an enlargement of the air sacs and accompanied by destruction of the walls of many of the sacs. This results in the conversion of many small air sacs into fewer large ones, with the inevitable decrease in surface area for oxygen and carbon dioxide transfer.

The kidneys are also injured in chronic cadmium poisoning. Diffuse extensive scar tissue develops between the tubules and subsequently destroys many glomeruli and tubules. Serum protein is then lost in the urine.

When administered to rats in dose levels comparable to that ingested by humans, cadmium produces high blood pressure, enlargement (hypertrophy) of the muscle of the heart (left ventricle), and hard patches (sclerotic plaques) in the very small arteries (arterioles) of the internal organs. (Communities in the United States with the highest environmental cadmium content also have the highest incidence of arteriosclerosis.) Another consequence of chronic cadmium exposure is an increased brittleness of bone, to an extent that the mere act of coughing may produce fractures of the ribs. Cadmium nitrate produces gene mutation in plants, but its effects, if any, on mammalian genes remain unknown.

MERCURY POISONING (MERCURIALISM)

The biological effects of excessive intake of mercury illustrate some important similarities and differences relative to cadmium. Mercury is unique in that it is the only metal that is a liquid at ordinary temperatures. Like cadmium, mercury is a relatively rare element, ranking near the bottom of the list of elements found in abundance on Earth.

Food grown under natural conditions contains traces of mercury. Five different studies, using a variety of methods from 1934 to the present, have shown that dietary meats contain on the average less than 0.01 ppm mercury. Fish generally have twice and vegetables have half as much (the commonly accepted limit for mercury in foods is 0.5 ppm). Some algae contain more than 100 times more mercury than the sea water in which they grow. Fish eating the algae concentrate the mercury, and predators that eat the fish in turn concentrate the mercury further.

Increasing prevalence of element. Man has evolved in an environment containing mercury, and throughout his history he has ingested plants, animals, and fish containing mercury. Through his evolution he has developed a tolerance, or possibly a need, for mercury as a trace constituent of his body. How has our urban technology affected this ecosystem?

First of all, mercury is being added to the North American environment at a rapid rate. Of greatest significance is the unmeasured increase of mercury to our atmosphere from the burning of fossil fuels. Coal and petroleum contain significant amounts of mercury; the mercury content of 36 American coals ranged from 0.10 to 33.0 ppm. In addition, organic mercurials are widely used as a fungicide and pesticide in agriculture. Other mercury compounds have been successfully used as a dressing for seed grains to prevent mildew, as a conditioner for lumber to prevent fungal discoloration, and in laundries to prevent garment mildew.

Not only is there more mercury in the environment, but evidence suggests that more of it is in a form that is most toxic to man. Liquid (metallic) mercury itself is not ordinarily toxic, since the body does not absorb mercury in this form from the digestive tract. However, many intoxications occur when metallic mercury is vaporized and inhaled. Figure 5 shows the effect of mercury vapor on lung tissue (it can be compared to the normal lung tissue in Fig. 3a). The tissue is from the lung of a 1-year-old child who died 6 days after exposure to the mercury vapor. The septa are swollen and thickened, and the air sacs contain fluid. Aside from such acute injuries to lung tissue, chronic low-level exposure to mercury vapor may produce symptoms or death because of its destruction of the nervous system.

Organic mercurials, too, which are readily assimilated, are becoming more prevalent. It has been proven that elemental, or inorganic, mercury, when released into the hydrosphere, is converted to an organic mercurial by the linkage of methyl or other carbon chains to the mercury by marine life. Repeated predation by animals in the food chain concentrates organic mercurials in their tissues, ultimately reaching toxic levels for man and animal.

Case histories. The increasing prevalence of organic mercurials has led to some serious episodes of intoxication of man. A case of direct ingestion of organic mercury occurred in Alamogordo, N. Mex., when several members of a family ate pork from a pig that had been inappropriately fed seed-wheat previously treated with a methyl mercury compound. Three children became seriously ill (but none died), whereas two other children who had not eaten pork were unaffected. The mother was pregnant at the time of onset of the symptoms in the three older children but she was symptom-free. The baby appeared normal at birth; however, soon thereafter this child developed convulsions and the characteristic symptom complex of chronic mercurialism, from which he seems to be gradually recovering.

Indirect chronic mercurial poisoning with a much higher morbidity and mortality rate occurred in Minamata and Niigata, Japan. The poisoning, called Minamata disease, was caused by discharge into Minamata Bay from an acetaldehyde and vinyl factory. The effluent contained large amounts of inorganic and organic mercury. There were 121 human cases of mercury poisoning recorded, with 46 deaths. About one-third of the afflicted were infants and children, some of whom had acquired mercury poisoning through the placenta prior to birth. The disease occurred mainly among fishermen and fish-eating families. Fish-eating animals such as cats, dogs, pigs, and seabirds were often affected, but herbivores such as rabbits, horses, and cows were not.

Symptoms. Chronic mercurialism presents five main symptoms: numbness, staggering gait, constriction of visual field ("tunnel vision"), garbled speech (dysarthria), and tremor. In addition, the children with Minamata disease exhibited emotional disturbances, and infants with congenital intoxication also suffered from impaired chewing or swallowing. Convulsions and mental retardation often occurred in the childhood cases. These symptoms of chronic poisoning are almost totally traceable to lesions in the central nervous system.

Autopsies of those who died soon after exposure revealed degenerative lesions in the liver, heart muscle fibers, and tubules of the kidneys. These

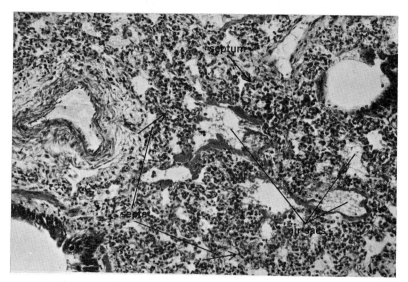

Fig. 5. Cells from lung of a child poisoned by mercury vapor. Arrows indicate swollen septa. The air sacs are filled with fluid.

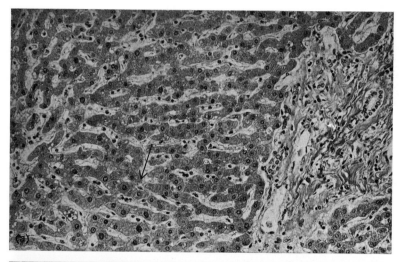

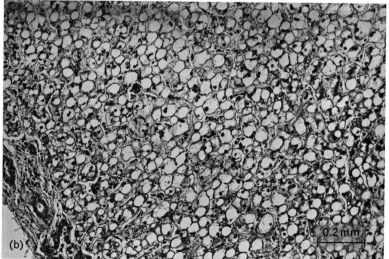

Fig. 6. Liver cells. (a) Normal cells. (b) Mercury-poisoned cells.

changes were similar to those seen with inorganic mercury poisoning. Normal human liver cells (Fig. 6a) are of uniform size and take up stain uniformly. Mercury-poisoned liver cells (Fig. 6b) accumulate fat, assume various sizes, and stain unevenly. This type of fatty degeneration of cells can be seen throughout the liver and is characteristic of other metal poisonings in addition to mercury.

Acute focal inflammation without ulceration was often seen in the stomach or duodenum. The autopsies also revealed the presence of a generalized edematous swelling of the brain and a decrease in blood cell production by the bone marrow.

Another important early lesion involved the blood vessels, especially those of the central nervous system. Blood vessels became cuffed with white blood cells, and in some there was degeneration of the vessel wall and capillary proliferation.

The lesions of chronic mercurialism are principally in the brain. In the autopsies, destruction of the nerve cells of the gray matter of the cerebral cortex was found in several regions, especially the visual cortex. Extensive necrosis of nerve cells in the visual area and an increase in the nonneuronal supporting cells had taken place. These lesions principally account for the "tunnel vision" of methyl mercury poisoning. Similar changes were seen in the cerebellum. An astounding destruction of the granular cell layer was found, as shown in Fig. 7 (at arrows). This layer contains nerve cells that receive stimuli from the muscles, tendons, and joints throughout the body and transmit them to the Purkinje cell processes in the molecular layer, where the many stimuli are correlated and integrated. The cerebellum is the area of the brain where position sense, balance, and fine muscle coordination are controlled. In most cerebellar pathologic processes the sensitive cells are the large Purkinje cells; mercury is the only known metallic toxin from an external source that results in the preferential destruction of the granular cell layer.

LEAD POISONING (PLUMBISM)

The toxicity of lead has been known since antiquity. Lead and lead-containing products are extremely important industrially and are of great health significance. Exposure of man may result from inhalation of fumes and dust in the smelting of lead, from the manufacture of insecticides, pottery, and storage batteries, or from contact with gasoline containing lead additives. Intoxication by the inhalation of lead fumes is a most serious mode of exposure. The ingestion of soluble lead compounds accidentally, or with suicidal intent, is another portal of entry of the body. Only tetraethyllead in gasolines can be absorbed through the intact skin. Traces of lead occur in the diet; small amounts are absorbed into the body when it is present in the food as a soluble salt. Lead is absorbed mainly through the small and large intestines.

Irrespective of the route of entry, lead is absorbed very slowly into the human body. Even this slow and constant chronic absorption is sufficient to produce lead poisoning because the rate of elimination of lead is even slower and a slight excess in intake may result in its accumulation in the body. Much of the lead is taken up by red blood cells and circulated throughout the body. Organic lead com-

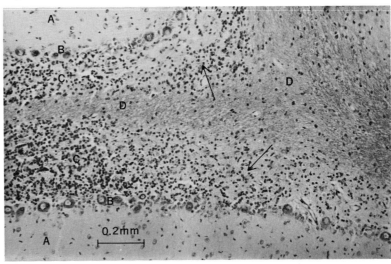

Key: A = molecular level C = granular cell level
B = Purkinje cells D = white matter

Fig. 7. Cells in cerebellum. Arrows show cells destroyed by methyl mercury.

pounds such as tetraethyllead become distributed throughout the soft tissues, with especially high concentrations in the liver and kidneys. Over a period of time, the lead may be redistributed, becoming deposited in bones, teeth, or brain. In bones, lead is immobilized and does not contribute to the general toxic symptoms of the patient. Organic lead compounds have an affinity for the central nervous system and produce lesions there.

Acute form. Acute plumbism is ordinarily seen as a result of the ingestion of inorganic soluble lead salts for suicidal, accidental, or abortion-inducing reasons. They produce a metallic taste, a dry burning sensation in the throat, cramps, retching, and persistent vomiting. The gastrointestinal tract is encrusted with the coagulated proteins of the necrotic mucosa, thereby hindering further absorption of the lead. Muscular spasms, numbness, and local palsy may appear.

Chronic form. Chronic lead poisoning is much more common. Two general patterns of symptoms relate to the gastrointestinal and nervous systems. One or the other may predominate in any particular patient; in general, the central nervous system changes, which predominate in children, are of greater significance. The abdominal symptoms in chronic lead poisoning are similar to those for acute cases, but are less severe: loss of appetite, a feeling of weakness and listlessness, headache, and muscular discomfort. Nausea and vomiting may result. Chronic excruciating abdominal pain, sometimes referred to as lead colic, may be the most distressing feature of plumbism.

An autopsy does not ordinarily reveal specific gross lesions, but a marked congestion and petechial hemorrhages are sometimes seen in the brain. Microscopic examination reveals inclusion bodies within cells of the kidney, brain, and liver. Focal necrosis of the liver is found in some cases. Lesions in the central nervous system are primarily vascular and consist of scattered hemorrhages, often in the perivascular tissue; degenerative and necrotic changes in the small vessels, surrounded by a zone of edema; and sometimes a fibrinous exudate.

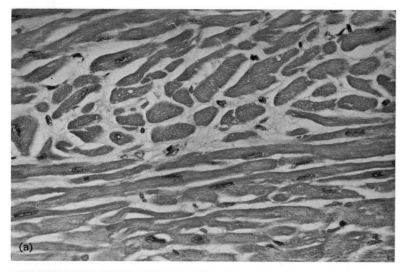
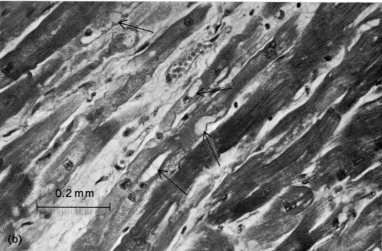

Fig. 8. Heart muscle fibers. (a) Normal fibers. (b) Fibers from a victim of beer drinker's syndrome. Arrows point to fatty vacuoles.

THALLIUM DISEASE (THALLOTOXICOSIS)

Since thallium was discovered more than 100 years ago, it has been responsible for many therapeutic, occupational, and accidental poisonings. During the first 50 years since its discovery, thallium sulfate gained general use as a treatment for syphilis, gonorrhea, gout, dysentery, and tuberculosis. It was subsequently discarded as a medicine because of its unpleasant side effects, principally the temporary loss of hair (depilation). Later, dermatologists utilized this feature as a method of removing unwanted hair.

Occurrence. By 1934 more than 700 poisonings with thallium were reported in the medical literature, with more than 90% of these the result of the use of thallium as a depilatory agent. In recent years hundreds of cases of thallotoxicosis have been related to its use as a rodenticide and insecticide. For example, a group of 31 Mexican workers in California were poisoned, six fatally, after eating tortillas made from a bag of thallium-treated barley intended for rodents.

Symptoms. The symptoms of thallotoxicosis vary extensively with the dosage, age of the patient, and the acuteness of the intoxication. Inflammation involving many nerves, loss of hair, gastrointestinal cramps, and emotional changes are the principal symptoms. For large doses, the first symptoms are gastrointestinal hemorrhage, gastroenteritis, a rapid heartbeat, and headache. Shortly thereafter abdominal pain, vomiting, and diarrhea may ensue. Neurologic symptoms appear after 2 or more days of exposure, with delirium hallucinations, convulsions, and coma occurring after severe poisoning, and death may follow in about a week due to respiratory paralysis.

When smaller doses are taken, a loss of muscle coordination (ataxia) and sensations of tingling or burning of the skin (paresthesia) may be the principal symptoms. These may be followed by weakness and atrophy of the muscles. Tremor, involuntary movements (choreic athetosis), and mental aberrations, with changes in the state of consciousness, may ensue.

Alopecia is one of the best known symptoms of chronic thallium poisoning. The hair loss begins about 10 days after the ingestion of thallium, and complete loss of hair may be reached in a month. Hair in the axillary and facial regions, including the inner one-third of the eyebrows, is usually spared. The evidence suggests thallium acts directly on the hair follicles. Cardiac symptoms—rapid heartbeat, irregular pulse, and high blood pressure, with angina-like pain—have been recorded. Heart muscle fiber degeneration accompanies these symptoms. Thallium affects the sweat and sebaceous glands, and the nails are white with transverse bands and may assume unusual shapes. A blue line may develop along the gums of an individual who has ingested thallium.

The turnover of thallium in the human body is extremely slow. Under normal circumstances the tissues contain trace amounts of the metal. The biological mechanism by which thallium produces its effects on the human body is not well understood, but as with other metals, it appears to block some enzyme systems. Autopsies on fatal human cases reveal the presence of thallium in all organs and tissues, with the highest concentrations in the kidneys, intestinal mucosa, thyroid, and testes. Fat, liver, and all types of nerve tissue are uniformly low in thallium content. The autopsies reveal bleeding at many points (punctate hemorrhages) in the gastric and upper intestinal mucosa, fatty change in the liver and kidney, and small punctate hemorrhages and focal necrosis of the surface layers of the adrenal glands. Congestion of the central nervous system blood vessels is evident, and there is some degree of cerebral edema.

Neurons show varying degrees of degenerative change, especially in the locomotor fiber pathways and associated centers. In the more chronic cases, the ganglion cells of the sensory and motor horns of the spinal cord degenerate, with chromatolysis, swelling, and fatty change. Examination of peripheral nerves shows marked degeneration in nerve cells, axons, and myelin sheath. Thallium poisoning is a serious problem not only because approximately 15% of those poisoned die, but also because there is persistent neurologic damage in over half the cases.

COBALT TOXICITY

Cobalt is a metal with a very industrially important property: It is immune to attack by air or water at ordinary temperatures and imparts this property to many alloys. Under natural circumstances cobalt does not produce a toxic syndrome in man. Cobalt in most naturally occurring foodstuffs is for the most part unabsorbed and is eliminated in the feces.

It takes an unusual set of circumstances to produce cobalt toxicity. Intravenously injected, large doses of cobalt have been observed to cause paralysis and enteritis, and sometimes death. When cobalt is injected into the bloodstream, the amount distributed throughout the tissues is very small; higher concentrations are present in the pancreas, liver, spleen, kidney, and bone. Elimination of cobalt is rapid.

Carcinogen. Researchers in England have shown that cobalt, like nickel and cadmium, produces a high incidence of cancer, namely rhabdomyosarcoma, when it is injected in powdered form into rats. A number of other metals, such as iron, copper, zinc, manganese, beryllium, and tungsten, are not carcinogenic under the same conditions. Cobalt has an inhibitory effect on cell oxidative metabolism and is toxic to connective-tissue cells grown in culture.

Beer drinker's syndrome. Another unusual kind of cobalt toxicity is the "beer drinker's syndrome." During the 1960s, some breweries in Nebraska and Quebec added cobalt to their beer to improve the stability of the foam. Subsequently, numerous fatalities occurred among people who consumed large quantities of beer. Following discontinuance of the cobalt additive, no new cases have been reported. Whether the cobalt acted independently or synergistically with other constituents of beer to enhance its toxicity is not known. The principal lesion was found in the heart; autopsies revealed enlarged, flabby hearts that were more than twice normal size. Microscopically the myocardial cells were vacuolated with numerous fat droplets. The microscopic view of heart muscle fibers from a normal heart is shown in Fig. 8a, and from a beer drinker's syndrome case in Fig. 8b. The fatty vacuoles (arrows) resulting from degenerative changes give a "moth-eaten" appearance to the fibers. Progressive degenerative changes in the myofibrils were also found in the autopsies, and complete necrosis and lysis of some heart muscle fibers were noted.

Cobalt chloride has been used as a therapeutic in the treatment of anemia, and dosages in excess of 100 mg per day for prolonged periods have been unattended by changes in the cardiac muscle. In contrast, the cobalt intake of the most avid beer drinker was calculated to be only 10–15 mg per day. This suggests that the production of the cardiac lesions was not due to cobalt alone. In addition to the myocardial lesions, the beer drinkers also frequently had gastrointestinal inflammation and extensive hemorrhagic necrosis of the liver. The production of similar lesions in experimental animals given cobalt in their water affirms the importance of cobalt in their genesis. However, the heart lesions are similar in appearance to those of beriberi, a thiamine vitamin deficiency disease, and other nutritional factors may have contributed to the beer drinker's syndrome.

Some experiments have linked cobalt intake to the development of hardening of the arteries (atherosclerosis).

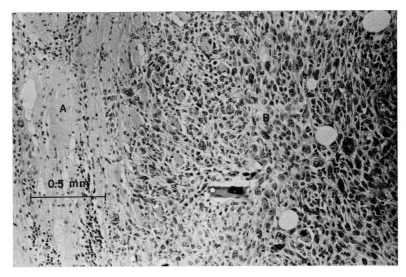

Fig. 9. Skeletal muscle fibers. Cells in region A are normal; those in region B are cancer cells.

NICKEL POISONING

Nickel is widely distributed throughout the Earth's crust and is a relatively plentiful element. It occurs in marine organisms, is present in the oceans, and is a common constituent of plant and animal tissues. The ordinary human diet contains 0.3–0.5 mg of nickel per day; diets rich in vegetables invariably contain much more nickel than those with foods from animal origin.

Depending on the dose, the organism involved, and the type of compound involved, nickel may be beneficial or toxic. It appears to be essential for the survival of some organisms.

Industrial hazard. The use of nickel in heavy industry has increased markedly over the last few decades, principally in the production of stainless steel and other alloys and in plating. Because of the excellent corrosion resistance of nickel and high-nickel alloys, these metals are used widely in the food processing industry. In fact foods can be contaminated with nickel during handling, processing, and cooking by utensils containing large quantities of nickel. Nickel is also used frequently as a catalyst; one of its most significant catalytic applications is in the hydrogenation of fat. Nickel carbonyl, $Ni(CO)_4$, is one of the most toxic nickel compounds and is a major industrial hazard. Lesser quantities of nickel are used in ceramics, as a gasoline additive, in fungicides, in storage batteries, and in pigments.

Nickel usually is not readily absorbed from the gastrointestinal tract except as nickel carbonyl. This compound has caused most of the acute toxicity of nickel. Recently it has been shown that nickel has a carcinogenic property and may be involved in hypersensitivity reactions. Figure 9 shows normal skeletal muscle fibers (region A) and rhabdomyosarcoma cells (region B) characteristic of the type produced by nickel, cobalt, or cadmium injection. Nickel dermatitis is reported with increasing frequency in industrial workers, especially nickel platers. It produces an allergic dermatitis on almost any skin area.

Nickel workers have approximately 150 times the cancer of the nasal passages and sinuses as the general population, and approximately five times

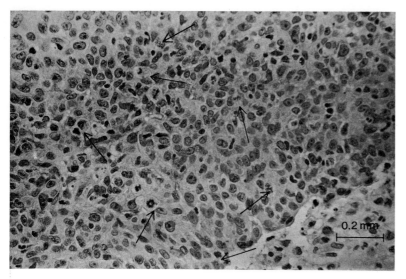

Fig. 10. Paranasal sinus carcinoma cells. Arrows indicate dividing cells.

the lung cancer. Figure 10 shows a paranasal sinus carcinoma. This type of neoplasm is seen with increasing frequency in industrial employees exposed to nickel carbonyl. The numerous dividing cells (arrows) suggest rapid growth. Several reports from Great Britain have established conclusively that inhaled nickel is a carcinogenic agent, and in Great Britain it is a compensable industrial disease.

Toxic action. The mechanism of toxic action of nickel is its capacity to inhibit oxidative enzyme systems. Within the body the main storage depots of nickel are the spinal cord, brain, lungs, and heart, with lesser amounts widely distributed throughout the organ systems. Acute poisoning causes headache, dizziness, nausea and vomiting, chest pain, tightness of the chest, dry cough with shortness of breath, rapid respiration, cyanosis, and extreme weaknesses. The lesions resulting from acute exposure are mainly in the lung and brain. In the lungs hemorrhage, collapse (atelectasis), and necrosis occur. Deposits of brown-black pigments are found in the lung phagocytic cells. In the brain extensive damage to blood vessels is seen. Lung cancers due to nickel exposure are usually of the squamous carcinoma variety and are indistinguishable microscopically from lung cancer of other causes.

CONCLUSION

As we have seen, some metals are extremely injurious to the body tissues, whereas others are less so or are nontoxic. Exposure to them therefore produces diverse patterns of pathologic change. Some metals may produce extensive degeneration and necrosis of the cells in a particular organ, whereas others may produce cancerous change or birth defects.

Because of the increasing incidence of such effects—and the inevitability of further additions of metals to the environment—some practical goals can be set: Man should learn more about the human biological effects of excessive metal exposure through research, and should manage the type and quantity of exposure to minimize its adverse effects.

[N. KARLE MOTTET]

Photographic Highlights

These photographs have been chosen for their scientific value and current relevance. Many result from advances in photographic and optical techniques as man extends his sensory awareness with the aid of the machine, and others are records of important natural phenomena and recent scientific discoveries.

Apollo 15 Lunar Rover on the Moon, with TV antenna in transmitting position. TV camera is below antenna. *(NASA)*

David R. Scott and the Lunar Rover at Hadley Rille. The astronauts touched down on the Moon on July 30, 1971, and spent nearly 67 hours there. *(NASA)*

Left. Scott holds sample-pickup tongs and camera on *Apollo 15* mission. *(NASA)*

Below. Rock-strewn crater near Hadley Delta at base of the Apennine Front. *(NASA)*

Above. Falcon lunar module. *(NASA)*

Right. James B. Irwin uses scoop to dig trench in seeking rock samples. *(NASA)*

Apollo 15 command module orbiting Moon. *(NASA)*

Splashdown of *Apollo 15* in the Pacific on Aug. 7, 1971. *(NASA)*

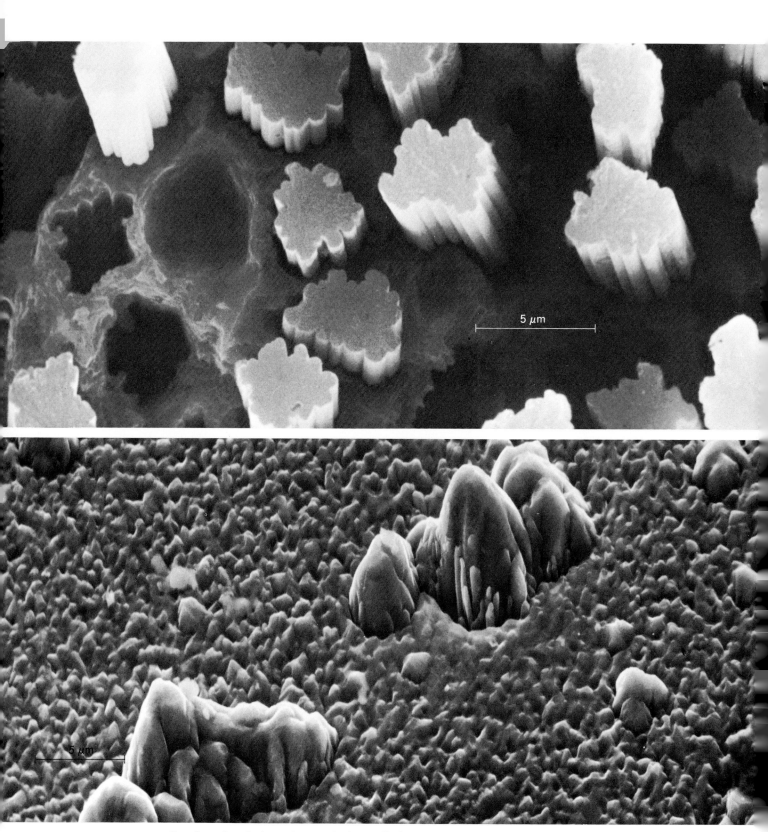

Top. Scanning electron micrograph of a tensile fracture of aluminum-silicone alloy composite with Union Carbide Thornel 50 graphite fibers. Fracture surface is shown.
(The Aerospace Corporation)

Bottom. Scanning electron micrograph of vacuum-evaporated Ti-6Al-4V alloy film, 0.002 in. thick, tilted at a 45° angle. (Etec Corporation)

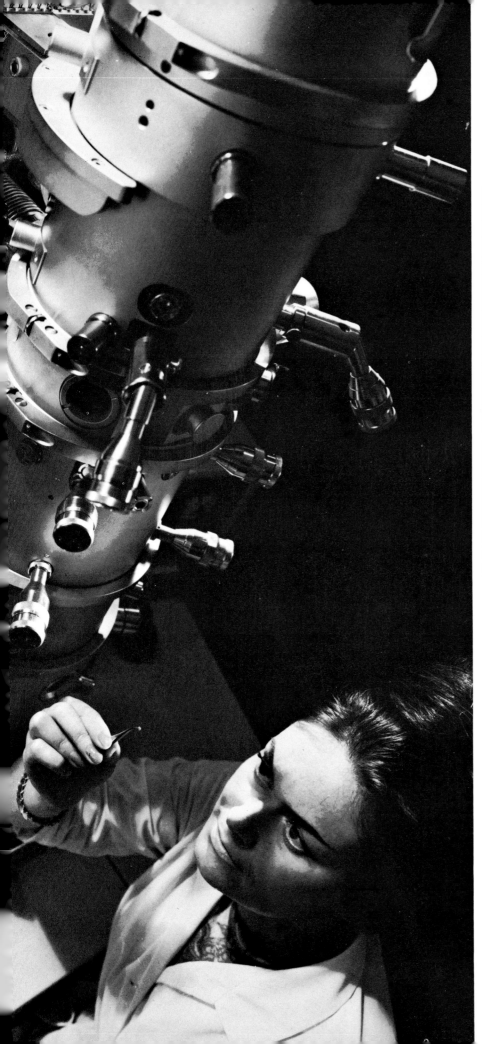

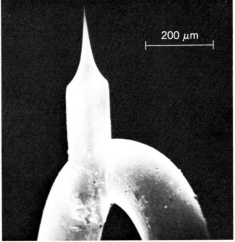

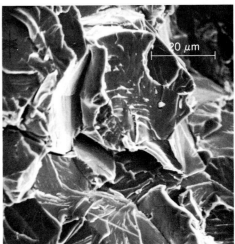

Top. Field-emission tip of CWIKSCAN/100 field-emission scanning electron microscope. The tip consists of an etched single crystal of tungsten welded to tungsten wire. *(American Optical Corporation)*

Bottom. Scanning electron micrograph of mild steel bolt which failed at subzero temperature. *(Industrial Research, June, 1971)*

Left. New Siemens Elmiskop 101 electron microscope. *(Industrial Research, June, 1971)*

Tennessee Valley Authority strip mine reclamation. Upper photo shows area before work was started. Lower photo shows the same area about 3 years later. Drainage and tree planting stabilized spoil banks and made road usable. *(Science, 170(3962):1119, 1970)*

This hailstone fell at Coffeyville, Kans., in September, 1970. Shown with a ruler and hen's egg for comparison, the hailstone may be a world record, being 44 cm in circumference and weighing 766 g. Previous record was 43 cm and 680 g. *(National Center for Atmospheric Research)*

Communications antenna designed to be boosted into space while folded up and then to unfold to the configuration shown here. *(Goodyear Aerospace Corporation)*

The geoduck *(Panope generosa)*, a giant clam on the West Coast, can weigh over 10 lb. All of the clam is edible and is considered a delicacy. The geoduck burrows 18–82 in. into soft mud below lowest tide level. It can be located by its occasional water spout and is best dug for with an open-ended 10-gallon can and a shovel. *(Washington State Fisheries Department)*

A laser cuts cloth with a precise, clean cut (right) compared with knife-cut strip (left).

Computerized laser fabric cutter, manufactured by Hughes Aircraft Company, in use at Genesco, Inc.

Laser-cut suit fabric rolls from cutting machine.
(Aviation Week & Space Technology, March 29, 1971)

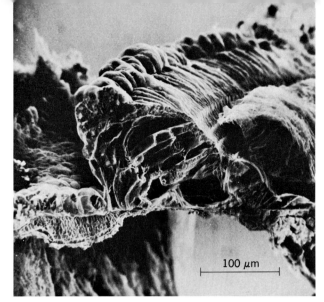

Left. Fracture through the organ of Corti of the guinea pig showing the acoustic papilla. Scanning electron micrographs of fractured freeze-dried specimens permit study of the interior structures of the cochlear sensory epithelium.

Below left. Surface pattern of the sensory epithelium of the normal organ of Corti is regular and geometric, with the hairs (stereocilia) of the sensory cells in characteristic W pattern.

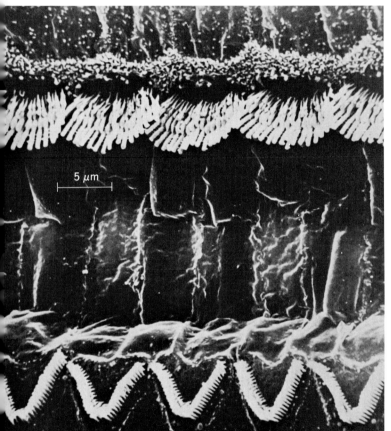

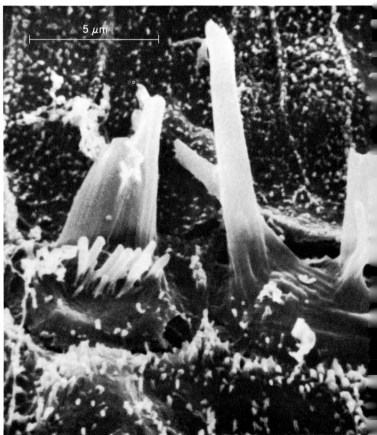

Above. Second coil of cochlea from cat exposed to pure tone (125 Hz at 152.5 dB for 4 hr). Note fusion of stereocilia and growth of giant hairs on two inner hair cells.

Apical coil from a guinea pig cochlea showing an area of complete degeneration of the organ of Corti due to exposure to a pure tone (125 Hz at 148 dB for 4 hr). (G. Bredberg et al., Scanning electron microscopy of the organ of Corti, Science, 170(3960): 861–863, 1970)

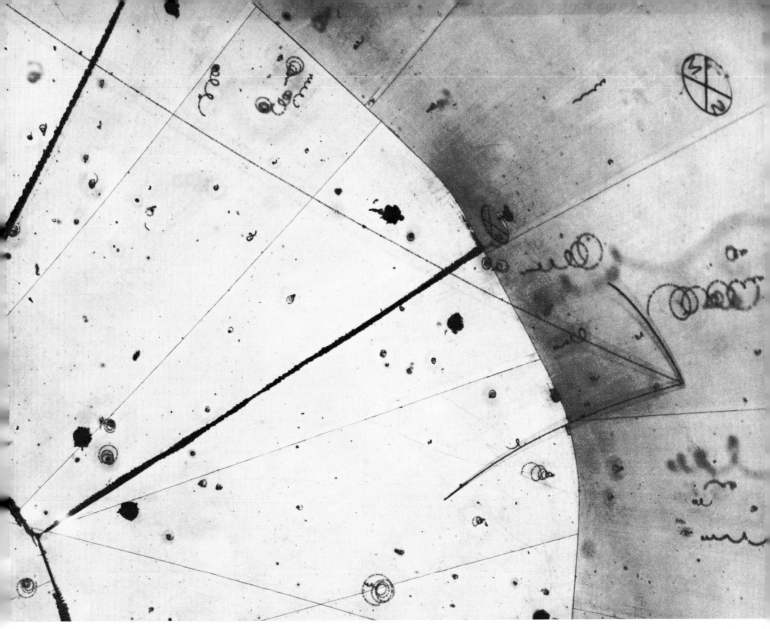

Photograph of first neutrino interaction to be observed in a hydrogen bubble chamber. Chargeless neutrino enters from right, leaving no track, and strikes a proton to produce three new particles. *(Argonne National Laboratory)*

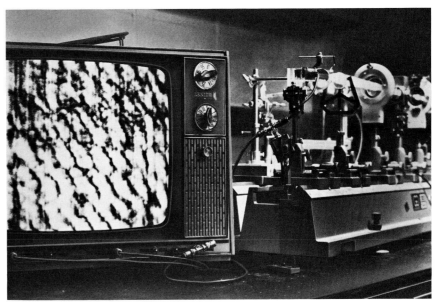

Individual cells of onion skin under an acoustic microscope. Sound waves sent through the onion skin are scanned by a laser beam. The deflected beam is converted into electrical signals and projected on the television monitor. *(Zenith Radio Corporation)*

Left. Completed 15,000-ton bottomless tank, higher than a 20-story building, at construction site. It was built for the Continental Oil Company as an underwater petroleum storage tank.

Below. Tank being towed by two ocean-going tugs to be submerged about 58 miles off the coast of Dubai in the Arabian Gulf.

Above. Tank releasing air during submerging process.

Right. Neck of tank shown just as the huge dome-shaped bottom was settling on the sea floor, 158 feet below the surface. *(Continental Oil Company)*

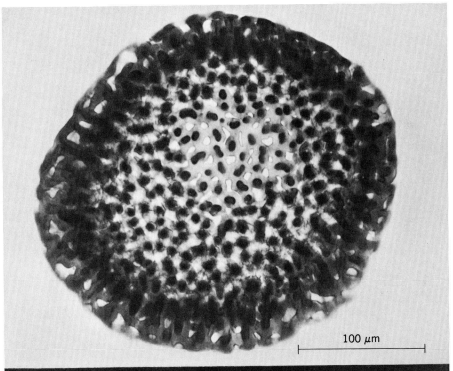
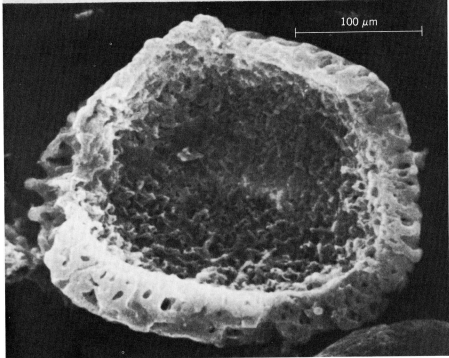
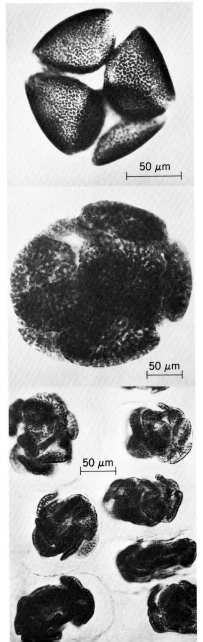

Distal face (top) and proximal face (bottom) of *Cymbopetalum odoratissimum* pollen grain. The pollen of the primitive angiosperm family Annonaceae is unique in its large size, in lacking exine over nearly 50% of the grain surface at maturity, and in having a proximalipolar aperture.

Right. Upper photo of tetrad of *Asteranthe asterias* shows proximal position of aperture. Annonaceae pollen is further distinguished by being in polyads, shown by middle photo of octad of *C. gracile*, which are compartmentalized individually within septate stamens, shown by lower photo of *Porcelia steinbachii*. (J. W. Walker, Unique type of angiosperm pollen from the family Annonaceae, Science, 172(3983): 565–567, 1971)

An 88-inch-diameter mirror blank, manufactured by Corning Glass Works, is the primary optical component in a reflector telescope at Mauna Kea Observatory, Hawaii.

Mauna Kea Observatory, near the top of an extinct 13,796-foot-high volcano, is the highest in the world.

Reflector telescope inside the Mauna Kea Observatory building.
(Corning Glass Works)

Views of a 138-inch-diameter glass disk undergoing final grinding and polishing. The 13-ton mirror blank was made for the new Italian National Observatory. *(Corning Glass Works)*

Wankel rotating internal combustion engine installed in a standard Cessna Cardinal 177. Liquid-cooled, 185-hp engine weighs about half as much as a conventional reciprocating engine of same horsepower. *(Aviation Week & Space Technology, 95(5):17, 1971)*

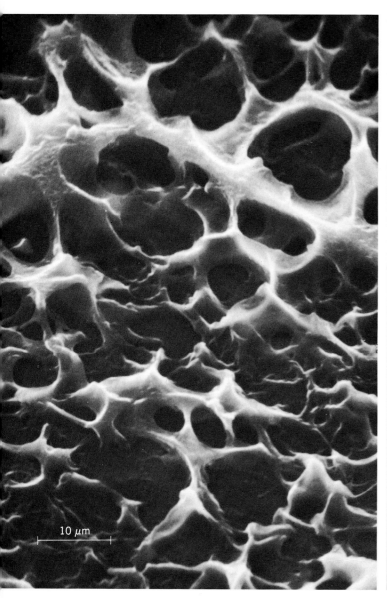
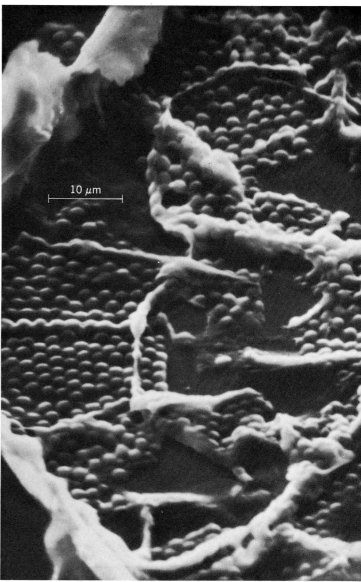

Scanning electron micrographs of corn endosperm sections from which the starch has been removed. Photo at right is of normal corn, in which the protein zein is stored in spherical bodies and is of relatively poor nutritive value. Photo at left is of a high-lysine corn variety, in which much of the protein is in a matrix of interstitial material and has a higher nutritive value. *(Northern Regional Research Laboratory, U.S. Department of Agriculture)*

Absorption of electromagnetic radiation

Absorption of electromagnetic radiation

Highly relativistic electrons coursing through the magnetic fields of synchrotrons or electron storage rings emit a continuum of electromagnetic radiation extending from the infrared to a maximum in the x-ray or soft x-ray region of the spectrum. Currently scientists are using this radiation to carry out high-resolution spectroscopic studies of matter in the vacuum ultraviolet (wavelengths from 105 nm down to 5 nm) where there is a lack of suitable laboratory light sources. Reflectivity measurements can be made at the longer wavelengths in the above range, and these have yielded many new spectral features, especially for wide-band gap materials. The transmission of thin films can also be observed, and this technique is especially suited to soft x-ray absorption spectroscopy at the shorter wavelengths near thresholds corresponding to inner shell excitation.

Soft x-ray absorption. Soft x-ray absorption spectroscopy is a technique which is complementary to x-ray emission spectroscopy. Whereas the x-ray emission technique gives information about the filled levels of valence bands of a solid, soft x-ray absorption yields information about the unoccupied states above the Fermi level in a metal or about the empty conduction bands in the case of an insulator. For a number of materials it is now

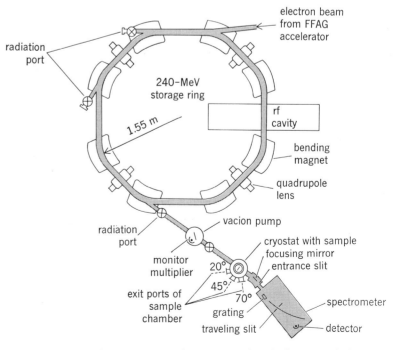

Fig. 1. Storage ring for extreme ultraviolet as set up for thin-film transmission measurements. (*From C. Gähwiller, F. C. Brown, and H. Fujita, Extreme ultraviolet spectroscopy with the use of a storage ring light source, Rev. Sci. Instr., 41:1275, 1970*)

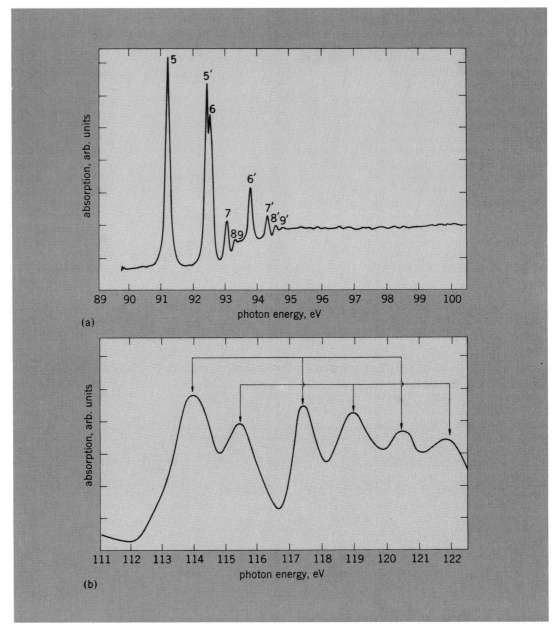

Fig. 2. Absorption as a function of energy. (a) Absorption of krypton gas in the vicinity of 95 eV. Two series of lines are apparent. (b) Absorption of the Rb+ ion (isoelectric with Kr) in the solid RbCl.

possible to understand the structure observed just above the soft x-ray thresholds in terms of the density of unoccupied states computed from band theory. Peaks in the observed spectra are to be associated with regions of momentum space where high state densities occur. In addition, some account has to be taken of matrix-element effects.

Figure 1 shows an outline of the 240-MeV electron storage ring at the University of Wisconsin, which is used by a number of groups as an ultraviolet light source. The storage ring itself consists of eight bending magnets and separate quadrupole lenses arranged so that electrons circulate in an ultrahigh vacuum chamber. A fixed-field alternating gradient (FFAG) accelerator produces a pulse of about 10^{10} 50-MeV electrons which are injected into the storage ring and then slowly accelerated by the electric field of a radio-frequency cavity to a final energy of 240 MeV. During this acceleration the magnetic fields of the storage ring are slowly increased until the maximum energy is reached, whereupon the field is held constant and the stored beam circulates in the ultrahigh (10^{-10} torr, where 133.32 torr = 1 N/m²) vacuum for many hours. During this time the FFAG accelerator is turned off and scientists can work next to the storage ring with little radiation hazard. The stored circulating beam current may be 10 mamp or larger, and synchrotron radiation losses are constantly replenished by means of power from the radio-frequency cavity.

For work in the extreme ultraviolet, synchrotron radiation emitted from a small segment of the electron orbit is focused by a grazing incidence mirror onto the entrance slits of a 2-m grazing incidence spectrometer. The entire optical system, including the spectrometer, is in a vacuum compatible with the very high vacuum of the storage ring. Thin-film samples are evaporated onto substrates of Formvar and then rotated into place in front of the entrance slits. A gas cell can also be positioned in the beam. The spectrometer employs photon counting and continuous scanning as a function of wavelength. All information is recorded in digital form and later analyzed by means of a computer. Corrections can be made for the response of the apparatus, stray light, the absorption of substrates and of gas-cell windows, and so forth.

Figure 2a shows the absorption as a function of energy for krypton gas in the vicinity of 95 eV. Two series of sharp absorption lines appear separated by 1.24 eV, the spin orbit splitting of the $3d^{10}$ core states involved. Krypton is a closed-shell rare-gas atom with an electronic configuration consisting of the argon core plus $3d^{10}4s^24p^6$. Each series of lines in the above spectrum corresponds to an excited electron bound to a hole in the d core. In other words, the final state electron configuration is $3d^94s^24p^6np$, where p is 5, 6, 7, 8, or 9 as indicated. The d shell may be left with inner quantum number $j = 5/2$, corresponding to the series indicated by unprimed numbers in Fig. 2, or with $j = 3/2$, corresponding to the primed series of Fig. 2.

Thin-film absorption. Figure 2b shows the d excitation spectra for Rb$^+$ in the ionic crystal RbCl. Rb$^+$ is isoelectronic with the krypton atom, and so one expects a d threshold similar to that of the rare gas but shifted to higher energy because of the higher atomic number of the Rb nucleus. It can be seen that this threshold begins at about 113 eV and consists of two series of very broad lines, or bands, separated by about 1 eV. The first, third, and fifth lines in the figure belong to the same series, whereas the second, fourth, and sixth bands are members of a second series shifted by about 1.0 eV, the known spin-orbit splitting of the Rb $3d$ shell. Although there is some resemblance, these broad lines in the solid are not excitation states in the same sense as for the rare-gas atom. A detailed comparison with the results of band theory for RbCl shows that they are due to maxima in the conduction-band density of states and thus correspond to specific regions in the Brillouin zone for the solid.

Figure 3 shows thin-film absorption spectra for RbCl and CsCl over a very wide range of photon energies. The Rb $3d$ threshold can be seen in the vicinity of 111 eV. Thereafter the absorption rises slowly to a new series of lines corresponding to excitation of the $L_{II,III}$ or $2p^6$ electrons on the chlorine just above 200 eV. Here again the spin-orbit splitting of the core levels and details of the conduction-band density of states appear in the spectrum.

In Fig. 3b one has a threshold corresponding to excitation of the $4d^{10}$ electrons on the cesium at about 80 eV. Here, however, the absorption rises much more rapidly beyond threshold to a large and very broad resonance extending from 80 to 140 eV. This is an atomic effect seen also in xenon gas, which is isoelectronic with cesium. This effect is due to transitions from the $4d$ shell to f like final

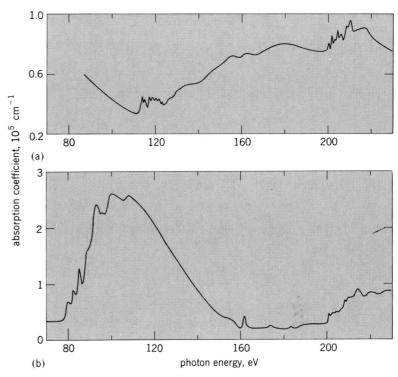

Fig. 3. Thin-film absorption spectra over a very wide range of energies in the extreme ultraviolet. (a) RbCl. (b) CsCl. (From F. C. Brown et al., Extreme ultraviolet spectra of ionic crystals, Phys. Rev., B2:2126, 1970)

states in the continuum. The threshold is suppressed but matrix elements increase greatly beyond threshold. Excitation of the cesium $4p$ shell can be seen at about 160 eV, followed by the chlorine $L_{II,III}$ threshold again at 200 eV. Again the details just above the various thresholds relate to the electronic band structure of the solid. The broad features and resonances far above threshold relate to atomic properties and the nature of atomic excitations above the ionization threshold. So far most observations in this part of the spectrum are rich and varied, and it can be said that one is just beginning to understand the phenomena involved within the principles of modern quantum mechanics.

For background information see ABSORPTION OF ELECTROMAGNETIC RADIATION; BAND THEORY OF SOLIDS; ELECTRON CONFIGURATION in the McGraw-Hill Encyclopedia of Science and Technology. [FREDERICK C. BROWN]

Bibliography: F. C. Brown et al., *Phys. Rev.*, B2: 2126, 1970; F. C. Brown et al., *Phys. Rev. Lett.*, 25: 927, 1970; U. Fano and J. W. Cooper, *Rev. Mod. Phys.*, 40:441, 1968; C. Gähwiller, F. C. Brown, and H. Fujita, *Rev. Sci. Instr.*, 41:1275, 1970; R. Haensel et al., *Phys. Rev. Lett.*, 25:208, 1970.

Agricultural meteorology

The recent emphasis on the weather phase of plant disease development has brought plant disease forecasting into the area of agricultural meteorology. Correlation of plant disease forecasts with weather forecasts and identification of the meteorological situations that predetermine the kind of weather favoring or hampering the development of a disease are subjects that require, and are receiving, the combined attention of plant pathologists and agricultural meteorologists.

Temperature-moisture relationship. Bacteria, fungi, and viruses are the principal pathogens that cause crop plant diseases. The meteorological factors of greatest importance in forecasting the incidence and development of plant diseases are probably temperature and moisture. Moisture that influences the growth and development of the plant pathogens may be rain, dew, fog, mist, relative humidity, or the guttation water that a plant pumps out onto its own leaf surfaces. Wind flow is another useful weather variable in crop plant disease development and forecasting.

Each plant pathogen has specific temperature-moisture requirements. Relating these requirements to the epiphytotics caused by these pathogens is the first step toward forecasting. Duration of temperature-moisture values is of critical importance, because if they are optimum for the pathogen growing in a large community of susceptible host plants in the field, the pathogen will produce an epiphytotic. J. R. Wallin in 1967 and Paul R. Miller in 1969 discussed the meteorological parameters of many different plant diseases.

P. M. A. Bourke in 1970 reviewed many individual research papers dealing with the use of weather information in the prediction of plant disease epiphytotics and stated that plant diseases for which forecasting with the aid of meteorological information can be of practical value to the grower are those which meet four basic requirements: (1) The disease is one which causes economically significant damage in terms of quantity or quality in the area concerned. (2) The disease is variable in seasonal impact—time of onset, speed of buildup, ultimate level—and an appreciable part of this variation can be attributed to weather factors, acting directly or indirectly. (3) Control measures, whether curative or preventative, are available and can be operated at economically acceptable cost. (4) Information is available from laboratory or other investigations of the nature of the weather dependence of the disease. Some important diseases meeting these four requirements will be discussed.

Temperature. Temperature is used in predicting the early appearance and severity of blue mold of tobacco. In the Southeastern United States warmer than normal January temperatures accurately forecast the appearance of the disease. Forecasts of bacterial wilt of corn are based upon "winter temperature indexes," that is, the sum of the monthly mean temperatures for December through February at specific locations. An index below 32°C indicates that the disease will be absent the following season. An index over 38°C indicates that the disease will be destructive. An index between 32 and 38°C indicates that moderate severity should be expected.

Temperature and rainfall. Incidence of downy mildew of lima bean is predicted after 8 consecutive days when the 5-day mean temperature is less than 26°C and the concurrent 10-day cumulative rainfall is 1.20 in. (30.48 mm) or more, provided that the current weather forecast calls for continued mildew-favorable weather. In the North Atlantic states, the critical appearance of potato late blight is forecast after 10 consecutive blight-favorable days with a weather forecast of continuing blight-favorable weather. A favorable day is defined as a day when the mean temperature for the 5 days ending that day is less than 26°C, coincident with the 10-day cumulative rainfall of 1.2 in. (30.48 mm) or more.

Temperature, relative humidity, and dew. In the North Central states the 90% relative humidity–temperature–time method of forecasting the incidence and spread of potato late blight was developed. A series of "infection values" were determined depending upon the coincidence of given average temperatures with relative humidity equal to or greater than 90% for a specific time in hours. In general, the first late blight symptom is expected to appear with a postemergence cumulative severity value of 18. After the disease becomes established, much lower values suffice to allow the pathogen to survive and be ready to spread with recurrence of higher values. The absence of any favorable period for 3 consecutive weeks indicates that fungicide application can be discontinued for the season.

Apple scab forecasting and warning services are conducted in Europe and the United States. The criteria are the duration of leaf wetness caused by rain, fog, dew, or mist and coincident favorable temperatures in the orchard, because infection depends upon the number of hours that the leaves are wet and the temperature during this period. A minimum of 10 hr of leaf wetness, with coincident temperatures of 16–24°C, is required for infection. The growers are advised of the time of ripening of the ascospores, the date when spore discharge was observed, and the occurrence of infection periods. Dew recorders are employed in several European countries to record duration of leaf wetness.

Dew is considered an important factor in downy mildew of grape epiphytotics in France and Italy, where mildew forecasting is based on the interrelation of meteorological factors and disease development. In Italy, if the dew point is higher than 12°C, the lowest temperature at which spore germination occurs, an outbreak of the vine mildew might be expected.

The leaf spot disease of peanuts is forecast on the basis of the number of hours the relative humidity remains at or above 95% and the minimum temperature during such periods. In general, rapid disease spread occurs when the relative humidity is equal to or exceeds 95% for 10 or more hours, coincident with temperatures equal to or exceeding 21°C. If these conditions persist for 2 or 3 successive days, disease increase is forecast.

Wind-flow data. The incidence of the barley yellow dwarf virus on small-grain crops in Iowa can be predicted from the occurrence and persistence of low-level jet winds for 24 hr or more from Texas and Oklahoma into Iowa. These winds transport the virus aphid vectors from their overwintering areas in the southern plains northward. After the occurrence of a 24-hr jet wind, the small-grain crops in the target area are searched immediately for the aphid vectors. Generally the symptoms of barley yellow dwarf disease can be predicted 21–30 days after the first vectors are noted.

The synoptic weather chart is an aid in disease forecasting. In coastal northwest Europe the major mechanism governing potato blight attacks is the frequency and duration of influxes of maritime tropical air. This warm, moist airmass, which originates in the tropical part of the Atlantic Ocean, cools as it moves toward Europe, so that it arrives with ideal conditions of temperature, air humidity, and precipitation not only for the potato disease but for the downy mildew of onions in the Netherlands. In the North Central United States the identifications of potato blight–favorable weather on the ground is made from 5-day mean surface and upper air maps. Southeasterly surface wind flow and southwesterly flow at 10,000 ft indicates blight-favorable weather. West-to-northwest wind flow at the surface and at 10,000 ft indicates blight-unfavorable weather. The typical model on a daily surface weather map indicative of late blight conditions in crop cover is a nearly stationary east-to-west front just south of the target area with a weak migratory low-pressure system along the front. On the ground the wind flow is from east to west, north of the front.

Basidiospore production and infection of pines with fusiform rust are favored when a surface high-pressure system is centered off the southeastern coast of the United States and causes a persistent wide current of maritime tropical air to move northward over the Gulf states. The prevalence and persistence of the same airmass were found to be related to outbreaks of the downy mildew of lima beans in New Jersey.

C. B. Gullach and Wallin in 1970 found the flow pattern of the upper air at the 500 mb (50,000 N/m^2) level to be the key to the surface weather pattern influencing the development of the fungus *Cercospora beticola* on sugarbeet. Stable longwave conditions of the upper air provide more persistent surface conditions unfavorable for leaf spot development. A train of shortwaves produces favorable weather conditions for leaf spot development. Favorable weather conditions on the ground are leaf-wetness periods of 48–72 hr, coincident with temperatures of 24–29°C. The identification of these wave conditions circulating in the Northern Hemisphere could be used in forecasting favorable or unfavorable conditions for sugarbeet leaf spot development on the ground.

Computer use in forecasting. In Germany a system for predicting potato late blight employs data on the four different stages in the life cycle of the fungus that are known to be related to different kinds of weather: (1) sporulation, requiring relative humidity equal to or greater than 90% for over 10 hr; (2) germination and infection at such humidities for over 4 hr; (3) mycelial growth, depending on temperature but independent of moisture; and (4) suppression of disease extension when relative humidities are less than 70%. The different weather sequences are weighted by multiplying them by empirically derived parameters in accordance with the air temperature, the parameter for the dry periods being negative. The current total weather rating is reckoned each week by adding the weekly values, starting from a known average date of emergence of early potatoes. The date on which the total weather rating reaches 150 is called the first critical date. Lower values indicate a negative forecast. Generally the first symptoms of the disease are seen 10–40 days after the first critical date. The date on which the total weather rating reaches 270 is called the second critical date. Growers are then advised to apply control measures immediately. The system is based upon weekly bulletins from about 50 standard weather reporting stations which codify the past 168 hourly observations of temperature, relative humidity, and rainfall. These data are fed into an electronic computer, which weighs the observations with the appropriate parameters and calculates the weekly weather rating to date and, if appropriate, the predicted times of occurrence, assuming average climatological conditions of the first and second critical dates.

Programming a computer to simulate plant disease was done in 1969 by P. E. Waggoner and J. G. Horsfall. The simulator, called EPIDEM, employs temperature, relative humidity, wind speed, sunniness, and wetness for each 3 hr of each day. Each 3 hr, it modulates the course of the following fungal stages according to the different affect of the weather factors upon them: formation of conidiophores, formation of spores, departure of spores on wind or rain, finding a host, germination of the spores, penetration of the host, incubation of the infection, and expansion of the lesion. This is a computer program written in FORTRAN IV. The simulator has not been applied to produce disease predictions for current use, but there seems no reason why it could not be supplied with up-to-date weather data in order to yield forecasts. EPIDEM is not readily adaptable to utilizing forecast weather. Currently the same investigators are experimenting with EPIMAY to simulate the serious southern corn leaf blight disease of corn.

For background information *see* AGRICULTURAL METEOROLOGY; PLANT DISEASE; PLANT DISEASE CONTROL in the McGraw-Hill Encyclopedia of Science and Technology. [J. R. WALLIN]

Bibliography: P. M. A. Bourke, *Annu. Rev. Phytopathol.*, 8:345–370, 1970; C. B. Gullach and J. R. Wallin, *Int. J. Biometeorol.*, 14:349–355, 1970; P. R. Miller, *Phytoprotection*, 50:81–94, 1969; P. E. Waggoner and J. G. Horsfall, *Bull. Conn. Agr. Exp. Sta.*, no. 698, 1969; J. R. Wallin, in *Ground Level Climatology*, American Association for the Advancement of Science, 1967.

Air pollution

In their combination of noxiousness and quantity the sulfur oxides (SO_2 and SO_3) are the most important of air pollutants, particularly in the light of the potential growth in their emissions (see table). Sulfur trioxide generally constitutes only 1–5% of the total emissions, depending on the nature of the combustion process producing the sulfur oxide emission. On entering the atmosphere it combines with water vapor to form a sulfuric acid aerosol, which is primarily responsible for the dense white stack plumes emitted from the smelting of sulfide ores and the combustion of high-sulfur fuels. However, sulfur dioxide presents a far more serious problem because of its quantity, its effects, and its difficulty of control.

Projections indicate that by the end of the century the potential emission of sulfur oxides in the United States could increase by threefold and that the emission for the world as a whole could approximately double. By far the greater part of both the current emission and the expected growth results from the combustion of sulfur-bearing coal and oil

Projected potential emissions of sulfur dioxide in the United States (no abatement assumed)

Source	Annual emission of sulfur dioxide, millions of tons*				
	1967	1970	1980	1990	2000
Power plant operation (coal and oil)†	15.0	20.0	41.1	62.0	94.5
Other combustion of coal	5.1	4.8	4.0	3.1	1.6
Combustion of petroleum products (excluding power plant oil)	2.8	3.4	3.9	4.3	5.1
Smelting of metallic ores	3.8	4.0	5.3	7.1	9.6
Petroleum refinery operation	2.1	2.4	4.0	6.5	10.5
Miscellaneous sources‡	2.0	2.0	2.6	3.4	4.5
Total	30.8	36.6	60.9	86.4	125.8

*One ton = 907.19 kg.
†Assumes use of breeder nuclear reactors.
‡Includes coke processing, sulfuric acid plants, coal refuse banks, refuse incineration, and pulp and paper manufacturing.
SOURCE: February, 1970, estimates by National Air Pollution Control Administration (transportation excluded).

fuels, primarily for the generation of electric power. Legal pressures for control of sulfur oxide emissions have increased rapidly during 1969–1971, particularly in the United States but in other industrialized countries as well.

Recovery of sulfur dioxide present at low concentrations in waste gases is inherently uneconomic by conventional standards. Increases in sulfur supplies and a sharp drop in sulfur prices since 1968 have made economic recovery from dilute gases even less favorable than previously. In 1970 the quantity of by-product sulfur recovered in the Western world for the first time exceeded the production of elemental sulfur by the Frasch process. The additional prospect of recovery of large quantities of pollutant sulfur reduces the potential opportunities for profitable recovery of by-product sulfur from any sources. The quantity of sulfur emitted as sulfur dioxide far exceeds the quantity used for all purposes.

Control of power plant emissions. Until 1969 there was widespread expectation that extensive application of sulfur dioxide removal processes to power plant flue gases was imminent. However, utility companies faced with the necessity of curtailing their sulfur oxides emissions have uniformly elected to switch to use of low-sulfur fuels (natural gas or low-sulfur oil or coal) rather than install stack gas treating systems. Although two full-scale limestone-injection wet-scrubbing systems installed in United States power plants were ostensibly commercial units, both are actually demonstration plants, and the process is not considered to be "commercially proven" at this time. Two more processes, the Monsanto Cat-Ox system and the Chemico-Basic system, are to be applied in full-scale demonstration plants with partial Federal financial support, and other demonstration projects are being planned. Nevertheless the utility companies evidently prefer to pay premium prices for low-sulfur fuels rather than accept the high capital investments and the technical risks associated with installation of the flue gas treating processes.

In power plant flue gases the concentration of sulfur dioxide is commonly in the range of 0.1 to 0.3%, and the values of the sulfur by-products potentially recoverable are small in relation to the capital and operating costs associated with their recovery. Profitable recovery of the by-product sulfur from flue gases is thus indicated to be virtually impossible regardless of the technical approach employed, despite unrealistic claims that occasionally have been made.

Because of the variability in the demand for electricity, power plants are themselves basically inefficient in their use of investment capital and operating labor, and stack gas treating systems attached to the power plants suffer from the same disadvantage. A large, new power plant handling base electrical loads may operate relatively steadily, but over a 30-year plant lifetime the average annual capacity factor is typically only 50–60%. This low percentage utilization greatly accentuates the influence of fixed charges on the economics of the gas treating system. Because of these considerations, utilities are currently less concerned with possibilities of economic return from stack gas treatment than they are with minimum overall cost, maximum reliability, and flexibility of the treatment system in coping with startups, shutdowns, and fluctuations in load.

Additional problems are encountered where flue gas treating systems must be fitted to existing power plants. Some proposed control systems cannot be installed without extensive alterations of the boiler system, and hence are suited only for application to new power plants. The space required for most of the systems is at least considerable, and frequently very large, and may exceed that available at power plants located in urban areas. The weight of the equipment may practically eliminate installation on power plant roofs. A growing appreciation of the foregoing problems is evidently now leading to much more active consideration of various techniques for production of low-sulfur fuels.

Small combustion sources. The problems of installing flue gas treating systems are accentuated at the smaller sources, such as small utility plants, industrial boiler and power plants and process and space heaters, and commercial space heaters. Although several of the simpler flue gas processes have been suggested for such applications, there are evidently few cases in which any course but substitution of low-sulfur fuels is being seriously considered. The cost of fuel is seldom more than a small fraction of the total cost of production of industrial products. Consequently the increased cost of low-sulfur fuels is usually of relatively less concern to industrial companies than are reliability, simplicity, and convenience.

SO₂ removal from flue gas. Currently there are probably no less than 100 proposed flue gas treating systems under study at stages ranging from laboratory experimentation to large demonstration plants. Many of these processes are essentially variations on a much smaller number of basic chemical processes. However, there are a number of radically different approaches among the basic processes.

Recently a number of novel processes, or novel adaptations of other technology, have been proposed. Some of these are notable for the complexity of the chemistry and mechanical arrangements that they would employ and offer little attraction where simplicity and economy are critical requirements.

For convenience, the flue gas treating processes can be divided into a small number of broad classes. Nonrecovery processes, generally based on absorption of sulfur dioxide by lime or limestone, are intended only to fix the sulfur in disposable waste materials. However, most of the processes or systems are intended to recover the sulfur dioxide in useful forms. Another division of recovery processes into categories separates the systems into dry, hot processes (generally operating at flue gas temperatures above 250°F, that is, 121°C) and processes that require that the flue gas be cooled and contacted with an absorbent or adsorbent at relatively low temperatures (usually under 150°F, that is, 65.5°C). The dry, hot processes generally employ adsorption or absorption of the sulfur dioxide by solids. The low-temperature processes usually employ aqueous solutions or slurries for absorption, or at least require water scrubbing for preliminary cooling and cleaning of the gas.

The by-products of the recovery systems are usually sulfuric acid, elemental sulfur, concentrated sulfur dioxide, or ammonium sulfate. Sulfur dioxide is seldom considered to be a practical final product but to be an intermediate in the production of sulfuric acid or elemental sulfur. Processes producing sulfuric acid directly from the flue gases yield dilute acid that would in most cases require concentration to produce an acceptable commercial product. Elemental sulfur is a highly desirable product because it can be easily stored or be shipped relatively cheaply. However, the reduction of the sulfur dioxide to elemental sulfur requires additional equipment and processing steps as well as an additional raw material to serve, directly or indirectly, as the reducing agent.

The presence of fly ash in the flue gas has presented many problems in flue gas treating systems, even in cases where the bulk of the ash has been removed by dust collectors upstream of the sulfur dioxide collection system. In a number of pilot and demonstration plant studies, the difficulties created by the ash have exceeded those originating in the sulfur dioxide treating systems themselves. For the systems intended to recover useful by-products, at least, it appears that high-efficiency removal of fly ash ahead of the recovery unit will be necessary.

Dry, hot collection systems. The development of dry, hot processes over the past decade was inspired primarily by concern with dispersion of the treated flue gas after it is discharged to the atmosphere. The earlier interest has now been diminished by the many serious problems encountered during development. Fly ash must often be removed from the flue gas at high temperature and with high efficiency to avoid interference with the sulfur dioxide collection process. Mass transfer is limited with the solid adsorbents and absorbents, and in comparison with systems using liquid absorbents, it has been difficult to attain high sulfur dioxide removal efficiency. With granular absorbents the amount of surface available for reaction is generally limited, and the fraction of the material utilized tends to be relatively low. The same problem remains to some degree even when the absorbent solids are finely divided. Loss of absorbent reactivity and porosity from chemical changes occurring during regeneration has been reported. Mechanical attrition of absorbent is one of the most serious of the difficulties encountered.

In the dry, hot collection systems the absorption or adsorption of the sulfur dioxide is generally accompanied by its oxidation to sulfate, the sulfites commonly being unstable at elevated temperatures. Chemical methods are therefore necessary for regeneration of the absorbent, whereas physical methods of regeneration can be used in some of the aqueous absorption systems in which sulfites are formed.

Several processes under test employ various forms of active carbon or semiactivated coke as adsorbents. Examples are the Reinluft and Bergbau-Forschung processes in West Germany, the Hitachi process in Japan, and the Westvaco process in the United States. Various forms of gas-solid contactors (fixed, moving, and fluidized beds) and methods of adsorbent regeneration (hot-gas stripping, washing) are used.

Lime and limestone have also been used as solid absorbents. In the simplest form of the process, finely ground limestone is injected into the boiler furnace and later removed from the flue gas by the fly ash collector. Tests conducted at a number of locations have indicated that it is generally impractical to attain removal of more than 40–50% of the sulfur dioxide by this method and that other adverse effects (such as overloading of the fly ash precipitator) may appear.

Although not a "dry" system, the Monsanto Cat-Ox process may be placed in this category because it involves no contacting of the flue gas with added water and produces relatively little cooling (40–60°F, that is, 20–35°C) of the exit gas below the usual levels of power plant stack gas temperatures. The Cat-Ox process is an adaptation of the contact process for sulfuric acid manufacture.

Broadly, it appears that the extra costs of the dry, hot systems may more than compensate for the cost of reheating the stack gases from aqueous scrubbing systems to insure dispersion, even assuming that reheating is required. However, no comprehensive cost analysis has been made to resolve the question.

Nonrecovery systems. The principal nonrecovery systems (both dry and wet processes) commonly employ limestone or lime (or sometimes dolomite) as the absorbent, with the resulting calcium sulfate and sulfite being discarded. The combination of injection of limestone into the boiler furnace and subsequent wet scrubbing of the flue gas for sulfur dioxide and fly ash removal has been receiving much attention. Operating problems in demonstration units have resulted from blockage of the scrubbers by fly ash and from scaling of surfaces by insoluble calcium compounds. Many of these problems appear to have resulted from deficiencies in the designs of the scrubbers and other system components. However, it is not yet certain that the injection of the limestone into the furnace is not responsible for some of the maintenance problems of both the scrubbers and the boiler furnaces. If the injection of the limestone into the furnace should have to be abandoned, either limestone or precalcined lime must then be added directly to the scrubber water. Separate calcination of the limestone, if required, will add to the costs of the process.

Nominally, at least, wet scrubbing with lime or limestone, with discard of the collected sulfur compounds, appears to be one of the cheapest of the sulfur dioxide control processes. However, the actual costs must include realistic charges for the lime or limestone and for disposal of the calcium sulfite and sulfate waste. The cost of waste disposal, in particular, may be generally underestimated. Attempts are being made to find uses for the waste, but if it must be processed in some manner, much of the presumed advantage of the nonrecovery process may be lost.

Other forms of nonrecovery processes under study use solutions of sodium hydroxide or of ammonia in the scrubber. The spent solution of sodium or ammonium sulfite is treated with lime to precipitate calcium sulfite and release the soluble base for recycle to the absorber. The principal advantages sought in such a system are elimination of handling insoluble materials in the scrubber, more efficient contacting of gas by solution rather than slurry, and reduction of the need for excess lime that is required where the lime is used directly as the absorbent.

Wet scrubbing recovery systems. Attempts have been made to absorb sulfur dioxide in water and oxidize it in solution directly to sulfuric acid. Such processes have failed because of the slow rate of oxidation of the dissolved sulfur dioxide and the difficulty of dissolving the gas in the acid solution.

Most of the absorption systems using various basic aqueous solutions or slurries as absorbents are intended to concentrate the sulfur dioxide from the flue gas stream in a cyclic process. The absorbents are commonly oxides, hydroxides, or salts of weak acids, such as carbonates or sulfites. During the absorption step sulfur dioxide is absorbed, with formation of a sulfite or bisulfite. In the succeeding regeneration step the sulfur dioxide is released at high concentration from the spent absorbent by physical or chemical means, and the absorbent base is recovered for recycle to the absorber.

The bases most commonly used or proposed for such processes are ammonia, sodium, and magnesium. Recovery systems based on ammonia have probably received more study than any others, and one, the Cominco process, has been in commercial use for some 35 years for treating dilute smelter gases and tail gases from sulfuric acid plants. In the Cominco process the ammonium bisulfite solution is treated with sulfuric acid, which releases the sulfur dioxide. The process is thus not fully cyclic but consumes ammonia and sulfuric acid and produces ammonium sulfate as a by-product.

All systems employing a sulfite-bisulfite cycle have a common problem in the oxidation of part of the sulfur dioxide to sulfate. The sulfate cannot be readily regenerated and must be purged from the system. If the sulfate cannot be used directly, it must either be disposed of as waste or else be subjected to additional chemical processing for recovery of the base.

Magnesium oxide can be used in slurry form as an absorbent for sulfur dioxide. The product, magnesium sulfite, which is also insoluble, can be separated from the water for regeneration. Heating the magnesium sulfite does not release sulfur dioxide without undesirable side reactions such as autoxidation to sulfate and thionates or sulfur. However, if the magnesium sulfite is calcined in the presence of excess carbon, the products are sulfur dioxide and magnesium oxide. The sulfur dioxide gas stream is rich enough to be fed to a conventional contact sulfuric acid plant, and the magnesium oxide is recycled to the absorption step. The Chemico-Basic process, which employs this cycle, proposes to carry out the regeneration and recovery operation at a central site, with shipment of spent absorbent from, and of fresh absorbent to, absorption facilities at a number of surrounding sites.

There is growing interest in a process that will yield elemental sulfur rather than sulfur dioxide or sulfuric acid. Such a process proposed by Chemico would employ (1) absorption of sulfur dioxide by sodium carbonate solution, (2) crystallization of the resulting sodium sulfite, (3) reduction of the sodium sulfite to sodium sulfide by producer gas, (4) stripping of the sodium sulfide with steam and carbon dioxide to release hydrogen sulfide and regenerate sodium carbonate, and (5) conversion of the hydrogen sulfide to elemental sulfur by the conventional Claus process.

Costs of flue gas treatment. The probable costs of removing sulfur dioxide from flue gas are poorly defined. A few independent attempts have been made to compare different systems on a common basis, but these estimates also suffer from unavoidable factors of ignorance and commonly employ unrealistically high power plant use factors. The estimated costs commonly lie in a range equivalent to an added electricity generating cost of about 0.5–1.5 mills/kwhr before allowance for by-product recovery credits, if any. This range, including as it does, estimates for radically different systems, is surprisingly narrow. Considering the factors of ignorance involved, it is questionable how significant variations of even twofold in the estimates may really be.

Assignment of by-product credits involves much uncertainty because of the fluctuations in the price of sulfur and the major influence of local market factors.

Pretreatment of fuels. The use of low-sulfur fuels avoids the many problems associated with flue gas treatment, although the supplies of natural low-sulfur fuels tend to be limited, undeveloped, or poorly located geographically. Interest in desulfurizing of fuels has therefore been growing. A number of processes have been developed for desulfurizing residual fuel oil, and a number of plants have already gone into service at petroleum refineries throughout the world. However, no comparable plants exist for pretreatment of coal.

Cleaning of coal can be used to remove a substantial part of the mineral content of coal, including part of the pyritic sulfur. However, work to date indicates that on the average it is not practical to remove more than half of the sulfur content of coal by such mechanical cleaning. About half the sulfur content consists of organic sulfur compounds forming part of the structure of the coal itself, and this can be removed only by chemical means.

Several processes now partly developed with

support from the U. S. government's Office of Coal Research or private companies offer the possibility of producing ash-free, low-sulfur fuels (solids, liquids, or gases) from coal. None of these newer processes are considered ready for immediate large-scale commercialization. From the standpoint of utility plant use, the preferred product would probably be a low-ash, low-sulfur solid fuel with a relatively low melting point, produced at large plants near the coal mines. The Pittsburgh and Midway Coal Mining Co. (PAMCO) process currently under development aims at production of such a material—"solvent-refined coal." The coal is dissolved in a coal-derived solvent, hydrogenated to remove sulfur, and filtered to remove ash. Removal of the solvent leaves a solid fuel.

Removal of sulfur from coal is most readily accomplished if the coal is gasified. New gasification processes are aimed at producing a gas equivalent to natural gas that could be manufactured in large plants and distributed by pipeline. However, individual power plants could employ existing coal gasification technology to generate producer gas for on-site use. Such a commercial producer gas generating system is being constructed by Lurgi for a power plant in West Germany, although desulfurization of the gas is not the objective sought and is not actually to be practiced.

On-site generation of producer gas by utility plants would lack some of the advantages of direct purchase of a low-sulfur fuel, but may be an alternative preferable to burning high-sulfur coal and then collecting fly ash and sulfur dioxide from the flue gas. It would use existing technology and the sulfur would be conveniently recovered in elemental form.

Cost estimates for control of sulfur oxide emissions by fuel pretreatment indicate that the costs may be roughly the same as the more optimistic estimated costs of flue gas treatment. However, the cost estimates for the two paths to control are generally not entirely comparable, since the depreciation charges and the required rates of return levied against utility plant equipment are lower than those usually assumed for fuel conversion plants.

New combustion processes. Some new combustion processes under study are intended to fix the sulfur in a nonvolatile form during the combustion process itself. The most notable process in this category is "fluidized bed" combustion. The boiler furnace contains a fluidized bed of limestone into which the coal is injected. As the coal burns, at least part of the sulfur oxides formed reacts with, and is absorbed by, the limestone, which is itself calcined. The potentials of this process are currently being investigated with prototype units and are still unresolved. Even if ultimately proved successful, fluidized bed combustion will be applicable only to new boiler furnaces built specifically to employ it.

Treatment of rich SO₂ gas streams. Gas streams containing high concentrations of sulfur dioxide (that is, 3.5–4% by volume and higher) can be treated economically by use of existing technology. The concentration of 3.5–4% represents the thermal balance point for the contact sulfuric acid process; at or above this concentration the heat released by the catalytic oxidation of the sulfur dioxide to sulfur trioxide can be exchanged to preheat the cooled, cleaned feed gas to the required reaction temperature; the plant is "autothermal" and requires no auxiliary heat. Gases containing less than about 3.5% of sulfur dioxide may be defined as "dilute" gases. The definitions of "rich" and "dilute" gases are thus made by reference to the economics of sulfur recovery.

The contact sulfuric acid process is by far the best developed process for recovering sulfur dioxide from rich gases. Where elemental sulfur rather than sulfuric acid is the desired product, the sulfur dioxide must be reduced with such agents as coke, natural gas, or producer gas. For reasonable economy, the sulfur dioxide should be present in the gas stream at very high concentrations (15% or more) and there should be little or no oxygen present. A stream of essentially pure sulfur dioxide is preferable but requires operation of a concentration system on the original sulfur dioxide–containing gas stream. Sulfur dioxide reduction plants have been operated commercially but are not well developed and standardized commercial units.

For concentration of sulfur dioxide from rich gases, the Asarco dimethylaniline (DMA) cyclic absorption process is available and in commercial use, although on only a limited scale. Compared with dimethylaniline, the absorbents used in processes intended for concentration of sulfur dioxide from dilute gases, such as flue gas, are relatively uneconomical for use on rich gases.

Noncombustion sources emissions control. Noncombustion sources of sulfur dioxide emissions may be defined as those in which the primary purpose of the operation is other than the burning of a fuel for the indirect transfer of heat, even though the sulfur may actually come primarily from the fuel. They include the roasting and smelting of sulfide ores, the tail gases from sulfuric acid plants, the tail gases from Claus sulfur plants, the refining of petroleum, the coking of coal, and a wide variety of direct-fired industrial processes in which the sulfur dioxide may come either from the fuel or the material being processed or both.

Smelters. Smelting of sulfide ores—primarily those of copper, lead, zinc, and nickel—is second only to combustion of fuels as a source of sulfur dioxide emissions. Major quantities of sulfur are also recovered from rich smelter gases, usually as sulfuric acid. However, large amounts of sulfur dioxide are still emitted from smelting operations yielding dilute gases, from which sulfur recovery is uneconomic. In some geographical areas, even rich gases are often discharged to the atmosphere because no adequate markets exist locally for sulfuric acid. In such areas, elemental sulfur would be a far preferable by-product, but smelters in the United States are currently continuing to install sulfuric acid plants, apparently because production of sulfuric acid—even at a financial loss—is the cheapest method immediately available for the control of the emissions. However, some new sulfur dioxide reduction plants are being constructed at smelters outside the United States.

The economical control of sources of dilute smelter gases actually depends on changes in smelting technology that will eliminate production of such gases and result in formation of only rich gases. New or improved smelting systems are beginning to be installed by smelting companies in the United States, as well as by smelters abroad, as the stringency of air pollution control regulations increases.

Sulfuric acid plants. The tail gases from contact sulfuric acid plants are themselves substantial sources of sulfur dioxide emission. The conversion of sulfur dioxide to sulfur trioxide is limited by the equilibrium of the reaction. The practical limit of conversion in conventional contact plants ranges up to about 98%, but the actual conversion in operating plants is usually lower. However, conversion can be increased to at least 99.5% by use of the "double-contact" (or "interpass-absorption") design. Conversion of the sulfur dioxide is effected in two stages. The sulfur trioxide formed in the first stage is absorbed, shifting the equilibrium of the oxidation reaction so that more of the residual sulfur dioxide can be converted in the second stage. For autothermal operation, the initial sulfur dioxide concentration in the gas must be at least 6–7%.

Existing conventional contact plants can be converted to double-contact plants, but use of a tail gas scrubbing system for collection of the residual sulfur dioxide is sometimes preferred. Scrubbing systems proposed for treatment of flue gases and other dilute gases may be employed. The Cominco, Lurgi Sulfacid, and Wellman-Lord processes are in commercial use for this purpose.

Claus sulfur plants. Claus sulfur plants are in greatly increasing use for conversion of the hydrogen sulfide in natural gas, petroleum refinery gases, and coke oven gas to elemental sulfur. The emission of sulfur dioxide in the tail gases from such plants has been restricted in part by increasing the number of catalytic reactor stages. However, the ultimate extent of conversion is limited by the equilibrium of the conversion reaction. The tail gas is commonly incinerated to convert residual hydrogen sulfide and any carbonyl sulfide and carbon disulfide to the less toxic and obnoxious sulfur dioxide. The sulfur dioxide may then be removed by use of one of the systems designed for treating dilute gases. One large commercial installation uses the Haldor Topsoe process, which is similar to the Monsanto Cat-Ox process.

Fuel gases. The technology for removal of hydrogen sulfide from various kinds of fuel gases such as natural gas, petroleum refinery gases, and coke oven gas is well developed and commercialized. Most of the systems employ a cyclic absorption-desorption process to separate and concentrate the hydrogen sulfide, which is then delivered to a Claus plant for conversion to elemental sulfur. Some alternative processes not only absorb the hydrogen sulfide but also convert it to elemental sulfur. The absorbing solution is also an oxidizing agent for the hydrogen sulfide. During the regeneration cycle, the spent (reduced) absorbent is reoxidized by blowing with air. The hydrogen sulfide is thus indirectly oxidized by air. Examples of such systems are the Giammarco-Ventrocoke and Stretford processes.

For background information *see* AIR-POLLUTION CONTROL; FLUIDIZATION OF SOLIDS; GAS ABSORPTION OPERATIONS; SULFUR in the McGraw-Hill Encyclopedia of Science and Technology.

[KONRAD T. SEMRAU]

Bibliography: National Academy of Engineering–National Research Council, *Abatement of Sulfur Oxide Emissions from Stationary Combustion Sources*, Rep. no. COPAC-2, 1970; P. F. H. Rudolph, *Amer. Chem. Soc. Div. Fuel Chem. Preprints*, 14(2):13, 1970; K. T. Semrau, *J. Metals*, 23: 41, 1971; A. V. Slack, G. G. McGlamery, and H. L. Falkenberry, *J. Air Pollut. Contr. Ass.*, 21:9, 1971.

Algae

The algae are a heterogeneous assemblage that include all organisms which possess the ability to carry out photosynthesis, except the more primitive photosynthetic bacteria and the more highly evolved archegoniate land plants. Colorless forms that are clearly related to pigmented photosynthetic taxa are also ranked among the algae.

The modern era of phylogenetic systems of algal classification began with the elegant studies of Adolph Pascher, who worked in Prague, Czechoslovakia, and published his results between 1914 and 1931. Pascher's system was first introduced to American readers by Gilbert Smith in his influential book *Cryptogamic Botany* in 1938, and has been adopted with minor modifications by most textbooks that have appeared since World War II.

During the 1960s an outpouring of new information, especially in the fields of comparative biochemistry, physiology, and cytology, has stimulated new interest in the taxonomy and phylogeny of the algae. The most recent classification to have gained wide acceptance is one proposed by the Danish phycologist Tyge Christensen in 1962 (see table). A chart based on Christensen's phylogenetic concepts is given in Fig. 1. His arrangement of the major algal divisions emphasizes the primitive condition of the nucleus and the absence of internal organelles in blue-green algae and bacteria (Procaryota) and their presence in all other algae (Eucaryota); the absence of flagella in Rhodophyta; and the presence of chlorophyll *b* as a principal accessory photosynthetic pigment in Chlorophyta and its absence in Chromophyta. Classes of algae are further separated on the basis of pigment composition, type of carbohydrate accumulated during photosynthesis, and the structure, number, and arrangement of flagella. Christensen's scheme includes several new classes that were not known to Pascher. The most important of these are Haptophyceae, which are golden-brown algae, and Prasinophyceae, which are green algae. These two classes were recognized as a result of the electron microscope studies of Irene Manton and Mary Parke in England. It is remarkable the extent to which new data acquired by means of the most sophisticated modern techniques have tended to confirm the taxonomic system which Pascher developed over 40 years ago.

Current information calls for some further changes in the classification of the major algal groups. The chief distinction of the scheme that will be used in this discussion is the recognition of the cryptomonads (Cryptophyta), the dinoflagellates (Dinophyta), and the euglenoids (Euglenophyta) as separate and distinct algal divisions, and the more restricted circumscription of Chromophyta to include only those classes of algae which have xanthophylls as their principal accessory photosynthetic pigments and β-1,3-linked carbo-

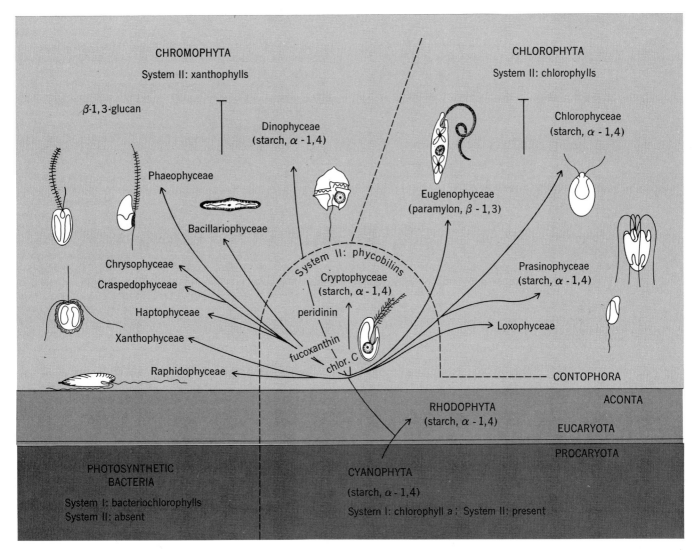

Fig. 1. Phylogenetic tree of the algae based on system of Tyge Christensen.

hydrates as storage products.

Research interest has centered recently on efforts to understand the evolution of fundamental life processes as they relate to systematics. Briefly these include (1) the biochemistry and physiology of the light-dependent steps in photosynthesis, (2) the mechanism of mitosis, (3) the structure and function of flagella and associated structures, (4) the process of cell wall formation, and (5) the origin of cell organelles and the transcription of genetic information regulating their activities.

Photosynthetic apparatus. According to the most widely held current model, bacterial photosynthesis involves one light reaction, whereas two independent light reactions are linked in series in the photosynthesis of all algae and higher plants. Chlorophyll a functions in both light reactions, while the accessory pigments found in algae are needed primarily to drive the second light reaction in which oxygen is evolved.

In Cyanophyta, Rhodophyta, and Cryptophyta the accessory pigments are either blue (phycocyanin) or red (phycoerythrin) and are associated with proteins. The phycobiliproteins of blue-green and red algae are structurally similar and have similar antigenic properties. They appear as discoid or spherical granules on the outer surface of saclike membranes, called thylakoids, that are present inside the chloroplast (Fig. 2). In contrast, the phycobiliproteins of Cryptophyta are small molecules having distinct antigenic properties. Moreover, they are incorporated within the chloroplast thylakoids (Fig. 3).

In the remaining algal classes the principal accessory photosynthetic pigment is either a xanthophyll or chlorophyll b. When chlorophyll b is present, as in Chlorophyta and Euglenophyta, the thylakoids are closely appressed, forming stacks. However, when xanthophylls predominate, the thylakoids tend to be grouped in threes with space between each thylakoid. Fucoxanthin is responsible for the characteristic brown color of Chromophyta. Exceptions are Xanthophyceae, in which two distinct esterified xanthophylls, heteroxanthin and vaucheriaxanthin, have been described, and the newly established class Eustigmatophyceae. Either peridinin or rarely fucoxanthin is found in photosynthetic dinoflagellates (Dinophyta).

Perhaps the most significant recent observation on the biochemistry of algal pigments is the discovery of carotenoids that contain acetylenic triple bonds between carbons 6 and 7 in Cryptophyta,

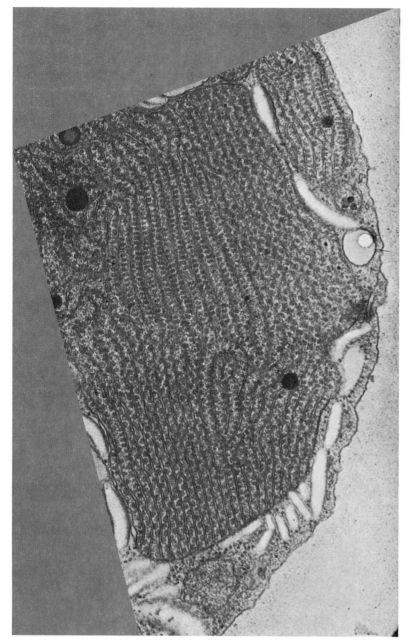

Fig. 2. Transverse section showing the arrangement of the granules (35 mm in diameter) which contain the phycobiliprotein pigments on the chloroplast thylakoids in *Porphyridium cruentum* (Rhodophyta). (Courtesy of M. R. Edwards, New York State Department of Health, and E. Gantt, Smithsonian Institution)

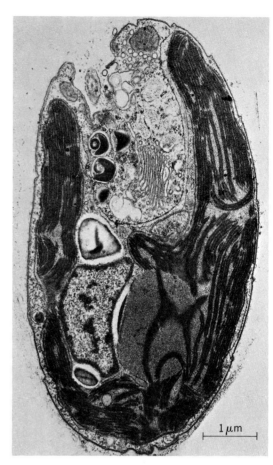

Fig. 3. Longitudinal section of a cell showing the location of the phycobiliprotein pigments within the chloroplast thylakoids in *Chroomonas* (Cryptophyta). (From E. Gantt, M. R. Edwards, and L. Provosoli, Chloroplast structure of the Cryptophyceae: Evidence for phycobiliproteins within intrathylakoidal spaces, J. Cell Biol., 48:280–290, 1971)

Chromophyta, Dinophyta, and Euglenophyta. It now appears that there are two series of xanthophylls present in algae which are capable of undergoing reversible light-dependent epoxidation and deepoxidation reactions. One series, which includes zeaxanthin and lutein, occurs in Rhodophyta and Chlorophyta. The other acetylenic series, which includes diatoxanthin and diadinoxanthin, is found in the remaining divisions of eukaryotic algae. The function of these carotenoid pigments is unknown. Violaxanthin is a major carotenoid pigment in both Phaeophyta and Chlorophyta. It is a diepoxide of a type that may have evolved independently in several different phylogenetic lines.

It has been recognized for many years that the chemical nature of the photosynthetic storage product is of major importance in the classification of algae. Two distinctly different kinds of carbohydrate reserves are known. In one (starches) the linkage between glucose units is α-1,4, and in the other (leucosins) the linkage is β-1,3. According to a recent study, the starch family has evolved through a loss of enzymatic capability leading to branching. The most primitive starch is the multiple-branched glycogens of blue-green algae. Red algae possess glycogen and simple-branched amylopectin, while green algae are the most specialized in lacking glycogen and possessing only amylopectin and straight-chained amylase. Cryptophyta and Dinophyta also accumulate starch.

The glycogen of Cyanophyta accumulates between the thylakoids. In Rhodophyta, Cryptophyta, and Dinophyta starch forms in granules that lie outside the chloroplast in the cytoplasm. In contrast, starch is formed within the stroma region of the chloroplast in Chlorophyta, commonly in association with a pyrenoid.

The principal food reserve in Euglenophyta is the β-1,3-linked carbohydrate known as paramylon. It has a distinct helical symmetry and is formed outside the chloroplast in association with the endoplasmic reticulum. Several names have

Three systems for classification of the algae

Pascher (1931)	Christensen (1962)	Hommersand (1970)

PLANTAE HOLOPLASTIDEAE
SCHIZOMYCOPHYTA
  Bacteria
  Cyanophyceae
	PROCARYOTA
	CYANOPHYTA
	  Cyanophyceae
		PROCARYOTA
		CYANOPHYTA
		  Cyanophyceae

PLANTAE EUPLASTIDEAE
CHRYSOPHYTA
  Heterokontae
    (=Xanthophyceae)
  Chrysophyceae
  Diatomaceae
    (=Bacillariophyceae)
	EUCARYOTA
	ACONTA
	RHODOPHYTA
	  Rhodophyceae
		EUCARYOTA
		RHODOPHYTA
		  Rhodophyceae
		
		CHLOROPHYTA
		  Chlorophyceae
		  Prasinophyceae

PHAEOPHYTA
  Phaeophyceae
	CONTOPHORA
	CHROMOPHYTA
	  Cryptophyceae
	  Dinophyceae
	  Rhaphidophyceae
	  Chrysophyceae
	  Haptophyceae
	  Craspedophyceae
	  Bacillariophyceae
	  Xanthophyceae
	  Phaeophyceae
		CHROMOPHYTA
		  Haptophyceae
		  Chrysophyceae
		  Bacillariophyceae
		  Phaeophyceae
		  Xanthophyceae
		  Eustigmatophyceae
		  Chloromonadophyceae?

PYRROPHYTA
  Cryptophyceae
  Desmokontae
  Dinophyceae		

EUGLENOPHYTA
  Euglenophyceae
	CHLOROPHYTA
	  Euglenophyceae
	  Loxophyceae
	  Prasinophyceae
	  Chlorophyceae
		CRYPTOPHYTA
		  Cryptophyceae
		
		DINOPHYTA (=PYRROPHYTA)
		  Dinophyceae

CHLOROPHYTA
  Chlorophyceae
  Conjugatae		

CHAROPHYTA
  Charophyceae
		EUGLENOPHYTA
		  Euglenophyceae

RHODOPHYTA
  Rhodophyceae
		GLAUCOPHYTA?
		  Glaucophyceae?

been applied to the β-1,3-linked carbohydrates of Chromophyta, including leucosin (Xanthophyceae), laminarin (Phaeophyceae), and chrysolaminarin (Chrysophyceae and Haptophyceae). Some studies indicate that these carbohydrates accumulate in a sac formed by protrusion of the outer plastid membrane through fenestrations in the periplastidial endoplasmic reticulum, often in association with stalked pyrenoids.

Mitotic apparatus. It is well known that the chromosomes of blue-green algae (Cyanophyta) are composed of deoxyribonucleic acid (DNA) that is not closely associated with protein. The strands of DNA are attached to the plasma membrane, as in bacteria, and are probably separated by membrane-dependent movements during cell division.

An intermediate type of mitosis has been described recently for dinoflagellates (Dinophyta). The nuclei are quite large and the chromosomes are made up of many parallel strands of DNA that are not associated with protein and are condensed and relationally coiled during interphase. Histone appears to be absent. Mitosis involves the invagination of the nuclear envelope, generating channels through the nucleus. The chromosomes replicate while still attached to the nuclear envelope and appear to separate by membrane flow. Microtubules are present only in the channels outside the nuclear envelope (Fig. 4).

The Euglenophyta possess large nuclei in which the chromosomes remain condensed during interphase. Here histone is associated with the DNA. The nuclear envelope remains closed during divi-

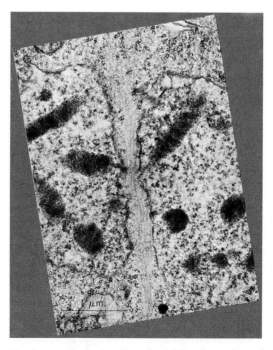

Fig. 4. Longitudinal section of one lobe of a nucleus of *Gymnodinium cohnii* (Dinophyceae) in a late stage of mitosis showing a bundle of microtubules passing through a channel outside the nucleus and two chromosomes attached to the nuclear envelope bounding the channel. (*From D. F. Kubai and H. Ris, Division in the dinoflagellate Gymnodinium cohnii (Schiller), J. Cell Biol., 40:508–528, 1969*)

sion and microtubules appear within the nucleus. The nucleolus (endosome) elongates along the division axis, and the movement of the chromosomes is thought to be autonomous, since there is no evidence of centromeres or of attachment between the microtubules and the chromosomes.

Centrioles appear to be absent in Rhodophyta, and their association with dividing nuclei is uncommon in Chlorophyta. Flagellar bases sometimes act as centrioles in Chrysophyceae and Xanthophyceae. Often the flagella may remain attached to the basal bodies and are seen extending from the poles of dividing nuclei (Fig. 5).

An unusual situation has been described in diatoms (Bacillariophyceae) in which the spindle precursor, consisting of a pair of parallel, electron-dense plates, forms the spindle outside the nuclear envelope (Fig. 6). After the spindle is fully formed, the nuclear envelope breaks down and the spindle sinks into the mass of chromosomes.

Neuromotor apparatus. The flagella together with the basal bodies and anchoring roots make up the neuromotor apparatus. All classes of algae except the Cyanophyta and Rhodophyta possess flagella having the same basic structure of nine pairs of microtubules arranged in a cylinder about two central, unpaired microtubules. Typically, motile algal cells possess two flagella, although a few species have only one through reduction and some have three, four, or more. Flagella may be naked, covered with scales, or may possess stiff hairs (mastigonemes) in unilateral or bilateral arrays. The origin of the mastigonemes has been elucidated in some important recent studies. They consist of one, two, or three protein fibers that are coiled in an alpha helix and originate in the perinuclear space or in vesicles produced by the evagination of the nuclear envelope (Fig. 7). It is not known just how the mastigonemes become fixed to the shaft of the flagellum, but they appear to attach to specific sites along the flagellar sheath.

In Cryptophyta one flagellum possesses a bilateral and one a unilateral array of mastigonemes. Potentially the same arrangement is found in some Dinophyta in which the trailing flagellum may have

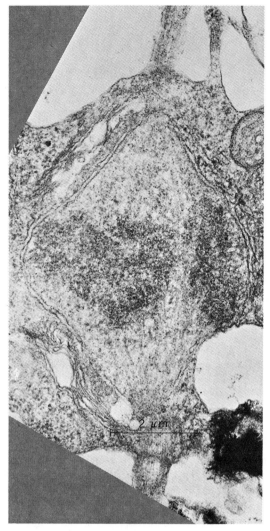

Fig. 5. Longitudinal section of a dividing nucleus in metaphase of mitosis from the antheridium of *Vaucheria litorea* (Xanthophyceae). Basal bodies bearing flagella are present at both poles. (*Courtesy of D. Ott, University of North Carolina*)

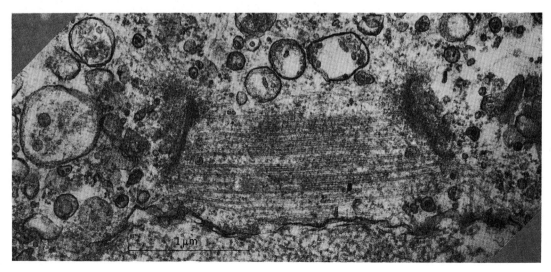

Fig. 6. Longitudinal section of a spindle appearing outside a spermatocyte nucleus in meiotic prophase in *Lithodesmium undulatum* (Bacillariophyceae). (*From I. Manton, K. Kowallik, and H. A. von Stosch, Development of spindle in Lithodesmium undulatum, J. Cell Sci., 5:271–298, 1969*)

a double row and the girdling flagellum a single row of mastigonemes. In Euglenophyceae the flagella possess a single row of fine hairs of uncertain origin. Most Chromophyta have one tinsel flagellum with two bilateral rows of mastigonemes and one naked whiplash flagellum. In the Haptophyceae of the Chromophyta the flagella are equal in length and smooth, and a third anterior filament, the haptonema, containing seven or fewer microtubules, may be present. In Prasinophyceae one, two, or four flagella are present which are usually covered by two or more layers of scales. In Chlorophyceae the flagella are equal in length and naked. Each group of flagellated algae has a characteristic swimming movement that is related to its flagellar structure.

Flagella are anchored by roots that consist either of bundles of microtubules or striated protein fibers or both. The roots may be anchored to the inner surface of the cell or to large cell organelles, such as the nucleus, chloroplast, or mitochondria. The number of roots and their ultrastructure and mode of attachment provide some important taxonomic characters that are useful in classification.

Cell wall formation. It is now well established that the great variety of types of cell walls, scales, plates, loricas, and frustules that invest the cells of the different classes of algae are produced within vesicles that originate from the nuclear envelope, the smooth or rough endoplasmic reticulum, or the dictyosomes (Golgi bodies). The Golgi apparatus is involved whenever carbohydrate components such as cellulose, mannan, or xylan are incorporated into the cell wall. Nowhere is this seen more strikingly than in the formation of organic scales in Haptophyceae. Here the scales have a basic plate-like framework of radiating and concentric cellulose fibers that are laid down within vesicles derived from the dictyosome (Fig. 8). Later, as in the marine coccolithophorids, calcium carbonate may be deposited in intricate patterns. Prasinophyceae are distinct in having both flagellar and body scales present in many species.

While the organic scales of Haptophyceae are formed within vesicles derived from dictyosomes, it has been demonstrated recently that the elaborate silicified scales of Chrysophyceae are produced within evaginated folds of the chloroplast envelope in conjunction with the surrounding periplastidial envelope which is derived from endoplasmic reticulum. Although it has not been proved, it is entirely possible that the silica frustules of diatoms (Bacillariophyceae), which overlap like the two halves of a petri dish, are formed in basically the same manner as the silicified scales found in Chrysophyceae.

The cell walls of many classes of algae contain cellulose as the principal fibrillar constituent. Xylan and mannan fibrillar components substitute for cellulose in some orders of Rhodophyceae and Chlorophyceae, while the most distinct cell walls among the algae are to be found in Cyanophyta, which, as in many bacteria, are composed of mucopeptides that are sensitive to the action of lysozyme and penicillin.

Origin of cell organelles. New information obtained during the last few years suggests that there is a close biochemical relationship between the inner membranes of chloroplasts and mitochon-

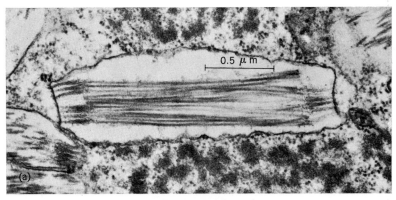
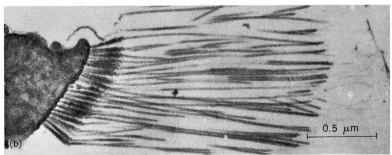

Fig. 7. Origin of flagellar hairs in the perinuclear space in *Olisthodiscus luteus* (Chrysophyceae). (*a*) Vesicle within a cell in which longitudinally sectioned elements of fibers lie parallel to one another, facing in opposite directions. (*b*) Longitudinal section through a clump of flagellar hairs showing the base, shaft, and terminal fiber regions. (*From G. F. Leedale, B. S. C. Leadbeater, and A. Massalski, The intracellular origin of flagellar hairs in the Chrysophyceae and Xanthophyceae, J. Cell Sci., 6:701–719, 1970*)

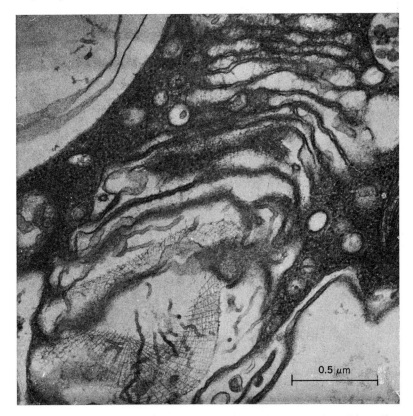

Fig. 8. Photomicrograph showing the role of the Golgi apparatus in wall formation in *Pleurochrysis scherffelii* (Haptophyceae). The concentric and radial arrangement of the cellulose microfibrils of a scale is seen in grazed section. (*From R. M. Brown, Jr., et al., Cellulosic wall component produced by the golgi apparatus of Pleurochrysis scherffelii, Science, 166(3907):894–896, November, 1969*)

dria present in eukaryotic organisms and the plasma membranes of bacterial or blue-green algal cells. The similarity extends from the structure and base composition of the DNAs and the molecular weights of several classes of ribonucleic acid (RNA) to the sensitivity of RNA polymerization and protein synthesis to inhibition by selective antibiotics. The chemical nature of membrane proteins and lipids are also shared features.

Chloroplasts and mitochondria may have acquired their distinctive properties through the invagination and differentiation of the plasma membrane of some ancestor among the Procaryota isolating segments of DNA. Thus it would be nuclear DNA and RNA that have changed in the course of evolution of eukaryotic organisms, while plastid and mitochondrial DNA and RNA have retained the properties of the ancestral nucleic acids. Alternatively it has been suggested that mitochondria and chloroplasts may have originated through the symbiotic association of prokaryotic bacteria or blue-green algae with an ancestor of the Protozoa. The algal component in such an association has been called a cyanelle. Such a mechanism could result in a polyphyletic origin of different major groups of algae. The symbiosis theory is riding a wave of popularity at the present time. Another possibility that leads to polyphyletic relationships among the algae is one in which an alga or a protozoan might "capture" the chloroplast of another species. Such harvested chloroplasts are functional and often long-lived in associations between the siphonaceous green alga *Codium* and the opisthobranch *Elysia*. Nonetheless, the chloroplasts do not appear to divide or maintain themselves autonomously. One difficulty with any of the symbiosis theories is that many or most important functions, such as chlorophyll synthesis, are known to be under the direct control of the nucleus, at least in those green algae and higher plants that have been studied. Whether the chloroplasts of other groups of algae are more nearly autonomous is not known.

The classes of algae most likely to represent symbiotic associations are (1) the anomalous group Glaucophyceae, some of which are flagellated but which have chloroplast characters of red algae; (2) Dinophyta, with their utterly distinct nuclei and mitotic mechanism, which have plastids having some of the characters of Chrysophyceae; (3) Euglenophyta, with their distinctive helical symmetry and mitosis, which have plastids that resemble those of green algae in having chlorophyll b, but otherwise have the pigmentation and other properties of Chrysophycean chloroplasts; and (4) Chloromonadophyceae, with large nuclei and dorsiventral symmetry, which have plastids like those of Xanthophyceae.

Phylogenetic relationships. Rhodophyta would appear to be intermediate between Cyanophyta and Chlorophyta, having plastid characters similar to Cyanophyta and a eukaryotic organization, a mitotic mechanism, and wall characters similar to Chlorophyta. Chromophyta all have similar chloroplast structure, xanthophylls as principal accessory photosynthetic pigments, and acetylenic-type carotenoid pigments and store β-1,3-linked carbohydrates or oil as food reserves. Except for Haptophyceae and Eustigmatophyceae, the flagella are of two types, one with a bilateral array of mastigonemes and the other naked, and often adjacent to an eyespot. Eustigmatophyceae are unusual in that it is the tinsel flagellum that is associated with an eyespot. The affinities of Cryptophyta are unclear, but the presence of chlorophyll c and flagella bearing mastigonemes suggest a distant relationship with Chromophyta. The distinctly different mitotic mechanisms found in Dinophyta and Euglenophyta set them apart from all other algae. They may represent very ancient relic groups or some kind of symbiotic association between distantly related organisms. A third possibility is that they are indeed highly specialized in having large polyploid nuclei and that the evidence has been misread as to the direction of their evolution.

As research on fundamental biological processes in the algae continue at an accelerated pace and data accumulate, one can expect a period of active speculation concerning the phylogenetic relationships of the algae. Some interesting proposals have been made and others are forthcoming in the next few years.

For background information *see* ALGAE; MICROORGANISMS in the McGraw-Hill Encyclopedia of Science and Technology.

[MAX H. HOMMERSAND]

Bibliography: J. R. Fredrick and R. M. Klein (eds.), *Ann. N.Y. Acad. Sci.*, no. 175, 1970; L. Margulis, *Origin of Eukaryotic Cells*, 1970; J. R. Rosowski and B. C. Parker (eds.), *Selected Papers in Phycology*, 1971; R. Y. Stanier, *J. Soc. Gen. Microbiol.*, no. 20, 1970.

Amplifier

Advances have taken place recently in the design and application of feedforward amplifiers. The term feedforward applies to correction systems in which the processing error imparted by an amplifier to the signal is sensed, amplified by one or more auxiliary amplifier stages, and then reinjected at some later point in the output chain in the proper phase to yield destructive interference. This designation of the error-correcting process is taken to distinguish it from feedback in which a portion of the signal is taken backward and injected in an earlier point in the amplifier chain.

Interestingly, feedforward predates feedback in that H. S. Black devised this process in 1924. For the most part feedforward held relatively minor status, reemerging only in the late 1940s with the independent work of B. McMillan and J. J. Zaalberg Van Zelst, who combined feedback and feedforward to achieve both redundancy and parameter desensitization. A summary and generalization of these and subsequent activities was published by J. J. Golembeski and colleagues in 1967.

The salient property to emerge from feedforward in very recent years is its total freedom from the stability restrictions of feedback, and as a consequence, its forceful applicability to situations generally inaccessible to feedback techniques. This applies particularly to the application of feedforward to both microwave traveling-wave tubes and to very wide baseband amplifiers. To appreciate more fully the vital differences between feedforward and feedback, a brief review of their basic features follows.

Basic features. Negative feedback, yielded by reintroducing a sampled output at the input, may be thought of as producing an endless train of suc-

Three-tone modulation product tabulation with and without feedforward correction*

A ($F_A = 14.5$)	B ($F_B = 15.2$)	C ($F_C = 16.6$)	Frequency (F), MHz ($F = AF_A + BF_B + CF_C$)	IMP†	IMP†
1	−2	1	0.7	79	100
−1	1	0	0.7		
−2	2	0	1.4	83	100
0	−1	1	1.4		
−1	0	1	2.1	78	100
−2	1	1	2.8	86	>110
0	−2	2	2.8		
−1	−1	2	3.5	87	>110
−2	0	2	4.2	88	>110
2	1	−2	11	88	>110
−1	2	−2	11.7	81	>110
2	0	−1	12.4	70	108
1	1	−1	13.1	68	107
2	−1	0	13.8	70	>110
0	2	−1	13.8		
−1	2	0	15.9	69	108
1	−1	1	15.9		
1	−2	2	17.3	66	108
−1	1	1	17.3		
−2	2	1	18	69	106
0	−1	2	18		
−1	0	2	18.7	69	>110
−2	1	2	19.4	78	>110

*A, B, and C are nondimensional numbers.
†IMP represents intermodulation products, in decibels, below output tone levels. Each tone level is at 5 mW.

cessively reduced and reshaped transformations of initial input events. If this sequence converges, it yields a frequency-dependent multiplicative factor, corresponding to a geometric sum, modifying the open loop gain. Clearly this sequence need not converge, particularly if the loop gain remains large at frequencies for which transit time corresponds to a phase change of 180°, shifting a degeneration at lower frequencies to regeneration. Exact criteria for stability are given variously by H. Nyquist, A. Hurwitz, and H. W. Bode. In particular, Bode showed that loop-gain reduction at higher frequencies generally should not exceed the order of 10 dB per octave and that moderately large loop gains may demand open loop cutoff frequencies possibly a decade or greater larger than the actual feedback amplifier bandwidth.

Feedforward operates in fundamentally different fashion. Since there is no return of output to input, correction is not a geometric sum but, instead, a single signal addition, or at most, a finite number of additions corresponding to the various auxiliary stages. These stages handle only the error portion of the original signal, and if there were no error there would be no modification of the main amplifier gain. This directly opposes the usage in feedback, where the entire signal is sampled and gain is multiplicatively modified, error or not. Further, stability considerations do not enter, nor are there any consequent demands for excess bandwidth capabilities of system components. Finally, as a most practical matter, the last auxiliary stage is the final arbiter of error. It removes the thermal noise of all prior stages and substitutes its own. The noise figure of this low-level stage may be made quite small, and it is that low noise figure which dominates the entire noise performance of the feedforward system.

Wideband developments. The key to large-bandwidth, low-error feedforward synthesis was shown by H. Seidel, H. R. Beurrier, and A. N. Friedman to relate to the synchronous timing of all paths, so that the signal and error portions retain constant relative phase relationships. This is the same recognition made earlier for distributed amplifiers. Proper design for timing requires directional wave sensing and well-matched components. Otherwise, small reflections or multiple reflection reincidence would produce unacceptable timing degradations. The couplers are broadband-matched four-ports and may be introduced in such fashion as neither to add substantial noise to the overall system nor to rob significant main amplifier power in the final error injection process.

The application of feedforward to a wide-baseband amplifier and to a microwave traveling-wave tube, indicated earlier, appeared in two publications by Seidel in 1971. The baseband amplifier assembly, devised purely to provide a benchmark for feedforward correction capabilities, is shown schematically in the illustration as a bootstrap correction system; namely, one in which the main and auxiliary amplifiers are identical. The input 10-dB power-splitting coupler favors the reference path in order to maintain a high signal-to-noise error determination. The joint operation of two central couplers provides this determination of error by both sampling and interfering the main amplifier output against the unamplified reference signal. The output coupler has a 10-dB coupling ratio, and while weakening the injection by the auxiliary amplifier, nevertheless it limits the amplifier transmission loss to a fraction of 1 dB.

The amplifiers used are those conventionally employed as distribution amplifiers within the Bell System L4 coaxial cable communication system. The table depicts distortion reduction under a three-tone modulation product test with and without the application of feedforward. Where measurable, distortion reduction over the 20-MHz band was of the order of 40 dB. It is worthy of note that

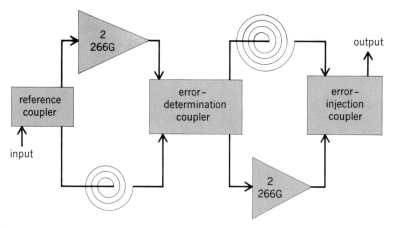

Diagram of feedforward correction system. Spiral represents a core of coaxial cable used as a delay line. Triangular box represents a cascade of two Western Electric 266G amplifiers.

these amplifiers had already been fed back to the limit of acceptable margin for stable operation.

Using the circuit shown in the illustration but adapted to microwave format, Seidel showed a similar feedforward capability for traveling-wave-tube amplifiers. The traveling-wave tube employed both for the main and auxiliary amplifiers is one used for the Bell System TD3 microwave relay, and the feedforward demonstration was operated over a 20-MHz channel between 4.080 and 4.100 GHz. Results yielded greater than 42 dB distortion reduction across the band. Long-term stability of this correction capability was obtained by the use of pilot tones providing sense correction to servo mechanisms which offset the effects of slow thermal and aging drifts.

A second auxiliary stage was also constructed of solid-state devices. It had far lower power capacity than the earlier stage, but the prior stabilization of the main amplifier limited its error range to lie within the capabilities of the second stage. With both stages operating, distortion products were unmeasurable with existing equipment.

Both the baseband and microwave demonstrations were performed for their impact on communications practices. To the present, cable amplifiers have been unable to use the available bandwidth residing with the cable because of the physical limitations of amplifiers. Amplifiers constructed to handle message power over these bandwidths suffer a constraint in cutoff frequency and, correspondingly, cannot provide sufficient feedback degeneration across the band to reduce crosstalk adequately. Both the experimental results and subsequent studies show great encouragement for the successful use here of feedforward.

The microwave demonstration may have impact on the form of modulation used in microwave relay systems. The lack of feedback at these frequencies has discouraged the use of linear amplitude transfer microwave devices. The intent, rather, has been to use these devices in a constant power mode so that amplitude compression effects would not enter. Consequently pulse or frequency modulations have been preferred although they assume large bandwidths per message channel compared to amplitude modulation. If other highly practical problems can be solved, the potential offered by feedforward is a possible tripling of the message-carrying capacity of microwave relay systems.

Not only does the microwave demonstration provide a benchmark for the frequency capabilities of feedforward correction, but it casts an interesting sidelight on the limitations of feedback. Transit time of the traveling-wave tube measures 13 sec, which also corresponds roughly to the duration of a 20-MHz bandwidth pulse. Were feedback attempted, the input pulse would be gone by the time the output pulse emerged, denying any possibility for joint interaction between the two pulses to improve transmission properties.

Frequency-dependent amplifiers. It is well known that feedback amplifiers may be given frequency dependence by shaping the feedback signal and yielding a frequency-dependent degeneration. M. O. Deighton, G. L. Miller, and E. H. Cooke-Yarborough achieved a similar capability for feedforward in their stablizing an amplifier with a single resistance-capacitance (RC) pole. They added an RC decay to the reference path so that this frequency dependence obtained uniformly throughout the error-sensing interferometer loop, and simply formed a factor modifying error determination. In general, a multipole dependence may be achieved by simple extension.

Bending the reference path to meet the amplifier shape, however, means a degradation to the signal-to-noise ratio in the reference path, and implies a higher noise figure in ultimate amplifier performance. A resolution proposed by Seidel in a 1970 patent was implemented as an adjunct activity in connection with his 1971 baseband amplifier study. Its relevance lay in the $\sqrt{f}$ neper characteristic required by cable repeater amplifiers. The technique employed here accepts the reference energy virtually intact, but samples the frequency-dependent gain of the amplifier with an inversely shaped coupler. The error, now measured with respect to a flat spectrum, is restored to its original frequency dependence in the main amplifier. Since the reference is undergraded and the error is measured within the context of a flat spectral weighting, there is no degradation to either noise figure or stabilization with frequency.

Term feedforward. The term feedforward has been used here to cover situations in which an unknown or an intractably complex fluctuation takes place in an amplifier or possibly other transfer device. The chemical industry has also laid claim to the same term to describe a control process in which a compensating change in processing later in a cycle is anticipated by earlier measurement of system parameters. In distinction to what has been described here, their use implies all input-output relations to be known and assumes control to be easily available. For whatever the title claims are worth, the term feedforward was used by Black in the same context as the use here.

For background information *see* AMPLIFIER; FEEDBACK CIRCUIT; TRAVELING-WAVE TUBE in the McGraw-Hill Encyclopedia of Science and Technology.   [H. SEIDEL]

Bibliography: H. S. Black, *Translating System*, U.S. Patent 1,686,792, Oct. 9, 1928; M. O. Deighton, E. H. Cooke-Yarborough, and G. L. Miller, in *Proceedings of the Northeast Regional Electrical*

Manufacturers Conference, Boston, 1964; J. J. Golembeski et al., *IEEE Trans. Circuit Theory*, CT14(1):69–74, 1967; B. McMillan, *Multiple Feedback Systems*, U.S. Patent 2,748,201, May 29, 1956; H. Seidel, *Bell Syst. Tech. J.*, in press; H. Seidel, *IEEE Trans. Commun. Tech.*, COM-19(3):320–325, 1971; H. Seidel, H. R. Beurrier, and A. N. Friedman, *Bell Syst. Tech. J.*, 47(5):651–722, May-June, 1968; J. J. Zaalberg Van Zelst, *Philips Tech. Rev.*, 9:25–32, 1947.

Animal virus

Picornavirus is a term of classification applied to a group of small (hence the prefix pico) viruses which infect vertebrate animals. These viruses are approximately spherical, some 30 nm in diameter, and composed of a ribonucleic acid (RNA) genome surrounded by a protein capsid. In recent years much progress has been made in elucidating the physical, chemical, and structural properties of picornaviruses, especially human poliovirus and the murine cardioviruses EMC-, ME-, and Mengovirus. This article will include a summary of the properties of these particular viruses because they have been the most thoroughly studied and because their properties can be considered as generally representative of all picornaviruses. Also included will be an outline of the known sequence of events in the process of assembly of these viruses in infected cells.

Background. Poliovirus and the cardioviruses have been adapted to replicate in cultured cells of mammalian origin. Virus particles for physical and chemical studies are produced by infecting monolayers or suspensions of such cells and allowing replication to proceed until large numbers (10^5 to 10^6 per cell) of progeny viruses have been produced and the cells lyse. Purification commonly involves enzymic digestion of cellular debris—care being taken that the enzymes used do not degrade the virions (virus particles) themselves, organic solvent extraction of cellular lipids, and low-speed centrifugation to remove debris followed by high-speed centrifugation to collect the virus. Final purification to obtain virions suitable for physical and chemical analysis is achieved by equilibrium centrifugation in density gradients of cesium salts or by column chromatography. It is now possible to routinely obtain milligram quantities of virus which is better than 99.9% pure as measured by the removal of radioactively labeled cellular material by these methods.

Virions thus prepared have been shown to contain only RNA and protein, in the ratio of about 70% protein to 30% RNA by weight. There is no evidence that lipid or carbohydrate are present.

Particle weight. Initial attempts to determine particle weight for the picornaviruses were hampered by the difficulties involved in obtaining sufficient quantities of highly purified material and resulted in the reporting of particle weights which ranged from 5.5×10^6 daltons for poliovirus to 10×10^6 daltons for EMC-virus. Improved purification procedures have largely overcome these difficulties, and accurate particle weights have been determined for Mengo- and EMC-viruses by hydrodynamic measurements.

Due to the presence of an RNA component, picornaviruses absorb ultraviolet light, an absorption maximum occurring at a wavelength of 260 nm. This property has permitted experimental determination of hydrodynamic parameters in analytical ultracentrifuges equipped with ultraviolet optics at virus concentrations low enough (50–100 μg virus/ml) to eliminate complicating particle-particle interactions. Sedimentation coefficients (150–160S) and diffusion coefficients (1.44×10^{-7} to 1.47×10^{-7} cm^2/sec) have been determined in this manner. Partial specific volumes of the virions ($\bar{v} = 0.68 - 0.70$ ml/g) have been calculated from chemical composition and confirmed hydrodynamically by measuring the rate of decrease in sedimentation coefficient with increasing ratios of D_2O to water in the solution being centrifuged. Once these three experimental parameters were obtained, the particle weight could be computed. A. T. H. Burness also used the ultraviolet optics combined with a photoelectric scanning system to determine EMC-virus particle weight by an equilibrium centrifugation method. Both approaches were in close agreement. The established particle weight for the cardioviruses is thus 8.3×10^6 to 8.6×10^6 daltons.

Further calculations based upon the hydrodynamic parameters indicate that the virion in solution is essentially spherical (frictional ratio ≃ 1.10), that its hydrated diameter is approximately 29.5 nm, and that its degree of hydration is of the order of 0.25 g water per gram of dry virus.

RNA component. The genetic material of the picornavirus particle is a single polynucleotide chain with a sedimentation coefficient of 35S. This RNA contains no unusual nitrogenous bases, and its adenine, guanine, cytosine, and uracil base components are present in approximately equimolar proportions.

Early experiments with polio- and Mengovirus had suggested that the molecular weight of the viral RNA was about 2×10^6 daltons, but more recently this value has been revised. N. Granboulan and M. Girard have examined RNA isolated from poliovirus in the electron microscope and have calculated its molecular weight, based upon measured contour length, to be 2.6×10^6 daltons. The identical value was obtained by P. D. Cooper, who compared the relative mobilities of poliovirus RNA and other RNAs of known molecular weight during electrophoresis in polyacrylamide gels. And Burness calculated a molecular weight of 2.7×10^6 for EMC-virus RNA from chemical analyses of RNA content and the particle weight of the virion. Thus, in summary, these picornaviruses have particle weights ≃ 8.4×10^6, their RNAs have molecular weights ≃ 2.6×10^6, and their total protein molecular weight (by subtraction) ≃ 5.8×10^6 daltons.

Protein component. The total protein isolated from the virus is quite ordinary in terms of amino acid composition. No unusual amino acid residues have been discovered. There is only a small amount of the sulfur-containing amino acids (a total of about 2.5 mole % of cysteine, cystine, and methionine) and a rather large amount of the imino acid proline (about 8 mole %).

Optical rotatory dispersion and circular dichroism studies have been carried out in order to determine the nature of the protein secondary structure in the living virus, and have revealed that the polypeptides in the virus capsid are largely random in

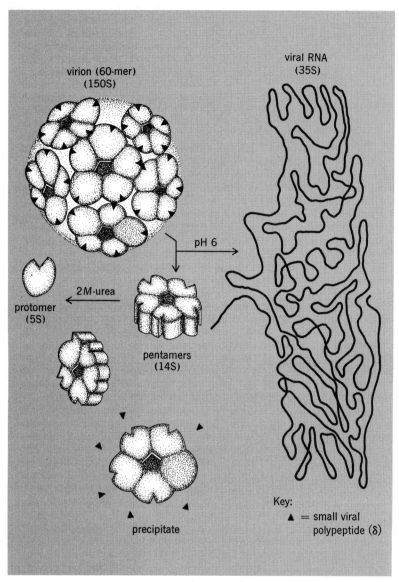

Fig. 1. Schematic representation of the degradation of ME-virus. (Redrawn from A. K. Dunker and R. R. Rueckert, Fragments generated by pH dissociation of ME-virus and their relation to the structure of the virus, J. Mol. Biol., 58:217, 1971)

present but also their molecular weights. Taking ME-virus as an example, there are five nonidentical polypeptide chain types which comprise the total capsid protein. These have been designated α (molecular weight = 33,000 daltons), β (30,000 daltons), γ (25,000), δ (10,000, and ϵ (41,000). Likewise the total proteins of polio-, EMC-, and Mengovirus have been shown to be made up of four or five nonidentical types of polypeptide chains. The logical next question to ask was how these chains are assembled to form the virus capsid.

Structure. Some 12 years ago J. T. Finch and A. Klug examined single crystals of poliovirus by x-ray diffraction and concluded that the virion had icosahedral (5:3:2) symmetry, and that the capsid was therefore likely to be composed of $60n$ asymmetric structural units ("capsomeres"). Further analysis of the precession photographs revealed an intensity modulation in the diffracted x-rays which was thought to correspond to a center-to-center subunit spacing of about 6 nm. Poliovirus was believed at that time to have a particle weight of 6.7×10^6 daltons and to contain 25% RNA. From these figures it was calculated that the poliovirus capsid was composed of 60 identical subunits, each of 6 nm in diameter and having a molecular weight of 80,000 daltons.

However, subsequent work by Finch and Klug on small, RNA-containing, icosahedral plant viruses made it clear that 5:3:2 symmetry and a 6-nm periodicity did not exclusively specify a $60n$ subunit particle. For example, turnip yellow mosaic virus (TYMV) was shown by electron microscopy to be composed of 32 rather than 60 capsomeres, and the total capsid protein was shown by chemical analysis to consist of 180 identical polypeptides. The TYMV capsid is assembled from these 180 polypeptides quasiequivalently spaced in 12 clusters of 5 chains and 20 clusters of 6 chains; the 6-nm periodicity in this case corresponded to the spacings between the 32 capsomeres. These results and the revised particle weight and RNA composition values for the animal picornaviruses left the x-ray data open to question; and new models were proposed which had 42 or 32 capsomeres, with the 32-capsomere configuration being the one which appeared most frequently in the literature.

Resolution of this controversy about the capsid architecture was delayed until just recently because of two sources of difficulty. The first stems from the fact that the picornavirus is extremely compact and is not penetrated by the negative stains which proved so useful in elucidating the capsomere arrangement in plant viruses such as TYMV and animal viruses of the adeno-, reo-, and herpesvirus groups. About all that one could conclude from electron microscopic examination of the picornaviruses was that they were approximately spherical in shape with anhydrous particle diameters of 27–28 nm. A second difficulty is the presence in the picornavirus capsid of several nonidentical polypeptides. Theories of virus structure have generally dealt with capsids containing a single type of polypeptide. It was therefore postulated that the picornavirus capsid might be composed of pentamers of one type of polypeptide and hexamers or tetramers or trimers of other types.

The solution has now been provided by A. K. Dunker and R. R. Rueckert, working with ME-vi-

intrachain organization. The calculated amount of α-helical structure was less than 5%. This is in contrast to the proteins of a rod-shaped bacterial virus, fd, which are more than 90% α-helical in configuration.

The total protein of the picornaviruses is comprised of not a single type of polypeptide but of several nonidentical types. This was first demonstrated when J. V. Maizel developed a procedure for electrophoresis of proteins in polyacrylamide gels in the presence of the detergent sodium dodecyl sulfate (SDS) and applied it to the examination of poliovirus protein. It was soon discovered that the rate of migration of a particular polypeptide during electrophoresis in SDS-polyacrylamide gels was directly proportional to the logarithm of its molecular weight and was independent of its particular amino acid composition. Thus one could dissociate a virus particle with SDS, subject it to electrophoresis in SDS-containing gels, and determine not only the number of different polypeptides

rus, and was based on the following information. (1) The virus can be degraded at pH 6 in the presence of chloride ions into soluble fragments which have a sedimentation coefficient of 14S, and these can further be degraded by urea into protomeric units which sediment at 5S. (2) The 14S fragment has a molecular weight of about 425,000 daltons, while that of the 5S protomer is about 86,000. The 14S fragment is therefore a pentamer of the 5S protomers. (3) The 5S protomer contains equimolar proportions of the viral polypeptides α, β, and γ, referred to previously, whose combined individual molecular weights total 88,000 daltons. Thus each protomeric subunit (capsomere) is a composite of one chain each of three of the viral polypeptides. (4) When the virion is dissociated by weak acid to 14S fragments, a precipitate also forms. Examination of the precipitate by SDS-polyacrylamide gel electrophoresis revealed the presence of α, β, and γ polypeptides, as well as the small polypeptide δ (molecular weight = 10,000) and a minor viral protein component ϵ (molecular weight = 41,000). Further examination of the precipitate showed that it contained aggregated δ polypeptides and distinctive 14S fragments composed of four normal protomers and one protomer containing one polypeptide each of α, γ, and ϵ. (5) Using the more accurate molecular weight determinations for the picornavirus particle and its RNA component which are now available, it was possible to compute the stoichiometry of the virus particle:

Molecular weight of 58 α, β, γ protomers
$= 4.988 \times 10^6$ daltons
Molecular weight of 2 α, ϵ, γ protomers
$= 0.182 \times 10^6$ daltons
Molecular weight of 60 δ polypeptides
$= 0.6 \times 10^6$ daltons
Total $= 5.76 \times 10^6$ daltons
Total molecular weight of capsid protein (virus particle weight minus RNA molecular weight)
$\simeq 5.8 \times 10^6$ daltons

Thus the picornavirus capsid is composed of 60 capsomeres (the 5S protomers) arranged in 12 clusters (the 14S pentamers). The exact location and functions of the δ polypeptides and the two "immature" protomers are unknown. The 5S protomer is composed of one chain each of the polypeptides α, β, and γ. The "immature" protomeric unit in precipitated 14S pentamers is composed one chain each of the polypeptides α, γ, and ϵ. There are 58 "mature" (α, β, γ) protomers and two "immature" (α, ϵ, γ) protomers in the virus capsid. It is thought that ϵ represents an uncleaved precursor of the β and δ polypeptides. This is shown schematically in Fig. 1. Electron micrographs of Mengovirus obtained during its degradation tend to confirm this model, at least as far as the assembly of the capsid from 12 of the 14S fragments is concerned (Fig. 2).

Assembly of virion. When the picornavirus particle enters a susceptible cell, it is disassembled into 14S fragments and the RNA is released. D. Baltimore has shown that the viral RNA is then translated by cellular enzymes and ribosomes into a single polypeptide of molecular weight in excess of 200,000, which is thought to represent the entire

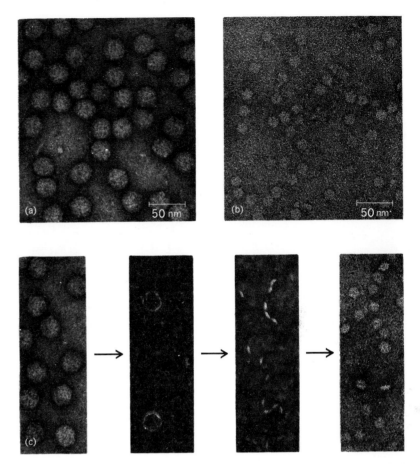

Fig. 2. Electron micrographs of Mengovirus and the degradation by treatment at pH 6 in the presence of chloride ions. (a) Whole virus. (b) 14S fragments. (c) Sequence of degradation. Preparations were negatively stained with heavy metal salts.

information content of the viral RNA. The large polypeptide is not stable and undergoes posttranslational cleavage to produce a number of smaller fragments. Some of these are structural proteins which will form the new capsids of progeny virions, and some probably perform enzymatic functions involved with production of more viral RNA, alteration of normal host cell functions, or final assembly of virus particles.

It seems that primary cleavages of the large original polypeptide give rise to three capsid structural proteins, these being designated α, γ, and ϵ in the case of ME- and EMC-viruses (the corresponding designations for poliovirus are VP1, VP3, and VP0). These three chains are assembled into protomers and then into 14S pentamers. The next stage is the formation of an empty capsid. In the case of poliovirus, the empty capsid is a stable structure sedimenting at 73S and can be isolated from infected cells. This is not so for the cardioviruses, where all attempts to obtain empty capsids from infected cells or by gentle degradation of virions have been unsuccessful. The next stage in the assembly of the poliovirus is the insertion of viral RNA into the capsid. It is probable that formation of the capsid occurs concomitantly with RNA packaging in the case of the cardioviruses, the RNA serving to stabilize the 14S pentamers in their capsid conformation. Maturation of the capsid also occurs during or after this process — maturation being defined as the cleavage of almost all of the ε proteins into β and δ proteins in the case of the cardioviruses, or the cleavage of almost all of the VP0 proteins into VP2 and VP4 proteins in the case of poliovirus. Cell lysis releases complete progeny virions which can then reinitiate the infectious cycle in other cells.

B. A. Phillips reported that 14S fragments obtained from poliovirus-infected cells will self-assemble in the test tube at appropriate concentrations to form stable 73S capsids with similar kinetics to those observed in living viruses. Studies are now underway to determine the polypeptide composition of these test-tube empty capsids and to further elucidate the maturation process.

For background information see ANIMAL VIRUS in the McGraw-Hill Encyclopedia of Science and Technology.   [DOUGLAS G. SCRABA]

Bibliography: A. T. H. Burness and F. W. Clothier, *J. Gen. Virol.*, 6:381, 1970; A. K. Dunker and R. R. Rueckert, *J. Mol. Biol.*, 58:217, 1971; M. F. Jacobson, J. Asso, and D. Baltimore, *J. Mol. Biol.*, 49:657, 1970; B. A. Phillips, *Virology*, 44:307, 1971.

Antigen

In the past few years studies of the immunological response have led to the characterization of the common antigens and antibodies involved in the response. Antigens are molecules that stimulate a specific immune response when introduced in a host. Essentially, two kinds of immunological response can be distinguished: humoral and cell mediated. The former is characterized by the production of immunoglobulin molecules, which react specifically with the antigen by which they are induced. These immunoglobulins (antibodies) are produced and secreted into the body fluids, predominantly by plasma cells. Cell-mediated immunity, on the other hand, is characterized by a lack of circulating antibodies. This response is mediated directly through an interaction between the antigen and specific receptors bound to the surface of the immunocompetent cells (lymphocytes).

The table gives the current classification of the more important human antigen polymorphic systems and the common antigens in each system.

Immunoglobulins. Immunoglobulins are serum proteins with antibody activity. Immunological investigations, combined with physicochemical studies, have led to the recognition of five different molecular classes of immunoglobulins (IgG, IgM, IgA, IgD, and IgE), their polypeptide constitution, and the various genetic polymorphisms among the classes.

As far as is known all the immunoglobulins have a basic structure consisting of two heavy and two light polypeptide chains, joined together by disulfide bonds. In a given antibody molecule the two heavy chains (molecular weight of 50,000–70,000) are identical and so are the two light chains (molecular weight of 20,000).

Light-chain polymorphism. Two kinds of light chain (κ and λ), common to all immunoglobulin classes, can be recognized with heteroantisera. Both types of light chains are composed of about 214 amino acids, of which the carboxy-terminal half of the light chains is essentially invariant in the sense that within one individual all molecules have the same sequence. In contrast, the amino-terminal half is somatically highly variable particularly in certain regions. This variability is undoubtedly related to the antibody function of the molecule.

Up to now, only one polymorphic system, Inv, has been described in detail for light chains. This system occurs on κ light chains and is due to point mutations at position 191 in the carboxy-terminal, somatically invariant half of the molecule. Inv was first detected using alloantibodies and is characterized by three alleles, Inv^1, Inv^{1-2}, and Inv^3. There is also evidence that an Inv^2 and a "silent" allele (Inv^-) occur at extremely low frequencies. Recently variation has been found in the invariant portion of λ chains as well. At position 190 either lysine or argenine may occur; the former determines antigen OZ(+) and the latter OZ(−). Both antigen types do occur in a single individual.

Heavy-chain polymorphism. The five classes of immunoglobulins are characterized by five different types of heavy chains, detected by means of appropriate heteroantibodies. The heavy chains

Principal human polymorphisms defined by immunological techniques

System	Antigens
TISSUE-FLUID ANTIGENS	
Immunoglobulin systems	
Inv	Inv(1), Inv(2), Inv(3)
Gm	Gm 1,2,3,4,7,9,18,20,22 (on IgG_1); GM 23 (on IgG_2); GM 5,6,10,11,12,13,14,16,21 (on IgG_3)
Isf	Isf(1)
Am	Am
β-Lipoprotein systems	
Ag	Ag(a)
Lp	Lp(a), Lp(x)
$α_2$-Globulin systems	
Gc	Gc(1), Gc(2)
Xm*	Xm(a)
CELL ANTIGENS	
Erythrocyte systems	
ABO	A_1, A_2, B, (H†)
MNSs	M, N, S, s
Rh	C, C^w, C^x, c, D, E, e
Kell	K, k, Kp(a), Kp(b), Js(a), Js(b)
Duffy	Fy(a), Fy(b)
Lutheran	Lu(a), Lu(b)
P	P_1, P_2, P*
Lewis	Le(a), Le(b)
Kidd	Jk(a), Jk(b)
Xg*	Xg(a)
I	I, i
Yt	Yt(a), Yt(b)
Diego	Di(a), Di(b)
Other minor systems	
Leukocyte systems	
HL-A	
LA (first) locus	H4-A1, H4-A2, HL-A3, HL-A9, HL-A10, HL-A11; 4–5 other antigens‡
FOUR (second) locus	HL-A5, HL-A7, HL-A8, HL-A12, HL-A13; 10–11 other antigens‡
NINE	9a
NA1	NA1
NB1	NB1
FIVE	5a, 5b

*Linked to the X chromosome.
†A genetically independent part of the system.
‡Not yet accepted by the WHO nomenclature committee.

are also composed of a somatically constant and a somatically variable region. The latter is approximately the same length as the variable region of the light chain. Together, these somatically variable regions of the light and heavy chains contribute to the formation of that portion of the molecules which recognizes and combines with antigen (antibody-combining site). Genetic polymorphism has been found in only two of the five immunoglobulin classes, IgG and IgA.

Within the IgG class, four subclasses have been distinguished (γG_1, γG_2, γG_3, and γG_4) according to differences in the heavy chains detected by special heteroantisera. Among these isotypic differences, allotypic differences have been discovered which define two polymorphisms, Gm and Isf. The Gm determinants are distributed throughout the heavy chain, predominantly in the invariant half; 23 Gm antigens have thus far been characterized. The table shows the distribution of these antigens among the IgG subclasses. The Isf system is defined by only one antigen, Isf(1), which is located in the invariable region of the heavy chain of the γG_1 subclass. Its frequency appears, in Caucasians, to be directly related to age.

Within the IgA class, two subclasses (IgA_1 and IgA_2) are isotypically distinguishable, by heavy-chain differences. A genetic polymorphism has been detected only in the IgA_2 subclass and is characterized by two alleles, Am^+ and Am^-.

The biological significance of the different classes and subclasses of immunoglobulins, the diseases characterized by their variation (deficiency syndromes, myeloma, and so on), and their use in preventive medicine and therapy were described by E. Merler in 1970.

The 1970 publication of K. A. Gally and G. M. Edelman exemplifies the way in which knowledge of structural and genetic properties of the immunoglobulins contributes to the proposal of new theories for the tantalizing puzzle of antibody diversity.

Beta lipoproteins. In normal human serum at least four different classes of lipoproteins are detectable by ultracentrifugation and electrophoresis. Of these four classes, only β-lipoprotein molecules are interesting from the genetic point of view. Serological studies can distinguish two separate genetic systems: Ag and Lp. Little is known about the function of β-lipoprotein molecules other than that they act as carriers for cholesterol and are associated with various enzyme activities, particularly esterases. β-Lipoproteins are made up of at least three different lipids which together constitute 75% of the molecule; 3% of the molecule is composed of carbohydrate and the remainder is protein.

The Ag system was first detected through the discovery of precipitating antibodies in the serum of a polytransfused patient. Five antigens have been defined up to now: Ag(x), Ag(y), Ag(a_1), Ag(z), and Ag(t). The results of family studies indicate that Ag(x) and Ag(y) are the products of allelic genes. The remaining antigens appear to be closely associated with them, since no recombination has been observed.

The Lp system is defined by two antigens, Lp(a) and Lp(x), which react with heteroantisera to give a line of precipitation in the Ouchterlony test. Lp(x) antigen, when present, is inherited with Lp(a). Thus the known phenotypes are Lp(a+, x−), Lp(a+, x+), and Lp(a−, x−). It has been shown that the Lp locus is not sex-linked and furthermore that Lp and Ag antigens are located on different β-lipoprotein molecules.

Alpha₂ globulins. A serum α_2-globulin, the function of which is still unknown, provides another system of antigens, the Gc system. This glycoprotein, which is synthesized by the liver, has been identified by means of immunoelectrophoresis using antibodies directed against human serum. Three major phenotypes are described: Gc 1-1, Gc 2-2, and Gc 2-1, coded for by a pair of autosomal alleles Gc^1 and Gc^2. Different phenotypic variants (Gc^z, Gc^x, Gc^y, and so on) appear sporadically in Negro and Caucasian populations. Among the Chippewa Indians, the variant allele Gc^{Chip} occurs with 0.10 frequency. Another unusual allele, Gc^{Ab}, has been found in Australian Aborigines, also with a frequency of 0.10. Chemical studies suggest the possibility that all Gc molecules are composed of at least two subunits, and the Gc 1 and Gc 2 chains differ by at least one amino acid. There appears to be close linkage between Gc and the albumin loci.

Up to now, the Xm system has been defined by only one antigen, Xm(a). This antigen, which is closely associated with, or a part of, α_2-macroglobulin, is detectable in the Ouchterlony test using a heteroantiserum. The Xm locus has proved to be on the X chromosome. Xm^a gene frequency in Caucasians is around 0.26 for males and 0.30 for females.

Erythrocyte systems. The importance of the knowledge of red-blood-cell antigens has been recognized for many years, chiefly in the area of blood transfusion. Recently, however, experimental skin grafting has demonstrated that ABO antigens also play a critical role in graft survival. Thus compatibility between donor and recipient has been accepted as a mandatory rule in organ transplantation. Detailed information about the chemical structure of some components of red-blood-cell surfaces and the interaction between the various genes involved in their production was published by W. H. Hildemann in 1970. *See* TRANSPLANTATION BIOLOGY.

Knowledge of the Ph system has not only permitted recognition of the fact that hemolytic disease of the newborn (erythroblastosis foetalis) is due to the aggression of the mother's antibodies against the fetal red blood cells, but has also permittted the development of efficient immunological procedures to prevent it. A complete description of erythrocyte systems, together with their legal and anthropological implications and their use in mapping human chromosomes, was published by R. R. Race and R. Sanger in 1968.

Leukocyte systems. In recent years organ transplantation has stimulated the study of the genetics of histocompatibility in man. Previous experimental work has demonstrated that leukocyte antigens mimic those found on other cells of the body. For this reason, as well as for their easy availability, leukocytes have been extensively used. The most common methods of study are cytotoxicity and agglutination, using alloantisera obtained either from pregnant women or from immunized volunteers. While five separate systems have now been well described, only one (HL-A) appears to play an important role in transplantation. *See* IMMUNOLOGY.

The HL-A system (human leukocyte, locus A) is considered to be the major histocompatibility system in man, comparable to the H-2 system of mice, Ag(B) system of rats, and H-1 system of rabbits. HL-A antigens are operationally defined using a complement-mediated cytotoxicity test performed with peripheral blood lymphocytes. These antigens are also well expressed on the surface of spleen, liver, kidney, lung, heart, and intestinal cells. Expression of HL-A antigens on spermatozoa is probably haploid, the genes from one chromosome being expressed on half the spermatozoa and those from the other chromosome on the remainder. No HL-A antigen can be detected on the surface of mature red blood cells.

The present concept about the HL-A system is outlined in the table. On one autosomal chromosome a complex region exists (HL-A) which is divisible into two loci, named locus LA (or first) and locus FOUR (or second).

The LA locus controls the production of one antigen from a mutually exclusive series of at least 10 members, only six of which are officially accepted by the World Health Organization (WHO) Nomenclature Committee (HL-A1, HL-A2, HL-A3, HL-A9, HL-A10, and HL-A11). The frequency of the genes coding for these antigens ranges in Caucasians from 0.26 (HL-A2) to 0.06 (HL-A10).

The FOUR locus has been defined up to now by 15 or 16 antigens, only five of which (HL-A5, HL-A7, HL-A8, HL-A12, and HL-A13) have been officially accepted. All genes corresponding to these antigens have low frequency. The most frequent is HL-A12 (0.16); the frequency of the others varies from 0.09 to 0.02.

Since close linkage exists between the LA and FOUR loci, a given LA determinant is almost always accompanied by the same FOUR determinant within a family. This chromosomal combination of one LA and one FOUR genetic determinant is called a haplotype and is inherited as a whole. Thus all individuals have two HL-A haplotypes, and because HL-A antigens are codominantly expressed, a maximum of four different antigens may exist on the surface of one cell. As the antigens of the HL-A region detected can only be a factor of all those existing, a minimal estimate leads to the theoretical expectation of 187 different haplotypes, 76,672 phenotypes, and 17,578 genotypes. However, some haplotypes (HL-A1,8 and HL-A3,7) appear to occur more often than would be expected from the gene frequencies. While the reason for such a "linkage disequilibrium" is unknown, a likely explanation is that there is a selective advantage in having these genes transmitted together.

The evidence that the HL-A region is composed of two separate loci is based on the following arguments: A recombination frequency of about 1% has been found between LA and FOUR loci, meaning that a distance of about 200 cistrons lies between these loci. LA antigens are biochemically separable from locus FOUR antigens, the former having a molecular weight of 35,000 and the latter a molecular weight of 60,000. The presence of a carbohydrate moiety on these molecules is still subject to debate. While cross-reacting antibodies have been described for antigens of the HL-A system, no cross reaction between products of the two loci has been observed. This suggests that LA and FOUR genes are derived from different ancestor genes.

The practical implications of the knowledge of the HL-A system are numerous. The fantastic polymorphism of the system suggests that in the near future it will become an important tool in paternity cases and in anthropology.

Recent reports have shown that the HL-A region is linked, on chromosome number 16, to the loci coding for phosphoglucomutase (PGM_3) and for haptoglobin (Hp). Should linkage between the HL-A region and any one hereditary disease be discovered, the probably haploid expression of the HL-A genes in spermatozoa may be utilized as a means of gametic selection. This may be accomplished by artificial insemination, using spermatozoa that have been treated in the test tube with cytotoxic antibodies directed against the HL-A antigen associated with the disease.

Experimental and clinical evidence has clearly demonstrated the dominant role played by the HL-A system (or by a system very close to it) in transplantation. Skin grafts between HL-A-identical twins have a mean survival time of 20 days, while grafts exchanged between HL-A-nonidentical twins have a mean survival time of only 13 days. Similarly, although kidney grafts between HL-A-identical twins are always successful, the percentage of successful grafts falls to 62% in sibs sharing one HL-A haplotype and to only 56% between sibs with no haplotypes in common. In unrelated individuals a correlation has been found between the number of HL-A incompatibilities and graft survival, but this correlation is not so impressive as in the case of grafts between sibs. Several hypotheses can be invoked for explaining these differences. Especially in skin grafting, which is a very sensitive technique for detection of transplantation antigens, a summation of weak histocompatibility antigens can obscure the role played by the HL-A system. Another possibility is that an unknown locus, closely linked to the HL-A system, plays a major role in transplantation. In this event, sibs with the same HL-A would also have the same hypothetical transplantation locus. On the other hand, unrelated individuals, although identical for HL-A, could have a different transplantation locus, resulting in rapid rejection of the graft.

Finally, recent evidence indicates that some HL-A antigens previously assumed to represent a single antigenic determinant may, in fact, be composed of at least two different determinants. Since some antisera commonly in use appear to recognize only one of these antigenic determinants, it is possible that the lack of correlation, found in some cases in unrelated persons, reflects inadequacies in present typing procedures.

The biological functions of the histocompatibility systems are not known. The complex polymorphism of HL-A is considered to be the result of selective advantage of the heterozygote with respect to various unknown selective mechanisms. It is well known that tumor cells induced by viruses and chemical carcinogens have new antigens on their surface. In view of this fact, it can be speculated that HL-A antigens become important markers for surveillance of the new cell variants. Tumor cells with the new foreign antigens are recognized as nonself and may be eliminated by the immunologic system of the body. Cross reaction has been found between some viral, bacterial, and histocompatibility antigens. This similarity has been invoked to explain why certain individuals are more susceptible to various bacterial and viral infections and certain virus-induced tumors. Particularly in the case of HL-A, it has been recently found that cross reaction exists between HL-A antigens and the M1 protein of streptococci. Correspondingly a high frequency of patients with haplotype HL-A2,X (X for unidentified gene at the FOUR locus) have glomerulonephritis, which probably results from streptococcal infection. Similarly, in 10 cases of lymphoblastic leukemia, six cases were observed carrying HL-A12 antigen, a frequency that is three times higher than that registered in normal individuals.

In this connection the extreme polymorphism of the HL-A system may have resulted from a process of selection which thus enables the species to combat the spread of those organisms which mimic some of its antigens.

The NINE system is specific for granulocytes; only one antigenic determinant, 9a, has thus far been described.

NA1 and NAB1 are two indepedent systems, each defined by a single antigen, expressed only on the neutrophilic granulocytes. A severe deficiency of neutrophilic granulocytes is frequently seen in the offspring of women with circulating antibodies against NA1.

The FIVE system is defined by two alleles, 5a and 5b, that can be detected only by agglutination.

None of these systems seem to be linked to the HL-A chromosome region.

For background information *see* ANTIGEN; ANTIGEN-ANTIBODY REACTION; HUMAN GENETICS in the McGraw-Hill Encyclopedia of Science and Technology. [VINCENZO MIGGIANO]

Bibliography: J. A. Gally and G. M. Edelman, *Nature*, 227:341, 1970; E. R. Giblett, *Genetic Markers in Human Blood*, 1969; W. H. Hildemann, *Immunogenetics*, 1970; F. Kissmeyer-Nielsen and E. Thorsby, *Human Transplantation Antigens*, 1970; E. Merler (ed.), *Immunoglobulins: Biological Aspects and Clinical Uses*, 1970; R. R. Race and R. Sanger, *Blood Groups in Man*, 1968.

Arteriosclerosis

Recent findings suggest that arteriosclerosis, a disease with high incidence in industrialized societies, is initiated by increased amounts of homocysteine formed from metabolism of diets rich in animal proteins. Arteriosclerosis produces death or disability by narrowing and occluding the arteries supplying vital organs, especially the heart, kidneys, and brain. Homocysteine initiates narrowing of arteries by stimulating growth of the cells lining the arteries and by increasing the sulfate content of arterial wall polysaccharides. The risk of arteriosclerosis is high in populations which consume a diet rich in animal protein, containing abundant methionine, the only metabolic source of homocysteine.

Homocysteine metabolism. Homocystinuria is an inherited disease caused by deficient activity

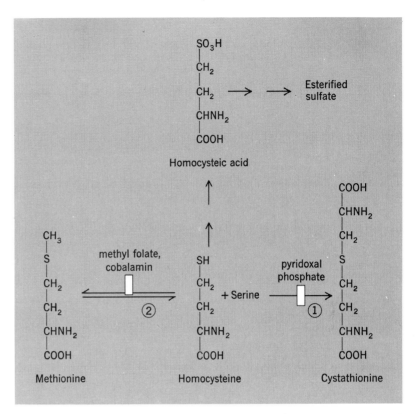

Fig. 1. Hereditary deficiencies of human enzymes catalyzing the conversion of homocysteine to cystathionine (reaction 1) or to methionine (reaction 2) lead to production of excess homocysteic acid, which results in increased synthesis of esterified sulfate.

of cystathionine synthetase, the pyridoxal phosphate–requiring enzyme which catalyzes the conversion of homocysteine and serine to cystathionine (enzyme 1 in Fig. 1). Children with homocystinuria are usually affected by excessive growth, dislocated lenses, mental retardation, and severe arteriosclerosis which frequently kills by occlusion of arteries to the vital organs. A child with a rare form of homocystinuria due to reduced activity of the folate and cobalamin enzyme catalyzing the methylation of homocysteine to methionine (enzyme 2 in Fig. 1) was discovered also to have died of severe arteriosclerosis. It was concluded that an increased amount of homocysteine, resulting from these inherited errors of metabolism, produced arteriosclerosis by a direct effect of homocysteine derivatives on the artery wall.

Methionine is converted to homocysteine in the liver by demethylation. Subsequently homocysteine is either converted to cystathionine by reaction with serine or it is oxidized to homocysteic acid. Homocysteic acid reacts in the laboratory with adenosinetriphosphate (ATP) and liver extracts to form phosphoadenosinephosphosulfate (PAPS), the precursor of sulfate ester groups of the proteoglycans (protein-polysaccharide polymers) of connective tissues (Fig. 1). In enzymatic disorders with excess production of homocysteine from methionine, increased amounts of proteoglycan macromolecules esterified with sulfate are produced in the walls of the arteries because of increased synthesis of homocysteic acid and PAPS from homocysteine. The oxidation of homocysteine to homocysteic acid is diminished in the

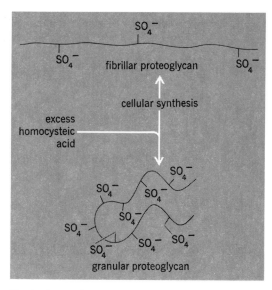

Fig. 2. Conversion of fibrillar to granular proteoglycan is produced by increased numbers of esterified sulfate groups, resulting from the effect of excess homocysteic acid on cellular metabolism.

liver of guinea pigs with scurvy, a disease produced by ascorbic acid deficiency in which the small blood vessels become fragile and bleed spontaneously. The resulting failure to produce normal amounts of sulfated proteoglycans in scurvy leads to thinning of blood vessel walls and hemorrhage, whereas in homocystinuria excess production of sulfated proteoglycans leads to thickening of artery walls and thrombosis.

Human cell cultures. Cells cultured from normal skin synthesize fibrillar, filamentous proteoglycan substances, but cells cultured from the skin of homocystinuric individuals form a similar sub-stance which assumes an abnormal, aggregated, granular configuration rather than the normal fibrillar form. Furthermore homocysteine in the culture medium of normal cells causes them to produce some granular proteoglycan in place of the fibrillar form. Homocysteine is an amino acid which is completely excluded from peptide linkage, and indeed the amino acid composition of the proteins of the homocystinuric cell cultures is free of homocysteine and identical to that of normal cell cultures. However, as shown in Fig. 2, increased homocysteic acid synthesis in the cell cultures, resulting from an enzyme block or exogenous homocysteine, leads to increased numbers of sulfate groups attached to the sugar residues of the proteoglycan, causing the substance to assume an irregular rather than fibrillar configuration. Some evidence suggests that proteoglycan with this abnormal configuration is less soluble and dispersible in the tissues, explaining why increased amounts of this substance accumulate in the arteries of children with homocystinuria.

Homocysteic acid and growth. Children with homocystinuria are usually taller than their normal siblings because of the effect of homocysteine derivatives on the epiphyseal cartilage of the skeleton. In 1957 W. D. Salmon and W. H. Daughaday discovered a heat-stable, dialyzable substance, which they called sulfation factor, bound to proteins of normal serum, which increases the binding of sulfate by hypophysectomized rat cartilage fragments incubated in the laboratory. Sulfation factor activity is reduced in serum from hypopituitary dwarfs, and only growth hormone therapy could restore the activity to normal. Recently homocysteic acid, the fully oxidized form of homocysteine, has been found to restore the sulfation factor activity of hypopituitary serum to normal. In addition, homocysteic acid administration increases the growth rate of young guinea pigs, showing that it has the growth-promoting properties expected for sulfation factor.

Significant human arteriosclerosis is unusual during the growing years, but early lesions begin to appear in the arteries of young men in the late teens and early twenties when growth has almost ceased. The process which caused growth of epiphyseal cartilage and bone in the early years increases the growth of cells lining the arteries in later years, thickening the artery wall and constricting its lumen. After a latent period of 10–20 years, the arteriosclerotic narrowing process has so progressed that symptoms begin to appear in susceptible individuals, usually in the form of angina pectoris and myocardial infarct.

Production of arteriosclerosis. For many years arteriosclerosis could only be produced in experimental animals by prolonged feeding of diets containing increased saturated fat and cholesterol. Homocysteine, administered subcutaneously every day in physiologic doses, has now been found to produce arteriosclerotic lesions in as little as 3 weeks. A section of the coronary artery of a rabbit given homocysteine injections is shown in Fig. 3. The marked thickening of the wall, which is due to increased numbers of cells, increased fibrous tissue, and fragmentation of elastic tissue (stained black), severely narrows the lumen. When cholesterol is added to the diet of rabbits on homo-

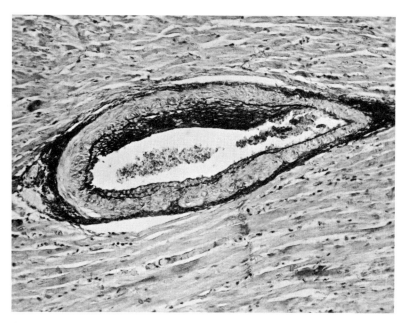

Fig. 3. Coronary artery of rabbit treated with homocysteine is narrowed because the walls are thickened by increased numbers of cells, increased fibrous tissue, and fragmented elastic tissue.

cysteine, lipid droplets rapidly accumulate in these fibrous arteriosclerotic plaques. With special staining techniques increased amounts of sulfated proteoglycan substances are found to be deposited both in these plaques and in the wall of the aorta.

Knowledge of the macromolecular changes in proteoglycan structure resulting from the effect of homocysteine derivatives on cell metabolism suggests a theoretical formulation on a molecular level of how the arteriosclerotic plaque develops. The initial events are cellular hyperplasia from the growth-promoting effect of homocysteic acid and deposition of sulfated proteoglycan in the wall of the artery. Deposition occurs because of excessive sulfation of the polysaccharide component of proteoglycan, conversion of the fibrillar to the granular form, and reduction of its solubility in tissue fluids. Elastin fibrils are fragmented because sulfated proteoglycans interfere with normal cross linking by desmosine molecules containing strongly basic pyridinium groups. Collagen deposition and fibrosis occur because sulfated proteoglycans increase the rate of tropocollagen polymerization. Low-density lipoproteins, fibrinogen, and calcium ions from filtered plasma are bound to the artery wall because insoluble complexes are formed with the sulfated proteoglycans. Complete occlusion may occur because of severe narrowing of the artery, or because the increased adhesiveness of platelets containing heavily sulfated proteoglycan compounds may initiate formation of a thrombus at a site of narrowing.

Prevention of arteriosclerosis. Pyridoxine deficiency produces mild arteriosclerosis in monkeys and rabbits, probably by reducing the conversion of homocysteine to cystathionine, a reaction requiring pyridoxal phosphate (Fig. 1). Some children with homocystinuria respond biochemically to large doses of pyridoxine, but the effect on their arteriosclerosis is not known. Experimental choline deficiency produces arteriosclerosis, and choline administration partially prevents experimental arteriosclerosis produced by cholesterol feeding, possibly by increasing the rate of methylation of homocysteine to methionine. Chronic liver disease, such as cirrhosis, slows the progression of arteriosclerosis, probably because the capacity of the liver to demethylate methionine to homocysteine is reduced in this disease.

Since methionine is the only metabolic precursor of homocysteine in the human diet, deliberate reduction in the amount of dietary animal protein, which contains more methionine than plant protein, may prove to be effective in preventing arteriosclerosis. The foods which are rich in methionine are also rich in cholesterol and saturated fats. These lipids have been implicated by many previous studies as a factor in the progression of arteriosclerosis. Changing to a predominantly vegetarian diet reduces the consumption of methionine and lowers elevated serum cholesterol concentration, but further study is needed to demonstrate a consequent reduction in the risk of arteriosclerosis.

For background information see ANEURYSMS; ANGINA PECTORIS; CHOLESTEROL; INFARCTION in the McGraw-Hill Encyclopedia of Science and Technology. [KILMER S. MC CULLY]

Bibliography: K. S. McCully, *Amer. J. Pathol.*, 56:111, 1969; K. S. McCully, *Amer. J. Pathol.*, 59:181, 1970; K. S. McCully, *Nature*, 231:391, 1971; K. S. McCully and B. D. Ragsdale, *Amer. J. Pathol.*, 61:1, 1970.

Astronomical instruments

Several large astronomical telescopes are under construction in various places throughout the world and will be in operation during this decade to help solve some of the "limit" type problems facing the astronomer of today. Six of these instruments have mirrors of 150 in. aperture range (3.81 m) and one of 236 in. aperture (the 6-m telescope), which will be the largest telescope in the world. In addition, the world's largest radio telescope was completed in the early part of 1971. The diameter of this instrument is over 328 ft (100 m).

Optical instruments. The 236-in. telescope project was started during the middle of the 1960s by the Soviet Union and, when completed, will be operated by the USSR Academy of Sciences. The location of the new observatory is in the Caucasus Mountains near Mount Pastukhov, not far from the village of Zelenchukskaya. This site was selected only after several years of extensive site survey in the mountains of Caucasia, the Crimea, eastern Siberia, and Soviet Central Asia. Elevation at the observatory site is 6830 ft (2083 m) above sea level. The metal tower of the Special Astrophysics Observatory, with portions of the telescope revealed, is shown in Fig. 1.

Fig. 1. View of the azimuth telescope under construction in the Soviet Union.

Fig. 2. The completed 100-m radio telescope near Bonn, West Germany.

The primary mirror weighs 42 tons (38,136 kg) and was cast of Pyrex-type, low-expansion borosilicate glass near Moscow at the Optical Glass Works. Cooling of the mirror blank to room temperature after casting required over 18 months to prevent strain in its cross section. The front surface of the blank was ground and polished to an f/4 paraboloidal figure with an accuracy of 1/1,000,000 to 2/1,000,000 in. To support this massive mirror and maintain the required optical figure, the blank was drilled on the back surface to produce 60 cavities for the support network to prevent the giant mirror from sagging under its own weight. When completed the mirror surface will be coated with aluminum a few millionths of an inch thick in a specially built vacuum chamber to provide the high-reflectance surface.

The optical arrangement of the telescope consists of a prime focus cage centrally located near the upper end of the 82-ft-long tube and a Nasmyth (sometimes referred to as broken Cassegrain) focus fed by a flat at 45° at the intersection of the axes. The focal ratio is f/4 with a 944-in. focal length for the prime focus and f/30 (7080-in. focal length) for the Nasmyth focus. A large high-dispersion spectrograph will be permanently mounted on one side of the giant fork at the Nasmyth focus. On the opposite side of the fork, three smaller spectrographs will be installed and operated by means of the flat mirror tilted through 90°. When using the prime focus, the observer rides inside the cage, as is the case with all large telescopes.

Historically all other large optical telescopes have been equatorially mounted, but an inclined polar axis for an instrument as large as the 236-in. telescope could introduce serious mechanical problems, since the moving parts of this telescope weight about 700 tons. Therefore the altazimuth mount was adopted for the design of the 236-in. instrument, incorporating oil pad bearings to reduce friction to an acceptable value. An altazimuth mount has two basic motions: a rotation about the horizon and a rotation permitting altitude adjustment. This type of mounting for a large telescope has several important engineering advantages. Loading is constant with rotation in azimuth; no change in elastic deformations will occur. Also, the mirror support system is much simpler than that for an equatorial mount, since deformations due to gravity are always in the same plane. The design has the disadvantage of requiring motion about three axes to track a star field to compensate for the rotation of the Earth, since derotation of the image plane is also necessary. In addition, the driving rates are constantly changing for a given object throughout the night. To handle this problem, a special digital computer was constructed to point the telescope at the desired object and track it. Data about atmospheric pressure and temperature are fed to the computer so that appropriate corrections for refraction can be made. The computer also senses elastic deformations in the telescope tube and controls the observatory dome automatically.

One serious disadvantage of the altazimuth mount is the "blind" cone near the zenith of about 2° in radius. This blind region is a consequence of very large acceleration requirements for tracking near the zenith.

Preinstallation assembly and test were done at the State Optical Instrument Plant in Leningrad in a specially built 138-ft-high bay area. Shipment from Leningrad to the observatory site in the Caucasus involved a trip from the Neva River by way of inland waterways to the city of Rostov-on-Don and a 310-mi journey by trailer truck to the mountain site. Installation of the instrument was started in early 1970; no completion date had been made public as of mid-1971.

The observatory dome is also the largest in the world, being 144 ft in diameter and 174 ft above the ground. This observatory, when completed, will become one of the world's major astronomical research institutions.

Construction of 150-in. telescopes are at the following locations: 158 in., at Kitt Peak National Observatory (KPNO), Arizona; 158 in., at Cerro Tololo, Chile (KPNO); 144 in., at La Silla, Chile; 150 in., at Siding Spring, Australia; 144 in. (proposed), site survey in process in France; and 138 in. (proposed), site survey in process in Italy.

Radio telescope. The largest steerable radio telescope in the world started astronomical observations early in 1971. The site of the giant paraboloidal reflector (Fig. 2) is in the Eifel Mountains, 49 km west of Bonn, West Germany. Design, construction, and operation were by the Max Planck Institute for Radio Astronomy in Bonn.

The diameter of 100 m (328 ft) surpasses the next largest fully steerable disk at Jodrell Bank, England, which is 250 ft. Design criteria also include observations of frequencies as high as 25 GHz (corresponding to 1.2 cm wavelength), as small an amount of antenna noise as possible in the centimeter range, and high pointing and steering accuracy of more or less 6 seconds of arc.

Important applications for this large instrument include line spectroscopy in the centimeter range of such molecular lines as OH, of H_2O at 1.35 cm, NH_3 at 1.26 cm, and H_2CO at 6.2 cm. Spectra of nonthermal sources in the centimeter, millimeter, and submillimeter range are also of great interest. See INTERSTELLAR MATTER.

In this same region variations in the spectra of high-velocity cosmic electrons might yield results concerning the generation and development of compact galaxies.

To operate the telescope at wavelengths as short as 1.2 cm with a design goal efficiency of 67%, the surface of the reflector must not deviate from the true paraboloid by more than 0.65 mm (root mean square). Constructing a surface of this size to such a high degree of precision is very difficult, particularly when one considers that the elastic deformations encountered when the reflector is tilted about the elevation axis are 20–30 times the required tolerance. Extensive computer design showed that back-ribbed design using 24 radial cantilevers, supported on one point, produces a surface of the required precision throughout the full travel in elevation. Some refocusing is necessary for different elevations since elastic deformations tend to change the effective focal length of the reflector surface.

The focal length of the 100-m reflector is very short, only 30 m (f/0.3), and the prime focus cabin is supported by a quadrupedal structure at the focal surface. A Gregorian concave secondary is interchangeable in the cabin to give an effective focal length of 364 m. This configuration yields the lowest antenna noise figure. The Gregorian usable field diameter is 20 minutes of arc; to take advantage of the full field, the secondary cabin can hold seven horn aerials and receiver systems at the same time. This multiple system permits operation on two or three different frequencies at the same time, thus saving background measuring time.

All the receivers employed use parametric preamplifiers refrigerated to 17K to reduce induced noise to a minimum. Receivers for frequencies of 1.42, 2.65, 4.95, and 10.5 GHz and a 100-channel spectrometer and an autocorrelation spectrograph are called for in the design.

The positioning and tracking of the telescope are computer-controlled; such variations as atmospheric refraction and elastic deformations in the disk must be corrected for, since no optical guiding or sighting is planned. The specifications of the 100-m telescope are listed in the table.

Construction began in February, 1968, with foundation work. Some portions of this project proved to be very difficult because of extreme alignments required for the azimuth track (1 mm

Specifications of the 100-m radio telescope in West Germany

Main reflector diameter	100 m	(328 ft)
Primary focal length	30 m	(98 ft)
Gregorian reflector channel	6.5 m	(21 ft)
Cone ring diameter	38 m	(125 ft)
Gregorian focal length	364 m	(1194 ft)
Azimuth track diameter	64 m	(210 ft)
Height of elevation axis above azimuth track	50 m	(164 ft)
Elevation axis length	45 m	(148 ft)
Elevation gear radius	28 m	(92 ft)
Azimuth travel	±360°	
Elevation travel	+5 to 94°	
Maximum azimuth speed	40° per minute	
Maximum elevation speed	20° per minute	
Positional accuracy	±6 seconds of arc	
Total weight of mounting (moving sections)	3200 metric tons	

on horizontal, and it must be circular within 0.2 mm).

The reflector disk was finished on May 5, 1970, and coarse adjustment of the reflector panels was started. The azimuth track was finally leveled and the precise positioning of the 2000 reflector panels across the reflector surface was started. Installation of electronic equipment ocurred in October, 1970, and the first tests ocurred near the end of 1970.

This instrument promises to be the world's most powerful, fully steerable radio telescope.

For background information see ASTRONOMICAL INSTRUMENTS; OBSERVATORY, ASTRONOMICAL; RADIO TELESCOPE in the McGraw-Hill Encyclopedia of Science and Technology.

[FRED G. O'CALLAGHAN]

Bibliography: L. Artsimovich, *Sov. Union*, p. 18, March, 1971; O. Hachenberg, *Sky Telesc.*, p. 338, December, 1970; H. A. Ingras, *Sky Telesc.*, p. 279, May, 1968; V. Lutsky, *Sky Telesc.*, p. 99, February, 1970.

Beaver

Among rodents, the Canadian beaver (*Castor canadensis*) is second in size only to the capybara of South America. The beaver, once widespread in North America, was a major inducement which favored the exploration and colonization of the continent. Not only the pelt of this mammal, but also its elaborate structures (dams, tunnels, canals, lodges, food storage piles, and so forth) have aroused the interest of man. Recent investigations have addressed themselves to the thermal problems encountered by the beaver in its range, from the Mackenzie Delta, which is north of the Arctic Circle, south to northeastern Mexico. Over this wide geographic range, the beaver must adjust to a variety of climatic conditions. It does so by a variety of anatomical, physiological, and behavioral adaptations.

Anatomy. While most of the beaver's surface is covered with fur with high insulative value, the foot-webs and tail are virtually naked surfaces which are available for the exchange of heat with the environment. Of these, the broad, flat, scaly tail is the most important.

The circulatory system of the tail retains heat when the animal is in the cold and acts as a radiator in warm surroundings. Blood enters the tail through a group of small arteries from the abdominal aorta. Immediately adjacent to the arteries are veins in antiparallel orientation, that is, the blood vessels are parallel but the blood flows in the arteries and veins are in opposite directions. Together these vessels comprise a vascular bundle, or rete mirabile. When found in appendages, this structure has been shown to function for heat retention in cold environments. While the blood may circulate the member's entire length, the heat contained in the blood of the arteries entering the limb is quickly passed to the cooler blood in the adjacent veins. Thus the flow of heat is shunted from artery to vein in this countercurrent exchange system at the base of the appendage so that the limb becomes cold and consequently heat loss from the appendage is reduced.

In warm environments the beaver utilizes the tail as a radiator, using arteries which emerge from the vascular bundle to supply an extensive subcutaneous network of veins with substantial amounts of blood to flow near the surface of the limb, where the elimination of heat is easily accomplished. In warm circumstances an alternate route for returning venous blood to the body is employed and the rete mirabile is by-passed and does not function to retain heat.

These vascular structures equip the beaver tail for an active role in the regulation of heat loss and thus in the maintenance of body temperature. At least in northern beavers, the tail must endure long periods at lowered temperature, with reduced blood supply. Consequently, most of the tail is composed of tissues with low metabolic requirements. The expanded lobes of the tail are constructed of adipose tissue subdivided with septae of connective tissues. In the central axis of the tail lies the vertebral column, with a few associated nerves and muscles and the vascular bundle ventral to the vertebrae. The fat stored in the tail represents a depot of energy reserves which may assist the beaver through the winter.

Actually the furless part of the beaver's tail represents about two-thirds of the appendage's length, since the basal third of the limb is not so broad and is invested with pelage. Measurements of the scaly part of the tail and comparison of an index reflecting scaly tail area with body length have revealed that as beavers grow, the tail area available as a radiator grows more rapidly than other parts of the animal. As a hypothetical spherical animal grows, its surface increases with the square of the radius, while its volume increases with radius cubed. Consequently, larger animals have less surface per unit of volume than do small ones, and larger animals of a homeothermic species (or larger species of homeotherms) have proportionately less surface for the elimination of excess heat. The beaver is more nearly spherical than many mammals; the tail represents the major deviation from sphericity. This rapidly growing appendage may be an important factor in the beaver's growth to body weights of 70 lb (31.8 kg) or more.

Physiology. Measurements of surface and internal temperatures of the beaver and its tail in a variety of environmental temperatures yield results which support the thermal interpretations of the limb's anatomy. At room temperature blood flow to the subcutaneous network is seen to increase and tail surface temperatures are elevated well above that of the surroundings. If heat dissipation from the tail is impaired with insulation, the animal's core temperature rises.

When the beaver is moved to a cold room, tail circulation is reduced and use of the rete mirabile allows tail temperature to slowly approach that of the environment, while the temperature of the body and its furred surfaces remain high and relatively constant. However, the beaver does not allow the tail to freeze. If impulse conduction in the nerves which control the tail circulation is blocked with anesthetic, then heat loss from the tail continues and the beaver's core temperature drops, in spite of vigorous shivering.

Measurement of tail heat loss into air and into water allows comparison of heat loss by these routes with the heat generated by the beaver's metabolism. In cold environments these measurements revealed that the beaver can confine heat loss to less than 2% of its resting heat production, while in warm environments tail heat loss into air was elevated to over 25% of the resting heat production. About 35% of the resting heat production could be eliminated by the tail if immersed in water. Thus physiological regulation by the tail plays an important role in the beaver's adjustment to varied environmental temperatures.

Behavior. In building a massive and conspicuous nest, or lodge, the beaver creates for itself a stable microclimate to which it may retreat at times of stress. Not only does the lodge provide protection of young and adults from predation and act as a focal arena for the maintenance of the beaver's social life, but the lodge acts as a buffer shielding its occupants from extremes in daily and seasonal temperature fluctuations. In fall the residents typically repair the lodge, adding mud and sticks to its exterior. This additional material, together with snow, makes the lodge an especially well-insulated retreat in winter, when the beaver's pond may be frozen over. The beavers enter the dry chamber of the lodge, which may be in the bank or in the center of the pond, through tunnels which open under water. A porous region at the top of the lodge is roofed with sticks alone, the mud being omitted. This "smokehole" provides ventilation to the nest chamber below. A submerged cache of branches is accumulated near the lodge and is consumed through the winter.

By these anatomical, physiological, and behavioral adjustments, the beaver is adapted for survival over a wide range of latitudes in North America.

For background information see BEAVER; HOMEOSTASIS in the McGraw-Hill Encyclopedia of Science and Technology. [RICHARD W. COLES]

Bibliography: M. Aleksiuk, *Ecology*, 51:264–270, 1970; M. Aleksiuk, *J. Mammalogy*, 51:145–148, 1970; R. W. Coles, *J. Mammalogy*, 51:424–425, 1970; L. Wilsson, *My Beaver Colony*, 1968.

Beryllium disease

The widespread use of beryllium in industry has been associated with numerous instances of illness, often severe and fatal, resulting from toxic exposure to the metal and its salts. Recent studies have elucidated the features of this illness. Although bodywide in its effects, the lung has been found to be the organ most commonly affected. Inhalation of toxic material in high concentration may result in a nonspecific acute form of the disease. The chronic form affects only a small proportion of workers and may occur after much less intense exposure; it develops after a latent period ranging from months to many years, and its effects range from mild respiratory distress to complete incapacitation and death. The chronic disease may also affect individuals indirectly exposed within the plant or living in its vicinity. Various clinical and pathological features suggest that immune mechanisms, as well as direct toxic effects, are involved in the pathogenesis of the chronic disease.

Beryllium disease—originally called berylliosis—was first described in Germany and the Soviet Union in the mid-1930s in workers in extraction plants where beryllium is separated from its ore (beryl or beryllium aluminum silicate). By the mid-1940s, when the disease was recognized in the United States, there were epidemics of some size in workers exposed in the extraction, alloy production, and fluorescent lamp and neon sign industries, as well as in laboratories engaged in research in atomic energy and the development of x-ray-tube windows. These epidemics ceased once the hazard was recognized and controls were instituted, starting in 1949. The fluorescent lamp industry discontinued the use of beryllium phosphors during the same year, with the result that all cases subsequently recognized in this industry are associated with earlier exposures. New cases continue to appear in other industries, however, and threaten to increase as new ways of using beryllium are being found by industry and government agencies concerned with military defense and space exploration. All compounds of beryllium used in industry, including the metal dust encountered in grinding and similar operations, have been associated with illness except for beryl itself, which is inert.

Acute form. Exposure to high levels of atmospheric contamination with beryllium compounds, usually through accidental overexposure, may result in severe inflammation of the upper respiratory tract or lungs. Massive exposure may result in a fulminating pneumonia appearing within 72 hr; exposure to lower doses more commonly produces a more insidious illness developing over a period of about 2 weeks. These patients develop an exhausting cough and marked difficulty in breathing, associated with such toxic symptoms as fatigue, lassitude, and loss of weight and appetite. Fever is usually mild or absent. Infiltration of the lung is seen by x-ray, and examination of the lungs of those patients who have died of the disease reveals severe congestion and outpouring of fluid into the air spaces (pulmonary edema), together with varying degrees of ulceration and inflammation of the walls of the bronchi. These changes are not specific for beryllium, however, and may also occur after exposure to other severely irritant gases and fumes. Similar lesions have been produced in experimental animals following inhalation of beryllium sulfate or fluoride mists.

Most patients with acute beryllium pneumonia recover within 1–6 months, but death may occur following intense overexposure as a result of severe pulmonary edema. About 10% of these patients subsequently develops the chronic disease, but this figure is probably low, since many mild acute forms have undoubtedly been unrecognized or unreported. Reexposure of the patient following recovery may also result in one or more subsequent acute attacks, although there is no clear evidence that one attack necessarily makes a patient more vulnerable to others.

Dermatitis and conjunctivitis may occur following exposure of skin or eyes to such soluble acid salts as beryllium sulfate or chloride. Accidental introduction of a crystal of such a compound into a crack in the skin may produce a chronic ulcer which resists healing until the crystal has been extruded or removed. Similar chronic ulcerations ("lamp lesions") were observed following accidental cuts on broken fluorescent lamps when beryllium phosphors were still being used.

Chronic form. After a latent period varying from a few months to more than 20 years following exposure, the chronic form of the disease may develop. Only a comparatively small proportion of exposed workers are affected, and most of these give no history of prior acute disease. Onset is usually insidious, with gradually developing difficulty in breathing (especially after exertion), cough, fatigue, malaise, loss of appetite, and often marked loss of weight. The course is highly variable; symptoms range from the very mild, in which the chest x-ray shows infiltration but the patient is aware of difficulty only after exertion, to the very severe, in which the patient is a respiratory cripple who can hardly move without distress and who requires almost continuous administration of oxygen for survival. Remissions and exacerbations are common. At least one-third of the known cases have died of their disease after periods ranging from a few months to more than 15 years. The use of steroids in treatment appears to have aided considerably in ameliorating symptoms and probably prolonging life, but there is no convincing evidence as yet that any individual who develops the chronic disease ever completely recovers.

In about 80% of patients, examination of the lung after death or sampled during life shows extensive infiltration of the lung tissue between the air sacs with lymphoid cells and tissue phagocytes (histiocytes), which interfere with gas exchange between air and blood and stiffen the lung, impeding normal ventilation. The development of fibrous tissue in more advanced cases exaggerates these phenomena, as well as the symptoms. Many of these patients also show prominent focal collections of histiocytes (granulomas) in association with the cellular infiltration. Large numbers of minute calcified bodies, often showing a concentric lamellation, may also be present. The remaining 20% of patients shows little diffuse cellular infiltration; pulmonary changes are confined to well-formed granulomas and varying degrees of fibrosis similar to those changes seen in sarcoidosis, a generally milder disease of unknown etiology and unrelated to beryllium exposure. These cases also resemble sarcoidosis in having a generally more benign and prolonged course than do those with extensive cellular infiltration and few or no granulomas.

Changes in the chronic disease are by no means

confined to the lungs, and lesions can also be demonstrated in other tissues, such as lymph nodes, liver, and spleen, all of which may become enlarged. In late stages the heart is often affected, usually secondary to changes in the lungs. Since beryllium is an enzyme poison capable of inhibiting alkaline phosphatase, and probably other calcium- or magnesium-activated enzymes as well, this, together with its close relation to these elements in the periodic table, probably accounts for the markedly disturbed calcium balance in some patients. This disturbed calcium balance is characterized by increased calcium loss in the urine, the formation of calcium stones in the kidney, and the numerous calcified bodies often seen in the lungs and occasionally in other tissues as well.

The mechanism whereby the chronic disease is produced is not yet known. The occurrence of the disease in only a small proportion of those exposed, the prolonged latent period, the occasional abnormalities in blood globulins usually associated with antibody production, the characteristic cellular infiltration in the lungs with granuloma formation, and many other features strongly indicate that immune mechanisms, especially those related to delayed hypersensitivity, are involved, in addition to direct toxic effects. The chronic disease has not yet been reproduced exactly in experimental animals, although focal granulomatous inflammation of the lungs has been observed. Bone tumors have been produced experimentally in rabbits, but neoplastic disease in human patients is rare.

"Neighborhood" cases. Chronic beryllium disease has been observed not only in those workers directly involved with beryllium or its compounds but also in those closely associated with these operations, such as nurses, janitors, watchmen, and office workers in the plants. Cases have also been observed in other members of the families of beryllium workers living nearby, especially those who have come in contact with dusty clothing brought home to be washed. Additional cases have been observed in unrelated families presumably exposed to atmospheric pollutants within about 0.5 mi of an extraction plant. These cases show no appreciable difference in severity and prognosis than do others, and their occurrence has been markedly reduced or eliminated following the introduction of more rigid safety measures in most plants.

Prevention. Continued recognition of the health hazards inherent in the use of beryllium and its compounds and the cooperation of most manufacturers in the use of adequate safety precautions have gone far toward reducing the incidence of this disease. The continued expansion of the industrial uses of beryllium in aircraft, spacecraft, missiles, lasers, computers, armor systems, electronics, and rocket fuel, however, represents an ever-present danger to the industrial and even the general environment which requires continued supervision and control. Occasional accidental overexposures will probably continue to occur, but all other forms of exposure are clearly preventable.

For background information see BERYLLIUM in the McGraw-Hill Encyclopedia of Science and Technology. [DAVID G. FREIMAN]

Bibliography: D. G. Freiman and H. L. Hardy, *Hum. Pathol.*, 1:25, 1970; J. D. Stoeckle et al., *Amer. J. Med.*, 46:545, 1969; H. E. Stokinger (ed.), *Beryllium: Its Industrial Hygiene Aspects*, 1966; L. B. Tepper, H. L. Hardy, and R. I. Chamberlin, *Toxicity of Beryllium Compounds*, 1961.

Brassin

Plants produce hormones which control their growth and behavior. Synthetically prepared growth-regulating substances have been used for many years to supplement these endogenous hormones, and thus they change to some extent the behavior of crop plants to suit the needs of man. But hormones produced by crop plants themselves so far have not been useful, mainly because they occur in extremely small amounts and are difficult to obtain. Tons of plant material are required to produce a few milligrams of hormone. However, some hormones made by plants eventually may be useful because sources are now known from which relatively large amounts of these endogenous regulators can be obtained. Furthermore, these hormones induce some responses that cannot be obtained with growth-regulating substances. Also of importance in the control of food crops is the fact that these hormones are present in plants consumed by humans and animals, an indication that these substances are relatively nontoxic. See MORPHACTIN.

Some hormones, brassins, that may eventually be useful were recently discovered in pollen of rape, a widely grown crop plant. The discovery and some characteristics of brassins are described here.

Detection. A new bioassay was devised to aid in the search for hormones in crop plants. The bioassay is based on the following principles: An entire plant is used rather than an explant, thus making possible the expression of a variety of responses by the test plant. The test compound is applied to an undeveloped portion of a stem (internode) just as this portion begins to grow rapidly, thus taking advantage of the extreme sensitivity of immature plant parts to the effects of hormones. An oily carrier is used. This carrier is suitable for all known hormones and also for any lipoidal compounds that may have growth-regulating properties. Brassins and similar hormones have been detected in rape and other crop plants with this bioassay.

Isolation. In searching for suitable sources of hormones in crop plants, it was found that cotton fibers contained them, but yields from this source were insufficient. It was also learned that corn pollen contains relatively large amounts of growth-accelerating substances, but the pollen of this wind-pollinated plant is difficult to obtain. Bees were used recently to facilitate collection of pollen for most of these hormone studies; 39 kinds of pollen were tested and 18 of these contained substances that accelerated plant growth in the assay used.

Pollen from the rape plant, *Brassica napus* L., proved to be most promising. Since rape plants are grown extensively to obtain oil, it is possible to collect large amounts (hundreds of pounds if needed) of pollen from this plant. Being rich in plant hormones, rape pollen represents a plentiful source of these substances.

Marked acceleration of stem growth resulted when crude extracts of rape pollen were tested. These extracts were made by simply soaking rape pollen in ether overnight. Chromatographic purification of the hormones obtained resulted in an oily product, 10 µg of which generally accelerated stem elongation by three- to fivefold, sometimes as much as tenfold (see illustration). Methods have been developed with which brassins can be purified using column and plate chromatography; 8–10 mg of the purified brassin complex is obtained from 60 g of rape pollen.

Pollen from plants other than rape may contain hormones that are similar to brassins. For example, pollen from alder, *Alnus glutinosa* (L.) Gaertn., contains hormones that behave like brassins chromatographically and induce plant responses typical of those induced by brassins.

Chemistry. The brassins, obtained as an oily product after chromatographic purification, were examined to determine their chemical structure. By use of various spectroscopic techniques, the following information has been obtained. The presence of isolated and conjugated-type double bonds in brassins is established since the ultraviolet spectrum showed very intense absorption bands at 207 and 226 nm. Brassins are made up of long-chain aliphatic esters, a conclusion reached on the basis of very strong infrared spectral bands at 2950–2850 cm^{-1} due to the CH stretching frequency in aliphatic carbon compounds present and a band at 1720 cm^{-1} for ester carbonyl groups. Further, brassins do not contain either hydroxyl or amino groups because there were no absorption bands in the 3600–3000 cm^{-1} region; the bands in this region are characteristic for X-H groupings, where X represents a heteroatom. Brassins are fatty acid esters with different types of unsaturation in the molecules. This conclusion is drawn because absorption signals in the nuclear magnetic resonance spectrum are evident at δ 0.9 for a terminal methyl group, δ 1.26 for methylene groups as in a fatty material, and between δ 1.5 and 2.5 ppm for different methyl, methylene, and methine groupings in the vicinity of double bonds. The spectrum also showed absorption due to olefinic protons at 5.3 ppm. Brassins appear to have a glyceride structure since absorption signals at 4.0 and 5.3 ppm are evident due to presence of methylene and methine groups, respectively, attached to oxygen atoms. Finally brassins contain several fat components with molecular weights ranging from 250 to 600. This conclusion is based on studies of brassins by mass spectrometry combined with gas chromatography which revealed several molecular ions. The fragmentation pattern suggests that all the fragments are derived from long-chain esters.

Separation of brassins by gel chromatography gave five compounds, each with hormonal activity. Saponification of brassins yielded fatty acids, some of which are linked to the glycerol moiety. These acids were converted into corresponding methyl esters, which were then analyzed by gas liquid chromatography. There are five major peaks on the chromatogram. Comparison of these peaks with those of standard methyl esters shows that the methyl esters in brassins have chain lengths of C_{16} to C_{24}. These results suggest that the brassin complex contains five compounds which are composed of fatty acids either in the form of esters or glycerides.

Accelerated stem growth induced in bean plant (right) with 10 µg of a fatty hormone complex (brassins) obtained from rape pollen, compared with that of an untreated bean plant (left).

Growth responses. Knowledge about the way that plants respond to brassins is very limited because at present only small amounts of these hormones are available. Both leaves and stems of bean plants respond to treatment with brassins. An internode treated with 10 µg of brassin complex usually begins to show an increase in growth rate, compared with that of an untreated plant, in about 24 hr. About 4 days after treatment, this same internode will be both longer and, in most cases, larger in diameter at the top than comparable parts of untreated plants (see illustration). Histological studies show that brassins induce both cell elongation and cell division within the same internode. This dual response (elongation plus division) is unique among the natural plant hormones. Stimulation of cell division is unusual, since with untreated plants growth of these internodes involves for the most part cell enlargement rather than cell division. This cell division in the stem usually involves parenchyma cells of the cortex, phloem, xylem, and pith. *See* XYLEM.

Applied to the stem, brassins also stimulate cell division in the pulvinus, petiole, and veins of bean leaves. When relatively large amounts of brassins are used (10 µg), leaves assume a distorted shape, and tissue arrangement, particularly that of the vascular tissues, is greatly modified when viewed histologically.

Responses to brassins occurred over a hundred-fold concentration range (1–100 µg), and the hormone complex was not toxic to bean plants even when applied directly and undiluted with the lanolin carrier.

For background information *see* PLANT HOR-
MONES in the McGraw-Hill Encyclopedia of Science and Technology.

[J. W. MITCHELL; N. MANDAVA; J. F. WORLEY]

Bibliography: J. W. Mitchell et al., *Nature*, 225:
1065–1066, 1970; J. W. Mitchell and G. A. Living-
ston, *Agriculture Handbook No. 336*, pp. 26–28,
U.S. Department of Agriculture, 1968.

Buildings

A new surge of activity in the construction of tall buildings has taken place within the last few years. Although there have been many advancements in building construction technology in general, spec-tacular achievements have been made in the de-sign and construction of ultrahigh-rise buildings. The construction of skyscrapers reached an earlier peak in the 1930s when the 102-story Empire State Building was built in New York, but most tall buildings built during that time used the tradi-tional column-and-beam frame construction, where stability depends on rigid connections between columns and beams at each floor, which results in a high "premium for height" as compared to other buildings.

The early development of high-rise buildings be-gan with structural steel framing. Reinforced concrete has since been economically and com-petitively used in a number of structures for both residential and commercial purposes. The newer high-rise buildings, ranging from 50 to 110 stories and now being built all over the United States, are the result of innovations and development of new structural systems.

Many of these innovations have been utilized and many of these structural systems have been developed since 1960.

The premium for height in tall buildings is mostly because of the use increased column and beam sizes to make buildings more rigid so that under wind load they will not sway beyond an acceptable limit. Excessive lateral sway per floor may cause serious recurring damage to partitions, ceilings, and other architectural details. In addi-tion, excessive sway may cause discomfort for occupants of the building due to perception of such motion. New structural systems in rein-forced concrete, as well as steel, take full advan-tage of the inherent potential stiffness of the total building and thereby do not require additional stiffening to limit the sway.

To establish an economic parameter for the premium for height, the design process should be viewed in two phases. The first phase involves de-signing the building for the floor loads (that is, gravity loads) only, without considering the effect of the lateral load—either seismic or wind. The columns, beams, and slabs are proportioned to carry only the normal floor loads. Because gravity loads cannot be reduced or overcome by any struc-tural manipulation, the first design phase sets the lower boundary of overall proportions, total quan-tity of materials, and cost. After phase one, the designer must consider the effects of lateral loads on the structure.

To provide adequate lateral resistance as well as to limit the lateral sway or to control the per-ception of lateral motion, the designer may have to increase sizes of columns or beams or both. The second phase, therefore, constitutes the final de-sign of the building and represents the upper boundary in terms of the proportions, quantity of material, and total economy. In a steel structure, for example, the economy can be fairly well de-fined in terms of the total average quantity of steel per square foot of total floor area of the building. Therefore, for a typical building frame, one can show graphically (Fig. 1) how an increase in the number of stories increases the gap between the lower bounds (phase one) and the upper bounds (phase two) in the design quantities. The gap be-tween the upper boundary and the lower bound-ary represents the premium for height for the tra-ditional column-and-beam frame. Within the last few years structural engineers have developed new structural systems with a view to eliminating this premium for height.

Systems in steel. Tall buildings in steel have de-veloped as a result of several types of structural innovation. The innovations have been applied to

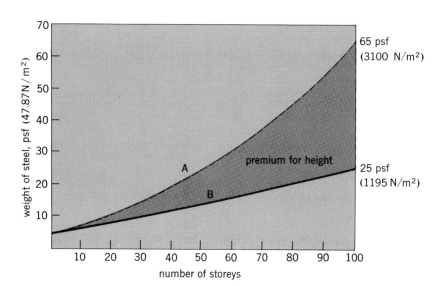
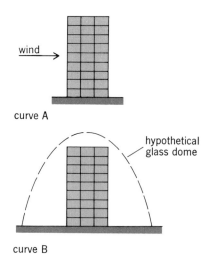

Fig. 1. Graphical relationship between design quantities of steel and building heights for a typical building frame. Curves A and B correspond to the boundary condi-tions indicated at the right.

Fig. 2. Belt trusses increase stiffness of the First Wisconsin Bank building, Milwaukee. (*ESTO*)

the construction of both office and apartment buildings.

Frames with rigid belt trusses. In order to tie the exterior columns of a frame structure to the interior vertical trusses, a system of rigid belt trusses at mid-height and at the top of the building has been developed. These horizontal belt trusses have been shown to increase the stiffness of the entire structure. The first building using a belt truss at mid-height and one at the top of the building is now being constructed in Melbourne, Australia. In the United States a good example of this system is the First Wisconsin Bank building under construction in Milwaukee (Fig. 2).

Framed tube. The maximum efficiency of the total structure of a tall building, both for strength and stiffness, to resist wind load can be achieved only if all column elements can be connected to each other in such a way that the entire building acts as a hollow tube or rigid box projecting out of the ground. There are many practical planning and architectural difficulties in tying all of the columns of a building together. However, exterior columns may be made to act together by various means within the architectural framework of rectangular windows. This can be done if exterior columns around the four walls of a building are tied together by spacing them as closely as possible so that these exterior columns connected at each floor with deep spandrel beams will simulate a rectangular or square hollow tube with perforated window openings. This particular structural system was probably used for the first time in the reinforced concrete 43-story DeWitt Chestnut apartment building in Chicago. The most significant use of this system has now been made in the twin structural steel towers of the 110-story World Trade Center building in New York.

Column diagonal truss tube. The exterior columns of a building can be spaced reasonably far apart and yet be made to work together as a tube by connecting them with diagonal members intersecting at the center line of the columns and beams. This simple yet extremely efficient system was used for the first time on the John Hancock Center in Chicago (Fig. 3). Because of the resulting tube or rigid box action in this system, only 29.7 lb (13.5 kg) of steel per square foot (0.09 m²) of floor area was used for the 100-story John Hancock Center, which is about the same as would have been required for a 40-story column-beam frame building.

Bundled tube. With the future need for larger and taller buildings, the framed tube or the column diagonal truss tube may be used in a bundled form to create larger tube envelopes while maintaining high efficiency. The 110-story Sears Roebuck Headquarters Building in Chicago, now under construction, will have 9 tubes, each 75 ft² (7 m²), bundled at the base of the building in three rows. Some of these individual tubes will terminate at different heights of the building, demonstrating the unlimited architectural possibilities of this latest structural concept. The Sears tower, at a height of 1450 ft (442 m), will be the world's tallest building when completed in 1973.

Systems in concrete. While tall buildings in steel had an early start, present development of tall buildings in reinforced concrete has prog-

Fig. 3. Exterior diagonal system is used in the John Hancock Center, Chicago. (*ESTO*)

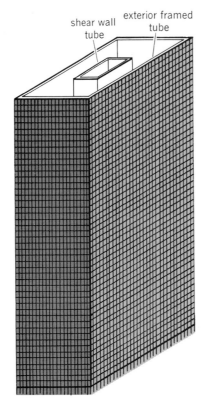

Fig. 4. Schematic sketch of tube-in-tube system.

ressed at a fast enough rate to provide a competitive challenge to structural steel systems for both office and apartment buildings.

Shear wall–frame interaction. A recent improvement in shear wall construction has been a tall concrete building with a central shear wall and an outer rigid frame of columns and beams and slabs, resulting in an efficient interaction between the two types of construction. The first design application of the shear wall–frame interaction was probably the 38-story Brunswick Building in Chicago, which was built without any appreciable premium for height.

Framed tube. As discussed earlier, the first framed-tube concept for tall buildings was used for the 43-story DeWitt Chestnut apartment building, which was built in reinforced concrete. The framed-tube concept is highly efficient for small floor areas where a central shear-wall core cannot be efficiently used. In the DeWitt Chestnut building, exterior columns were spaced at 5 ft 6 in. (1.67 m) centers and interior columns used as needed to support the 8-in.-thick (203 mm) flat-plate concrete slabs. The closely spaced column systems in reinforced concrete do not generally require any special premium in construction costs. They, therefore, have a distinct advantage in eliminating the use of hung curtain walls by using the closely spaced columns as the supporting structure for the window panes.

Tube in tube. The most recent system in reinforced concrete for office buildings combines the traditional shear-wall construction with an exterior framed tube. The system consists of an outer framed tube of very closely spaced columns and an interior rigid shear-wall tube enclosing the central service area. This system (Fig. 4), known as the tube-in-tube system, made it possible to design the world's present tallest (714 ft or 217.6 m) reinforced concrete building (the 52-story One Shell Plaza building in Houston) for the unit price of a traditional shear-wall structure of only 35 stories.

New systems combining both concrete and steel have also been developed, the latest of which is the composite system developed by Skidmore, Owings & Merrill in which an exterior closely spaced framed tube in concrete envelopes an interior steel framing, thereby combining the advantages of both reinforced concrete and structural steel systems. The 52-story One Shell Square building in New Orleans and the 40-story Union Station Building in Chicago, presently under construction, are based on this system.

For background information *see* BUILDINGS; REINFORCED CONCRETE; STRUCTURAL ANALYSIS in the McGraw-Hill Encyclopedia of Science and Technology. [FAZLUR R. KHAN]

Bibliography: F. R. Khan, *Civil Eng.*, October, 1967; F. R. Khan, in *Proceedings of the ACI Annual Convention*, April, 1970; F. R. Khan, *Inland Architect*, July, 1967.

Buoy

Studies of the ocean environment frequently require that research ships involved in making measurements be far removed from the sensing instrumentation being employed so as not to adversely affect observations. For example, when hydrophones are suspended over the side of a ship for making acoustic measurements of ambient noise in the open ocean, the noise radiated by the measurement ship could seriously contaminate the ambient noise field and thus render the observations invalid. A ship's movements can produce erroneous results in measurements because of motions imparted to the sensing instrumentation by the umbilical connection between the sensor and the ship.

A number of different types of buoys, or buoy systems, capable of deploying a variety of sensors for oceanographic and underwater acoustic measurements have been developed to overcome the difficulties mentioned above. Some approaches which permit both meteorological and oceanographic observations employ surface buoys which are moored in water depths as great as 4570 m (2500 fathoms) so as to maintain a fixed geographic position. With such buoys, sensors can be located above the ocean surface, at the air-water interface, and in the water column at almost any depth. A preferred approach for cases where only in-water measurements are to be made is to use completely submerged tethered or free buoys. Complete submergence produces freedom from any surface excitation. Free buoys, or floats, designed to stabilize at a given depth have been developed which employ materials in configuration that are less compressible than sea water. Such a body will gain buoyancy as it sinks and if its excess weight at the surface is small, it can, at some depth, become neutrally buoyant and sink no farther. Buoys of this type have been fitted with equipment which sends out acoustic signals. A ship can track these buoys for a number of days and make direct measurements of current at depth which are free from the uncertainties inherent in current meters on an anchored ship. Aluminum alloys have been employed in making these buoys. These alloys are available in convenient cylindrical form of various lengths. In addition to having sufficiently low compressibility, these floats must provide sufficient additional buoyancy to carry desired instrumentation and yet must not collapse at the greatest working depth. Perhaps the most versatile buoy developed to date is the AUTOBUOY. This device is a free-diving, self-controlled, programmable device capable of diving to a maximum depth of 6100 m (20,000 ft.)

AUTOBUOY operational characteristics. This buoy was developed by Lear Siegler, Inc., with support from the United States government, and is shown in outline form in Fig. 1. The buoy weighs 386 kg (850 lb), is cylindrical in shape, and has a diameter of 53 cm (21 in.) and a length of 2.7 m (9 ft). In use it assumes a vertical attitude with the radio-frequency (rf) antenna (beacon) uppermost. The ends of the buoy are tapered, with one end finned to assure a stable vertical trajectory in the water. Deployed from an oceanographic ship the buoy descends rapidly to an initial maximum depth. At this point the 22.7-kg (50-lb) lead descent weight used to achieve quick descent is jettisoned. The lead weight provides a descent velocity of approximately 1.2 m/sec (4 ft/sec). After jettisoning the weight, depth control of the buoy is accomplished by the alternate valving to sea of perchloroethylene and mineral spirits. These fluids have, respectively, a specific gravity greater than and less than sea water. Buoyancy is thus adjusted to a neutral value in response to pressure and velocity error signals derived from self-contained instrumentation. Simultaneous with the jettisoning of the descent weight, the two hydrophones are released from their storage chambers and deploy to the full length of their electrical connecting cables, 2.7 m (9 ft) and 11.0 m (39 ft), respectively.

Release of the heavy and light fluids is achieved without pumping by the opening of values and following the fluids to gravitate out of their respective tanks, thus minimizing the possibility of noise generation. As fluid is released it is replaced by sea water which enters through a permanently opened vent tube located in each tank.

The design of the buoy incorporates several safety features to abort a dive and cause the buoy to return to the surface if a malfunction should occur. These features include a leak detector, explosive cutters for releasing the hydrophones, and a magnesium valve which after corroding for approximately 72 hr will dump fluid ballast. This frees the hydrophones, thus compensating for negative buoyancy which may have resulted from flooding of the instrumentation housing.

AUTOBUOY programming. Programming of the buoy is accomplished by means of a number of digital switches. These provide for initial descent to the deepest depth, ascent to as many as four shallower depths (in sequence), and final ascent to the surface. At each of the five selected depths, various prolonged hovering periods may be chosen with individually selectable data-acquisition programs. Figure 2 is a block diagram of the AUTOBUOY control and sensor systems.

The buoy monitors water temperature, depth

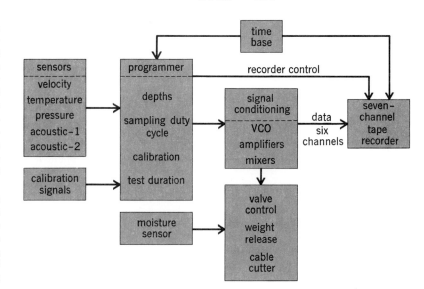

Fig. 1. Schematic view of AUTOBUOY.

(pressure), two channels of acoustic information, and velocity of the unit through the water. AUTOBUOY can be programmed to make observations over a period up to 48 hr. Data are recorded on magnetic tape for a time period up to 160 min and a frequency response from 10 to 5000 Hz. Recording can be continuous or intermittent. Sampling intervals can be selected to cover descent and ascent modes as well as the five hovering modes. The number of samples at any depth can vary from 1 to 24. The programmer also permits injecting calibration signals on the magnetic tape to permit deriving absolute acoustic sound pressure levels. A self-powered Accutron provides the basic timing frequency of 360 Hz for the entire system.

To provide a wide dynamic range for acoustic signal monitoring, each of the hydrophones occupies two channels of the seven-track tape recorder. The recording channels of the two channels differ by 30 dB apart. Multiplexing is employed for recording the remaining four channels of information on the other three tracks of the tape recorder.

AUTOBUOY announces its return to the surface by initiating both an rf signal and a flashing xenon light. These features are provided in a recovery package which is completely independent of the other systems. A pressure switch keeps the recovery package inactivated below the water surface, while an internal gravity switch permits oper-

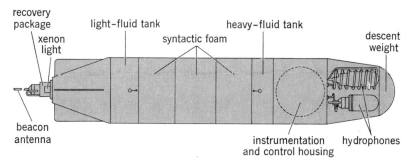

Fig. 2. Functional block diagram of AUTOBUOY control and sensor systems.

ation only in a vertical operating position.

Deployment of AUTOBUOY is normally scheduled so that it returns to the surface a few hours before daybreak. A radio direction finder is used for detecting the buoy's return to the surface and vectoring the mother ship to it. The flashing xenon light permits final identification and location. With the coming of daybreak, the buoy is retrieved aboard the mother ship.

The buoy has been deployed in the Atlantic and Pacific oceans and Mediterranean Sea and has been used successfully at varying depths down to 5200 m (17,000 ft).

For background information *see* BUOY in the McGraw-Hill Encyclopedia of Science and Technology.   [RAYMOND W. HASSE, JR.]

Bibliography: R. W. Hasse, Jr., *IEEE 2d Annu. Int. Geosci. Electron. Symp.*, April, 1970; T. Rossby and D. Webb, *Deep-Sea Res.*, vol. 17, 1970.

Cell division

Recent studies of cell division have been concerned with meiotic and somatic chromosome pairing, deoxyribonucleic acid (DNA) replication, and repetition in heterochromatin during chromosome replication.

Meiotic and somatic chromosome pairing. Chromosomes display a strong tendency to associate with one another in a variety of ways. Most commonly the associations are pairwise and highly specific in nature, occurring between identical chromosomes (homologs) or their multiples. Associations of a less specific nature taken place between nonhomologs or their parts. Pairing may be initiated between condensed homologs, such as following karyogamy in certain fungi, or between condensed nonhomologs, such as at mitotic metaphase in the somatic and gonial cells of *Drosophila melanogaster* and in the oocyte of that species, as inferred from genetic studies. Whether pairing ever begins between greatly extended chromosomes remains problematical. The molecular forces, long-range or otherwise, responsible for the initiation of pairing are unknown.

Functionally, pairing is a prerequisite for the exchange of parts between homologs, for reduction in chromosome number during gametogenesis, and conceivably for the execution of metabolic processes. Pairing which leads to exchange must be extremely intimate to provide for the precision of the process; it is most easily thought of as base pairing between complementary, single strands of DNA, one from each of two homologous chromosomes. Less specificity and intimacy are required for the pairing that precedes chromosome reduction, for here the degree of discrimination need normally proceed only to the gross chromosomal level. Thus, in *Drosophila* females, pairing for segregation (distributive pairing) displays no requirement for homology, but rather uses size similarity as the criterion for chromosome recognition. This is an example of karyotypic control of meiotic behavior.

Meiotic pairing. This section will discuss exchange pairing and distributive pairing taking place during meiosis.

Although the molecular events surrounding exchange pairing remain elusive, the identification of the stage when it begins seems an approachable goal. Early cytological literature contains descriptions of homologous associations initiated in the final gonial division stages or during premeiotic interphase. These data were largely ignored in the widespread commitment to C. D. Darlington's precocity model of pairing. Deriving from the sequence of H. de Winiwater, which assumes that chromosomes first become cytologically visible as fine, unpaired threads at leptotene, establish initial contact at zygotene, and attain a stable, paired state at pachytene, Darlington further proposed that precocious entrance into prophase in an unreplicated and hence unsaturated state was responsible for homogolous pairing and saturation at zygotene-pachytene. He speculated that, upon chromosome replication during pachytene, saturation between sister chromatids replaced that between homologs, and a substituted repulsion between homologs led to the breakage and non-sister-strand rejoining required for recombination. Demonstrations that meiotic DNA replication preceded zygotene-pachytene invalidated the precocity theory but had the net effect of uncoupling replication from recombination and retaining the latter as a pachytene event. Evidence favoring the traditional sequence, which demands that homologous pairing begin at zygotene, is not sufficiently compelling to eliminate alternatives.

Recently, efforts employing light microscopy, electron microscopy, and a combination of autoradiographic and genetic techniques have attempted to resolve the problem. Cytological investigations with a broad spectrum of plants, reviewed by S. M. Stack and W. V. Brown, have confirmed observations of early workers that homologs are often paired during premeiotic interphase. Electron microscopy has revealed the presence of a tripartite structure (the synaptinemal complex) associated with meiotic chromosomes. Apparently restricted to germ cells, the complex is thought by some to represent the point-to-point pairing of homologs preparing for or undergoing exchange. Originally its duration was said to coincide with the "synaptic stages" (that is, zygotene and pachytene). However, subsequent studies have disclosed single complexes preceding "synapsis" at interphase or leptotene and following "synapsis" at diakinesis and metaphase, as well as multicomplexes at premeiotic interphase and in spermatids. This and the presence of single complexes in haploids of maize and tomato, in the single X of XO genotypes, and in certain achiasmate forms, together with failure to detect the complex in strains of *D. anassae* males in which crossing-over occurs, weaken previous assumptions about the essential role of the synaptinemal complex in exchange pairing or exchange.

A different approach uses the final DNA replication as a fixed point in a cell undergoing meiosis (meiocyte) which can be marked by label. Alterations in exchange frequency, notably by temperature, can be temporally related to the replication period. By simultaneously exposing a meiocyte to tritiated thymidine and heat treatment, it becomes possible to determine whether the temperature effect on exchange precedes, coincides with, or follows the time of DNA replication; since pairing is required for exchange, information concerning pairing initiation is thus indirectly obtained. Initial

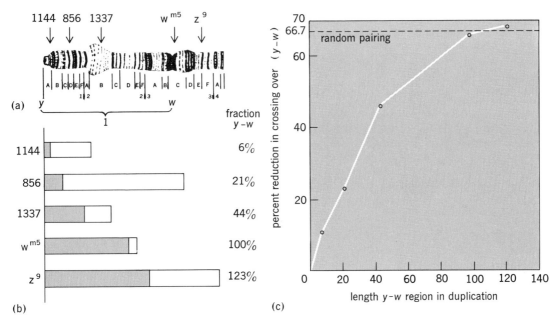

Fig. 1. Correlation between extent of homology and reduction in exchange in region between yellow (y) and white (w) with different X-duplications. (a) Euchromatic breakpoint at the tip of the X salivary gland chromosome for each duplication used. (b) Five duplications in a, with percent of region between y and w shown in gray; portion in white reflects relative amount of heterochromatin in the duplications, but the fraction within each duplication has been reduced for illustrative purposes. (c) Reduction in crossing-over from y to w as a function of the extent of euchromatin from y to w carried by the duplication. Apparent inability of different heterochromatic contents to affect values suggests that heterochromatin does not undergo exchange pairing. (Modified from R. F. Grell, Pairing at the chromosomal level, J. Cell. Physiol., 70(1):119, 1967)

studies with adult *Drosophila* females indicated that the maximal heat-induced increase in exchange was roughly coincident with DNA replication. Later studies with immature females, which permit a more precise resolution of events, demonstrated that the temperature-sensitive period was partially, if not entirely, coextensive with DNA synthesis. Parallel studies with maize, *Sphaerocarpus*, and *Chlamydomonas* have identified premeiotic interphase or DNA synthesis as the temperature-sensitive period, while related experiments with *Ornithogalum* have similarly localized the cold-sensitive period for univalent induction. Additional analysis of the *Drosophila* data has disclosed that heat induces drastic alterations in chromosome interference which appear independent of effects on exchange. As defined by H. J. Muller, positive interference is the tendency of one crossover to interfere with the occurrence of another crossover in its vicinity. If, as generally assumed, positive interference arises through the configurations homologs assume during pairing, changes in interference, like changes in exchange frequency, are most easily interpreted as direct effects. Interestingly it is found that maximal heat effects on interference occur in the final oogonial and earliest oocyte stages.

G. Pontecorvo and R. H. Pritchard, early iconoclasts with respect to the traditional sequence, suggested on the basis of the clustering of exchanges they found in *Aspergillus* that cytologically visible pairing at pachytene might represent a mechanical device for segregation, with crossing-over occurring much earlier, perhaps coincident with chromosome replication. Instead of parasynapsis along the entire chromosome, they visualized short, discrete "effective pairing sites" arising through chance contact but within which the probability of exchange would be extremely high. An altered version, suggested by R. F. Grell in 1965, proposed that a rough alignment of homologs preceding the establishment of intimate, discontinuous sites during interphase provides ways to avoid interlocking of nonhomologs and to position sites for positive interference. In *Drosophila*, tests of the two versions with free duplications (triplicated regions) favor the latter (Fig. 1a and b). Chance contact of small regions predicts that a third copy should triple the probability of contact and, depending on whether the duplication partici-

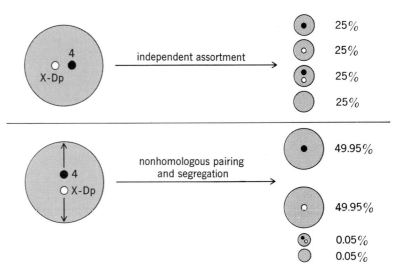

Fig. 2. Expected distribution with independent assortment of two nonhomologous chromosomes to the progeny (above), and observed distribution (below). (From R. F. Grell, Distributive pairing in man?, Ann. Génét., in press)

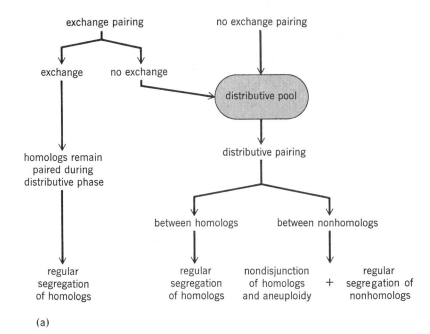

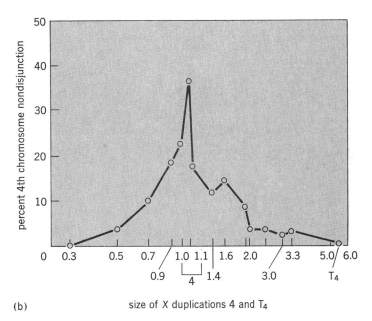

Fig. 3. Distributive model for meiosis in *Drosophila melanogaster* females with mode of operation. (*a*) Model. (*b*) Induction of distributive nondisjunction of fourth chromosomes by X-duplications. Highest nondisjunction value (37%) is reached with a duplication closest in size to chromosome 4, and lowest with duplications farthest in size from the 4. (*Modified from R. F. Grell, Pairing at the chromosomal level, J. Cell Physiol., 70(1):119, 1967*)

At present, several interpretations of the time and manner of exchange pairing are tenable, so that commitment to any one seems premature.

A second type of pairing, concerned solely with the segregation process, has been recently uncovered in the *Drosophila* oocyte and characterized through genetic techniques. This pairing probably corresponds to that observed by E. B. Wilson in certain Hemiptera where condensed microchromosomes, as they moved onto the spindle, united in pairs and then segregated at anaphase 1. Recognition of its occurrence in *Drosophila* came by way of the finding that two nonhomologs, when present without mates, displayed a strong tendency to segregate. For example, a female trisomic for both a sex chromosome (XXY) and an autosome (triplo-4) should, through independent assortment, produce four equal classes of progeny, namely, the double trisomic, two kinds of single trisomics, and the normal diploid. Instead, over 90% were single trisomics. Significantly the number with the Y and no fourth was precisely equivalent to the number with the fourth and no Y, indicating that these aneuploid types represented the two reciprocal products resulting from the meiotic segregation of the Y and fourth. A more extreme example is given in Fig. 2. Pairing and segregation of nonhomologs, in a manner characteristic of homologs, violate a basic tenet of neo-Mendelian genetics.

Analysis of the requirements for nonhomologous pairing disclosed that the chromosomes involved were invariably noncrossovers. Furthermore it was found that when one member of a pair of homologs segregated from a nonhomolog, the total frequency of exchange between the homologs remained unaltered. This meant the homolog had had the opportunity to undergo exchange, and that failing to do so it was later able to associate with, and segregate from, a nonhomolog. The distributive model (Fig. 3*a*) therefore recognizes two distinct phases during meiosis, each characterized by a special type of pairing. The first type, called exchange, precedes exchange and requires homology; the second, called distributive, follows exchange and requires noncrossover chromosomes. Chromosomes available for distributive pairing are considered to compose a pool, within which a size criterion operates to permit nonhomologous pairing but generally favors homologous pairing, thus serving as a secondary mechanism for proper segregation. The dot-like fourth chromosomes of *Drosophila*, between which crossing-over is very rare, are dependent on this segregational device. While their uniquely small size normally ensures a successful outcome, it is possible to disrupt their regular segregation through a nonhomolog of proper size (Fig. 3*b*). By introducing different numbers of chromosomes of varying size into the pool, a set of precise rules governing their behavior has begun to emerge. There is evidence to suggest that distributive pairing occurs in a variety of forms, including man.

Somatic pairing. Somatic pairing encompasses a wide spectrum of associations ranging from greater than random proximity to close fusion. Although popularly associated with Diptera, somatic pairing was first described in the gonia of many animals and in the somatic cells of plants. Recently, with more refined techniques for analyzing spatial relationships, additional cases have

pates in exchange or not, should triple exchange values or leave them unaltered. Parasynapsis assumes that maximal contact is achieved with two copies, so that the corresponding predictions are for unaltered exchange or a 67% reduction. Figure 1*c* shows that reductions reaching the predicted 67% are obtained. Additional support for a spatial constraint upon exchange pairing in *Drosophila* comes from studies of a single duplication attached at different locations in the genome or free. Again, exchange is markedly reduced in the triplicated region, with the extent of reduction depending on the proximity of the duplication to its homologous region, being greatest (95%) when closest to it.

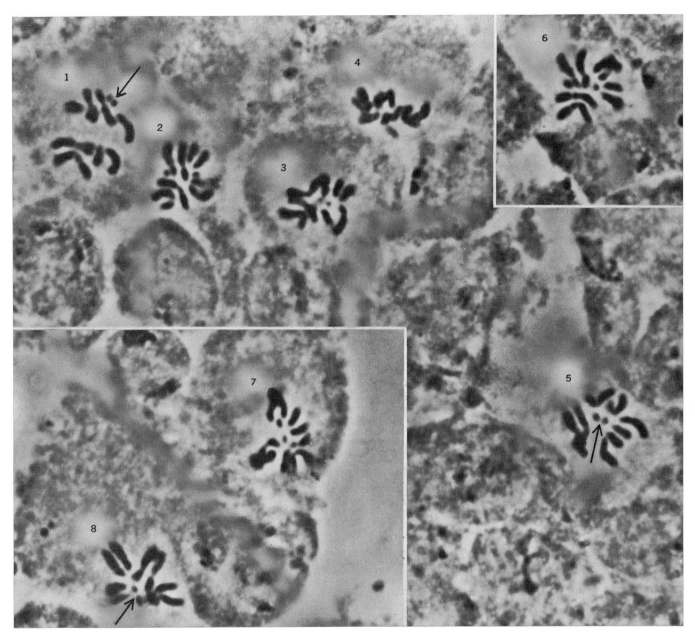

Fig. 4. Nonhomologous pairing between an X-duplication and fourth chromosome in a cyst of eight oogonial cells at final premeiotic metaphase. Arrows in nuclei 1, 5, and 8 indicate paired nonhomologs. (From R. F. Grell, Pairing at the chromosomal level, J. Cell. Physiol., 70(1):119, 1967)

been reported—including those for maize, wheat, and possibly humans. Recognition of somatic pairing is optimal with condensed metaphase chromosomes, but descriptions appear for all mitotic stages and for interphase. Detection at interphase requires the presence of a haploid number of chromatin bodies or close positioning of distinctive, satellited chromosomes in the interphase nucleus. Observations of homologous pairing during interphase suggest that this may be the rule and that the degree of its retention accounts for the extent of pairing typically observed during mitotic stages. If the Drosophila salivary gland nucleus is correctly interpreted as interphase, then pairing is not only intimate but extremely precise, since it occurs band-for-band. By comparison, mitotic pairing is less intimate, for homologs rarely touch, and is less specific, since inversion heterozygotes exhibit good alignment, indicating large segments nonhomologously paired. Even less specificity is seen in the mitotic pairing between nonhomologs recently reported in the oogonial cells, as well as in brain cells, of male and female Drosophila (Fig. 4). In these cases, as with distributive pairing in the oocyte, a size-dependency is evident.

What functions might somatic pairing serve? Two kinds are conceivable. The first is a metabolic function in processes such as transcription, translation, regulation, or organization of chromosomal products. Evidence for a metabolic function is meager, resting on a single case in Drosophila where rearrangements that disrupt somatic pairing between homologs carrying different mutant alleles cause an enhancement of the mutant phenotype, and on a single case in maize where physical proximity of alleles appears to affect the ability of

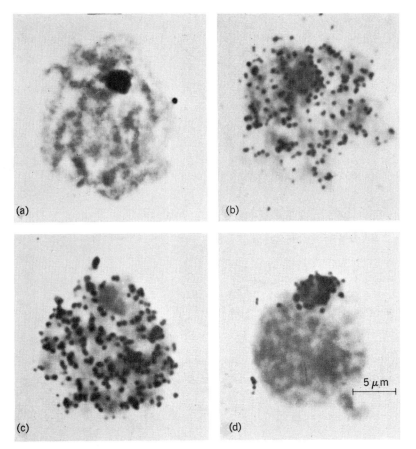

Fig. 5. *Melanoplus differentialis* spermatocytes at zygotene-pachytene labeled with H³-thymidine. (a) Unlabeled nucleus showing the heterochromatic sex chromosome (at 12 o'clock, in the four nuclei) surrounded by the euchromatic autosomes. (b) Fully labeled nucleus. (c) Nucleus labeled in the autosomes but not in the sex chromosome. (d) Nucleus showing the sex chromosome labeled and the autosomes unlabeled. Feulgen stain. (From A. Lima-de-Faria, Differential uptake of tritiated thymidine into hetero- and euchromatin in Melanoplus and Secale, J. Biophys. Biochem. Cytol., 6: 457, 1959)

one to alter the functioning of the other. Second, a genetic function is implied in situations where mitotic crossing-over is known to occur. Mitotic exchange is well known in *Drosophila*; there is some evidence for it in birds and rodents; but it is in certain fungi that it assumes a preeminent role in recombination, supplementing or even replacing the sexual cycle.

In conclusion, aside from the rules governing chromosome behavior during the distributive phase, knowledge of pairing is still largely in the descriptive stage. New biochemical and electron microscopical approaches to chromosome structure and function may soon provide much needed insight into the subject.   [RHODA F. GRELL]

Chromosome replication. This section will discuss DNA replication and DNA repetition in heterochromatin. It will cover the following aspects of this subject: (1) the relationship between heterochromatin and late replication, (2) the rule of chromosome replication derived from research with plants, insects, and mammals, (3) DNA replication in relation to differentiation, (4) the relation between late ending and late beginning of DNA synthesis, (5) the replicons of human chromosomes, (6) gene amplification in heterochromatin, and (7) DNA redundancy in heterochromatin.

Heterochromatin and late replication. Heterochromatin, as originally described by E. Heitz in 1928, represents a differential behavior of the chromosome phenotype. Heterochromatic regions are characterized in interphase and early prophase nuclei as "chromosome regions that form massive blocks of Feulgen positive material."

Until 1959 heterochromatin was considered to be an elusive chromosome material difficult to approach experimentally at the molecular level. Work with phosphorus-32 and tritium had not then produced any relevant information about this chromosomal material.

The first experimental demonstration that heterochromatin synthesizes its DNA later than the euchromatic regions of the chromosome complement was obtained in the grasshopper *Melanoplus differentialis*. In the male of this species the X (sex) chromosome at prophase of meiosis appears as a large heterochromatic mass. The autosomes are chiefly euchromatic. The injection of tritiated thymidine into the body cavity revealed that in the spermatocytes, at zygotene and pachytene, the heterochromatic X chromosome and the autosomes were labeled at different times. In the testicular tubules the spermatocytes are grouped in cysts which are synchronized at a given meiotic stage. The cysts move along the tubule as meiosis progresses, so that it is possible to tell from their relative position which have developed furthest. In sections of this material, the spermatocytes of the cysts which were at a late stage in development were labeled only in the heterochromatic X chromosome, whereas cysts at an earlier stage of development were labeled only in the autosomes. This result established that heterochromatin in this species, and in this tissue, replicated later than euchromatin (Fig. 5).

Since then many workers have studied the same phenomenon in plant, animal, and human material, confirming this result.

Four species of plants have been found to exhibit this phenomenon: *Secale cereale, Fritillaria lanceolata, Vicia faba,* and *Hordeum vulgare*. Four species of insects show also late replication in heterochromatin: *Melanoplus differentialis, Bombyx mori, Pseudococcus obscurus,* and *Drosophila melanogaster*. *Drosophila, Melanoplus,* and *Pseudococcus* are classical examples of organisms with heterochromatin.

Mammals, including man, provide most examples on the late replication of heterochromatin. Man is the species in which the relation between the metaphase and interphase pictures has been best established because of the occurence of individuals with multiple X chromosomes. The number of chromocenters (Barr bodies) seen at interphase in diploid cells is equal to the number of X chromosomes minus one (Harnden's rule). Thus in man the autoradiographic picture of the X chromosome at mitotic metaphase can be considered to furnish reliable information about the late DNA replication of heterochromatin. Studies of DNA replication in individuals with multiple sex chromosomes have confirmed the relation betweeen the number of Barr bodies seen at interphase and the number of late-replicating X chromosomes present at metaphase.

The Chinese hamster (*Cricetulus griseus*) has X and Y chromosomes which form distinct chromocenters at interphase. These were found to replicate later than the euchromatic regions of the autosomes in interphase nuclei and at metaphase. A similar situation is found in the Syrian or golden hamster (*Mesocricetus auratus*) and in the guinea pig (*Cavia cobaya*).

The sex chromosomes of the field vole (*Microtus agrestis*) are among the largest known in mammals, and their heterochromatin was also found to replicate later than the euchromatin.

In every mammalian species included in Table 1, the autoradiographic picture and the heterochromatin were studied in the same tissue at interphase, disclosing the late replication of the chromocentral heterochromatin.

Rule of chromosome replication. The results for the 17 species of plants, insects, and mammals in Table 1 allow the formulation of the following rule of chromosome replication: When tritiated thymidine is incorporated into the cells of a given tissue in which there are massive Feulgen-positive chromosome regions at interphase or early prophase, the heterochromatic regions replicate their DNA later than the euchromatic segments. This takes place when the autoradiographic picture is studied in the same tissue and at the same stage in which heterochromatin is displayed. This phenomenon occurs irrespective of the evolutionary position of the organism and of the tissue in which heterochromatin is observed.

It is not a corollary of the DNA rule, formulated above, that any chromosome or chromosome region that is late replicating is heterochromatic. Thus, within the chromosomes of various species, different regions replicate at different times, showing no indication of being heterochromatic. When heterochromatin is not present, other factors not yet apparent determine the sequence of replication of various segments within a chromosome and between chromosomes.

It may be stated at present that what determines the presence of heterochromatin determines the late replication of a chromosome or chromosome segment. In other terms, what changes the condensation cycle of the chromosome at the same time shifts its timing of DNA replication. However, what determines the late replication of a chromosome or chromosome region may not necessarily determine the onset of heterochromatin.

DNA replication and differentiation. If a chromosome or chromosome segment is heterochromatic in a given tissue and replicates its DNA later than the euchromatin, it does not necessarily follow that it replicates its DNA later in other tissues where it may be euchromatic. Demonstration of this phenomenon was obtained in *Melanoplus*. At prophase of meiosis, in spermatocytes, the X chromosome is heterochromatic in the males and replicates its DNA later than the autosomes. The same chromosome is, however, euchromatic or slightly negatively heteropycnotic in the spermatogonia, and in the spermatogonia the incorporation of tritiated thymidine takes place at the same time as in the autosomes.

Two other examples of this phenomenon are found, in the female of *Mesocricetus auratus* and in *Cricetulus griseus*. The three cases demonstrate (1) that there are controlling mechanisms affecting the chromosomes of higher organisms at the molecular level, which shift the time of DNA replication of given chromosomes or chromosome segments from one period to another of the S (synthetic) phase; (2) that this phenomenon is tissue-specific, that is, it is correlated with the differentiation of the organism; and (3) that the shift occurs, at least in one case, in connection with the heterochromatization of the chromosome.

Late ending and beginning of DNA synthesis. The question arises whether chromosomes which finish their DNA synthesis after all the others also initiate their DNA synthesis later than the other chromosomes. As a rule, late-replicating chromosomes start later than the other chromosomes, but the difference at the beginning of the S period does not seem to be as dramatic as at the end. In the spermatocytes of *Melanoplus*, where this phenomenon was studied in detail, the heterochromatic X chromosome starts only slightly later than the euchromatic autosomes, but at the end of the S period the X goes on replicating for an appreciable period of time after the autosomes have finished. In man, where the phenomenon has been studied in several laboratories and through the use of individuals with chromosome complements containing multiple Xs, it is well established that the heterochromatic X chromosome begins and ends its replication later than the other chromosomes.

Replicons of human chromosomes. When DNA synthesis was studied in human leukocytes in 1961, a complex pattern of chromosome replication was found. Since then the contributions from many different laboratories have established the following major points: (1) The asynchrony of replication between the chromosomes of the same cell, as well as between regions of a chromosome, has been generally well demonstrated. (2) One of the X chromosomes of the human female has been found to be late in replication. (3) Homologs, other than the X, seem to have essentially the same labeling features but at times they are out of phase in their

Table 1. Species where heterochromatin has been found to replicate later than euchromatin

Plants	Insects	Mammals
Secale cereale	*Melanoplus differentialis*	*Cricetulus griseus*
Fritillaria lanceolata	*Bombyx mori*	*Chinchilla lanigera*
Vicia faba	*Pseudococcus obscurus*	*Microtus agrestis*
Hordeum vulgare	*Drosophila melanogaster*	*Mesocricetus auratus*
		Cavia cobaya
		Canis familiaris
		Donkey–Grévy's zebra hybrid
		Mus musculus
		Homo sapiens

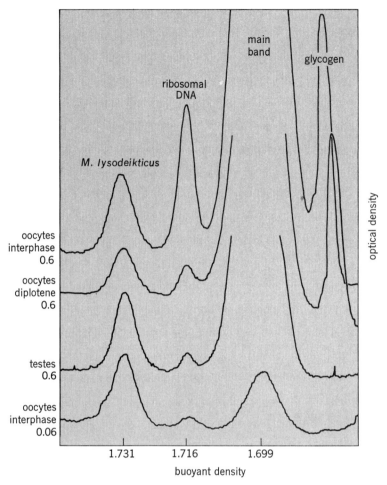

Fig. 6. Analytical centrifugation of the DNA of different tissues of *Acheta*. Optical density 0.6 of DNA from oocytes at interphase of meiosis is compared with the same amount of DNA from oocytes at diplotene and from testes. No other satellite is present between the main band and the marker DNA from *Micrococcus lysodeikticus*. The amount of DNA contained in the satellite of ovaries at interphase, where the DNA body is largest, is 18 times that found in the satellite of the testes, where no DNA body of comparable size is present. (From A. Lima-de-Faria et al., Amplification of ribosomal cistrons in the heterochromatin of Acheta, *Genetics*, 61(1):145, 1969)

strongly condensed, and the length of 1 μm in a metaphase chromosome corresponds to several micrometers in the same chromosome at mitotic prophase or pachytene. What are seen in mitotic chromosomes are major replicons, which represent sequences of a large number of minor replicons, found at the level of the DNA molecule, and which happen to synthesize their DNA with a given degree of synchrony.

Gene amplification in heterochromatin. The results of the study of DNA synthesis with H^3-autoradiography have improved the cytogenetic knowledge of heterochromatin by bringing the experimental approach to the level of DNA replication, but the results have not furnished data on the genetic constitution of this material.

Demonstration of gene amplification for ribosomal cistrons in heterochromatin was obtained in 1969, following a biochemical analysis of the DNA body of the house cricket (*Acheta domesticus*). Ribosomal cistrons are the genes that code for ribosomal ribonucleic acid (RNA). Gene amplification is defined as the occurence of extra copies of a gene in a chromosome segment of a given tissue. The evidence in *Acheta* is the following: (1) There is a DNA satellite in the analytical centrifuge which has a calculated guanine-cytosine (GC) content of 56%. (2) The ribosomal RNA of *Acheta* has a GC content of about 56%. (3) The hybridization peak between ribosomal RNA and DNA of *Acheta* occurs at the position of the DNA satellite. (4) The DNA satellite of ovaries at interphase of meiosis, where the DNA body is largest, represents 14% of the total ovarian DNA, whereas the DNA satellite of testes, where no DNA body of comparable size is present, represents 0.8% of the total testes DNA. (5) The amount of satellite DNA in the ovaries at interphase is 18 times higher than in the testes (Fig. 6). (6) The DNA/RNA hybridization tests at the saturation level reveal that the DNA of ribosomal cistrons is about 18 times as much in the ovaries at interphase than in the testes.

The same situation occurs in the DNA body of two other insect species, the water beetle (*Dytiscus*) and *Colymbetes*. Gene amplification has been thoroughly investigated in amphibians, where preferential synthesis and amplification of the DNA in the oocyte is evident. This gene amplification can now be seen to be also localized in the heterochromatin of the oocyte nucleus.

DNA redundancy in heterochromatin. DNA redundancy occurs when extra copies of a given DNA sequence are present in the chromosomes of a species. These extra copies may or may not occur in the form of a DNA satellite and may be spread throughout one or more chromosomes of the complement.

DNA redundancy has recently been demonstrated to occur in heterochromatin. The best studied species is the mouse (*Mus musculus*), where a DNA satellite containing 10^6 copies and 10% of the total DNA is localized in the heterochromatin of most of the chromosome arms (at sites adjacent to the kinetochore).

Moreover it has been shown that the heterochromatin fraction of guinea pig nuclei (Table 2) and of calf nuclei is particularly rich in satellite DNA.

Thus heterochromatin contains both amplified and redundant copies. Since heterochromatin re-

replication. This results in a differential labeling of the homologs in the X chromosomes and in the autosomes. (4) Characteristic patterns of replication have been found for those chromosomes that can be identified with accuracy and for given pairs within a group. (5) Within each chromosome, there are segments which replicate at a different time from others. (6) These segments are well delimited. (7) The segments can be followed from cell to cell, being present at the same site of a given chromosome and having the same size. (8) These segments are of different sizes in different chromosomes. They are as large as a whole arm, or they may be as small as 1 μm, the limit of resolution for tritium labeling. (9) There is no indication of a given sequence of synthesis within each chromosome or within each arm. Evidence that the kinetochore or the telomeres are determinants in the replication sequence is not available. (10) Each chromosome finishes replication at a site or region not related to the other chromosomes.

It should be noted that the tritium picture obtained in a mitotic metaphase chromosome is quite a crude one. At this stage the chromosome is

Table 2. Species where repetitive DNA has been localized in heterochromatin

Gene amplification	Heterochromatin	Biochemical evidence	Multiple DNA sequences	Heterochromatin	Biochemical evidence
Acheta domesticus	Nucleolar organizing heterochromatin; DNA body	DNA satellite; DNA-RNA hybridization (on filters)	*Mus musculus*	Proximal regions of chromosome arms	DNA satellite; fractionation; DNA-DNA hybridization (in situ)
			Cavia cobaya	Chromocenters	DNA satellites; fractionation
Xenopus laevis	Heterochromatic "cap" at pachytene	DNA satellite; DNA-RNA hybridization (on filters and in situ)	*Drosophila melanogaster*	Heterochromatin in giant chromosomes	DNA satellites; DNA-RNA hybridization (in situ)
Dytiscus marginalis	Heterochromatic DNA body	DNA satellite; DNA-RNA hybridization (on filters)	*Microtus agrestis*	Chromocenters	DNA-RNA hybridization (in situ)

plicates later than euchromatin, the question arises whether the DNA satellites that contain the extra copies are also found to be late-replicating.

In synchronized cell cultures of the mouse, the cells replicate their satellite DNA preferentially later in the DNA synthetic period, while the other parts of the genome are replicated early.

The information so far available agrees in demonstrating that the DNA of heterochromatin contains extra copies of ribosomal genes, as well as other DNA sequences of unknown function, and that these DNA sequences replicate later than the others of the chromosome complement.

For background information *see* CHROMOSOME; DEOXYRIBONUCLEIC ACID (DNA); RIBOSOMES in the McGraw-Hill Encyclopedia of Science and Technology.   [A. LIMA-DE-FARIA]

Bibliography: R. F. Grell, *Chromosoma*, 31:434, 1970; R. F. Grell, in E. Caspari and A. Ravin (eds.), *Genetic Organization*, vol. 1, 1969; A. Lima-de-Faria, in A. Lima-de-Faria (ed.), *Handbook of Molecular Cytology*, 1969; G. Pontecorvo, *Trends in Genetic Analysis*, 1958; S. M. Stack and W. V. Brown, *Nature*, 222:1275, 1969; P. M. B. Walker, in J. A. V. Butler and D. Noble (eds.), *Progress in Biophysics and Molecular Biology*, vol. 22, in press; J. J. Yunis and W. G. Yasmineh, in E. J. DuPraw (ed.), *Advances in Cell and Molecular Biology*, 1971.

Cell surface

The cell surface has been viewed as a direct or indirect mediator of cellular growth control. It has been hypothesized that the surface membrane may be responsible for such properties of tumor growth as metastasis, loss of contact inhibition, and invasive growth. Many attempts have been made to correlate chemical changes of the surface membrane with alterations in growth control. Recently, through the use of plant agglutinins which specifically bind to various carbohydrates on the cell surface and produce cell agglutination, K. D. Noonan and M. M. Burger observed changes of the cell surface which may play a role in cell growth.

Plant agglutinin role. A number of different agglutinins with specificities for different cell surface carbohydrates have been isolated from plant material, the best known so far being wheat germ agglutinin, concanavalin A (conA), and soybean agglutinin. Through the use of these plant agglutinins it has been demonstrated that tissue culture cells transformed by DNA viruses, exposed to chemical carcinogens, or selected as spontaneous transformants agglutinate upon exposure to the agglutinin, while normal tissue culture cells agglutinate much more poorly. This suggested that a glycoprotein not exposed on the normal cell surface was exposed on the transformed cell surface. Agglutination patterns following infection with RNA viruses (viruses containing RNA in place of DNA) are not yet clear.

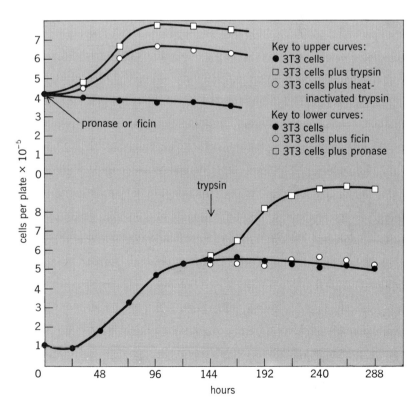

Fig. 1. Stimulation of overgrowth by proteases. At arrows, pronase was added to a final concentration of 0.005% and trypsin to 0.007%. Recent data indicate that exposure of 3T3 cells to pronase for only 5 min gives the same result.

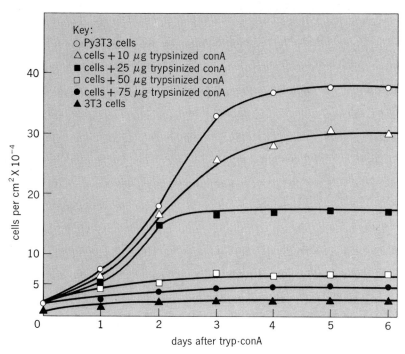

Fig. 2. Effect of varying concentrations of trypsinized ConA on growth of Py3T3 cells.

Normal and transformed cell growth. One of the most striking qualities of many normal tissue cells is that once the cells grow to a population density at which the surface of the tissue culture flask is covered with cells one cell layer deep, the cells stop growing. Transformed cells, on the other hand, do not stop at the monolayer but continue to grow until they have piled up several cell layers deep, that is, they do not exhibit the phenomenon of contact inhibition of growth.

Agglutinin binding sites. Using cells transformed in the laboratory, the change in the surface membrane monitored with the plant agglutinins has, in many cases, been correlated with the loss of contact inhibition. Cell lines demonstrating little contact inhibition of growth agglutinate extremely well, whereas the cell lines demonstrating good contact inhibition agglutinate poorly, suggesting that the availability of the agglutinin receptor sites on the cell surface parallels the increase in saturation density and loss of growth control of a cell line. In fact, it was shown that if one selected for cells which had reverted back from a transformed to a normal growth pattern, there was a concomitant decrease in agglutinability. Further evidence for the involvement of the agglutinin site in transformation comes from work with W. Eckhart and R. Dulbecco using a DNA virus which has a specific viral function that is temperature-sensitive. At 37°C the virus infects and transforms the cell, while at 39°C the virus infects but does not transform the cell. It was demonstrated that if the cells were infected and grown at 37°C, they would behave like transformed cells and would agglutinate. If the same cells were shifted to grow at 39°C, they would behave like normal cells and would not agglutinate. This suggests that exposure of the agglutinin site is part of the transforming process and that some continuous viral function is necessary for the configurational change observed in the membrane.

Isolation of the partially purified wheat germ agglutinin (WGA) site from normal mouse fibroblasts (3T3) and polyoma-infected mouse cells (Py3T3) has shown that an approximately equal number of WGA binding sites are present in the normal and transformed cell membrane. This suggests that the appearance of the sites following transformation is not the result of synthesis of new sites but, rather, a structural rearrangement of the surface membrane to expose the WGA sites. Presently this is still an assumption and will have to await a careful analysis of the carbohydrate structure in the glycolipid and glycoprotein receptors from transformed and normal cell membranes.

Cell exposure to protease. Better support for this hypothesis comes from work in which normal mouse fibroblasts were exposed to a protease (trypsin). After such exposure it was found that these cells agglutinate to the same degree as Py3T3 cells. The agglutination pattern seen with the protease-treated 3T3 cells appears to be both quantitatively and qualitatively equivalent to Py3T3 agglutination.

It was possible to show that the agglutinin receptor sites seem to be important in growth control through two different experiments. In the first experiment 3T3 cells were grown to confluency and then briefly treated with a protease, such as pronase, ficin, or trypsin, thereby producing the surface configuration normally present on a transformed cell surface. The protease was prevented from acting anywhere but at the surface by binding it to beads too large to penetrate the cell. As can be seen in Fig. 1, a burst of cell division follows the protease pulse, suggesting that the surface configuration plays a role in the control of cell division under crowded or confluent conditions and that a simple rearrangement of the surface of a normal cell causes it to grow like a transformed cell.

Monovalent agglutinin. In the opposite experiment the agglutinin receptor sites of a transformed cell are covered. The agglutinin used in this experiment was a monovalent rather than a divalent agglutinin, that is, a molecule which binds to the cell

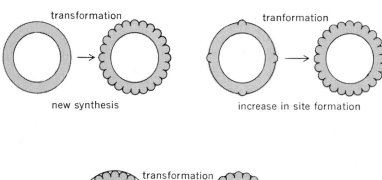

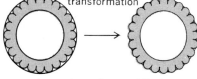

Fig. 3. Summary of possible membrane changes in the process of transformation from a normal to a malignant cell.

surface with its one remaining active site but which cannot cause agglutination because it lacks a second active site. Divalent agglutinins produce cell death. Monovalent agglutinins produce a cell surface change which leads to reestablishment of contact inhibition (Fig. 2). If the monovalent agglutinin is then removed through the use of carbohydrates specific for the agglutinin and which act as haptens, the cells resume growth. That this effect is due to the specific covering of the agglutinin receptor sites can be shown by the fact that a nonspecific protein, such as bovine serum albumin, when added to the growth medium, also adsorbs to the cell surface but does not produce growth control. Only carbohydrates specific for the release of the agglutinin from the cell surface allow the cells to resume the growth pattern of transformed cells.

It has recently been shown that a transitory exposure of the agglutinin receptor sites occurs during part of the cell cycle of normal cells. As a working hypothesis, it has been postulated that such configurational changes of the cell membrane may be an integral part of the normal cell cycle.

Molecular mechanism for site exposure. A number of molecular mechanisms may be responsible for the exposure of the agglutinin receptor sites following transformation. Figure 3 suggests some of the different ways in which agglutinin sites buried in the cell membrane could come to appear on the surface following transformation. In all these models the number of sites on the surface of the normal cell would be less than those on the transformed cell. A second possibility is that the number of sites on the surface of the normal cell is the same as on the transformed cell but that agglutination would be prevented due to surface charge differences or a nonfavorable localization of sites on the cell surface.

Therefore, whatever the molecular mechanism of rearrangement, it is known that configurational changes in the cell surface do occur following transformation and that these changes are important in neoplastic growth and, in fact, may be important in normal and embryonic growth.

For background information see AGGLUTININ; DEOXYRIBONUCLEIC ACID (DNA); ONCOLOGY; RIBONUCLEIC ACID (RNA) in the McGraw-Hill Encyclopedia of Science and Technology.

[KENNETH D. NOONAN; MAX M. BURGER]

Bibliography: M. M. Burger, in *International Conference on Biological Membranes, Stresa*, sect. 2, ch. 5, p. 107, 1970; M. M. Burger, *Nature*, 227: 170, 1970; M. M. Burger and K. D. Noonan, *Nature*, 228:512, 1970.

Charge-coupled device

The charge-coupled device (CCD) is a relatively new class of semiconductor devices which can be used in information-handling applications. The information is represented by charge which is stored in potential wells created at the surface of the semiconductor. This charge is then moved from one position to another by proper manipulation of the potential wells. Inputs and outputs to create and detect the charge complete this shift-register-type device. Analogy can be made with other shift-register-type devices in which, for example, the information is in the form of magnetic domains, as in disk files and magnetic tapes, or in the form of currents, as in some electronic shift-register circuits.

Several characteristics are apparent from this brief description. First, there is no inherent gain in the charge-coupled device, which implies a degradation of a signal fed through the register. Second, the potential well in the semiconductor is formed by reverse biasing the material, so that the storage time of the information is limited by the reverse-bias leakage currents, making it a dynamic shift register. Third, the stored charge can vary continuously from zero to some maximum value, so that the device is basically analog. See SEMICONDUCTOR.

Transfer of charge. A schematic diagram of a simple charge-coupled device element is shown in Fig. 1. This shows a cutaway view of a slice of semiconductor on which there is an insulating layer. On this insulating layer is a linear array of metal plates which form capacitors with the semiconductor substrate as the other plate. Leads to these capacitors are brought in over the thick oxide portion shown in the rear of the diagram. Negative voltages are applied to these leads to drive the negatively charged conduction electrons away from the surface of the *n*-type semiconductor. The boundary between the region depleted of electrons and the bulk is indicated by the broken line.

In Fig. 1*a* a voltage ($V_2 > V_1$) more negative than the adjacent ones is applied to the center electrode so that a potential minimum for positive charge (holes) is created under this electrode. Storage of holes at the surface between the semiconductor and the insulator is indicated by the positive signs. To transfer this charge to the adjacent electrode, a more negative voltage ($V_3 > V_2$) is applied to it, causing the positive charge to flow to the new po-

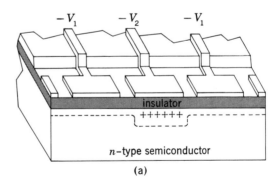

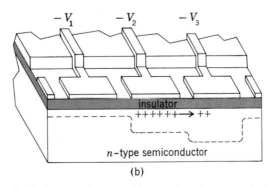

Fig. 1. Charge storage and transfer in a charge-coupled device. (*a*) Storage. (*b*) Transfer.

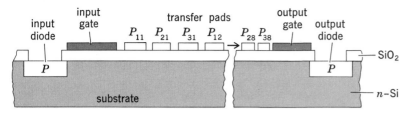

Fig. 2. Cross section of a charge-coupled device showing input and output diodes.

tential minimum, as is shown in Fig. 1b. After a time sufficient to allow most of the charge to transfer over, the cell is reset by reducing V_2 to V_1 and V_3 to V_2, thereby moving the potential well and its stored charge one physical location. The process can now be repeated to transfer to the next plate. For a linear array, every third plate would be electrically connected together, forming a three-phase shift register. Other structures, such as a two-phase device which uses asymmetrical electrodes, have been made or proposed but all utilize the charge-shifting feature.

The most common method for injecting charge at the beginning of a line and detecting it at the end is shown in Fig. 2. A p-type region which acts as a source of holes (input diode) is shown at the left end of an array of plates in which every third plate P_{11}, P_{12}, and so on is connected together to form a three-phase shift register. The input signal is applied to the input gate which is used to form a conduction path between the diode and the first storage plate (P_{11}) much in the same way as charge is gated from source to drain in an insulated-gate field-effect transistor. The output diode on the right end is reverse biased through an output resistance so that any charge transferred from the last storage site via the output gate to the diode flows in the external circuit and is detected as an output current.

Limitations. The principal limitation to this device is the lack of complete transfer of charge from one plate to the next. Two effects contribute to this effect. One is that it takes a finite time for the charge to flow from one plate to the next under the forces of drift and diffusion, and the other is the fact that "traps" for charge exist at the interface between the insulator and the semiconductor in any practical system. The charge which is left behind does not remain but is added to the charge packets which follow. The net effect is a "smearing" of the signal; that is, large packets of charge become smaller and small ones larger. Since a larger percentage of charge is left behind as the transfer rate increases, this represents a frequency limitation to the device. Using current fabrication techniques, $\approx 99.9\%$ of the charge can be transferred per cycle at a 10-MHz frequency and operation in the hundreds-of-megahertz range is feasible.

The low-frequency limit of the device is determined by the reverse-bias leakage currents in the semiconductor. Electron-hole pairs will be thermally created in the region depleted of electrons (Fig. 1) and the holes (positive charge) will collect under the electrodes. Given enough time, every plate will collect a maximum amount of charge and the information will be lost. In a practical device, the time for this maximum amount of charge to collect is approximately 1 sec.

Both of these limitations dictate that a long shift register be made by transferring a certain number of steps, regenerating the signal, transferring again, regenerating, and so on. A simple regenerator can be made by applying the charge to the gate of an insulated-gate transistor which in turn drives the input gate of the next register. For digital applications, a typical scheme is to transfer about 20 bits between regenerators, with a total of about 1000 bits on a single semiconductor chip. Typical operating parameters use 5–10 V on the plates and dissipate about 1 μW of power per bit at 1 MHz. The output current is about 1 μA. These numbers assume an area of about 10^{-6} cm^2 per plate with silicon as the semiconductor and silicon dioxide as the insulator.

Applications. There are several potential applications for these devices. The most obvious is a simple digital shift register. If the information from the output of the shift register is fed back to the input and recirculated, the device can be used as a serial memory. The small size, and resulting low cost, makes this mode of operation attractive for bulk-storage applications. Since the amount of charge stored can vary continuously from zero to some maximum value, the shift register can be used as an analog delay line. The ability to vary the delay by varying the drive frequency is particularly useful in some applications. For buffering applications, the information can be read in at one rate and read out at another.

The device has also been used to make an imaging device. In this mode of operation, potential wells are created under the plates and light is allowed to fall on the device. The light makes electron-hole pairs in the semiconductor and a number of holes collect under the plates which is proportional to the intensity of the incident light. After allowing a certain integration time, the information is shifted out and the output current replicates the optical image incident on the device. Both one- and two-dimensional devices can be made. A final use for these devices is in digital information processing. Devices similar to regenerators can be fabricated on the semiconductor chip which perform logical operations on information being carried on separate shift registers. The simplest operation is to transfer the charge from each of two registers to a single register, thereby doing a simple form of addition. Many more complicated interactions between registers can be constructed both serially and in parallel.

For background information see COMPUTER MEMORY; INTEGRATED CIRCUITS; SEMICONDUCTOR in the McGraw-Hill Encyclopedia of Science and Technology.   [GEORGE E. SMITH]

Bibliography: G. F. Amelio, M. F. Tompsett, and G. E. Smith, *Bell Syst. Tech. J.*, 49:593, 1970; W. S. Boyle and G. E. Smith, *Bell Syst. Tech. J.*, 49:587, 1970; M. F. Tompsett, G. F. Amelio, and G. E. Smith, *Appl. Phys. Lett.*, 17:111, 1970.

Cholera

Reports on recent research in cholera indicate that enterotoxin produced by *Vibrio comma* acts on the mucosa of the small bowel, causing it to produce the large quantities of fluid seen in typical cholera diarrhea.

Cholera is an extremely severe epidemic diarrheal disease occurring exclusively in the human and caused by the gram-negative bacterium *V. comma*. The disease, now in its seventh widespread epidemic (1961 through 1971), occurs predominately in Southeast Asia. The last cases in the United States occurred during the Civil War. Cholera was present in Europe as late as 1923. Interest in the international spread of this disease has been rekindled by the rapid means of travel presently available from Southeast Asia to other parts of the world. The infection is initially waterborne; however, it is further spread by food contaminated with feces from people with inapparent infections. Fortunately the disease is not highly contagious. Observation of rules of simple hygiene is thought to protect hospital personnel from developing the disease.

This infectious disease is characterized by the sudden onset of vomiting with copious and incapacitating diarrhea. While the vomiting lasts only 1 or 2 days the diarrhea persists for 5 to 7 days. During this time the patient may lose an amount equal to his body wieght in diarrheal effluent. The diarrheal fluid becomes progressively clearer, containing flecks of mucus, and has been compared in its appearance to "rice water." The choleraic stool contains very little protein and is similar in its electrolyte content to plasma, although it contains a greater quantity of bicarbonate and potassium. The copious diarrhea results in a syndrome of dehydration, acidosis, and shock.

For years it has been known that the organism did not invade the intestinal epithelium but remained in the lumen of the gastrointestinal tract. In the early 1960s an enterotoxin produced by the viable organism was isolated which appears to be entirely responsible for the findings in this disease. This toxin is separate and distinct from the endotoxin of the organisms' cell wall. The organisms are present throughout the entire gastrointestinal tract. However, exceedingly large numbers are found in the small bowel, an area which is usually bacteriologically sterile. It is the action of the toxin on this area of the gut that is responsible for the diarrhea. Treatment is directed at replacing the losses of water and electrolytes by intravenous infusions. With early and adequate intravenous therapy, no deaths or serious complications should occur. The prompt administration of oral antibiotics significantly shortens the duration of the disease by killing the organism within the gastrointestinal tract.

Toxin. A similar enterotoxin is produced by various strains of the vibrio. This single toxin is referred to by a variety of names, such as choleragen, skin permeability factor, vascular permeability factor, type-2 cholera toxin, cholera enterotoxin, and cholera exotoxin. The most highly purified form of this toxin is a protein polypeptide with a molecular weight of 90,000 which is composed of repeated subunits of a molecule that contains no carbohydrate or lipid. This thermolabile toxin is unique in that it appears to cause two distinct responses in the experimental animal. When instilled into the lumen of the bowel, it causes diarrhea, and when instilled into the skin, in minute doses, it causes a delayed increase in permeability of the small vessels of the skin.

Experimental models. Studies directed toward a better understanding of the pathogenesis of cholera were greatly facilitated by the development of three reliable experimental models which utilized the small intestine of the infant rabbit, the adult rabbit, and the dog. All the models require the instillation of the organism or its toxin into the lumen of the gastrointestinal tract.

Concepts of pathogenesis. Although it was initially thought that the bacterium evoked its damage by causing the mucosal epithelial surface to slough (mucosal denudation), recent studies of human and animal material by light and electron microscopy failed to demonstrate any consistent morphologic lesion.

Vasculature. One concept of the pathogenesis of cholera holds that the major area of action of the toxin is to the vasculature supplying the mucosa of the small bowel. This concept found support from the demonstration of the skin permeability factor characteristics of the toxin. However, it has been convincingly demonstrated that when the toxin is instilled in the lumen of the small intestine, there is no change in the permeability of these vessels for large-sized molecules such as albumin. Patients with cholera have very little protein in their stool, and markers injected into their vasculature do not appear in the stool. Another proof that the toxin's action is not primarily of vascular origin is seen when the blood flow to the choleraic small bowel is mechanically decreased. The small bowel continues to produce effluent, even when the blood flow in the artery supplying it has been reduced to 30% of its original flow rate. It therefore appears that the toxin effects on the vasculature of the small bowel are secondary.

Hypersecretion. Others thought that the active transport of sodium from the gut lumen into the cell was interrupted by the vibrio or its toxin (inhibition of the sodium pump). With the use of biophysical measurements to assay the active transport of sodium (the Ussing chamber), no inhibition of the active transport of sodium was found during the first few minutes that the bowel was studied in the preparation. Studies over much longer periods of time indicate that there may be a gradual diminution in the active transport of sodium.

It is now generally held that the toxin's major action is on mucosal ion transport, causing the small bowel to "hypersecrete" its usual product, an isotonic solution containing sodium, potassium, and increasing amounts of bicarbonate and lessening amounts of chloride as production approaches the ileal area of the small bowel. The exact mechanism through which this occurs is still unclear. Studies using the isolated small bowel indicate that the primary response of the ileal mucosa to the toxin is an active secretion of chloride. Studies in the dog indicate that both bicarbonate and, to a lesser extent, chloride are actively secreted. No alteration of small bowel absorption of a simple sugar (glucose) or an amino acid (glycine) is present in patients that are actively purging, indicating that the mechanism of injury is quite specific. Recently it has been postulated that the toxin acts through stimulation of an intermediary compound such as 3'5'-adenosinemonophosphate (cyclic AMP). Cyclic AMP is thought to be the intracellular trigger that causes the cell to respond to the stimulus of hormones or other extracellular chemical regulators. Increased levels of this substance have been found in choleraic mucosa. The role of cyclic AMP in ion transfer is yet to be ascertained.

The area of the mucosa responsible for fluid production is not definitely known. Cycloheximide, a potent inhibitor of protein synthesis, will prevent effluent production when administered prior to the instillation of toxin. It is speculated that a protein synthetic step, probably in the mucosal crypts, is interrupted, prohibiting fluid production.

Action on nonintestinal tissues. Cholera toxin has a wide variety of actions on tissues outside the gastrointestinal tract. Because of the expense and difficulty of preparing the toxin, studies on these tissues are useful because much smaller doses are required to elicit their response. These studies are particularly helpful in understanding enterotoxin-antitoxin activities and alterations in cellular metabolism.

The alteration in permeability of vessels of the skin has been mentioned. Edema of the rat footpad is produced 2–4 hr after instillation into this area. Cholera enterotoxin causes an increased rate of lipolysis from rat epididymal fat cells and an enhanced rate of glycogenolysis in platelets and liver. It is speculated that the latter effects are mediated through a direct increase in cyclic AMP in these tissues.

Immunity. While the vibrio remains in essence outside the body, it still is capable of eliciting serum antibodies. In humans, resistance to cholera infection correlates best with serum vibriocidal antibody titer. In experimental animals, resistance to oral infection is correlated with antibodies present in the intestinal juices of previously challenged animals. This coproantibody, probably IgA, is produced in large quantities during the infection. Its clinical significance is not yet apparent. Cholera vaccination is no longer required in the United States for travelers returning from endemic areas. This is in part due to the effectiveness of appropriate therapy, and in part to the ineffectiveness of the currently used cholera vaccines composed of killed organism. Studies are directed toward the production of an appropriate toxoid.

For background information see CHOLERA VIBRIO; TOXIN, BACTERIAL in the McGraw-Hill Encyclopedia of Science and Technology.

[H. T. NORRIS]

Bibliography: C. C. J. Carpenter, Jr., *Amer. J. Med.*, 50:1–7, 1971; M. Field, *New Engl. J. Med.*, 284:1137–1144, 1971; G. F. Grady and G. T. Keusch, *New Engl. J. Med.*, 285:831–841, 891–900, 1971; W. B. Greenough III et al., in G. B. Jerzy-Glass (ed.), *Progress in Gastroenterology*, vol. 2, 1970; W. L. Moore, Jr., et al., *J. Clin. Invest.*, 50:312–318, 1971; H. T. Norris and G. Majno, *Amer. J. Pathol.*, 53:263–279, 1968; N. F. Pierce, W. B. Greenough III, and C. C. J. Carpenter, Jr., *Bacteriol. Rev.*, 35:1–13, 1971.

Chromosome

In recent years the term chromosome has been evolving from its traditional meaning to include the genetic apparatus of viruses, bacteria, and sometimes even such cellular organelles as mitochondria and chloroplasts. The essential structural and informational component is deoxyribonucleic acid (DNA) in all of these forms. Although some of the simpler viruses have ribonucleic acid (RNA) which codes and transmits their genetic information, these linear polymers appear to be small enough to qualify as macromolecules in the usual sense rather than chromosomes. The same might also be said for the DNA of some of the smaller viruses, but in any case the term chromosome is restricted in this article to DNA-containing structures. Recent investigations have attempted to elucidate the occurrence and possible function of repeated segments in chromosomes and the details of structure and arrangement of DNA in chromosomes.

Classification. The simpler and presumably more primitive life forms such as viruses, bacteria, blue-green algae, and similar organisms are classified as prokaryotes (literally "before nuclei"). These organisms are characterized by a single chromosome per genome (set of genes characteristic of a species). The chromosome consists of one continuous piece of DNA, either in the form of an open-ended strand or, in many cases, a closed strand referred to as a circle, although it is flexible and may take any shape. These chromosomes lack the histones typical of traditional chromosomes of higher cells. The DNA forms complexes with other types of proteins and basic substances which serve to neutralize the charge of the nucleic acid.

Prokaryotes lack the nuclear membrane which separates nucleus from cytoplasm in cells of higher organisms. They also lack nucleoli and mitochondria. Nucleoli are dense bodies found in nearly all nuclei and represent sites where ribosome assembly begins. Ribosomes are small bodies which bind to certain types of RNA molecules and translate their coded information (sequence of nucleotides) into sequences of amino acids, or polypeptides, which are the principal functional macromolecules of cells, that is, enzymes and structural components for membranes, fibrous elements, and so on. In spite of the absence of the nuclear membrane and nucleoli, bacteria are efficient both in the synthesis of ribosomes and their utilization in protein synthesis. See DEOXYRIBONUCLEIC ACID (DNA).

In cells of eukaryotes (all organisms with nuclei), the assembly of proteins, that is, the translation of coded information and the production of the coded messages (transcription of RNA chains from the DNA of chromosomes), is separated in space and time. RNA is transcribed from DNA in the chromosomes and then transported by means not yet understood to the cytoplasm (region of the cell outside the nucleus). Translation occurs in the cytoplasm on ribosomes whose assembly began in the nucleus, but which only become active, and therefore presumably finished, after or during transport to the cytoplasm.

Prokaryotes are known to by-pass these apparent impediments. In these forms the translation of the messages begins before the chains are released from the chromosome during their synthesis or growth. The advantages or necessity for a nuclear membrane remain obscure, although it is assumed that some considerable advantage accrued from the possession of this organelle. No higher cell has ever dispensed with a separation of the cytoplasm from the nucleus and its chromosomes.

To summarize, the chromosomes of eukaryotes contain DNA in common with those of prokaryotes; but the chromosomes of eukaryotes are isolated from the cytoplasm by a nuclear membrane, except during the middle stages of mitosis. The chromosomes typically contain histones and nonhistone proteins, as well as some RNA.

Repetitious DNA. One of the most puzzling features of eukaryotic chromosomes is the highly versatile ability to evolve genomes with repetitious DNAs. Typically prokaryotes contain chromosomes with one set of almost all genes; each segment of a chromosome is unique in the sense that a long chain of nucleotides in which the same sequence of components occurs repeatedly is rare, if not entirely absent. On the other hand, eukaryotes have highly reiterated sequences, particularly in vertebrates and higher plants. For example, mammals have nearly 1,000,000 copies of a sequence 300–400 nucleotides long. Whether the fit is perfect in reannealing (reassociation of separated chains put into solution under appropriate conditions) is difficult to determine; but reannealing of separated complementary chains has been the method used to get this information. The exact length and number of copies are uncertain, since one parameter varies in relation to the other. For example, 500,000 copies 600–800 nucleotides long would probably fit the data as well as the first estimate given above. The occurrence and possible function of these repeated or nearly repeated segments is a matter of current debate and active investigation. The present evidence indicates that such repeated sequences are not translatable messages for ribosomes, but they nevertheless seem to be transcribed and the copies are broken down within the nucleus. In addition to the highly reiterated sequences, a large class of moderately redundant sequences appear to exist in the chromosomes. The remaining 30–70% of the DNA in chromosomes is thought by some investigators to be composed of unique sequences similar to the situation in prokaryotes.

Circular chromosomes. In the bacterium *Escherichia coli*, fertilization was discovered in the 1950s. Sexuality in bacteria and related lower organisms was previously thought to be absent. However, conjugation was discovered in which a tube is formed between cells and the chromosome or part of one chromosome is transferred from the male cell into the female (F^-) cell. For particular strains of male cells called Hfr (high frequency recombinants), the transfer in one strain begins at a certain gene and the other genes follow in sequence as if on a string. The time required for a gene to be transferred is a measure of its distance from the first gene to enter. However, different strains of Hfr were found to vary in the first gene to enter the F^- cell. When many strains were studied, the genes of all could be arranged in a characteristic order as if the chromosome were a circle which in different strains could open and begin entry at different but characteristic points for each strain. A circular chromosome was postulated, and in later years so was the physical structure; long double chains of DNA without ends have been demonstrated by autoradiography and electron microscopy in both bacteria and a number of viruses. Some viruses have open chains which can form a circle during their growth or replicative stage, while some small viral DNAs appear always as closed chains or circles.

Lengths of DNA. The length of the DNA in chromosomes can be estimated by determining the mass of DNA per cell or per chromosome. In some of the forms with shorter genomes, such as the DNA-containing viruses, the length has been rather accurately measured in electron micrographs, where an image of the DNA can be viewed directly. Even in some bacteria, very long chains have been seen in electron microscopic preparations; but the total length was first visualized by associated radioactivity. For example, tritiated thymidine is an intermediate utilized exclusively for the synthesis of DNA. One or more of the hydrogen atoms can be replaced by the radioactive isotope of hydrogen, tritium (H^3). When sufficient H^3 thymidine is incorporated into the chromosomes of a bacterial cell (*E. coli*) and the cell is very gently lysed to avoid breakage of the long strands, the DNA can be allowed to untangle and float out on a filter disk. There it is covered with photographic emulsion, which records the sites of radioactive atoms as small black silver grains. Since these lie along the DNA and shows its position, the chromosome's shape and size (a circle more than 1000 μm in contour length) can be seen with the regular light microscope by the distribution of silver grains.

More recently electron microscope pictures have demonstrated the chromosome of *Mycoplasma hominis* to be a circular piece of DNA about 500 μm long. *M. hominis*, sometimes called pleuropneumonia-like organism (PPLO), is usually a parasitic bacteroid inside higher cells; but by special methods it can be grown on artificial media in the laboratory. The size of its chromosome indicates that it has about one-half the genes characteristic of a free living bacterium such as *E. coli* or *Bacillus subtilis*.

Besides the regular, single chromosomes which comprise most of the genetic material of bacteria, small extrachromosomal DNAs have been found which are also circular. These elements were first identified by their genetic effects, and only in recent years have they been identified as morphological entities. These elements may carry sex factors and genes regulating various metabolic pathways in the cells; a very interesting one confers resistance to six different antibodies. In fact, the extrachromosomal elements are small chromosomes, or plasmids, of various sizes and genetic content. Most, if not all, of them can be incorporated into the regular chromosomes by crossing-over between the two rings. Likewise certain virus chromosomes can become a part of the host's chromosome by a similar process of integration by crossing-over. The virus genomes and presumably the plasmids can be released from the chromosomes under certain conditions by appropriate nucleases which induce release by crossing-over similar to that which originally led to integration.

Linear. Circular DNA is largely absent from chromosomes of higher cells, although fragments of chromosomes can sometimes be induced to join ends and form a circle. For circle formation of this type, complementary segments of single chains must extend from each end of the fragment. Such circles can be broken by heating to a temperature at which the base pairing of these single chains is eliminated. For many years the structure and ar-

Fig. 1. Metaphase chromosomes of *Bellevalia romana*, member of lily family.

rangement of DNA in chromosomes of eukaryotes have been debated, and evidence has been pursued to settle this still-unanswered question. If all of the DNA in some of the larger chromosomes were in one double chain, it would be several feet long. In the chromosomes there may be one continuous piece of DNA; but it is always folded into a very small package. The first coiling or folding forms a strand of DNA-protein about 23 nm in diameter in which the DNA is folded so that it is 100–50 times as long as this relatively large fiber. The straight double chain DNA is only 2 nm in diameter. The DNA-protein fibers are looped or folded into a rod-shaped chromosome. At early stages of division, each chromosome is already composed of two rods called chromatids (the two new chromosomes will be formed from these) which are usually attached side-by-side with a small number of twists (Fig. 1). The cell whose chromosomes are shown in Fig. 1 was treated with colchicine, a drug which destroys the spindle of the normal division and allows the chromosomes to be flattened into one plane for photography.

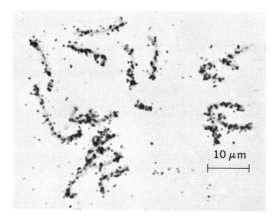

Fig. 2. Separated chromatids in an early anaphase from a root cell of *Bellevalia* grown in a solution containing thymidine labeled with tritium. Photograph shows chromosomes and the autoradiograph superimposed.

Fig. 3. Chromosomes in late metaphase from a cell which grew in tritiated thymidine for the same time as the cell shown in Fig. 2. After that, it was allowed time to replicate one more time before making the autoradiograph.

The integrity of the DNA chains in chromosomes was demonstrated some years ago by a technique called autoradiography in experiments designed to test the hypothesis of DNA replication originally proposed by J. D. Watson and F. H. C. Crick. The radioactive isotope of hydrogen, tritium, is substituted for a hydrogen in the small molecule thymidine, which, as mentioned before, is utilized in cells exclusively for DNA synthesis. When cells are grown in the presence of thymidine, only the chromosome (DNA) becomes radioactive. At the subsequent division, cells placed in contact with photographic emulsion yield autoradiographs. The decay of radioactive hydrogen emits beta particles with very low energy. These are stopped close to the chromosomes and thereby activate silver bromide grains, which when developed yield black silver grains in the photographic film. These can be examined with the high-power light microscope to determine which parts of the chromosome contain the tritium. Since thymidine is only incorporated when new DNA is built, the original, unlabeled DNA can be distinguished from the new, labeled DNA. The experiments revealed that each chromatid at the subsequent division was labeled (Fig. 2); but after one additional replication cycle, the labeled DNA chains were segregated into separate chromatids. The experiment indicated that the whole unit of DNA, more than a foot long in these large chromosomes, remained as one piece and segregated as a unit in subsequent divisions. The original, unlabeled chromosome, like the DNA, was composed of two chains which separated during replication so that each chromatid at the subsequent division contained one labeled and one unlabeled chain. After one more replication in the absence of the labeled thymidine, these two chains, a labeled one and an unlabeled one, separated and formed chromatids with the newly built chains, one with tritium, revealed by black grains, and one free of tritium (Fig. 3). In Fig. 3 whole chromatids are labeled or unlabeled, but occasionally a breakage and rejoining of labeled and unlabeled segments has also occurred during the 2 days' growth after the incorporation of tritiated thymidine. Subsequent experiments proved that DNA indeed does replicate as originally proposed by Watson and Crick. The two chains separate, and each serves as a pattern or template for the assembly of a complementary chain. The result is two identical double chains which contain the same sequence of components (nucleotides) and therefore code the same genetic information.

The details of structure and arrangement of DNA in chromosomes are still not solved, but the whole DNA per chromosome behaves in reproduction as if composed of one long double helical chain. However, when cells are very gently and carefully lysed (dissolved with detergents and salt solutions), the DNA of eukaryotic chromosomes dissociates into many segments about 100 μm long. Since the genes are only a fraction of a micrometer in length, other subdivisions and various punctuation marks, that is, variations in nucleotide sequences different from those which code amino acids, must also be a part of the DNA in chromosomes. What these may be is still unknown.

For background information see CHROMOSOME; DEOXYRIBONUCLEIC ACID (DNA) in the McGraw-Hill Encyclopedia of Science and Technology.

[J. HERBERT TAYLOR]

Clay

Recent research appears to indicate that the physical and chemical properties of clays and of clay minerals as they appear in the natural environment or as they are used in industrial processes are ruled largely by the properties of their surfaces. Clays and clay minerals are found in soils and in sediments. In soils the clay fraction is usually, but somewhat arbitrarily, defined as that fraction containing particles with diameters less than 2 μm. In soils and sediments the clay particles are mixed with coarser particles (sands and other products of the weathering of rocks). The surface area developed by clays is quite appreciable because of the small dimension of the clay crystallites. As the specific surface area (that is, the area developed by one gram of material) is roughly inversely proportional to the diameter of the crystallites, it is not surprising to obtain values for the area between a few square meters per gram and several hundred square meters per gram.

Consequently in soils and clay sediments the surface properties are determined by the clay fraction.

From the aspect of the extent of surfaces, clays and clay minerals may be divided into two classes. First, there are clay minerals with only an external surface, corresponding to the external crystalline planes of the lattice. Kaolinite, dickite, nacrite (1.1 lattice), and hydrous micas, such as illite or hydrobiotite (2.1 lattice), belong to this first class.

Second, there are clay minerals that have both an internal and an external surface. The internal surface corresponds to that in pores inside the crystal lattice. This class may be divided into two subclasses: the nonswelling porous clay minerals, such as attapulgite (a fibrous clay), and the swelling clay minerals, such as montmorillonite, beidellite, and vermiculite (1.2 lattice). In the latter subclass the porous volume is determined by the volume of the adsorbed materials.

The surface activity of clays may be considered to consist of an intensive factor, the surface activity per unit surface multipled by an extensive factor, the extent of the surface area. The specific surface area is of the order of 10 m²/g for kaolinite, 100 m²/g for hydrous micas, and 700–800 m²/g for montmorillonite. Lattice defects, such as isomorphous substitutions of silicon by aluminum within the tetrahedral layer, give rise to electrical charges which are balanced by exchangeable actions.

The surface activity is therefore a complex function determined by the nature of the surface as well as the exchangeable cations.

Nature of surface atoms. The surface of a clay mineral is covered either by hydroxyl (OH) groups or by oxygen atoms. Because of their polar nature, the OH groups are particularly active in sharing their hydrogen atoms with adsorbed molecules bearing a hydrogen acceptor group. This chemical process is called hydrogen bond formation. It may be illustrated very simply by Eqs. (1) and (2).

$$]OH + :O\begin{matrix}H\\ \backslash\\ \diagdown\\ H\end{matrix} \rightarrow]OH \ldots O\begin{matrix}H\\ \diagup\\ \diagdown\\ H\end{matrix} \quad (1)$$

$$]OH + :N\!\!-\!\!H \rightarrow]OH \ldots N\!\!-\!\!H \quad (2)$$
with CH_3 and H substituents on N.

The hydrogen bond energy is of the order of 4–8 kcal/mole. Thus the hydrogen bond formation may be an important factor in the surface activity. At first sight, the surface oxygen atoms might also be involved, as in Eq. (3). The orientation of the orbitals containing the lone pairs of electrons is such that this bond does not exist to the same extent.

$$]O: + HO\!\!-\!\!H \rightarrow]O \ldots HO\!\!-\!\!H \quad (3)$$

A fraction of the external surface of kaolinite (and also of dickite and nacrite) is covered by OH groups (Fig. 1). However, because of the limited

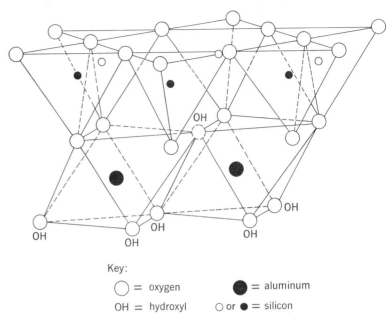

Fig. 1. Schematic structure of a sheet of kaolinite.

extension of the crystal lattice, and consequently the presence of broken bonds on cleavage faces parallel to the c axis, these planes are also probably covered mainly by OH groups. The origin of these OH groups may be explained very easily by considering the reaction shown in Eq. (4), occurring in the hydrated medium where the clay minerals are formed. The OH bond to silicon (the silanol group) has a stronger acid character than that bond to aluminum.

$$\begin{matrix}Si\\ \diagdown\\ O + H_2O \rightarrow \\ \diagup\\ Al\end{matrix}\quad \begin{matrix}]Si\!\!-\!\!OH\\ \\ ]Al\!\!-\!\!OH\end{matrix} \quad (4)$$

The OH surface groups not only play an important role in the adsorption processes but also in the first steps of transformations undergone by clays when heated progressively to high temperature (ceramic formation). As the temperature increases, the OH groups belonging to two adjacent particles may form a water molecule and an oxygen bridge. This initiates the sintering of the porous mass, resulting in a decrease of the available surface area.

The reactivity of OH groups in the interlamellar space of the kaolinite minerals may expand the lattice by intercalating organic molecules through the formation of hydrogen bonds. The intercalation process may be obtained, for example, with urea and its derivatives and with formamide and its derivatives. Accordingly the spacing between the layers is increased by the thickness of the intercalated molecules. The total surface area of kaolinite is then added to the internal pores, and it is of the order of magnitude observed for the swelling clay minerals.

Role of exchangeable cations. Exchangeable cations have a high affinity for molecules with lone pairs of electrons. This affinity results in a high hydration energy (>10 kcal/mole). In the adsorption process of water on clay surfaces, it may thus be expected that the first molecules form a hydra-

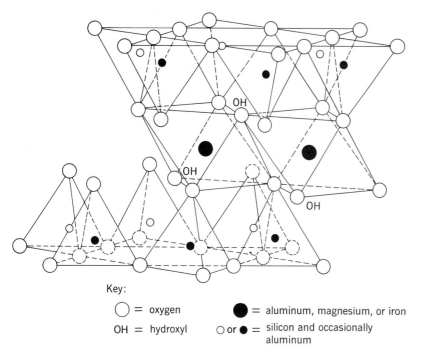

Fig. 2. Schematic structure of a sheet of montmorillonite. The exchangeable cations are located on the two planes shown.

tion shell around the exchangeable cations. These molecules are near a high polarizing field and have a degree of dissociation much higher than normal. It has been suggested that in the first or the first two layers of water adsorbed on the internal surface of montmorillonite (Fig. 2), the degree of dissociation, namely, the ratio of the surface concentration $[H^+]/[H_2O]$, is of the order of 10^{-2}, instead of $\sim 10^{-8}$ in the liquid water at 25°C.

Because of the high hydration energy, the last traces of water are almost impossible to remove, even under severe pressure and temperature conditions. Since the residual water molecules are strongly affected by the polarizing field of the exchangeable cations, the partially hydrated clay surfaces have an appreciable acid character.

This high acidity may greatly affect organic molecules adsorbed on clays. A pesticide such as triazine is protonated when in contact with the clay. These processes must be taken into account in the evaluation of the residual effects of this class of compounds in the environment.

The water molecules in the hydration sphere of cations are also ordered to some extent in regular geometric configurations (for instance, octahedral) and thus the nature of the cations rules the extent of the swelling of expansible lattices.

Exchangeable cations may also interact with inorganic or organic molecules in processes similar to that described for water, forming coordination complexes, for example. The stability of these complexes depends on the chemical characters of both partners.

In the evaluation of the catalytic properties of clay surfaces, the effects of the exchangeable cations and the acidic character of the residual water molecules are probably the most important factors to be considered. For instance, in clay sediments where oil reservoirs have been formed from buried organic materials, the catalytic effects due to the nature of the clay surface probably superimpose their actions upon the biological degradation.

For background information see CLAY; CLAY MINERALS; HYDROGEN BOND in the McGraw-Hill Encyclopedia of Science and Technology.

[J. J. FRIPIAT]

Bibliography: G. W. Brindley, in *Proceedings of the Hispano-Belgian Meeting on Clay Minerals*, 1970; M. Cruz, A. Laycock, and J. L. White, in *Proceedings of the International Clay Conference 1969, Tokyo*, 1969; M. Cruz, A. Laycock, and J. L. White, in *Proceedings of the Hispano-Belgian Meeting on Clay Minerals*, 1970; J. J. Fripiat, *Bull. Groupe Fr. Argiles*, 1970.

Coffee

More than 300 diseases attack coffee, but of all the parasites the most destructive is the leaf rust *Hemileia vastatrix*. It is a fungus, a native of northeastern Africa, and during 1869–1900 it spread to many other parts of the eastern tropics. Very recently the disease has been found in South America. Control measures that have been judged practical consist of spraying and the breeding and substitution of rust-resistant varieties of coffee.

The symptoms of this rust are yellow-orange to reddish leaf spots (Fig. 1). Each spot bears at least 150,000 spores. Infection causes spectacular defoliation which, if not controlled, results in crop failure. Years ago, country-wide occurrences were systematically recorded, and in some cases crop losses were estimated. A few examples of the average losses are as follows: Uganda 30%; in the wild Ethiopian forest harvests 12–80%; Kenya 25%; the Mysore and Coorg states of India 50–100%; Philippines 90%; and Ceylon 95%.

Origin and spread of rust. Rust made its first impact on cultivated coffee, the aromatic *Coffea arabica*, in Ceylon. In 1869 rust was found on four or five trees, and in a little more than two decades most of the millions of plantation trees succumbed to the disease. As a result, the country, which had

Fig. 1. Photograph of Arabica coffee seedling about 2 months after artificial inoculation on undersides of lower leaves with spores of rust *Hemileia vastatrix*. Diseased leaves dropped in 4–6 weeks after the picture was taken.

been noted for its wealth, became bankrupt.

At about this time Java planters had redoubled their efforts at cultivation, and soon this large island became a good source of aromatic Arabian. However, by 1876 the rust had spread to Java, and in some 12 years the specialized coffee agriculture was being devastated. Fortunately, some growers saw that coffee grew slowly but with less rust in plots on very cool and dry hillsides. They planted more in these areas and it brought premium prices, since the harvest was much reduced. On the better lands the coffee growers replanted with Robusta coffee, *Coffea canephora*, of much poorer quality but tolerant and resistant to rust.

A small amount of Arabica was grown in India but, as in the other cases, the coffee business faltered when *Hemileia* rust occurred. Planters began to use resistant varieties at the turn of the century. About then the first spraying with Bordeaux mixture had been started in Ceylon. However, because of primitive methods and high costs, it did not save the crop. With newer equipment and more experience, Bordeaux mixture was used successfully in nearby southern India. Spraying is still a common practice there.

In Uganda and Kenya coffee growers produced good Arabica but with difficulty. There were severe dry seasons and much rust. Kenya would have lost all its coffee crop except for the redesigning of pruning and mulching practices and the use of copper spray compounds.

Coffee growing expanded in Africa to the farthest areas on the west coast. At first the fields of Arabica were rust-free, but in a few years rust appeared. As the rust spread from east to west, it was seen in one season to "jump" at least 600 mi (nearly 1000 km). From 1952 to 1963 it was in serious amounts in the west coast countries of Ghana, Dahomey, Liberia, the Guineas, Angola, and the far west offshore islands of São Tomé and Fernando Poo. Crops became enriched sources of rust spores, and it was almost inevitable that spores would be carried across the Atlantic to the American tropics.

Coffee (almost all Arabica) is of paramount importance in tropical America. It has been grown there, rust free, since the 16th century. Brazil, with its great stretches of suitable lands with literally billions of Arabica trees, is the largest producer of this crop anywhere and of all time. It deserves mentioning that the lack of rust led to the success in growing the aromatic coffee bean in Brazil.

On Jan. 17, 1970, the pathologist A. G. Medeiros reported that typical *Hemileia* rust spots had been found on coffee in the Brazilian state of Bahia. The news had an unsettling effect on the world coffee trade.

No one can ever be sure how *Hemileia vastatrix* spores came to Brazil. It is probable that they were picked up by winds from Africa and, by themselves or attached to insects and other particles, were carried to Brazil. Most of the storms and airflow to eastern South America come from Africa. Airplane flights may also have had some part in it, since there is regular traffic between large cities on the west coast of Africa and the east coast of Brazil. In any case it is only about 1400 mi (some 2200 km) across the Atlantic between rust-diseased western African crops and plantations near Brazilian eastern shores.

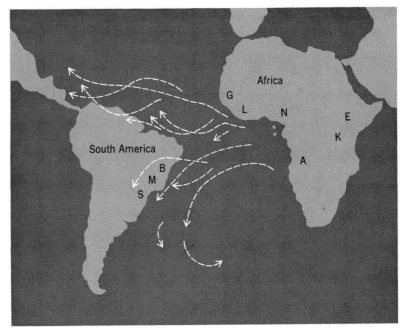

Key

--→ storm tracks and direction of storm

A = Angola

B = Brazil

E = Ethiopian region where both Arabica coffee and *H. vastatrix* originated

G = Guineas

K = Kenya

L = Liberia

M = Minas Gerais, state of Brazil (unmarked, next to the coast, is the smaller state of Espírito Santo)

N = Nigeria, below are islands of Fernando Poo (upper) and São Tomé (lower)

S = São Paulo, state of Brazil

Fig. 2. Sketch map of some locations where *Hemileia vastatrix* rust occurs on coffee. Other areas in Africa are affected with coffee rust.

Control measures. Rust is now fully established in some places and scattered over a wide area in tropical America. As of July, 1971, it was in the states of Bahia, Espírito Santo, Minas Gerais, and São Paulo (Fig. 2) and spreading. When rust was first encountered, unsuccessful attempts were made to extirpate the disease. But the disease occurred over a wide area with difficult terrain, and conditions were impossible for the kind of total eradication program needed. Another program was tried: setting aside of a sanitation corridor to keep rust from moving into highly productive and rust-free areas. But the disease continued to jump the corridor. The corridor was moved many times, but it was not feasible to keep the rust contained in the large area where it was first discovered.

The only possibilities to control the rust in Brazil are by chemicals and substituting resistant varieties of coffee. Special techniques of spraying are being developed. In extensive *fazenda* plantings, where land surfaces are very undulating and sharply broken by hills and valleys, airplanes have limited usefulness. The problems are in loading enough spray, penetrating the thickset foliage, and the danger to pilots who fly over the rough terrain.

Brazilian coffee is thickly planted. There are about 1000 official trees to a hectare (a "tree" is actually about four so arranged that their trunks are in actual contact at the *pes* or *cova*); and a heavily leaved "skirt" of branches is encouraged around the bottom half of the multiple tree to protect the bases against frosts and the Sun in the dry season. It is possible for a worker to walk between rows with a machine-driven knapsack sprayer on his back and hand-direct the spray mist. In countries such as Kenya and parts of India, plantings are arranged for the use of sprays. Trees are placed singly and kept pruned; there is space in every other row for tractor drawn spray equipment with fixed nozzles.

Thus far in Brazil, the best spray technique has been to use knapsack equipment and start just before the rainy season begins and then repeat applications every 4 weeks. Where a supply of skilled labor is at hand and when at most only five applications will be needed and where the *fazenda* owner secures a sufficient income from his plantation, spraying will probably succeed. Certain organic spray chemicals, such as Ziram and Zineb, have given hopeful results but others, such as Benlate or Difolatan, seem of little value. Inorganic sprays with zinc and copper compounds are more satisfactory. Copper oxychlorides and related compounds bring good results. Basing dilutions with compounds having 35–50% copper, the liquid is put on so that about 6 kg of copper will be on a hectare of trees. Of all materials used, Bordeaux mixture at 1.0 or 1.5% by volume is still the best, with 350–500 l applied to a hectare of trees.

Rust-resistant trees. For the long pull, the substitution of rust-resistant trees is of utmost importance. Planters are well aware that in many parts of Brazil, and in other tropical countries, spraying is not always a feasible addition to their farming practices.

Fortunately, since 1951–1952 basic and detailed studies have been conducted in Portugal by B. d'Oliveira, the world leader on *H. vastatrix* research. He has also worked in Africa and recently in Brazil. He and coworkers have found that the parasite has at least 26 races, and breeding lines of coffee trees have been developed with both special and general resistances.

Interestingly, supplies of the rust in Brazil when tested on coffees in Oeras, Portugal, have all proved to be race number two (II), the most common. There is no telling how soon others with different virulences will appear in the American tropics, but they are expected. A wide range of resistance is needed, along with horticulturally acceptable tree types and beans of good aroma having commercial acceptance by coffee processors, roasters, and manufacturers. Progress takes time with the many-faceted problem of breeding a tree crop such as coffee.

Years of arduous breeding experiments have resulted in the developing of lines (varieties) with genetic constitutions that give them various rust resistances. The Brazilian geneticist A. de Carvalho and coworkers, in collaboration with d'Oliveira in Portugal, have been steadily advancing toward the breeding of good resistant Arabicas. These will eventually replace the billions of trees presently grown.

There are some outstanding examples of coffees resistant to probably all races of rust: Geisha, Kawisari, Hibrido de Timor, Brazil 387, and another Brazil progeny that is not designated. There are other types of varying degrees of resistance. By good fortune, numbers of trees of some of these are in botanical garden plantings in Brazil and other tropical American countries. These are now being multiplied to be grown in plantings that will be mother sources.

It is safe to say that all is not known about the growth habits and commercial values of the rust varieties in Latin America. Probably the most intense study of this has been done by personnel of the Ministry of Agriculture in Costa Rica and by P. G. Sylvain of the Instituto Interamericano in that country.

For background information *see* BREEDING, PLANT; COFFEE; PLANT DISEASE in the McGraw-Hill Encyclopedia of Science and Technology.

[FREDERICK L. WELLMAN]

Bibliography: A. J. Bettencourt and A. Carvalho, *Bragantia*, 27(4):35–68, 1968; G. M. Chaves et al., *Rev. Seiva*, vol. 30, December, 1970; F. L. Wellman, *Plant Dis. Rep.*, 54(7):539–541, 1970.

Collenchyma

Recent ultrastructural investigations have shown that, irrespective of the type of collenchyma recognized in earlier anatomical studies, the pattern of cell wall organization is constant. Other work has been concerned with the possible changes in wall organization during growth, the cytology of the tissue, and the ultrastructure of lignified collenchyma.

Cell wall organization. The wall thickening is usually uneven. In early studies it was reported that some types of collenchyma showed a transverse heterogeneity of composition such that

Fig. 1. Electron micrograph of transverse section of thin area of stained collenchyma cell wall in *Rumex conglomeratus*. Reduction in thickness is most marked in lamellae showing a longitudinal orientation of cellulose microfibrils. All lamellae appear to be continuous through the thinner portion of the wall. (*From S. C. Chafe, The fine structure of the collenchyma cell wall, Planta, 90:12–21, 1970*)

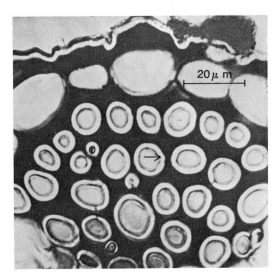

Fig. 2. A transverse unstained section of lignified collenchyma of *Eryngium rostratum* photographed in an interference microscope with the collenchymatous wall at compensation point (black). The lignified wall appears white (arrow). (*From A. B. Wardrop, The structure of the cell wall in lignified collenchyma of Eryngium species (Umbelliferae), Aust. J. Bot., 17:299–240, 1969*)

pectin-rich layers alternated with layers poor in pectin. S. C. Chafe has reported investigations on four types of collenchyma generally accepted as being anatomically distinct on the basis of the appearance of their cell walls when viewed in cross section: angular (the wall thickening prominent at cell corners), lacunate (intercellular spaces adjacent to which the wall is most prominently thickened), lamellar (thickening mainly on the tangential walls), and annular (wall thickening uniform or almost so). It was shown that for each type there is an outer limiting structure—probably corresponding to the first-formed meristematic wall in which the cellulose microfibril orientation appeared random on its outer surface. The thickening consisted of alternate lamellae in which the cellulose microfibrils were arranged either transverse or parallel to the longitudinal cell axis. In the thicker areas of the wall, the layers consisting of longitudinal microfibrils were wider than those consisting of transversely oriented microfibrils. The thinner areas of the wall reflected a reduction in thickness of all layers but particularly a reduction in thickness of layers with longitudinally oriented microfibrils (Fig. 1).

The wall of lignified collenchyma consists of a typical unlignified collenchyma wall as described above plus an additional lignified layer (shown by the arrow in Fig. 2). The unlignified part of the wall has the same organization as that already described (Fig. 1), whereas the lignified layer consists of successive lamellae of helically oriented microfibrils making an angle of about 45° with the cell axis. The presence of lignin in the lignified layer was shown by staining reactions, ultraviolet fluorescence, and ultraviolet absorption. This layer differs markedly in its optical and physical properties from the unlignified regions. Thus the path difference of the lignified layer as determined by interference microscopy was tenfold that of the unlignified collenchyma wall. On drying, the collenchyma wall underwent shrinkage to 0.9% of its original thickness compared with a value of 66% for the lignified layer.

The distribution of pectic substances, using the hydroxylamine–ferric chloride reaction of R. M. Reeve as well as the reaction with ruthenium red, was studied by Chafe. At the optical level, discontinuous distribution of material stained by these agents was observed in only three of nine genera studied. In species in which lamellation was observed at the electron microscope level, the hydroxylamine–ferric chloride stain was most intense in those lamellae which were oriented parallel to the cell axis (Fig. 3).

Growth and formation of cell wall. The ontogeny of collenchyma has been summarized by K. Esau. Early investigations showed that considerable extension of collenchyma cells takes place after the onset of wall thickening. This was confirmed in the study by A. B. Wardrop of leaf collenchyma of *Eryngium rostratum*. It was shown that in a glucose–indoleacetic acid solution the extension of leaf segments containing thin-walled parenchyma was over twice that of segments containing thick-walled collenchyma. Cells containing a lignified layer showed no extension. The structural changes in the cell wall which accompany growth of these cells have been discussed by Wardrop. It can be seen in Fig. 4 (arrows) that the lamellae of transversely oriented microfibrils were less apparent in radial sections of the cell the greater their distance from the lumen. It was suggested that this may reflect some form of modified multinet growth as postulated by P. A. Roelofsen and A. L. Houwink. On this basis, in collenchyma it might be expected that, during longitudinal growth of the cells, the microfibrils of the first-formed (outermost) lamellae would undergo great-

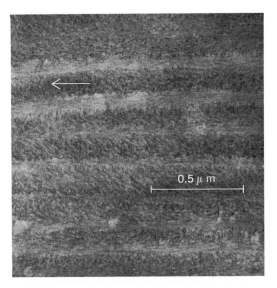

Fig. 3. Electron micrograph of a transverse section of a collenchyma cell wall in *Petasites fragrans*, stained by the Reeve method, showing a lamellation of pectic substances. Those lamellae having the higher pectin content appear to be those in which the orientation of the cellulose microfibrils is longitudinal (arrow). (*From S. C. Chafe, The fine structure of the collenchyma cell wall, Planta, 90:12–21, 1970*)

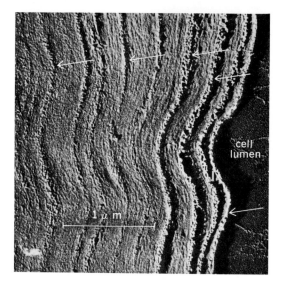

Fig. 4. An electron micrograph of a platinum-palladium shadowcast radial longitudinal section of the collenchyma wall of *Eryngium rostratum*. Arrows indicate lamellae in which microfibrils are transversely oriented. These lamellae appear less prominent the greater their distance from the cell lumen. Transverse undulations in the wall are also obvious. (*From A. B. Wardrop, The structure of the cell wall in lignified collenchyma of Eryngium species (Umbelliferae), Aust. J. Bot., 17:229–240, 1969*)

est reorientation toward the longitudinal axis and at any stage the innermost lamellae would have undergone least reorientation toward the longitudinal direction and would, as shown in Fig. 4, appear more prominent when the cell was cut in longitudinal section.

An alternative concept is that the forces causing elongation of the collenchyma cells do not originate within the cells themselves but that the cells are caused to elongate by the growth of cells adjacent to them (passive growth). This concept would be consistent with the known phenomenon apparent in Fig. 4 that, on isolation, collenchyma cells develop transverse undulations due to the release of growth stresses.

On either hypothesis, however, it is apparent that if reorientation of microfibrils occurs, then relative movement between them must occur. The mechanisms of growth of collenchyma cells remain unknown, as do the mechanisms by which the cell deposits lamellae of microfibrils alternately of transverse and longitudinal orientation.

In cell wall studies generally, it has been pointed out that, in view of the length of the cellulose molecule and the fact that in many cells the orientation of the molecular aggregates reflects the cell form, it is reasonable to suggest that only large organelles can act as templates for the synthesis of cellulose molecules or govern their orientation. The endoplasmic reticulum, the plasmalemma, and the cytoplasmic microtubules have been suggested as possible organelles involved in this function.

In a study of the collenchyma of *Apium graveolens*, in the wall of which the typical organization had been established, Chafe and Wardrop observed that, in general, the orientation of microfibrils in the wall was parallel to the orientation of microtubules in the cytoplasm, although in some instances the orientation of the microtubules did not parallel either of the directions of microfibril orientation in the cell wall. In this study the arrangement of particles on or in the plasmalemma, which have been demonstrated by the freeze-etching method and which have been suggested to have the function of being involved in the synthesis and orientation of microfibrils in the wall, was also studied. The distribution of plasmalemma particles was found to be uniform, rather than random, and unrelated to microfibril orientation in the wall. Accordingly it was concluded that, so far as the evidence goes, the microtubules are the most likely organelles governing the orientation of microfibrils in the cell wall.

Cytology of collenchyma. In unpublished studies of the cytoplasmic organization of collenchyma, Chafe has recorded the presence of cellular organelles known to be present in most eukaryotic cells and noted electron microscopic evidence indicative of secretion of vesicles, probably of Golgi origin, into the cell wall. This was also demonstrated in the studies of J. C. Roland. In all these investigations the progress of cell development was characterized by the progressive vacuolation of the cells and the development, at an early stage of differentiation, of organelles characteristic of senescence, such as myelin bodies. Whether this sequence of changes is indicative of "passive growth" of the cells requires further study. Roland has drawn attention to the differences in nature of calcium deposits in the cytoplasm and cell wall by means of histochemical and electron-probe analytical studies. The significance of these investigations remains to be assessed.

Morphogenesis. Early anatomical studies demonstrated that the extent of development of collenchyma in petioles could be influenced by mechanical stress such that sustained stress increased the extent of collenchyma formed. Roland and M. Bessoles have shown both qualitatively and quantitatively that the removal of a leaf influences the development of collenchyma in the subadjacent internode and also the incorporation of xylose-l-C^{14} into the differentiating tissue. These observations may serve as a starting point for the integration of anatomical and physiological studies of the tissue.

For background information see CELL WALLS (PLANT); COLLENCHYMA in the McGraw-Hill Encyclopedia of Science and Technology.

[A. B. WARDROP]

Bibliography: S. C. Chafe, *Planta*, 90:12–21, 1970; S. C. Chafe and A. B. Wardrop, *Planta*, 92: 12–24, 1970; J. C. Roland, *C. R. Acad. Sci.*, 267: 589–592, 1968; J. C. Roland, *J. Microsc.*, 6:399–412, 1967; J. C. Roland and M. Bessoles, *C. R. Acad. Sci.*, 269:894–897, 1969; A. B. Wardrop, *Aust. J. Bot.*, 17:299–240, 1969.

Coordination chemistry

Some of the most important recent developments in the field of coordination chemistry relate to reactions of coordination compounds. These developments include studies leading to new insights into the mechanisms of such well-know reactions of coordination compounds as substitution and electron transfer, as well as discoveries and applications of new types of reactions. One such class

Characteristic coordination numbers of low-spin transition-metal complexes

Coordination number	Complexes	Electron configuration	Total number of valence electrons
8	$Mo(CN)_8^{3-}$, $Mo(CN)_8^{4-}$	d^1, d^2	17, 18
6	$M(CN)_6^{3-}$ (M = Cr, Mn, Fe, Co)	d^3, d^4, d^5, d^6	15–18
5	$Co(CN)_5^{3-}$, $Ni(CN)_5^{3-}$	d^7, d^8	17, 18
4 (square planar)	$Ni(CN)_4^{2-}$	d^8	16
4 (tetrahedral)	$Cu(CN)_4^{3-}$, $Ni(CO)_4$, $Pt(PPh_3)_4$	d^{10}	18
3	$Pt(PPh_3)_3$	d^{10}	16
2	$Ag(CN)_2^-$, $Pt(PPh_3)_2$	d^{10}	14

of reactions, which is the principal subject of this article, is oxidative addition. This class of reactions is characterized by considerable chemical novelty and by important synthetic and catalytic applications.

The term "oxidative addition" has come to be used to designate a rather widespread class of reactions, generally of low-spin transition metal complexes, in which oxidation (that is, an increase in the oxidation number of the metal) is accompanied by an increase in the coordination number. Illustrative of the reactions encompassed by this designation are the examples depicted by Eqs. (1–3).

$$2Co^{II}(CN)_5^{3-} + CH_3I \rightarrow$$
$$Co^{III}(CN)_5I^{3-} + Co^{III}(CN)_5CH_3^{3-} \quad (1)$$

$$Ir^I(CO)(PPh_3)_2Cl + H_2 \rightarrow$$
$$Ir^{III}(CO)(PPh_3)_2H_2Cl \quad (2)$$

$$Pt^0(PPh_3)_2 + Ph_3SnCl \rightarrow$$
$$Pt^{II}(PPh_3)_2(SnPh_3)Cl \quad (3)$$

The occurrence of such oxidative addition reactions is related to the pattern of characteristic coordination numbers of low-spin transition metal complexes depicted in the table. Reflected in this pattern is an inverse dependence of the preferred coordination number upon the d-electron population of the transition metal atom, a trend which becomes especially pronounced as the filling of the d subshell approaches completion. This trend reflects the constraints of the well-known "18 electron" (or "noble gas") rule, according to which stable configurations in such complexes are restricted to those in which the total number of valence electrons (comprising the d electrons of the metal and the σ-bonding electron pairs donated by each of the ligands) does not exceed 18. Closed-shell configurations corresponding to 18 valence electrons tend to be particularly stable and widespread (especially for complexes with good π-acceptor ligands, such as carbonyls) but configurations having fewer valence electrons, for example, square planar d^8 complexes, also commonly occur.

One of the consequences of the trends described above is that the oxidations of certain classes of complexes (notably those with nearly filled d shells, that is, d^7–d^{10}) are accompanied by increases in the preferred coordination numbers of the metal atoms and hence by the incorporation of additional ligands into their coordination shells. The ligands required to complete the coordination shells may in certain cases be derived from the oxidant itself, and indeed such complexes are especially effective as reductants for molecular oxidants which undergo dissociative reduction to yield anionic ligands $(X-Y + e^- \rightarrow X^{\cdot} + Y^-;$ $X-Y + 2e^- \rightarrow X^- + Y^-)$.

Three such classes of reactions which have attracted particular attention in recent years are those exemplified by Eqs. (1–3), namely, the oxidative addition reactions of five-coordinate d^7, four-coordinate d^8, and two-coordinate d^{10} complexes. In each of these cases the driving force for reaction is associated with the increase in stability in going from the initial open-shell configuration (containing 17, 16, and 14 valence electrons, respectively) to the closed-shell configuration of the product. The range of molecules which undergo oxidative addition to such complexes is very extensive and encompasses hydrogen, the halogens, hydrogen halides, water, and hydrogen peroxide, as well as various organic halides and metal halides which add through carbon-halogen and metal-halogen bonds, respectively. The oxidative addition of organic molecules through either carbon-carbon or carbon-hydrogen bonds is rare, although a few such cases are known.

Mechanisms. The principal studies on the mechanisms of oxidative addition reactions of d^7 complexes, exemplified by Eq. (1), have been concerned with $Co(CN)_5^{3-}$. Most of these reactions proceed through stepwise free radical mechanisms such as that depicted by Eqs. (4–6), where Eq. (4)

$$Co(CN)_5^{3-} + CH_3I \rightarrow Co(CN)_5I^{3-} + CH_3 \quad (4)$$
$$Co(CN)_5^{3-} + CH_3 \rightarrow Co(CN)_5CH_3^{3-} \quad (5)$$
$$\overline{2Co(CN)_5^{3-} + CH_3I \rightarrow Co(CN)_5I^{3-}}$$
$$+ Co(CN)_5CH_3^{3-} \quad (6)$$

determines the rate. Other low-spin cobalt(II) complexes including bis(dimethylglyoximato)cobalt(II) and cobalt(II) Schiff's base complexes undergo similar reactions. Interest in the study of such oxidative addition reactions of low-spin cobalt complexes, notably reactions leading to organocobalt products, is enhanced by certain parallels with corresponding reactions of vitamin B_{12} derivatives and consequently by their possible relevance as vitamin B_{12} model systems.

The most extensive studies on the oxidative addition reactions of d^8 complexes have been made on square planar iridium(I) complexes of the type $Ir(CO)L_2Y$ (Y = Cl, Br, I, and so on; L = PPh_3, PPh_2Me, $PPhMe_2$, and so on). Such complexes oxidatively add a variety of molecules, including H_2 and organic halides (RX = CH_3—I, $C_6H_5CH_2$—Br, and so on). The addition of H_2, as in Eq. (7a), is a concerted process, involving a relatively nonpolar transition state and leading to a cis adduct. On the other hand, the oxidative addition of alkyl or benzyl halides, as in Eq. (7b), appears to proceed through an S_N2-type attack, with at least partial halide displacement, through a highly unsymmetrical transition state, resulting, in certain cases at least, in trans addition.

The decarbonylation of acyl halides and aldehydes, RCZO(Z = H or halogen) by tris(triphenylphosphine)chlororhodium(I) (= Rh^IL_3Cl) probably proceeds through the oxidative-addition mechanism depicted by Eq. (8).

$$Rh^IL_3Cl + R\overset{O}{C}Z \longrightarrow L_3ClRh\overset{\overset{O}{\|}\overset{}{CR}}{\underset{Z}{|}} \xrightarrow{-L}$$

$$L_2ClRh\overset{\overset{R}{|}}{\underset{\underset{Z}{|}}{-CO}} \longrightarrow Rh^IL_2(CO)Cl + RZ \qquad (8)$$

The best characterized oxidative addition reactions of d^{10} complexes are probably those of platinum(0), notably $Pt(PPh_3)_3$. This complex reacts with a variety of molecules (X—Y = CH_3—I, $C_6H_5CH_2$—Br, Ph_3Sn—Cl) to give adducts of the type $Pt^{II}(PPh_3)_2(X)(Y)$. Several such reactions have been studied and shown to proceed through mechanisms of the type depicted in Eq. (9), in which oxidative addition actually occurs to $Pt(PPh_3)_2$ which is formed by dissociation of the parent complex.

$$Pt^0(PPh_3)_3 \xrightleftharpoons{-PPh_3} Pt^0(PPh_3)_2 \xrightarrow{X-Y} Pt^{II}(PPh_3)_2(X)(Y) \qquad (9)$$

Synthetic applications. Oxidative addition reactions afford widely applicable routes to the synthesis of a variety of organo- and hydrido-transition metal compounds, as well as compounds containing metal-metal bonds. The syntheses of $Co(CN)_5CH_3^{3-}$, $Ir(CO)(PPh_3)_2H_2Cl$, and $Pt(PPh_3)_2$-$(SnPh_3)Cl$, depicted by Eqs. (1), (2), and (3), respectively, illustrate these applications.

Catalytic applications. The ability of certain transition metal complexes to dissociate molecular hydrogen through oxidative addition permits such complexes to function as catalysts for the hydrogenation of olefins and other substrates. Among the general requirements for such catalytic activity are the ability of the transition metal hydride, formed through oxidative addition of H_2, to transfer the coordinated hydrogen ligands to the substrate and thereby regenerate the original complex to complete the catalytic cycle. The known homogeneous hydrogenation catalysts for which these conditions apparently are fulfilled (although the mechanisms in many cases are still incompletely understood) include several d^7, d^8, and d^{10} complexes, notably of Co(II), Co(I), Rh(I), Ir(I), Pt(II), and Fe(0).

Among d^7 complexes, the outstanding example of a hydrogenation catalyst is undoubtedly $Co(CN)_5^{3-}$. This complex is catalytically active for the hydrogenation of a variety of organic substrates, including olefinic compounds containing conjugated (but not isolated) double bonds. Thus butadiene, styrene, and cinnamic acid are selectively hydrogenated in aqueous (or aqueous-alcoholic) solutions, in the presence of $Co(CN)_5^{3-}$, to butene, ethylbenzene, and phenylpropionic acid, respectively. A general mechanistic scheme which accommodates all these reactions is depicted by Eqs. (10) and (11), where X is a conjugating group such as —CH=CH_2, —C_6H_5, —CN, or —COOH. According to this scheme, the primary reaction in each case is the oxidative addition of H_2 to $Co(CN)_5^{3-}$ to form $Co(CN)_5H^{3-}$, followed by reaction of $Co(CN)_5H^{3-}$ with the conjugated olefin. The latter reaction leads either (through Co—H addition across the C=C bond) to an organocobalt adduct, as in Eq. (11a), or (through H atom transfer) to a free radical, as in Eq. (11b), followed in each case by further reaction with $Co(CN)_5H^{3-}$ to complete the catalytic hydrogenation cycle and regenerate $Co(CN)_5^{3-}$. The former mechanism apparently applies to the hydrogenation of butadiene, and the latter to cinnamic acid.

$$2Co(CNO)_5^{3-} + H_2 \rightarrow 2Co(CN)_5H^{3-} \qquad (10)$$

Oxidative addition of H_2 also appears to be a feature of the mechanisms of hydrogenation with catalysts of d^8 configuration. The mechanism of one such reaction which has been at least partially elucidated, namely that involving $Ir(CO)(PPh_3)_2Cl$ as catalyst, is shown in the figure. The versatile homogeneous hydrogenation catalyst $Rh(PPh_3)_3Cl$,

Proposed mechanism of the $Ir(CO)[P(C_6H_5)_3]_2Cl$-catalyzed hydrogenation of maleic acid in dimethylacetamide (S = solvent).

$$Co(CN)_5H^{3-} + CH_2=CHX \begin{cases} \rightarrow Co(CN)_5CHXCH_3{}^{3-} \xrightarrow{Co(CN)_5H^{3-}} CH_3CH_2X + 2Co(CN)_5{}^{3-} & (11a) \\ \quad \updownarrow \\ \rightarrow Co(CN)_5{}^{3-} + \dot{C}HXCH_3 \xrightarrow{Co(CN)_5H^{3-}} CH_3CH_2X + 2Co(CN)_5{}^{3-} & (11b) \end{cases}$$

discovered and studied by G. Wilkinson, probably functions through a similar mechanism.

In addition to hydrogenation, there are several other reactions of olefins catalyzed by d^8 complexes that involve transition metal hydride intermediates, formed by oxidative addition either of H_2 or of hydrogen-containing molecules such as hydrogen halides or silanes. Such reactions include the hydroformylation of olefins catalyzed by cobalt(I) and rhodium(I) complexes; the hydrosilation of olefins catalyzed by cobalt(I) and platinum(II) complexes; and the isomerization of olefins, as well as various olefin-to-olefin addition reactions, catalyzed by complexes of rhodium(I).

Oxidative addition steps are also involved in the catalysis by transition metal compounds of certain reactions involving the formation or breaking of carbon-carbon bonds, the concerted pathways of which are "thermally forbidden" according to the Woodward-Hoffmann rules of orbital symmetry conservation. Such reactions include the 1,2-cycloadditions of olefins to form cyclobutane rings, and the reverse cyclo-reversion reactions. It has been found that certain such reactions, for example, the valence isomerizations of highly strained cyclobutane-containing compounds to the corresponding dienes (quadricyclene → norbornadiene; hexamethylprismane → hexamethyldewarbenzene), are catalyzed by transition metal complexes, notably of rhodium(I). At least one such reaction, the valence isomerization of cubane to syn-tricyclooctadiene, as in Eq. (12), catalyzed by

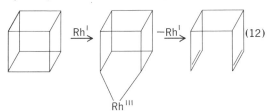

rhodium(I) complexes of the type $[Rh(diene)Cl]_2$ (where diene = norbornadiene, syn-tricyclooctadiene, and so on), has been shown to proceed through a stepwise mechanism involving oxidative addition of the cubane through a carbon-carbon bond.

For background information see COORDINATION CHEMISTRY; WOODWARD-HOFFMAN RULE in the McGraw-Hill Encyclopedia of Science and Technology. [JACK HALPERN]

Bibliography: J. P. Collman, *Accounts Chem. Res.*, 1:136, 1968; R. Cramer, *Accounts Chem. Res.*, 1:186, 1968; J. Halpern, *Accounts Chem. Res.*, 3:386, 1970; J. Halpern, *Advan. Chem. Ser.*, 70:1, 1968.

Copper chemistry

For over a century copper has played an important role in promoting organic reactions such as the well-known Ullman biaryl synthesis, the Ullman condensation to make aromatic ethers, and the Sandmeyer reaction to make aromatic halides. Only recently has organocopper chemistry been broadly investigated and become important to synthetic organic chemists. This has been made possible through development of techniques for handling air- and heat-sensitive compounds and the discovery of stable, soluble, easily handled organocopper derivatives.

Organocopper compounds contain an organic group such as an alkyl, vinyl, aryl, or acetylenic group directly bound to copper by a carbon-copper bond. Almost all stable organocopper compounds have copper in the +1 or cuprous oxidation state and are simply represented as RCu. Work with organocoppers was undoubtedly discouraged by the properties of the few early examples—copper acetylides, methylcopper, and phenylcopper—many of which are explosive, insoluble, and thermally unstable at room temperature, react with oxygen in air, and hydrolyze with moisture. The problems of solubility, thermal stability, and explosive nature have been solved by new structures, complexes, and solvents, while the air and moisture reactivity is controlled by careful technique and the use of inert-gas-atmosphere glove boxes.

Fluorinated organocoppers have been particularly valuable in studying organocopper chemistry for they are unusually thermally stable, are soluble in many common organic solvents such as ether and benzene, can be safely isolated as pure dry solids, and hence are the best characterized organocoppers.

Preparation. Fluorinated organocoppers are prepared by reacting a Grignard or organolithium with a cuprous salt. With soluble organocoppers generated from Grignards, the unwanted magnesium halide by-product is removed as its insoluble dioxane complex as in Eq. (1). The solvent then is

$$RMgBr + CuBr \longrightarrow RCu + MgBr_2 \xrightarrow{Dioxane} MgBr_2 \cdot dioxane + RCu \quad (1)$$

removed leaving the solid organocopper.

Methods of lesser importance are decarboxylation of cuprous salts of carboxylic acids, organocopper–organic halide exchange, copper metal–organic halide reaction and copper-catalyzed decomposition of diazo compounds. Some of the recently prepared fluorinated organocoppers are listed below.

Perfluoroheptylcopper	$CF_3(CF_2)_6Cu$
Hexafluoropropyl-1,3-dicopper	$Cu(CF_2)_3Cu$
3-Phenylhexafluoropropylcopper	$C_6H_5(CF_2)_3Cu$
Perfluoro-*t*-butylcopper	$(CF_3)_3CCu$
m- and *p*-Fluorophenylcopper	FC_6H_4Cu
o, *m*, and *p*-(Trifluoromethyl)phenylcopper	$CF_3C_6H_4Cu$
2,3,5,6-Tetrafluorophenylcopper	$4\text{-}HC_6F_4Cu$
Pentafluorophenylcopper	C_6F_5Cu

Structure. Fluorinated arylcoppers are discrete aggregates. Pentafluorophenylcopper and *o*-(trifluoromethyl)phenylcopper were shown by mass spectroscopy and cryoscopic molecular weights

COPPER CHEMISTRY

(a) $\left(\begin{array}{c} \text{Cu} \\ \text{CF}_3 \end{array} \right)_4$

(b) $\left(\begin{array}{c} \text{Cu} \\ \text{CF}_3 \end{array} \right)_8$

Fluorinated aryl copper aggregates. (a) Tetramer. (b) Octamer.

to be tetramers. o-(Trifluoromethyl)phenylcopper has a particularly clean mass spectrum with a strong parent ion for the tetramer, weak peaks (1%) for trimer and dimer, and a very weak peak (0.1%) for monomer. These tetramers are remarkably stable, existing as discrete tetramers at 190°C under high vacuum as well as in benzene solution at 5°C.

A subtle change in structure to m-(trifluoromethyl)phenylcopper gives rise to an octamer, according to molecular weight and kinetic studies. Structural formulas for the tetramer and octamer are shown in the figure.

Apparently, bulky ortho substituents are accommodated better by smaller aggregates, analogous to the situation in organolithiums. Presumably the tetramers have central tetrahedral clusters of copper as in $[CuS_2CN(C_2H_5)_2]_4$ and the octamer has a central cube of copper as in $Cu_8[S_2CC(CN)_6]^{4-}$. The copper clusters probably are held together by bridging aryl groups.

Hydrolysis. Organocoppers react with proton sources to form the protonated ligand and cuprous salts. Water gives cuprous oxide, as in Eq. (2), hy-

$$2C_6F_5Cu + H_2O \rightarrow 2C_6F_5H + Cu_2O \quad (2)$$

drogen halides give cuprous halides, carboxylic acids give cuprous carboxylates, and alcohols slowly give cuprous alkoxides. Hydrolysis is greatly inhibited in dimethylsulfoxide solvent.

Oxidation. Organocoppers are readily oxidized by oxygen, as in Eq. (3), nitrobenzene, benzoyl peroxide, bromine, and cupric halides. Usually the organic group is coupled in high yield and cuprous or cupric salts are formed, depending on the amount and type of oxidant used.

$$2C_6F_5Cu + O_2 \rightarrow C_6F_5{-}C_6F_5 + 2CuO \quad (3)$$

Thermal decomposition. Alkylcoppers are the least stable thermally and decompose to many types of products, depending on their structure. n-Alkylcoppers decompose below 0°C, while perfluoroalkylcoppers are stable at 80–150°C.

Vinyl and arylcoppers decompose very cleanly to coupled products. The m-(trifluoromethyl)phenylcopper octamer is unusual in that it loses its aryl groups pairwise, forming a stable deep-green intermediate containing copper in the zero oxidation state, as in Eq. (4).

$$[m\text{-}(CF_3)C_6H_5]_8Cu_8 \rightarrow$$
$$(m\text{-}(CF_3)C_6H_4)_2 + [m\text{-}(CF_3)C_6H_4]_6Cu_8 \quad (4)$$

In all cases pyrolysis eventually produces deposits of copper metal.

Coupling with organic halides. Organocoppers show an unusually high reactivity toward carbon-halogen bonds. Reactivity is in the order I>Br≫Cl, and fluorides do not react. Carbonyl halides and allyl halides are the most reactive, and aromatic halides react quite cleanly. Some examples are shown in Eqs. (5), (6), and (7).

$$C_6F_5Cu + C_6H_5I \rightarrow C_6F_5{-}C_6H_5 + CuI \quad (5)$$

$$C_6F_5Cu + CH_3COCl \rightarrow C_6F_5COCH_3 + CuCl \quad (6)$$

$$C_6F_5Cu + CBr_2{=}CHBr \rightarrow$$
$$C_6F_5C{\equiv}CC_6F_5 + CuBr \quad (7)$$

Since cuprous halide is formed in the coupling reaction, Grignards or organolithiums, plus a catalytic amount of cuprous halide, may be substituted for the organocopper where these other organometallics do not normally react with the organic halide.

Other reactions. Organocoppers add very rapidly and specifically 1,4 to α,β-unsaturated carbonyl compounds. They promote the rapid loss of nitrogen from diazoalkanes, first inserting (in a formal sense) the carbene in the carbon-copper bond and then catalyzing carbene reactions. In some cases they react with diazonium ions to form azobenzenes. Allylic and propargyl acetates couple like organic halides. Some epoxides and strained hydrocarbons are opened by organocoppers.

Other organocopper reagents. Pure organocoppers have been very valuable in understanding organocopper structure and chemistry. However, many pure organocoppers are not stable, soluble, or reactive enough for practical use.

Complexes with strong ligands, particularly tributylphosphine and triphenylphosphine, greatly improve the solubility and thermal stability of alkylcoppers without impairing their reactivity toward organic substrates.

Likewise complexation of organocoppers with an equivalent of organolithium forms a stable soluble "ate" complex. These reagents, particularly lithium dimethylcuprate, may be the most valuable form of organocoppers. Also, they probably have enhanced nucleophilic reactivity, that is, they react better with organic halides than regular organocoppers.

For many purposes mixtures of Grignards or organolithiums with cuprous salts are effective.

Summary. Organocoppers have proven to be powerful reagents for selective carbon-carbon bond formation. Oxidative and thermal coupling, coupling with organic halides, and conjugate addition are becoming widely used synthetic methods. The reactivity pattern of organocoppers toward functional groups is quite different from, and complementary with, that of Grignards and organolithiums, making organocoppers quite valuable to the synthetic chemist. Some recent examples of complex molecules synthesized with the aid of organocoppers are trans,trans-farnesol and dl-C_{18} cecropia moth juvenile hormone.

For background information see COPPER; GRIGNARD REACTION; ORGANOMETALLIC COMPOUND in the McGraw-Hill Encyclopedia of Science and Technology.

[ALLAN CAIRNCROSS]

Bibliography: A. Cairncross and W. A. Sheppard, J. Amer. Chem. Soc., 93:247, 1971; A. Cairncross, H. Omura, and W. A. Sheppard, J. Amer. Chem. Soc., 93:248, 1971; E. J. Corey et al., J. Amer. Chem. Soc., 90:5618, 1968; A. E. Jukes, S. S. Dua, and H. Gilman, J. Organometal. Chem., 24:791, 1970; V. C. R. McLoughlin and J. Thrower, Tetrahedron, 25:5921, 1969; G. M. Whitesides, C. P. Casey, and J. K. Krieger, J. Amer. Chem. Soc., 93:1379, 1971.

Corn

Prior to 1970, Southern corn blight was a comparatively minor disease in the United States. In 1970, however, the disease developed in epiphytotic proportions and caused a greater production loss on a single crop in a single year than any similar event in American agricultural history. This was due mainly to the widespread distribution of a new and highly pathogenic strain of the fungus causing the disease. The new strain of the fungus was uniquely adapted to, and highly destructive of, corn hybrids which carry the Texas (cms-T) source of cytoplasm for male sterility. This cytoplasm, from one plant in 1944, was multiplied and used by 1970 in the production of about 70–80% of the United States corn hybrids. By the end of the growing season the new fungus strain had spread across most of the corn-producing areas in the eastern half of the United States. Except where dry weather prevailed, climatic conditions were favorable for disease development throughout this vast area. Losses were catastrophic in the Southern states and substantial in many Corn Belt fields. It was estimated that American corn production was reduced by about 10%. The disease also seriously reduced seed supplies for the 1971 corn crop.

History of blight and disease. Southern corn blight has been known in the United States since first described in Florida in 1925. The disease is essentially worldwide in distribution. It is widespread and a major disease in tropical and subtropical corn-growing areas of the world. Up to and including 1969, the disease was of minor importance in the United States because most inbred lines and hybrids produced from them were resistant.

Southern blight reduces both the yield and quality of the crop. Losses in any individual field from infection by the old strain of the fungus rarely exceeded 30% but losses from the new strain were as high as 100% when infections were early and severe. Grain from infected plants is lower in test weight.

Although the grain may be infected with the fungus, it is not toxic to animals.

Cause of disease. Southern corn blight is caused by a fungus known in the conidial, or asexual, stage as *Helminthosporium maydis* Nisikado and Miyake (Fig. 1). The perfect, or sexual, stage has the name *Cochliobolus heterostrophus* Drechsler. Ascospores are produced during the sexual stage, which is rarely seen in nature but can easily be produced in the laboratory.

The fungus persists in crop refuse between corn-growing seasons. It produces spores (conidia) on this refuse. The spores are airborne and some land on corn leaves, where they germinate in the presence of water and penetrate the leaf. Some infections are directly through the cuticle, while others are through stomatal openings. Once inside the leaf the fungus hyphae (threadlike mold growths) spread through the cells, killing them. After the leaf tissue is killed and if kept moist for 10 or more hours, the fungus produces thousands of new conidia on the surface of this dead tissue. These conidia are easily dislodged and blown through the air to initiate new infections. The entire cycle from infection to reproduction can take place in as little as 60 hr if conditions are favorable. Hence an epiphytotic can be explosive.

At least two distinct races of *H. maydis* are known. These have been named race T and race O through work at the Illinois Agricultural Experiment Station. Race T was first recognized in the early spring of 1970 from greenhouse and laboratory studies of the fungus isolated from infected corn leaves collected in a west-central Illinois

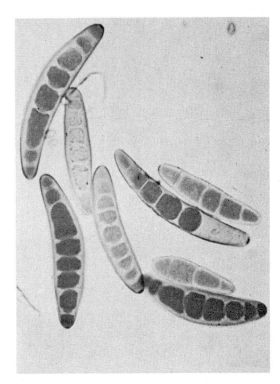

Fig. 1. Conidia of *Helminthosporium maydis* race T.

cornfield in September, 1969. The Illinois research group was well familiar with the old race O, which was the common race prior to 1969. They have continued to study the two races.

The most important distinction between race T and race O is in their specificity to different corn cytoplasm types. Race T is unique in that it is pathogenic only to corn plants that have the cms-T cytoplasm for male sterility. Race T can also attack other plants with certain forms of male sterility but these forms are now recognized as being similar, if not identical, to the cms-T cytoplasm. Although plants with other cytoplasms can be infected as seedlings, they show little or no evidence of infection at older plant stages in the field until late in the season. Race O shows no specificity to different plant cytoplasms and can attack a wider range of plant genotypes and cytoplasms than can race T.

The two races differ in toxin production. Race T produces a highly specific pathotoxin in infected plants and in laboratory cultures. This toxin shows the same specificity to plant cytoplasms as does the organism itself. The pathotoxin is a metabolic byproduct of the fungus, is dialyzable and thermostable, and is probably a low-moleculer-weight compound. It has a paper chromatography Rf value between 0.85 and 1.00 in upper phase *n*-butanol, acetic acid, and water (4:1:5; volume/volume/volume) and gives a negative reaction with ninhydrin. Race O toxin, while similar to race T toxin in some respects, is distinctly different. Race O pathotoxin is produced in lower amounts, is nonspecific to plant cytoplasms, has a substantially lower Rf value, and gives a positive reaction with ninhydrin.

Race T pathotoxin, when applied to roots from germinating seeds, inhibits the growth of susceptible types but not of resistant types. It has been

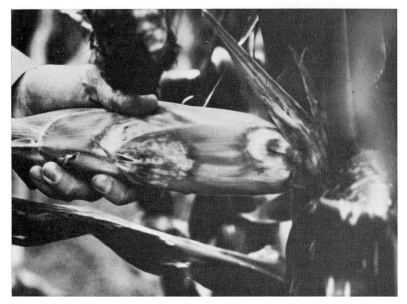

Fig. 2. *Helminthosporium maydis* race T husk infection.

used in seed-testing laboratories to determine the proportion of resistant and susceptible seeds in a mixture. This root bioassay test takes only a few days. In more elaborate studies, cell-free mitochondrial preparations from susceptible *cms-T* cytoplasm plants were inactivated by the pathotoxin from race T, whereas the mitochondria from resistant plants were unaffected. This pathotoxin plays an important part in disease damage to susceptible corn plants.

The two races differ in plant parts infected.

Fig. 3. *Helminthosporium maydis* race T ear infection.

Race O usually infects only the leaf blade, where it produces lesions or dead areas that are about 1/4 in. wide and 1/2 to 3/4 in. long. The leaf veins limit the lateral spread of the lesion so that it tends to be parallel sided. The lesions are tan in color and sometimes have a pigmented border. Race T attacks the leaf, leaf sheath, husk (Fig. 2), shank, ear (Fig. 3), and sometimes the stalk tissue of susceptible plants. The fungus can infect the seed and from there cause seedling blight. The fungus can be carried on the seed but is not seed transmitted in the sense that infected plants with leaf infection arise from infected seed. The lesions on susceptible corn infected with race T are spindle shaped, are about 1/2 to $1\frac{1}{2}$ in. long and about 1/4 to 1/2 in. wide, and usually have a distinct yellowish-green (chlorotic) border. Later the lesions often have a dark reddish-brown border. Rather commonly the lesions merge, and the entire leaf dies a few days after the individual lesions form. Similar infections occur on the leaf sheath and husk. Early infections on the husk can penetrate the various husk layers and infect the side of the ear. Ear rot is more frequently initiated through the silk end penetrating to the ear tip.

Race T has a higher reproductive rate than has race O. Consequently race T can spread rapidly in the field when environmental conditions are favorable.

In past years, race O has been limited by climate and temperature to the warmer part of the United States. Only rarely have temperatures been warm enough during a growing season to enable the fungus to spread into the Corn Belt. Race T, however, does not seem to be limited to any geographic region of the United States. *See* PLANT GROWTH.

Crosses made in the laboratory between Illinois isolates of race O and rate T show that the specific pathotoxin production and specific pathogenicity of race T to plants having *cms-T* cytoplasm are inherited as a single genetic factor. Other genetic factors determine the amount of pathotoxin produced and the degree of pathogenicity expressed.

The origin of race T is not known. Although first identified from infected corn plants in Illinois, there is ample evidence to show that it was widely distributed in 1969 through the central portion of the eastern half of the United States from Pennsylvania to Iowa. It is now known that races of *H. maydis* with similar specificity to *cms-T* cytoplasm have occurred on corn and other grass hosts in the United States since 1955. Similar isolates occur elsewhere in the world. In fact, the first observation that plants with *cms-T* cytoplasm were more susceptible to *H. maydis* than other plants was made in the Philippine Islands in 1961.

Race T was identified in Florida during the winter and early spring of 1970, causing losses in epiphytotic proportion in this area. Spores from these fields apparently were carried from field to field northward by winds and were first deposited throughout the Southern states. Here local weather conditions were favorable for disease development. The fungus then spread northward up the Mississippi River and surrounding areas on into Wisconsin and part of Minnesota. Losses were most severe in southern and western Illinois and eastern Iowa. Considerable damage also occurred in portions of states near the Ohio River Valley. Another northward path of the fungus moved along the eastern coastal areas into Pennsylvania and the New England states.

Disease resistance. Disease resistance is the most feasible control for Southern corn blight. In areas where both races occur, resistance to both races is needed. In a major portion of the Corn Belt and northward where race O had not been known to cause disease in past years, resistance to only race T is needed. If new races appear in the future, resistance will be needed in those areas where they are adapted.

Resistance to the two races has been studied at the Illinois Agricultural Experiment Station.

Resistance to race O is based on genetic factors in the cell nucleus. Plant reaction is quantitative in expression. In the field, plants highly resistant to race O show few, if any, lesions. Plants that are highly susceptible have many lesions on them and may die prematurely. A continuous range in plant reaction between these two extremes is seen. Genetic studies involving segregating populations obtained from crosses of resistant and susceptible inbred lines have shown that several genetic factors are involved in resistance. This type of inheritance is known as polygenic, meaning many genes. Statistical studies of the data show that most of the genes act in an additive manner, with little evidence for dominance or gene interaction.

In Nigeria a source of resistance to *H. maydis* has been identified that produces chlorotic lesions when plants are infected by the fungus. The fungus produces few spores on these lesions in contrast to many spores produced on lesions of susceptible plants. Genetic studies in Nigeria reveal that the chlorotic-lesion form of resistance is due to the presence of two linked recessive genes in the plant. Although unadapted to the Corn Belt, this form of resistance has been transferred by breeding to more adapted types. In Illinois this source of resistance is highly resistant to race O, and segregates with the chlorotic-lesion type of resistance have been observed in greenhouse studies. Investigations now under way will determine the usefulness of this form of resistance to strains of *H. maydis* in the United States.

Resistance to race T, on the other hand, is based both on factors in the cytoplasm and genetic factors in the nucleus. The factors in the cytoplasm are the most important and override the factors in the nucleus. Many sources of normal (not male-sterile) cytoplasm give high resistance to race T (Fig. 4). So far all cytoplasms for male sterility proven to be different from *cms-T* react the same as normal cytoplasm in being highly resistant to race T. Several of these, belonging mostly to the *cms-C* and *cms-S* group of cytoplasms, were released to the public by the Illinois Agricultural Experiment Station and the U.S. Department of Agriculture.

Nuclear factors for resistance to race T are expressed when they are acting within *cms-T* cytoplasm. In all probability the same nuclear genes for resistance to race O give partial resistance to race T. The resistance is expressed as small restricted lesions, less leaf area blighted, less sensitivity to race T pathotoxin, and ability to yield well even though infected with race T. An extensive

Fig. 4. Resistant reaction to Southern corn blight expressed by plant with normal cytoplasm (top leaf) compared with susceptible reaction expressed by plant with *cms-T* cytoplasm for male sterility (bottom leaf).

search is being made for nuclear genes that may restore resistance to *cms-T* cytoplasm to the level of normal cytoplasm much as the nuclear fertility restoring genes restore full pollen production to plants with *cms-T* cytoplasm.

Comparison with other diseases. The history of Southern corn blight seems similar to another event in American agriculture. In the Corn Belt several oat varieties with crown rust and smut resistance derived from the oat variety Victoria were developed. Since they were also superior in yielding ability and had strong straw, they soon became widely grown. In 1945, within about 3 years after they had been released for production, a new disease known as Victoria blight and caused by *H. victoriae* Meehan and Murphy became destructive; the varieties had to be replaced. A pathotoxin was also important in disease development. In oats, however, the susceptibility was due to a dominant gene closely associated with the genes for resistance to crown rust. Varieties selected for rust resistance were also uniformly susceptible to Victoria blight. It is believed that the fungus *H. victoriae* existed for some time before the disease-susceptible oat varieties were grown.

Control. Although less effective than disease-resistant varieties, other control measures can be used. It is generally recognized that corn plants grown under stress are more damaged by leaf diseases than are more vigorously growing plants. Hence special care should be followed in the culture of the crop. Adequate soil fertility and insect and weed control are needed to be achieved. The plant population should not exceed the stress level of the hybrid. Clean plowing and good seedbed preparation are believed to be helpful. Since the disease usually becomes more destructive during the later part of the growing season, early planting of susceptible hybrids might enable them to achieve a major portion of their growth before the disease develops and hence escape the effects of disease. In 1970 early planted hybrids were frequently less damaged than later planted ones. Chemicals in the form of protectant sprays are believed to be effective in disease control if applied adequately and often enough. Not enough testing has been done to prescribe an adequate spray control program. Since repeated applications may be necessary during the growing season and since these are relatively expensive, corn growers need to carefully consider costs and potential return benefits before electing to use a fungicide spray program.

The obvious and most effective control is the use of disease-resistant hybrids. The American seed industry is a responsible one and is moving rapidly in this direction. Because additional races of *H. maydis* other than those now recognized may exist or appear in the future, continual research is needed on pathogen variation and host resistance to protect the most valuable crop in the United States from diseases of this type.

For background information see CORN; PLANT DISEASE: PLANT DISEASE CONTROL in the McGraw-Hill Encyclopedia of Science and Technology.

[ARTHUR L. HOOKER]

Bibliography: A. L. Hooker, D. R. Smith, S. M. Lim, and J. B. Beckett, *Plant Disease Rep.*, vol. 54, 1970; A. L. Hooker, D. R. Smith, S. M. Lim, and M. D. Musson, *Plant Disease Rep.*, vol. 54, 1970; D. R. Smith, A. L. Hooker, and S. M. Lim, *Plant Disease Rep.*, vol. 54, 1970; L. A. Tatum, *Science*, vol. 171, 1971; A. J. Ullstrup, *Plant Disease Rep.*, vol. 54, 1970.

Cosmic ray

Recent studies on cosmic rays have elucidated three areas. First, from cosmic ray research scientists have gained insight into the nature of the strong interactions of elementary particles above 100 GeV (10^{11} eV). Second, experiments conducted to study cosmic ray muons have disclosed a zenith angle dependence of the muon intensity. Third, studies have determined the importance and effect of interfaces such as air-water on the flux and energy of the cosmic ray–produced neutron.

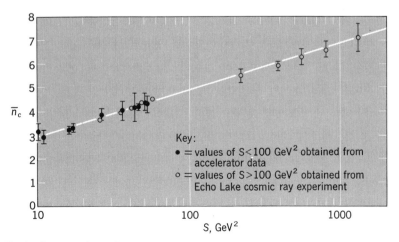

Fig. 1. Average charged prong multiplicity $\bar{n}_c$ from proton-proton collisions plotted against S, the square of the total center-of-mass energy (numerically equal to approximately twice the laboratory proton energy).

Cosmic rays above 100 GeV. Cosmic rays have always provided a window into the nature of elementary particles and their interactions at energies beyond those available in the laboratory. Now, as physicists open the energy range of laboratory experiments above 100 GeV, the knowledge of elementary particles and their interactions so far gained from cosmic ray studies may be reviewed.

Each step in man's understanding of the microscopic structure of matter has seen the study of some unit building block in nature reveal an internal structure of yet smaller, perhaps more fundamental constituents. Thus compounds were found to be composed of molecules whose constituents are atoms; each atom contains a nucleus surrounded by a cloud of electrons. The nucleus in turn is composed of nucleons, protons and neutrons, which are now found to be members of a large family of "elementary" particles. Over the past two decades physicists have identified a number of mesons and baryons and have come to understand how they are related to each other as members of several families or multiplets of elementary particles. It is only natural to wonder what the internal structure of elementary particles might be; what factors in their composition determine not only the known static properties of the many elementary particles but determine their interactions and the character of the various particle debris resulting from their collisions. See ELEMENTARY PARTICLE.

In order to discover clues to the internal structure of these particles, it is necessary to conduct experiments at higher energies than the 10–30 GeV energies at which most of the existing data have been accumulated. There are two reasons for this: First, the quantum mechanical nature of these particles dictates that the spatial resolution on the structure studied in a particular interaction is related to the momentum transferred, so that the higher the energy, the smaller the structural details which may be explored. Second, it appears that at very high energies many particles are produced per collision and that the number and distribution in momentum, angle, and particle type are probably a consequence of the basic structure of the particles.

Prior to 1971 all the data on the interactions of elementary particles above 100 GeV had come from cosmic ray research. Now the 200–500 GeV synchrotron of the National Accelerator Laboratory in Illinois and the 1600 GeV (equivalent laboratory energy) intersecting storage ring system at CERN in Geneva, Switzerland, are being completed, so that soon much more quantitative data on strong interactions at very high energies will be available. It is hence an appropriate time to review what has been learned on the nature of strong interactions above 100 GeV from cosmic rays. See PARTICLE ACCELERATOR.

Recent data. The most recent, quantitative cosmic ray data in this area is from an experiment at Echo Lake, Colo., at 3218 m elevation, carried out by a group from the universities of Michigan and Wisconsin. In this experiment a 2000-liter liquid-hydrogen target was employed to study the interactions of cosmic ray protons with the free protons of the hydrogen target. Large spark chambers above and below the target were photographed for determination of the tracks of protons entering and reaction products leaving the target. The energy of the incident proton, whether or not it interacted in the target, was totally absorbed in a 90,000-kg stack of iron plates; signals from 10 scintillation counters interspersed with the iron plates provided quantitative sampling of this energy deposition and permitted a measurement of the energy of each event.

During a 6-month period more than 50,000 protons of at least 100 GeV were detected in this apparatus and provided a data sample from which several characteristics of high-energy interactions were learned. The average multiplicity of charged particles from the proton-proton interaction is seen in Fig. 1 to increase as the logarithm of the incident energy. Power-law dependence had been the basis of earlier cosmic ray theories and was suggested by some earlier experiments with complex nuclear targets. The distribution in numbers of charged particles produced in proton-proton collisions within a particular energy interval duplicates a Poisson distribution in pairs of particles.

The angular distributions of particles as observed in this experiment and elsewhere are peaked both forward and behind the center of mass system. However, if the distribution of transverse momenta is peaked about small values and is relatively independent of longitudinal momentum (as is seen qualitatively in other cosmic ray experiments and from precise data below 30 GeV), then these angular distributions indicate a maximum in the secondary particle distributions near zero longitudinal momentum. This may be seen when the distributions of particles are plotted in the variable log tan θ (Fig. 2), where θ is the angle between the secondary particle and the incident particle direction. The variable, log tan θ, is approximately proportional to the particle center-of-mass longitudinal momentum.

Other cosmic ray experiments, including measurements in nuclear emulsions and indirect evidence from cosmic ray muons, confirm that the large majority of produced secondary particles are pi mesons, or pions. These may be produced directly, or may be the decay products of very short-lived resonant states or of more massive elementary particles.

A group from the Lebedev Institute in the Soviet Union, under the leadership of N. A. Dobrotin, has studied cosmic ray interactions at a mountain-top laboratory using a cloud chamber in a magnetic field and targets of carbon and of lithium hydride. They have made direct momentum measurements of the slower produced secondaries and have ascertained the fraction of the interaction energy carried away by gamma rays, which in turn are presumed to be from the decay of neutral pions. Their results in the same energy range generally agree with the results and interpretations of the Echo Lake experiment, although the techniques are complementary in many important respects.

An essential variable in the proton-proton interaction has been the total cross section. Indirect cosmic ray data from extensive air showers and nuclear emulsions have long indicated that the total inelastic cross section does not vary over a fac-

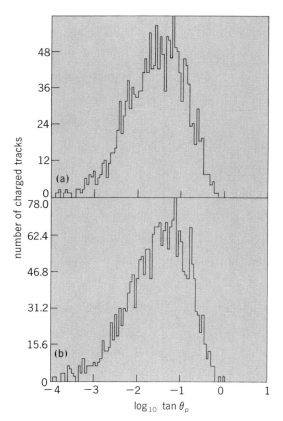

Fig. 2. The angular distributions of secondary charged particles from proton-proton collisions from the Echo Lake cosmic ray experiment for two incident energy intervals. The angles are expressed in terms of the variable $\log_{10} \tan \theta_p$, where θ_p is the projected angle relative to the incident proton direction: (a) 132 events at incident energy of 291 GeV; (b) 192 events at 203 GeV.

tor of 2 up to energies of hundreds of thousands of GeV, albeit with sizeable uncertainties. Now the Echo Lake experiment has provided direct data on proton-proton reactions, giving a nearly constant cross section up to 800 GeV. Other data from Echo Lake and from Dobrotin's group have given a constant cross section for protons on iron with much greater precision over the same energy range. It must also be noted that another Soviet group, under the direction of N. L. Grigorov, has conducted an ambitious series of satellite experiments, from which he concluded that the proton carbon inelastic cross section increases slowly from 20 to 600 GeV by 20% or more. G. Yodh has reached similar conclusions on the proton nitrogen cross section from studying the variation of measured cosmic ray flux with elevation in the Earth's atmosphere.

Models of particle interactions. The data thus far available do not solve the problems of the internal structure of the nucleon (or more generally of elementary particles). However, they do lend support to some proposed models of interaction and cast doubt on others. For example, one possible mechanism for particle interaction suggested that each particle was excited in the interaction and subsequently shook off, or radiated, mesons in its own on-going center-of-mass system. This would not account for the maximum in secondary particle production near zero longitudinal momentum in

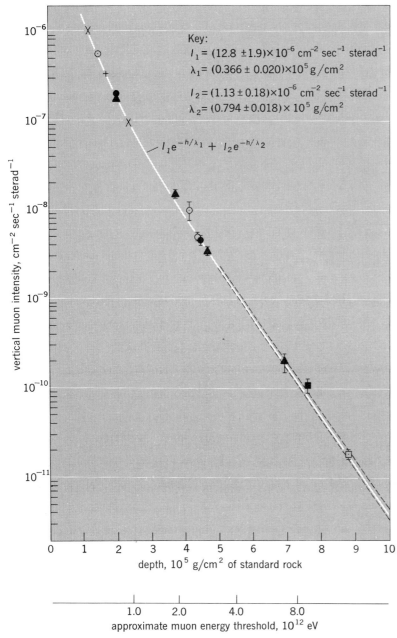

Fig. 3. The vertical cosmic ray muon intensity I as a function of depth h. Broken curves indicate the statistical error in the solid curve fitted to the data.

the overall center-of-mass system at several hundred GeV, and therefore it is probably that this mechanism is not dominant at high energy.

Another early notion developed from older cosmic ray data suggested that the collision of particles produced globules of excited matter, or "fireballs," which decayed into mesons. Again newer data, such as from the recent Echo Lake experiment, do not require fireballs for their interpretation.

On the other hand, a straightforward extension of the virtual one-particle-exchange peripheral mechanism (which has fitted data well at 2–10 GeV) to a more complex virtual particle exchange does seem to give a reasonable agreement with the cosmic ray data above 100 GeV. This multi-peripheral model suggests that there may be a chain of virtual exchanged mesons, with other mesons being produced or radiated at each vertex along the chain. In this model, as the energy increases more links are produced, so that the radiated particles from internal links have energy distributions nearly independent of the total energy. This then results in the logarithmic increase in average multiplicity with energy, as observed experimentally. The subenergies of the produced particles and momentum transfers in each exchange link are taken from parameters which fit lower energy data.

A more statistical, or thermodynamic, model, proposed by Rolf Hagedorn, can also reproduce the broad features of the data. Here it is considered that in a collision the overlapping regions of hadronic matter are heated to a temperature equivalent to about 160 MeV and that produced particles are boiled off much as photons are radiated from a blackbody.

Richard P. Feynman has proposed that the interacting protons may be composed of small pointlike components, or "partons," as has been suggested by high-energy inelastic electron-proton scattering. The produced secondary mesons from a proton-proton collision could then be understood as the appropriate combinations of partons. This model also gives good qualitative agreement with the data.

These successful models all support the notion that the general character of very-high-energy collisions may be understood on the basis of a multiplicity of low-energy reactions.

The present data and the refinement of the theories are still insufficient to decide among the models; perhaps there are elements of truth in several of the theories. The coming experiments at the new accelerators should greatly help in clarifying this picture. Meanwhile, cosmic ray data at even higher energies continue to hint at new frontiers in knowledge—or ignorance—of fundamental particles. Some experiments suggest that the average transverse momentum may increase slowly with energy. Intermediate vector bosons, magnetic monopoles, or quarks might be produced at very high energies (although in the case of quarks, for every experiment reporting their existence there seem to be two with negative results). It seems that nature has an inexhaustible store of surprises and mysteries and that new frontiers always appear just as the older ones yield to conquest. In this quest cosmic rays continue to provide a glimpse into the future for the particle physicist. [LAWRENCE W. JONES]

Cosmic ray muons. Muons have two distinctive properties which result in their peculiar value as cosmic ray research tools. First, they exhibit remarkably little interaction with other matter. They find atomic nuclei quite transparent, interacting only by virtue of the fact that they are charged. Even their electromagnetic interactions are of less importance than for electrons because of the muon's greater mass. Cosmic ray muons are consequently capable of penetrating to great depths underground. The final depth reached by a given muon depends upon the energy of its cosmic ray parent, and it also "remembers" the direction of the parent. Deep underground experiments may therefore be performed to study the properties of rare high-energy cosmic ray events, with the overburden automatically eliminating low-energy events which are thousands of times more common at sea level. Such experiments also provide an interesting "bootstrap," since most of the information about muon interactions at high energies has come from these experiments. Second, the comparatively short lifetime of muons (2.2 μsec) ensures that their point of origin was near the Earth, even for the most energetic particles studied. This means that they originate in the atmosphere and that the production processes can be observed in detail. The results so far have disclosed a zenith angle dependence of the muon intensity which cannot be understood in terms of known processes.

Vertical muon intensity. Experimental measurements of the vertical cosmic ray muon intensity at depths greater than 1.5×10^5 g/cm² are shown in Fig. 3. (In ordinary rock 10^5 g/cm² is about 420 m.) Some early data of questionable dependability have been deleted, and two new deep points have been added. As indicated by the energy scale, most of the muons observed at the extreme depths originally had energies of about 10^{13} eV. The particular form of the curve fitted through the data is not significant; any of several smooth four-parameter functions fit just as well. However, it is noteworthy that at depths greater than about 4×10^5 g/cm² the second term in the expression for the curve dominates and the behavior of the intensity is that of a pure exponential.

On the basis of other cosmic ray experiments, it is thought that the intensity of primary cosmic rays is proportional to a power of the energy ($I \propto E^{-\gamma}$). Moreover, the average rate of energy loss in matter for a muon with energy E should be given by Eq. (1), where a and b are only mildly energy dependent.

$$dE/dh = a + bE \qquad (1)$$

Electromagnetic processes are responsible for a and most of b, and about 10% of b is presumably due to nuclear interactions. When the power law spectrum and this form for the average energy loss are assumed, together with fluctuations in the rate of energy loss, the observed exponential behavior at great depths may be deduced. The data thus demonstrate that muon interactions show no drastic departure from expectation at energies up to about 10^{13} eV. At the same time, it must be pointed out that such a compressed semilogarithmic plot is misleading: If the muon nuclear interaction cross section were different by a factor of two or so or if the new production process suggested by University of Utah experiments exists, the departure of the data from theory would not be especially evident.

Zenith angle dependence. According to the accepted theory, almost all cosmic ray muons are the decay products of π- and K-mesons. These mesons are known to be copiously produced when primary cosmic rays (protons or occasionally heavier nuclei) strike atomic nuclei in the atmosphere. The mesons normally undergo further nuclear collisions themselves, but decay with some small probability (a few percent at 10^{12} eV). If the particles are moving at large angles with respect to the vertical, they spend a longer time in rarefied atmosphere and decay is more probable, compared to interaction, than if they are heading straight down. It is not difficult to demonstrate that the intensity becomes proportional to the secant of the zenith angle at high energies.

Several years ago, experimenters at the University of Utah reported measurements which indicated that the intensity did not rise as steeply with zenith angle as would be expected from this model. They recently published a new analysis based upon about 10 times more data; the main result is greatly strengthened evidence for the anomaly. At any given slant depth it is possible to represent adequately the intensity as a function of angle as in Eq. (2). Here $G(h, \theta)$ is the ratio of slant to vertical

$$I(h,\theta) = I_{\pi K}(h) G(h,\theta) + I_X(h) \qquad (2)$$

intensity as predicted by the theory discussed above and differs but little from sec θ. At each of several independent depth cuts between 3×10^5 and 5×10^5 g/cm², they found the isotropic component to be statistically different from zero by more than three standard deviations.

An example of their data is shown in Fig. 4. The Utah data are represented by the solid points, and the vertical intensity from Fig. 3 is indicated by the bar at sec $\theta = 1$. Relative normalization was carried out by making independent vertical intensity measurements at the Utah site. The broken curve is the prediction based upon the vertical intensity and the π-K decay theory discussed above, while the solid curve is a best fit obtained by admitting the isotropic component. As a function of depth, this component appears to rise to a maximum contribution of about 0.5 of the vertical intensity at 4×10^5 g/cm² and thereafter decreases in importance.

So far, independent experiments have not convincingly corroborated or disproved these results.

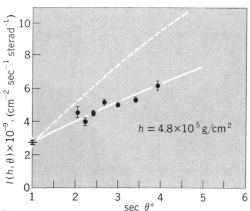

Key:

--- prediction based on the vertical intensity and the πK theory

—— best fit obtained by admitting the isotropic component

Fig. 4. The cosmic ray muon intensity as a function of zenith angle θ^* at fixed slant depth $h = 4.8 \times 10^5$ g/cm², as measured by the Utah group. Data within $\pm 0.4 \times 10^5$ g/cm² of the central depth have been combined. An isotropic component equal to 0.61 ± 0.15 of the vertical intensity was obtained when fits were made to the Utah data above, and 0.50 ± 0.05 was obtained when the vertical intensity from Fig. 3 was also used.

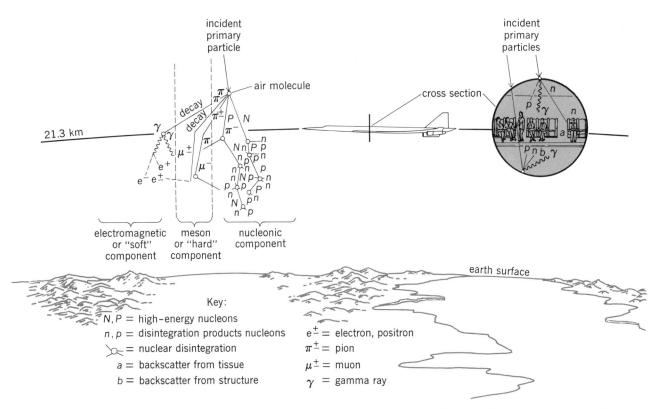

Fig. 5. Pictorial representation of cosmic ray interaction with the atmosphere at supersonic transport (SST) levels. (C. A. Randall, Jr., ed., Extra-Terrestrial Matter, Northern Illinois University Press, 1969)

Theoretical attempts to understand them have been equally unsatisfactory. The most credible theories have supposed the existence of very heavy ($m_x \simeq 45 \times 10^9$ eV) particles which decay quickly to muons. Much remains to be done.

Direction isotropy. Most muons observed at depths greater than 0.4×10^5 g/cm² originate from cosmic rays with energies in excess of 0.5×10^{12} eV, an energy sufficiently large that the particles have not been significantly deflected by the magnetic fields of the solar system. Careful experiments can be done at this depth to search for directional anisotropies in the cosmic ray flux. No such anisotropy exists down to the 1% level. On the other hand, an anisotropy at about the 0.1% level should exist if cosmic rays are produced in the Galaxy and are diffusing outward. Such would be the case if they originate in supernovae or pul-

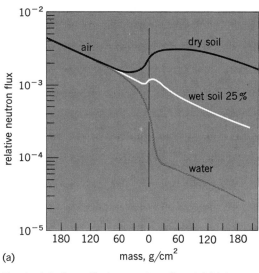
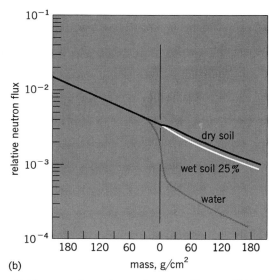

Fig. 6. Interface effect on neutron flux at (a) intermediate (0.465–0.215 eV) energy and (b) high (4.0–2.5 MeV) energy (boundary at 0 g/cm²). (Argonne National Laboratory)

sars. An important test concerning the origin of a significant fraction of the energy in the universe is thus suggested, and experiments are presently in the planning stage. [DONALD E. GROOM]

Cosmic ray neutrons. When primary high-energy galactic cosmic rays impinge on the atmosphere (or any material in space) at high energies, nucleonic cascades are formed (Fig. 5). In the first encounter, high-energy disruptions of nuclei occur, with the formation of high-energy neutrons. Later these react to produce still more neutrons through the mechanism of the evaporation stars. All the neutrons are slowed down by collision and eventually captured, in general producing new isotopic types, such as carbon-14 or tritium. Recent investigations have been concerned with the effect of interfaces such as air-soil or air-water on the flux and energy spectrum of the cosmic ray–produced neutron. Another area which has been studied most recently is the production of neutrons by primary cosmic ray particles incident on supersonic aircraft and space capsules. An important goal in such studies is the determination of the potential radiation exposure dose to astronauts and passengers when traveling at high altitudes.

Effects at interfaces. Years ago, measurements and calculations concerned with cosmic ray neutrons dealt primarily with the interactions of galactic primary charged particles with the N_2 and O_2 of the atmosphere. The variation with atmosphere depth of the number-energy distribution of the secondary neutrons was the subject of much debate and was only recently well defined. These early investigations largely ignored the interface questions. Very recently major questions have been raised and answered, such as: What happens to the neutron flux when soil is nearby as well as air? What happens over water? How is the flux and energy spectrum of neutrons changed in the vicinity of the top of the atmosphere, or at the Moon's surface, where one side of the interface is a vacuum?

A variety of detectors—liquid scintillators, proportional counters, nuclear emulsions, etched-track detectors, and thermoluminescent crystals—has been employed to measure the environmental neutron flux from thermal (room-temperature) energies to more than 10 MeV. These measurements have been carried out below ground, at ground level, on a lake surface, on the top of a ranger tower, at altitudes of 20,000 and 70,000 ft (6.1 and 21.3 km), and on an orbiting satellite. In many cases results have been compared with calculations based on the established programs for computing the slowing down of neutrons in the shielding of reactors and accelerators. Figures 6 and 7 show how strongly the interface effect depends on the energies of the neutrons. Since the abscissa is in gm cm^{-2}, the curves begin at the left at about 10,000–15,000 ft (3–4.6 km) above sea level. At thermal energies the effect is enormously magnified because the nitrogen component of the air has a large capture cross section for low-energy neutrons and thus the flux in the air at these energies is less than the flux at high energies. After the remaining neutrons enter the Earth, a large thermal contribution boils back out of the Earth. The same phenomenon ("albedo") arises at the top of the Earth's atmosphere and at the Moon's surface. Table 1 lists data developed recently by different investigators. (The data are far from precise and thus the totals appear to be less than the sums of the parts.) Note the relatively large modifications in the intensities of slow neutrons resulting from the proximity of soil or water.

Neutron exposure to SST travelers. The radiation dose absorbed by man resulting from cosmic ray neutron exposures has occupied the attention of

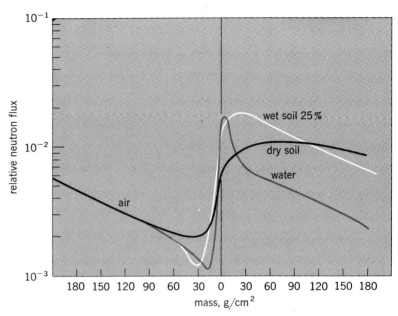

Fig. 7. Interface effect on neutron flux at near thermal energies, 0.215–0.025 eV (boundary at 0 g/cm²). (*Argonne National Laboratory*)

Table 1. Recent cosmic ray neutron flux measurements (neutron/cm²-sec)

Altitude, km	Depth, g/cm²*	Slow (<0.1 MeV)	Fast (>1 MeV)		0–20 MeV	Remarks
0.150	1000	0.002	0.0016	(1–10)		Water-air interface
0.200	1000	0.013	0.0038	(1–10)	0.008	Soil-air interface
0.235	1000	0.00033				Top of 35-m tower
4.7	600				0.32	Balloon flight
6.0	450		0.25	(1–10)		DC-6 aircraft
25	100				4.4	Ballon flight
32	9		0.15	(3–10)		Ballon flight
66	0.5				0.42	Rocket (albedo)
555	0		0.001	(40–200)		Orbiting Solar Observatory

*Approximate "depth" of the atmosphere; for example, 1000 g/cm² corresponds to about sea level.

Table 2. Cosmic ray neutron annual doses

Altitude, km	Depth, g/cm²	Slow dose, rads	Fast dose, rads	Total Dose, rads	Total Dose, rem*	Remarks
0–0.2	1000	0.6×10^{-3}	0.26×10^{-3}	0.5×10^{-3}	3.3×10^{-3}	Land-air interface
0.15	1000	0.1×10^{-3}	0.1×10^{-3}			Water-air interface
6	450		16×10^{-3}			DC-6 aircraft
12	190			2.6		Balloon
18	150			5		Balloon
20	125				4	Balloon (SST altitudes)†
66	0.5			36×10^{-3}		Albedo (leakage)
15,000	0			4×10^{-3}		Albedo (leakage)

*Here rem = Roentgen equivalent man = absorbed dose in "rad" × relative biological effectiveness (RBE) factor; for a 1-MeV neutron, RBE = 10; for a 1-MeV photon, RBE = 1.

†0.5 rem is the estimated annual dose to SST crew members and passengers flying 500 hr per year.

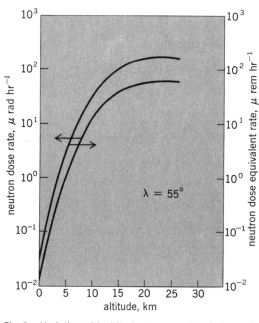

Fig. 8. Variation with altitude above sea level of cosmic ray neutron dose rates. (*U.S. Atomic Energy Commission*)

Fig. 9. Cosmic ray neutron detector package placed on a bridge to measure air-water interface effect. (*Argonne National Laboratory*)

many investigators in recent years. Table 2 summarizes the present knowledge. Although other supersonic transport (SST) problems are more serious than those of radiation exposure, a great deal of effort has been made to investigate the exposure to cosmic radiation experienced by passengers and crew at 65,000–75,000 ft (20 km). The problem is very complex due to the different dependencies in tissue response to radiation, the large variations in cosmic ray intensity, and the effects of the Earth's magnetic field. Nevertheless, a reasonable estimate is under 5 millirem for the "dose equivalent" obtained by a passenger for a single New York to London trip. For the galactic cosmic ray contribution, when one considers the longer flight time, there is essentially little difference between flying in an ordinary jet at 12 km and the SST at 18–20 km (Fig. 8). During a solar flare, of course, the SST, would presumably have to take evasive action and drop altitude.

Neutron exposure to astronauts. The natural environment of radiation in space consists of a number of components—one of which is the galactic radiation whose estimated absorbed dose rate is well below 100 millirad per 24 hr. As flight durations increase, however, monitoring of the low-dose-rate components due to cosmic rays will become more important. Furthermore, the biological effects of this radiation have hitherto not included the activation possibility resulting from the secondary neutrons which become thermalized. For example, it is now possible to estimate the potential activation of the Apollo lunar astronauts.

The flux and energy distribution of neutrons generated by cosmic rays in lunar material can be inferred from knowledge of neutrons generated by cosmic rays in the Earth's atmosphere and at the air-ground interface. A number of factors should enhance the neutron flux at the Moon's surface. The total flux should be greater because of the vastly higher cosmic ray intensity resulting from the lack of a shielding atmosphere and magnetic field. Also, the higher average atomic mass of lunar materials should result in a greater neutron yield and reduced neutron moderation so that more of the neutrons can leak out into space.

The thermal or slow neutrons should be relatively even greater because of the lack of nitrogen to act as a sink (for the production of C^{14}). Furthermore, the Moon's gravity should trap most thermal neutrons (especially on the night side, where the Moon's escape velocity of 2.5 km/sec is certainly greater than the mean neutron speed). Finally, the neutron decay ($\tau = 13$ min) should restrict the slow neutrons to within about one or two moon radii of the lunar surface.

The mean free path for cosmic ray protons in the Earth's atmosphere is about 145 g/cm^2. Thus the atmosphere (1 kg/cm^2) reduces the incident cosmic ray proton flux by a factor of about 10^3 (Fig. 9). One would therefore expect the thermal neutron flux at the Moon's surface to be greater than that on the Earth by the factors: 1000 (lack of atmosphere) × 2 (lack of magnetism) × 2 (atomic mass yield) × 2 (gravity turn around), that is, greater by a factor of 8000. If the hydrogen content of the lunar surface is significant, the leakage of thermals would be even greater.

The thermal neutron flux at an air-land interface (at sea level) has been measured by its $N^{14}(n,p)C^{14}$ reaction to be about 10^{-2} N/cm^2/sec. Thus at the Moon's surface it is estimated to be about 100 N/cm^2/sec for thermal neutrons.

As far as can be determined, none of the neutron leakage due to galactic protons striking the lunar surface is severe enough to produce significant dosage problems, at least as compared with the risk of solar flares. However, the thermal neutron cross section (varying as $1/V$) will be very great, because of gravity turnaround, at or near the altitude of low-flying astronauts, for example, 100-mi-high (160 km) orbits. On the cold night side, for example, the astronauts' tissue sodium could well be activated to Na^{24}. It would be very informative for the future assessment of risk to personnel on the Moon to test lunar astronauts with a whole-body counter immediately upon splashdown. *See* SPACE FLIGHT.

For background information *see* ALBEDO; COSMIC RAYS; RADIATION INJURY (BIOLOGY); SCATTERING EXPERIMENTS, NUCLEAR in the McGraw-Hill Encyclopedia of Science and Technology.

[JACOB KASTNER]

Bibliography: H. E. Bergeson et al., *Phys. Rev. Lett.*, 27:160, 1971; F. Hoyle, *Galaxies, Nuclei, and Quasars*, 1965; L. W. Jones et al., in *Proceedings of the 6th Interamerican Seminar on Cosmic Rays, La Paz, Bolivia*, 3:651, July, 1970; L. W. Jones et al., *Phys. Rev. Lett.*, 25:1679, 1970; J. Kastner et al., *IEEE Trans. Nucl. Sci.*, 17:144, 1970; M. R. Krushnaswamy et al., *Proc. Roy Soc. London Ser. A*, 323:511, 1971; D. E. Lyon in *Proceedings of the Symposium on High Energy Interactions and Multiparticle Production*, Argonne National Laboratory ANL/HEP 7107, November, 1970; D. E. Lyon et al., *Phys. Rev. Lett.*, 26:728, 1971; B. S. Meyer et al., *Phys. Rev.*, D1:2229, 1970; K. O'Brien et al., *USAEC Health and Safety Report 228*, May, 1970; C. A. Randall, Jr. (ed.), *Extra-Terrestrial Matter*, 1969; E. A. Warman (ed.), *National Symposium on Radiation in Space, Las Vegas, March, 1971*, NASA-SP Proc., in press.

Critical point

Experimental data of increasing precision, together with new analyses of older data, have in recent years confirmed the fact that the van der Waals and similar equations of state do not give a correct description of the thermodynamic properties of a fluid near its liquid-vapor critical point. The "homogeneous function," or "scaling-law," equation of state, first proposed by B. Widom, is in

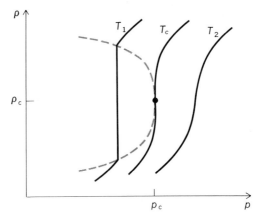

Fig. 1. Isotherms near the critical point of a pure fluid: $T_1 < T_c < T_2$.

much better agreement with experiment for fluids and ferromagnets near their Curie points.

At a first-order phase transition certain thermodynamic variables, typically density and entropy (per mole), are discontinuous; that is, they have different values in the coexisting phases. If these discontinuities go to zero continuously as a function of a suitable parameter, for example, the temperature, and if a further change in this parameter results in the complete absence of any phase transition, the point at which the first-order phase transition vanishes is called a critical point. Critical points are observed in certain fluid mixtures and in some solid-state transformations (for example, ferromagnetism) as well as in pure fluids. *See* EQUILIBRIUM, PHASE.

Critical point indices. Isotherms in the density ρ, pressure p plane in the vicinity of a typical liquid-vapor critical point in a pure fluid are shown schematically in Fig. 1. For temperatures $T < T_c$ (the subscript c denotes the value of a quantity at the critical point), each isotherm shows a discontinuity in density which occurs at the corresponding vapor pressure. The endpoints of these discontinuities form the coexistence curve, shown by a broken line in Fig. 1.

For T close to T_c, it is found that the discontinuity follows approximately a power law, varying as $(T_c - T)^\beta$ (see table).

The critical isotherm, $T = T_c$, is continuous with an infinite slope at the critical point, with the pressure difference $|p - p_c|$ varying as $|\rho - \rho_c|^\delta$ close to the critical point. The compressibility $K_T = \rho^{-1}(\partial \rho / \partial p)_T$ evaluated at the critical density diverges as $(T - T_c)^{-\gamma}$ as T approaches T_c from above. For temperatures below T_c, the compressibility evaluated separately in the liquid or vapor phase is assumed to diverge as $(T_c - T)^{-\gamma'}$ near the critical point. Recent experiments indicate that the heat capacity at constant volume, C_V, probably

Values of critical indices

Index	α	β	γ	δ
Classical theory	0	1/2	1	3
Real fluids	0.05 ± 0.05	0.355 ± 0.01	1.25 ± 0.1	4.5 ± 0.2
Ising model (3 dim.)	0.12 ± 0.02	0.31 ± 0.01	1.25 ± 0.01	5.0 ± 0.15

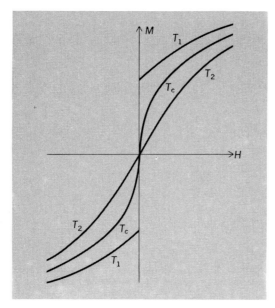

Fig. 2. Isotherms near the critical (Curie) point of a ferromagnet: $T_1 < T_c < T_2$.

diverges to infinity near the critical point. For a density equal to ρ_c, it is assumed C_V diverges as $(T-T_c)^{-\alpha}$ for $T > T_c$ and as $(T_c - T)^{-\alpha'}$ for $T > T_c$ (that is, for the heat capacity of a mixture of liquid and vapor in a container of fixed volume with average density given by ρ_c).

The isotherms for a ferromagnet in the magnetization M, field H plane for temperatures near the critical (or Curie) temperature are shown schematically in Fig. 2. It is assumed that hysteresis can be neglected and that H is the internal field, obtained by correcting the external magnetic field for sample demagnetization. The situation is analogous to that in a fluid, with M playing the role of $\rho - \rho_c$ and H the analog of p (except that the discontinuity always occurs at $H=0$ for $T < T_c$). The critical indices α, α', β, and so on are defined for a ferromagnet in complete analogy with the fluid. Thus, for example, the susceptibility $\chi = (\partial M/\partial H)_T$ evaluated at $H=0$ is assumed to diverge as $(T-T_c)^{-\gamma}$ for $T > T_c$.

Breakdown of classical theory. Experiments have shown that a number of pure fluids (for instance, CO_2, He, and Xe) have very similar values for their critical indices. These values are given in the table, together with the predictions of "classical" equations of state, such as van der Waals, Dieterici, and Redlich-Kwong. (Since it appears that α' is equal to α and γ' to γ within experimental error, α' and γ' have been omitted from the table.) The discrepancies between experiment and classical theory may appear, at first sight, to be small, even though significantly larger than the experimental error (except for α). They are, nonetheless, very significant. The underlying assumption in the classical theories is that the Helmholtz free energy may be expanded in a power series in $\rho - \rho_c$, with coefficients which are smooth function functions of the temperature. This assumption leads very naturally to $\beta = 1/2$ (that is, a parabolic coexistence curve in the ρ, T plane), $\gamma = 1$, and so on, independent of other details of the equation of state. The experimental evidence, on the other hand, indicates that a power series expansion is probably not possible and that an alternative approach is needed.

The indices obtained experimentally for ferromagnets do not differ a great deal from those of fluids, though in some cases the heat capacity seems to rise to a finite cusp, rather than diverging to infinity, at T_c. Again there is a discrepancy between experiment and the classical theories (such as the mean-field theory) which predict $\beta = 1/2$, $\gamma = 1$, and so on.

The third row of the table gives values of indices obtained by special numerical techniques for a particular statistical model, the Ising model (in three dimensions). The values are closer to experiment than is the case for classical theories, though some significant discrepancies remain.

Homogeneous or scaling functions. The basic geometrical idea of a scaling-law equation of state can be explained by reference to Fig. 1. One supposes that all the isotherms near the critical point for $T < T_c$ can be superimposed on top of each other by stretching ("scaling") the ρ and p axes by amounts which depend on the temperature, while simultaneously shifting the p scale so that the different curves coincide at $\rho = \rho_c$. Similarly the isotherms for $T > T_c$ can be superimposed on one another.

For certain technical reasons, it is customary in practice to use the chemical potential μ in place of the variable p (though the essential idea of the method remains intact if μ is replaced by p in the following formulas). The scaling hypothesis can be expressed mathematically as Eq. (1) for $T < T_c$,

$$\Delta\rho/|t|^\beta = f_-(\Delta\mu/|t|^{\beta\delta}) \qquad (1)$$

where $\Delta\rho$ is $\rho - \rho_c$; $\Delta\mu$ is $\mu - \mu_x(T)$, with $\mu_x(T)$ the value of μ for the coexisting phases at the temperature in question; t is $(T-T_c)/T_c$; and f_- is a universal function valid for all isotherms with T less than, but near to, T_c. The factors $|t|^\beta$ and $|t|^{\beta\delta}$ are "scale factors" indicating the amount by which ρ and μ axes are to be stretched for a given temperature. For temperatures $T > T_c$ one can write Eq. (2),

$$\Delta\rho/t^\beta = f_+(\Delta\mu/t^{\beta\delta}) \qquad (2)$$

where μ_x becomes the value of μ at $\rho = \rho_c$ and f_+ is again a universal function for isotherms with T greater than, but near to, T_c.

There are numerous other ways to express the functional relationship embodied in Eqs. (1) and (2). One of the most useful (for certain purposes) is the parametric form proposed by P. Schofield. The scaling-law equation for a ferromagnet is obtained by replacing $\Delta\rho$ by M and $\Delta\mu$ by H in Eqs. (1) and (2).

The scaling law is more flexible than the classical theory in that the two indices β and δ (or another pair, such as β and γ) can be chosen arbitrarily. The remaining indices are then fixed in terms of these by the relations (3).

$$\begin{aligned}\alpha &= \alpha' = 2 - \beta(1+\delta) \\ \gamma &= \gamma' = \beta(\delta - 1)\end{aligned} \qquad (3)$$

M. Vicentini-Missoni and collaborators have shown that, with suitable choices for β, δ, and the functions f_+ and f_-, the thermodynamic data for several fluids and ferromagnets in the vicinity of the critical point (for fluids this is roughly the range

$-0.01 \leq t \leq 0.03$, $|\Delta\rho| \leq 0.3\,\rho_c$) can be fitted to the functional form of Eqs. (1) and (2). In a few cases the relations (3) have been checked by an independent analysis, and the discrepancies are within, or not seriously outside, the experimental errors.

Additional topics. Discussions of the microscopic (statistical) basis of scaling laws, the scattering of light and neutrons and the behavior of correlation functions near the critical point, transport properties in the critical region, and a more extensive discussion of critical point thermodynamics will be found in the references given below.

For background information *see* EQUILIBRIUM, PHASE; THERMODYNAMIC PRINCIPLES; VAN DER WAALS EQUATION in the McGraw-Hill Encyclopedia of Science and Technology.

[ROBERT B. GRIFFITHS]

Bibliography: C. Domb and M. S. Green (eds.), *Phase Transitions and Critical Phenomena*, 1972; M. E. Fisher, *Rep. Progr. Phys.*, 30:615, 1967; P. Heller, *Rep. Progr. Phys.*, 30:731, 1967; H. E. Stanley, *Introduction to Phase Transitions and Critical Phenomena*, 1971.

Density

Significant advances have been made recently in the determination and application of the density of solutions. The rapid and convenient evaluation of the density property, in which small volumes (<1 ml) can be employed with sufficient accuracy (10^{-5} g/ml or better), will be of great importance to science and technology. If the change in density could be measured quickly, such as during the course of chemical processes, in flow systems, and as the pressure P and temperature T are varied, a unique and powerful tool would become available which could lead to new insights. In order to understand the behavior and biological role of dissolved macromolecules, such as proteins, nucleic acids, and other biopolymers (which often are obtained pure in very small amounts), the need for advanced density methods has long been evident.

With adequate methodology, however, the scope of applications becomes much broader, because the density reflects the volume V peculiar to a given chemical system of specified mass. Thus the change in volume ΔV during a reaction bears an obvious relationship to structural changes in the molecules and the forces between them; that is, the volume change from the breaking of chemical bonds and formation of new ones, or simply an alteration in the arrangement of molecules with respect to one another, hardly ever results in exactly a zero volume change. ΔV, however, is often so small that it goes unnoticed and it is seldom measured, despite the fact that it is an obligatory quantity in the overall driving force or free energy G of a reaction. (For any process at constant temperature and pressure, the change in G is given by the thermodynamic relation $\Delta G = \Delta E + P\Delta V - T\Delta S$, where E is the internal energy and S is the entropy.)

Density determinations also can be used to evaluate the partial volume change contributed by each component in a reacting mixture. This gives substantially more information on a system, because the partial specific volume $\bar{v}$ of the various components may change appreciably even though ΔV for the overall process is very small or zero. (A component is a defined material which can be added to a mixture independently.) The partial specific volume of a component i in a solution is defined as the change in total volume in milliliters on varying the mass in grams g of the component while holding all other masses g_j constant at a given temperature and pressure. Thus v_i is represented by Eq. (1), whereas the total volume V of any solution is the sum of $\bar{v}$ of each component times its mass in grams, or $V = \Sigma \bar{v}_i g_i$.

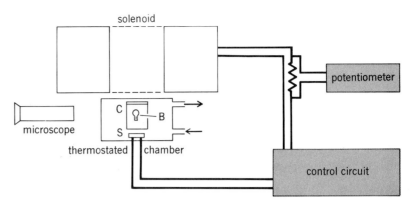

Fig. 1. Schematic diagram of the basic components for a magnetic densimeter. For this design, the buoy (B) is fashioned to sink in the liquids of interest and is pulled up and held at a precise height by the air-core solenoid above the cell (C). The height sensor for the buoy is a pickup coil (S), in this case, whose impedance changes depending on the proximity of the ferromagnetic slug within B.

$$\bar{v}_i = \left(\frac{\delta V}{\delta g_i}\right)_{T,P,g_j} \qquad (j \neq i) \qquad (1)$$

Methods. Two methods have emerged which currently hold the most promise of meeting the requirements outlined above.

Magnetic densimetry. The most tested new method, now called magnetic densimetry, which was pioneered at the University of Virginia by J. W. Beams and associates, utilizes the magnetic suspension approach. This method has continued to receive vigorous efforts in order to increase its power, convenience, and versatility. Figure 1 is a schematic diagram of the principle. A compact instrument, utilizing solid-state components and optical sensing of the height of a magnetically suspended buoy within the solution has now been developed by J. P. Senter. This model (Fig. 2) has been applied to a variety of experimental problems in biochemistry. A somewhat similar design, developed by R. Goodrich and coworkers, employs a pickup coil to sense the position of the buoy as described earlier by Beams and coworkers.

The magnetic densimeter is a device whereby a tiny (<10 mm³) ferromagnetic cylinder (such as hydrogen-annealed permalloy, about 80% nickel and 20% iron), encased in a glass jacket, is held at a precise height within a solution by virtue of a solenoid controlled by a servo system in circuit with a height sensor. The glass jacket and ferromagnetic material are called a buoy or float. The solenoid induces a magnetic moment M at the buoy which is proportional to the electric current I to the solenoid if the ferromagnetic body is truly "soft" magnetically (that is, free of hysteresis and memory). The total magnetic force on the buoy is the product of this moment and the field gradient dH/dz, where H is the magnetic intensity and z is the distance

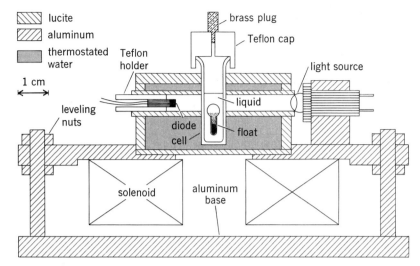

Fig. 2. Magnetic densimeter with optical sensing. (*From J. P. Senter, Magnetic densimeter utilizing optical sensing, Rev. Sci. Instr., 40:334, 1969*)

ence between the density of the buoy and the medium increases.

A complete measurement of a sample (Fig. 2) requires only 5 min, including its insertion and removal; this includes the temperature equilibration, which requires about 3 min since the sample volume is small (for this purpose, a thermostated brass block around the cell has been found more suitable). The float or buoy in the thermostated cell, containing 0.2–0.3 ml of liquid, is pulled down to a chosen height in the microscopic field by the air-core solenoid. The position of the float is maintained by automatic current adjustment to the solenoid through the amount of light reaching the photodiode receptor in circuit with the solid-state servo system. A precision resistor in series with the solenoid allows the voltage drop across the resistor to be measured with a differential voltmeter; the voltage drop is directly proportional to the solenoid current. A change of 0.1 mV corresponds to approximately 2×10^{-6} g/ml difference in the density of the liquid. A sensitivity of ± 0.01 mV may be achieved with close temperature control.

As is evident, this approach readily lends itself to a variety of experimental protocols, and the density may be determined as a function of a number of variables which were not feasible to manipulate heretofore. For example, a pressure bomb may be constructed around the cell so that the density, and hence the compressibility, is determined as the pressure is increased, as was shown by P. F. Fahey and coworkers. Also, temperature variations are easily effected because of the small volumes employed (but reference densities at these temperatures must be known); and flow systems, as well as the control of gas mixtures above the solution, can be adapted.

Where light is deleterious or must be controlled, a pair of sensing coils above and below the cell may be used to fix the position of the buoy reproducibly instead of by optical means. One of the most important advantages of magnetic densimetry is in the field of kinetics because the buoy is maintained at a fixed height in the solution while the density is changing; hence a time-chart record of the voltage can be used for kinetic analysis of various reactions.

A recent advance, due to Beams, in the magnetic suspension approach is that the change in the moment M can be canceled (by introducing two Helmholtz coils, one above and one below the cell) so that the change in magnetic force (dH/dz) on the buoy is a linear function of the current to these coils; with this modification, the same buoy can be used at any range of densities by utilizing the main solenoid at reproducible settings for the principal supporting force and the crossed coils for the differential measurements.

Another very recent advance is that remote drive coils surrounding the cell have been introduced, by Beams and M. G. Hodgins, in order to apply a uniform torque on the buoy; the rotation time (~ 1 rpm) has been found to be proportional to the viscosity with an accuracy of about 2 parts in 10^4. This new method of obtaining the viscosity circumvents the theoretical and practical problems associated with forcing liquids through small capillaries and eliminates the need for applying density and kinetic corrections. Thus both the density and viscosity of a solution can be deter-

from the center of the solenoid. The field gradient varies with z and is also proportional to the current. Thus the total magnetic force at a particular distance z in the solution which compensates for the difference in the opposing forces of gravity (downward) and buoyancy (upward) exerted by the medium, through Archimedes' principle, is $M(dH/dz)$. The magnetic force, under proper conditions, is directly proportional to the square of the current, which can be measured very accurately. Thus $M(dH/dz) = kI^2$, where k is a constant. If the buoyant force on the buoy is sufficient to make it float on the liquids of interest, the force generated by the solenoid must be downward to add to the force of gravity, as indicated in Fig. 2, where the solenoid is below the cell containing the solution and buoy. The equation relating these forces is as in Eq. (2), where V_B is the volume of the buoy, g is

$$M\left(\frac{dH}{dz}\right) = gV_B\rho - gV_B\rho_B = gV_B(\rho - \rho_B) \quad (2)$$

the acceleration of gravity, and ρ and ρ_B are the densities of the solution and the buoy, respectively.

In some designs the solenoid is placed above the cell, in which case the buoy is fashioned to sink in the liquids to be measured as in Fig. 1 (ρ and ρ_B in the equation are then interchanged). It is convenient for many purposes, however, to have the top of the cell unobstructed. When the height of the buoy is reproduced precisely ($\pm 10^{-4}$ cm), as by the use of a microscope, the magnetic force, and hence the current, is a strict function of the density. The buoy is held stationary by the servo circuit which responds to the light sensor or to the impedance of a pickup coil to control the power to the solenoid.

By means of a precision resistor and an accurate differential voltmeter, the measurements consist simply of reading or recording the voltage, which is a quadratic function of the density. A calibrated line obtained with standard solutions of known density is prepared for a given buoy from which to calculate the densities of unknown solutions. A particular buoy can cover a range in density of about 0.03 g/ml with good accuracy; with present instruments, the sensitivity drops off as the differ-

mined concurrently on the same small sample.

Oscillator method. The second method utilizes a mechanical oscillator principle devised by O. Kratky and coworkers. This method is a more recent development and its versatility remains to be documented; however, it probably will be useful for a variety of purposes. Although the method has not received as much attention from experimenters as has the magnetic approach, under certain conditions it appears to be as convenient, accurate, and almost as rapid. By the oscillator approach, the density of a sample is related to the change in resonance frequency f of a laterally vibrating tube (made of special glass or quartz). This frequency is inversely proportional to the square root of the mass m of the tube and its contents. Thus Eq. (3) is

$$f = \frac{1}{2\pi}\sqrt{\frac{Y}{m}} \quad (3)$$

formed, where Y is the elasticity coefficient.

By calibrating the tube with media of known density at a given temperature, the density of unknown solutions may be determined if the volumes are strictly identical (an overfilling procedure is utilized to ensure a constant volume). In effect, the method resembles the classical pycnometer, except that the mass is determined more quickly by applying vibrational forces than by direct weighing. In order to calculate the density of an unknown, the square of the oscillation period T, which is the reciprocal of the resonance frequency, is related to the constants A and B obtained from the calibration with known standards. Accordingly, Eq. (4) is formed.

$$\rho = AT^2 - B \quad (4)$$

Advanced electronic techniques have been applied in order to reduce the time of vibration and to measure accurately the oscillation period T. Temperature studies appear to be feasible and a modification for varying the pressure has been reported, albeit with lower accuracy in the density values. The effect of the mechanical agitation on reactive solutions has not yet received attention. Maximum accuracy ($\sim 2 \times 10^{-6}$ g/ml) is achieved when the difference in density $\Delta\rho$ between the standard, or standards, and unknown is less than 0.05 g/ml; the relation to be employed for this purpose is Eq. (5).

$$\Delta\rho = (T_s^2 - T^2)/A \quad (5)$$

In commercial instruments, such as the one manufactured by Anton Paar, K.B., of Graz, Austria, sample volumes as small as 0.7 ml may be used. T may be measured after about 20 sec; however, several measurements are required to be certain of temperature equilibration. The total time for a complete measurement depends primarily on the filling procedure and temperature equilibration protocol used and on the time required for cleaning the tube before insertion and measurement of a known sample, which must accompany the measurement of the unknown for best accuracy.

Applications. Heretofore the principal purpose of carrying out highly precise density determinations on solutions was to evaluate the partial specific volume $\bar{v}$ of a dissolved substance. For example, the value of $\bar{v}$ of dissolved macromolecules is essential when studying their sedimentation or x-ray scattering behavior. With the advent of rapid, micro methods, a number of other applications of the density property of solutions becomes feasible. Theoretical guides have recently been outlined by D. W. Kupke for applying density (1) to the routine determination of the volume change during reactions and of the partial volume change of each component; (2) to the assessment of preferential solvation of proteins, including the solvent spaces within viruses; (3) as a unique titration method; and (4) to the determination of highly precise compositions of two- and three-component mixtures.

In addition, the density is a generally good indicator of whether a system has attained equilibrium. Future applications include automatic product monitoring, kinetic analysis, pressure-volume-temperature studies on the equation of state of liquids and gases, and perturbation-relaxation experiments. Finally, statistical thermodynamic analyses on the structure of solutions may be furthered by having available easily and accurately obtained values of the density property.

The applications of density rest on its definition, the mass per unit of volume. This is a fundamental, intensive property of any macroscopic phase. For solutions, it is convenient to define the density ρ in grams per milliliter; that is, the density is the sum of the masses in grams g of N number of components added together per milliliter of solution. Thus Eq. (6) may be formed, where the volume V is

$$\rho = \frac{\sum_{i=1}^{N} g_i}{V} \quad (6)$$

in milliliters. Since the grams per milliliter of a component i is its concentration c_i, the density of a solution becomes simply Σc_i. The volume of a solution in terms of c_i leads to the useful relation $\Sigma \bar{v}_i c_i = 1$. By determining the density of weighed amounts of solutions before and after mixing, the change in the total volume ΔV for the reaction is obtained. By obvious extensions of the principle, it is possible to determine the change in the partial volume for each component.

Some headway has already been made in relating total volume and partial volume changes to chemical events. The extensive and detailed work which is still needed with model systems to further define the chemical changes in terms of volume is now feasible through the recent advances in densimetry previously described. The partial specific volume of a component may be determined at any concentration from the derivative when the density is plotted against the concentration of the component. The derivative or tangent $(d\rho/dc_i)_m$, where m refers to constant molality of all other components, is related to $\bar{v}_i$ by Eq. (7).

$$\left(\frac{d\rho}{dc_i}\right)_m = \frac{1 - \bar{v}_i \rho}{1 - \bar{v}_i c_i} \quad (7)$$

At comparatively low concentrations, the curve is usually linear (especially for macromolecules), and $\bar{v}_i$ is obtained more simply from the average slope $(\Delta\rho/c_i)_m$ by Eq. (8), where ρ^0 is the density in

$$\bar{v}_i^0 = \frac{1}{\rho^0}\left[1 - \left(\frac{\Delta\rho}{c_i}\right)_m^0\right] \quad (8)$$

the absence of component i and superscript zero refers to values at vanishing c_i.

With the new methods available, it becomes

convenient to obtain the partial volume spectrum over the entire range of concentration when ρ is plotted against c_i. The theoretical value of such spectra cannot yet be fully assessed, but some examples include transitions in molecular conformation, binding between different molecules, self-association of macromolecules, and secondary interactions such as the amount of mutual exclusion between water and small solutes in regions around and within macromolecules. Binding or titration phenomena have been observed as a sharp break in the density-concentration curve after the component being added has fully interacted with another species within the mixture. Density has become perhaps the most convenient and accurate means of assessing preferential interaction of macromolecules for components of the solvent medium. Invariably proteins exhibit a preference for either water or some dissolved solute, such as salt, saccharides, or denaturing agents — the preference changing often with the conditions employed (namely composition, temperature, and pressure).

Hence by density it has become possible to determine the net hydration of proteins and viruses by using inert solutes, such as sucrose. For example, when a virus has been dialyzed in various sucrose-water mixtures, a constant amount of preferential hydration ξ_h (grams of water per gram virus in excess of that defined by the solvent or dialyzate composition) has been observed. This is evidence that sucrose cannot penetrate the interior spaces. When the virus loses its nucleic acid, solutes such as sucrose can then penetrate. The grams of excess water ξ_h on the virus side of the dialysis membrane can be utilized, along with other data, to give the size of the interior spaces and the outside diameter of the virus. The quantity ξ_h is obtained simply by determining the density of the macromolecule solution ρ and the dialyzate ρ^0, which values are then combined with the ones for $\bar{v}$ and c of the macromolecule, in Eq. (9),

$$\xi_h = \frac{\left(\frac{\rho - \rho^0}{c}\right) - (1 - \bar{v}\rho^0)}{1 - \bar{v}_h^0 \rho^0} \quad (9)$$

where $\bar{v}_h^0$ is the partial specific volume of water in the dialyzate obtained from density-composition tables for the particular two-component solvent medium.

If ξ_h is negative, the amount of sucrose or other diffusible component in the medium is then preferred over water, and this amount is calculated simply by substituting the partial specific volume of the former component for $\bar{v}_h^0$ in the equation. The variation of ξ_h with composition of the solvent medium leads to information on the amount of both water and the dissolved solute (or reactant) which exclude each other (relative to the composition of the pure solvent) in the presence of the macromolecule. In this way it has become possible to evaluate the simultaneous interaction of proteins with both water and dissolved solutes, such as occurs in all living tissues.

For background information see DENSITY; DENSITY MEASUREMENT in the McGraw-Hill Encyclopedia of Science and Technology.

[DONALD W. KUPKE]

Bibliography: J. W. Beams, *Rev. Sci. Instr.*, 40:167, 1969; J. W. Beams and M. G. Hodgins, *Bull. Amer. Phys. Soc.*, ser. II, 15:189, 1970; P. F. Fahey, D. W. Kupke, and J. W. Beams, *Proc. Nat. Acad. Sci. U.S.*, 63:548, 1969; R. Goodrich et al., *Anal. Biochem.*, 28:25, 1969; O. Kratky, H. Leopold, and H. Stabinger, *Z. Angew. Phys.*, 27:273, 1969; D. W. Kupke, in S. J. Leach (ed.), *Physical Principles and Techniques of Protein Chemistry*, pt. C, 1971; J. P. Senter, *Rev. Sci. Instr.*, 40:334, 1969.

Deoxyribonucleic acid (DNA)

A new mode of information transfer in biological systems, from ribonucleic acid (RNA) to deoxyribonucleic acid (DNA), appears to have been established by work done in 1970. Whereas most viruses with RNA genomes replicate their nucleic acid through an RNA intermediate, the RNA tumor viruses and their relatives appear to replicate their nucleic acid through a DNA intermediate, that is, they replicate RNA $\rightarrow$ DNA $\rightarrow$ DNA $\rightarrow$ RNA. This discovery has immediate consequences for molecular biology and for cancer research. It may have further consequences for embryology and gene therapy.

RNA tumor viruses were discovered by V. Ellerman and O. Bang in 1908 and Peyton Rous in 1911. Rous found a chicken sarcoma that could be transferred by a cell-free filtrate. The virus he isolated is called the Rous sarcoma virus and is the prototype for this class of viruses (Fig. 1). Since these early discoveries, viruses similar to Rous sarcoma virus have been found to cause leukemias and sarcomas in other birds and in mice, hamsters, cats, and dogs; related viruses cause mammary carcinomas in mice. Other related viruses have been isolated from tumors in snakes, rats, and monkeys, but have not yet been shown to cause neoplasia. Because of this widespread distribution, these viruses are considered to be important as possible etiologic agents of "spontaneous" tumors of animals and, some people think, of man.

RNA-directed DNA polymerase. Since the late 1950s these viruses have been known to differ from most other viruses in the nature of their interaction with the host cell. They do not kill infected cells, but are continually produced by multiplying cells. In 1964, as a result of experiments involving inhibitors of nucleic acid synthesis and involving nucleic acid hybridization, H. Temin proposed that the regularly inherited intracellular form of these viruses, the provirus, was DNA and that the virus replication involved RNA to DNA information transfer. This hypothesis appeared to contradict what was generally assumed to be the "central dogma of molecular biology" and was at first not taken too seriously. However, in 1970 Satoshi Mizutani and Temin and, simultaneously, David Baltimore announced the discovery in the virus particle, or virion, of RNA tumor viruses of an RNA-directed DNA polymerase. This polymerase is located in the core of the virion and transcribes the sequences of the viral RNA into double-stranded DNA. The endogenous virion polymerase system appears first to make an RNA-DNA hybrid and then to use this hybrid molecule as a template for synthesis of double-stranded DNA (Fig. 2). In addition to the polymerase, Mizutani and Temin found that the virion of Rous sarcoma virus contains a series of other enzymes involved in DNA metabolism: DNA endonuclease, DNA exonuclease, and DNA ligase. The action of these enzymes could lead to integration of the DNA copy of the viral RNA into the host cell chromosome. The virion therefore appears to have all of the machinery required for the establishment of the provirus and may not need cellular enzymes for this step.

Because of these enzymes, the virion of RNA tumor viruses may become an important reagent for use in possible genetic therapy. It is easier to prepare purified specific messenger RNAs than it is to purify genes. If such a specific messenger RNA could be placed in the virion proteins of an RNA tumor virus, the messenger RNA might be able to transfer its information into DNA and so establish a new gene in a cell to counteract or replace a bad or inactive gene.

RNA-directed DNA polymerase in virions. Similar RNA- and DNA-directed DNA polymerases have now been found in several other viruses. All of the viruses previously classified as RNA tumor viruses have this RNA-directed DNA polymerase. There is, however, one exception, RSVα. H. Hanafusa and T. Hanafusa found that this defective variant of Rous sarcoma virus does not have a DNA polymerase. This result suggests that the polymerase is essential for infection by Rous sarcoma virus. Some other viruses, for example, Visna and primate syncytium-forming virus, have also been found to have this polymerase. These viruses previously were not considered related to tumors but were classified as slow or latent viruses. The finding that they contain a DNA polymerase raises the possibility that this RNA → DNA → DNA → RNA mode of replication may be important for latent infections with nontumor viruses. Once an RNA virus inserted its genome as DNA into a cell's genome, the viral genome would not have to be expressed for its own maintenance and replication. Cellular enzymes would take care of these functions. Only when the virus was activated to reproduce its RNA would the activity of the viral genes be necessary. And, since the viral information could be transferred from DNA to RNA and then packaged in a virion, no special machinery for excision of the viral DNA would be needed to obtain infectious virus. The presence of an RNA-directed DNA polymerase in a virion may also mean that it is potentially oncogenic. To test this hypothesis, it will be necessary to investigate many different RNA viruses for DNA polymerase and then to use many different systems to test for possible oncogenicity.

Other viruses structurally similar to the RNA tumor viruses, that is, having a membrane and much RNA—the myxoviruses, paramyxoviruses, and rhabdoviruses—do not have a DNA polymerase in their virions. They have an RNA-directed RNA polymerase.

RNA-directed DNA polymerase in cells. In addition to the virions of the RNA tumor viruses, there have been many reports of RNA-directed DNA polymerase in cells. In some cases, for example, cells infected with RNA tumor viruses, the polymerase system in the cells is probably a precursor of the virion. In other cases, the cells may have contained a latent or passenger rousvirus (an RNA virus with a DNA intermediate), so that the RNA-directed DNA polymerase was viral and not cellular. In still other cases, the RNA-directed DNA polymerase activity may have been caused by a normally DNA-dependent DNA polymerase which under experimental conditions was able to use RNA as a template. This possibility was originally suggested by S. Lee-Huang and L. F. Cavalieri, who showed that *Escherichia coli* DNA polymerase could use RNA as a template.

All these possibilities make it difficult to say whether there is in uninfected cells a DNA polymerase which, in its normal functioning, uses RNA as a template. Temin proposed that such a polymerase is present in some uninfected cells and that it functions in normal differentiation—the pro-

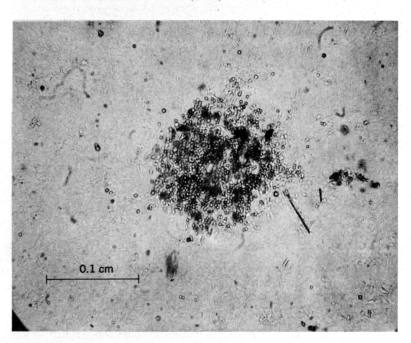

Fig. 1. A focus of chicken cells converted by infection with Rous sarcoma virus.

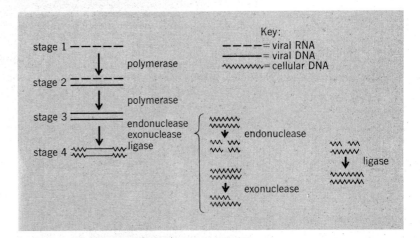

Fig. 2. Diagram of possible steps in formation of DNA provirus of rousvirus. Viral RNA (stage 1) acts as a template for the virion DNA polymerase to form a viral RNA-DNA hybrid molecule (stage 2). Hybrid (stage 2) acts as a further template for the virion DNA polymerase to form viral double-stranded DNA (stage 3). Also shown is how the virion DNA endonuclease cuts DNA in the middle; how the virion DNA exonuclease cuts DNA from an end; and how the virion polynucleotide ligase joins cut sections of DNA. Action of these three enzymes could serve to integrate viral DNA with cellular DNA (stage 4).

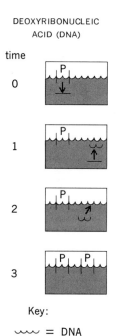

Fig. 3. Possible role of RNA-directed DNA polymerase in gene amplification.

tovirus hypothesis. Such RNA to DNA transfer of information, using such an RNA-directed DNA polymerase, could be involved in gene amplification and in differentiating the genomes of somatic cells. Figure 3 presents a model of how such a polymerase might act for gene amplification. One cell is shown at different times. Time 0, RNA is transcribed from protovirus region of DNA. Time 1, DNA is transcribed from this RNA. Time 2, this DNA is integrated into cellular DNA. Time 3, the cell has two regions of protovirus DNA. Types of transfer other than that shown in Fig. 3 could lead to duplications at other places in the cell genome and could transfer information from one cell to another. If RNA were transcribed from regions between two protoviruses, new sequences might result.

With all of these possibilities to consider, it is not clear whether this RNA-directed DNA polymerase will be useful in determining the etiology of "spontaneous" tumors. Merely finding RNA-directed DNA polymerase is not definitive proof that the tumor was caused by a virus. However, careful characterization of the polymerase activity might determine whether it was related to a virion polymerase.

The discovery of the RNA-directed DNA polymerase and consequent validation of the DNA provirus hypothesis lends support to theories of the importance in neoplasia of vertical transmission, that is, from parent to offspring, of viruses or related genetic elements. In one of these theories, the oncogene theory, R. J. Huebner suggested that all cells contain in an inactive state the information for an RNA tumor virus and that neoplasia results when this information is activated. An alternative theory, based on the protovirus hypothesis, states that germ-line cells do not contain the genes for neoplastic transformation, but that these genes are created during development by misevolution of protovirus elements.

Much more research will have to be carried out to see whether either of these theories is correct and to determine the general importance of the discovery of information transfer from RNA to DNA.

For background information see DEOXYRIBONUCLEIC ACID (DNA); RIBONUCLEIC ACID (RNA); TUMOR VIRUSES in the McGraw-Hill Encyclopedia of Science and Technology. [HOWARD M. TEMIN]

Bibliography: S. Mizutani et al., *Nature New Biol.*, 230:232, 1971; H. M. Temin, *Annu. Rev. Microbiol.*, vol. 25, 1971; H. M. Temin, *Perspect. Biol. Med.*, 14:11, 1970.

Dinosaur

Two areas of research on dinosaurs have brought forth significant new theories. First, the classification scheme for carnivorous dinosaurs has been reevaluated. This was a result of new discoveries concerning the carnivores *Deinonychus* and *Deinocheirus*. Second, new evidence suggests that brontosaurs were terrestrial and not aquatic feeders.

Carnivorous dinosaurs. Traditional classifications of the carnivorous dinosaurs (order Saurischia, suborder Theropoda) usually recognize two major kinds: large, sometimes gigantic creatures such as *Allosaurus*, *Gorgosaurus*, and *Tyrannosaurus* (infraorder Carnosauria), and the small and usually lightly built animals such as *Ornitholestes*, *Coelurus*, *Compsognathus*, and *Struthiomimus* (infraorder Coelurosauria). A variety of anatomical features have long been recognized by paleontologists as apparently diagnostic of each of these two infraorders, but size differences are the most conspicuous, as well as the most frequently applied, criteria for assignment of particular species of carnivorous dinosaurs to one or other of these two categories. Small-sized specimens invariably have been referred to the coelurosaurs, and large specimens usually have been assigned to the carnosaurs. Two recent discoveries, one in the United States and the other in Mongolia, provide new evidence that indicate this two-fold subdivision of the Theropoda may be incorrect.

Deinonychus. The American discovery, made in 1964 by a Yale University expedition to Montana, unearthed a small carnivore (now called *Deinonychus*) that in life stood 3–4 ft (0.9–1.2 m) tall, measured about 8 ft (2.4 m) in length, and probably weighed only about 150 lb (68 kg) (Fig. 1). Study of the skeletal remains of *Deinonychus* has established that this animal must have been a most extraordinary and highly specialized predator, quite unlike any previously known carnivorous dinosaur. One of the most remarkable features of *Deinonychus* was the presence of a large and very sharp, sicklelike claw on the inside toe of each hind foot (Fig. 2). This claw was more than twice as large as the other claws of the foot and was much sharper and more strongly curved. In addition, the joints of the sickle-bearing toe were unusual in that they permitted the toe, and the sickle claw, to be bent upward, raising the claw well off the ground. The joints of the other toes, as with all normal walking toes, provided very little freedom for such elevation (extension). The shape of this claw, and the unusual joints of that toe, clearly show that this structure was not adapted for ordinary walking and must have been used for killing prey. But the most extraordinary fact is that these slashing claws were on the feet—not the hands—of an obligatory bipedal animal, that is, one that could walk only on its hindlimbs. This means that *Deinonychus* must have leaped about from one foot to the other while slashing out at its prey with the free foot. This adaptation is reminiscent of the living cassowary that fights with its feet and has long and very sharp spikes on the inside toes of both feet, with which it has been said to kill men and even tigers. Analogy with the cassowary implies that *Deinonychus* was both a rapid runner and a very agile animal. *Deinonychus* may even have been able to leap at its victims slashing out simultaneously with both hind feet.

The above interpretations of *Deinonychus* are quite different from those that have been postulated for most other theropods, but credence is given to this reconstruction by a series of unique structures in the tail of *Deinonychus*. The entire tail of this animal was enclosed in bundles of long, thin, parallel, bony rods that extended from near the base of the tail to the tail tip. At any given position on the tail, there are as many as 60 to 80 bony rods along the sides, top, and bottom of the vertebrae. These rods are believed to be ossified tendons that

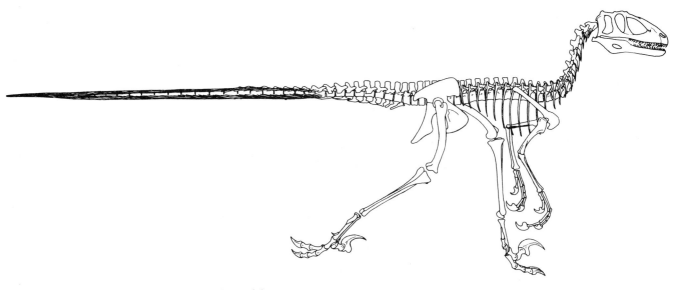

Fig. 1. Reconstruction of skeleton of *Deinonychus antirrhopus*.

originally were connected to powerful muscles at the base of the tail (Fig. 3). In fact, the preserved pattern of these ossified tendons is nearly identical to that of (unossified) tendons in the tails of modern lemurs, cats, and dogs that connect the caudal vertebrae with the powerful muscles that raise and lower the tail. Because these tendons were apparently ossified in *Deinonychus*, contraction of the attached muscles would have locked the tail vertebrae into a rigid series with a unified moment of inertia. The net result is a dynamic stabilizer or active counterbalance comparable to a tightrope walker's balancing pole.

In addition to these unusual adaptations, *Deinonychus* was also characterized by very long and powerful forearms with long, raptorial-clawed hands. These, combined with the unusual tail and specialized feet, clearly establish that *Deinonychus* must have been a very active and agile predator. Since the discovery of *Deinonychus* in the Early Cretaceous rocks of Montana, fragmentary remains of several closely related dinosaurs have been recognized in previous collections from the western United States, Canada, and Mongolia,

Fig. 2. Foot of *Deinonychus antirrhopus* with the "sickle-clawed" second toe.

Fig. 3. Section of the tail bones of *Deinonychus antirrhopus* showing the multiple ossified tendons.

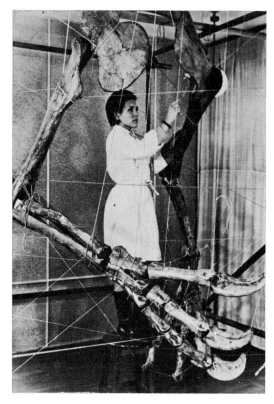

Fig. 4. Forearms of *Deinocheirus mirificus*. (Photograph courtesy of Z. Kielan-Jaworowska, Warsaw)

showing that a modest radiation of small, "sickle-footed," predaceous dinosaurs occurred during the Cretaceous, perhaps on a worldwide basis.

Because of its small size, *Deinonychus* would normally be referred to the Coelurosauria, but a number of anatomical features indicate close affinity to the large carnosaurs; its disproportionately large skull, a short neck with short cervical vertebrae, certain features of the foot, and carnosaurian hindlimb proportions. On the other hand, other anatomical characters resemble those of so-called coelurosaurs: very long forelimbs and hands, the anatomy of the hand, the tail vertebrae, and the morphology of the lower jaw. This peculiar mosaic of carnosaurian and coelurosaurian anatomy makes assignment to either category unsatisfactory. E. H. Colbert and D. A. Russell have attempted to resolve this dilemma by proposing a third infraorder of theropods—the Deinonychosauria—just for *Deinonychus* and its relatives. This is one possible solution, but there is another alternative.

Deinocheirus. The Mongolian discovery was made by Z. Kielan-Jaworowska of the Polish Academy of Sciences during an expedition to the Gobi Desert in 1965. This discovery consisted of both forelimbs and hands of an enormous carnivorous dinosaur, which was subsequently named *Deinocheirus*. Unfortunately all of the rest of the skeleton had eroded away before the discovery. But what was left was startling, to say the least. The claws on the three fingers of each hand measured up to 12 in. (30.5 cm) in length, and the total length of each forearm and hand was slightly less than 9 ft (2.7 m) (Fig. 4). On the basis of the enormous size of these forelimbs, which are almost three times larger than those of the next longest theropod forelimb, this animal has been classified as a carnosaur. Paradoxically, however, the forelimbs and hands of all known "carnosaurs" are disproportionately short, relative to other body dimensions, whereas in all "coelurosaurs" they are relatively long. The extremely short forelimbs of *Gorgosaurus*, *Tarbosaurus*, and *Tyrannosaurus* are examples. This carnosaurian trait is in sharp contrast to the greatly elongated forelimbs of *Deinocheirus*. The forelimb morphology in *Deinocheirus* —long slender humerus, radius, and ulna, long slender form of the digits, and especially the presence of three (rather than two) digits—is strikingly similar to the forelimb morphology in most coelurosaurs, despite the enormous size. Of even greater significance, however, is the construction of the wrist in *Deinocheirus*, in which all three metacarpals are of subequal length. This condition is

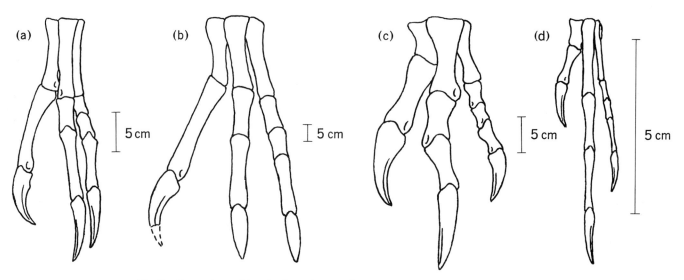

Fig. 5. Comparison of hands of carnivorous dinosaurs to illustrate the similarity of hand anatomy in (a) *Struthiomimus* and (b) *Deinocheirus* as contrasted with that of (c) *Allosaurus* and (d) *Ornitholestes*. All are drawn to unit length to eliminate differences due to size.

known in only one theropod family—the Struthiomimidae—a coelurosaurian family. All other carnivorous dinosaurs, coelurosaur and carnosaur alike, have a first metacarpal that is much shorter than the second and third metacarpal bones. This evidence, together with the overall forelimb proportions, seems to indicate a close relationship between *Deinocheirus* and *Struthiomimus*, but a struthiomimid larger than *Tyrannosaurus* is contrary to all traditional concepts of theropod classification (Fig. 5).

Deinocheirus and *Deinonychus* appear to violate some of the traditional criteria that have been used for subdivision of the carnivorous dinosaurs. In view of the "hybrid" nature of *Deinonychus* anatomy and the combination of gigantic size with coelurosaurian anatomy in *Deinocheirus*, it is possible that "carnosaurs" and "coelurosaurs" may not have existed as distinct groups of carnivorous dinosaurs. Doubt is cast on whether these man-made categories are real.

Whatever significance *Deinonychus* and *Deinocheirus* may have for the formal classification of carnivorous dinosaurs, their discoveries have revealed the former existence of two previously unknown, very peculiar, and highly specialized evolutionary lineages of dinosaurian predators.

[JOHN H. OSTROM]

Brontosaurs. The greatest dinosaurs and the largest land animals of any age were the brontosaurs, comprising the suborder Sauropoda of the order Saurischia. Traditional restorations show brontosaurs as semiaquatic herbivores, wading in swamps and feeding on soft water plants. But a biomechanical reevaluation of brontosaur skeletons demonstrates that sauropods were fully terrestrial browsers with a complex suite of adaptations for feeding high among the forest canopies.

Terrestrial adaptations. Brontosaur external nares are usually very large and high on the skull. The elevated narial position supposedly enabled the animals to breathe with only the top of the skull exposed above water. However, many terrestrial tetrapods have elevated nares—monitor lizards, ground iguanas, elephants, and several families of fossil elephantlike mammals. Moreover in truly aquatic reptiles olfaction is of little importance and the nasal capsule and external nares are small. The large nares in brontosaurs indicate the presence in life of a spacious capsule and a well-developed sense of smell, as in monitors and ground iguanas.

Brontosaurs lacked grinding or shearing molariform teeth. But their incisiform teeth were stout and often show severe wear indicative of coarse food, not water plants. Living crocodilians have powerful gastric mills with pebbles embedded in the muscular stomach walls. Gastrolith concentrations have been found within sauropod skeletons, and comminution of tough food may have occurred in a gastric mill.

In nearly all semiaquatic tetrapods, from otters to hippos, the short-limbed configuration is complemented by very broad round thoraxes. In brontosaurs and elephants, however, the thorax is deep and slab-sided anteriorly (Fig 6a).

Large, aquatic tetrapod herbivores—hippos and the extinct hippo-like rhinos—invariably have short, relatively weak limbs flexed somewhat at

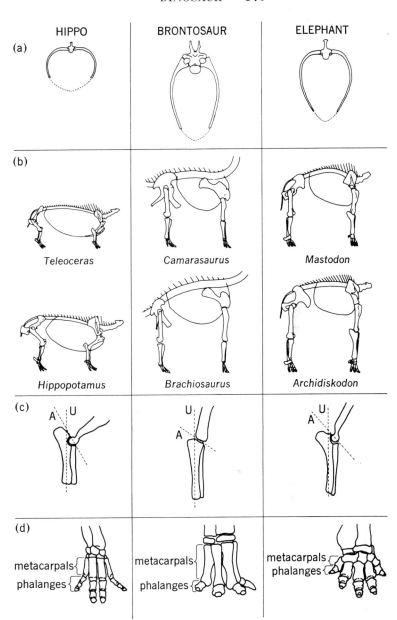

Fig. 6. Comparison of features of the hippo, brontosaur, and elephant. (a) Cross sections of thorax of hippo (*Hippopotamus*), brontosaur (*Diplodocus*), and elephant (*Loxodonta*). (b) Outlines of right side of skeletons drawn to same acetabulum-to-shoulder length of hippo-like rhino (*Teleoceras*) and hippo (*Hippopotamus*), brontosaurs (*Camarasaurus* and *Brachiosaurus*), and elephants (*Mastodon* and *Archidiskodon*). Full length of brontosaur necks are not shown. (c) Sections through elbow to show orientation of articular surfaces of hippo (*Hippopotamus*), brontosaur (*Camarasaurus*), and elephant (*Elephas*). A = axis of elbow facets on humerus; U = axis of facets on the radius and ulna. (d) Right forefeet of hippo (*Hippopotamus*), brontosaur (*Diplodocus*), and elephant (*Mastodon*).

knee and elbow (Fig. 6b). Elephant limbs are markedly different—long, columnar, with little flexure at the joints. Sauropod limbs were long and columnar and closely resembled those of elephants in reduction of joint flexure (Fig. 6c). The long, straight limbs of brontosaurs and elephants are adaptations for large body size and extensive terrestrial activity in feeding and migrating.

The toes of hippos and swamp-dwelling antelope are long and spreading to provide support on soft ground. Elephant toes are short and encased in a broad cushioning pad, and elephants can negotiate

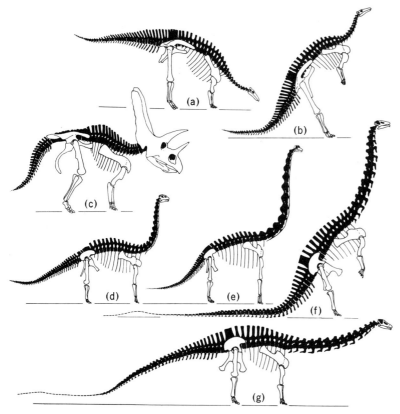

Fig. 7. Skeletons of dinosaurs drawn to the same acetabulum-to-shoulder length, with vertebral column in black. (a, b) *Stegosaurus* (bipedal and quadrupedal). (c) Horned dinosaur (*Pentaceratops*). (d) *Haplocanthosaurus*. (e) *Brachiosaurus*. (f, g) *Apatosaurus* (bipedal and quadrupedal).

marshy ground only with difficulty. Brontosaur toes were exceptionally abbreviated (Fig. 6d) and fossil footprints show the presence of a footpad like that of an elephant. Brontosaurs probably avoided soft ground much as elephants do today.

The tail of swimming reptiles is distinctive—tall and flat-sided with long neural spines throughout its length. The brontosaur tail resembled that of terrestrial lizards—the neural spines rapidly decreasing in height posteriorly (Fig. 7d–g).

Brontosaurs supposedly took to the swamps to escape predators. But the carnivorous dinosaurs were probably more agile in and around water than were the sauropods. The toes of the carnivores were long and spreading like those of the living large ground birds rheas and cassowaries. These birds move with ease over soft terrain and swim well with powerful kicks of the hindlimbs. Fossil trackways demonstrate that brontosaurs traveled in herds with the adults shielding the juveniles. On land a herd could protect itself with the lashing tails and crushing feet of the adults. In water or soft ground brontosaurs would have bogged down and have been at a serious disadvantage compared to the carnivores.

The necks of hippos and hippo-like herbivores are invariably short, adapted to the browsing and grazing of plants near ground level on the river bottoms and along the shore. Long necks are characteristic of high browsers, such as camels, giraffes, okapis, and gerenuks, which have a great vertical feeding range. Elephants have the equivalent of a long neck in their proboscis which extends their vertical feeding range to giraffe levels. Brontosaurs had extremely long necks which must have been associated with high browsing, not hippo-like feeding habits.

Two approaches to high browsing. Sauropods are most diverse in the Jurassic. In the two best known genera, *Apatosaurus* (=*Brontosaurus*) and *Diplodocus*, the tail was very long and heavy, the forelimbs were short, the number of dorsal vertebrae was reduced, the hip-to-shoulder distance was reduced relative to limb length, and the neural spines were tallest over the hips (Fig. 7f, g). Long heavy tails, shortened back regions, and neural spine patterns of this sort are characteristic of bipedal dinosaurs and large ground birds, and the short forelimbs of *Apatosaurus* and *Diplodocus* have been interpreted as evidence that these genera had bipedal ancestors. However, apatosaurs and diplodocines are not primitive sauropods; in most features of skull and vertebrae they are the most specialized brontosaur genera. In the most primitive Jurassic sauropods, *Cetiosaurus* from Europe and *Haplocanthosaurus* (Fig. 7d), the tail was relatively short and light, the back region was long, the forelimbs were not shortened, and the neural spines were low over the hips and highest over the mid-back. This pattern of neural spines is characteristic of typical quadrupedal dinosaurs, such as ceratopsians (Fig. 7c). The common ancestors for all sauropods have been identified as the Late Triassic melanorosaurs, known from South Africa and South America. A. J. Charig and J. Attridge have shown that melanorosaurs were quadrupeds and that sauropods were consistently quadrupedal all through the early stages of their evolution.

Haplocanthosaurus and *Cetiosaurus* had long necks and relatively high shoulder regions, which would have permitted them to browse high among Jurassic conifers and long-trunked cycads. *Brachiosaurus* (Fig. 7e) represents a modification of the primitive haplocanthosaur pattern—the forelimbs were lengthened, raising the shoulders above hip level, and the neck was exceptionally long. *Brachiosaurus* is often credited with being the heaviest dinosaur, but large individuals of *Apatosaurus*, *Camarasaurus*, and even *Haplocanthosaurus* were probably equally robust. In summary, the cetiosaur-haplocanthosaur-brachiosaur branch of the brontosaurs represents a basically quadrupedal lineage with a trend to increase the vertical feeding range by forelimb and neck elongation.

If sauropods were primitively quadrupedal, what is the significance of bipedal features in *Diplodocus* and *Apatosaurus*? The trend for forelimb shortening in these and related specialized brontosaurs is puzzling since it would have decreased the shoulder height and reduced the vertical reach of the long neck. Strong tails and tall sacral neural spines are present in kangaroos, which can stand upright on a tripod composed of the two hindlimbs and tail. Ground sloths and some primitive ungulates (*Barylambda*) also had stout tails and probably fed in an upright, kangaroolike posture. Probably the bipedal features in *Diplodocus* and its allies are related to a trend toward a bipedal feeding posture. The heavy tail would serve as a counterbalance and third hind leg, the neural spines over the sacrum would improve the transmission of the entire body weight to the hind limbs, and the shortening of the back and forelimbs would reduce the weight in front of the hips. By employing an upright feeding posture (Fig. 7*f, g*), *Apatosaurus* and *Diplodocus* could feed at a 50–60 ft (15.2–18.3 m) level, beyond the reach of even *Brachiosaurus*. *Camarasaurus*, the commonest sauropod of the North American Jurassic, was intermediate between primitive cetiosaur-like brontosaurs and the diplodocines and apatosaurs in structural features and was probably intermediate in feeding habits as well, employing upright feeding posture less often than in the specialized bipedal feeders. *Stegosaurus* (Fig. 7*a, b*), a large ornithischian dinosaur found in association with Jurassic sauropods, had shortened forelimbs, a stout tail, and a neural spine pattern like that of apatosaurs. Stegosaurs were probably also bipedal high browsers.

In conclusion, all brontosaurs were terrestrial high browsers, with a great vertical feeding range provided by long neck and tall shoulders in the *Haplocanthosaurus-Brachiosaurus* line, and an even greater vertical range in the *Camarasaurus-Apatosaurus* line provided by upright, bipedal posture.

For background information *see* DINOSAUR in the McGraw-Hill Encyclopedia of Science and Technology.

[ROBERT T. BAKKER]

Bibliography: R. T. Bakker, *Discovery*, 3:11, 1968; R. T. Bakker, *Evolution*, September, 1971; R. T. Bakker, *Nature*, 229:172–174, 1971; A. J. Charig, J. Attridge, and A. W. Crompton, *Proc. Linn. Soc. Lond.*, 176:197–221, 1965; E. H. Colbert and D. A. Russell, *Novitates Zool.*, 2380:1–49, 1969; H. Osmolska and E. Roniewicz, *Palaeontol. Pol.*, 21:5–19, 1970; J. H. Ostrom, *Peabody Mus. Nat. Hist. Bull.*, 30:1–165, 1969; J. H. Ostrom, *Peabody Mus Nat. Hist. Postilla*, 128:1–17, 1969.

Earth tides

In recent years, gravimeters have been used extensively to study the elastic properties of the Earth by measuring its response to the gravitational forces from the Sun and Moon. These studies have been particularly fruitful as a result of the use of the superconducting gravimeter, a new type that yields data with less error.

The response of the Earth to these tidal forces from the Sun and Moon is usually expressed as the ratio of the observed variations in the force of gravity to the variations which would be observed on a rigid Earth. This ratio can be as large as 1.3 as a consequence of several different effects. The deformation of the Earth adds a term of approximately 0.16 due to the redistribution of mass and the variation of the distance from surface to center of the Earth. The perturbation from other effects, such as the loading of the Earth by ocean tides, can add or subtract a term as large as 0.2. These perturbations, as well as instrumental noise, have placed serious limitations on the determination of the elastic properties of the Earth from earth-tide measurements. However, studies of seismic disturbances and excitations of the normal modes of the Earth have made it possible to predict the tidal deformation of the Earth with an uncertainty which is smaller than that of the direct observations. Thus sufficiently precise measurements of the gravity tides can be used to study the perturbing effects. At coastal locations, such as La Jolla, Calif., the ocean tides are dominant.

Interestingly, tides with a frequency of less than 1 cycle per day have been measured only at the South Pole by L. B. Slichter and at Strasbourg, France, by P. Lecolazet and L. Steinmetz. In general, instrumental noise increases toward zero frequency, making the observation of these small, slowly varying gravity changes extremely difficult.

Superconducting gravimeter. The superconducting gravimeter consists of a 2.54-cm-diameter superconducting sphere levitated in the magnetic field of a pair of current-carrying superconducting coils designed to produce a small force gradient in the vertical direction. The vertical position of the sphere, which depends on the vertical force on it, is detected by a capacitance displacement transducer. The same capacitor plates which detect the ball position are also used to apply a feedback force to the ball which counterbalances the variations in the force of gravity and holds it in a nearly fixed position. Thus the variations in the force on the sphere are observed by measuring the feedback voltage which is required to maintain the ball at the capacitance bridge null point.

The low-noise, low-drift feature of the instrument is a result of the inherent stability of the "persistent" currents in a superconductor. The magnets are constructed in such a way that current can be trapped in the coils by shorting them to form closed superconducting loops. The induced currents in the sphere which result in its levitation, of course, also flow entirely in the superconductor and thus will not decay. The zero drift of the instrument in its present configuration corresponds to a drift in the support-magnet current of 2.5 parts in 10^{10} per day.

The most serious disturbance to the data occurs when the liquid-helium Dewar is refilled every 6–8 days, causing large temperature and pressure fluctuations. The plot of the signal in Fig. 1 contains a typical helium-transfer disturbance. The large spike can be removed in the analysis. These spikes are sometimes accompanied by small permanent offsets, which can also be removed. However, both spikes and offsets contribute non-Gaussian noise which is responsible for at least part of the noise and drift on the data. The bandpass filter shows an oscillation with a period of 20 min following the transfer. The oscillation appears to be associated with a thermal instability of the helium liquid and gas in the Dewar. It damps out in about 8 hr and cannot be reexcited by external mechanical disturbance. The data at tidal frequencies are not noticeably affected by this oscillation. The instrument is located in the basement of the five-story physics-chemistry building at the University of California at San Diego, where ball displacements are measured at frequencies greater than 10 cycles per second, which correspond to direct-current accelerations of 1 milligal. The quality of the record under these conditions is impressive, but the building must certainly contribute noise at low frequencies also. The amount of noise contributed by the instrument itself is not known. However, better temperature control should eliminate the 20-min mode which appears after helium transfers.

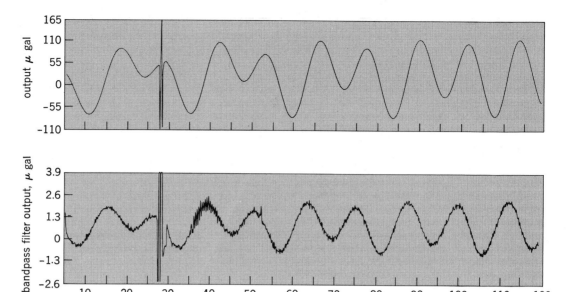

Fig. 1. A 5-day plot of the output of the gravimeter as it was stored by the data-logging system. The upper plot is the output filtered by a three-pole Butterworth filter with a gain of 7.07 and cutoff frequency of 0.18 cycle per minute. The lower plot is the output of a bandpass filter with a gain of 1 for direct current and 100 for frequencies between 1 and 30 cycles per hour. The large spike after $t = 25$ hr is due to liquid-helium refilling.

Data. The plot of Fig. 2 shows the results of a Fourier analysis of run 2 (44 days of data). Of particular interest is the 50- to 60-dB signal to noise on the principal diurnal and semidiurnal tidal contituents, prominent fortnightly and terdiurnal tides, and the low noise level at low frequency. Uncertainties in the amplitudes of the tidal lines may be estimated by assuming that the contribution to the Fourier spectrum from noise varies smoothly with frequency and may be extended monotonically through the tidal peaks. At most, the noise may change the amplitude δ of the tidal constituent X by $\pm \delta(X)/S(X)$ and the phase by $\pm 1/S(X)$ radians.

The term $S(X)$ represents the signal-to-noise ratio of the Fourier coefficient of gravity tide constituent X. The error limits of the table have been computed by this method. Systematic errors are not reflected in these uncertainties and may be present due to variations in overall gain, temperature and barometric pressure influences, time base inaccuracies, and uncertainties in the tiltmeter corrections of the instrument. The term $\delta_o(X)$ is defined as a vector corresponding to the measured tidal constituent X which has an amplitude $A(X)/B(X)$. Its phase is equal to the phase of the observed tidal constituent [of amplitude $A(X)$] minus the phase of

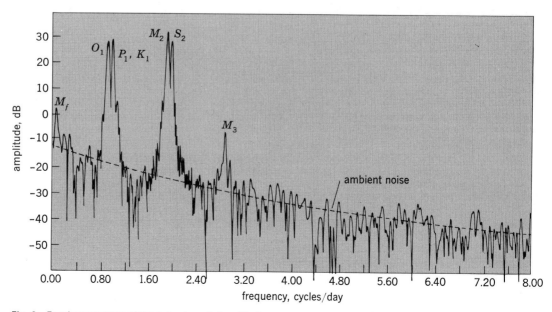

Fig. 2. Fourier spectrum of the data of run 2. Amplitude 0 dB corresponds to approximately 1.4 (microgals)², or a power spectral density of 2×10^{-3} (microgals)²/cycles per hour.

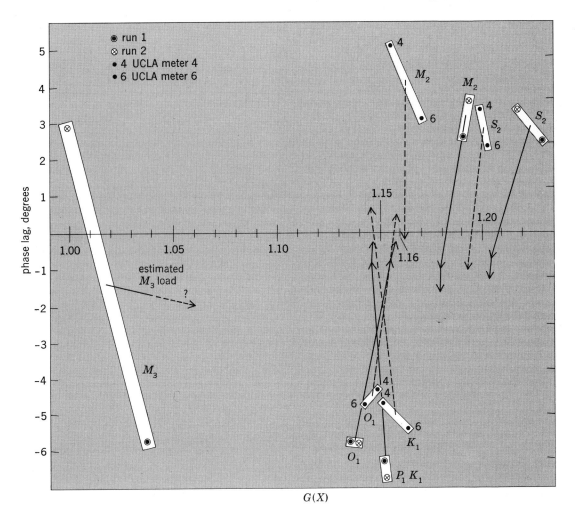

Fig. 3. Vector plot of La Jolla and UCLA gravity tide amplitudes and their load corrections. The data points represent the tip of the tidal amplitude vectors, $\delta_o(X)$. Load vectors are solid for La Jolla and broken for UCLA data. The small vector added to the La Jolla load correction represents the correction for the direct gravitational attraction of the ocean.

the rigid-Earth tidal constituent [of amplitude $B(X)$]. Using this convention, a phase lag is positive. The table is a summary of these results for the O_1, P_1K_1, M_2, S_2, M_3, and M_f tide components. Other periodic phenomena with the same frequency as the tidal constituents add vectorially to the tidal amplitudes, so a load vector $L(X)$ is defined. The phase of $L(X)$ is measured relative to the rigid-Earth tide. The corrected tidal amplitude is formed by performing the vector subtraction $\delta_c(X) = \delta_o(X) - L(X)$. To a precision of perhaps 1%, the most likely amplitudes of $\delta(X)$ are 1.16 for the M_f, O_1, M_2, and S_2 tides; 1.14 for the K_1; and 1.07 for the M_3. The Q of the lowest order Earth modes places upper limits on the contribution to the phase shift from the solid Earth of 0.2° for the semidiurnal constituents. This is much smaller than that due to the oceans and so is negligible here.

W. E. Farrell has computed the loading effect of the ocean within a radius of 2215 km from La Jolla. He used a flat-Earth approximation to the model of G. Backus and F. Gilbert which has its density and elastic parameters adjusted for the best fit to selected normal mode frequencies and seismic travel time data. The ocean tides in this region have been measured extensively at coastal and offshore (deep-sea) stations by W. Munk, F. Snodgrass, and M. Wimbush. Other gravity tide measurements have been made in southern California by Slichter at the University of California at Los Angeles (UCLA) and J. C. Harrison at Glendora.

The corrections to the UCLA data are approximated by using the La Jolla amplitude for the ocean load vectors, but with the diurnal and semidiurnal phases increased by 3 and 11°, respectively. This takes into account the phase lag in the offshore ocean tides between La Jolla and Los Angeles. The load corrections and corrected tidal amplitudes are summarized in the vector plot of Fig. 3.

In the absence of an absolute amplitude calibration, it was decided to normalize the tidal amplitudes so that $[|\delta_c(K_1)| + |\delta_c(O_1)|]/2$ is equal to the corresponding quantity computed from the UCLA data. The diurnal constituents were chosen to determine the normalization rather than semidiurnal constituents because of the smaller scatter in the data at each station and the greater simplicity of the K_1 offshore ocean tide distribution.

From Fig. 3, the difference in the ocean-load-corrected O_1 and K_1 amplitudes, $|\delta_c(O_1)| - |\delta_c(K_1)|$,

Tidal amplitudes and their phase lags measured at La Jolla (lat. 32.87°N, long. 117.27°W). Run 1 began Nov. 1, 1969, and lasted 52 days; run 2 began Feb. 5, 1970, and lasted 44 days

Tide*	Run 1		Run 2	
	$|\delta_o(X)|$	Phase lag, degrees	$|\delta_o(X)|$	Phase lag, degrees
O_1	1.1407 ± 0.004	-5.00 ± 0.23	1.1427 ± 0.004	-5.02 ± 0.23
P_1, K_1	1.1576 ± 0.003	-5.41 ± 0.18	1.1590 ± 0.003	-5.74 ± 0.18
M_2	1.1920 ± 0.0016	2.30 ± 0.04	1.1955 ± 0.0016	2.98 ± 0.04
S_2	1.2311 ± 0.0026	1.98 ± 0.26	1.2191 ± 0.0020	2.72 ± 0.26
M_3	1.042 ± 0.12	-5.49 ± 4.5	1.003 ± 0.12	2.90 ± 4.5
M_3^1	1.041 ± 0.19	-0.98 ± 7	1.097 ± 0.19	1.95 ± 7
M_3^2	1.150 ± 0.19	-9.41 ± 7	1.075 ± 0.19	9.00 ± 7
M_3^3	0.890 ± 0.19	2.16 ± 7	0.974 ± 0.19	6.79 ± 7
M_3^4	1.081 ± 0.19	-1.42 ± 7	—	—
M_f	1.42 ± 0.25	13.5 ± 10	1.24 ± 0.22	-13.8 ± 10

*M_3^n is the M_3 tide computed by dividing runs 1 and 2 into four and three parts, respectively, and Fourier analyzing each segment separately. Data are normalized so that average of amplitudes of O_1 and P_1K_1 tides agree with corresponding values from UCLA after ocean load corrections are performed.

is 0.013 for the UCLA data and 0.011 for the La Jolla data. This is to be compared to the value of 0.022 predicted by M. S. Molodenski by considering the dynamical effect of the liquid core. The corrected M_2 and S_2 tidal amplitudes disagree with those measured at UCLA by as much as 2%. Of course, if the normalization was chosen to make the M_2 amplitudes agree, the O_1 and K_1 tides would then disagree. In any event, the corrected La Jolla tidal amplitudes are not consistent with the predictions of the Earth models. The discussion which follows is an an attempt to identify possible sources of this disagreement.

Sources of data inconsistencies. The accuracy of the ocean load calculation by Farrell is determined by the accuracy to which the offshore ocean tides are known and the suitability of the Green's function (response to a delta-function pressure source) which is convolved with the ocean tide to produce the ocean load vector. Farrell's load computations considered only the ocean within 2215 km of La Jolla. Just beyond this distance, the ocean acts to decrease the M_2 load and increase the K_1 load. Futhermore, the tides of the world's oceans produce a global distortion which could amount to a significant fraction of the total effect. Contributions to the La Jolla load from the ocean at given radii add nearly in phase for both the K_1 and M_2 tides. Thus, for the offshore ocean tide amplitudes used by Farrell, the phases of the load vectors are only weakly dependent on the Green's function used to compute them. However, the M_2 ocean tide contains an amphidrome about 2000 km offshore, which results in cancellation of the load due to water on either side of it. The contribution to the load at La Jolla from water in the vicinity of the amphidrome is significant, so that the phase of the M_2 load is probably sensitive to its location and to the position of the station along the coastline. Agreement would be achieved with the UCLA data if the lag in the M_2 load there was increased by 10°. A proper convolution for gravity stations in Los Angeles would clarify this question.

The S_2 tide occurs at a frequency of twice the solar day. Most of the periodic barometric pressure and external temperature variation occurs at this frequency and influences the S_2 gravity tide at the 1% level. The consistent relationship between M_2 and S_2 amplitudes measured at UCLA and La Jolla 7 years apart indicates that perturbations of the S_2 tide are fairly constant and that it should be possible to identify the source of those perturbations and perform accurate corrections for them.

The amplitude of the M_f tide varies considerably between run 1 and run 2 but remains within the limits set by the background noise and uncertainty in temperature correction. The phases disagree by an excessive amount, possibly due to the large number of offsets during run 1. The amplitude dependence of the M_f tide on latitude (λ) of the station is $\sin^2 \lambda - 1/3$. La Jolla is near to the zero of this factor. Its amplitude is 17 times smaller than at the pole and 5 times smaller than at a northern United States location, such as Seattle. One expects both the local and global M_f ocean tide to have a larger relative effect near to the node ($\lambda = 35°16'$), but the error limits of the data reported here are too large to make a definite statement about the amplitude and phase of this quantity. If the M_f ocean tide could be computed for the world's oceans, an accurate measurement of the M_f gravity tide could provide a relatively direct test of a global ocean load theory, as well as a test of a low-frequency ocean tide theory.

The M_3 tide amplitudes, computed by dividing the runs into 2-week intervals, vary considerably (see table). The phase shifts for the complete analyses of runs 1 and 2 are of opposite sign, although they agree with one another within the relatively coarse error limits of the data. The M_3 load vector shown in Fig. 3 was crudely estimated by using the phase of the M_3 ocean tide at La Jolla. Because of the probable complexity of the offshore M_3 ocean tide, this is a very crude estimate. The M_3 ocean tide is not well enough known to make a more accurate determination, but it can be said that the direction of the estimated load vector is consistent with the measured M_3 gravity tide.

Conclusions. The results of the ocean load corrections of gravity tide data taken at La Jolla by W. A. Prothero, Jr., and at UCLA by Slichter indicate that meaningful restrictions on acceptable deep ocean tide amplitudes can be made. Accurate gravity tide measurements at coastal and island stations around the Pacific Ocean could provide boundary conditions for theories of the ocean tides in much the same way that observed frequencies of the normal modes of the Earth place discrete boundary conditions on acceptable Earth models, as suggested by J. T. Kuo and Farrell. Gravity tide measurements "feel" the ocean at great distances in contrast to coastal ocean tide measurements, which measure the ocean height at a point. The expense and effort in making land-based gravity measurements is much less than that involved in deep-sea work, and this approach to the deep ocean tide problem could be extremely fruitful.

The high precision with which the superconducting gravimeter is able to measure the gravity tides can be combined with the increased understanding of the solid Earth to study the offshore ocean tides and other "indirect" tidal effects. Further improvement in the instrument could result in the continuous observation of tectonic uplift, very long period tides, seasonal gravity changes, and other low-level gravity changes and accelerations for direct-current to seismic frequencies. *See* TECTONOPHYSICS.

For background information *see* EARTH; EARTH TIDES; TERRESTRIAL GRAVITATION in the McGraw-Hill Encyclopedia of Science and Technology.

[WILLIAM A. PROTHERO, JR.]

Bibliography: W. E. Farrell, *Gravity Tides*, thesis, University of California at San Diego, 1970; P. Melchior, in R. Dejaiffe (ed.), *6th Symp. Int. Marees Terr.*, A(9):10 and 20, Strasbourg, Sept. 15–16, 1969; W. Munk et al., *Geophys. Fluid Dyn.*, 1:161–235, 1970; W. A. Prothero and J. M. Goodkind, *Rev. Sci. Instr.*, 39:1257–1262, 1968.

Earth-resource satellites

A national experimental program is being implemented by the United States to provide for Earth resources surveys and environmental monitoring. The satellites constitute an essential part of this experiment. The other major portions of the program include measurements made from aircraft, ground investigations, the development of data-analysis methods, and techniques for using information extracted from data. The objective is to make it possible to take actions that will be beneficial to society. The National Aeronautics and Space Administration (NASA) is responsible for obtaining requirements and providing for the instrument development, spacecraft, major portions of the aircraft capability, and data-processing system. Other agencies of the Federal government participating in the program are the Department of Agriculture, the Department of Commerce, the Department of the Interior, the Navy, and the Army Corps of Engineers. Many other agencies have potential interest in the program and are expected to take a more active role in the future. The interest of all these agencies is to make use of data that can now be obtained rapidly and economically to improve the present methods of resource and environmental evaluation and to provide for new functions that cannot be accomplished with present techniques. *See* ENVIRONMENTAL ENGINEERING.

Science of remote sensing. Remote sensing can be defined as the science of determining properties of materials without coming into physical contact with the object being observed. The forms of remote sensing of immediate concern to the Earth resources surveys are measurements within the electromagnetic spectrum that allow the objects to be identified.

Visual observation. The most familiar form of remote sensing is visual observation. The human eye is capable of observing colors and distributions of color and interpreting this information to reach conclusions such as what is being observed and its condition. The human eye is very limited in the portion of the electromagnetic spectrum where it can detect energy. The human eye is not capable of seeing reflected or emitted energy in the ultraviolet and at shorter wavelengths or of seeing reflected or emitted energy in the infrared or microwave portions of the electromagnetic spectrum. Instruments have been developed that are capable of detecting these types of energy and of measuring the distribution of each color over the field of view that can be observed by the instrument. These devices provide a capability for extending visual observations, documenting observations for future reference, and providing quantitative data for the full scene rather than qualitative data that are observed by the human eye.

Camera. The camera is a very important remote-sensing instrument. Most photographic methods record data in the portion of the spectrum where the human eye is sensitive. This coincidence is not always necessary. The development of camouflage-detection film extended the range of the camera into the near infrared. This extension in spectral range made it possible to separate healthy vegetation from vegetation that was under stress or from artificial materials that were indistinguishable in the visual range. The reason this is possible is that healthy vegetation reflects a larger amount of near-infrared energy than unhealthy vegetation or artificial materials.

Radiometers. Other detectors have been built that extend the spectral range beyond what is practical for cameras and photographic film. These sensors fall in the general class of instruments known as radiometers. A radiometer is an instrument that has the capability of measuring the reflected or emitted radiation from an object. This radiation may be measured in a single wavelength, or it may be divided into several color or wavelength bands and each band measured independently. In this way it is possible to measure the amount of blue, green, red, near-infrared, thermal infrared, or heat energy and even the amount of radio-wave energy coming from an object. A radiometer can be made to scan a scene in a similar way that a television receiver scans the image of an object being observed. The radiometer can then separate the spectral components for each element (spot) of the scene. It is then possible to reconstruct the scene using any portions of the spectrum that are desired. It is also possible to use electronic computing methods to determine what the objects in the scene represent and their condition if their spectral signatures are known or parts of the scene can be positively identified and used as a reference. Scanning radiometers are commonly called multispectral scanners.

Previous experience. Methods of remote sensing have been developed that use aircraft and even balloons to carry the instruments. Aerial photographic surveys have become very important in providing the basic information needed to construct maps, measure crop acreage, determine terrain types, interpret geologic features, and provide a wide variety of other information needs. Most aerial photography is obtained on black-and-white film. More recent experience using color aerial photography has demonstrated that the color of the scene, as well as its shape, can be used in the interpretation. In addition, radiometers that operate in the infrared portion of the spectrum are used for a number of Earth resources applications. These sensors measure surface temperature changes. These types of measurements are important in agriculture, hydrology, pollution investigations, and determining the dynamics of large water bodies. More recently, experimental multispectral scanners have been used in aircraft surveys.

Microwave sensors have been used on aircraft. These sensors may be either active or passive. The active microwave systems operate very similarly to conventional radar. A signal is sent out from a source and is reflected back to the receiver. Passive systems depend on microwave energy from sources that are independent from the receiver. The other sources may be unrelated to the object being observed, in which case the receiver measures the reflected energy, or the object being observed may be the source, in which case the receiver detects the emitted energy.

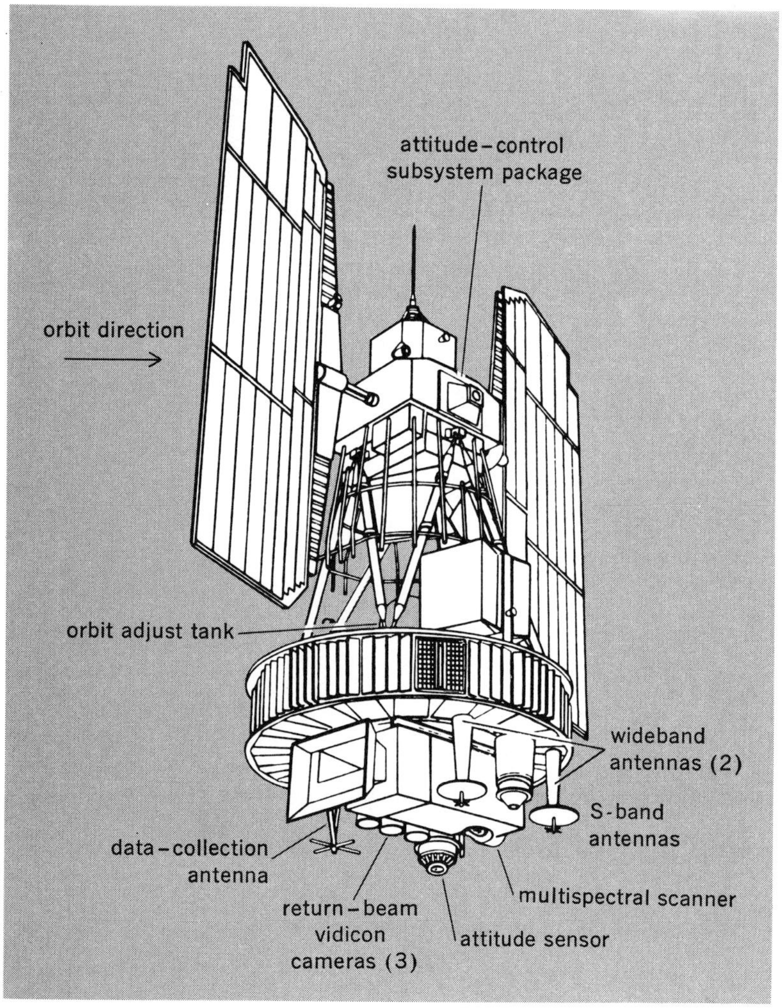

The meteorological satellites have demonstrated the feasibility of using space platforms for remote sensing. The experimental meteorological satellites have provided the basis for operational systems of meteorological satellites that now constitute a very important portion of the world's weather prediction capability. Similar technology can be applied to viewing the surface of the Earth. The technology has to be advanced for this purpose. It is necessary to observe smaller objects on the surface and to separate the electromagnetic spectrum into smaller units (narrower color bands) to do effective Earth resources surveys. The basic concept, however, of repetitive coverage in observation of changing and even static phenomena on the surface of the Earth makes it possible to extend remote sensing capabilities so that large areas, possibly global, can be observed on a repetitive basis. The solution to major resource problems of the future may well depend on such observations.

Experimental data have been obtained from the manned space flights conducted by the United States. The Gemini and Apollo programs have been of special importance in providing these data. Photographs taken during these space flights have been used extensively by resource scientists to determine their utility and value. The unique characteristics that can be obtained from space are: The whole scene of the camera covers a very large region and the observations are made under uniform lighting conditions; the angle from the vertical needed to observe a large scene is small, so that distortions due to surface elevation changes are minimal; the spacecraft moves at a very rapid rate, so that adjacent pictures along the orbit track have almost the same illumination, and by selection of appropriate orbits, it is possible to observe the same scenes repetitively at the same local times.

A major disadvantage in using aerial photographs to study large surface areas is that many photographs must be assembled into a mosaic. It is not possible for an aircraft to cover large areas without obtaining data at several Sun angles and illumination conditions. This causes tonal changes in the photographs, so that the mosaics are difficult to interpret. Major features are often obscured by these tonal changes. Distortions due to relief are large in aerial photographs. These distortions also make it difficult to produce high-quality mosaics. However, it is these distortions that make it possible to use aerial photographs in making topographic maps.

Experience using space-acquired and aerial photographs clearly indicates that both data-acquisition methods are important for Earth resources surveys. These two data-acquisition methods are complementary, and both methods depend on the collection of data at some locations on the surface. The balance between the effort required on the ground, the amount of aircraft coverage required, and how these can be directed by the space-acquired data will require additional experience and system studies.

Planned space flight programs. Two Earth Resources Technology Satellites (*ERTS A* and *B*)

Fig. 1. Earth Resources Technology Satellite showing location of major system components.

are being constructed. Figure 1 shows the modified Nimbus-type spacecraft that will be used for these missions. Plans call for the launch of *ERTS A* in spring, 1972, and the launch of *ERTS B* about a year later. These satellites will be equipped with two types of multispectral remote sensors. The *ERTS A* spacecraft will carry three high-resolution television-type cameras in the green, red, and near-infrared color bands. There will also be a multispectral scanner on *ERTS A*, which will observe the surface features in the green, red, and two bands of the near infrared. In addition, this spacecraft will carry a data-collection system, so that information from instruments on the surface can be relayed through the satellite to the ground tracking stations for correlation with observations made by the satellite.

These satellites will be in near-polar orbit and will pass over each latitude of the Earth's surface at the same local time. Plans call for *ERTS A* to cross the Equator at 9:30 a.m. local time on each orbit. The orbit of the satellite is planned so that it will repeat its coverage of any given location every 18 days. This repeated coverage will make it possible to compare observations of the same scenes made 18 days apart with only small changes in Sun angle. The scene observed by the television cameras will be 100 × 100 naut mi. The multispectral scanner will observe a strip 100 naut mi wide that will correspond to the frame width observed by the television cameras.

The Skylab Program includes a major experimental system for Earth resources surveys. The presence of man in Skylab, the availability of power, and the ability to carry heavy instruments into orbit make it possible to carry out much more complicated experiments than would be practical in unmanned satellites. The Earth Resources Experiment Package planned for Skylab includes six film cameras that will photograph the Earth in six different color bands or combinations of colors, a radiometer that can be pointed at a target by the astronaut to obtain very detailed spectral measurements, a multispectral scanner that divides the spectrum in 13 different bands, a microwave radiometer and scatterometer that functions both as a passive and an active microwave instrument, and a lower frequency passive microwave radiometer. The microwave radiometer-scatterometer is also equipped to be used as a radar altimeter. Figure 2 shows the arrangement of these instruments in Skylab. Plans call for the launch of *Skylab A* in 1973. It should be possible to make detailed comparisons between the repetitive coverage capability of ERTS with its multispectral sensors and the larger number of more complex instruments on Skylab for determining the relative utility of various techniques for resources surveys.

Aircraft will be used extensively in obtaining data prior to the launch of *ERTS A* and *Skylab A* so that more will be known concerning spectral signatures and repetitive coverage to make the data from these space experiments more valuable at an early date. The aircraft will also be used extensively during the flights of *ERTS A* and *Skylab A* to provide additional information on sampled regions and to provide additional time sequence data.

Benefits. The benefits that may be derived from Earth resources surveys depend on the ability to

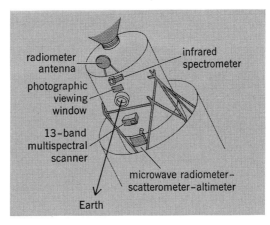

Fig. 2. *Skylab A* Earth Resources Experiment Package mounted in multiple-docking adaptor.

improve efficiency in data acquisition in terms of cost or time and the contribution these data can make to reaching better management decisions and taking actions that benefit society. The use of resources information often depends on rapid data acquisition, analysis, and interpretation. Most benefits depend on using this information to take actions that provide the direct benefit. The Earth Resources Survey Program offers potential for making it possible to take such actions in a timely manner and on the basis of sufficient information to provide assurance that the correct actions were taken. This capability also provides for observing surface phenomena after actions are taken to document the results of the action.

For background information *see* RADIOMETRY; SATELLITES, SCIENTIFIC; TERRAIN SENSING, REMOTE in the McGraw-Hill Encyclopedia of Science and Technology. [JOHN M. DE NOYER]

Bibliography: *Ecological Surveys from Space*, NASA SP-230, 1970; *Remote Sensing with Special Reference to Agriculture and Forestry*, National Research Council–National Academy of Sciences, 1970.

Electric power system, interconnecting

The history of electric power systems is uniquely characterized by three facts: exponential rise in its demand, reduced cost per kilowatt-hour, and increased dependence on its reliability. The reduction in cost was due to improved device designs and efficiencies, systems integration and interconnections, and resulting economy of scale. However, as unit sizes increase and more and stronger ties are built, systems are exposed to more disturbances and disturbance effects propagate further. Thus, even as sufficient reserves are being planned for high reliability, the ever-changing operating conditions, tightening environment constraints, increasing incidents of disturbances, and extended reaches of their effects increase the burden on system operators and make it necessary to develop sophisticated methods for system security control.

Estimates for next two decades. The demand for electric energy has been doubling every 7–10 years and will probably increase at even a faster pace as both new and presently nonelectric markets open up. In the United States this demand

Benefits of pooling power systems

Features	Savings, %	Added cost, %
Lower generation reserve	48	
Larger generating units	35	
Generation at economic sites	5	
Transmission system		5
Fuel expenses	12	
Manning	5	

will require annual net additions to generating capacity ranging upward from 25,900 MW in 1970 to 37,000 MW in 1975, 42,000 MW in 1980, and 63,000 MW in 1985. The mix, over these 15 years, will shift from 63% fossil fuel–steam and less than 9% nuclear for capacity added in 1970 to 40% fossil fuel–steam and 45% nuclear in 1985. The cost for added generation in 1970 was about 6×10^9 and will step sharply upward to about 8.5×10^9 in 1975, 10.4×10^9 in 1980, and 15×10^9 in 1985, in 1970 dollars. Transmission expenditures will rise, too, since these massive generation additions will require substantial links to load areas and higher capacity internal and interregional interconnections. These expenditures will ease upward from about 2×10^9 in 1970 to 3.5×10^9 in 1975, 3.9×10^9 in 1980, and 6×10^9 in 1985. For 1971–1975 overhead transmission is being added at a rate of nearly 11,000 mi per year. Of this total, 63% will operate in the 115–230 kV range and 37% in the 345–765 kV range. In terms of transmission capability, however, the extrahigh-voltage (EHV) lines (over 230 kV) will add 26,634 gigawatt-miles (GW-mi) against only 9093 for 115–230 kV lines. Underground transmission will average 273 mi per year, 89% in the 115–230 kV range and 11% at 345 kV of combined capacity of 165 GW-mi; underground transmission will be about 0.4% of that for overhead.

Future transmission technology. Extensions of high-voltage technology, without need of radical breakthroughs, indicate that ultrahigh-voltage (UHV) transmissions and equipment are feasible at least up to 1500 kV and that levels above 1000 kV can be foreseen within about 10 years in the United States. However, one of the main difficulties to be solved before such voltage levels became practical has to do with audible noise associated with the higher gradients. Other means for transmission of large quanities of power are high-voltage direct current (HVDC), cryogenic cables, and compressed-gas cables. Future choices among these alternatives for bulk power transmission will depend on complete economic comparisons taking into account all engineering, operational, security, and environment aspects. Also, increased strength of transmission links is making it necessary to develop methods to improve system stability, such as faster relay protection, new concepts for local backup breaker failure, more rapid breakers, and single-pole tripping and reclosing. While EHV is fulfilling the basic requirements of integrated regional networks, HVDC can provide high-capacity interregional links where considerations of different operating requirements, frequency regulations, reserve capacities, and load-time diversities may be met by the unique stability characteristic of direct current. This is due to the very rapid power control possibility of direct-current transmission which provides strong damping to disturbances within the power systems.

Power pools. As systems grow and the number of generating units and plants increases and the transmission network expands, the possibility of attaining higher levels of bulk power system reliability also increases. Higher reliability is also attained through the use of properly designed and coordinated interconnections among separate systems. Today all of the continental United States (except parts of Texas) and parts of Canada are interconnected into one gigantic power grid known as the North American Power Systems Interconnection. However, most of the power systems are planned and operated as regional groups called power pools. Each individual power system within such a pool operates technically and economically independently, but is contractually tied to the other pool members in respect to certain generation and scheduling features. Beside the higher levels of reliability attained through pooling and area coordination, the economic gains are considerable. For example, in the case of a 6000–8000 MW typical power pool, estimates of economic savings are of the order of from 8×10^6 to 10×10^6 annually. The breakdown of such savings is shown in the table.

Reliability and security. Increased interconnection capacity among power systems reduces the acceptable generation reserve of each of the separate systems. In most companies the loss-of-load probability (LOLP) is used to measure the reliability of electric service, and it is based on applications of probability theory to unit outage statistics and load forecasts. Computation of LOLP and joint planning are used to avoid having one system count on the other to have enough reserve. LOLP decreases (that is, reliability increases) with increased interconnection between two areas until a saturation level is reached which depends on the amount of reserve, unit sizes, and annual load shape in each area. Any increase in interconnection capacity beyond saturation level will not result in a corresponding reserve in the level of system reliability.

Traditionally systems were planned to withstand all reasonably probable contingencies, and for most of the time operators did not have to worry about the possible effects of unscheduled outages. Operators' normal security functions were to maintain adequate generating capacity on-line and to ensure that system variables, such as line flows and station voltages, remain within the limits specified by planners. However, stronger interconnections, larger generating units, and the rate of system growth are spreading the transient effects of sudden disturbances and increasing the responsibilities of operators for system security.

System security is concerned with service continuity at standard frequency and voltage levels. The system is said to be insecure if a contingency would result in overloading of some system components, abnormal voltage levels at some stations, change of system frequency, or system instability, even if there were adequate capacity as indicated by some reliability index.

Regional and national coordination. The blackout of Nov. 9, 1965, brought considerable attention to the electric power industry and started a chain reaction affecting planning, operation, and control procedures. Also regional and national coordina-

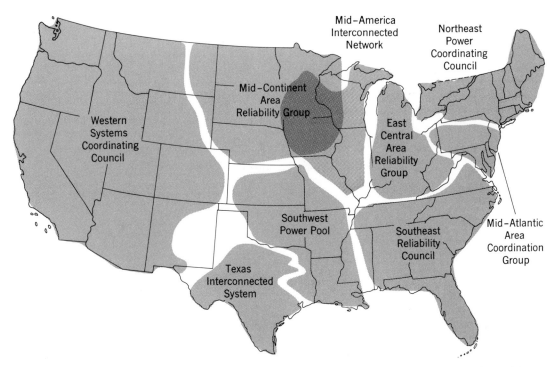

Map of the continental United States showing the nine regional reliability organizations involved in coordination and operation of electric power generation and transmission facilities.

tion councils were organized by the industry; various committees were established and reliability bills were introduced affecting the industry regulation on both state and Federal levels. The following is a brief summary of these efforts.

1. Regional reliability organizations. The whole United States is now organized into nine such groups (see illustration). In general, these groups were not formed to engage in making plans, but rather "to further augment reliability of the parties' bulk power supply through coordination of the parties' planning and operation of their generation and transmission facilities."

2. The National Electric Reliability Council (NERC). This council was established in June, 1968, and continues to be an effective body for the collection, unification, and dissemination of various reliability criteria while retaining planning and operating responsibilities at the individual utility system level.

3. The North American Power System Interconnection Committee (NAPSIC). This committee, which was organized by the industry years ago, was reorganized in 1968 with the formation of NERC. It is responsible for operating criteria and now operates as an advisory group to the technical advisory committee of NERC.

4. Government regulation. Public service commissions continue to express a great amount of interest in the reliability of electric service in their states. The Federal Power Commission (FPC) did not obtain the degree of active participation in the planning of electric facilities which it had requested; however, its role is strengthening through the issuance of Order 383 Docket R-362 in April, 1970. Appendix A of that order requires annual reports to the FPC regarding 10-year generation and transmission plans (including statements of criteria and adequacy), supporting load-flow computer studies, and other information related to scheduling and reliability from each of the nine regions or councils.

All these regional and interregional coordination efforts are imposing a tremendous work load on power system engineers. A recent Edison Electric Institute survey showed that the industry has conducted over 14,000 computer studies for coordination purposes alone in 1969, costing over $3,500,000.

Computer control centers. It is estimated that by the middle of the 1970s about 20 utilities in the United States will have installed computer control centers to provide improvements in both the areas of system operation and system security. Economic consideration such as fuel and start-up costs, which are a function of the operational phase of these centers, become secondary under emergency operating conditions. The security-related functions of the control centers have the primary goal of reducing the probability that system-wide emergencies ever arise. In all cases, although the functions carried by the computer are being expanded, the decision-making role still lies with the human system operator.

The several functions of the computer control center include data collection and computation, monitoring, alarming, data logging and display, load and frequency control, economic dispatch, operation planning and scheduling, security analysis, recording of prefault and postfault telemetered data, and communication with other centers and computers. Modern control centers can provide all these functions by the proper combination of equipment (hardware) and programming (software).

New control systems. All the computer control center functions are still limited mainly to the role of assisting the power system operators during the normal operating state, that is, they are of the preventive control type. Continuing research and emerging concepts of security control (l) will produce more effective methods for preventive control functions; (2) will develop and introduce means for trouble detection during emergency operating conditions; and (3) will develop methods to assist in (and perhaps automate) emergency and restorative control actions to optimally return the power system to its normal operating state.

One good example of a new control system is being designed and built for Cleveland Electric Illuminating Company and is scheduled for initial service in late 1972. This control system is planned to improve the security of the power system through the implementation of several innovations in the areas of security analysis, trouble analysis, automatic circuit restoration, generalized operation load flow, interconnection modeling, exact economic dispatch, man-machine interfaces and display, and operator training.

Improved system reliability. Today it is theoretically feasible to plan, construct, and operate a power system which would be immune from uncontrolled, widespread cascading power interruptions, even under the most adverse set of contingency conditions. This goal can be achieved by following the basic principles of bulk power supply planning. These basic principles are well known and are the basis for the simulated testing criteria recommended by reliability coordination groups such as the East Central Area Reliability Group (ECAR). However, due to obvious economic reasons, the testing criteria are chosen to meet a specific reliability level (close to, but never, 100%) within which widespread system cascading can be avoided. Also, due to increasing interconnections and uncertainty in both demand growth and starting date of new capacity, the responsibility of operators for system security is increasing. In other words, the incidents of critical and emergency operating conditions are going to increase; hence improved reliability will require improved operating criteria and procedures (security control) on top of planning criteria and improved device and system designs. Research areas for improved power system reliability cover a wide spectrum: (1) basic understanding of power devices and systems, (2) improved planning and design ideas, (3) security control, and (4) implementation requirements of security control.

Security control is the proper integration of automatic and manual controls to the maintenance of electric power service continuity under all conditions of operation. The concept is now developing and is approaching enough precision for mathematical formulation and analysis for eventual computer-aided or computer-directed implementation.

Security control assumes several objectives and relies on varied control actions that depend on the different operating states of the power system: normal, emergency, and restoratives states.

Normal operating state is when all customer demand is satisfied at standard voltage and frequency. The control objective in this state is to preserve (hence it is preventive) the normal state at minimum operating cost.

Emergency operating state comes about if one or more components are overloaded, if frequency is changing toward a value at which load or generation is dropped, or if the system is in the process of losing synchronism. In this state the control objectives become the relief of the system distress and preventing further system degradation while satisfying maximum demand.

Restorative operating state is the state following system relief from emergency if some customer loads have been lost (otherwise the system would have returned to normal state). The objective in this case is to restore 100% demand in minimum time.

Developments in the area of trouble detection and emergency and restorative control are presently of lower priority, and important contributions are not expected before the middle of the 1970s.

For background information see ELECTRIC DISTRIBUTION SYSTEMS; ELECTRIC POWER SYSTEMS in the McGraw-Hill Encyclopedia of Science and Technology.

[A. H. EL-ABIAD]

Bibliography: T. E. Dy Liacco, B. F. Wirtz, and D. A. Wheeler, in *Proceedings of the 7th Power Industry Computer Applications Conference*, Boston, May 24–26, 1971; A. H. El-Abiad, in *Proceedings of the 3d Annual Southeastern Symposium on System Theory*, 1:Cl, Apr. 5–6, 1971; *Elec. World*, 174(6):35–50, Sept. 15, 1970; O. I. Elgerd, *Electric Energy Systems Theory: An Introduction*, 1971.

Electrical utility industry

Adding system capacity at the unprecedented rate of about 32,000,000 kW during 1971, while power consumption was growing at considerably under its recent 8.0% per year, enabled United States electrical utilities to supply the summer peak loads with a minimum of service interruptions and voltage reductions. This capacity increase, breaking the previous record of 27,004,000 kW added in 1970, boosted the reserve margin to 20.5% of the summer peak—a more comfortable margin for unplanned generation outages than the 16.6% and 18.6% of recent years. This recovery to a more normal reserve margin can be attributed to a sustained high level of utility construction during a year when industrial growth stood still and residential use of electricity gained a fourth less than its average annual percentage increase for the previous decade.

As the United States economy resumes its growth pattern, however, power system capacity must be added at an ever-increasing rate to supply the needs of industry and the growing population. Hence some 37,000,000 kW of generating capacity already under construction must be completed during 1972—half of it before June—to enable the electrical utilities to supply the hot-weather loads of 1972 with a reasonable margin.

Paralleling this construction of power-generating facilities was the completion during 1971 of additional transmission circuits to convey the power to areas where it was needed, and distribution facilities to connect some 1,500,000 new customers and to bolster the capacity serving more than 72,000,000 existing customers. Thus upward of 11,000 mi of overhead transmission circuits—about 65% for operation at 230,000 V or higher—was placed in service, as well as 80 mi of underground transmission cable. Distribution facilities were expanded by about 1000 substations containing 32,000,000 kVA of transformer capacity to supply 45,000 mi of distribution feeders and 35,000,000 kVA of distribution transformers.

Money outlay. Expenditures during 1971 for building this additional system capacity totaled somewhat over $15,000,000,000. Of this, about 56% was for generating plants, 16% for transmission, 24% for distribution, and 4% for a variety of miscellaneous facilities. These expenditures, already about 15% above those of 1970, will jump another 20% or so in 1972, according to *Electrical World*'s 22d Annual Electrical Industry Forecast (see table).

Part of this sharp rise in expenditures must be attributed to the additional cost of adding cooling towers or ponds for delaying the discharge of cooling water at steam power stations that otherwise would have circulated their cooling water from rivers or lakes. Another factor has been the cost of converting existing stations designed to burn readily available coal to permit the substitution of low-sulfur coal or fuel oil; such a change requires different burners, extensive modification of the boilers, and frequently a complete replacement of the fly-ash collecting system. All too often, the power station's output and efficiency are reduced by the conversion, and the new fuel, selected to meet the recent and rapidly tightening rulings of environmental agencies, costs as much as 50% more than the fuel the station was designed to burn.

The most unfortunate aspect of these additional costs is that they stem from the forced adoption of methods that are still in the development phase to satisfy hypothetical tolerances for stack emissions and water temperatures for which there are as yet no proven values. A further complication is that tolerances that were in effect when the power station was designed have too often been superseded by more stringent ones before the station is ready for operation.

Rate schedules for customer bills were never set to cover such sharp increases in expenditures. Consequently many electric utilities have been forced to apply to their respective regulatory commissions for rate increases. This movement began in 1966, the first year in which a sharp rise occurred in expenditures, and climbed rapidly to an annual rate increase of $426,000,000 in 1970 and nearly $500,000,000 in 1971. For other utilities, a fuel-adjustment factor in their approved rate schedules provided automatic increases. As a result, the long-term decline of about 2% each year in the average price per kilowatt-hour sold to residential customers stopped at 2.09¢ for 1969. The following year, 1970, brought a minimal upturn, but 1971 will bring a 3% jump to 2.17¢, and the trend is expected to continue upward until at least 1975. These increases, while they have been described here in terms of the cost for residential service, are having an equivalent effect on all classes of customers.

Generation methods. No new methods for generating electricity came into use in 1971, but large funds were set up to accelerate the research and development of two new methods. One of these, the liquid metal–cooled fast-breeder reactor (LMFBR) nuclear plant, is slated for final design and proof testing in a moderate-sized demonstration plant by the late 1970s with joint funding by the U.S. Atomic Energy Commission (AEC) and the entire electric power industry. The other method, magnetohydrodynamic (MHD) generation of electricity from coal, received about $2,000,000,000 from the U.S. Office of Coal Research, a group of investor-owned electric utilities, and Avco Corp. to develop plans for a demonstration plant. The MHD process, hailed as the solution to pollution problems, can meet air-quality specifications only with the inclusion of advanced antipollution devices which, when developed, may well find applications in conventional steam power plants.

Another generation method, rarely used by electric utilities until recently and now considered almost a standard part of many power systems, is the gas-turbine plant. This method got off to a slow start because it was originally offered in ratings of about 5000 kW and this was deemed too small to cover a large need for generating capacity. But the size range was extended to 25,000 kW per unit a few years ago and then to 100,000 kW or more by grouping standard modules under a single control system. With this expansion, gas turbines became attractive to many electrical utilities as power sources for supplying peak loads and for emergencies. They also provide the quickest way of adding generating capacity when it becomes apparent that other generating facilities will not be completed on schedule. These attributes, in various combinations, led to the installation of 6,700,000 kW of gas-turbine generating capacity in 1971, nearly 21% of the total capacity added during the year.

Gas turbines require no cooling water, since they exhaust their spent gas directly to the atmosphere. This is one reason why their efficiency has always been poorer than that of conventional steam plants. But the heat remaining in this exhaust gas can be used to improve the efficiency of a steam plant, generally by discharging the hot gas into the combustion chamber of a conventional

United States electric power industry statistics for 1971*

Parameter	Amount	Increase over 1970, %
Generating capacity, kW ($\times 10^3$)		
Total	370,000	8.8
Hydro	56,900	3.4
Fossil-fueled	273,800	5.8
Nuclear	12,800	98.0
Gas turbine and internal combustion	26,500	33.8
Energy production, kwhr ($\times 10^6$)	1,616,700	5.0
Energy sales, kwhr ($\times 10^6$)		
Total	1,461,700	5.1
Residential	482,000	7.7
Commercial	339,200	8.4
Industrial	577,600	0.9
Other	62,900	7.9
Revenue, total ($\times 10^6$)	$24,400	10.3
Capital expenditures, total ($\times 10^6$)	$15,092	15.0
Customers ($\times 10^3$)		
Residential	65,200	1.8
Total	73,800	1.7
Residential usage, kwhr (average)	7,480	5.8
Residential bill, ¢/kwhr (average)	2.17	3.3

SOURCE: *Electrical World*, Sept. 15, 1971, and extrapolations from Edison Electric Institute data.

boiler, where it adds to the heat produced by the boiler fuel. This concept is not new; it was used in a few so-called "combined cycle" units in the 1960s, but it appears to have gained substantially in 1971 with the development of at least four balanced designs involving a high degree of factory prefabrication for combined outputs on the order of 200,000 kW. Electric utilities ordered a number of these combined units for early installation as intermediate-load-factor or "cycling" units, a service for which they offer better overall economy than gas turbines alone or steam plants designed for cycling operation.

Transmission circuit expansion. Overhead transmission circuits for operation at 765,000 V ac were extended 649 mi in 1971, for a United States total approaching 1000 mi. Included was a link from northern Indiana into Illinois, where the link ties into the Commonwealth Edison Co. and provides an interconnection capable of transferring upward of 2,500,000 kW in either direction. Hence, when the planned total mileage of 765,000-V circuits is completed, there will be a continuous high-capacity system operating at that voltage from western Virginia through the East North Central Region to Illinois.

Existing 500,000-V systems were expanded even more. As a result, the Atlantic Coast States from New Jersey to Georgia and west to Texas and Arkansas are linked by transmission circuits each capable of carrying about 1,200,000 kW. In addition, a pair of 500,000-V circuits completed last year between the Columbia River and the Los Angeles area is carrying the regional interchange of power for which the world's largest high-voltage dc system was completed in 1970 only to be knocked out in 1971 by the San Fernando Valley earthquake. Damage to the terminal station there was extensive, and full restoration is not expected before 1973.

In addition, the Consolidated Edison Co. 345,000-V link to the 500,000-V Pennsylvania–New Jersey–Maryland interconnection, cited by the Federal Power Commission several years ago as the most vitally needed interconnection in the United States, will be completed in time to help Con Edison carry next summer's peak loads. This line was delayed for several years by a variety of interveners, and its construction is on a compromise route finally accepted as the least objectionable. The new line will assume additional importance in the years ahead as the only feasible way to get power out of a large new generating station being built jointly by Con Edison and Orange & Rockland Utilities Inc. on the west bank of the Hudson River at Bowline Point. Before the plant comes on-line, however, the 345,000-V system will be extended with another line across northeastern New Jersey to the Public Service Electric & Gas Co. Hudson Generating Station, thence by way of 345,000-V cable under the Hudson River, across Manhattan, and under the East River to Con Edison's system backbone in Brooklyn.

Thus the interconnections among seven of the nine regional reliability organizations comprising the National Electric Reliability Council were strengthened materially during 1971. About the only remaining weak links are between the westernmost of the seven regions and the Pacific Coast States, and between the southern regions and Texas. Of these, the former are linked by several lower voltage interconnections that are often used but are too weak for reliable operation. *See* ELECTRIC POWER SYSTEMS, INTERCONNECTING:

Problems with environment. All this progress might be interpreted as proving that the electrical utility industry has overcome the difficulties of recent years. But this would be far from correct. On the nuclear power program, for example, interveners are still prolonging the public hearings and delaying the granting of permits to build new plants and of licenses to operate them when they have been completed. The recent creation of the U.S. Environmental Protection Agency, and its early rulings on cooling-water discharges, gave interveners a new ally and, with the agency's support, they won a 1971 court decision under which the AEC has ordered the reopening of argument on all permits issued since Jan. 1, 1970. As a result, 25,000,000 kW or more of nuclear power plants currently under construction may suffer yet another delay.

During 1971, also, staged tests of emergency cooling systems for nuclear reactors were interpreted as casting doubt on the safety of nearly half of the nuclear power units in service or under construction in the United States. This development spurred the AEC to order a review of each individual unit's nuclear system design, but none of them has yet been found unsafe.

For fossil-fueled power plants, attacks focus on their stack emissions and fuel sources, as well as on their cooling-water discharges. One result of these attacks, and regulations inspired by their proponents, has been a scramble for low-sulfur coal and oil. But suitable coal is in short supply east of the Mississippi River, so that many utilities applied for exemptions to the quota for imports of low-sulfur oil from Africa and Asia because domestic oils cannot satisfy the sulfur limits.

Other utilities turned to the low-sulfur coals of the Rocky Mountain States, which, because of the long rail haul, in some cases doubled their fuel costs. The alternative of building generating stations in the coalfields and transmitting the power by way of extrahigh-voltage lines was adopted by a group of Southwestern utilities collaborating as West Associates. Several members of this group built two 1,500,000-kW generating stations, one in New Mexico and the other in Nevada, only to face environmentalist attacks on their coal sources. The coal for one is obtained by strip mining on the Navajo Indian reservation. However, the coal has a high ash content, and it has been claimed that the plant is discharging 300 tons of particulates into the air each day. Accordingly the U.S. Department of the Interior has called a moratorium on three additional stations in the design or planning stage.

Water is scarce in the vicinity of these Rocky Mountain coalfields, and all of the new stations have had to use cooling towers or artificial lakes. Even these usually require allotments of water to replace evaporation losses to the arid air. Hence there has been pressure to develop nonevaporative cooling systems. One small station in eastern Wyoming was recently put in service with an air-cooled condenser for the exhaust steam, and a unit of about 200,000 kW was authorized in 1971. This is about the limit for air-cooled condensers, however, and several consulting firms are investigating the possible adoption of nonevaporative or dry cooling towers, as developed in Eastern Europe, for larger generating units.

There is reason to believe, too, that several antipollution methods now on trial at a number of generating stations will permit the burning of readily available coal without violating the new regulations. Some of these methods depend upon injecting dry reagents into the flame to capture the sulfur compounds, while others pass the stack gases through a solution of reagents. In some cases the sulfur is removed as an insoluble compound, while others convert it to saleable sulfuric acid. At this point it is still uncertain how efficient these processes can be made and what complications they may introduce, but it is very certain that they will significantly increase the cost of generating electricity.

Hydroelectric power. Hydroelectric power additions in 1971 were confined generally to adding generating units at existing Columbia River stations to use the additional water flow provided by building three huge storage reservoirs in British Columbia. This improvement came about as a result of a treaty negotiated several years ago with Canada under which a group of utilities in the Northwestern States paid British Columbia for its participation in the plan. Meanwhile environmentalists have intervened in the arguments that have blocked construction of the 1,750,000-kW High Mountain Sheep project on the Snake River since 1964.

Pumped-storage hydro plants, which function for power systems much as storage batteries do for smaller installations, are under construction in at least four states, with some 1,800,000 kW scheduled for service in 1972. They function by drawing power from other sources during light-load periods, usually nights and weekends, to pump water from a river or lake to a reservoir at a higher elevation, where it is stored until the power system needs additional generation. Then the water is permitted to flow back to its original source, passing through a hydraulic-powered generating unit on the way. The largest installation of this type yet proposed, Con Edison's 2,000,000-kW project on the Hudson River near Cornwall, N.Y., remains in doubt after 8 years of controversy. No pumped-storage plants of any consequence were completed in 1971.

Underground transmission circuits. Interest continues for putting transmission circuits underground, although there seems to be growing recognition that the cost of doing this with today's technology is exorbitant at the voltage levels used for major lines. Research continues, under direction of the Electric Research Council, at the Westinghouse Waltz Mill test station to develop cables for 500,000- and 765,000-V service, as well as better cables for lower voltages. A growing faction in the industry advocates forced-cooled or refrigerated gas-insulated intermediate-voltage cable in place of high-voltage cable for high-capacity circuits.

Substations, too, have come under fire for their unsightliness. Designers have, in recent years, redoubled their efforts to improve the appearance, in many cases by strictly limiting the exposed height of structures and surrounding the station with architecturally attractive walls or landscaped plantings. Several utilities turned in 1971 to the so-called "mini sub" concept in which gas-insulated components make for extreme compactness and minimum structural exposures.

Because suitable technology was in hand and the overhead plant was highly visible, several state commissions made underground construction mandatory for new distribution. Many utilities in other states negotiated more or less voluntary agreements for doing so. The result is apparent in the mileage of distribution cable reported for *Electrical World*'s 1971 Construction Survey—21,260 mi in 1971 and a gain of 10% or more projected for 1972—and in the types of distribution transformer capacity installed—36.3% suitable for underground distribution in 1971 and a rise each year through 1977.

For background information see ELECTRIC POWER SYSTEMS; NUCLEAR POWER in the McGraw-Hill Encyclopedia of Science and Technology. [LEONARD M. OLMSTED]

Bibliography: *Elec. World*, 175(6):39–70, Mar. 15, 1971; L. M. Olmsted, *Elec. World*, 176(6):41–56, Sept. 15, 1971; L. M. Olmsted, *Elec. World*, vol. 176, no. 9, Nov. 1, 1971; *The 1970 National Power Survey*, Federal Power Commission, pt. 1 and 2, 1971.

Electrodiagnosis

Medical signals taken each year are becoming so numerous that manual means of analysis are becoming impractical. Conventional analysis requires a significant amount of the physician's scarce time. Further, the cost of analysis is rising as a result of increases in the physician's office expense overhead in conjunction with his being overburdened with routine tasks from the increased volume. However, it has been found that the printouts from automatic analysis improve and simplify record keeping. Since simplified data-acquisition machines can be used when computer input is involved, the technician's time can be used more effectively.

In addition, automated medical signals provide a more suitable data base for research. The advantage over manual methods resides in the fact that the computer is more objective than humans and is highly consistent.

Electrocardiogram (ECG) analysis by computer has begun to pass from the research phase to a validated, standardized, cost-effective technique with defined quality of interpretations. Since the electrocardiogram was the first major signal selected for automation, it can be used as a model for the field of electrodiagnosis.

Several ECG computer programs have been or are in development from many groups. Furthermore, several hardware system components for ECG analysis have been developed and introduced commercially as a result of efforts by the U.S. Public Health Service's former Medical Systems Development Laboratory (MSDL). That organization, concerned with development of an information system in medicine to serve as a model for the total informational needs of practitioners, coordinated the efforts of many groups and fitted them into an overall integrated system design to secure the inherent advantages of automatic analysis.

An orthogonal ECG lead program has been under development at the Washington, D.C., Veterans Administration. A Mayo Clinic program is under development for orthogonal leads in clinical usage. New York City's Mt. Sinai Hospital has developed a system incorporating interpretation of criteria. This system is available. Other programs are being evolved.

In dealing with automated electrocardiographic analysis one must consider the basic system, from data acquisition to storage and retrieval. A central point is the stability of the software. The design for the best system to service both physicians and hospital-based services at the requisite cost and quality standards requires in-depth knowledge of the capabilities and limitations of available hardware/software combinations.

Various users have different requirements for ECG analysis which may require modification and extension of the basic system design. Data input from various classes of ECGs (stress, resting diagnostic, screening, or resting) presents problems due to the need for comparison of tracings, arrhythmic data bases, and long-term storage requirements. The geographical location of physicians is another pertinent point in considering computerized information systems.

The 12-lead electrocardiogram is recommended as the best technique for current clinical use since it is standardized and also a very sensitive and commonly utilized procedure. A 3-lead orthogonal electrocardiogram may be a good substitute, but lead application is more difficult. At this date practitioners do not know the criteria for orthogonal leads as well as the 12-lead. The alternative types of electrocardiograms, which are less sensitive and less specific are 6 or 1 conventional-limb leads. These generally miss about 15% of the ECG abnormalities. The specific abnormalities missed depend on the lead configuration used, as well as age and body configuration of the subject's test.

The known drawbacks of electrocardiography are dependence on (1) the population being tested and relevance of the criteria used on their tracings, (2) the subject's biologic variations, (3) machine reliability factors, and (4) variability of the observer interpretation. None the less the electrocardiogram is a valid and useful clinical test.

It should be mentioned that the absence of atypical, borderline, or abnormal findings in electrocardiographic data alone is not necessarily the same as the interpretation of normality made by an electrocardiographer, who might with additional patient data also call certain atypical or borderline findings normal.

Approach to computerization. Any automated system requires evaluation. The system must be evaluated for repeatability, signal fidelity, noise, equipment stability, precision, and cost. Quality control techniques require careful methodological documentation that must be included in the design. The MSDL-developed software has been translated into several computer languages designed to run on a wide variety of commercial hardware systems. The practicality of using several of these systems on a regional network basis is being tested. Standardization and quality control for these must be designed. Evaluation of the uses of the various programs can now be carried out to determine how they may complement each other for better patient care.

A system for resting diagnostic or screening, stress, and monitoring electrocardiogram information should have the following basic subsystems: (1) signal acquisition, (2) transmission, (3) preprocessing, (4) processing, (5) display, and (6) storage and retrieval.

The recordings (that is, paper tracing) produced by conventional manual ECG machines could be specially manipulated to serve as input to a preprocessing system. Optical scanning techniques could be used to produce copy to be digitized for input to processing systems, but technical lag has delayed these efforts, and the alternative routes seem to be better choices for compatibility with other than medical data processing.

There are two general types of acquisition units: standard single- and simultaneous-lead machines. Adaptations can be designed for computer compatibility. Simultaneous-lead machines either transmit an analog signal by telephone or record the ECG signal on either analog or digital magnetic tape. There are both portable and mobile units available. Digital units are directly compatible with processing software. Analog acquisition units undergo preprocessing, that is, analog-to-digital conversion. The choice from the group of acquisition units with acceptable technical quality and performance characteristics now depends primarily on economic considerations. The principal determinants of this are "turn-around" time requirements, volume of ECGs, operational efficiency, simplicity, and unit costs. Rapid turn-around time ("real time") is desirable but not required in some applications.

The area of stress and continuous ECG has special requirements for identification of events as well as elimination of noise artifacts. Wave onset and baseline determination are necessarily more complex in this field. Experimental programs have incorporated procedures to handle some of these difficulties. More frequent measures of specific data points have been tried as well as time-normalization procedures. Measurement and criteria are still nonstandard in these areas, and thorough consideration of benefits and limitations of certain techniques must be undertaken. Online monitoring has been successful. Astronaut monitoring, coronary care, animal testing, and anesthesia monitoring are some examples of recent uses. Astronauts were monitored by computer throughout the *Gemini 7* through *11* missions.

Transmission and preprocessing. The first telephone transmissions for computer use were from a doctor's office in 1961. The situation still resembles that of current community physician needs. ECGs transmitted by public health nurses with portable ECG machines taken to the patients' homes have been investigated to determine the feasibility of the system. The results (good and bad) have been helpful in research, development, and trials. Other tests have included telephone transmission over long distance by way of private and public lines from many points in the United States and throughout the world by various facilities such as satellites and radio links. These tests were not satisfactory. These transmissions have included long-term routine day-to-day operations, emergency room, and special screening systems, as well as single demonstrations.

The requirements of the processing software, in part, indicate some of the preprocessing hardware. Most of the current processing software requires 500 samples per second. Since the requirements over the next decade may change, design must allow for alternative sampling rates.

The preprocessing function requirements differ depending upon the input, direct telephone or tape playback and demodulation of signal, amplification, filtering, and analog-to-digital conversion. Specifications still need to be developed for various interfaces. Recent usage indicates that the time has come to reevaluate filter characteristics and their effect on signal degradation, tape playback problems of alignment, speed, and demodulation, as well as analog-to-digital conversion procedures and equipment.

Processing. The major sections of an ECG processing software package are pattern recognition and interpretation. The pattern-recognition portion of the software generally contains logic for data condensation and data combination. The interpretation software allows for interaction of criteria and measurements. The design of the software must be oriented for modifications based on future evaluations and on research requirements. Detachable subroutines for various peripheral units and subsystems must be considered.

The ultimate system design must be such that any diagnostic criteria required can be easily incorporated into the system using stable pattern analysis techniques. Most programs contain empiric constants which require changing if different input data specifications are used. The MSDL program computes the derivative of the ECG wave, performs a selective "averaging," locates a fiducial point and makes limited rhythm analysis (which needs to be augmented), selects an appropriate complex, locates and measures the wave amplitude and durations, formats and prints these measures, computes the frontal plane axis (which can be augmented to be three-dimensional), determines the interpretation to be made, and classifies the ECG as normal, borderline, atypical, or abnormal. Provisions are made for bad data, rejection of bad data, and redundancy checks to compensate for rejected bad data.

Any system such as the one for automated ECG analysis is only as good as its documentation. The success of any system depends on its ability to be used by others. The MSDL computer documentation has resulted in widespread use of their program and its conversion for several types of computers. Processing has been successfully carried on by both large and small computer systems. These operate in both online and batch-processing modes. Similarly, engineering specifications were developed and allowed duplication and improvement in the whole system.

Other medical signal analysis work has been conducted. The forced vital capacity signal for respiratory evaluation has been extensively tested. The procedures to analyze this type of monotonically increasing wave differ from those used in ECG analysis only in that the curve is much simpler. A phonocardiographic program has been designed which required a much higher data rate and the use of an ECG as a timing signal. A tonography feasibility study is an example of work in large-scale data reduction. Other studies include pattern recognition and measurement programs for electroencephalography, dye dilution, and electrophoretic waveforms. A generalized system design for electrodiagnosis should consider the possibility of several signals.

Display. The clinician has to be taught how to interpret an acceptable report from an automated system. This, of course, requires time and effort. If it is not in a format acceptable to the clinician, its use is limited. Much time must be spent in developing report outputs acceptable for routine clinical use. Educational guidance must be included in future systems-development efforts.

The statistical analysis of the data and of the population groupings stratified in various epidemiologically significant ways is an area of importance. The first study of multidimensional analysis of the ECG was made in the early 1960s. Although still not possible for current routine immediate use, such techniques will be essential by the late 1970s, and modes of incorporating them into processing systems must be studied.

The display subsystem must provide formats suitable for (1) hard copy of the ECG report for use by the submitting physician, hospital, clinic, and staff member; (2) reproduction of the tracing for quality control of the processing system and for duplicate and retrieval purposes (possibly computer-generated); (3) a computer-storable form (such as digital tape) for retrieval and comparison purposes.

In addition, in many situations, computer-generated voice reports might be extremely easy to use and be useful.

Storage and retrieval. These systems can range from very simple indexing storage and retrieval systems to large online sophisticated systems. Both the economics and statistics are significant to the overall system. Both measurement and interpretive codes can be stored on appropriate media within a computer system. Software specifications for economical record retrieval must be developed after full consideration with physicians of the needs and uses for the data. Comparison techniques must be incorporated into the retrieval system.

For background information see CARDIAC ELECTROPHYSIOLOGY; DATA-PROCESSING SYSTEMS; ELECTRODIAGNOSIS in the McGraw-Hill Encyclopedia of Science and Technology.

[CESAR A. CACERES]

Bibliography: C. A. Caceres and L. S. Dreifus (eds.), *Clinical Electrocardiography and Computers*, 1970.

Electron diffraction, high-energy

High-energy electron diffraction (HEED) is one of the most important methods for the study of the surfaces of crystalline materials, of thin films (thickness from $\sim 10^{-8}$ to $\sim 10^{-4}$ cm), and of small particles (diameter from $\sim 10^{-7}$ to $\sim 10^{-4}$ cm). The most significant recent developments in the experimental methods are the use of ultrahigh-vacuum (UHV) instruments (pressure $\sim 10^{-10}$ torr, where 1 torr $= 133.32$ newtons/m²), espe-

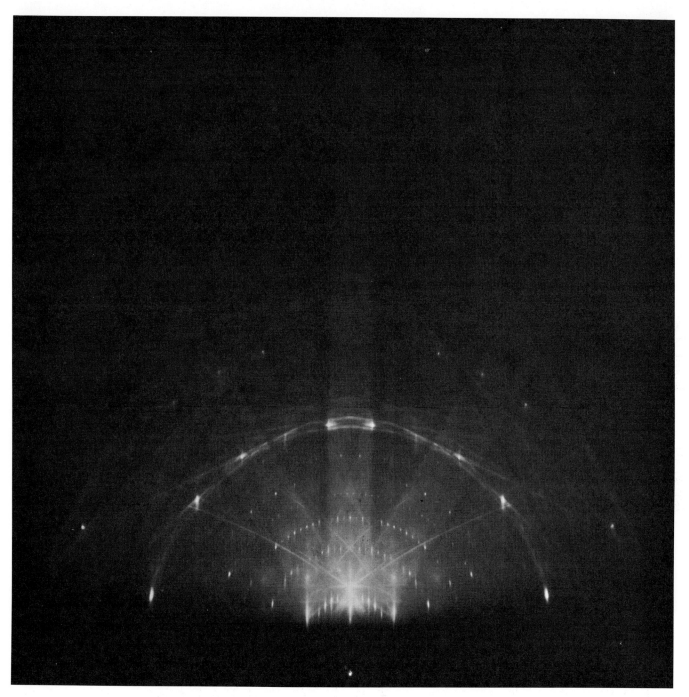

Fig. 1 RHEED pattern from a flat (111) surface of a silicon single crystal. (*Varian Associates*)

cially in conjunction with other analytical tools; the perfection of scanning electron diffraction instruments, partially with energy filtering; and the development of high-voltage, high-resolution diffraction electron microscopes and of special illumination systems for these instruments allowing high-resolution microscopy with diffracted electrons. Significant progress has also been made in the theory of HEED: Dynamical effects and the influence of absorption can now be taken into account quantitatively.

Reflection HEED. UHV instruments have become easily available recently, which has made it possible to study the structure of surfaces under clean conditions by reflection HEED (RHEED). Many experiments have confirmed the theoretical expectation that RHEED has a sensitivity comparable to that of low-energy electron diffraction (LEED). Therefore, similar to LEED, RHEED can be used for the determination of the lateral arrangement of the atoms in the topmost layers of the surface, including the structure of adsorbed layers. Although it is more convenient to deduce the periodicity of the atomic arrangement parallel to the surface from LEED patterns than from RHEED patterns, LEED frequently becomes inapplicable when the surface is rough. This occurs usually in the later stages of corrosion or in precipitation, for example, of silicon carbide on silicon, when small crystals grow on the surface. In such investigations RHEED is far superior to LEED. Figure 1 shows a RHEED pattern of a flat silicon (111) single crystal surface, a sample for which LEED is equally well suited. It is a Kikuchi diagram characterized by bands and lines of varying intensity onto which the short streaks of the Laue diagram of the reconstructed surface layer are superimposed. Figure 2 is the RHEED pattern of a rough surface of a polycrystalline evaporated film of CaF_2 with strong fiber texture. In this sample the surface consists of many small facets distributed at random and inclined against the mean surface so that no LEED pattern could be obtained.

Like LEED, RHEED gives little information on the chemical nature of the atoms which produce the diffraction pattern. Therefore RHEED has been combined with analytical tools, such as Auger electron spectroscopy or x-ray emission spectroscopy, in the same system. A second limitation inherent to both RHEED and LEED is the difficulty of determining the atomic positions normal to the surface. This is a consequence of dynamical effects and of absorption. These limitations have led to a number of controversial interpretations, such as the question of reconstruction upon gas adsorption or of the nature of reconstructed "clean" surfaces with superstructure. Nevertheless both techniques have given valuable information on corrosion, precipitation, adsorption, condensation, film growth, and other surface and interface phenomena which could not be obtained by other methods.

Scanning HEED. In scanning HEED (SHEED) the diffracted electrons are not recorded on photographic film but measured directly electronically with sensitive detectors. By moving the detector across the diffraction pattern or by deflecting the diffracted electrons across a stationary detector ("scanning"), the intensity distribution in the diffraction pattern can be displayed quantitatively on an XY recorder. If an energy filter is put in front of the detector, the inelastically scattered electrons which are usually not taken into account in the quantitative intensity evaluation may be filtered out so that only the elastically scattered electrons are measured. This technique is particularly useful in transmission through polycrystalline samples which produce a ring pattern (Debye-Scherrer diagram) but has also been used in single crystal samples producing a spot pattern (Laue diagram) and even in reflection diffraction from surfaces. The main application of SHEED is in the study of processes which are accompanied by changes of the intensity distribution, such as the growth of thin films and annealing and corrosion processes.

The technological importance of thin film and interface devices has led to an upsurge of thin film growth studies by conventional transmission HEED (THEED), usually combined with transmission microscopy. Information obtained this way has been mainly on the orientation of the crystallites comprising the film. Figure 3 shows an example, the diffraction pattern of an epitaxial Au film on a single crystal PbS layer.

Diffraction electron microscopy. The third significant progress is in an area in which HEED is only an auxiliary tool, that is, in the transmission diffraction electron microscopy of crystalline specimens. In such specimens the main contrast mechanism is the diffraction contrast. To understand

Fig. 2. RHEED pattern from an evaporated calcium fluoride film.

it, knowledge of the diffraction pattern and of diffraction theory is necessary. In recent years high-voltage electron microscopes (with electron energies from ~0.15 to ~1.5 MeV) have become increasingly available. As a consequence, samples with thicknesses of the order 10^{-4} cm can now be studied, as compared to the former upper thickness limit of about 10^{-5} cm. This has significance for metallurgy and the physics of metals and semiconductors: It is now possible to study the structure, distribution, and behavior of imperfections in metals, alloys, and semiconductors in the bulk with little influence from the boundaries of the material. Other electron-microscope-technique im-

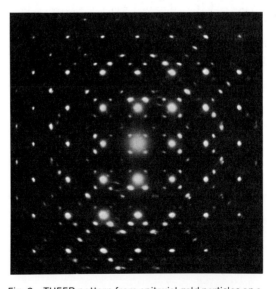

Fig. 3. THEED pattern from epitaxial gold particles on a lead sulphide single crystal film. (*Courtesy of A. K. Green*)

provements which rely heavily on diffraction are the use of illumination systems producing tilted or conical illumination of the specimen and of objective apertures which transmit only diffracted electrons. With these techniques it has become possible to produce, by diffraction contrast, images of crystal planes 1.5×10^{-8} cm apart.

The developments in diffraction electron microscopy just described have stimulated considerable effort to understand quantitatively the processes involved in electron diffraction. As a result, much progress has been made in this field recently. The theoretical problem is as follows: When an electron beam traverses the sample for a considerable distance ($\sim 10^{-6}$ cm in the case of electrons with energies of the order of 10^4 to 10^5 eV), the beam is attenuated strongly by diffraction (elastic scattering) and by losing energy to the sample (inelastic scattering). As a consequence of elastic scattering, the incident beam is attenuated to such an extent that the diffracted beams equal it in intensity and are diffracted again, partially in the direction of the incident beam: The various beams in the crystal are coupled with each other by diffraction (dynamical coupling). Thus the assumption, basic to conventional structure analysis with x-rays and to the kinematical theory of diffraction, that the diffracted beams are weak as compared to the incident beam is not fulfilled. Rather the much more complicated dynamical theory has to be used. In recent years this theory has been developed to a high degree of sophistication, so that quantitative structure determinations with HEED and quantitative contrast analysis of diffraction electron micrographs have become possible. In these quantitative studies it is necessary to include the influence of inelastic scattering. Inelastic scattering reduces the number of electrons which have the same energy as the incident electrons (absorption).

A quantitative theory of electron diffraction thus has to take into account both dynamical coupling and absorption, whether it be LEED, RHEED, SHEED, or THEED. Further development of the dynamical theory, including absorption, should increase the usefulness of these methods considerably.

For background information *see* AUGER EFFECT; ELECTRON DIFFRACTION; ELECTRON SPECTROSCOPY; MICROSCOPE, ELECTRON; X-RAY CRYSTALLOGRAPHY; X-RAY DIFFRACTION in the McGraw-Hill Encyclopedia of Science and Technology.

[E. BAUER]

Bibliography: E. Bauer, in R. F. Bunshah (ed.), *Techniques of Metals Research*, vol. 2, pt. 2, 1969; P. J. Estrup and E. G. McRae, *Surface Sci.*, 25:1, 1971; C. W. Grigson, *Advan. Electron. Electron Phys.*, suppl. no. 4, p. 139, 1968; G. Thomas and W. L. Bell, in R. F. Bunshah (ed.), *Techniques of Metals Research*, vol. 2B, 1970.

Elementary particle

Insights have been gained recently from two areas in the study of elementary particle resonances. First, the study of exotic resonances has led to new thoughts on the quark model for the strongly interacting particles. Second, new investigations of coupled and overlapping resonances have produced a better understanding of the strong interac-

Quantum numbers for two properties of quarks

Type of quark	Electric charge (in units of the proton charge)	Hypercharge = Strangeness + Baryon number (in units of the proton hypercharge)
p	+2/3	+1/3
n	−1/3	+1/3
λ	−1/3	−2/3

tions producing these resonances.

Exotic resonances. The subject of exotic resonances has developed in the past few years in connection with a particular model of the strongly interacting particles, namely that in which these particles are composed of more fundamental constituents called quarks. Exotic resonances are resonances whose quantum numbers can not be obtained by combining the quarks in the simplest way. The simplest way in the quark model to form a baryon is from three quarks ($3q$); the simplest way to form a meson is from a quark and an antiquark ($q\bar{q}$). More complicated states for the baryons can be imagined, such as $4q\bar{q}$ and $5q2\bar{q}$, and likewise for the mesons, such as $2q2\bar{q}$ and $3q3\bar{q}$. These more complicated states have been given the name "exotic." If there were experimental evidence for exotic states, the quark model of mesons and baryons would lose some of its attractiveness. On the other hand, if exotic states can be shown to be absent, one's confidence in the quark model as a fruitful model is increased. At the present time there is no clear-cut experimental evidence for exotic resonances in spite of considerable ingenious efforts to discover them. *See* PARTICLE ACCELERATOR.

In the table are listed two properties of quarks, electric charge and hypercharge, that are important in determining what quantum numbers one expects for exotic resonances. The quantum numbers of the antiquarks are obtained by reversing the signs of both the charge and the hypercharge. The quark model pictures baryons as composites of three quarks, and mesons as quark and antiquark pairs. Thus one can form, for example, a +2 electric charge state from $p + p + p$ or a −2 hypercharge as $\lambda + \lambda + \lambda$. But a +2 charge state or −2 hypercharge state of the mesons cannot be formed from a combination of q and $\bar{q}$. Such meson states are exotic. As another example, a baryon state of hypercharge +2 is exotic. There are other restrictions that can be imposed on the allowed charge and hypercharge states; the states which have the forbidden values of charge and hypercharge are sometimes called exotic resonances of the first kind. For the mesons, and the mesons only, there is another category of exotics in which certain combinations of the spin (J), parity (P), and charge conjugation quantum number (C) are not allowed, independent of the value of charge or hypercharge. If the spin and parity of the meson resonance is in sequence $0^+, 1^-, 2^+, 3^-, \ldots$, then the q-$\bar{q}$ model requires $C = P$. Therefore states with J^{PC}, 0^{+-}, 1^{-+}, 2^{+-} are not allowed and are all called exotic mesons of the second kind. Baryons cannot be eigenstates of charge conjugation, and there is no restriction on their spin and parity from the $3q$ model.

Information on the presence or absence of exot-

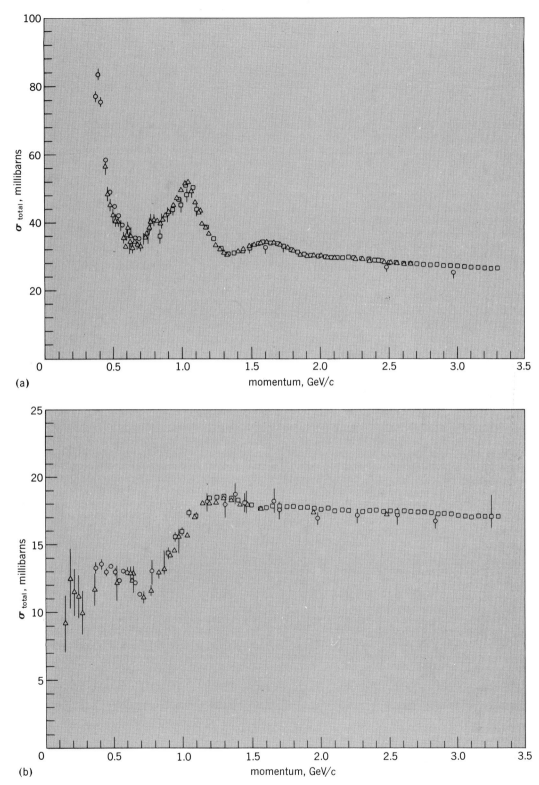

Fig. 1. Total-cross-section data as a function of incident K momenta. (a) K⁻ proton. (b) K⁺ proton.

ics comes from three basic types of experiments: formation experiments (the resonance is formed at rest in the center-of-mass system); production experiments (the resonance is produced with other particles and is therefore moving in the center-of-mass system); and reactions in which the presence of an exotic resonance as an exchange particle would show a peak in the angular distribution. The identification of resonances in data from formation and production experiments is more straightforward than in data from angular distribution measurements.

Formation experiments. Formation data from total-cross-section measurements and studies of two-body scattering have yielded in the past 15 years extensive evidence for many nonexotic resonant states of the baryons. The method of phase shift analysis for treating the two-body scattering data is well established and has given resonance parameters, mass, width, isospin, spin, and parity for most of the know nucleon and hyperon resonances.

Figure 1 shows a collection of data on resonance formation as seen in K^--proton and K^+-proton total cross sections plotted as a function of incident K momenta from 0 to 3.5 GeV/c. The K^-p system has charge 0 and hypercharge 0 (strangeness -1) and is therefore not exotic. The bumps in the K^-p total cross section indicate strong resonance formation, and there are more than 10 resonances known in the region below 2 GeV/c incident K^- momentum. In contrast to the K^-p data, the K^+p data are relatively flat except for a rise around 1 GeV/c K momentum and a subsequent slight fall. The K^+p system has hypercharge +2 (strangeness +1) and is therefore exotic. The top of the curve in the K^+p data occurs at a center-of-mass energy of 1.91 GeV. This is in the isospin $T=1$ channel. There is also a similar structure in the $T=0$ channel obtained from K^+d total-cross-section measurements located at a mass value of 1.87 GeV. These structures cannot be interpreted as indicating the presence of exotic resonances without supporting data from phase shift analyses of relevant two-body scattering processes, for example, $K^+p \rightarrow K^+p$ and $K^0p \rightarrow K^+n$. At the present time all that can be said is that if there are exotic K^+p resonances, they are created much more weakly (by more than a factor of 10) than K^-p resonances in the corresponding mass region. The scattering data are consistent with the interpretation of no exotic resonance formation.

Other combinations of meson projectiles and nucleon targets, π^+p, π^+n, π^-p, and π^-n, have nonexotic quantum numbers and their total cross sections show strong resonance behavior similar to the K^-p data of Fig. 1. In contrast, nucleon-nucleon total cross sections show behavior quite similar to the K^+p data, and there is no evidence for resonant nucleon-nucleon states if one does not consider the deuteron to be a resonance. Strictly speaking, nucleon-nucleon resonances are not exotic; however, they will be absent if the only allowed resonances of strongly interacting particles are states of three quarks and states of quark-antiquark pairs.

Production studies. The second category of experiments to search for possible exotics is production studies. Over the last 10 years, most of the information on meson resonances and cascade resonances (baryon resonances with hypercharge = -1) has come from production experiments, principally from measurements using liquid-hydrogen bubble chambers in a magnetic field. These experiments have produced scores of resonances with nonexotic quantum numbers but no well-established exotic states. Typical production cross sections for meson resonances in the intermediate momentum region of 2 to 5 GeV/c in excess of 100 μbarns are common, whereas cross-section limits for possible exotic states are generally set at less than 10 μbarns. These limits on exotic production come from a study of mass plots of various combinations of final-state particles. Resonance production would be indicated by the presence of bumps on a more or less smooth background. One way to increase the sensitivity of this type of experiment is to increase the number of events.

There are currently two groups working on experiments which can push the limits on exotic production to the level of about a microbarn. These experiments, one at Argonne National Laboratory and the other at CERN, employ electronic techniques for particle detection rather than bubble chambers. One of the particular reactions studied by these experiments is $\pi^+ + p \rightarrow \pi^+ + \pi^+ + n$, and resonant behavior in the $\pi^+\pi^+$ (charge +2) system is sought. Nevertheless structure in the $\pi^+\pi^+$ mass spectrum would not be unambigious evidence for exotics. Careful analysis of the final-state particles would be required. In general, because there are more particles in the final state, the job of establishing resonance quantum numbers (spin and parity, for example) of a signal seen in production is more difficult than to establish the spin and parity of a resonance in a formation experiment.

Peaks in angular distribution. A third method of detecting exotics is by a study of reactions in which the quantum numbers exchanged between target and projectile are exotic. For example, the $\pi^+p \rightarrow \pi^-\Delta^{++}$, in which the π^- is observed near the π^+ direction, involves the exchange of two units of electric charge and is therefore an exotic exchange reaction. When there is a single well-established particle that can be exchanged between target and projectile, the angular distribution shows a characteristic peak at small scattering angles. Until 1969 there were no reactions involving exotic exchange which showed this behavior, and this was taken to be further evidence against the existence of exotics. Since 1969 a number of exotic exchange reactions have shown peaks at small scattering angles. Evidence of this sort is, however, weaker than evidence from either production or formation experiments. The experimental situation regarding exotic mesons based on scattering information has recently been reviewed by John Rosner of the University of Minnesota and he concluded that "two-meson exchange seems to be the most likely explanation of the data" rather than a single exotic particle exchange.

[RICHARD C. LAMB]

Coupled and overlapping resonances. Modern elementary particle experiments can detect interference effects in mass distributions, which can result when two (or more) different elementary particles (resonances) have nearly the same mass. In the language of quantum mechanics, the state functions of these two particles have overlapping components so that transitions from one state to the other occur.

The word "resonance" will be used interchangeably with particle. Historically it was used to describe an intermediate or excited state produced when target particles, for example, protons or neutrons, were struck by energetic mesons or photons. All these particles (except the proton and electron, which are stable) can occur in nature with varying mass values distributed about some most probable value known as the mass of the particle. The distance in energy units between half the most probable values on either side of the most probable value is called the width (Γ). When resonances lie close enough in mass for interference and mixing to occur, their mass distributions will overlap and their most probable mass values will be separated by very few widths.

If the individual resonances themselves are well studied entities through independent experiments, then the interference, or mixing, effects can be used to gain important knowledge about the production process itself. Such information is extremely useful because the strong interactions responsible for producing these particles are poorly understood. Alternatively, these interference effects can yield information, through unusual distortions of conventional mass shapes, about particles which would otherwise remain undiscovered. In fact, the mass distributions can be so different as to make them worthy of study for their own sake.

K-Mesons. Many of the important ideas in part developed or follow from attempts in the past 10 years to describe the decay of the two neutral K-mesons. These are produced by strong interactions as either K^0- or $\bar{K}^0$-mesons; however, their decay is caused by weak interactions and proceeds from linear combinations of K^0 and $\bar{K}^0$ called K_S and K_L (S = short life and L = long life), the former having a width over 600 times that of K_L ($\Gamma_S = 620 \Gamma_L$). The separation between their most probable values (masses M_S and M_L) is half of Γ_S (that is, $M_L - M_S = 0.47 \Gamma_S$), so that the mass distributions overlap significantly. The Bell-Steinberger relation gives the amount of overlap between the state functions themselves (that is, $\int \Psi_L^* \Psi_S dV \equiv \langle K_L | K_S \rangle$) in terms of all possible ways one state (K_S, for example) can be connected to the other state (K_L) through any intermediate state. It gives the restrictions due to unitarity (a way of saying that probability must be conserved) for transitions involving the overlapping resonances. In mathematical form, Eq. (1) applies,

$$\langle K_L | K_S \rangle [\tfrac{1}{2}(\Gamma_S + \Gamma_L) - i(M_L - M_S)]$$
$$= \sum_F \langle F | T | K_L \rangle^* \langle F | T | K_S \rangle \quad (1)$$

where $i = \sqrt{-1}$ and T stands for the transition operator connecting the K-meson states to possible decay products F which are summed over. The fact that the overlap integral $\langle K_L | K_S \rangle$ is not zero indicates transitions are connecting the two states. In this case, the symmetry given by charge conjugation times parity, believed to be equivalent to time reversal, is being violated; under strict conservation, $\langle K_L | K_S \rangle = 0$ and then K_L does not decay into two pi mesons, or pions—only K_S does.

Electromagnetic mixing. In 1969 careful study of the two pion decay mass distribution of the ρ-meson, a resonance with intrinsic spin one (in $\hbar$ units), isotopic spin one ($I = 1$), a mass of 765 MeV, and a width $\Gamma_\rho = 125$ MeV, revealed a deep, narrow dip at the mass of the ω-meson, a resonance with spin one, isotopic spin zero ($I = 0$), a mass of 784 MeV, and width $\Gamma_\omega = 12$ MeV, which decays into three pions. Both ω- and ρ-mesons had been studied in detail before but never with such good mass resolution. The existence of this interference distortion of the ρ mass spectrum indicates nonzero overlap of the ω- and ρ-meson state functions ($\langle \omega | \rho \rangle \neq 0$) and that isotopic spin symmetry is violated. In other words, electromag-

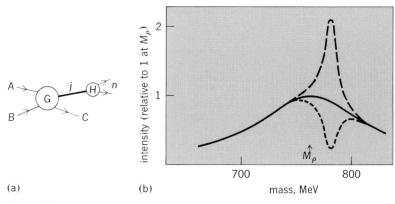

Fig. 2. Representations of resonances. (*a*) Schematic diagram depicting the reaction $AB \rightarrow jC$. The resonance j is produced with an amplitude G and decays into the final particles n with the amplitude H. (*b*) Mass distribution curves for the ρ-meson (solid line) calculated from Eq. (3), and for the ρ-ω system for constructive (long dashes) and destructive (short dashes) interference calculated from the absolute value squared of Eq. (4). The dashed curves are obtained with $\delta G_\omega / G_\rho = -2i$ and $+2i$, respectively, and correspond generally to what is expected for the $\pi\pi$ mass in $\pi^+ p \rightarrow \pi^+\pi^- \Delta^{++}$ and $\pi^- p \rightarrow \pi^+\pi^- n$ (or $\pi^+\pi^-\Delta^0$), respectively.

netic forces which violate isotopic spin symmetry mix or couple these $I = 0$, $I = 1$ particles together. Equivalently, in application of Eq. (1) K_S and K_L may be replaced by ρ and ω, respectively, and the intermediate states F are reached from the ρ (or ω) state by electromagnetic transitions.

One can depict the production of a resonance j of mass M_j and width Γ_j, as in Fig. 2a, for a reaction $A + B \rightarrow j + C$. Particles A and B collide (time increases to the right) with amplitude G for producing the resonance j and some other particle C ($|G|^2$ is proportional to the probability that j will be produced). The amplitude for j to disintegrate into decay products n is H. For the overall process the amplitude is $T_{AB \rightarrow jC}$, as in Eq. (2),

$$T_{AB \rightarrow jC} = H \frac{1}{M - M_j + i\Gamma_j/2} G \quad (2)$$

where M is the mass with which j is actually produced and M_j is the most probable mass value. The absolute value squared $|T_{AB \rightarrow jC}|^2$ gives the probability of finding j with mass M, as in Eq. (3).

$$|T_{AB \rightarrow jC}|^2 = |HG|^2 [(M - M_j)^2 + \Gamma_j^2/4]^{-1} \quad (3)$$

The factor in square brackets gives the famous Breit-Wigner mass distribution, which is shown plotted in Fig. 2b for $j = \rho$. Clearly the peak value is at M_ρ and the full width at half maximum is Γ_ρ. In a typical experiment (done at Berkeley) $AB \rightarrow jC$ corresponds to $\pi^+ p \rightarrow \rho^0 \Delta^{++}$, with ρ^0 subsequently decaying into $\pi^+ + \pi^-$.

When two mesons mix (or couple) as do the ρ- and ω-mesons, Fig. 2a and its mathematical expression Eq. (2) still apply if G is reinterpreted to contain two terms for production of the two particles (G_ρ and G_ω, below) and H likewise includes decay terms for each particle (H_ρ and H_ω being the amplitude for each to decay to $\pi\pi$). In Eq. (2), the factor between H and G is often called the propagator because from it one can calculate the probability that j will go or propagate from production to a particular decay point. It must be generalized so that the masses and widths of the two particles are each included. A convenient way of doing this is to replace $M_j - i\Gamma_j/2$ by a mass matrix; for the ρ-ω interference situation, Eq. (2) becomes Eq. (4) (in matrix form, with spin projection dependence suppressed throughout), where δ measures the strength of all possible ways the ω-meson can turn into a ρ-meson. This expression can be expanded (by doing the matrix multiplication) into four terms each somewhat like Eq. (2). In forming the absolute value squared $|T_{AB \to (\rho,\omega)C}|^2$ to get the probability for observing the ρ-ω system at the mass value M, one must multiply each of these four terms by the complex conjugate of all four so that numerous cross or interference terms contribute. For example, the numerator of one such term is $H_\rho^* G_\rho^* H_\rho \delta G_\omega = \delta |H_\rho|^2 G_\rho^* G_\omega$, from which it is clear that the magnitude and particularly the sign depend on the production amplitude of the ρ-meson times that for the ω-meson. Thus, compared with the single-resonance result of Eq. (3), there will be interference contributions which add or subtract as a function of M, illustrated by the dashed lines for production amplitudes with relative phase 90 and $-90°$ in Fig. 2b. With the masses, widths, coupling (δ), and decay parameters known, the ρ-ω interference studies yield quite direct information on the mechanism by which these particles were produced in strong interactions (about which physicists need better understanding). A large number of different reactions have been studied and compared with the relative production phase expected from Regge theory (a phenomenologically developed, after-the-fact description of high-energy reactions). Quite remarkably, the majority of these relative phases were predicted before confirmatory experiments were done, indicating considerable progress.

A_2-Meson. It appeared that a remarkable example of overlapping resonances was provided by the A_2-meson of mass 1300 MeV, width 90 MeV, spin 2, and isotopic spin 1, when the CERN missing mass (MM) group published their results in 1967. In this experiment the meson is not observed directly; energy and momentum conservation are used to calculate the mass of the meson as missing mass from measurements of the momentum and angle of the recoiling proton in the reaction $\pi^- + p \to (MM)^- + p$. These MM data in the A_2 region are shown as the histogram in Fig. 3; a deep dip at $\sim$1300 MeV splits this mass distribution into two symmetric peaks. Subsequent analyses for the spin and parity quantum numbers indicated identical values for each peak. This and other evidence imply that a relation similar to Eq. (4) with two mesons ($A_2^{(1)}$ and $A_2^{(2)}$ in the following) should be used in analysis of these measurements. As a result, this mass region in recent years has been one of the most studied and most intriguing in elementary particle physics.

An interesting way of explaining the dip follows from requiring that the overlap between the two particle states be maximal, $\langle A_2^{(2)} | A_2^{(1)} \rangle = 1$. Then, in Eq. (4) $A_2^{(1)} \to \rho$ and $A_2^{(2)} \to \omega$; also, in Eq. (1) $A_2^{(1)} \to K_S$ and $A_2^{(2)} \to K_L$. With the restrictions implied by Eq. (1) imposed on $H_{A_2(1)}$ and $H_{A_2(2)}$ and also on $G_{A_2(1)}$ and $G_{A_2(2)}$ (for example, $H_{A_2(1)} = \langle F | T | A_a^{(1)} \rangle$, and so on), Eq. (4) reduces to the two-parameter form, relation (5). The mass distribution will be relation (6), which gives an excellent fit to

$$T_{AB \to (\rho,\omega)C} = (H_\rho, H_\omega) \begin{pmatrix} M - M_\rho + i\Gamma_\rho/2 & \delta \\ \delta & M - M_\omega + i\Gamma_\omega/2 \end{pmatrix}^{-1} \begin{pmatrix} G_\rho \\ G_\omega \end{pmatrix} \quad (4)$$

$$T_{\pi^- p \to (A_2^{(1)}, A_2^{(2)})p} \propto (M - M_0)/(M - M_0 + i\tfrac{1}{2}\gamma)^2 \quad (5)$$

$$|T_{\pi^- p \to (A_2^{(1)}, A_2^{(2)})p}|^2 \propto (M - M_0)^2/[(M - M_0)^2 + \gamma^2/4]^2 \quad (6)$$

the CERN data with $M_0 = 1300$ MeV and $\gamma = 28$ MeV (Fig. 3). This form has the fewest possible parameters to which Eq. (4) can be reduced and is often called the "dipole" resonance shape. More generally, when the overlap integral is not zero, the system of two particles can be called a double resonance.

The most recent experiments have caused considerable controversy. A repetition of the CERN experiment (same reaction, same energy) in the United States by a group from Northeastern University and the State University of New York at Stony Brook failed to find structure! These data are shown by the error flags in Fig. 3. This second

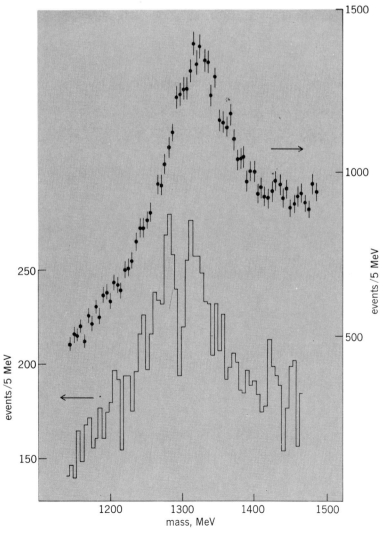

Fig. 3. Mass distributions found by the CERN missing mass group (histogram) and by the Northeastern University – State University of New York collaboration (upper) for the A_2 meson.

experiment had poorer resolution than the CERN experiment but more events, which the authors feel should compensate. The authors of the European work, however, strongly believe this is not so. Another experiment is needed to solve this dilemma; it appears that interference between the A_2(system?) and the f^0-meson may provide the answer.

For background information see ELEMENTARY PARTICLE; MESON; QUANTUM THEORY, NONRELATIVISTIC; QUANTUM THEORY, RELATIVISTIC; QUARKS; REGGE POLE; SYMMETRY LAWS (PHYSICS) in the McGraw-Hill Encyclopedia of Science and Technology. [KENNETH E. LASSILA]

Bibliography: C. Baltay and A. H. Rosenfeld (eds.), *Experimental Meson Spectroscopy*, 1970; A. Donnachie and E. Gabathuler (eds.), *Vector Meson Production and Omega-Rho Interference*, Science Research Council, Daresbury, England, 1970; G. Fox and C. Chiu (eds.), *Conference on the Phenomenology of Particle Physics*, Caltech, 1971; G. Goldhaber, in *Hyperon Resonances—70*, Duke University, 1970; P. K. Kabir, *Springer Tracts Mod. Phys.*, 52:91, 1970; H. J. Lipkin, Resonance physics, *Proceedings of the Lund International Conference on Elementary Particles*, 1969; J. Rosner, Review of exotic mesons, *1970 Philadelphia Meson Spectroscopy Conference*, 1970.

Elements (chemistry)

An interesting development in the theory of new elements has been the recent theoretical calculations which indicate that there may be several regions of increased nuclear stability far beyond the present limits of the periodic table. The elements falling within the bounds of these regions of nuclear stability are referred to as the superheavy elements. Efforts are now underway to produce these elements synthetically; simultaneously, corresponding efforts are being exerted to search for some of these elements in nature.

For some time nuclei have been known to possess shell structure for the neutrons and protons, the particles which form the basic ingredients of the nucleus, in a manner analogous to the electron orbitals that determine atomic structure. Closed shells containing 28, 50, and 82 neutrons or protons have been found to impart extra stability to a nucleus in a manner similar to the special stability of atoms associated with 8, 18, or 32 electrons. The different numbers result from different rules governing the interaction of nucleons as opposed to electrons. In an earlier period in the development of nuclear theory, it had reasonably been assumed that since 126 was the next closed neutron shell beyond 82, in an analogous fashion the next closed proton shell would also occur at $Z = 126$. However, calculations by several theorists almost uniformly conclude that the next closed proton shell occurs at $Z = 114$, with $Z = 126$ a less likely prospect. Similar calculations predict closed neutron shells at $N = 184$ and 196. Even further away, closed shells at $Z = 164$ and $N = 318$ have been predicted by several groups.

Thus the nuclide 114^{298} is expected to have a closed neutron and proton shell with relatively large stabilization energies. It should be noted that the stabilizing influence of a closed shell extends over a range of neutrons and protons. Hence 114^{298}

Table 1. Calculated alpha and spontaneous-fission half-lives of some superheavy elements

Atomic number (Z)	Neutron number (N)			
	180	182	184	186
108	$10s^*$	10^2y	10^8y	10^8y
	$10^4y\dagger$	10^8y	$10^{13}y$	10^6y
110	10min	10^4y	$10^{10}y$	$10^{10}y$
	10^2y	10^6y	10^8y	10^2y
112	10d	10^6y	$10^{13}y$	$10^{13}y$
	1y	10^3y	10^4y	1y
114	10^2y	10^9y	$10^{16}y$	$10^{15}y$
	10d	1y	10y	1d
116	1d	10^5y	$10^{11}y$	$10^{11}y$
	½min	10s	½min	0.1s

*Upper value for each nuclide is the spontaneous-fission half-life.
†Lower value for each nuclide is the alpha half-life.

is predicted to be the center of a rather large region of exceptional nuclear stability with a similar region possibly existing in the vicinity of 164^{482}.

Limiting half-lives. Accepting the possibility of relatively stable doubly-closed-shell superheavy nuclei, what would the expected alpha and spontaneous-fission half-lives be in these regions? These are the truly limiting half-lives as far as detection of new elements are concerned, since these half-lives can be much less than 10^{-9} sec, whereas the half-lives for beta and electron-capture decay do not become much shorter than 0.1–0.01 sec.

Since most doubly-closed-shell nuclei have even numbers of neutrons and protons and since calculations are simpler for such nuclei, the spontaneous-fission and alpha half-lives have been computed for even-even superheavy elements. Table 1 lists some half-lives of elements in the region of element 114 as calculated by S. Nilsson. It should be noted that nuclei with an odd number of neutrons or protons or both are known to have longer alpha-decay half-lives, and in the case of spontaneous fission, much longer half-lives than neighboring even-even nuclei.

As can be seen from Table 1 in the region of $Z = 114$, some nuclei are predicted to have very long alpha and spontaneous-fission half-lives, with the longest overall half-life falling in the region of element 110, mass 294. Although calculations of this type may be in error by orders of magnitude, the trend is nevertheless clear that relatively stable nuclides should exist in the islands of stability. Nuclides with half-lives of the order of seconds or even microseconds are readily detected with modern instruments, so that there can be quite a large margin of error in the predictions without materially effecting the chances of detecting the superheavy elements.

Possible syntheses. The known elements extend through element 105. Those beyond uranium (element 92) and through fermium (element 100) are best prepared in high-flux reactors. It appears that elements beyond fermium will not be formed in such reactors, since the buildup of elements by neutron capture terminates at Fm^{258}, an isotope that decays by spontaneous fission with an extremely short half-life. The source of elements beyond 100 has been, and seems likely to continue to be, accelerator irradiations. In order to reach the desired superheavy nuclei in the islands of

Conventional form of periodic table showing predicted locations of new elements.

stability, new accelerators must be built. The heavy-ion accelerators in existence cannot develop useful beams of projectiles above $Z \cong 18$ (argon) or lighter. As the mass of projectiles is increased, the cross section for producing a given heavy element decreases rapidly as a result of fission competition. Thus the neutron-deficient isotopes that can be produced with existing accelerators fall short of the regions of stability. However, recent considerations indicate that if much heavier ions with adequate energy are used, they will be able to form the superheavy nuclides in the region of interest.

Table 2 shows some types of nuclear reactions that might be used to synthesize isotopes of element 114 near or in the region of stability. Reactions 1 and 2 represent conventional types of heavy-ion nuclear reactions employing neutron-rich targets and projectiles, but yielding isotopes of element 114 that are neutron-deficient with respect to the region of enhanced stability and thus having short half-lives. The product in reaction 2 may well have a half-life long enough to be detectable. Reactions 3 and 4 are illustrative of nuclear reactions in which the product nucleus is formed with low excitation energy, and perhaps with a large cross section, since the competitive fission reaction would be suppressed. However, existing combinations of projectiles and targets produce isotopes of element 114 which probably have too short a spontaneous fission half-life to allow detection. Perhaps unusual transfer reactions such as reactions 5 and 6 might yield neutron-rich isotopes of element 114 in or near the desired region, but they may prove to have formation cross sections that are too small.

Currently reactions such as 7 and 8 appear to offer the most hope for getting the desired doubly-closed-shell isotope of element 114. These reactions are in essence fission reactions, in which the superheavy elements occur as fission products. Extrapolation of results obtained by United States and Soviet scientists, using carbon, neon, and argon as projectiles to induce fission in uranium targets, indicate that reactions 7 and 8 may be orders of magnitude more favorable than most of the types listed in Table 2.

Reaction 9 represents an "overshoot" where the primary product decays by alpha-particle emission into element 114. It should be emphasized that the nuclear reactions in Table 2 represent only selected kinds of interactions between complex nuclei; many other possibilities exist.

Electronic structure and chemistry. From the foregoing discussion it appears that several types of nuclear reactions may be used to produce superheavy elements, some of which are expected to have long alpha and spontaneous-fission half-lives. In particular, the fission-type reactions, which appear to be a promising technique, may produce new elements accompanied by a wide distribution

Table 2. Possible heavy-ion reactions leading to element 114

1. $_{94}Pu^{244} + _{20}Ca^{48} \rightarrow 114^{288} + 4n$
2. $_{96}Cm^{248} + _{20}Ca^{48} \rightarrow 114^{290} + 2n + _{2}He^{4}$
3. $_{50}Sn^{124} + _{64}Gd^{160} \rightarrow 114^{284}$
4. $_{54}Xe^{136} + _{60}Nd^{150} \rightarrow 114^{286}$
5. $_{64}Gd^{160} + _{70}Yb^{176} \rightarrow 114^{298} + 20p + 18n$
6. $_{48}Cd^{116} + _{92}U^{238} \rightarrow 114^{298} + _{26}Fe^{56}$
7. $_{64}Cd^{160} + _{92}U^{238} \rightarrow 114^{298} + _{42}Mo^{98} + 2n$
8. $_{92}U^{238} + _{92}U^{238} \rightarrow 114^{298} + _{70}Yb^{174} + 4n$
9. $_{20}Ca^{48} + _{98}Cf^{252} \rightarrow 4n + 118^{296}$
   $118^{296} \xrightarrow{\alpha} 116^{292} \xrightarrow{\alpha} 114^{288}$
10. $_{94}Pu^{244} + _{20}Ca^{52*} \rightarrow 114^{294} + 2n$

*$_{20}Ca^{52}$ may be a product formed in certain types of high-energy nuclear reactions.

of other elements as fission products. Thus, to identify the new elements, it may be necessary to separate chemically each one from a mixture of other elements. To predict the chemical behavior of the superheavy elements and thus provide the necessary clues to devise separation procedures requires a consideration of the probable position of the new elements in the periodic table.

The positions of the elements in the periodic table depend on their electronic structure. An excellent discussion of this problem has been given by Glenn T. Seaborg, and the expected locations of the new elements in the periodic table, as predicted by Seaborg, are shown in the illustration. This chart is based on various Hartree-Fock-type calculations made by groups at the Los Alamos Scientific Laboratory, Northwestern University, Argonne National Laboratory, and Oak Ridge National Laboratory. It should be possible to develop chemical separations based on the known chemistry of the lighter homologs of the superheavy elements. The preparation of the superheavy elements presents the exciting prospect of exploring the chemistry of a whole new group of elements and of probing the electronic structure of the atoms in regions well beyond the present periodic table.

For background information see ELEMENTS (CHEMISTRY); NUCLEAR REACTION; NUCLEAR STRUCTURE in the McGraw-Hill Encyclopedia of Science and Technology. [PAUL R. FIELDS]

Bibliography: J. Gruman et al., *Z. Physik*, 228: 371, 1969; S. G. Nilsson et al., *Nucl. Phys.* A131:1, 1969; G. T. Seaborg, *Annu. Rev. Nucl. Sci.*, 18:68, 1968; G. T. Seaborg et al., *The Transuranium Elements: The Mendeleev Centennial*, 13th Robert A. Welch Foundation Conference on Chemical Research, November, 1969.

Emphysema

Emphysema is a common and important disorder of the lungs characterized by an enlargement of the gas-exchanging part of the lung which is accompanied by, or perhaps is a consequence of, destruction of the walls of the air spaces. Reports of recent research indicate that the final common pathway producing various types of emphysema may be an unbalance in the protein metabolism of the lung.

Classification. Classification may be based on appearance (morphology) or clinical and functional criteria. Morphological classification is based on the way in which the gas-exchanging portion (acinus) of the lung is affected. The acinus is the normal respiratory unit and consists of, in series, respiratory bronchioles, alveolar ducts, and alveolar sacs. These are the air passages which contain air sacs or alveoli in their walls. Four morphological types of emphysema occur, depending on what part of the acinus is predominantly involved; these are termed centrilobular, panlobular, paraseptal, and irregular emphysema. When the first part of the acinus is destroyed, the respiratory bronchioles enlarge and become confluent; this is centrilobular or centriacinar emphysema, the most common sort. The peripheral part may be predominantly involved, in which instance the alveolar sacs will be selectively enlarged, producing paraseptal or periacinar emphysema. Uniform destruction and enlargement of the entire acinus is termed panlobular or panacinar emphysema, while irregular involvement is called irregular or scar emphysema.

Clinical classification depends on the features emphysema produces in patients affected. Recognized types of emphysema include obstructive pulmonary emphysema, familial emphysema, congenital lobar emphysema, unilateral pulmonary emphysema, and bullous disease of the lung. Thus there are several forms of emphysema, each of which may have its own cause, effect, and outcome. Two clinical types are responsible for the great majority of symptomatic patients: obstructive emphysema and familial emphysema.

Obstructive pulmonary emphysema. This condition is due to widespread severe emphysema, usually centrilobular or panlobular or both. Characteristically airflow is impeded during expiration, and hence the term obstructive. A rapid and startling increase in deaths and disability due to obstructive pulmonary emphysema has occurred in the United States in the last 20 years, with the death rate doubling every 6 years. Men are affected seven times more commonly than women, and the peak age incidence is in the sixth and seventh decade. The most important cause of this form of emphysema is heavy cigarette smoking. However, the occurrence of emphysema in light smokers, its absence in many heavy cigarette smokers, and the wide variation in incidence and severity of emphysema from country to country indicate that other factors, such as individual susceptibility, are important. Certain occupations, notably coal mining, have high emphysema-incidence rates. It is likely that the environmental factors of urban dwelling also play a part.

The chief symptom of a patient with obstructive emphysema is shortness of breath; this is often insidious in onset. Chronic bronchitis (chronic cough and sputum) and repeated lung infections are important associated disorders in the majority of cases, and death and disability are often due to these complications. Heart failure may also result from emphysema and associated chronic bronchitis.

Mechanical properties of the lung are affected in obstructive emphysema. The elasticity of the lung is diminished, and this affects both the maximum force that can be applied to the lungs and the proper patency of the airways. This means in particular that maximum airflow out of the lung during expiration is greatly diminished. In addition, destruction of lung tissue alters the flow of blood through the lungs, so that the transfer of gas from the air spaces to the circulating blood is impaired.

While no treatment can restore normal function to the parts of the lung already destroyed, medical management is of considerable value. In particular, chronic bronchitis and lung infections can be controlled and treated. Graduated physical exercise, breathing exercises, and controlled oxygen therapy can produce remarkable subjective improvement.

Familial emphysema. Severe emphysema may be associated with a deficiency in the blood of a specific protein named alpha-1-antitrypsin. This deficiency is genetically determined and inherited in a complex way. One homozygous state (ZZ type) is associated with almost complete absence of circulating alpha-1-antitrypsin. These homozygotes differ from patients with obstructive pulmonary emphysema in that the former develop symptoms at an earlier age and often do not have bronchitis. ZZ homozygotes account for more than 60% of patients under the age of 40 who have severe emphysema and 1–5% of all patients with disabling emphysema. Not all patients with absent alpha-1-antitrypsin will develop emphysema, however. Abstinence from cigarette smoking delays, and may obviate, the onset of emphysema, and affected women are less prone to develop emphysema than men. Morphologically the emphysema is panlobular in type, and characteristically the lower parts of the lungs are much more affected than the upper.

Heterozygotes for alpha-1-antitrypsin deficiency are much commoner than homozygotes. One particular type (MZ) is responsible for the carrier state, and this type makes up 3–5% of the population. MZ heterozygotes have levels of antitrypsin in their blood which are intermediate between normals and ZZ homozygotes. Heterozygotes who develop emphysema have the usual features of patients with chronic obstructive emphysema rather than the features of familial emphysema.

Pathogenesis. While it is clear that cigarette smoking and alpha-1-antitrypsin deficiency are implicated in the great majority of patients with symptomatic emphysema, the exact pathogenesis of emphysema in these cases is uncertain.

The frequent association of chronic bronchitis with emphysema was responsible for the view that bronchitis produced emphysema. It was thought that bronchial infections obstructed airways which caused overexpansion of air spaces with subsequent loss of tissue. However, chronic bronchitis is presently considered less important as a cause of emphysema. Their frequent coexistence is explained by the importance of cigarette smoking in causing both conditions and to the predisposition to infection of the airways in emphysema.

Several recent observations have emphasized the importance of protein metabolism of the lung in the pathogenesis of emphysema. In animals emphysema can be produced experimentally by injecting papain, an enzyme which digests protein, into the lung. While the exact biological function of alpha-1-antitrypsin is not known, it can inhibit the action of trypsin, another proteolytic enzyme. The occurrence of emphysema in association with absence of this antitryptic factor suggests an alteration in the balance between synthesis and breakdown of protein in the lung. DL-Penicillamine (β,β-dimethylcysteine) produced emphysema in another experiment. This compound interferes with the cross-linkages of elastin and collagen, the two most important structural proteins of the lung. One hypothesis to explain the relationship of smoking to emphysema is that cigarette smoke irritates the lung and produces mild inflammation in the air spaces. The white cells which are attracted by the inflammation release their enzymes, some of which digest protein. The balance between synthesis and breakdown is upset and emphysema occurs. Patients who lack alpha-1-antitrypsin have diminished resistance to these proteolytic enzymes and are thus predisposed to emphysema.

For background information *see* EMPHYSEMA; LUNG DISORDERS in the McGraw-Hill Encyclopedia of Science and Technology.

[WILLIAM M. THURLBECK]

Bibliography: D. V. Bates, R. V. Christie, and P. T. Macklem, *Respiratory Function in Disease*, 2d ed., 1971; M. F. Fagerhol and C. B. Laurell, *Progr. Med. Genet.*, 7:96–111, 1970; B. L. Gordon, R. A. Carleton, and L. P. Faber, *Clinical Cardiopulmonary Physiology*, 1969.

Environmental engineering

Everything man does affects the environment in which he lives to some degree. Environmental engineering is the discipline which evaluates these effects and develops controls to minimize environmental degradation.

In the 1960s the United States became acutely aware of the deterioration of its air, water, and land. The roots of the problem lie in the rapid growth of the national population and the industrial development of natural resources which has given Americans the highest standard of living in the world. Since 1964 there has been enactment of national, state, and local legislation directed toward the preservation of these resources.

The technology which provided society with all the necessities and luxuries of life is now expected to continue providing these services without degradation to the environment. How well this goal is attained will depend in great measure on the environmental engineers who must cope with the enormous challenges presented by society.

It is the feeling of industry and governmental agencies that the ultimate goals should be the design of processes and systems which need minimal treatment for pollution control and the ultimate recycling of all wastes for reuse. This philosophy is both logical and necessary in a society that is rapidly depleting its nonrenewable natural resources.

Governmental policy. State and local governments have the prime responsibility for the adoption, administration, and enforcement of suitable air, water, and land quality standards. Legislation has been or is being enacted by the state and Federal governments.

The Federal Wilderness Act of 1964 has been enacted, followed by passage of the Clean Air and Water Act of 1965 and the Clean Water Restoration Act of 1966. These were in turn followed by the Air Quality Act of 1967 and attendant state and local regulations. To assure that the national legislation was effectively administered and enforced, the National Air Pollution Control Administration was created under the Department of Health, Education, and Welfare and the Federal Water Quality Administration became an integral part of the Department of the Interior.

In December, 1970, by presidential order, the Environmental Protection Agency (EPA) was formed. Under Public Law 91-604, which extended the Clean Air and Air Quality acts, this agency now embodies under one administrator the responsibility for setting standards and compliance timetables for air and water quality improvement. This agency will also administer grants to state and local governments for construction of waste-water treatment facilities and for air and water pollution control.

Industrial policy. Through its environmental engineers, industry is directing much of its capital and technological resources to the correction and maintenance of improved environmental quality. The solution of the myriad complex environmental problems requires the skills and experience of persons knowledgeable in health, sanitation, physics, chemistry, biology, meteorology, engineering, and many other fields.

Each air and water problem has its own unique approach and solution. Liquid wastes can generally be treated chemically or physically or a combination of the two for removal of contaminants with the expectation that the liquid can be recycled. Air or gaseous contaminants can be removed by scrubbing, filtration, absorption, or adsorption and the clean gas discharged into the atmosphere. The removed contaminants, either dry or in solution, must be handled wisely or a new water or air pollution problem may result.

Industries that extract natural resources from the earth, and in so doing disturb the surface, are being called on to reclaim and restore land for reuse. Many companies are making reclamation a part of planning and operation. It seems more practical and economical to pursue this course rather than to leave reclamation until the project is completed.

Air quality management. The air contaminants which pervade the environment are many and emanate from multiple sources. A sizable portion of these contaminants are produced by nature, as witnessed by dust that is carried by high winds across desert areas, pollens and spores from vegetation, and gases such as sulfur dioxide and hydrogen sulfide from volcanic activity and the biological destruction of vegetation and animal matter.

The greatest burden of man-made atmospheric pollutants are carbon monoxide, hydrocarbons, nitrogen oxides, and particulates. These are generated by the burning of fuels to power cars and public transportation systems. It has been conservatively estimated that 60% of all major pollutants in the United States come from this source alone.

Industry contributes approximately 17% of the total air pollutants, with utility power plants following closely with 14%. The major pollutants from these sources are sulfur dioxide, particulates, hydrocarbons, and nitrogen oxides.

California has set the pace in the promulgation of standards to reduce internal-combustion-engine emissions. The Federal government and heavily populated states and cities are rapidly developing standards for control of carbon monoxide, hydrocarbons, and nitrogen oxides.

Many states have ambient air and emission standards directed primarily toward the control of industrial and utility power plant pollution sources. The general trend in particulate control is to limit the emissions from a process stack to a specified weight per hour based on the total material weight processed. Process weights become extremely large in steel and cement plants and in large nonferrous smelters. The degree of control necessary in such plants can approach 100% of all particulate matter in the stack. Retention equipment can become massive both in physical size and cost. The equipment may include high-energy venturi scrubbers, fabric arresters, and electrostatic precipitators. Each application must be evaluated so that the selected equipment will provide the retention efficiency desired. *See* POWER PLANT.

Sulfur oxide retention and control present the greatest challenge to industrial environmental engineers. Ambient air standards are extremely low, and the emission standards adopted by some states are beyond the present state of the control art for weak sulfur oxide gas streams. The states of Arizona, Montana, Nevada, and Washington have adopted standards requiring nonferrous smelters to limit sulfur emissions to 10% of the total sulfur present in the concentrate treated. Most copper smelters and all utility power plants have large-volume, weak sulfur oxide effluent gas streams. To scrub these large, weak gas volumes with limestone slurries or caustics is extremely expensive, requires prohibitively large equipment, and presents waste-water problems of enormous magnitude. However, gas streams containing high concentrations of sulfur dioxide gas can be treated more economically to obtain such by-products as elemental sulfur, liquid sulfur dioxide, sulfuric acid, calcium sulfate, and ammonium sulfate. *See* AIR POLLUTION.

The task of upgrading weak smelter gas streams to produce products with no existing market has led to extensive research into other methods of producing copper. A number of mining companies are now piloting hydrometallurgical methods to produce electrolytic-grade copper from ores by chemical means, thus eliminating the smelting step. Liquid ion exchange, followed by electrowinning, is also being used more extensively for the heap leaching of low-grade copper. This method produces a very pure grade of copper without the emission of sulfur dioxide to the atmosphere.

Water quality management. Recently, by presidential executive order, the EPA and its office of water quality has assumed the Federal responsibility in water pollution control. Also the Army Corps of Engineers, under the Navigable Streams Refuse Act of 1899, has initiated regulation of waste effluents by requiring permits to discharge wastes or treated effluents to navigable waters. It remains to be seen if this duplication of effort will continue since EPA officials have also requested that the states develop liquid waste effluent standards. Standards can only be arbitrary and, in some cases, punitive without epidemiological investigation of receiving-water biota. Some mining industry environmental engineers can be expected to take issue with the philosophy of wasting a portion of the assimilative capacity of receiving waters by not utilizing part of this natural phenomenon and thus producing finished products at a lower cost to the consumer.

The state of the art of treatment of waste waters containing metals has barely advanced beyond neutralization and chemical precipitation. As water quality standards continue to become more stringent, especially concerning dissolved solids, it is apparent that chemical treatment of metals wastes will be unacceptable. Chemical treatment simply substitues one compound for another and thus does not reduce total disolved solids in a treated effluent.

Some segments of the mining industry are considering physical treatment methods or desalination techniques such as evaporation, reverse osmosis, and electrodialysis. These techniques require vast quantities of energy and are prohibitive in both capital and operational costs. Continued experimentation will undoubtedly improve the benefit-cost ratio.

For the most part the mining industry tends toward a policy of complete recycle of water. Some metallurgical processes are amenable to reuse of process waters with only minimal treatment, such as removal of suspended and settleable solids supplemented by a low total dissolved-solids content. Other processes require higher quality water with lower dissolved-solids content. Even this, however, generally requires removal of only a portion of the solids from a waste stream in order to maintain an acceptable process water quality. Many of these solids can also be reclaimed, thus recovering values now lost. This approach is not only logical but economical. If receiving-water quality criteria are more stringent than process-water requirements, it is impractical to comply with those standards and then waste the water to the nearest surface drainage. In the more arid parts of the world recycle of water has become a practice by necessity; in water-rich areas it will become practice indirectly by governmental decree.

Land reclamation. It has been stated that strategic mineral development is the most productive use of land because of the great values that are received by the nation from such small land areas. Open-pit and strip mining have recently been criticized for their impact on the surrounding ecosystems. Some preservationists and conservationists feel that the Earth's surface should not be disturbed by either exploration or by mining activities. Mining companies must reverse the spoiler image that has been created in the past, especially in the large-scale stripping of coal in the eastern and mid-central states. Most states now have regulations that require that all stripped land be reclaimed. Strip mining of coal has often produced an attendant acid drainage problem.

Ecosystem studies are now being conducted by many mining companies during exploration and prior to the commencement of mining operations. These studies lead to the effective planning for the most desirable method of mining that will least disturb the environment and yet lend itself to later reclamation.

The Federal Public Land Law Review Commission has recently recommended that those who use public lands and resources should be required to conduct their activities in a manner that avoids or minimizes adverse environmental impacts. These users should also be held responsible for restoring areas to an acceptable standard upon completion of their activities. This recommendation will certainly affect future mining land laws.

A number of states in which large-scale surface mining is conducted have already enacted land reclamation laws. Many states are presently considering such regulations, and undoubtedly all states will soon regulate the mining and restoration of its land.

Mining companies, too, have taken the initiative in planning operations so as to limit the adverse impact on the environment. Notable examples of this forward-looking and concerned approach are programs initiated and carried on by American Cyanamid Co., American Metal Climax, Inc., The Anaconda Co., Bethlehem Steel Co., and Peabody Coal Co.

Typical of these industrial programs is that of The Anaconda Co. which undertook the development of the Twin Buttes Copper Mine in the desert area near Tucson, Ariz. In excess of 240,000,000 tons of alluvium were removed from this open-pit site before production could begin. This material and future waste will be used to form large dikes that will impound tailings. The dikes are terraced and planted with vegetation indigenous to the area and, when completed, will be 1000 ft wide at the base and 250 ft wide at the apex, with a maximum height of 230 ft. These dikes take on the appearance of mesas and blend into the desert landscape.

Moreover, the American Cyanamid Co. has found restoration of Florida's phosphate-mined lands can best be accomplished by reclaiming the major portions of those areas immediately and simultaneously during mining operations.

Land reclamation is made an integral part of mine planning by advanced consideration and decisions regarding what the area should look like upon completion of mining activity. By systematically forming eventual lakes and by distributing stripped wastes into previously mined cuts, grading the area and restoring it to desirable land become relatively easy. Many of these reclaimed areas have become useful as parks, recreational areas, wildlife sanctuaries, and agricultural and residential development sites.

Conclusion. Aroused public interest and the increased awareness of industrial firms coupled with governmental action have provided a base for environmental improvements. While steps are now being implemented toward this goal, each of these segments must recognize the particular problems of the other. No extensive progress can be accomplished without mutual cooperation and understanding. The public, industry, and government will have to act sensibly and reasonably to overcome a long-existing problem which has no single immediate solution.

For background information *see* ATMOSPHERIC POLLUTION; ECOLOGY; ECOLOGY, APPLIED; WATER POLLUTION in the McGraw-Hill Encyclopedia of Science and Technology.

[LEWIS N. BLAIR; JOHN C. SPINDLER; WALTER H. UNGER]

Bibliography: Cleaning Our Environment: The Chemical Basis for Action, American Chemical Society Report, 1969; J. A. Danielson (ed.), *Air Pollution Engineering Manual*, Public Health Service Publ. no. 999-AP-40, 1967; J. E. McKee and H. W. Wolf (eds.), *1963 State of California Water Quality Criteria*, California Water Quality Board and U.S. Public Health Service, 2d ed.,

1963; E. H. Peplow, Jr., Western mining's land restoration efforts, *Mining Congr. J.*, vol. 56, no. 9, September, 1970.

Epidermis (plant)

The epidermis and the cuticle which lies over it are the first defense of the plant against a hostile environment. They have always helped to defend the plant against desiccation and the attacks of pathogens and insects. In modern agriculture they may also serve as the first line of defense against toxic sprays and environmental pollution.

All the aerial surfaces, and some others, of a plant are covered by a protective skin, the cuticle. Even the apical dome of the shoot tip, overlain by the close-fitting leaf primordia, already possesses a recognizable cuticle.

The cuticle lies over, and is normally intimately connected to, the epidermal cell wall. The incrustation of epidermal wall components is cutinization, and the adcrustation of the cuticle proper onto the cutinized layer is cuticularization. The deposition by the plant of materials beyond the cuticle layer may be refered to as epicuticular development.

Abnormal distributions of cuticle. Although cuticles are normally formed upon the outer surfaces of epidermal cells, they may also be found within the substomatal cavity; at the base of the glands of salt-secreting plants; as a lining to stylar canals; below the water-absorbing scales of the Bromeliaceae; and as inner cuticles during the development of integuments into seed coats in some seeds. The old idea that the cuticle could be likened to the drying of a varnish in the air can therefore no longer be held.

Epidermal cell. The epidermis may comprise one or more layers of cells; these layers are usually morphologically and physiologically distinct from the ground tissue within the leaf. Most are basically tabular in shape, but a few are specialized in relation to their position upon the leaf, such as stomata, hydathodes, and trichomes. Little work has been done on the fine structure of epidermal cells themselves, but as Fig. 1 shows, they are not normally so extensively vacuolated as their mesophyll neighbors and sometimes, again unlike their neighbors, they contain large areas of granular osmiophilic material. They often contain leucoplasts and, occasionally, poorly developed chloroplasts with a little starch. They contain normal mitochondria, endoplasmic reticulum, Golgi bodies, and ribosomes, but very little is known about where the many components external to the epidermal wall are synthesized or how they are moved there.

Frequently the outer tangential wall of the epidermal cell is strikingly thicker than all its other walls. The walls of epidermal cells are thought to have a normal architecture and to consist of cellulose embedded in a pectin and hemicellulose matrix. They can be considered to be thickened primary walls similar in some ways to collenchyma. Occasionally, as in *Apium* petioles, the inner tangential wall may also be thickened. Sometimes a lamellation is seen in the outer epidermal wall similar to that of collenchyma and, as Fig. 2 shows, these lamellations, of which they may be up to 50

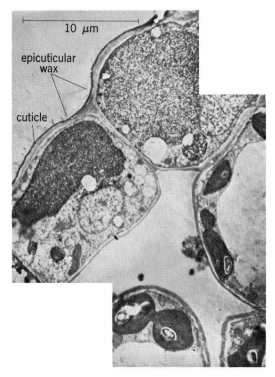

Fig. 1. Mosaic of electron micrographs of a section through leaf epidermal cells of *Eucalyptus papuana*. Epicuticular wax and the cuticle can be seen covering the epidermal cells. Note the large osmiophilic areas and the small ill-developed chloroplasts in the epidermal cells. (*From J. T. Martin and B. Juniper, The Cuticles of Plants, Edward Arnold, 1970; courtesy of N. D. Hallam*)

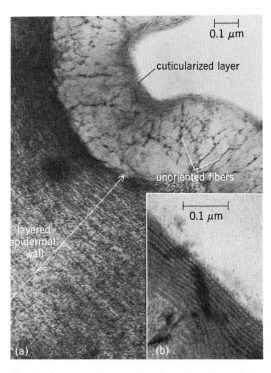

Fig. 2. Electron micrographs of sections through the epidermal wall and cuticle of *Apium graveolens*. (a) Layered epidermal wall and cuticularized layer. (b) The apparent lamellar structure of the outer layer of the cuticle shown in higher magnification.

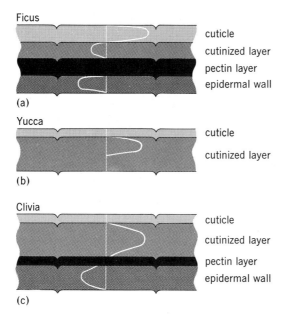

Fig. 3. Distribution and optical properties of different components of three cuticles. Vertical line indicates the position of minimal optical acitivity; deviation of the line to either side represents positive or negative birefringence. (a) Ficus epidermal wall is strongly birefringent and the "pectin" layer is weakly birefringent, but of the opposite sign. Cutinized layer shows the same direction of birefringence as the cellulose wall, but cuticle is anomalous in being birefringent in the opposite direction. (b) Yucca has a conventional isotropic cuticle, no obvious pectin layer, and a weakly birefringent cutinized layer. (c) Clivia has the most common structure. The cuticle proper is isotropic, the cutinized layer is birefringent, the pectin layer is isotropic, and the epidermal wall is birefringent at opposite sign to the cutinized layer. (From J. T. Martin and B. E. Juniper, The Cuticles of Plants, Edward Arnold, 1970)

in some plants, become progressively thinner as the cuticle is approached. These lamellations probably represent cellulose-rich layers with a predominantly longitudinal microfibril orientation, interspersed with cellulose-poor regions in which the microfibril orientation is random. Lamellae appear when any primary wall is increasing in thickness and elongating at the same time and will be most marked in structures with a predominantly axial growth, such as petioles and monocotyledonous leaves.

Figure 3 shows that the epidermal wall is variable. In most cases the innermost layer is a normal cellulose-rich wall, though Yucca (Fig. 3b) is an exception. Usually, but not invariably, there lies outside the cellulose region a layer of pectin or pectinlike material. Its presence may account for the ease with which intact cuticles can be isolated from some plants, as reported by P. J. Holloway and E. A. Baker in 1968, and its absence for the impossibility of removing the cuticle from other plants. Since the epidermal wall is expanding rapidly as the leaf extends, there must be a continuous augmentation of this pectin layer from below. The means whereby it reaches this region and the form in which it does so are unknown. However, another explanation for the inability to remove cuticles intact may be that cuticular pegs penetrate far down the anticlinal walls

(Fig. 4) and continue so far that they may be penetrated by the plasmodesmata joining together epidermal cells.

Presumably the cutinized layer consists of cellulose and matrix material, but with the addition of greater or lesser quantities of the polymer cutin in the cutinized layer (Fig. 3). Fibers which might be cellulose (Fig. 2) extend up as threads or sometimes tufts into the cutinized layer. In some *Eucalyptus* species and *Ilex integra* these tufts are so well marked that they can be seen under polarized light in the light microscope, according to M. Z. Hülsbruck in 1966. However, what is thought to be cellulose is not oriented in any obvious way and cannot be held responsible for the birefringence observed in the cutinized layer shown in Fig. 3. Cutin cystoliths are commonly found in this layer.

Cuticularized layer. Regardless of the variability of the epidermal wall and the cutinized layer below, there exists in higher plants and in many other plants as well a coherent cuticle, more or less unbroken over all the aerial parts of the plant except at such apertures as stomata and salt glands. The cuticle varies greatly in thickness from one part of the plant to another. There is no correlation between the thickness of the cuticle and the cellulose layers beneath. For example, in *Prunus laurocerasus* the cuticle is 0.7 µm and the cellulose layers are 6.0 µm thick. In *Olea lanceolata*, on the other hand, the cuticle is 13.5 µm and the cellulose layers are 1.2 µm thick. As a general rule, the adaxial (upper) cuticle is thicker than abaxial (lower) cuticle.

The chemistry of cutin, the principal constituent of the cuticle, has been extensively studied, particularly by J. T. Martin and coworkers at Long Ashton Research Station, England. Cutin is a polymolecular network of carboyxlic and hydroxycarboxylic acids connected through ester, ether, and possibly peroxide bridges. Of the acids present, 9,10,18-trihydroxyoctadecanoic acid and 10,16-

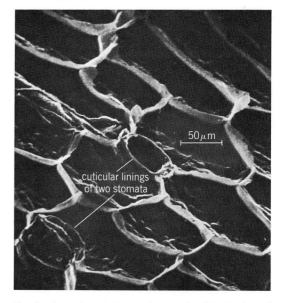

Fig. 4. Scanning electron micrograph of the inside of the isolated cuticle of a leaf of *Aloe keayi*. Lighter areas are the ridges of cuticle which in the whole leaf extend down the anticlinal walls. (Courtesy of L. E. Newton)

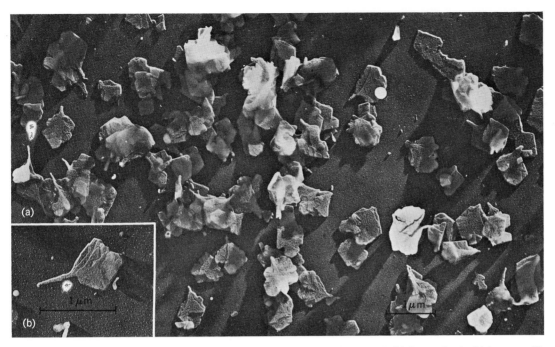

Fig. 5. Electron micrograph of the scales from the inside of the pitcher of *Nepenthes rufescens*. (a) Scales washed from the feet of a fly which had attempted to scale the pitcher wall. (b) One scale at a higher magnification. Scales probably project from the epidermal wall, supported at the thinner end.

dihydroxyhexadecanoic acid are common in many cutins, but many other long-chain acids have been found in small amounts.

Many other compounds are known to occur in cuticles, such as phenolics, tannins, and sometimes simple sugars. The cuticularized layer is sometimes homogeneous in appearance under the electron microscope and sometimes banded (Fig. 2b). The bands appear to represent layers of material of greater and lesser osmiophilic density anastomosing within the cuticle.

Beyond the cuticle itself may lie a variety of compounds, of which the most common are the heterogeneous waxes. Wax is present on about half of all the cuticles so far examined. Some plants have a prominent waxy bloom. This is due to the reflection and scattering of light on the surface by waxy crystals whose dimensions are close to the wavelength of light. The bloom of a pea leaf is due to snowflakelike waxy forms about 1 μm across. There is very little definite information, although many theories, as to how the waxes leave the epidermal cells and come to form, often in bizarre crystalline shapes, on the leaf surface. In some plants their function is obscure, and in others, such as the waxy, easily broken scales of the pitcher plant (*Nepenthes*) (Fig. 5), they prevent the escape of trapped insects. They bend at the junction of the "blade" and "handle" (Fig. 5b) and fold over to form a more or less complete covering held about 0.5 μm away from the epidermal surface. In most plants they cause the rejection of water or other liquids applied to plant surfaces and hence are important in modern agriculture. Apart from the waxes, crystalline deposits of such substances as ursolic acid, sugars, and flavones may also occur.

Almost all the modern techniques available to the plant research worker have been applied to the modern study of plant surfaces.

For background information *see* EPIDERMIS (PLANT); STEM (BOTANY) in the McGraw-Hill Encyclopedia of Science and Technology.

[BARRIE JUNIPER; GUY COX]

Bibliography: E. A. Baker and J. T. Martin, *Ann. Appl. Biol.*, 60:313, 1967; K. Esau, *Plant Anatomy*, 1965; P. J. Holloway and E. A. Baker, *Plant Physiol.*, vol. 43, 1968; M. Z. Hülsbruck, *Pflanzenphysiol.*, 55:181, 1966.

Equilibrium, phase

Over the past two decades there has been extensive research to determine the thermodynamic properties and phase stability of metallic, ceramic, and polymeric materials. This research has generated many data on phase equilibria over a wide range of composition, temperature, and pressure. It has been found that these data can be managed most efficiently by means of phase diagrams. However, practical utilization of this knowledge often requires information on metastable phases in addition to stable equilibria. Such additional information has been obtained, for example, in the recent studies of spinodal decomposition within miscibility gaps. Nevertheless, the most complete and extensive picture of a system consists of an explicit representation of the free energy of each of the competing phases within the system as a function of temperature, composition, and pressure. Such a description implies that the equilibrium phase diagram can be constructed and that the relative stability of metastable phases, as well as common thermodynamic properties such as heats of formation, vapor pressures, and heats of transformation and fusion, can be ascertained.

In the past, collection of experimental data relevant to these areas has been fragmented into two separate parts consisting of collations of phase diagrams on the one hand and compilation of thermochemical data on the other. The recent development of computer techniques dealing with the general problem leads to a natural merger of these activities. The advantages stemming from such a combination include an opportunity for feedback between "fundamentalists" (who are interested in first-principle calculations of phase stability) and "practitioners" (who are interested in detailed information concerning stable and metastable phases). See INTERMETALLIC COMPOUND.

First-principle calculations. Depending upon the input required to describe the system of interest, the current studies fall into two categories. The first method, first-principle calculations, requires extensive thermochemical data for the observed stable phases based on heats of formation and either vapor pressure or specific heat or both as a function of temperature and composition. The data can be cast into mathematical form in order to describe the excess free energy of mixing of the solution phases involved. Such mathematical data can be employed along with the equations for equilibrium (lowest common tangents) to calculate the phase diagram. Such calculations have been performed with very good results for a number of metallic and ceramic systems where extensive thermo-chemical data are available. In several of these cases the calculated "equilibrium diagram" represents a higher level of accuracy than the phase diagram determined by conventional methods. This often happens with solid-phase transformations when nucleation and diffusion effects limit the attainment of equilibrium in a practical way.

The limitations to this method of coupling thermochemical and phase equilibria data include the need for extensive thermochemical information on the phases of interest and the fact that this approach provides little information concerning metastable or unstable phases. Moreover, extension of the features of binary systems into ternary or multicomponent cases is restricted to specific cases which include miscibility gaps, solid/liquid or solid/solid phase equilibria between components exhibiting the same stable crystal forms, or eutectics (or eutectoids) with little or no solubility in the low-temperature phases. Thus this method is not readily applied to cases where extensive solubility exists in systems between component partners exhibiting different stable crystal structures. This latter class consists of a large number of systems of considerable interest which can only be treated under the first method when

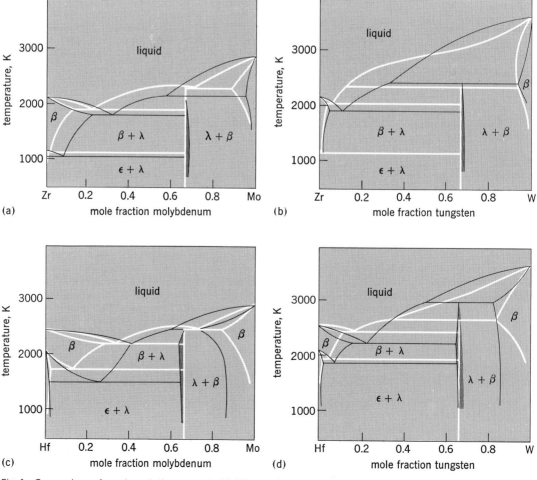

Fig. 1. Comparison of regular solution computed (white lines) and observed (black lines) phase diagrams for the (a) Zr-Mo system, (b) Zr-W system, (c) Hf-Mo system, and (d) Hf-W system.

extensive thermochemical data are available.

Practical approach. By contrast, a second approach in which little or no prior information is required has been developed and applied to more than 100 binary metallic systems. This method is based on simple descriptions of the solutions and compound phases requiring a few parameters for complete specification. Thus solution phases are approximated by the regular solution model, while compound phases are defined at fixed stoichiometries. Explicit procedures have been developed to calculate the regular solution interaction parameters for the solution phases covering the face-centered cubic (fcc), body-centered cubic (bcc), and hexagonal close packed (hcp), and liquid forms in all of the binary systems comprising second- and third-row transition metals. Additional procedures have been provided to cover titanium base binary systems. In addition, Laves and $AuCu_3$-type phases formed in transition-metal binary systems have been explicitly described. Although the above-mentioned simple description (regular solutions – line compounds) can be readily expanded in order to encompass more complex behavior and approach real systems more closely, the present idealization offers a method for dealing with metastable and unstable phases and extension into ternary and higher order systems.

Computer techniques. Computer programs have been developed for carrying out the phase-diagram calculation for the binary and ternary cases. These programs are based on equilibration of chemical potentials (or the lowest common tangent rule) and feature an examination of phase competition in order to establish the most stable phases. Sample binary phase-diagram calculations as shown in Fig. 1 have been performed in which the competition between four solution phases and two compound phases were examined simultaneously as a function of temperature and composition.

The ternary computer programs consider solution and compound phase interactions in addition to dealing with isolated miscibility gaps for ternary systems where the binary components exhibit complete solubility. An isothermal section of the Zr-Ta-W system is shown in Fig. 2 as an example.

This description permits calculation of the heat of formation and vapor pressures of components as a function of composition and temperature in the systems of interest. The description also can predict the occurrence of congruently vaporizing compositions in binary or multicomponent systems based on combinations of second- and third-row transition metals.

The procedures employed in calculating the regular solution and compound phase parameters reflect size factors, solubility parameters, and electronegativity and structural energy terms which are characteristic of the binary system. Calculation of ternary or higher order systems is accomplished by summing the pair-wise binary interaction parameters which are explicitly defined.

The keystone of this procedure is quantitative definition of the lattice stability of pure metals which describes the relative stability of the fcc, bcc, and hcp forms as a function of temperature. These differences, which are essential in carrying

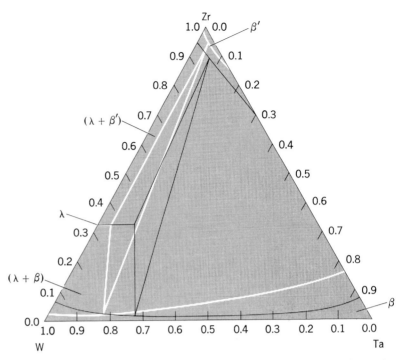

Fig. 2. Comparison of calculated (white lines) and observed (black lines) phase relations in the Zr-Ta-W system at 1873K.

out complete and explicit phase-diagram calculations for binary and multicomponent metallic systems, have proven difficult to calculate from first principles owing to their small magnitude (0.001–0.02 eV/atom). In spite of recent efforts to calculate these differences by means of pseudopotential techniques, suitable values for the transition metals have not been derived. However, some progress has been made by combining experimental information on polymorphic transitions at 1 atm and high-pressure transitions. These observations for several metals (iron, manganese, titanium, zirconium) have been coupled with thermochemical data on observations of phase equilibria to derive an empirically based description of the lattice stability of pure metals which reflects zero-point energy differences and vibrational, magnetic, and electronic specific heat contributions. Thus a quantitative expression of the lattice-stability parameters (entropy and enthalpy differences) for the fcc, bcc, and hcp forms of 30 metals has been obtained. This information includes the melting temperatures and the entropy of fusion of the unstable forms, which can be employed to compute melting trajectories, and eutectic and peritectic behavior in binary and higher order systems in a completely explicit manner.

The development of the computer techniques, when coupled with the above-mentioned methods for computing lattice-stability, interaction, and compound parameters, provides a framework for dealing with multicomponent systems to achieve a logical means of data management. Although this approach has been developed for describing metallic systems, the general method can be readily applied to ceramic and polymeric systems.

For background information see ALLOY STRUCTURE; EQUILIBRIUM, PHASE; SOLID SOLUTION in the McGraw-Hill Encyclopedia of Science and Technology. [LARRY KAUFMAN]

Bibliography: A. Alper (ed.), *Phase Diagrams*, 3 vols., 1970; L. Kaufman and H. Bernstein, *Computer Calculation of Phase Diagrams*, 1970; O. Kubaschewski and W. Slough, *Progr. Mater. Sci.*, 14:1, 1969; A. Reisman, *Phase Equilibria*, 1970; P. S. Rudman, J. Stringer, and R. I. Jaffee (eds.), *Phase Stability in Metals and Alloys*, 1967.

Faunal extinction

Since life first appeared on the Earth some 3,000,000,000 years ago or more, it alternately has experienced periods of extensive evolutionary diversification and periods of widespread faunal extinctions that affected wholly unrelated groups in widely different habitats. The disappearance of these once-thriving faunas was not merely the result of their evolution into somewhat different descendents, nor were the groups that became extinct immediately replaced by others that were better adapted. Paleontologists have long sought an explanation for these major "faunal crises," such as occurred near the end of the Devonian some 350,000,000 years ago, the end of the Permian about 225,000,000 years ago, and the end of the Cretaceous about 70,000,000 years ago.

Theories. Those theories that have been proposed in explanation of the extinction of one group or another, for example, the Paleozoic trilobites and the Mesozoic ammonites and dinosaurs, not only were inapplicable to some groups that disappeared simultaneously but failed to explain the persistence of others unscathed. Many of the theories postulated major changes in the physical environment, ranging from climatic extremes, such as the continental glaciation of the Pleistocene ice ages, to bursts of energy from a supernova, collision with comets, major fluctuations in solar radiation, increase or decrease of atmospheric oxygen, extensive volcanism, mountain building, fluctuations in sea level that restricted or expanded the area of the continental shelves, an excess or deficiency of metallic trace elements, hypersaline or brackish water oceans, increased influx of cosmic rays at loss of the Earth's magnetic field during polarity reversals, or the results of the movement of continental plates by sea-floor spreading. For

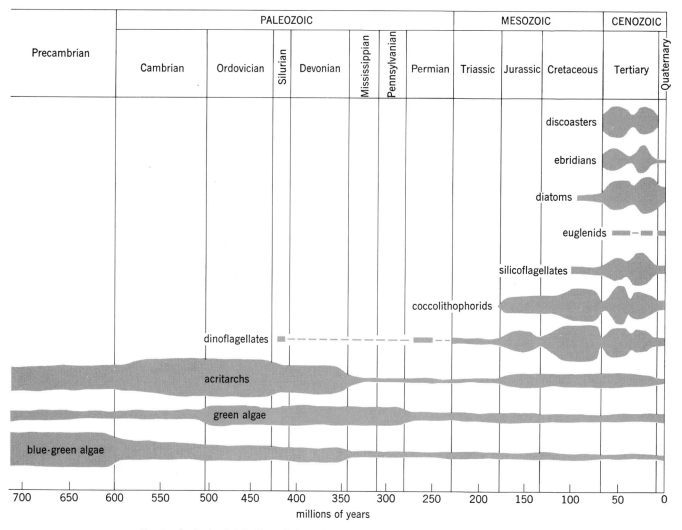

Fig. 1. Geologic distribution of phytoplankton groups having an important fossil record, showing their variation in abundance, relative scarcity during the late Paleozoic, and expansion in abundance and diversity during the Mesozoic and Cenozoic. (From H. Tappan and A. R. Loeblich, Jr., Geobiologic Implications of Fossil Phytoplankton Evolution and Time-Space Distribution, Geol. Soc. Amer. Spec. Pap. no. 127, 1971)

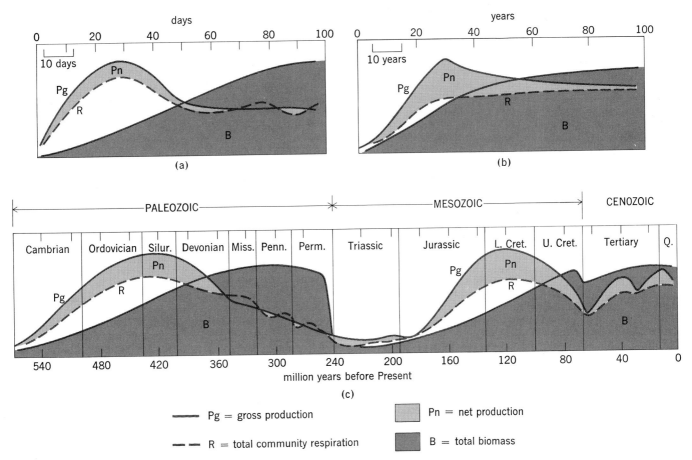

Fig. 2. Patterns of ecologic succession. (a) Laboratory microcosm succession, occupying a matter of days; (b) forest succession, requiring a century (based on E. P. Odum, Science, 164:264, 1969). (c) Marine evolutionary succession. The general pattern of succession is similar in various ecosystems, differing in size and duration. The hypothetical curves of marine evolutionary succession are for the global ecosystem. The mature, or climax, stage is delicately adjusted and susceptible to disruption; worldwide disruptions occurred at the end of the Paleozoic, the end of the Mesozoic, and on a lesser scale with the Tertiary. (From H. Tappan, Microplankton, ecological succession, and evolution, Symposium North American Paleontological Convention, Chicago, 1969, Proc. Part H, 1971)

some of these suggestions no possible means of verification exists; the timing of major volcanism and glaciation is not closely correlative with extinctions; and the selectivity yet universality of extinctions was inexplicable by excess cosmic-ray influx, increased solar ultraviolet, comet collisions, or other unique events. The major periods of extinction were in fact not instantaneous but represented a gradual though effective decrease in the number of species and higher categories of organisms. Hence these catastrophic explanations seem unnecessary as well as lacking in proof.

Biological explanations also were suggested, such as a hypothetical racial senescence, gigantism, or the depletion of genetic vigor (all unsupported by any biologic evidence), excessive competition or predation by newly evolved or newly immigrated species, epidemic disease, or the disappearance of favored food sources and resultant breakdown of food chains.

H. Tappan recently suggested that quantitative and qualitative changes in the oceanic phytoplankton (Fig. 1) are indicated by its fossil record. In Fig. 1 blue-green algae, green algae, acritarchs (fossil cysts of unknown exact affinities), dinoflagellates, and euglenids are represented as fossils by their highly resistant organic walls; coccolithophorids and discoasters by calcareous plates that covered the tiny flagellate cells; silicoflagellates and ebridians by their opaline silica internal skeletons; and diatoms by their siliceous bivalved frustules. These past changes in phytoplankton productivity would have affected not only all the oceanic food chains based on it but, because of the importance of phytoplankton in both carbon fixation and oxygen production, may well have changed the nature of the atmosphere sufficiently to have also affected terrestrial ecosystems. As these fluctuations in phytoplankton productivity resulted from various changes in the extent of lands and seas, elevation of continents, climatic regime, and oceanic circulation, they form a connecting link between changes in the Earth's physical features and the evolution of its inhabitants.

Photosynthesis influence. All life on Earth ultimately is dependent upon plant photosynthesis (use of the Sun's light energy to split water molecules and reduce carbon dioxide to produce organic compounds). The energy thus stored by this primary production is used by the plants in their growth and respiration, by grazing animals feeding on the plants, and by predators feeding on the herbivores. Free oxygen is a by-product of plant photosynthesis, and atmospheric oxygen represents the excess accumulation resulting from photosynthesis over geologic time. Microscopic floating marine plants, the phytoplankton, are by far the most important primary producers, having available the greatest area (71% of the Earth's surface is covered by oceans). Because of its astronomical abundance and extremely rapid turnover, the tiny phytoplankton annually produces between 50 and 90% of the world's total output of organic matter. During much of past geologic time, when seas were even more extensive and before a diversified land flora had evolved or when land areas were covered by extensive glaciers, phytoplankton relatively was even more important.

Ecologic succession. All ecosystems consist of the physical environment and the organisms living in and interacting with it. Tending to develop a maximum biomass for the amount of energy flow, they show many common characteristics in a regular progression termed ecologic succession. Developmental stages of this succession are similar for widely different types of ecosystems although progressing at different rates. All show a close relationship to the primary productivity of the particular system, the relative stability of the environment, and the time elapsing since its establishment. Large-scale fluctuations in productivity through geologic time similarly influenced the nature of the coexisting ecosystems; as a result, evolutionary succession may be considered analogous to the different stages of community development that characterize an ecologic succession (Fig. 2). In Fig. 2 all patterns of ecological succession show a similar low early productivity and small biomass with a rapid increase to a maximum in gross production and its utilization indicated as community respiration. Productivity then declines and becomes equalled by its utilization, while biomass continues to increase to the climax stage of succession. Of course, ecologic succession proceeds by the gradual development and replacement of faunas and floras by selective immigration, emigration, or elimination of preexisting species and thus differs greatly in scale from the evolutionary succession made possible only by chance mutations or genetic recombinations variously favored or rejected by the selection pressures of the environment. Comparison of local fossil assemblages or comparison of the total biota of the Earth at earlier geologic times with present regions of high or low productivity or with the early and late ecologic successional stages makes evident the many similarities of the large-scale fluctuations to those of shorter duration and suggests that their interrelationships are equally predictable.

In an early stage of ecologic succession, the community of plants and animals is strongly affected by the physical environment; thus the ratios of gross primary production to total community respiration and to community biomass are high; the total organic matter and the biomass supported by the energy flow within the community are low; food chains are short, mostly grazing and nonselective; organisms are small, unspecialized, and of simple morphology; life cycles are short, offspring numerous, and individual growth rapid, with strong fluctuations in abundance; and nutrient conservation is poor, species diversity per area low, and spatial organization poor. In the marine environment such an ecosystem is found in areas of upwelling. In such areas deeper waters, rich in dissolved nutrients, rise toward the surface, where they are available within depths penetrable by sunlight to which the phytoplankton is restricted. Turnover is rapid and primary production is high, although only a fraction of the available energy is used. In the geologic record many of these characteristics are found in the evolutionary development of the early Paleozoic or early Mesozoic. Selection pressures favored population abundance and highly plastic phenotypes, and food chains were nonselective, with many relatively inefficient, sessile suspension-feeding invertebrates. Rapid evolution of one group of species after another was a consequence of species abundance, short life spans, and great individual variability.

In the later or mature stages of ecologic succession, the physical conditions of the environment are less severe and the community is characterized by biological interactions and accomodations. The available energy is more completely utilized, and community respiration tends to approach or equal the primary production. The larger biomass is also greater in proportion to the energy flow through the system; food chains are many-linked complex webs, with detritus feeders using the final remnants of organic matter; and nutrients are largely tied up within the biomass and transferred in a closed cycle. Species diversity is high and spatial distribution well organized. Organisms show a greater size range, with highly ornate morphology, increased strength and motility, delayed maturity, and long complex life cycles with fewer but better protected offspring. The increased level of interactions between species involves predation, competition, mutualism, symbiosis, and parasitism. Such features characterize areas of stratified seas in modern oceans where the nutrient supply is limited to that recycled efficiently from one organism to another and every possible bit of energy is utilized by long and complexly interwoven food chains.

On an evolutionary scale these conditions characterize the late Paleozoic or Late Cretaceous. Diversity was at a maximum and the biomass was large, much of it consisting of large and ornate species represented by small populations and thus reduced variability. Many animals developed specialized methods or structures for increased efficiency in food gathering; indiscriminate filter feeders were replaced by actively motile grazers and predators or by deposit feeders. Thus in the late Paleozoic the more active starfish and echinoids replaced sessile crinoids, cystoids, and blastoids as the dominant echinoderms; detritus-feeding or siphonate bivalves replaced less efficient suspension-feeding brachiopods; and many groups showed phyletic size increase. Highly organized communities included complex reefs of corals, sponges, and bryozoans. The greater competition for limited resources resulted in the selective extinction of any less well adapted or less efficient species.

In its final or mature stage, an ecosystem is so delicately adjusted for maximum use of all resources that it is also highly susceptible to disruption. During the middle Paleozoic the influx of a dissolved nutrient supply to the ocean was lessened by its retention for the first time in the newly developing land plant biomass. Recycling of nutrients in the oceanic reservoir progressively slowed due to burial of organic matter in sediments, as in the Devonian-Mississippian black shales or the Carboniferous coals and petroleum. Phytoplankton was extremely rare and ecosystems increasingly adapted to more efficient recycling or to the use of terrestrial detritus as an energy source. Invertebrate plankton larval stages would have been particularly affected by the reduction of phytoplankton, and successive extinctions of the most sensitive and least adaptable taxa followed. The Permian marine extinctions affected the sessile, suspension-feeding benthos and higher carnivores most severely. Yet, when productivity again increased in the early Mesozoic, selection for efficiency became less important than rapidity of growth; this was reversed again in the latest stages of evolutionary ecologic succession in the Late Cretaceous. Marine extinctions again followed the same pattern as in the Paleozoic, although different taxa were involved. Cenozoic radiations and extinctions followed two such cycles, although in lesser extremes, one characterizing the Paleogene and the other the Neogene.

Animal and plant assemblages adapted for high productivity or physical limiting conditions differ distinctly from those in biologically accomodated communities where the increased biomass must share, cycle, and reuse the limited resources. These differences are both geographic and temporal, with seasonal successional fluctuations and longer term ones related to climatic change or modified current patterns. The longer term geologic fluctuations also were related to global geographic features. Rejuvenated continents of high elevation affect wind patterns and climatic conditions, and the greater latitudinal temperature gradient produces strong oceanic currents and prominent upwelling zones where nutrients are recycled. Such conditions in the geologic past had the high productivity and cosmopolitan species of a youthful ecosystem. Continents worn down by erosion and covered by extensive epicontinental seas had more moderate climates and thermally stratified seas. Organic matter sank to the sea floor without recycling and nutrients were buried in the sediments or trapped in the deeper oceanic waters. The resultant reduction in productivity caused strong selection pressures for efficiency, with many extinctions. In the eventual breakdown of the system, those organisms most dependent on phytoplankton abundance were caught in a final wave of extinctions.

As seas advanced over the continents, the global ecosystem remained youthful due to its continued expansion. Once maximum inundation was attained, evolutionary succession proceeded on the grand scale. It progressed more rapidly as seas retreated and formerly separated assemblages were crowded together, increasing both the biomass and diversity per given area and allowing all the biologic interactions of the late stages of an ecologic succession. The effects of crowding or expanding ecosystems resulted in more youthful terrestrial ecosystems coinciding with mature oceanic ones, so that extinctions were not always simultaneous. During the major faunal crises already mentioned, the greatly decreased phytoplankton productivity also slowed the rate of replenishment of atmospheric oxygen, as little organic carbon was buried in the sediments (for example, in the Permo-Triassic). The reduced oxygen pressures may have resulted in selective extinctions among terrestrial animals, although terrestrial floras continued to expand.

Thus the major changes in phytoplankton primary production of the geologic past are suggested to form a connecting link between the successive changes in the configuration of lands and seas and the patterns of evolution, radiation, and extinction of animals and plants. Faunal extinctions in the past, as now, resulted from the many varied interactions between species whose pattern was set by the nature of the physical environment. The extinctions were triggered by the changing primary productivity of the system.

For background information see ECOSYSTEM; FAUNAL EXTINCTION; MARINE ECOSYSTEM; PHYTOPLANKTON in the McGraw-Hill Encyclopedia of Science and Technology. [HELEN TAPPAN]

Bibliography: H. Tappan, Microplankton, ecological succession, and evolution, *Symposium North American Paleontological Convention, Chicago, 1969, Proc. Part H*, 1971; H. Tappan, *Palaeogeogr. Palaeoclimatol. Palaeoecol.*, vol. 4, 1968; H. Tappan, *Palaeogeogr. Palaeoclimatol. Palaeoecol.*, vol. 8, 1970; H. Tappan and A. R. Loeblich, Jr., *Geobiologic Implications of Fossil Phytoplankton Evolution and Space-Time Distribution*, Geol. Soc. Amer. Spec. Pap. no. 127, 1971.

Feather (bird)

Apart from performing aerodynamical, hydrodynamical, and heat-preserving functions, feathers also serve as a highly sophisticated water-repelling structure in both aquatic and terrestrial birds. Recent studies have revealed that effective waterproofing properties result from the optimal balance of some structural parameters of the feathers and resistance to water penetration. There also appears to be a close evolutionary relationship between this balance and the behavioral patterns of several bird families.

Repellency and feather structure. Water birds, particularly ducks, were generally regarded as having attained perfection in water repellency, and this quality was usually attributed to the superior properties of the uropygial gland oil. However, chemical analysis has failed to show the presence of any constituents in the gland oil that could explain the unusually effective water repellency of

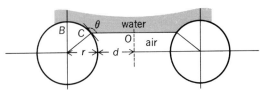

Fig. 1. Schematic diagram of cross section of two barbs with their axes perpendicular to the plane of the paper (barbules not shown). Here r is the radius of the (cylindrical) barbs with their axes $2(r+d)$ apart. Furthermore, $f_s = (\text{arc } BC)/(r+d)$ and $f_a = (CO)/(r+d)$.

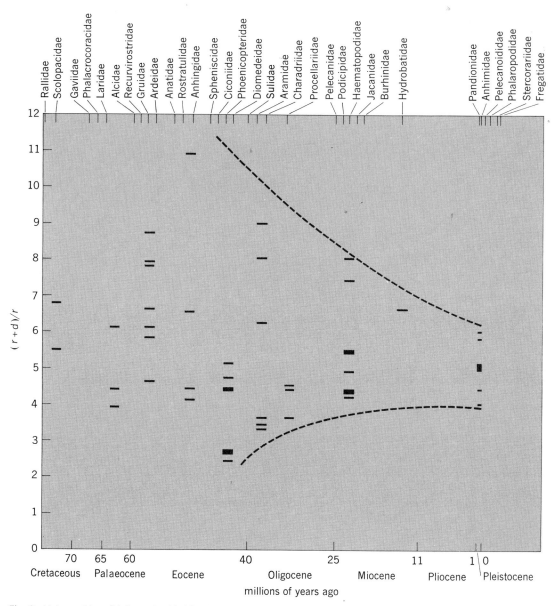

Fig. 2. Values of $(r+d)/r$ for water-bird feathers plotted against geological time interval of earliest fossil record for each particular avian family. Values were measured for dorsal surface near rachis of the feather.

feathers. Moreover, when gland oil is spread on smooth surfaces, contact angles of 90 and 60° are measured for the advancing and receding water drop, respectively, which are essentially the same as found for any waxy surface. Thus both chemical and physical data suggest that factors other than the properties of gland oil are responsible for the water-shedding qualities of feathers.

Studies on the wettability of porous surfaces have shown that the water repellency of a surface with air entrapped in the interface is greatly enhanced by some structural parameters peculiar to the surface. These principles are also applicable to feathers. In fact, microscopic investigation has revealed that the substructure of feathers conforms closely to the theoretical requirements of optimal water repellency. The effective contact angle Θ_A, which causes the drop to pearl and roll off, is related to the true contact angle Θ by Eq. (1),

$$\cos \Theta_A = f_s \cos \Theta - f_a \qquad (1)$$

where f_s is the area of solid-water interface and f_a that of air-water interface per unit apparent surface area. This equation shows that Θ_A will always be larger than Θ when an air-water interface is formed. When drops of water under zero hydrostatic pressure rest on a feather surface, a flat air-water interface will touch the barbs under a contact angle Θ, as depicted schematically in Fig. 1. Elementary calculations on this model show that Eqs. (2) and (3) may be written.

$$f_s = (\pi - \Theta) r/(r+d) \qquad (2)$$
$$f_a = 1 - r \sin \Theta/(r+d) \qquad (3)$$

It is seen that the contribution of the feather structure to the values of f_s and f_a is determined by the ratio $(r+d)/r$ only and not by the absolute values of r and d. Large values for this ratio mean large f_a and small f_s values, increasing the apparent contact angle in the manner described by Eq. (1). Large effective contact angles have much significance for the movement of water drops on feather surfaces. The drops move much as drops of mercury move on glass. Excellent waterproofing will be maintained provided that the drops do not penetrate the feathers on impact and that the weight of the bird does not force water between the barbs and barbules. This latter will be prevented if the distance between the barbs is sufficiently small. The combination of large $(r+d)/r$ and small d values can only be realized if r is very small. In practice, however, the barb diameters are rarely smaller than 30 μm. Thus the range of $(r+d)/r$ values that is associated with outstanding waterproofing qualities is restricted; small $(r+d)/r$ values will insufficiently increase the effective contact angle, whereas too large values will be associated with poor resistance to water penetration. In Fig. 2 the parameter $(r+d)/r$ of different families is placed in the geological time interval that corresponds to the earliest fossil finding of the family. It is seen that the widest range occurs in the middle and upper Eocene, when many birds existed whose anatomical characteristics are closely related to those of their present-day descendants. Fossil records of more recent periods involve families with more narrowly spaced $(r+d)/r$ values, which ultimately narrow down for families of Pleistocene origin to the range of about 4 to 6, which corresponds to structures of optimal waterproofing properties. Thus there appears to exist a phylogenetic tendency toward increasingly enhanced water repellency and resistance to water penetration with the course of time.

Several behavioral patterns may have evolved under the selective pressure of poor water repellency or poor resistance to water penetration. For instance, the underwater feeding habits of Anhingidae are probably assisted by extensive penetration, since too much buoyancy would prove to be a disadvantage. Waders and shore birds with their high $(r+d)/r$ values appear to be well equipped to shed water drops continously, but in general are less suited for swimming. The opposite appears to be true for the fully aquatic families, such as Gaviidae, Phalacrocoracidae, Alcidae, and Spheniscidae. Terrestrial birds show large $(r+d)/r$ values, which suggests that they have evolved in their habitat under the pressure of water repellency only, without the necessity to prevent water penetration.

Rehabilitation of oiled seabirds. The presence of gland oil on the plumage is essential for maintaining a finite contact angle Θ and flexibility of the feather components. However, after coating with commercial oil, the feather structure is destroyed and the water-repelling function annihilated. Flight is impaired or lost and the risk of poisoning, followed by fungal infections of the intestinal organs, is very real. Detergents used to wash oiled seabirds also remove the feather wax. Due to the critical role of wax in maintaining water repellency and heat insulation, no seabird can be returned to its natural environment until the wax has been replaced and its feather structure restored. Spraying a substitute wax after treatment with a detergent may lead to overdoses, with results similar to the original oiled plumage. Promising results have been obtained using Larodan 127, which waxes and cleans at the same time, as some car waxes do. The preparation consists of a dispersion of hydrophylic lipid crystals in water with synthetic wax included in the hydrophobic regions of the lipid crystal matrix. Larodan 127 has been tried on a large scale in Scandinavia with success. Birds could be returned to their habitat within 2 weeks after treatment. See FLIGHT.

For background information see FEATHER (BIRD); INTERFACE OF PHASES; SURFACE TENSION in the McGraw-Hill Encyclopedia of Science and Technology.

[A. M. RIJKE]

Bibliography: K. Larsson and G. Odham, *Mar. Pollut. Bull.*, 1:122, 1970; G. Odham and E. Stenhagen, *Acc. Chem. Res.*, 4:121, 1971; A. M. Rijke, *J. Exp. Biol.*, 52:469, 1970.

Flight

Since antiquity man has been enchanted and mystified by the flight of birds. In the Bible (Proverbs XXX) the writer lists "the way of an eagle in the air" first among the things that are "too wonderful" for him. Some of the mystery of bird flight has gone with the understanding of aerodynamics; nevertheless only the broader principles are known and few of the details of the natural flight process are understood, in the sense that the reasons for much of the wing structure, for the complicated wing strokes, and for flight pattern are not clear. However, recent theoretical analyses have suggested that the V formation adopted by migrating birds gives very significant reductions in flight power requirements.

Flapping flight fundamentals. Classical unsteady aerodynamics gives an understanding of the fundamentals of flapping flight. In essence, the bird's wing lifts in the same manner as an airplane wing, although the magnitude of the lift varies, being greater on the downstroke because of the increased angle of attack due to the vertical motion. The bird can twist its wings to change their

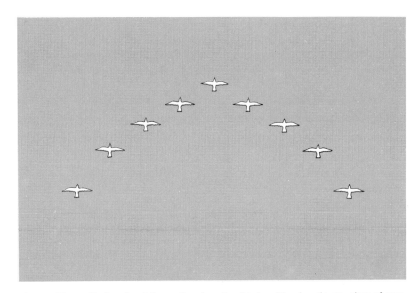

Fig. 1. Theoretical optimal formation for nine birds with wing-tip spacing of one quarter of wingspan.

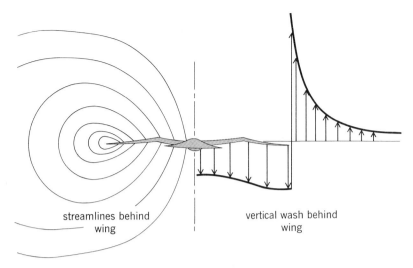

Fig. 2. Steady streamlines and downwash behind a bird.

angle, thus minimizing this variation; however, there are little quantitative data on this subject.

The flapping motion produces a thrust which propels the bird and overcomes the skin friction resistance of the air. Theory shows that this thrust is created by the forward rotation of the lift vector during the downstroke of the wing, according to the Kutta-Joukowski law, which states that lift is approximately perpendicular to the relative wind. On the upstroke, the lift is rotated backward, producing a rearward force. The wings are articulated to minimize this effect, and develop a net forward thrust. A mathematical analysis of the flight of birds, insects, and ornithopters, by D. Kuchemann, has shown that under ideal conditions a flapping wing is an extremely efficient thrust producer.

The bird must expend energy to fly—this appears physically in the muscular work exerted by the shoulder and elbow muscles during the downstroke, while the upstroke is apparently relatively effortless. Recently experiments by V. A. Tucker on a live bird flying in a wind tunnel have measured the energetics of natural flight.

V flight formation. The remarkable nonstop range flown in bird migration has always been a mystery, along with the deeper puzzle of navigation on oceanic and continental passages. It has long been surmised that the characteristic V formation adopted by many species is related to an increase in flight efficiency. In 1914, C. Weiselsberger published a simplified analysis showing that V formation gave a small aerodynamic advantage. Recent calculations by P. B. S. Lissaman and C. A. Shollenberger have shed more light on formation flight of birds. Using theoretical aerodynamics, aided by a computer, it appears that a significant power reduction can be achieved by close formation flight and that the V formation is optimal. For example, a formation of 25 birds can theoretically achieve a range of about 70% more than that of a single bird. According to these computations, a nine-bird formation should be arranged as shown in Fig. 1, which is drawn to scale and is consistent with that observed in nature.

The explanation of this effect lies in the general theory of lift. Any lifting wing, natural or mechanical, creates a downflow behind it—this downward momentum of the air being responsible for the lift. At the tip, however, the air flows from the lower to the upper surface, as shown by the streamlines in Fig. 2. Beyond the wing there is an appreciable upwash and this favors a neighboring bird, which experiences an apparent upcurrent. In a line-abreast formation each bird benefits from the others, reducing the flight power required. With closer spacing the advantage increases, although if a bird flies behind another, the effect becomes unfavorable.

The saving in drag can most conveniently be expressed as the ratio of the induced drag of a bird flying in formation to that in solo flight ($1/e$). Assuming the bird consumes fuel at a rate proportional to its flapping power requirements, the range of a formation is increased by a factor of $e^{1/2}$, while the optimal formation flight speed is lower, by a factor, $e^{-1/4}$. Values for $1/e$ are shown in Fig. 3. Appreciable savings occur only for very close wing-tip spacing, and while the larger formations are more effective, the saving does not increase much for more than a score of birds. Drag savings occur in any formation where no bird is behind another, but in general the birds do not share equally in the power reduction. In line-abreast formation the center birds experience favorable upwash from their neighbors on both sides and have approximately twice the drag savings of the tip birds. However, the flow fields are more intense behind a bird than ahead, so that in a V formation it is possible for each bird to share the drag equally. Here the tip birds experience a field from one

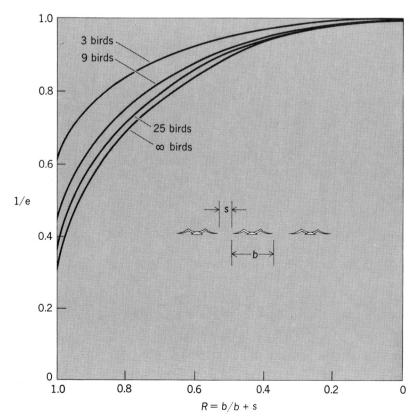

Fig. 3. Drag reduction for formation flight showing the effect of wing-tip spacing and number of birds.

side only but the upwash is fully developed since it is created by the birds ahead. At the V apex the bird is in two partially developed fields. Thus this is the proper formation for equipartition of flight power. Also, if a bird flies ahead of its optimal position, it requires more flight power and this natural mechanism causes it to fall back into line. Thus the bird can feel when it is in the right position. Again, since optimal wing lift distribution is almost independent of the formation selected, the bird will always fly so that it "feels the same." These two effects constitute a plausible explanation of how the optimal formation is adopted.

The brown pelican utilizes a very regular V formation, even for short foraging flights. Possibly the sensory mechanisms described above make this a natural means of holding formation without continuous visual checking, so that the bird can devote its entire attention downward to the ocean in its search for food. Apparently the group is a more efficient food gather than the set of individuals, since each fish is subject to the strikes of several birds.

For background information see AERODYNAMIC FORCE; FLIGHT in the McGraw-Hill Encyclopedia of Science and Technology.

[P. B. S. LISSAMAN]

Bibliography: D. Kuchemann and J. Weber, *Aerodynamics of Propulsion*, 1953; P. B. S. Lissaman and C. A. Shollenberger, *Science*, 168:1003–1005, 1970; V. A. Tucker, *Sci. Amer.*, 220:70–76, May, 1969.

Food

A few years ago the threat of massive famine loomed on the planet, particularly in the densely populated poor countries of Asia. Today that threat is less acute thanks mainly to a historic breakthrough by plant breeders in the development of new, high-yielding varieties of wheat and rice for tropical and subtropical regions. These new strains are at the heart of the modern agricultural revolution in the hungry countries, popularly known as the "green revolution." The new wheats and rices are not just marginally better than traditional or indigenous varieties but actually double yields, with proper management and sufficient water and fertilizer.

Fertilizer responses. The key to productivity of the new varieties is a remarkable feat of biological engineering that greatly enhances their responsiveness to fertilizer. Traditional strains of wheat and rice, particularly those of the tropics, are tall and thin-strawed. When fertilizer use exceeded 40 lb per acre (18 kg per 4050 m^2), the tall, thin straw, so necessary in the past for survival against weeds, heavy rains, and floods, would not support the heavier yield of grain and would fall down, or "lodge," causing heavy losses. Plant breeders redesigned the wheat and rice plants, producing dwarf plants with short, stiff straw that stands up under the weight of heavier yields. Yields from the new dwarf plants increase until nitrogen application reaches 120 lb per acre (54.5 kg per 4050 m^2). The new cereals not only respond to much larger quantities of fertilizer but also use fertilizer more efficiently. Thus a given level of production can be reached using far less fertilizer with the new seeds.

Aseasonal effect. In addition to being more responsive to fertilizer, the new strains are aseasonal, that is, not very sensitive to daylength (photoperiod). Their aseasonality makes them adaptable to a wide range of geophysical locations and seasons of the year. Many of the new varieties also mature early. The aseasonality and early maturity of new strains are opening new possibilities for multiple cropping and year-round farming.

Heretofore, most technological advances in human food production were geared to temperate-zone conditions or to plantation agriculture that benefited those living in tropical countries only residually at best. The new high-yielding wheats and rices are the product of the first systematic effort to develop agricultural technology designed specifically to take advantage of the unique growing conditions of the tropics and subtropics, particularly their wealth of solar energy, and to improve the lot of millions who live in desperate material poverty.

The new wheats came first. They were developed in Mexico by the Rockefeller Foundation, in collaboration with the Mexican government, in a pioneering program begun in the early 1940s. Norman Borlaug, director of the Rockefeller Foundation wheat-breeding program and winner of the 1970 Nobel Peace Prize for his role in creating the remarkable new Mexican wheats, used wheat germ plasm from widely scattered parts of the world—Japan, the United States, Australia, and Colombia. In addition, he alternated growing sites, raising a summer crop just south of the United States border and a winter crop near Mexico City, some 800 mi (1300 km) away. The two sites differed in daylength and other environmental factors. Given the cosmopolitan ancestry of the seeds, Borlaug was able to produce a dwarf wheat variety that was remarkably adaptable to a wide range of growing conditions. Mexican dwarf wheats flourish today in latitudes near the Equator, where days are uniformly short, as well as at higher latitudes, in Turkey, for example, where daylength varies greatly by season. This adaptability is new.

Buoyed by the success of the Mexican wheats and keenly aware that most of the world's poor eat rice, the Rockefeller and Ford foundations joined forces in 1962 to establish the International Rice Research Institute at Los Banos in the Philippines. The objective was to produce dwarf rice strains with advantages comparable to those of the Mexican wheats. The institute assembled some 10,000 strains of rice from every corner of the world and began to crossbreed them. Success came early when a tall, vigorous variety from Indonesia, called Peta, was combined with a dwarf rice from Taiwan, called Deo-geo-woo-gen, to produce IR-8, the first of a series of "miracle" dwarf rices for the tropics.

As a result of their high-yield capacity, the new grains have spread rapidly in the poor countries where ecological conditions are suited to their use. In the 1964–1965 crop year, only 200 acres (0.8 km^2) in Asia were planted to the new varieties, and that largely for experimental and industrial purposes; the next year there were 41,000 acres, then 4×10^6, then 16×10^6, and then 31×10^6; in 1969–1970 the total reached 44×10^6 acres (178.060 km^2). This is close to one-tenth the grain land in Asia.

Almost one-tenth of India's cereal acreage now has the new strains.

The rapid spread of the new seeds has been facilitated by their low cost. The poor countries have been able to import the new wheats by the shipload at prices only marginally higher than the world market prices. Since many countries were already importing cereals, the new technology was essentially free, sparing them large research and development costs. And because the seeds could be imported in huge quantities, the time required to multiply new seed was drastically condensed. Pakistan, for example, imported 42,000 tons (38×10^6 kg) of new Mexican wheat seeds in 1967, enough to plant more than 10^6 acres (4050 km²). Harvest from this crop provided enough to cover all of Pakistan's wheat land, collapsing into 2 years a process normally taking many years.

But perhaps even more important than the actual tonnage of dwarf wheats and rices imported are the prototype they represent and the genetic raw materials they provide. Already local plant breeders are refining and modifying the prototype to suit specific local growing conditions and consumer tastes, crossbreeding imported and local strains to create new varieties with desired characteristics.

The green revolution is not a universal phenomenon in the poor countries and is limited largely to wheat and rice crops. In some countries, particularly those of Latin America and sub-Saharan Africa, landholding arrangements form the main barrier to agricultural development. But the chief constraint to the spread of the new varieties in areas where they are being used is water: the supply of water for wheat, and the supply and control of water for rice. The dwarf wheats yield best under high rainfall or irrigated conditions. The dwarf rices do not perform well in conditions of natural flooding or in rain-fed fields where they may be submerged for some time.

Plant breeders are working to develop wheats that will yield abundantly under low rainfall conditions and high-yield rices that are more tolerant of flooding. Farmers and governments are expanding and intensifying irrigation and flood-control systems. Efforts are also underway to develop high-yielding varieties of corn, sorghum, millets, potatoes, and legumes.

Even with constraints on the spread of the new seeds, unprecedented gains in cereal production have been realized in more and more of the developing countries. India's production of wheat, expanding much faster than that of other cereals, increased by 80% between 1966 and 1970. In one of the most spectacular advances in cereal production ever recorded, West Pakistan increased its wheat harvest nearly 60% between 1967 and 1969, making West Pakistan a net exporter of both wheat and rice. The Philippines ended half a century of dependence on rice imports in 1968 and has become a rice exporter. Ceylon's rice crop increased 26% between 1967 and 1970. Mexico, once importing one-third of its wheat needs, is today exporting wheat, rice, and corn. Among other countries that are beginning to benefit from the new seeds are Turkey, Afghanistan, Burma, Indonesia, Iran, Laos, Malaysia, Nepal, South Vietnam, Morocco, and Tunisia.

The green revolution has arrested the deteriorating food situation in some of the most populous countries of Asia—India, Pakistan, Indonesia, and the Philippines. Still in its early stages, it has also had a pronounced impact on cereal production per person in four important countries: Mexico, India, Pakistan, and Ceylon (see table).

The word revolution is frequently used and greatly abused, but no other term adequately describes the effects of the new seeds on the poor countries where they are being used. Rapid increases in cereal production are but one aspect of the agricultural breakthrough. Agricultural scientists have achieved a technological breakthrough that foreshadows widespread changes in economic, social, and political orders. The new varieties may be to the agricultural revolution in the poor countries what the steam engine was to the industrial revolution in Europe.

Second-generation problems. The new varieties not only double the level of production over the levels of local varieties, but in doing so they often triple or quadruple profits. Millions of farmers have suddenly found themselves earning incomes that they had not dreamed were possible. The aspirations of millions of others, especially rural laborers without land, are being aroused in the process. As the new varieties and the new technologies associated with them spread, they introduce rapid and sweeping changes, creating a wave of expectation throughout society and placing great pressure on the existing social order and political systems. Distributing the benefits of the new varieties equitably and in ways that do not excessively exacerbate tensions between landless laborers and other farmers, between city dwellers and rural people, and between regional groups is a paramount challenge issuing from the green revolution.

The woefully inadequate state of marketing systems in countries where new varieties are being planted is the most immediate obstacle to the agricultural revolution. With the new technologies, farmers' marketable surpluses of cereals have increased far faster proportionately than production. A farmer accustomed to marketing a fifth of his wheat harvest finds that his marketable surplus triples when his crop suddenly increases 40%. Even after retaining more for home consumption, as many are doing, farmers who have doubled their output with the new seeds are increasing their

Annual production of selected cereals in countries using new seeds, in pounds per person of total population*

Year	India (wheat)	Pakistan (wheat)	Ceylon (rice)	Mexico (all cereals)
1960	53	87	201	495
1961	55	83	196	496
1962	59	87	213	525
1963	51	86	218	546
1964	46	83	213	611
1965	56	90	150	639
1966	46	71	188	649
1967	49	80	216	655
1968	76	116	247	680
1969	80	121	—	—

*1 lb = 0.454 kg.
SOURCE: U.S. Department of Agriculture.

marketable surpluses severalfold.

The green revolution found some countries with marketing systems oriented to a considerable extent toward handling imported grain. Now domestic marketable surpluses are threatening to overwhelm all components of the marketing system—storage, transport, grading and processing operations, and the local market intelligence system. Storage facilities, for example, are so inadequate that great amounts of grain have to be stored in open fields or in public buildings, such as schoolhouses. In West Pakistan, land planted to the new IR-8 rice rose from 10,000 to nearly 10^6 acres (40.5 to nearly 4050 km^2) in 1 year (1967–1968). West Pakistan suddenly found itself with an exportable surplus of rice, but without the processing, transport, and pricing facilities needed to handle an export trade efficiently.

A major and increasing threat to the green revolution is the availability of foreign markets. Introduction of the new seeds in tropical and subtropical regions, with their greater abundance of solar energy and year-round growing temperatures, is strengthening the competitive position of the poor countries. But while many tropical and subtropical countries are beginning to produce exportable surpluses of cereals, many of the rich industrial countries, such as Japan and the Common Market countries of Western Europe, are pursuing protectionist policies. As the roster of poor countries with exportable surpluses of wheat, rice, and feedgrains lengthens, there will be increasingly strong political pressure on the rich countries to limit domestic agricultural production and open their markets to competition. If the poor countries cannot gain access to the markets of the rich countries, their overall development will be thwarted.

Since high-yielding varieties exist that are adapted to almost every ecologic zone in the tropics and subtropics, there is no agronomic reason why people in any less-developed country should be deprived of the benefits of the green revolution. Its spread depends primarily on the commitment of political leadership, not on technology. How rapidly the green revolution progresses also depends on the extent of financial and technical assistance from the rich countries. Many poor countries, for example, are not able to finance imports of as much fertilizer as they need.

However exciting and encouraging in the short run, the green revolution should not reduce concern about the threat of uncontrolled human fertility. As Borlaug pointed out, the agricultural breakthrough is clearly not an ultimate solution to the food-population problem. But by laying the specter of imminent famine to rest, at least temporarily, the green revolution has bought urgently needed time to develop the technologies, strategies, and will to stabilize global population growth.

Stresses on Earth's ecosystem. Concern about population growth, and particularly the environmental consequences of man's burgeoning quest for food, is rising among observers. Prior to the invention of agriculture some 10,000 to 12,000 years ago, the Earth could not support more than 10×10^6 people, fewer than live in Afghanistan, London, or Ohio today. Since then, man has intervened extensively in the natural system, increasing the Earth's food-producing capacity several hundredfold. The human population has climbed to more than 3.5×10^9 people, although an estimated two-thirds are not well nourished. And man is currently adding more than 70×10^6 to his number annually. The net effect of continued population growth and an almost universal desire for better diets is increasing agricultural pressure on the Earth's ecosystem.

Some of the adverse effects of man's efforts to expand food production to keep up with population growth are beginning to come into view. For example, even at the current level of food production, agriculture is contributing to the eutrophication of hundreds of lakes, rivers, and streams throughout North America, Europe, and increasingly in the poor countries where fertilizer usage is beginning to climb. Usage of chemical fertilizers has climbed steadily and rapidly in the 20th century, particularly since World War II, as food demands mounted and frontiers disappeared. The use of chemical fertilizers today accounts for easily one-fourth of the world's food supply. In 1970 the world's farmers were adding some 70×10^6 tons (63.5×10^9 kg) of fertilizer to the Earth's soil, and usage is expected to double or triple in the decades ahead. Water runoff from agricultural land, carrying large quantities of nutrients, raises the nutrient content of freshwater bodies and contributes to the excessive growth of some forms of algae and eventual decline of other aquatic plants and animals. One result of eutrophication is the reduction in numbers of edible fish. *See* SOIL.

Man's agricultural activities are also endangering species by destroying habitats and by introducing chemical pesticides, including the well-known DDT, dieldrin, and mercury compounds, into the environment. Wild elephants in Ceylon now number no more than 2500, less than half the elephant population of 20 years ago. Their sources of subsistence are diminishing steadily as their forest and jungle habitat is cleared to produce food for the island's human population, which now doubles every 23 years.

Pravda reported that the reckless use of chemical pesticides in agriculture is decimating many forms of wildlife in the Soviet Union, causing many species to become "zoological rarities." The Soviet duck-hunting season was canceled entirely in 1970 because of the diminishing flocks of wild ducks.

The Department of the Interior maintains a list of species endangered within the United States, totaling 101 species in 1970. One worldwide list of endangered species, though obviously far from complete, now includes 275 species of mammals and 300 species of birds threatened with extinction. No one knows how many species of fish are threatened. What will the situation be 10 or 20 years hence if present trends continue?

The relevant question in any effort to project the food situation into the future is no longer simply can enough food be produced to feed a certain number of people, but what are the environmental consequences of attempting to do so in a finite ecosystem?

In light of what is now known about the environmental consequences of man's quest for food, there is increasing agreement that man's numbers must be stabilized within the next decade or two, not at some point in the next century. There is also mounting concern because the time bought by the green revolution to achieve a breakthrough in controlling human fertility is short and is not being very well used. The green revolution has been underway for several years already, but little progress has been made in slowing population growth. Many observers are calling for a reordering of global, national, and personal priorities, elevating family planning and population control to a much higher place.

For background information see BREEDING (PLANT); FERTILIZING; FOOD in the McGraw-Hill Encyclopedia of Science and Technology.

[GAIL W. FINSTERBUSH]

Bibliography: L. R. Brown, *Seeds of Change*, 1970; D. Dalrymple, *Imports of Plantings of High-Yielding Varieties of Wheat and Rice in the Less Developed Nations*, Foreign Economics Development Service, 1971; C. R. Wharton, Jr., *Foreign Affairs*, 47 (3):464–476, April, 1969.

Galaxy clusters

The tendency for galaxies to be associated in aggregates ranging from groups of a few galaxies to enormous clusters of tens of thousands to perhaps hundreds of thousands is perhaps the single most significant feature of the visible universe. Although the nonuniform distribution of galaxies was recognized by the Herschels long before their true nature as vast assemblages of stars was known and although the existence of a few great clusters was known to E. Hubble in the 1930s, it has only been in the past two decades that astronomers have realized that grouping and clustering are the general rule and not the occasional exception. In fact, recent studies reveal that 80% of all galaxies are associated with some group or cluster. This article discusses some of the fundamental properties of the clustering phenomenon that have been elucidated recently. In addition, the interrelationships between these various characteristic properties are examined.

Characteristic properties. The following observed properties have been discussed in reports recently.

Dimensions of clusters. The dimensions of cluster cells range from approximately 0.1 megaparsec (1 Mpc = 1,000,000 parsecs = 3,260,000 light-years) for small groups of several galaxies to enormous clusters whose dimensions are characteristically on the order of 50 to 100 Mpc. This was reported by G. de Vaucouleurs in 1971. Thus comparative sizes of clustering cells take on values over a range of approximately 1000 to 1. The upper end of this scale is made uncertain not so much by the ability of astronomers to count galaxies but the difficulty in perceiving clustering on a scale which begins to approach, within an order of magnitude, the size of the observable universe.

Number of galaxies in cluster. The number of galaxies in clusters or groups ranges from as few as two to as many as several tens of thousands. Typically, an average cluster may contain several hundred galaxies, although effects of distance make absolute number determinations very difficult.

Galaxy distribution within cluster. Two fundamentally different types of distributions are recognized. One is characterized by apparent spherical symmetry with a strong central concentration of galaxies and a number density of galaxies exhibiting a uniform approximately logarithmic decline with increasing distance from the center of the cluster. The second basic type of distribution exhibits one or more "nuclear" regions of varying degrees of concentration (frequently with little or no concentration), with the remaining galaxies being more or less randomly distributed throughout the cluster volume. This second type of cluster often exhibits a noticeable degree of flattening or elongation which, as pointed out by de Vaucouleurs in 1970, suggests a possible rotation of the cluster.

Galaxy type according to form. In studies of clusters of galaxies, classification systems dependent on form with relatively few parameters have been adopted, since forms of galaxies in remote clusters are difficult to perceive in detail. Either the Hubble system (basically E, S0, Sa, Sb, Sc, Ir, with a separate SB sequence for barred spirals) or the Yerkes system, based on spectral properties of the nuclei of relatively nearby galaxies (kE, kD, kS, gS, fS, aS, aI, also with a separate SB sequence for barred spirals), may be utilized effectively for form classification of galaxies in clusters out to intermediate distances.

The morphological system proposed by de Vaucouleurs in 1959 may also be applied, utilizing the basic form parameters and omitting the detailed parameters which are not amenable to observation

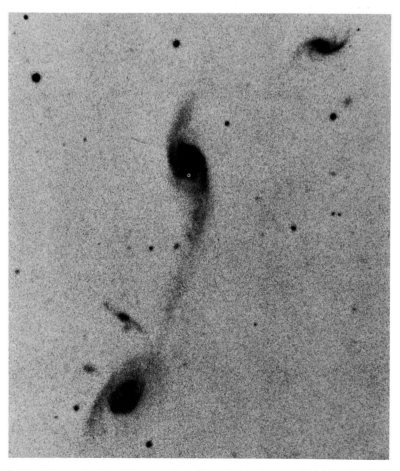

Fig. 1. An example of an interconnecting filament of luminous material (mostly stars) between two galaxies. (*From H. C. Arp, Atlas of Peculiar Galaxies, California Institute of Technology, 1966*)

with present techniques at intermediate cluster distances. Regardless of the specific galaxy form-classification system chosen, however, the following general classification of clusters is clearly evidenced.

First of all, galaxies themselves may be subdivided into two basic categories, galaxies containing little or no observable dust (the elliptical and lenticular galaxies), and those which contain moderate to large amounts of dust (the spiral and irregular galaxies). Accordingly a definite tendency exists for clusters either to consist mostly of dust-poor (E and So) galaxies, or to be composed of a rather large percentage of dusty galaxies (Sa, Sb, Sc, Ir) with a smaller percentage of E and So galaxies.

Intergalactic medium in clusters. The intergalactic medium may also be divided into two types: luminous material and nonluminous material. For either case the material is difficult to observe, but it appears probable that most of the observable luminous material is in the form of stars. The detailed nature of the nonluminous material which may be present is unknown, although work reported by F. Zwicky in 1962 seems to establish the presence of an exceedingly tenuous "atmosphere" of dust in association with intermediate to large clusters.

Subclustering and superclustering. The property of some clusters to be composed of smaller groups of galaxies is referred to as subclustering. On a larger scale, should a group of clusters be associated in a single comprehensive aggregate, this grouping would be called a supercluster or a second-order cluster. The concept of a cluster of second-order clusters is referred to as a third-order cluster, and so on. Although there can be no doubt that very small and very large clusters do exist, the physical reality of any actual hierarchy of clustering has not yet been rigorously established, or denied.

The fact that so little can be said about each isolated property is indicative of the large degree of interdependence of the cluster properties. The nature of the interdependence, one of the most rewarding areas of research on clusters of galaxies, will be discussed next.

Relationships between properties. At the small end of the scale are clusters consisting of pairs or small groups of three to four galaxies. The average size of these clusters is about 0.2 Mpc and may range up to 0.5 Mpc. Two fairly typical examples of small groups such as these are the Milky Way galaxy–Magellanic Clouds system of three galaxies and the M 31, M 32, M 33, NGC 147, NGC 185, and NGC 205 group. It should be noted that the Milky Way–Magellanic Clouds system is accompanied by no less than nine dwarf galaxies, many of which are of very low surface brightness and could rather easily go undetected at distances even as near as M 31 with its attendant companions. Thus, even within the Local Group of galaxies, composed of the Milky Way galaxy and its companions and nine additional galaxies (including the newly discovered Maffei 1 and 2 galaxies), effects of distance on the detectability of faint galaxies are significant. Since larger clusters will be found at a great distance (more distant than the galaxies of the Local Group), raw number counts of galaxies in clusters as a function of cluster size will not include the fainter,

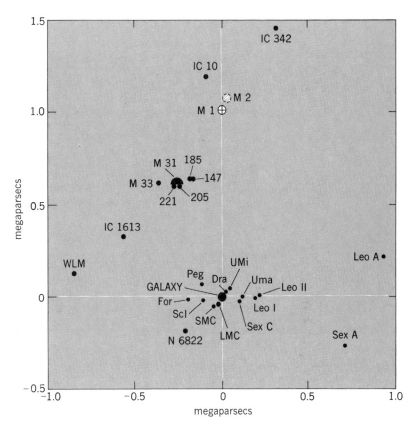

Fig. 2. The Local Group of galaxies projected onto the supergalactic plane. Note the subclustering around the Galaxy and around M 31. There is evidence for a chain of galaxies stretching from WLM through IC 342. (*From G. de Vaucouleurs, Distribution of galaxies and clusters of galaxies, Publ. Astron. Soc. Pac., 83:113, 1971*)

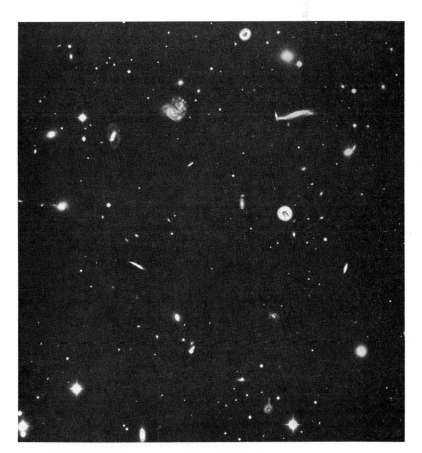

Fig. 3. Hercules Cluster of galaxies. (*Hale Observatories*)

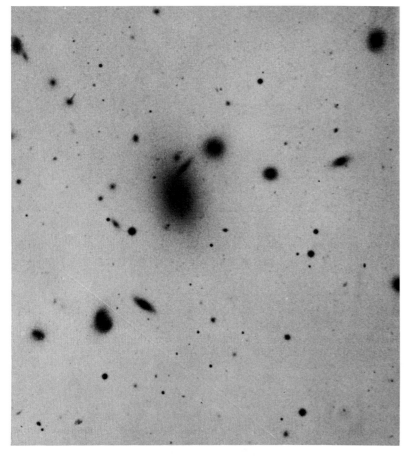

Fig. 4. Central portion of the Coma Cluster of galaxies. (*Hale Observatories*)

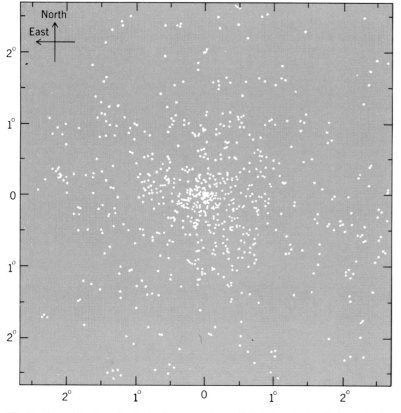

Fig. 5. Coma Cluster of galaxies from counts by F. Zwicky with the Palomar 18-in. Schmidt telescope. (*From G. de Vaucouleurs, Distribution of galaxies and clusters of galaxies, Publ. Astron. Soc. Pac., 83:129, 1971*)

undetected galaxies. Only by considering groups and clusters at similar distances can accurate values be obtained for the relative number of galaxies as a function of cluster size. Unfortunately the true total number of galaxies in any cluster, at distances sufficient for such comparative counts, can only be estimated on the basis of those few faint galaxies observable in the Local Group, in the vicinity of the Milky Way galaxy. It should be noted that the discovery of such low-surface-brightness galaxies in the vicinity of the Andromeda Galaxy (M 31), would add considerably to the statistical knowledge of the number of faint galaxies in clusters containing groups of galaxies similar to the Local Group.

Small subgroups frequently consist of spiral galaxies and their companions, usually small spiral or irregular galaxies. Small groups consisting entirely of elliptical galaxies are not uncommon, however, and these groups frequently take the form of chains of galaxies. H. C. Arp in 1967 and 1968 pointed out that many groups of galaxies are oriented in straight lines, a phenomenon which is evident in a number of systems illustrated in B. A. Vorontsov-Vel'jaminov's *Atlas and Catalog of Interacting Galaxies*, published in 1959.

Luminous intergalactic material. It is among the pairs and multiplets of galaxies that the presence of luminous intergalactic material is most conspicuous, frequently taking the form of bridges or filaments extending between adjacent galaxies. This phenomenon is illustrated in the small group of galaxies shown in Fig. 1. In the case of chains of elliptical galaxies, the faint outer portions of the galaxies are often seen to overlap. Recently Arp and collaborators have shown that giant elliptical galaxies have much larger tenuous outer regions than had previously been generally realized, although Zwicky had stated earlier that in large clusters containing giant elliptical galaxies, such as the Coma Cluster, luminous material probably composed of stars and pygmy galaxies is tenuously distributed throughout the intergalactic regions between the giant galaxies.

Stability of systems. The Local Group, mapped schematically in Fig. 2, is an example of a group or small cluster of galaxies with a characteristic size on the order of 2 Mpc and which contains usually from 25 to 100 galaxies. These clusters are composed predominantly of either spiral and irregular galaxies or elliptical and lenticular galaxies. Most of the small clusters are irregular in structure and are composed of spiral and irregular galaxies. The irregular distribution of galaxies in these clusters suggests that these systems are not dynamically relaxed; hence it has been suggested that there is no reason to assume that they are stable or bound systems. Nevertheless, studies of relatively nearby galaxies by H. G. Corwin in 1967, studies which were based on galaxies in de Vaucouleurs's *Reference Catalog of Bright Galaxies* (brighter than the thirteenth magnitude), indicate that not less than 80–85% of the more than 2500 galaxies examined were members of some subgroup, group, or small cluster of galaxies. Since the majority of the galaxies in the sample were spiral, one may infer that if the small clusters composed mainly of spiral galaxies are not bound systems, either the time scale for dispersion is long compared to the time since formation of the cluster galaxies or the dispersed galaxies have assumed a distribution which is sufficiently nonrandom as to mimic a cluster distribution. The latter possibility is highly unlikely from the statistics, while the former is physically unlikely on the basis of the calculated time scales. According to C. N. Limber in 1962, these time scales are on the order of 5×10^7 to 10^9 years. He suggests that this is too short a period for the development of the very old stellar populations found in many of these galaxies. Thus, despite the unrelaxed appearance of the loose clusters of spiral and irregular galaxies, there is some evidence for considering these to be gravitationally bound systems.

The same arguments hold true for larger clusters consisting mainly of spiral galaxies that are about 10 Mpc in size. An example is the Hercules Cluster of galaxies (Fig. 3), which contains both basic types of galaxies and exhibits a considerable degree of subclustering. These clusters may contain about 10^4 galaxies, not counting low-surface-brightness or faint dwarf galaxies. Another example is the Virgo Cluster of galaxies of mixed type. In this full-scale category of clusters of galaxies another type of cluster appears, the circularly symmetric globular cluster of galaxies, such as the Coma Cluster (Figs. 4 and 5). Such clusters contain mostly the dust-free galaxies. The concentration of these dust-free galaxies toward the center of the cluster is also particularly pronounced. The apparent spherical symmetry of these clusters, with the brighter more massive galaxies located near the center of the cluster, is highly indicative of a gravitationally bound system with the kinetic and potential energies of the system in a state of equilibrium. Zwicky in 1962 advanced evidence on the basis of diminished counts of extremely distant clusters in his *Catalog of Galaxies and of Clusters of Galaxies* that some form of intergalactic obscuration is present in both the Virgo and the Coma clusters. He concluded that fluctuations in the distribution of first-order clusters are due to intergalactic absorption and not to any second-order clustering. Nonetheless, a number of investigators have found strong evidence for second-order clustering. It may well be that the validity of Zwicky's argument will ultimately depend on observation of any third-order clustering.

Supercluster size. Second-order clustering, or superclustering, exhibits a characteristic dimension on the order of 40–50 Mpc. All types of galaxies are found in these cluster complexes, of which the Local Supercluster reported on by de Vaucouleurs in 1953 and 1958 is an excellent example. The distribution of galaxies of the Local Supercluster in the northern galactic hemisphere is shown in Fig. 6. The form of second-order clusters, whose existence must for the most part be established on the basis of statistical analyses of galaxy and galaxy cluster counts, is usually flattened, indicative of possible systemic rotation. Both Zwicky and G. O. Abell published in 1958 comprehensive catalogs of rich clusters of galaxies. The analyses of these catalog data provide the basis for the investigation of second-order clustering. In 1969 T. Kiang and W. C. Saslaw advanced evidence, on the basis of Abell's catalog, for clustering at levels higher than second order with characteristic dimensions as large as 200 Mpc.

These data must be interpreted within the context of all the other available data. In particular, it

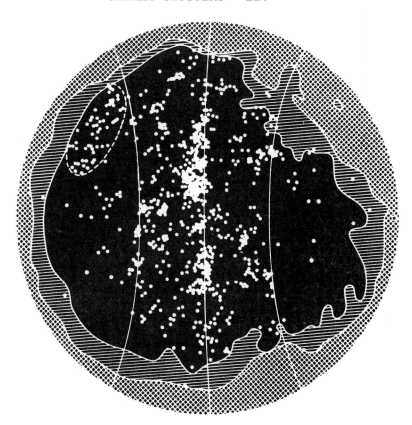

Fig. 6. Plot of nearby galaxies showing clustering toward the plane of the Local Supercluster. The view is a "fish-eye" perspective of the northern galactic hemisphere; the bright haze around the periphery is due to stars in the Galaxy. (*From G. de Vaucouleurs, Distribution of galaxies and clusters of galaxies, Publ. Astron. Soc. Pac., 83:131, 1971*)

appears that the larger the cluster (order $n+1$), the fewer are the clusters of order n which comprise it. Furthermore, as the cluster sizes increase the amount of intercluster space decreases, until the point is reached where essentially no room is left for higher order clusters to manifest themselves within the scale of the observable universe.

It appears that these conditions are being rapidly approached in the case of third-order clustering, so that higher order clustering with characteristic dimensions greater than 100 to 200 Mpc simply may not exist.

Interpretations. In conclusion it may be pointed out that evidence exists for either of two interpretations of the hierarchy of clustering: (1) The characteristic dimensions for clustering follow a smooth unbroken relationship which admits of clustering at any dimension, with no particular set of preferred dimensions; or (2) the characteristic dimensions for clustering have certain preferred values which are more or less universal. These two possibilities are shown in Fig. 7. It was suggested by de Vaucouleurs in 1971 that particular characteristic dimensions may indeed hold for small groups, with the tendency for specific characteristic dimension becoming progressively less pronounced as the clusters increase in size, until finally no specific characteristic size exists at all for clusters of perhaps third order or greater.

In any case it is evident that the phenomenon of clustering of galaxies is a very real and significant aspect of the observable universe, an aspect which must be adequately explained by any theory which

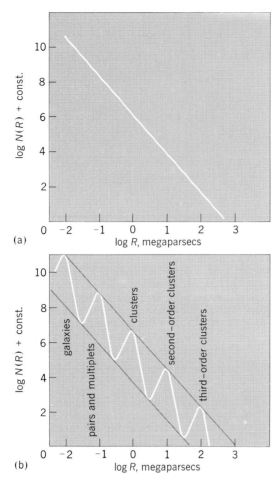

Fig. 7. Idealized frequency functions of characteristic lengths of galaxy clusters in two possible models of hierarchical clustering. (a) Clustering of galaxies occurs on all scales with no preferred sizes; the number density of clumps per unit volume decreases smoothly with increasing radius R. (b) Clustering of galaxies occurs on all scales but with relative maxima near a series of preferred characteristic radii corresponding to galaxies, small groups, clusters, and higher order clusters. (From G. de Vaucouleurs, Distribution of galaxies and clusters of galaxies, Publ. Astron. Soc. Pac., 83:140, 1971)

purports to describe either the present state of the universe or the nature of its origin and subsequent evolution.

For background information see GALAXY; GALAXY, EXTERNAL; MAGELLANIC CLOUDS in the McGraw-Hill Encyclopedia of Science and Technology. [JAMES D. WRAY]

Bibliography: G. O. Abell, Astrophys. J. Suppl., 3:211, 1958; H. C. Arp, Atlas of Peculiar Galaxies, 1966; T. Kiang and W. C. Saslaw, Mon. Notic. Roy. Astron. Soc., 143:129, 1969; D. N. Limber, in G. C. McVitte (ed.), Problems of Extra-Galactic Research: International Astronomical Union Symposium no. 15, 1962; G. de Vaucouleurs, Publ. Astron. Soc. Pac., 83:113, 1971; B. A. Vorontsov-Vel'jaminov, Atlas and Catalog of Interacting Galaxies, 1959; F. Zwicky, in G. C. McVitte (ed.), Problems of Extra-Galactic Research: International Astronomical Union Symposium no. 15, 1962; F. Zwicky et al., Catalog of Galaxies and of Clusters of Galaxies, 6 vols., 1961–1968.

Geodesy

Two major problems still confront geodesists today, both of which involve extremely accurate measurements: (1) determination of geodetic positions at sea (three-dimensional coordinates) and (2) determination of the geoid (an equipotential surface in the Earth's gravity field coinciding everywhere with mean sea level). The solution of these problems requires accurate knowledge of certain factors associated with the dynamic ocean environment that have limited the accuracy of geodetic measurement obtainable with conventional geodetic techniques. Examples of these factors are open-ocean tides, changing sea state, and underwater acoustic refraction. Another important factor is the motion (or instability) of the ships or floating platforms from which the measurements are made. See BUOY.

Many of the details involved in the solution of these major problems are interrelated. For example, although the required three-dimensional coordinates of positions can be determined without knowledge of the geoid, for practical purposes an accurately defined geoid is needed, particularly for determining geodetic heights, mapping, Earth-gravity modeling, and ocean-dynamics computations. Conversely, for determination of the geoid, accurate knowledge of the three-dimensional coordinates of several points is required to provide the necessary scale information.

Significant advances have been made within the last few years toward accurate determination of geodetic positions at sea. New techniques have been developed for handling ocean-surface to ocean-bottom measurements. Two recent experiments were aimed at establishing geodetic positions (or control points) at sea, marked by bottom-mounted acoustic transponders. A third experiment involved establishing geodetic control and also determining the geoid using satellites and other shipboard measurements. These techniques are now being explored for use in solving other problems associated with solid-earth physics and ocean dynamics.

Three-dimensional coordinates. The determination of the three-dimensional coordinates of geodetic positions at sea involves (1) accurate determination of ocean-surface positions with respect to ocean-bottom markers and (2) determination of the absolute position or relative positon or both in a chosen geodetic datum. Several techniques involving satellites and surface measurements have been developed for determining the three-dimensional coordinates of points on land. Positional accuracy with respect to the geocenter of ±5 to ±15 m have been obtained for several land control points in a unified world geodetic system. Some of these techniques can be applied to the establishment of similar control points in the ocean with comparable accuracies, provided that high accuracy can first be achieved in reducing the measurement from the ocean surface to the bottom markers which define ocean stations. This can be accomplished by reducing acoustic measurements made from different ship positions to ocean-bottom-mounted acoustic transponders or hydrophones. In practice, at least three transponders are used to permit later recovery of ship positions.

Much recent research has been focused on the development of new techniques that will permit better than 1 m precision in depth determination and a few meters in the horizontal coordinates.

Very recently three marine geodetic experiments designed specifically for geodetic research have been conducted in the Pacific Ocean, in the Bahamas, and over the Puerto Rico Trench. In the Pacific experiment a standard point error of ±16 to ±18 m was achieved in determining the horizontal geodetic coordinates of a marine geodetic control point about 200 km from shore at water depths of 2000 m. In the Bahamas experiment, standard errors of ±6 to ±7 m were achieved in determining the horizontal geodetic coordinates and ±2 to ±3 m in determining the depth. Most important, however, was the achievement of ±1 m standard deviation in recovery of ship heights above the transponders using independent ranging measurements.

Ocean-surface positions. Three newly developed techniques for relating ocean-surface positions to ocean-bottom markers and their specific applications follow.

Modified line crossing. Conventional line crossing is a technique that has been extensively used for determination of the geometry and orientation of transponder arrays. In its various versions, the technique presumes accurate knowledge of (1) the transponder depths, (2) ship speed and heading, which must also be constant, and (3) the point of closest approach for each transponder, at which the slant range from the ship to the transponder is perpendicular to the ship's track. These assumptions present many difficulties that render the technique very inefficient and produce inconsistent results. The transponder array location by coplanar ranges (TALCOR) technique, which is a modified line-crossing approach, eliminates the need for making assumptions required with the conventional technique. It permits analytically determining the depth of each transponder and the geometry of the transponder array. The technique involves a three-dimensional weighted-least-squares solution, utilizing near-coplanar slant ranges in the various vicinities where the ship crosses the vertical plane containing any two transponders. It is amenable to statistical analysis of the observations and the adjusted parameters. It can also furnish geodetic orientation of the array if the geodetic coordinates (from satellites or other electronic positioning systems) of at least two adequately separated ship positions and the corresponding slant ranges to at least two transponders are determined.

TLSP. Transponder location by surface positioning (TLSP) involves a weighted-least-squares solution for determining simultaneously the three-dimensional coordinates of ocean-bottom acoustic transponders using acoustic slant ranges and surface ship coordinates. It is based on the principle of intersection of a minimum of three slant ranges from three known and noncolinear points to an unknown point (Fig. 1). The technique includes statistical analysis for determining adequate weighting criteria before the adjustment. The adjustment is performed in a three-dimensional cartesian coordinate system whose primary plane is the mean equator and primary axis is parallel to the mean rotational axis of the Earth. The TLSP

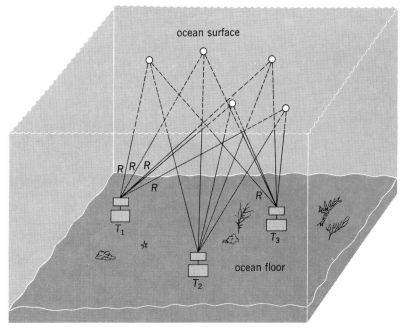

Key:
○ = Surface ship position (known coordinates)
T = Ocean bottom acoustic transponders (unknown coordinates)
R = Acoustic slant ranges

Fig. 1. Principle of intersection in TLSP technique.

computer program incorporates a ray-tracing subroutine that corrects for acoustic refraction as it computes the acoustic slant ranges from the curved-path travel times using velocity-of-sound profile data for the area. The advantage is that depth is determined by a least-squares adjustment in contrast to, for example, line crossing in which depth to the transponder is measured and then held fixed throughout the computation.

Horizon coordinates technique. The application of the horizon coordinates technique for adjustment of marine geodetic measurements was developed at Battelle to avoid the use of estimated geoidal heights as with TLSP. It solves directly for depth of the transponders and permits more flexibility for introducing constraints in a generalized least-squares solution with parameter weighting. The adjustment is performed in a three-dimensional coordinate system whose primary planes are parallel to the local horizon and the local meridian.

The previous two techniques, TALCOR and TLSP, were primarily three-parameter solutions. A five-parameter solution has been developed, treating as free variables the three coordinates of the transponders, the acoustic system bias, and the sound velocity. The ill-conditioned normal equations resulting from the direct solution has been remedied by introducing certain constraints regarding depth and velocity of sound through weighting of the parameters.

Geodetic coordinates. Some of the techniques developed for determining the geodetic coordinates of surface positions for land use that can be applied at sea are satellite techniques, surface-based techniques, and airborne techniques. In addition, the very-long-baseline interferometry (VLBI) technique is potentially applicable.

Satellite techniques. The two satellite techniques most recently experimented with involve Doppler and C-band radar ranging systems. They are applicable anywhere in the oceans. Most limitations on accuracy are imposed by errors in orbital data, particularly in remote ocean areas where no tracking stations are available. In the future, satellite altimetry, drag-free satellites, and satellite-to-satellite tracking are expected to help improve orbital accuracy.

Position accuracies of ±10 to ±50 m are now possible using Doppler satellite techniques or shipboard C-band radar ranging. Improved accuracies are also expected from future DOD satellite techniques. Studies are underway on the use of present laser-ranging systems with the stabilized shipboard C-band radar system. Accuracies better than ±10 m in determining the coordinates of marine geodetic points can be expected with a combined radar and laser system.

Surface-based techniques. The most useful techniques in this category involve the use of electronic positioning systems, such as Lorac, Raydist, Lambda, and Autotape. Electronic and inertial types of systems have been experimented with, and are widely used to provide horizontal geodetic coordinates "directly." The electronic positioning systems that have sufficient accuracies for geodetic positioning are limited to use about 200 to 300 km from shore. The inertial systems are affected by drift, and update information is required from external positioning systems. Also, since they are aligned with the local vertical, they are highly correlated with the geoid and the associated deflection of the vertical which are not yet known accurately at sea.

Airborne techniques. The most common use of airborne techniques is for independent distance measurements to establish trilateration networks which connect points of known and unknown coordinates. The adjustment of this network furnishes the unknown coordinates. Several systems, such as Shoran, Hiran, Shiran, and Lorac, can be used.

Shiran is perhaps the most sophisticated and accurate of these systems; however, it has not been used in experiments at sea. On the other hand, the Lorac system was used in a Pacific experiment with the line-crossing technique; standard point errors of about ±16 to ±18 m in determining the latitude and longitudes of the marine control point resulted. Airborne techniques are useful for measuring distances up to 800 km.

VLBI technique. One promising technique for the future involves the application of VLBI technique for the establishment of geodetic points at sea. The technique could utilize either stellar or satellite sources, and it has potential for yielding better than ±10 m accuracy at sea.

Geoid determination. Accurate determination of the geoid and geodetic positions at sea is extremely important for geodesy, oceanography, satellite orbital computation, space research, and other scientific purposes, as well as for national defense. A new approach to determination of the geoid has been under consideration in recent years. This involves equipping a special satellite to measure the height of the sea surface, using microwave altimetry techniques. Experiments will be conducted with two altimeter systems, one on Skylab in 1973 and the other on the *GEOS-C* satellite in 1973 or 1974. Eventually it will be possible with the satellite altimeter to determine an accurate global geoid at sea.

Methods by which a geoidal surface can be determined at sea include (1) astrogeodesy, which involves determination of astronomic coordinates at corresponding points, using surface-based techniques; (2) astrosatellite, which involves determination of astronomic coordinates and absolute geodetic coordinates at corresponding points, using satellite positions at sea; (3) gravimetry, which involves global gravity measurements which can determine only the shape of the geoid; linear and angular geodetic measurements are required to furnish correct scale; (4) astrogravimetry, which involves determination of geodetic and astronomic coordinates and gravity measurements; and (5) satellite, which involves either analysis of satellite orbit perturbation in the Earth's gravity field or determination of the geoid using satellite altimetry.

The methods involving astronomic observations require a positional accuracy of at least ±1 arc second. Current research is aimed at this accuracy goal. Present equipment accuracy is ±10 to ±20 arc seconds. Only the method of using satellite altimetry for determining the geoid will be discussed here because geoid determination at sea from satellite altimetry promises to be the most economic and accurate technique.

Two radar altimeter experiments are being planned for demonstrating the technical feasibility of the hardware for measuring accurately the height of sea surface relative to the satellite. Most of the research to date on satellite altimetry has involved theoretical studies of application techniques, hardware specification and design, and simulated aircraft tests. In addition, a NASA marine geodetic experiment was conducted in 1970 over the Puerto Rico Trench. This experiment was designed to provide data for assessing the feasibility of in-orbit calibration of a satellite altimeter in

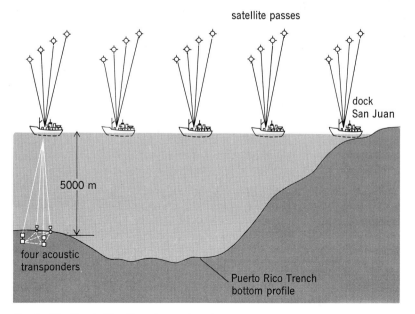

Fig. 2. The Puerto Rico Trench experiment.

advance of the *GEOS-C* launching and to provide marine geodetic control and geoidal profiles. The latter could be used as a basis, or "ground truth," for evaluating the capability of satellite altimetry for determining the geoid and the validity of the proposed calibration concept.

One aspect of the experiment involved determination of the heights of the ship relative to satellite orbit, and thus the heights of the sea surface at a series of points across the Puerto Rico Trench, utilizing the orbital dynamics of the *GEOS-2* satellite as determined from C-band radar and other shipboard systems (Fig. 2). The experiment involved the use of the Apollo ship *Vanguard* and its instrumentation systems, the *GEOS-2* satellite, the Navy Transit satellites, a land-based C-band radar tracking network, and four underwater acoustic transponders. The shipboard systems included a C-band radar system, Doppler SRN-9 receiver, inertial/star tracker system, and bathymetric navigation system. A geoidal traverse about 200 km long, consisting of eight stations across the Puerto Rico Trench, was made, using the shipboard systems for measurements. The data are being analyzed. Two geoidal profiles will be computed for comparison. One profile will be based on the measurements made with the C-band radar to the *GEOS-2* satellite, and the other on the astrogeodetic measurements made with the inertial/star tracker and the Doppler satellite data.

International symposium. In 1971, for the first time, the International Association of Geodesy (IAG) and the International Association of Physical Sciences of the Oceans (IAPSO) sponsored a 1-day symposium on marine geodesy. The symposium was held in August, 1971, during the International Association of Geodesy and Geophysics 15th General Assembly Meeting in Moscow. Speakers described the progress made in marine geodesy and the problems associated with establishment of geodetic positions at sea and with determination of the geoid by satellite altimetry. Also described were the application of marine geodetic techniques for solving such ocean-related problems as open-ocean tidal measurement, tsunami prediction, sea-state determination, and ocean-spreading measurements.

For background information *see* GEODESY; TERRESTRIAL GRAVITATION in the McGraw-Hill Encyclopedia of Science and Technology.

[A. GEORGE MOURAD]

Bibliography: A. G. Mourad, *EOS*, vol. 51, no. 12, December, 1970; A. G. Mourad, J. H. Holdahl, and N. A. Frazier, in *Marine Geodesy*, Marine Technology Society, 1970.

Gravitational collapse

After a lifetime which can be measured in billions of years, a star begins to lose its thermal pressure and to contract under the pull of gravity. The culmination of this gravitational collapse might be the formation of a white dwarf, or a supernova explosion and the formation of a neutron star. Recently interest has centered on a third possibility: continued collapse and the formation of a general-relativistic "black hole." Theorists are probing the properties of black holes by means of Albert Einstein's general relativity theory, while astronomers are looking for observable phenomena that might be associated with them.

Supernovae and black holes. A low-mass star dies quietly and becomes a white dwarf. But white dwarfs have a maximum mass of about 1.5 solar masses, so that the gravitational collapse of a large star must be halted in a more dramatic fashion. It is known that a star can undergo a violent supernova explosion which throws much of the star's mass far into space. As early as 1934, Fritz Zwicky proposed that a supernova explosion might be accompanied by the implosion of the stellar core and the formation of a neutron star, an extremely dense star whose matter is mostly in the form of neutrons. The success of neutron star models in explaining pulsars, and especially the discovery of a pulsar near the center of a supernova remnant (the Crab Nebula), is a proof of the validity of some of the basic ideas about supernovae. *See* NEUTRON STAR.

It is important for astrophysicists to understand the supernova process in order to determine whether all high-mass stars become supernovae and to predict what sort of remnant will be left behind after a supernova explosion. Calculations by Stirling Colgate and Richard White imply that an intense neutrino flux from nuclear reactions in the stellar core is responsible for blowing off the outer layers of the star. James Wilson, however, has recently performed computer studies of gravitational collapse which indicate that the neutrinos from the core pass through the outer layers without much interaction. The mass-ejection mechanism and the details of the supernova process must be considered an open and controversial question. Wilson's work does suggest that heavy stars may not shed mass as easily as was previously believed.

The gravitational collapse of a large star continues beyond the neutron-star stage if there is no rapid mass loss or if the stellar core left after a supernova explosion is too massive or is imploding inward too quickly to become a neutron star. When the star has shrunken down to a radius of the order of the Schwarzschild radius (R_{Sch} = 3.0 km × mass of star/solar mass), then the predictions of Newton's theory of gravitation are not correct, even qualitatively. In particular, according to general relativity, R_{Sch} marks a point of no return for the collapse. Once the star's radius shrinks through R_{Sch}, no matter or light or information can ever again radiate out. The star leaves a "black hole" in space.

Detection of black holes. Since black holes are inherently nonluminous and of very small radius (on the order of kilometers), their detection is a difficult problem for astronomers. Evidence for their existence will certainly have to be indirect. Y. B. Zel'dovich and O. H. Guseynov have suggested that black holes might be present in known binary-star systems where only the light from one star is seen. (In such systems the presence of a nonluminous second star, as well as an estimate of its mass, can be inferred from the varying Doppler shift from the primary, luminous star.) V. L. Trimble and K. S. Thorne have considered systematically all such binaries and have found that in all cases there exist explanations less

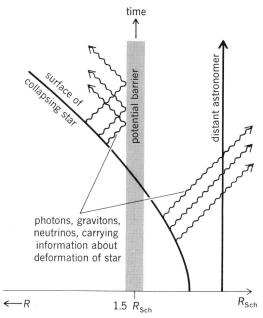

The propagation of information about a star's deformations during late stages of general-relativistic gravitational collapse.

exotic than the black-hole hypothesis.

Recently A. G. W. Cameron has argued that there is a black hole in the strange eclipsing binary system ε Aurigae, which contains an unseen component heavier than 10 solar masses. When ε Aurigae eclipses every 27 years, its light is diminished by half but its spectral distribution is unchanged. This suggests that the light from the luminous primary is being obscured by either an opaque body partially blocking the view, or by a semitransparent cloud of dust whose particles are much larger than the wavelength of light. (A black hole by itself would be far too small to block a significant fraction of the light.)

Cameron's model of ε Aurigae postulates that the obscuring body is a large ($\sim 2 \times 10^9$ km) disk of relatively small mass (a small fraction of a solar mass) containing relatively large (~ 0.5 cm) grains of dust. The disk surrounds, and is held together by, a massive (~ 23 solar masses) black hole. Although this claim of the detection of a black hole received much public attention, the interpretation is rather controversial; most astrophysicists believe the observational evidence is better explained by a model with an opaque gaseous disk surrounding an ordinary star.

A black-hole process that may give rise to observable phenomena is mass accretion. As a star moves through a gaseous region of space, its gravity draws in some of the gas. In the case of a neutron star or black hole the in-falling gas will be accelerated to high velocity and heated by compression to temperatures so high that x-rays and γ-rays can be emitted. The x-ray and γ-ray spectra from a black hole should differ from those of a neutron star because gas falling into a black hole never impacts on a stellar surface. Theorists are now trying to get a better understanding of these complicated phenomena and are awaiting improvements in the technology of x-ray astronomy.

The gravitational collapse in which a black hole is born produces a large flux of gravitational radiation. The gravitational waves which Joseph Weber at the University of Maryland seems to be detecting are almost certainly associated with violent gravitational collapse. One interesting possibility is that the waves are produced when a star is drawn into an enormous black hole created long ago by the gravitational collapse of matter at the center of the Milky Way galaxy. *See* GRAVITATION.

Asymmetric gravitational collapse. The equations of general relativity prohibit the fields outside a nonrotating black hole from showing any small asymmetries. A gravitational quadrupole field or an electric dipole field, for instance, cannot exist around a nonrotating black hole. John A. Wheeler has conjectured that this is true in a very general sense: A black hole (rotating or nonrotating) is characterized soley by its mass, angular momentum, and electric charge. As Wheeler put it: "A black hole has no hair." Recent work by Werner Israel, Brandon Carter, and others has all but completely confirmed this.

The fate of any "hair" (for example, an electric dipole) during collapse has been an enigma. According to general relativity theory, the material in a star experiences no strong forces as the star's surface collapses through the Schwarzschild radius. Thus forces should not develop to smooth away an asymmetry (for example, to redistribute the electrons in a star and neutralize an electric dipole) and the star should form a black hole with asymmetries. This dilemma caused some theorists to abandon the qualitative features of gravitational collapse and the formation of black holes as idiosyncracies of perfect spherical symmetry. They believed that forces develop, due to the small deformations in any real star, which either halt the collapse or greatly change its nature.

Scientists now know that the gravitational collapse of a star with small asymmetries leads to the formation of a black hole just as in the case of a star with no "hair." The star retains its asymmetries as it collapses through the Schwarzschild radius, but no manifestation of these asymmetries can be detected outside the black hole that is formed. The mechanism for this is shown in the illustration. In this diagram the scale for the radial (R) coordinate is distorted in such a way that photons, gravitons, and neutrinos move on 45° lines. This requires shifting the Schwarzschild radius an infinite distance to the left in the diagram. Electromagnetic and gravitational waves carrying information about the magnitude of the collapsing star's deformations are radiated outward, but are pulled back in and collapse with the star. Mathematically the waves are reflected off a potential barrier created by the curvature of space-time around 1.5 R_{Sch}. All radiatable information (gravitational multipoles, lepton number, and so forth) suffers the same fate. The only pieces of information that can penetrate the potential barrier and reach out from the black hole, to be detected by a distant astronomer, are nonradiatable quantities: the electric monopole (the charge), the gravitational monopole (the mass), and a fixed combination of higher gravitational multipoles corresponding to the star's angular momentum.

The nature of the late stages of gravitational collapse of highly nonspherical bodies is a more difficult question, on which work has just begun. Studies of the collapse of long cylinders, the only highly nonspherical configuration for which results are known, show that they do not form black holes. During their collapse, a point of no return analogous to the Schwarzschild radius does not develop.

For background information *see* GRAVITATIONAL COLLAPSE; PULSAR; X-RAY ASTRONOMY in the McGraw-Hill Encyclopedia of Science and Technology. [RICHARD H. PRICE]

Bibliography: A. G. W. Cameron, *Nature*, 229: 178, 1971; R. Ruffini and J. A. Wheeler, *Phys. Today*, 24:30, 1971; K. S. Thorne, *Comments Astrophys. Space Phys.*, 2:191, 1970; J. R. Wilson, *Astrophys. J.*, 163:209, 1971.

Great Lakes

The Great Lakes today, as they have been since man first used them for his purposes, form the heart of the North American continent. They provide water for municipal, industrial, agricultural, and recreational purposes for about 35,000,000 people living around their shores in the United States and Canada. By the end of the century, the population around the Great Lakes may be doubled and the use of their water quadrupled. Much concern is being shown at this time for the present and continuing quality of water in the lakes, and conflicts regarding the use of water for various purposes are already common. It has become obvious that major changes will have to be made in current water-use and waste-disposal practices if the high quality of water in the Great Lakes is to be maintained with increased use by a growing population. *See* ENVIRONMENTAL ENGINEERING.

Interest in long-range problems of lake management are concentrated on water qualtiy or pollution. Two principal factors are involved: first, the physical (and chemical and biological) characteristics of the resources; and second, the demands or stress put on the system. Part of the stress is natural—annual and seasonal fluctuations in climate, for example—and part is man-made. Important man-made stresses include diversion of water into or out of the Great Lakes and the addition of large quantities of organic and inorganic pollutants. *See* WASTES, AGRICULTURAL.

The five Great Lakes form one of the largest reservoirs of fresh water on the face of the Earth. They have a total area of about 95,200 mi^2, slightly greater than that of New York and Pennsylvania combined, and contain nearly 5,800 mi^3 of fresh water. This is about one-fifth of all the fresh water (not ice) in the world.

Moreover, the Great Lakes occupy almost one-third of their drainage basin. This is an unusually high proportion; most lakes occupy one-tenth or less. One result is that their lake levels and discharges are more constant than those of most lakes. The maximum difference between maximum high and minimum low on any lake since 1860 is about 6½ ft. The maximum flow of the outlet river is only about three times its minimum, in contrast to other large North American rivers which have maximum-minimum flow ratios of as much as 60:1. This stability makes the lakes dependable, but it also makes management more difficult, because the boundaries within which management must operate are already closely limited by the physical environment.

In addition to supplying water for the surrounding region and acting as a sink for its wastes, the Great Lakes also provide many recreational facilities, fish, and waterways for commercial and industrial ships. There are about 77,000 acres of parks and beaches along the lower lakes alone. More than 200,000,000 net tons of bulk cargo have moved annually in the Great Lakes in recent years, and the Lake Erie fish catch is worth about $5,000,000.

IJC report on pollution. After a 6-year study of the pollution of Lake Erie, Lake Ontario, and the international section of the St. Lawrence River, the International Joint Commission (IJC) found that the waters of the lower Great Lakes are being polluted seriously by both countries to their own and each other's detriment. The IJC recommended that both Canada and the United States adopt specific water-quality objectives and enter into agreement on programs, measures, and schedules to achieve them.

The general objective of the recommendation is that the waters of the lower lakes should be free from substances attributable to municipal, industrial, or other discharges that (1) adversely affect aquatic life and waterfowl, (2) are unsightly or deleterious, (3) create a nuisance, (4) are toxic or harmful to human, animal, or aquatic life, and (5) create nuisance growths of aquatic weed and algae.

Specific quality objectives were set for the "receiving waters except in the restricted mixing zones at outfalls." Total coliform should not exceed 1000/100 ml and fecal coliform should not exceed 200/100 ml; dissolved oxygen in connecting channels and upper lake waters should not be less than 6.0 mg/l; temperature changes should be held within limits that would not adversely affect any local or general use; there should be no objectionable taste or odor (phenols not to exceed a monthly average of 1 mg/l, pH to remain between 6.7 and 8.5); iron should be less than 0.3 mg/l; phosphorus loading (as a constituent of various organic and inorganic compounds) to Lake Erie should be less than 0.39 g/m^2/year and to Lake Ontario, less than 0.17 g/m^2/year; and radioactivity is to be limited to less than 1000 picocuries/l for gross beta activity, 3 picocuries/l for radium-226, and 10 picocuries/l for strontium-90.

The report touched on soil erosion and the adverse effects of increased sediment due to agricultural and construction practices but it made no specific recommendation for sediment limits.

The IJC urged immediate remedial measures, including reduction of phosphorus content of detergents and the prompt implementation of municipal and industrial waste treatment. The cost of the remedial treatment facilities is estimated, in 1968 U.S. dollars, to be $211,000,000 in Canada and $1,373,000,000 in the United States. The recommended phosphorus and other quality standards are disputed by some industries and scientists, largely on the basis of the buffering effect of minerals absorbed and stored by bottom sediments.

The current methods for cleaning oil spills on the Great Lakes, according to the IJC, are "primitive and inadequate," and it urged the adoption of a formal plan for international cooperation for prevention and cleaning.

The IJC made two other significant recommendations. It recommended that the IJC be authorized to investigate pollution in the remaining boundary waters of the Great Lakes system, and that it be given the authority, responsibility, and means for coordinating and ensuring the necessary surveillance and monitoring of water quality and the effectiveness of pollution abatement programs.

Implementation of the latter recommendation will provide for the first time a continuing joint United States–Canadian source of information basic to rational lake management.

One of the difficulties faced by lake managers is the intricacy of institutional arrangements concerned with the Great Lakes. Involved are two nations, which include one province (Ontario) and eight states (Minnesota, Wisconsin, Michigan, Illinois, Indiana, Ohio, Pennsylvania, and New York). There are two formal (treaty) international bodies, the IJC and the Great Lakes Fishery Commission. On the United States side alone, seven Federal departments and at least a dozen major agencies are involved in Great Lakes studies, investigations, and regulation, not to mention a multitude of state, university, and local government groups. On the United States side, two coordinating bodies, the Great Lakes Basin Commission (a Federal group) and the Great Lakes Commission (a state compact group), are effecting some integration of the diverse and often conflicting interests involved in managing the waters of the Great Lakes.

Recent developments. Reports of high mercury content in fish, water, and bottom muds have led to the restriction and banning of commercial fishing, more stringent waste-disposal controls, and more specific water-quality standards. Mercury levels in walleye pike in Lake Erie and near Detroit have been measured as high as 3.6 mg/l. Near Detroit, mercury levels in sediments were as high as 86 mg/l near a chlor-alkali industrial plant; they decreased downstream to 0.4 mg/l and less where the Detroit River enters Lake Erie. Mercury levels of more than 1 mg/l in fish and as much as 560 mg/l in bottom muds resulted in the banning of fishing in the Wisconsin River. Much of the source of mercury is believed to be waste from industrial plants along the shores of the Great Lakes and their tributaries.

Mercury and most of its salts are insoluble in water. However, mercury may be oxidized in aerobic parts of lakes, and there is a good correlation between mercury and organic carbon in the top layers of sediment. Microorganisms on the sediments ingest the mercuric ions, transforming them to forms which may be either volatilized into the atmosphere or incorporated into the plant and animal food chain. As larger fish prey on smaller, the mercury accumulates, eventually to the point it may become dangerous to man. So far there have been no reports of deaths due to people eating fish with high mercury content in the Great Lakes area.

Since 1678 Niagara Falls has retreated about 1/4 mi, and in recent years fallen rock has accumulated at the base of the American Falls to the extent that it marred the spectacle of an uninterrupted veil of water.

In a dramatic demonstration of man's local control over nature, the fall of water over the American Falls, which carry about 10% of the water over Niagara Falls, was stopped. At an estimated cost of about $1,500,000 the flow was diverted to the Canadian side and the extent of rock fracturing and deterioration was studied under virtually dry conditions. Jointing and permeability examinations and rock stress measurements are providing a basis for planning a program to stabilize the escarpment. In this way the Niagara River and Lake Erie, with their valuable shore installations, will remain as they are, and the American Falls will continue to delight the public where installations have been built to view them.

Although attention is being concentrated on the quality problems, it is gradually being accepted that an understanding of the physical hydrology of the Great Lakes is basic to the management of both the quantity and quality of their waters. The first effort to investigate the physical aspects (including related biological and chemical phenomena) of the hydrology of a whole large lake is being undertaken by the International Field Year for the Great Lakes (IFYGL). Since 1966, planning and feasibility studies for the investigation of Lake Ontario and its basin have been progressing under the guidance of a binational steering committee set up by the Canadian and United States National Committees for the International Hydrological Decade. Field operations will begin in 1972 and will be coordinated or conducted mainly by the Canada Centre for Inland Waters and the U.S. National Oceanographic and Atmospheric Administration. These two agencies will be assisted by many other Federal, state and provincial, and university institutions in both the United States and Canada.

The overall objective of IFYGL is to provide a framework of integrated information, by way of analytical and mathematical models, that will make it possible for management to predict (or to improve the prediction of) the effects of its policies and practices on all aspects of the physical environment of the Great Lakes and man's use of it. Specific programs are concerned with the energy balance; atmospheric water balance (evaporation, precipitation, movement, and distribution); terrestrial water budget (inflow and outflow balances of both surface and ground water); water movement (circulation, seasonal turnovers, wind effects, and dispersion); and nutrient cycle (sources and distribution of chemical and biological materials).

It is hoped that IFYGL will not only provide valuable data on the water regimen of Lake Ontario and its basin but will also be a pilot model for more sophisticated future studies on the other Great Lakes and other large lakes elsewhere.

For background information *see* EUTROPHICATION, CULTURAL; SEWAGE DISPOSAL; WATER POLLUTION in the McGraw-Hill Encyclopedia of Science and Technology. [L. A. HEINDL]

Bibliography: Great Lakes Basin Commission, The Laurentian Great Lakes of North America, *Organization of Economic Cooperation and Development Proceedings of Symposium on Large Lakes and Impoundments*, 1970; International Joint Commission, Canada and United States, *In-*

terim Report on the Regulation of Great Lakes Levels, 1968; International Joint Commission, Canada and United States, *Pollution of Lake Erie, Lake Ontario, and the International Section of the St. Lawrence River*, 1970; S. S. Philbrick, Horizontal configuration and the rate of erosion of Niagara Falls, *Geol. Soc. Amer. Bull.*, 81:3723–3732, 1970.

Growth hormone

In December, 1970, two California scientists published a short communication summarizing their work on the synthesis of a protein possessing growth-promoting and lactogenic activities. This publication opens the possibility for unlimited supply of synthetic products having human growth hormone (HGH) activities for both clinical and experimental investigations.

The earlier studies of H. M. Evans, J. A. Long, P. E. Smith, and others provided very convincing evidence that the pituitary gland secretes a hormone possessing growth-promoting activity in experimental animals. The final proof for the existence of growth hormone (also called somatotropin) was furnished when it was isolated in a highly purified form and characterized as a protein of molecular weight 45,000 and isoelectric point pH 6.8 from bovine pituitaries by C. H. Li and H. M. Evans in 1944. When the bovine hormone was repeatedly tested in man and found to have no growth-promoting activity, the hypothesis was advanced that the bovine and human hormones are chemically and immunologically different. In 1956 Li and H.

Table 1. Some physicochemical properties of HGH

Molecular weight	21,500
Isoelectric point, pH	4.9
Sedimentation coefficient, $S_{20,w}$	2.18
Diffusion coefficient, $D_{20} \times 10^7$	8.88
Specific rotation, $[\alpha]_D^{25}$(0.1 M acetic acid)	$-39°$
Ellipticity $[\Theta]$, at 221 nm	19,700
α-Helix content, %	55
pK_a of tyrosine residues	10.8
Absorptivity, $E_{1cm}^{0.1}$ at 277 nm	0.931

Papkoff isolated growth hormone from human and monkey pituitaries in highly purified form and showed that the primate hormones are chemically distinct from bovine growth hormone. In addition, it was also demonstrated that human and monkey growth hormones are active as anabolic agents in man. Thus the isolation of HGH opened up new vistas in clinical studies. It is now well known that HGH is highly active in growth promotion in humans; especially striking is the metabolic effect of this hormone on young children with hypopituitary function.

Chemistry of HGH. HGH is a protein of molecular weight 21,500 and isoelectric point pH 4.9; some physicochemical properties of the hormone are presented in Table 1. The amino acid sequence of HGH was first proposed in 1966 by Li, W.-K. Liu, and J. S. Dixon and recently revised as shown in Fig. 1. It may be noted that the hormone consists of a single polypeptide chain with the single tryptophan residue at position 85 and the two disulfide

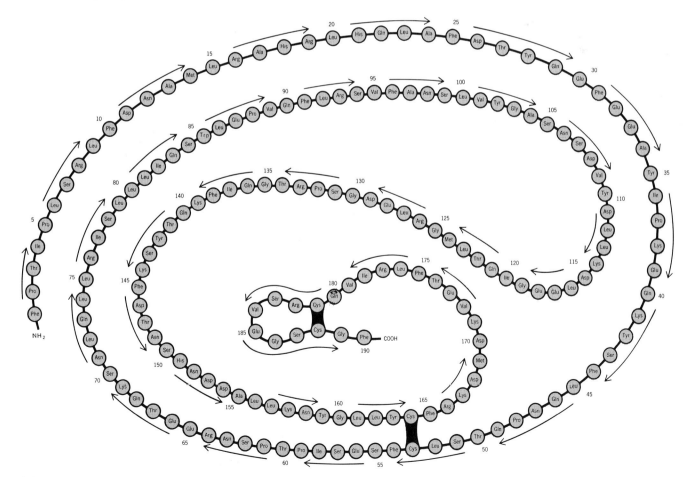

Fig. 1. The complete amino acid sequence of the HGH molecule.

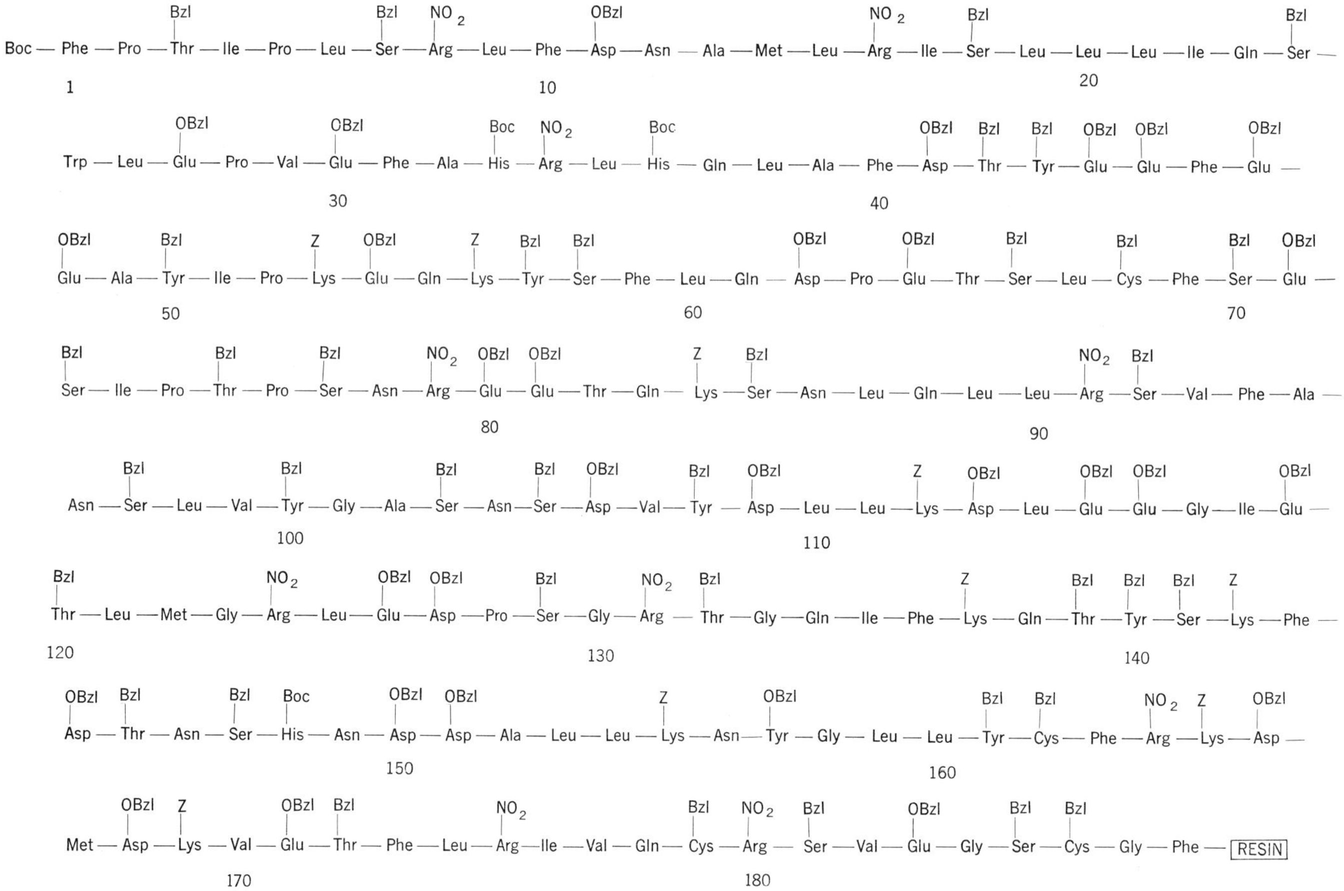

```
      Bzl              Bzl  NO₂         OBzl            NO₂     Bzl                           Bzl
Boc—Phe—Pro—Thr—Ile—Pro—Leu—Ser—Arg—Leu—Phe—Asp—Asn—Ala—Met—Leu—Arg—Ile—Ser—Leu—Leu—Leu—Ile—Gln—Ser—
1                               10                                           20
           OBzl         OBzl       Boc  NO₂     Boc                 OBzl  Bzl  Bzl  OBzl OBzl  Bzl
Trp—Leu—Glu—Pro—Val—Glu—Phe—Ala—His—Arg—Leu—His—Gln—Leu—Ala—Phe—Asp—Thr—Tyr—Glu—Glu—Phe—Glu—
                     30                              40
 OBzl  Bzl        Z    OBzl   Z   Bzl  Bzl              OBzl   OBzl      Bzl       Bzl  Bzl  OBzl
Glu—Ala—Tyr—Ile—Pro—Lys—Glu—Gln—Lys—Tyr—Ser—Phe—Leu—Gln—Asp—Pro—Glu—Thr—Ser—Leu—Cys—Phe—Ser—Glu—
        50                               60                                70
 Bzl       Bzl    Bzl       NO₂  OBzl OBzl      Z    Bzl                  NO₂  Bzl
Ser—Ile—Pro—Thr—Pro—Ser—Asn—Arg—Glu—Glu—Thr—Gln—Lys—Ser—Asn—Leu—Gln—Leu—Leu—Arg—Ser—Val—Phe—Ala—
                            80                                90
           Bzl       Bzl       Bzl       Bzl  OBzl   Bzl OBzl        Z    OBzl  OBzl      OBzl
Asn—Ser—Leu—Val—Tyr—Gly—Ala—Ser—Asn—Ser—Asp—Val—Tyr—Asp—Leu—Leu—Lys—Asp—Leu—Glu—Glu—Gly—Ile—Glu—
                   100                           110
 Bzl        NO₂   OBzl OBzl    Bzl    NO₂  Bzl             Z        Bzl  Bzl  Bzl   Z
Thr—Leu—Met—Gly—Arg—Leu—Glu—Asp—Pro—Ser—Gly—Arg—Thr—Gly—Gln—Ile—Phe—Lys—Gln—Thr—Tyr—Ser—Lys—Phe—
120                               130                                 140
 OBzl  Bzl   Bzl  Boc  OBzl OBzl            Z    OBzl            Bzl  Bzl      NO₂  Z   OBzl
Asp—Thr—Asn—Ser—His—Asn—Asp—Asp—Ala—Leu—Leu—Lys—Asn—Tyr—Gly—Leu—Leu—Tyr—Cys—Phe—Arg—Lys—Asp—
             150                                         160
      OBzl  Z    OBzl Bzl        NO₂            Bzl  NO₂ Bzl      OBzl     Bzl      Bzl
Met—Asp—Lys—Val—Glu—Thr—Phe—Leu—Arg—Ile—Val—Gln—Cys—Arg—Ser—Val—Glu—Gly—Ser—Cys—Gly—Phe—|RESIN|
            170                            180
```

Fig. 2. Fully protected HGH-like protein-resin. Boc = *tert*-butyloxycarbonyl; Bz = benzyl; Z = carbobenzoxy.

bridges forming two loops: one between residues 53 and 164 and the other between residues 181 and 188.

Biological properties. As its name implies, the most readily observed effect of growth hormone following its administration to either normal or hypophysectomized animals is enhancement of body weight and length, an effect representing a selective activity of growth hormone. The hormone is also known to influence the growth of bony tissue. Whereas growth hormone is not responsible for skeletal maturation, it is involved in the process of osteogenesis. In addition, there is now much evidence to the effect that in experimental animals growth hormone accelerates the mobilization and oxidation of depot fat. Recent data show that growth hormone lowers blood cholesterol levels in hypophysectomized rats. In addition, growth hormone elicits signs of diabetes and brings about various changes in carbohydrate metabolism in the dog.

An important role of growth hormone in the body is that of biologic synergist, reinforcing the activity of other glandular secretion. For example, growth hormone enhances the effect of pituitary interstitial cell–stimulating hormone (ICSH or luteinizing hormone, LH) in increasing ventral prostate weight in hypophysectomized-castrated male rats, and in increasing uterine weight in female rats whose pituitaries, adrenals, and ovaries have been removed. It acts in synergism not only with ICSH in restoring the organs of male and female rats but also with testosterone in males and with 17-β-estradiol in females.

In addition to its growth-promoting activity, HGH exhibits the various activities previously demonstrated by ovine lactogenic hormone: It promotes pigeon crop sac growth by either local or systemic tests; it synergizes with the ovarian hormone to induce ductal and alveolar mammary growth; and it induces localized milk secretion. In 1967, W. R. Lyons demonstrated that HGH causes an increase of milk secretion in lactating women. Together with the fact that the synthetic HGH-like protein exhibits both growth-promoting and lactogenic activities, it is beyond any doubt that HGH has intrinsic lactogenic activities.

Table 2 presents some biological properties of bovine growth hormone (BGH) and HGH.

Synthesis. The synthesis of a protein having HGH activities was accomplished by Li and D. Yamashiro in 1970 using the solid-phase procedure of R. B. Merrifield. The synthesis began with the attachment of the carboxy-terminal amino acid, phenylalanine, to the resin support and continued with the stepwise addition of protected amino acids until the entire amino acid chain was assembled on the resin (Fig. 2). Briefly, it was initiated by esterification of *tert*-butyloxycarbonyl-protected phenylalanine to the chloromethylated 1% cross-linked polystyrene-divinylbenzene resin. The *tert*-butyloxycarbonyl-protected group (that is, Boc) was subsequently removed by reaction of the protected phenylalanine-resin ester with triflu-

Table 2. Some biological properties of BGH and HGH	
Hormone	Properties
BGH	Growth-promoting activity in experimental animals
	Stimulation in the text tube of amino acid incorporation in the diaphragm of hypophysectomized rats
	Reduction in carcass fat in the rat
	Inhibition of fatty acid synthesis in the rat liver
	Increase glycogen in the fasted rat heart
	Lower plasma cholesterol of hypophysectomized rats
	Insulin antagonism in hypophysectomized dog
	Enhancement of the effect of testosterone
	Synergist to the effect of estrone
HGH	Growth-promoting activity in men and experimental animals
	Calcium retention in men
	Stimulation of RNA polymerase activity in hypophysectomized rat livers
	Elevation of free fatty acid in plasma of normal and hypophysectomized human subjects
	Lactogenic activity in pigeons, rabbits, and men

oroacetic acid in methylene chloride. Thereafter Boc-protected glycine was coupled to the resin-supported phenylalanine residue using dicyclohexylcarbodiimide as the coupling agent. After removal of the Boc protective group from the glycine residue with the same trifluoroacetic acid deblocking agent, the resulting free amino group was ready for coupling to the next protected amino acid, cysteine. The stepwise coupling of the successive amino acid residues of the sequence shown in Figure 2 was continued as described.

All amino acids were protected on the α-amino position with the Boc group and the following side chain blocking groups were used: Asp (OBzl), Glu (OBzl), Cys (Bzl), Ser (Bzl), Thr (Bzl), Tyr (Bzl), Lys (Z), Arg (NO_2), His (Im-Boc). The coupling reactions were carried out with dicyclohexycarbodiimide in the manner described above, except that the asparagine and glutamine were coupled by means of their nitrophenyl esters. After the tryptophan residue was incorporated, dithiothreitol was added to the trifluoroacetic acid deblocking agent, and this combination was used in all subsequent removals of the Boc protective group in order to deblock and preserve the tryptophan residue.

The foregoing procedure yielded a protected polypeptide in the sequence shown in Fig. 2. In order to remove all protecting groups from the synthetic polypeptide and to liberate it from the supporting resin, it was reacted with hydrogen fluoride and then with sodium in liquid ammonia. Next the deblocked polypeptide was autooxidized in air under controlled conditions. The oxidized polypeptide was isolated by lyophilization and was desalted on Sephadex. It was then subjected to repeated gel filtration on Sephadex G-100 in 20% acetic acid until a fraction was isolated which traveled in the column as a single peak at a maximum comparable to the curve of natural HGH.

Spectrophotometric measurements on the synthetic protein indicated a tyrosine-tryptophan ratio of 7.5 as compared to the known value of 8. Amino acid analysis of an acid hydrolysate gave: $Lys_{12.8}$-$His_{1.8}Arg_{9.7}Asp_{25.5}Thr_{9.2}Ser_{17.8}Glu_{22.0}Pro_{5.9}Gly_{9.9}$-$Ala_{7.8}Cys_{4.3}Val_{9.2}Met_{1.5}Ile_{7.2}Leu_{30.1}Tyr_{3.6}Phe_{12.4}$. These values were comparable with the analysis of HGH treated with HF and $Na-NH_3$: $Lys_{9.9}His_{2.5}$-$Arg_{10.0}Asp_{23.3}Thr_{9.9}Ser_{17.1}Glu_{28.7}Pro_{8.1}Gly_{9.3}Ala_{7.1}$-$Cys_{3.1}Val_{7.6}Met_{1.2}Ile_{6.7}Leu_{24.7}Tyr_{6.4}Phe_{11.9}$.

The synthetic product was found to react immunologically with the rabbit antiserum to HGH as revealed by the agar diffusion test. When the synthetic product was assayed by the rat tibia and pigeon crop sac tests, it gave approximately 10% growth-promoting potency and 5% lactogenic activity in comparison with that of natural HGH. Although the purity of the synthetic product has not been established, it exhibits to a significant degree those activities associated with HGH. These data provide evidence that the hormone does possess intrinsic lactogenic activity. Moreover, the accomplishment demonstrates that synthesis of a protein of molecular weight 21,500 is not impossible.

For background information *see* HORMONE; PITUITARY GLAND (VERTEBRATE) in the McGraw-Hill Encyclopedia of Science and Technology.

[CHOH HAO LI]

Bibliography: *Ann. N.Y. Acad. Sci.*, vol. 148, Feb. 5, 1968; *Growth Hormone: Proceedings of the International Symposium, September 11–13, 1967, Milan, Italy*, Int. Congr. Ser. no. 158, Excerpta Medica Foundation, Amsterdam, 1968; C. H. Li and D. Yamashiro, *J. Amer. Chem. Soc.*, 92:7608–7609, 1970.

Hadronic atom

Hadronic atoms are hydrogen-like systems that consist of a strongly interacting particle (hadron) bound in the Coulomb field and in orbit around any ordinary nucleus. The kinds of hadronic atoms that have been made and the years in which they were first identified include pionic (1952), kaonic (1966), Σ^--hyperonic (1968), and antiprotonic (1970). They were made by stopping beams of negatively charged hadrons in suitable targets of various elements, for example, potassium, zinc, or lead. The lifetime of these atoms is of the order of 10^{-12} sec, but this is long enough to identify them and study their characteristics by means of their x-ray spectra. They are available for study only in the beams of particle accelerators. Pionic atoms can be made by synchrocyclotrons and linear accelerators in the 500-MeV range. The others can be generated only at accelerators where the energies are greater than about 6 GeV. *See* PARTICLE ACCELERATOR.

The hadronic atoms are smaller in size than their electronic counterparts by the ratio of electron to hadron mass. For example, in pionic calcium, atomic number $(Z) = 20$, the Bohr radius of the ground state is about 10 fermis (1 F = 10^{-13} cm), and in ordinary calcium it is about 2500 F. Thus the atomic electrons are practically not involved in the hadronic atoms and the equations of the hydrogen atom are applicable. The close approach of the hadrons to their host nuclei suggests that hadron-nucleon and hadron-nucleus forces will be in evidence, and this is one of the motivations for studying these new types of atoms.

X-ray emissions. Negative hadrons are captured into orbits of principal quantum number

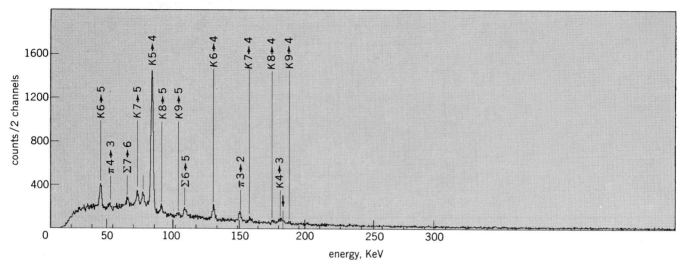

X-ray spectrum resulting from kaons stopped in chlorine. The transition of closest approach to the nucleus, $n=4$ to $n=3$, is broadened and displaced from its Coulomb energy. Lines from Σ^- hyperonic atoms are seen at 65.5 and 108.7 keV. The pionic lines come from decay products. The origin of the line at 78 keV is unknown. (*Lawrence Radiation Laboratory*)

$n \sim 100$ by the attraction of the positively charged nuclei. As the hadrons fall through successively smaller Bohr orbits, electrons are ejected from the cloud of atomic electrons (Auger effect). When a hadron has reached about the same radius as that of the electronic atom ground state, x-ray emission becomes the dominant method for the system to shed its excitation energy. X-rays, whose energy increases with each successive jump, are emitted until the hadron reaches the ground state ($n=1$) of the hadronic atom or is absorbed by the nucleus in a strong interaction.

The lines of the spectra of special interest are due to transitions between the lowest quantum levels because the hadrons are then closest to the nuclei and the nuclear forces perturb the orbits. The effects expected and observed are that some of the lines are slightly broadened (energy indefinite) and the average transition energy is different from that predicted solely by Coulomb effects.

The series of x-ray lines generally cuts off rather abruptly at $n>1$. However, in light pionic atoms the ground state is reached. Kaons, Σ^-, and antiprotons ($\bar{p}$) can probably reach the ground state only when the nucleus is singly charged. In kaonic chlorine atoms, for example, the series ends at $n=3$, and in kaonic lead the series ends at $n=7$.

Experimentally the hadrons are generated by a beam of protons incident on a metallic target. A secondary beam is used to transport the particles to the target in which the atoms are to be made and studied. The arrival of a hadron is signaled by a set of scintillation counters as it is slowed down by passage through a moderator of carbon or beryllium. The thickness of the moderator is adjusted so that a maximum number of hadrons stops in the target under investigation. The targets are usually sheets of metal or disk-shaped boxes of powder or liquid. The x-ray detectors are semiconductors of silicon or germanium. Efficiencies for detecting an x-ray that comes from within the target are around 5×10^{-3}, including factors of solid angle and target self-absorption. The energy resolution of the detectors is of paramount importance. Currently in use are detectors whose line widths at half maximum height are around 500 eV for energies below 30 keV and about 1 keV for energies from 50 to several hundred keV. The illustration is an example of a kaonic x-ray spectrum of chlorine obtained by a lithium-drifted germanium detector. The lines are labeled according to their hadronic transitions. The intensity of the lines average about 0.3 x-ray per stopped kaon for the principal lines ($\Delta n = -1$). About 5,000,000 kaons were stopped to obtain the spectrum.

The interpretation of the x-ray spectra of pionic atoms is complicated by the necessity for the pion to react with two nucleons rather than one nucleon alone, as in Eq. (1), where N stands for either a proton or a neutron. The two-nucleon final state is required for the reaction to conserve momentum.

$$\pi + N + N \rightarrow N + N + \text{Kinetic energy} \qquad (1)$$

Through the use of the line width and energy shift data of all the measurements on many elements throughout the periodic table, a calculation was made to determine whether pionic x-rays would show a difference between the root means square (rms) radii of the distributions of protons and neutrons in the nuclei. The result was that no significant difference was seen.

Targets of chemical compounds have been employed to determine the probability that the pions land on nucleus Z_1 (or nucleus Z_2). It had been predicted that the probability of formation of a pionic atom would be proportional to Z. The intensities of the x-ray lines indicate that in general the number of pionic atoms found are proportional to Z, but there are exceptions in some oxides.

The pions' decision is made in the outer regions of the atoms where the energies are those governing chemical effects. A more comprehensive study of the "chemistry" of pionic atoms could lead to increased understanding of some solid-state structures.

Surfaces of nuclei. Kaonic atoms were expected to yield valuable information concerning the surface of nuclei because a kaon reacts very strongly with either a single neutron or a single proton, as in Eq. (2).

$$K^- + N \rightarrow \pi + \text{Hyperon} \qquad (2)$$

On the basis of theory it had been predicted that more neutrons than protons would be found on the surfaces of nuclei. One of the first interpretations of the behavior of the series of x-ray lines of various elements ranging from $Z = 3$ through $Z = 92$ suggested that the neutron dominance of nuclear surfaces was verified. The reasoning went along the following lines: Electron scattering experiments measure to a high accuracy the charge (proton) distributions in nuclei. If one assumed the same distribution was applicable to neutrons and if one used the known kaon-nucleon interaction strengths, it would be possible to predict at which Z the x-ray series of the kaonic atoms would terminate. The experiments indicated a higher capture rate than expected, and this led to the idea that nuclear matter (neutrons) was encountered by the kaons in a region above the conventional nuclear surface. It was later pointed out that a resonance between K^- and proton, $Y^*(0)1405$, would probably enhance the affinity of kaons for protons on the nuclear outskirts, and this could account for the increased capture rate observed experimentally. The precision of nuclear matter distributions and kaon-nucleon interaction strengths are presently not sufficiently accurate to make a definitive calculation. However, a preliminary theoretical investigation indicates that the resonance enhances the kaon capture by protons but leaves some margin for capture on neutrons at a nuclear radius larger than the conventional charge radius. See ELEMENTARY PARTICLE.

Some kaonic x-ray lines are shifted from their Coulomb energies and broadened, as are some of the pionic lines; in mid-1971 the $K^- + P$ resonance prevented a definitive interpretation of the experimental results.

Σ^--hyperonic atoms. When kaons are absorbed by nucleons, about 20% of the hyperons produced are Σ^--particles. In the light elements most of the Σ^--hyperons are ejected from the nucleus in which they are generated. Some of them are captured by target nuclei; Σ^--hyperonic atoms are formed and emit characteristic x-rays, just as do the kaonic atoms. Weak x-ray lines due to the hyperonic atoms are found along with the kaonic x-ray lines (see illustration). The hyperonic lines are of special interest because they are actually doublets due to the magnetic moment of the Σ^-. By mid-1971 no Σ^- x-ray lines had been observed with high-Z targets, where the spacing between the lines would be greatest. The doublet splitting in light elements is beyond the resolution of present detectors. A measurement of the splitting would yield the value of the Σ^- magnetic moment, which has not been determined. Perhaps Σ^- will turn out to be a suitable probe of the nucleus. No resonances are known that would complicate the interpretation of the x ray spectra, but future experiments depend upon the availability of more intense beams of kaons.

Antiproton atoms. Antiproton atoms are the latest in the series of hadronic atoms to be observed. Their x-ray spectra are very similar to the kaonic x-ray spectra. Their x-ray series appears to terminate at a higher n than does the corresponding kaonic series. Antiprotonic x-ray lines are doublets due to the magnetic moment of $\bar{p}$, which is expected to be the same as the p magnetic moment. In this case the splitting is almost observable with present detectors and beams. An accurate comparison of the p and $\bar{p}$ magnetic moments would be a test of CPT (charge conjugation, parity, and time reversal) symmetry theory.

There are two more hadrons with lifetimes long enough to be candidates for hadronic atom formation: the negative xi (Ξ^-) and the negative omega (Ω^-), but the tracks of only about 10,000 Ξ^- and a few dozen Ω^- have been observed.

For background information see AUGER EFFECT; ELEMENTARY PARTICLE; HADRON in the McGraw-Hill Encyclopedia of Science and Technology.

[CLYDE E. WIEGAND]

Bibliography: A. Bamberger et al., *Phys. Rev. Lett.*, 33B:233, 1970; E. H. S. Burhop, in E. H. S. Burhop (ed.), *High Energy Physics*, vol. 3, 1969; C. E. Wiegand, *Phys. Rev. Lett.*, 22:1235, 1969.

Hibernation

Recent work on the neural control of hibernation has shown that ablation of those hypothalamic areas that cause disturbances in food intake, temperature regulation, and sleep patterns in homeotherms (animals that maintain a constant body temperature) leads to similar disturbances in hibernators. During hibernation the amplitude of the electrical activity of the brain decreases at least 90% and the animals no longer show cycles of brain waves corresponding to sleep and wakefulness. Some species of hibernators have no spontaneous cortical activity during hibernation at all, although they may remain dormant for weeks at a time. All hibernators arouse periodically throughout the winter. The cause of these spontaneous arousals is unknown. It appears that they are necessary for maintaining life because animals that are prevented from awakening die at about the time they normally would have aroused. Aside from spontaneous arousals, hibernating animals can be awakened by handling, by intraperitoneal injections of various substances, and by intrahypothalamic injections of biogenic amines. While the animals are hibernating, many of their physiological systems remain under precise neural control, but one current theory holds that regulation of body temperature is not among them.

Seasonal behavior. Seasonal hibernators such as ground squirrels, chipmunks, woodchucks, and hedgehogs go through large yearly fluctuations in various homeostatic behaviors such as food intake, body temperature, and sleep and wakefulness. In summer and early fall seasonal hibernators eat a great deal and become extremely fat. They maintain a homeothermic temperature of approximately 37°C and show normal daily rhythms of sleep and waking. In late fall and early winter they eat little or nothing and lose weight. Their body temperature falls, either all at once or in a series of increasingly larger drops, with a return to 37°C in between, until they reach their final hibernating level which may be as low as 2°C. They arouse briefly between bouts of hibernation, their body temperature returns to 37°C, and then they reenter hibernation, repeating this cycle until the following spring.

Brain activity. Electroencephalographic recording of cortical brain waves indicates that there are large species differences with respect to the activity of this part of the brain. Continuous spontaneous cortical activity during deep hibernation was seen in marmots and one species of ground squirrel. In hamsters, hedgehogs, and several other species of ground squirrel there is electrocortical silence for long periods until the animal's temperature rises above a certain point (which may be species-specific or vary with experimental conditions). In such animals the cortex may help in controlling the necessary adjustments for hibernation but may no longer be important once the internal state of the animal reaches equilibrium. Because hamsters arousing from hibernation seemed highly emotional, P. Chatfield and C. Lyman suggested that the limbic system (which controls emotionality in homeotherms) was an important neural structure for arousal. Indeed, researchers have found the first signs of electrical activity during arousal to occur in subcortical limbic structures, such as the hypothalamus, and in the olfactory bulbs.

Hypothalamic controls. As far as is known, there is no special "hibernating nucleus" in the brains of mammals that hibernate. Corresponding areas of the brains of hibernators and homeotherms subserve the same functions. Ground squirrels with lesions in the ventromedial nucleus of the hypothalamus overeat and become obese, just as do rats, cats, and humans with similar damage. However, this does not destroy their yearly cycle of weight changes; they continue to gain weight in the summer and fall and lose it in the winter, but at a higher, obese level. Lesions in the lateral hypothalamic area, adjacent to the ventromedial nucleus, lead to aphagia and adipsia for varying periods of time and the animals must be tube-fed a nourishing liquid diet or they will starve to death before they recover their ability to regulate their weight. Posterior hypothalamic lesions lead to somnolence and lethargy. Lesions in the anterior hypothalamus and adjacent preoptic area cause thermoregulatory deficits.

Brain-damaged ground squirrels made hypothermic took 2–6 times as long to regain normal body temperature as normal controls. Heating and cooling this area changed the metabolic rate and body temperature of ground squirrels in the same direction as in nonhibernating mammals. When ground squirrels with ventromedial or preoptic lesions were placed in a cold room in winter, they hibernated normally but died at about the time they should have awakened. This observation implies that the periodic arousals are necessary to maintain the life of the animals and are not simply a preprogrammed device to enable the animal to check on its environment to see if it is time to arouse for the rest of the season.

Given that they cannot arouse, why do they die? More generally, why do normal hibernators wake up every few weeks instead of hibernating through the winter, which would certainly conserve more energy? (It has been estimated that a hibernator uses as much energy to arouse as it uses in 10 days of hibernation.)

They may wake up because of the accumulation of a toxic metabolite or to replenish some nutrient that has been used up. They do not arouse to eat, drink, or defecate and may not even get out of the nest. They always urinate, but the urine probably accumulated during the course of arousal, not while they were hibernating. No change has been found in the levels of nonprotein nitrogen, blood urea, creatine, tissue water content, glucose, or other substances measured at various intervals of hibernation. All that can be said at this point is that the periodic arousals are necessary for the life of the animal and are timed rather closely, because ground squirrels that could not arouse because of hypothalamic damage died very shortly after they would have awakened normally.

Premature arousals. Until very recently nothing but handling or otherwise disturbing a dormant animal was known to cause arousal before the normal end of a bout of hibernation. The search for substances that lead to premature arousal is based on the supposition that anything that arouses an animal from hibernation may be the trigger for natural arousals. Many substances injected intraperitoneally into hibernating ground squirrels cause arousal, including epinephrine and all the trophic hormones except prolactin, adenosine, adenine nucleotides, and cyclic adenosinemonophosphate (AMP). Most of these compounds stimulate the activity of the enzyme adenyl cyclase which catalyzes the conversion of adenosinetriphosphate (ATP) to cyclic AMP; it is an appealing hypothesis that cyclic AMP, a chemical involved in the production of energy in the body, may play an essential role involved in spontaneous arousals which require a tremendous amount of energy. However, it is still too early to assume this for various reasons. The main ones are the unanswered questions of whether these substances which induce arousal pass the blood-brain barrier, and if so, how they stimulate nervous tissue to initiate arousal.

Another line of investigation involves injecting neural transmitter agents directly into the brain. When very small amounts of the biogenic amines norepinephrine and serotonin were injected intrahypothalamically into the preoptic area of hibernating ground squirrels, they evoked complete arousal from hibernation within 20 min. When these amines were injected into nonhibernating animals, brain temperatures rose about 2°C. Injections of the carrier solutions without the amines had no effect on brain temperature or arousal. It is still too early to tell how the amines cause arousal. A variety of sensory stimuli, such as a pinch on the tail, can cause arousal. Either the amines initiate arousal by activating thermoregulatory pathways or they mimic the arousing action of sensory stimuli.

Regulation during hibernation. An interesting current theory, put forth by T. Hammel, holds that animals have no functional thermoregulatory mechanisms below a certain body temperature. During hibernation, then, body temperature regulation is solely an equilibrium between heat loss and heat production, which is in turn a function of the ambient temperature and the heat transfer between body core to skin and skin to environment. The animal arouses, by this theory, when its thermodetectors in the hypothalamus are reactivated by the still unknown signal. Heating and cooling the hypothalamus above or below the equilibrium temperature should not activate any thermoregulatory responses.

One fact that argues against this theory is the

well-known ability of hibernators to increase their metabolic rates without arousing when the ambient temperature drops slightly below 0°C and thereby maintain a larger differential between body temperature and temperature of the environment. This reaction probably originates in the central nervous system, because peripheral cold receptors are not activated until stimulated by temperatures below $-12°C$.

Even if body temperature is not regulated during hibernation, there are other complex regulations going on which illustrate that tissues of mammals that hibernate are specially adapted to function remarkably well at temperatures near 0°C. A hibernator will increase its respiratory rate in response to an increase in inhaled CO_2. In hibernation the pH of the blood remains constant and the blood pressure remains high, indicating maintained vascular tone probably under the control of the sympathetic nervous system. Behaviorally the animal will change its position in the nest without awakening when disturbed, and it responds to sounds by turning its head in the direction of the sound stimulus. Furthermore it appears that there is actually a hyperresponsiveness in the spinal cord to changes in temperature. Placing a small heated or cooled disk on the flanks of hibernating ground squirrels led to larger and more prolonged muscle action potentials the lower the animal's body temperature was.

Ground squirrels, even when they are not hibernating, are more resistant to hypothermia and to anoxia produced by cooling below the physiological limit for breathing and cardiac activity. For instance, 100% of ground squirrels can be reanimated after 3 hr of complete cardiac arrest at body temperatures of 0°C, whereas the time limit for rats is 60–70 min. Ground squirrels rewarm themselves from artificial hypothermia with higher rates of heat production (O_2 consumption) than rats. Measures of energy utilization of the brain have shown that ground squirrels have a larger initial amount of available glycogen and energy-rich phosphates than do rats, and also can tolerate much higher accumulation of brain lactic acid, the end product of glycolytic metabolism. It appears that the tolerance-limiting factor for life at low body temperatures, just as for asphyxiation when there is an interrupted supply of oxygen, is the failure of energy-yielding processes. The brains of hibernators, then, seem to have special metabolic adaptations which confer greater resistance to cold than is found in homeotherms.

For background information see ADENYLIC ACID, CYCLIC; HIBERNATION in the McGraw-Hill Encyclopedia of Science and Technology.

[EVELYN SATINOFF]

Bibliography: R. K. Andjus, *Symp. Soc. Exp. Biol.*, 23:351, 1969; H. T. Hammel et al., *Physiol. Zool.*, 41:341, 1968; C. P. Lyman and R. C. O'Brien, *Symp. Soc. Exp. Biol.*, 23:489, 1969; E. Satinoff, in E. Stellar and J. Sprague (eds.), *Progress in Physiological Psychology*, vol. 3, 1970.

High-pressure physics

Over the past three decades very high pressure has developed from being rather a curiosity to become an effective, significant, integral tool for investigating the structure and behavior of solids. Earlier work was confined largely to establishing the existence and boundaries of first-order phase transitions involving the rearrangement of the atoms or ions in the lattice (changes in crystal structure). More recently high pressure has been most effective in elucidating the electronic behavior of solids, and this article is restricted to this topic, with emphasis on the improved understanding of semiconductors, of magnetic and ferroelectric materials, and of electronic rearrangements at high pressure.

This limitation forces the omission of two areas where there is a high level of activity: the development, improvement, and testing of equations of state for solids, and the measurement and analysis of atomic motion in solids. In both of these fields there is continuing significant and fruitful research. However, there have been no important breakthroughs in very recent years.

Hydrostatic pressure range. There are two aspects in which high pressure contributes to the understanding of solids. Theories developed to describe phenomena at atmospheric pressure very frequently contain the interatomic distance (or volume) as a basic parameter. The effective testing of a theory, or the selection between competing theories, then demands that the phenomenon be studied as a function of pressure. These studies can usually be carried out in the range of pressures up to 10–15 kbar (1 kbar = 987 atm or 100 N/m^2), where it is easily practical to use hydrostatic pressure media (liquids or gases). In this range it has been possible to apply almost all the tools of modern solid-state physics effectively.

The most important early application was to the understanding of the structure of the conduction band of semiconducting solids important in transition technology, such as silicon, germanium, and compounds of similar structure, for example, gallium arsenide. There has been recent work extending these techniques to other semiconductors and refining the understanding of the role of impurities. High pressure has also been very effective in improving the understanding of magnetic fields and of ferromagnetism and antiferromagnetism in metals alloys and compounds. A third area where there has been recent significant work in the hydrostatic pressure range is in the study of ferroelectricity and ferroelectric critical points. Another area of investigation involves measurement of the change in shape of the Fermi surface (the surface in energy space which represents the chemical potential of the electrons) with compression. These changes and, in particular, changes in the intersections of the Fermi surface with the Brillouin zone boundaries are very important in understanding changing electrical behavior in solids. See CRITICAL POINT.

For the past several years, one of the most active areas of high-pressure research primarily concerned with elucidating moderate-pressure phenomena has been measurements concerned with the Mott transition. The fact that pure NiO is an insulator, although one might expect metallic conduction due to the partially filled band arising from the nickel 3d orbitals, has long been a puzzle. Sir Nevill Mott has developed a theory of insulator-metal transitions which indicates that there is a critical correlation distance for electrons below which the system will be a metal, while at larger electron-electron distances it will be an insulator.

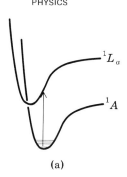
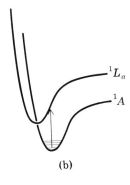
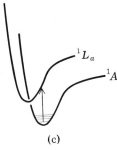
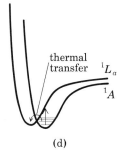

Potential energy diagrams showing the change in optical transitions for two hydrocarbons. (a) Anthracene at 1 atm. (b) Anthracene at high pressure. (c) Pentacene at 1 atm. (d) Pentacene at high pressure.

There have been many experimental efforts to test the theory, very largely through study of transition-metal oxides. Many such systems exhibit the gross features of the Mott transition but differ radically in numerous details. In a series of recent studies at Bell Telephone Laboratories, it has been shown that the system V_2O_3-Cr_2O_3, which forms a series of mixed crystals, exhibits almost all the features requisite for a Mott transition. The phase boundaries between paramagnetic insulator, antiferromagnetic insulator, and metal have been mapped in some detail. There are still some questions as to whether the transition involves phonons, whereas a true Mott transition involves only electron-electron interaction. Nevertheless this study is especially valuable because it illustrates very well the relationship between change of chemical composition and change of volume (pressure) in establishing the electronic properties of solids.

As Mott himself has indicated, the Mott transition may be only one of several ways in which an insulator may become a metal. (For example, iodine appears to transform from insulator to metal at high pressure with no discontinuities in physical properties.) However, both Mott's analysis and the experiments which have been generated to test it have yielded a great deal of important information on high pressure and electronic structure.

Very high pressure. The second aspect of high-pressure studies of electronic structure is a more exploratory one. There exist numerous phenomena at high pressure, the explanation of which can greatly extend the understanding of the structure of matter. The notion of the electronic transition, especially the recent understanding of the generality of the phenomenon, is such a development.

The basic effect of pressure is to increase overlap between adjacent orbitals. There are a number of consequences of this increased overlap. In the simplest order there is a delocalization of electrons—a broadening of energy bands. Ultimately this can lead from insulator to semiconductor to metal, as has been observed for iodine. In the next order there may be a relative shift in the energy of different orbitals. Since orbitals of different quantum number may differ in radial extent, orbital shape (angular momentum), and diffuseness or compressibility, one can expect this to be a common event. An important example involves the changes in band structure of semiconductors noted above. See JAHN-TELLER EFFECT.

Under a wide variety of circumstances there may be an excited state of different electronic characteristics not too far in energy above the ground state. The relative displacement of orbital energies with pressure may be sufficient to give the system a new ground state or to alter greatly the characteristics of the ground state by mixing of configurations. This event is called an electronic transition. It can occur discontinuously at a given pressure or continuously over a range of pressures. Electronic transitions have important consequences for physics and chemistry, but they may also have implications for geophysics and biology. Electronic transitions may occur in metals or insulators. In metals they are largely deduced from discontinuities in electrical resistance and in volume, together with determination of structure by x-ray diffraction. In insulators they may be deduced from measurements of electrical resistance, Mössbauer resonance, optical absorption, or irreversible chemical consequences. The mechanism of the transitions in insulators has largely been elucidated through optical studies at high pressure.

Electronic transitions were first discovered in metals. Over 20 years ago P. W. Bridgman discovered a volume discontinuity and resistance cusp in cesium metal. F. Sternheimer attributed this to the transfer of the $6s$ electron to a $5d$ orbital (really to a change of that character in the conduction band). This explanation appears to be correct. More recently a second cusp was observed near 135 kbars (13,500 N/m²) which may involve the introduction of $4f$ character to the conduction band. A resistance discontinuity in rubidium near 145 kbar (14,500 N/m²) is attributed to the $5s \to 4d$ electron promotion, and a similar event in potassium which occurs at 260 kbar (26,000 N/m²) at 77K and disappears near 250K is possibly the $4s \to 3d$ transition disappearing in a critical point.

In the same period when cesium was first studied, a discontinuity in resistance and volume was observed in cerium at a few kilobars pressure (1 kbar = 100 N/m²). This has been studied in detail and is associated with a $4f \to 5d$ or $4f$-conduction band transition. It disappears at a critical point with increasing temperature. Resistance discontinuities and cusps observed at higher pressure in a variety of rare-earth metals are undoubtedly also electronic transitions. Related transitions have also been observed in rare-earth monotellurides.

The alkaline-earth atoms have filled electronic shells. They are metals because of the complex structure of the conductor band, involving a mixture of s, p, and d orbitals. At high pressure calcium and strontium become semiconductors due to a demixing of orbitals, until a first-order phase transition rearranges the structure and causes a return to the metallic state. Similar behavior has been observed in the analogous metal ytterbium.

The observation of electronic transitions in insulating solids is a more recent phenomenon, observed in two classes of systems; in crystalline aromatic hydrocarbons and their charge-transfer complexes, and in ionic and covalent compounds of iron. Two types of electronic transitions are observed in the organic crystals, and three in compounds of iron.

Optical absorption studies of a series of crystalline aromatic hydrocarbons showed that the difference in energy between the ground state and the first excited state decreased by the order of an electron ~150 kbar (15,000 N/m²). Those compounds with excited states initially at less than about 2 eV reacted at high pressure to form novel polymeric compounds. Those with higher initial energy excited states did not. On the other hand, many hydrocarbons which do not react in the pure state form electron donor-acceptor complexes with such acceptors as iodine. These have low-lying excited states which decrease in energy with increasing pressure. Most of these compounds react at high pressure; sometimes the product involves the acceptor, and sometimes not.

The behavior can be understood in terms of the schematic diagram (see illustration) for two hydrocarbons where the potential energies of the ground

(^1A) and excited states (^1L_a) are plotted against some vibrational coordinate. Optical transitions are represented vertically on such a diagram because they are very rapid compared with vibrational motion (the Franck-Condon principle). For pentacene, a modest decrease in the energy of the optical transition is sufficient to bring the excited state in the range for thermal transfer of electrons, which process is not subject to the Franck-Condon principle. This transfer creates a new reactive ground state for the system. In the electron donor-acceptor complexes the function of the acceptor is to provide a low-energy excited state so that thermal transfer at high pressure can put the donor into a reactive configuration. Because of the much larger difference in initial ground state energies, this process does not occur in anthracene. Analogous diagrams describe other materials.

Iron is a ubiquitous element. Compounds of iron are important in biology and geophysics, as well as in physics and chemistry. High-pressure studies of iron, in the range to 200 kbar (20,000 N/m²), involve Mössbauer resonance and optical absorption. As a result of interaction between the $3d$ electrons of the iron and the surrounding ions or atoms (the ligands), three types of electronic transitions are observed. Pressure tends to increase the intensity of field due to the ligands, thus increasing the splitting among the $3d$ orbitals due to this field. The increased potential energy necessary to distribute electrons among these $3d$ orbitals in the normal high-spin configuration may ultimately overcome the interelectronic repulsion and result in a transition to the low-spin state (for ferrous ion, a transition from paramagnetic to diamagnetic behavior). Many compounds of ferrous iron with organic molecules are diamagnetic (low spin) probably because the iron bonds strongly by donating electrons to empty ligand orbitals. At high pressure these ligand orbitals may tend to be occupied by ligand electrons, by a transition similar to that discussed above for pentacene. This reduces the bonding to the iron and results in a diamagnetic-to-paramagnetic transition. Finally, with pressure there is a tendency for the iron $3d$ orbitals to decrease in energy relative to the filled orbitals of the ligand. There is a resulting thermal transfer of electrons from ligand to metal, that is, ferric iron reduces to ferrous ion at sufficiently high pressure in many compounds.

It would appear that similar processes may well occur in compounds of other transition-metal ions, such as copper, as well as in compounds of the rare earths. Certainly one of the most fruitful areas for high-pressure research lies on this border where physics, chemistry, biology, and geophysics overlap.

For background information *see* BAND THEORY OF SOLIDS; FRANCK-CONDON PRINCIPLE; HIGH-PRESSURE PHYSICS; SEMICONDUCTOR in the McGraw-Hill Encyclopedia of Science and Technology.

[H. G. DRICKAMER]

Bibliography: D. Bloch (ed.), *Propriétés physiques des solides sous pression*, 1970; R. S. Bradley (ed.), *Advan. High Pressure Res.*, 3:1–38, 41–47, 155–240, 1969; H. G. Drickamer, *Comments Solid State Phys.*, 3:53, 1970; N. F. Mott, *Comments Solid State Phys.*, 2:183, 1970.

Highway engineering

Along with a growing public concern over the effects of engineered works on man's environment, an increasing demand has been placed on highway engineers to develop methods and procedures to effectively study and forecast those effects resulting from a particular highway location. No longer can engineers be satisfied with a route selection which is superior when judged solely on the basis of least cost and primary purpose of providing access from one area to another in a safe and efficient manner. The engineer's task has been further expanded with a clear-cut declaration by the Federal government to include in his investigation a study into the socioeconomic and environmental effects of all new highways.

To answer this challenge a two-pronged attack has been launched. The first, to develop new procedures to produce basic engineering plans more efficiently, thereby freeing the engineer for more detailed investigations into the socioeconomic and environmental effects of new highways. The second, to develop a new methodology which will quantify in rational terms the socioeconomic and environmental effects of new highways in order to assist in the decision-making process. Recent advances in computer applications and aerial photogrammetry have assisted the engineer in realizing these goals.

COGO. Since the coordinate geometry (COGO) system was first introduced in 1961 by C. L. Miller of the Massachusetts Institute of Technology (MIT), this programming system designed to solve geometric problems has found wide acceptance in the profession and has undergone several updates designed to make it more flexible and efficient. The original system freed the user from the process of undertaking tedious and time-consuming computation by hand or by using standardized computer programs for specific highway problems such as ramp alignments, mainline alignments, and spiral curves.

The COGO system was based on the theory that all highway alignment problems could essentially be reduced to specific problems of intersections of straight lines, intersections of curves and lines, intersections of curves, computations to determine distance, bearing, azimuth, and areas, and so on. Verb packages were developed to solve each of these problems with appropriate language to communicate with the computer. The engineer, therefore, had only to formulate his problem in an orderly fashion in terms of the various verbs in order to solve all alignment problems.

In 1968, ICES-COGO-I was introduced. Whereas the original COGO system was based on the coordinated point (x, y values) as the basic input/output format, this new system was file-oriented and allowed the engineer to communicate with the computer in terms of lines, courses, curves, and chains (general term for parcels, alignments, and so on) in addition to coordinated points. This system represented an increase in the scope of the original COGO, further decreased engineering time required for computations, and allowed for the additional flexibility of computer plotting of the results.

In 1970, MIT announced the development of

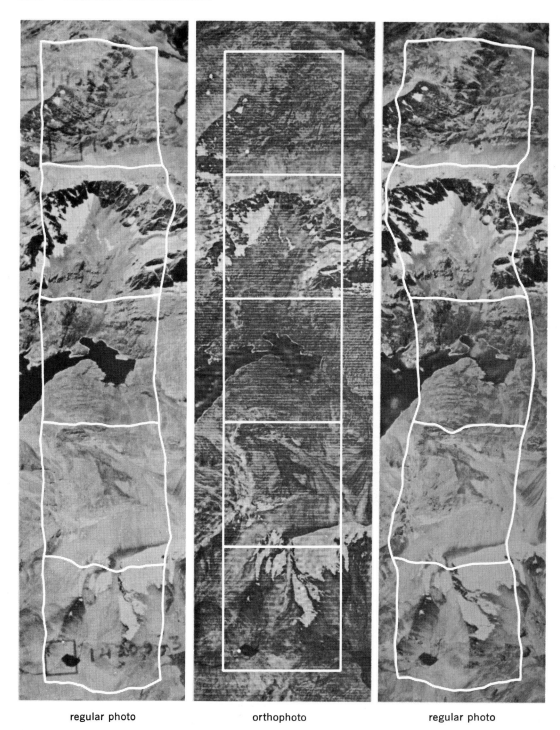

regular photo   orthophoto   regular photo

Fig. 1. Five square kilometers as they appear in regular photos of mountainous terrain show gross errors of scale and angle that occur locally but are corrected in the orthophoto, which is like a planimetric map.

URBAN COGO, which allows for additional flexibility by handling networks and blocks. In the words of the MIT *URBAN COGO Users' Guide*, "the network capability allows street networks and utility networks to be handled as named objects which can be defined, stored, displayed, manipulated and processed. Work on graphical capability to input, display, and plot such objects has been initiated."

URBAN COGO promises to give the engineer additional capability to interact directly with the computer, studying many more conditions and possible vertical and horizontal configurations than he was previously capable of and performing them in a shorter period of time.

Orthophotographs. Most major highway design projects begin with an accurate topographic base map showing all natural and man-made details, including roads, buildings, drainage channels, vegetation, and fence lines, and depicting ground elevations with contours (a line connecting all points of equal elevation). These maps are prepared by either conventional ground or aerial photogrammetric methods.

Orthophotographs (Fig. 1), on the other hand, are produced optically by manual methods or electronically by automatically scanning narrow strips of terrain and projecting the detail from one of the original photographs of a stereoscopic pair onto a new negative in perfectly joined strips fully corrected in scale and plan position. The orthophoto produced in this manner, although similar in appearance to an aerial photograph, is in fact quite different. Whereas an aerial photograph of rolling or mountainous terrain will have inherent scale and angular distortions, the orthophoto is true to scale; shapes, angles, and distances are correct. What has been produced, in effect, is a photographic planimetric map equivalent in all respects to a planimetric map produced by more conventional ground or aerial photogrammetric methods.

This ability to produce orthophotos has recently been combined with the ability to digitize terrain information. The system consists of the automatic photogrammetric plotter functioning on-line with a computer. Using standard aerial photography as original material, digital numerical terrain data (x, y, and z value of any number of ground points) can be extracted simultaneously with an orthophoto. The digitized terrain data can then be used in computer programs to obtain original ground plots for profile development, original ground cross sections, slope stake plots for right-of-way acquisition, and earthwork quantities.

The orthophoto, like any good map, allows the engineer to lay out a proposed highway with accurate scale, direction, and curvature. It also allows existing survey and map information such as property and contour data to be precisely correlated with full ground detail. Hence, at an early stage of a highway location study, the facts and physical limiting factors can be drawn together by the engineer not just qualitatively but quantitatively as well, resulting in better route selection.

Highway esthetics. In recent years it has become increasingly apparent that conventional design procedures, which depend on the engineer's ability to envision his design in perspective based on plan and profile studies, should be broadened to include more detailed three-dimensional studies. Highways designed by using accepted vertical and horizontal criteria provided adequate traffic service but sometimes failed to give the driver confidence nor did they in some cases fit the topography when viewed from a distance. New tools and procedures have been devised to assist the engineer to view the design in three dimensions, thus achieving one which is not only functional but esthetically pleasing.

One approach to obtain a three-dimensional view of a proposed highway has been to construct a model to scale (Fig. 2). This method has been particularly successful when studying complex interchanges which normally occupy a comparatively small land area yet by their nature involve complex relationships between roadways at various levels. These models have been constructed in some cases simply by cutting profiles from plain or corrugated cardboard mounted on boards to fit the horizontal alignment. Where a representation of the surrounding terrain is required to determine whether the interchange is in harmony with the topography, more elaborate models constructed from contours cut from a single styrofoam sheet have been constructed.

Photographs taken from various vantage points surrounding the model assist the engineer to study the interchange from the viewpoint of a pedestrian viewing the area surrounding the interchange. Similarly, photographs taken by various types of periscope or movie camera devices allow the engineer to view portions of the interchange as seen by the driver. The model, therefore, not only affords the engineer the opportunity to view his design as a combination of vertical and horizontal alignments, but to study through observations and photographic techniques its relationship to the surrounding terrain and view the roadway as seen through the driver's eyes. The disadvantages to using models are that they are generally expensive to construct and require many man-hours of effort to complete.

Another approach has been to draw perspective views of the highway. Traditionally this has been accomplished by preparing individual one-point or oblique perspectives depicting the relationship of the roadway to the surrounding terrain at key locations of interest to the engineer. During the last few years research has been conducted into the feasibility of using the computer to prepare perspective views of the roadway. Computer programs have been developed that transform three-dimensional terrain coordinate information into perspective drawings at varying intervals and

Fig. 2. Model of the approaches to Throgs Neck Bridge, New York City.

heights above the roadway. These perspective views can then be plotted by the computer on flatbed or roll-type plotters or displayed on a cathode-ray tube. The plotted perspective views can then be photographed in sequence to obtain movies of the proposed highway.

A further refinement can be obtained by the stereoscopic projection of a pair of perspective drawings of the same object, seen from two slightly different positions. When viewed simultaneously, one picture with each eye, or with the aid of a stereoscope, the engineer can see the roadway in three dimensions. These latest developments are promising and indicate that in the near future another technique will be available to highway engineers to determine whether or not a proposed design is in harmony with the surrounding topography.

For background information *see* HIGHWAY ENGINEERING; LAND-USE PLANNING; PHOTOGRAMMETRY in the McGraw-Hill Encyclopedia of Science and Technology. [GEORGE G. ALEXANDRIDIS]

Bibliography: R. O'Connell, Applying stereomat orthophotographs to highway route location, *Consult. Eng.*, 33(5):119–122, November, 1969; R. A. Park, N. J. Rowan, and E. E. Walton, A computer technique for perspective plotting of roadways, *Geometric Design: Photogrammetry and Aerial Surveys*, Highway Research Record no. 232, pp. 29–45, 1968; R. G. Porter, Models for highway design: Some construction and photographic techniques, *Photogrammetry and Aerial Surveys*, Highway Research Record no. 270, pp. 25–35, 1969; *URBAN COGO Users' Guide*, Urban Systems Laboratory, MIT, June, 1970.

Homeostasis

Homeostasis is the ability of an organism to maintain its internal physical state and chemical composition with relative constancy over a wide range of external environmental conditions so that survival is ensured. Since living things are composed predominantly of water, perhaps the greatest challenge to their homeostatic capabilities occurs at temperatures below 0°C. At subzero temperatures the organism must either maintain its body fluids in a liquid state or face solidification by freezing. Recent progress in the study of freezing resistance has elucidated two unusual physiochemical mechanisms which enable marine teleost fishes to live at subzero temperatures without freezing. These mechanisms include the ability to survive in a supercooled state and the formation of an antifreeze to lower the freezing point of the body fluids.

Supercooling. Since the freezing point of sea water (−1.9°C) is approximately 1°C lower than the freezing point of the blood of most marine teleosts (−0.8°C), the possibility exists that these fish may encounter temperatures which would freeze their blood. Recent studies have demonstrated that virtually all arctic, antarctic, and temperate marine teleosts avoid freezing at subzero temperatures by existing in a supercooled state. For example, the common killifish (*Fundulus heteroclitus*) of the Atlantic coast of the United States has blood that should freeze at −0.8°C. However, the killifish can, in fact, swim about without freezing in sea water cooled to −1.5°C. Similarly the blood of the antarctic fish *Notothenia neglecta* has a freezing point of −1.08°C, but this fish nevertheless remains unfrozen in sea water at −1.9°C.

Since supercooling is a physically unstable state, it has been suggested that supercooled fish release into their blood some substance or substances which stabilize the supercooled blood, thus reducing the probability of freezing. Recent attempts to discover the chemical identity of the "supercoolant" present in the blood of killifish and *N. neglecta* have implicated the simple sugar glucose as the compound responsible.

When the blood serum chemistry of supercooled killifish at −1.5°C is compared with that of killifish acclimated to 20°C (Fig. 1), it can be seen that most serum constituents show only minor changes. Serum glucose concentrations, however, increase by three- to sixfold in supercooled killifish, depending on the season of the year. Winter fish exposed to the subzero cold have almost twice as much glucose in their blood as summer fish maintained under the same conditions. Moreover, when kept at −1.5°C, a winter fish can be expected to survive for an average of 63 days whereas a summer fish can be expected to survive for an average of only 28 days. There is a seasonal cycle in the ability of the killifish to tolerate subzero temperatures which is related to the amount of blood sugar accumulated.

Killifish acclimated to temperatures as low as

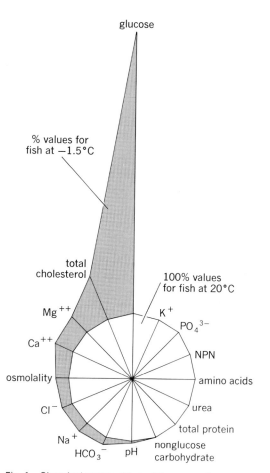

Fig. 1. Chemical composition of the serum from supercooled killifish (*Fundulus heteroclitus*) acclimated to −1.5°C expressed as the percentage change from the serum chemistry of killifish acclimated to 20°C.

2°C show practically normal blood glucose concentrations. Only when temperatures drop below 0°C does the dramatic increase in serum glucose levels occur. This increase in serum glucose concentrations at subzero temperatures results from a depletion of liver glycogen. At warmer temperatures liver glycogen is continually being replenished by breakdown products of the ingested food. However, at subzero temperatures the killifish refuses to eat, so that liver glycogen reserves are eventually exhausted. When liver glycogen is totally used up, serum glucose can no longer be maintained at high concentrations. Eventually serum glucose levels fall to normal, causing the fish to die. It is characteristic that all fish living in the subzero cold show a marked hyperglycemia and livers rich with glycogen, whereas those fish dying under the same conditions have low concentrations of serum glucose and totally depleted livers. However, if glucose is added daily to the water in which the killifish live, this sugar is taken into their bodies, elevating blood glucose concentrations. Such treatment with exogenous glucose prolongs the survival of killifish at −1.5°C since it prevents the decline of serum glucose levels.

On the basis of these studies, it has been suggested that glucose is the substance in the blood of the killifish that permits it to survive in a supercooled state. Similarly studies on the blood chemistry of the supercooled antarctic fish *N. neglecta* have demonstrated that concentrations of reducing sugars (predominantly glucose) are 10 times higher than those in fish living in the temperate zone. Other organisms which live at very low temperatures that have been investigated chemically are insects and plants. Terrestrial insects make extensive use of supercooling as a mechanism to avoid freezing during extremely cold winters. Chemical analyses of the tissues and hemolymph of supercooled insects often reveal high concentrations of either glycerol, sorbitol, trehalose (insect blood sugar), fructose, mannitol, or glucose (Fig. 2). In addition, practically all frost-tolerant plants exhibit increases in cellular sugar concentrations in winter. The protective sugar in plants is usually sucrose, but raffinose, stachyose, glucose, fructose, mannitol, sorbitol, and glycerol are also found to increase in many cold-hardened plants (Fig. 2). Therefore, in all organisms adequately studied, the ability to survive subzero temperatures in a supercooled state is accompanied by the accumulation of sugars or polyhydric alcohol derivatives of sugars in the body fluids and tissues.

Physiochemical studies of supercooled aqueous solutions have demonstrated that the presence of small amounts of glucose, glycerol, or other organic compounds slows down the rate of ice crystal growth when the solution is seeded with a crystal of ice. Thus it is reasonable to suspect that such substances may stabilize the supercooled state by preventing spontaneous nucleation and subsequent freezing. If one examines the chemical structure of all solutes known to promote supercooling in organisms (Fig. 2), one notices that all contain numerous hydroxyl (—OH) groups which readily form hydrogen bonds. Since hydrogen bonding is needed for water molecules to arrange themselves in the lattice structure requisite for the formation of ice crystals, it has been suggested

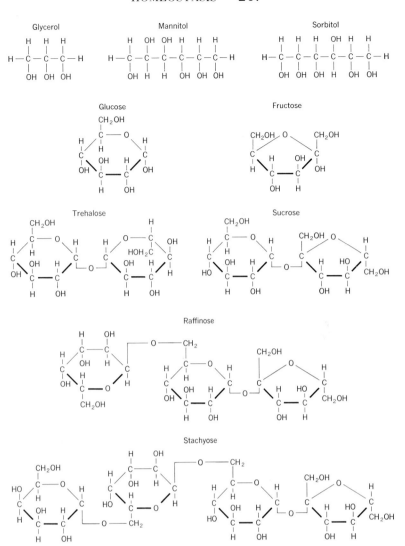

Fig. 2. Naturally occurring compounds associated with an organism's ability to withstand subzero temperatures in a supercooled state. Such compounds may be termed cryoprotective.

that the numerous hydroxyl groups on cryoprotective compounds might somehow interfere with water-to-water hydrogen bonding to inhibit the formation of ice nuclei. Equally plausible is the notion that such substances promote supercooling by poisoning ice crystal growth surfaces.

Antifreeze formation. Although most marine teleosts survive subzero temperatures in a supercooled state, a few species of antarctic fish in the genus *Trematomus* do not. These fish produce an "antifreeze" which lowers the freezing point of the blood. For example, the blood of *T. borchgrevinki* has a freezing point of −2.07°C, whereas the blood of a typical marine teleost has a freezing point of −0.8°C. Since the freezing point of the blood of *T. borchgrevinki* (−2.07°C) is slightly lower than the freezing point of sea water (−1.9°C), this fish is resistant to freezing even when swimming in its native habitat of McMurdo Sound, Antarctica, where the water is laden with crystals of ice.

Until quite recently the chemical identity of the antifreeze in *Trematomus* was unknown. Sodium chloride is the most abundant constituent of the blood of temperate marine teleosts, accounting for approximately 80% of the freezing point depression. However, in *Trematomus* sodium chloride is responsible for less than 50% of the serum freezing point depression. Clearly substances other than sodium chloride account for a large portion of the freezing point depression in *Trematomus*. These antifreeze substances have recently been identified as glycoproteins with relatively high molecular weights of 10,500 to 21,500 g. The protein backbone of the glycoproteins is composed of only two kinds of amino acids, alanine and threonine, whereas the carbohydrate side chains consist mostly of galactose and *N*-acetylgalactosamine. These glycoproteins have many unique properties, such as complete solubility in most protein-precipitating agents and a resistance to heat denaturation.

Another peculiarity of these glycoproteins is their ability to depress the freezing point of water more than expected on the basis of the number of particles present in solution. Theoretically these large glycoprotein molecules should not appreciably lower the freezing point of the serum of *Trematomus* at the concentrations normally present in the blood. In fact, they account for approximately 30% of the freezing point depression, markedly lowering the serum freezing point by $-0.7°C$. Moreover, if the glycoproteins are digested with an enzyme that cleaves peptide bonds, there is a complete inactivation of their ability to depress the freezing point. Thus the freezing-point-depressant activity is a conformational property of the glycoproteins: Fragments of the glycoproteins cannot perform the function of the intact molecules.

The ability of these glycoproteins to lower the freezing point of water to a considerably greater extent than is theoretically possible indicates that these substances are somehow interfering with the structure of large volumes of water. It has been suggested that the glycoproteins might be expanded, long-chain molecules in which the hydroxyl groups of the sugar residues bind large amounts of water in such a way as to make the water unavailable for freezing. Just as hydroxyl groups are important in maintaining the supercooled state in some fish, they are equally important in lowering the serum freezing point of other fish. Hydroxyl groups (present predominantly on carbohydrates) and hydrogen bonding (for restructuring water) appear to be emerging as important concepts in understanding basic aspects of freezing resistance in fish.

For background information *see* HIBERNATION; HOMEOSTASIS; HYPOTHERMIA in the McGraw-Hill Encyclopedia of Science and Technology.

[BRUCE L. UMMINGER]

Bibliography: A. L. DeVries, S. K. Komatsu, and R. E. Feeney, *J. Biol. Chem.*, 245:2901–2908, 1970; R. N. Smith, in M. W. Holdgate (ed.), *Antarctic Ecology*, vol. 1, 1970; B. L. Umminger, *Biol. Bull.*, 139:574–579, 1970; B. L. Umminger, *J. Exp. Zool.*, 173:159–174, 1970.

Hypofluorous acid

Hypofluorous acid, HOF, is the only known oxyacid of fluorine. It is a volatile, reactive compound of limited stability at room temperature. Although a number of inorganic and organic hypofluorites have been known for some time, the parent acid has long resisted discovery. Reports of the preparation of fluorine oxyacids appeared in the early 1930s, but these were soon discredited as being due to chlorine impurity. The reaction of fluorine with water has frequently been observed to yield small quantities of oxidizing substances, but it is only recently that HOF has been identified and isolated as a product of this reaction.

Synthesis. In 1968 P. N. Noble and G. C. Pimentel incorporated fluorine and water into a solid nitrogen matrix at 14–20K. They photolyzed the matrix with ultraviolet light and observed the appearance of new infrared absorption bands, some of which they identified as belonging to HOF. They expected, however, that this compound would be too unstable to isolate. In 1971 M. H. Studier and E. H. Appelman isolated milligram quantities of HOF formed by passing fluorine over water at 0°C. The fluorine was subsequently passed through traps cooled to -50 and $-79°C$ to remove water and HF, and the product was finally collected in a trap cooled with liquid oxygen to $-183°C$. In this way some 5–10 mg of HOF could be prepared from about 1 g of F_2.

The method of Studier and Appelman succeeds by trapping the HOF at $-183°C$ before it can react further. Thus if the fluorine stream is too slow or if the fluorine is bubbled through water, the HOF reacts with the water and cannot be isolated. On the other hand, fluorine itself does not react very rapidly with water, and if the gas flow is too rapid, most of the fluorine will escape reaction. An ideal preparative system would use a rapid flow and would recycle the unreacted fluorine.

Properties. Hypofluorous acid is a white solid which melts at about $-117°C$ to a colorless liquid. Its vapor pressure is less than 1 mm Hg (133.3 N/m^2) at $-79°C$ and about 5 mm Hg (666.6 N/m^2) at $-64°C$.

At room temperature, HOF decomposes to HF and O_2. In a Kel-F vessel (a vessel of halocarbon plastic resistant to fluorine and reactive fluorine compounds), the half-time of decomposition is usually 5–15 min, although some samples have occasionally shown much slower decomposition. The decomposition rate is probably dependent on the presence of catalytic impurities.

Hypofluorous acid reacts rapidly with water to yield HF, H_2O_2, and O_2. It also oxidizes many substances in aqueous solution; for example, Ag^+ is oxidized to Ag^{2+}, and in alkaline solution bromate is oxidized to perbromate. This last reaction puts HOF with F_2 and XeF_2 as the only known oxidants capable of forming perbromate. *See* XENON COMPOUNDS.

The reaction of hypofluorous acid with aqueous iodide solutions has been utilized for the analysis of HOF, as in Eq. (1).

$$HOF + 3I^- \rightarrow I_3^- + F^- + OH^- \quad (1)$$

It appears that HOF is the intermediate that is responsible for the oxidizing power of fluorine in aqueous solution, and one may imagine the reaction of fluorine with water to take place by a scheme in which the slow initial step is the dissolution of the fluorine in the water, as in Eq. (2),

$$F_2(g) \rightarrow F_2(aq) \quad (2)$$

followed by a rapid disproportionation, as in

Eq. (3). This disproportionation probably goes to

$$F_2(aq) + H_2O \rightarrow HOF + HF \qquad (3)$$

completion, unlike the analogous reactions of chlorine, bromine, and iodine. The HOF formed in the disproportionation, unless it is quickly removed, reacts further, as in Eqs. (4) and (5). If

$$HOF + H_2O \rightarrow H_2O_2 + HF \qquad (4)$$
$$HOF + H_2O_2 \rightarrow HF + H_2O + O_2 \qquad (5)$$

oxidizable substances are present in the solution, the HOF may attack them in preference to the water.

The relation between HOF and OF_2, oxygen difluoride, is somewhat mysterious. Although OF_2 is formally the anhydride of HOF, the reaction of OF_2 with water does not yield HOF, nor does HOF yield OF_2 either by itself or in the presence of water or alkaline solutions. Oxygen difluoride is formed as a by-product in the reaction of fluorine with water, but it is not clear just how this takes place.

The infrared spectrum of hypofluorous acid has been studied in a matrix obtained by codepositing the HOF with solid N_2 at 8K. The fundamental frequencies observed are 886 cm^{-1} for the O-F stretching vibration, 3537 cm^{-1} for the O-H stretching vibration, and 1359 cm^{-1} for the bending vibration. The two stretching frequencies are close to those of F_2 and the OH radical, respectively, suggesting that the F and OH groups in HOF are both nearly neutral. The dimensions of the HOF molecule have been determined from its microwave spectrum. The O-H bond length is 0.0964 ± 0.001 nm, the O-F bond length is 0.1442 ± 0.0001 nm, and the H-O-F angle is $97.2 + 0.6°$.

For background information see FLUORINE; HYDROGEN FLUORIDE; INFRARED SPECTROSCOPY in the McGraw-Hill Encyclopedia of Science and Technology.

[EVAN H. APPELMAN]

Bibliography: J. A. Goleb et al., *Spectrochim. Acta*, in press; P. N. Noble and G. C. Pimentel, *Spectrochim. Acta*, 24A:797, 1968; M. H. Studier and E. H. Appelman, *J. Amer. Chem. Soc.*, 93: 2349, 1971.

Immunology

Recent studies have demonstrated that the initiation of immunological reactions requires interactions between several diverse types of cells. Among the cells involved are two different types of lymphocytes, one originating in the thymus and the other in the bone marrow, both cells performing their function in effector organs such as spleen and lymph nodes, where the antibodies are made. The two types, morphologically quite similar, differ from each other in that the thymus-derived (T) cell, although it recognizes antigens in a highly specific manner, does not synthesize and secrete detectable amounts of antibody, while the bone-marrow-derived (B) cell does synthesize and release antibody, but only when acting in concert with, or in response to, T cells in the presence of antigen. See ANTIGEN.

Developmental aspects of formation. The number of cells that can respond at a given time to a single, specific antigen is quite small, in most instances on the order of 1 in 10,000 to 1 in 100,000 cells. To understand not only how the T and B cells differ from each other, but to learn how within the population of T and B cells a given cell becomes that rare cell that reacts to a specific antigen at all, are key tasks of current immunological research.

The critical events determining the special properties of antigen-reactive T cells or antibody-forming B cells must occur early in the development of the embryo. For example, in man, a fetus in midgestation can already perform the many specific reactions that characterize the immune response. How do the different lymphocytes arise? Why are there diverse types? And why and how do the lymphocytes gain the peculiar ability to respond to just one of what must be thousands of different antigenic stimuli? The embryo must be studied to find answers to these questions.

Origin of lymphocytes. The first area in the embryo to become a blood-forming tissue is the embryonic yolk sac. Required as a source of red blood cells, the yolk sac has been shown also to contain cells that can be induced to differentiate into lymphocytes under appropriate conditions. Experiments involving transplantation and special marker systems indicate that during ontogeny (embryonic development) some of the yolk sac cells migrate away from the yolk sac itself, settling in the thymus, while others migrate to the embryonic liver. It is at this time that the dichotomy between the B and T cell populations probably first becomes manifest.

The embryonic liver replaces the yolk sac as a functional organ for blood cell formation, functioning in this capacity for some time before the bone marrow takes over as the major organ for elaboration of blood cells. However, by the time the embryo is completely formed, the bone marrow is the only major source of blood cells, and it is the marrow that continues to supply these cells throughout the course of adult life. It is the bone marrow, then, that ultimately provides the B cell population which participates in immune reactions. The origin of bone marrow itself is not fully understood, but marrow is generally believed to be formed largely by cells migrating from the embryonic liver. Thus the pattern of development seems clear: The yolk sac sends precursor cells to both the thymus and the embryonic liver. In the thymus, T cell populations of lymphocytes are formed. The embryonic liver forms B cells, and moreover passes stem cells on to the bone marrow to provide the continuity in the production of B cells.

The effector organs that actually are directly involved in antibody formation, however, are not the yolk sac, embryonic liver or bone marrow, or the thymus. Rather, antibody formation takes place in such organs as the spleen and lymph nodes. These organs contain macrophages and connective tissue, as well as the T and B cells which they sequester from the circulation. By the time the B and T cells are released into the circulation to find their way to the spleen and lymph nodes, they have already acquired their B or T properties and the specificity required to recognize and respond to specific antigen; therefore this last phase of construction of an antibody-forming organ, although important to the organism, probably occurs too late in the sequence of events of differentiation to provide much insight into the nature

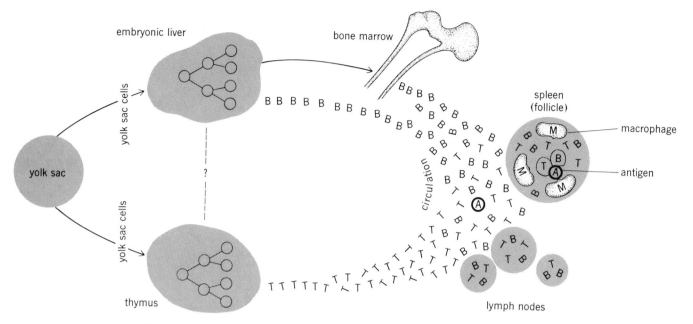

Sequence of events in differentiation of antibody-forming cells. Final synthesis of antibody involves collaboration between macrophages from spleen, B cells from liver and bone marrow, and T cells from thymus.

and diversity of antibody-forming systems (see illustration).

Differences between T and B cells. In the thymus the migratory (yolk-sac-derived) stem cell is subject to developmental influences of the thymus mesenchyme and epithelium, which act inductively to control the development and proliferation of these immigrant cells. This inductive control, characteristic of embryonic development in general, is not well understood. Under this control the stem cells undergo a precise sequence of divisions and differentiations, emerging as unique cells with specific and characteristic biological properties, such as surface antigens (TL, θ), and capable of recognizing specific antigens or cells as foreign.

In the embryonic liver and bone marrow the migratory stem cells (again, yolk-sac-derived) differentiate into B-type cells. Here, too, inductive influences must play a role, but little information is available on how the environment of the bone marrow or embryonic liver favors B-type development. Again, the end result is remarkable in terms of specificity, with the surface of the B cell unique and different (antigens LyA, LyB, and so on), with specific surface recognition sites for antigen, and with the capacity upon stimulation to secrete gamma globulin (antibody).

Origin of variability. Theories have been recently proposed which suggest (1) that initially all T and B cell precursors have the potential to make all antibodies but become restricted by the same type of controlled events as those that govern the restrictive development characteristic of all differentiation, for example, of muscle and cartilage; (2) that while restrictive differentiation may determine that a cell become a B or T cell, specific antigen recognition and antibody synthesis results from random mutations, chromosome rearrangements, or other spurious genetic alterations of the cell; (3) that a combination of these two processes occurs, involving first the creation of cells with certain restrictions, with refinement and modifications then being superimposed as the result of further mutation and selection; and (4) that T and B cells are determined by a series of inductive events including embryonic interactions and other means of information transfer. Of considerable interest is the fact that none of the current theories of immunity consider the antigen itself as playing a major role in the generation of antibody diversity, although subsequent cellular events, such as differentiation of plasma cells and cell proliferation, are doubtlessly influenced by the specific antigen under study. All the theories require extensive cellular selection following the generation of diverse immunocompetent cells, and assume the clonal proliferation of antibody-forming cells subsequent to selection.

Immunological tolerance. Fundamental to an understanding of immunity is the fact that during embryogenesis a developing individual learns to recognize self from not-self. Failure of this self-tolerance leads to autoimmunity and is obviously not compatible with survival. Since the Nobel-prize-winning experiments of the 1950s, it has been clear that presentation of foreign material to an embryo leads to the acceptance of that material as nonforeign, or self, resulting in tolerance. Numerous subsequent experiments have shown that tolerance in adults can also be achieved, although with more difficulty, and usually by using modified antigens or by employing antigenic stimulation in the presence of immunosuppressive drugs which destroy responding cells.

Applying the concept of tolerance to the B and T systems described above, it has been established that neither B nor T cells from tolerant individuals can participate in immune reactions against those antigens against which these cells have been tolerized. This argues strongly for the idea that the maturational steps leading to specificity must occur either before divergence of T and B stem cells of the yolk sac, or, more likely, in parallel during maturation of cells in the thymus, liver, or bone marrow.

Tissue culture studies. With the development in 1965 of organ culture methods permitting the steps involved in antibody formation to be studied outside of the animal and with the development shortly thereafter of methods of cell cultures, involving the explantation of large numbers of individual cells, much of the work on antibody-producing systems has been performed using these newer procedures. A number of recent observations, however, indicate clearly that, even in tissue culture, cell interactions must occur if antibody production is to ensue. In general, follicular structure is maintained or reestablished, and the immune reactions require collaboration between B and T cells acting in the environment of macrophages or "fibroblast" cells which form the matrix of the culture systems. While the precise cell requirements are still not defined, increasing attention is being placed on requirements for structure involving multicellular organization of diverse cell types.

Evolutionary considerations. Although the old notion that "ontogeny repeats phylogeny" (that is, the embryo recapitulates the evolutionary history of the species) is not considered tenable, still much validity exists for a phylogenetic approach to help understand developmental sequences and functional alterations. Although earlier it had been thought that the evolution of immunological capacity occurred gradually, with cellular immunity arising late in the evolution of cyclostomes and immunoglobulin (antibody) production occurring with elasmobranchs, it now appears that all vertebrates can mount typical immune reactions, including graft rejection, delayed hypersensitivity, and immunoglobulin production. Only the details appear to have undergone changes, these appearing primarily in terms of size and assembly of immunoglobulin molecules. Moreover, even such unrelated organisms as earthworms, whose evolutionary divergence from the vertebrate line probably occurred as early as coelenterates, can reject skin grafts, manifest immunological memory, and become immunologically tolerant by appropriate treatment. Moreover, annelids produce at least partially specific secretory chemicals, although these have not yet been linked to immunoglobulin-like materials. These evolutionary findings suggest that the immune system plays an integral role in survival and encourages the concept that specificity, as seen in the immune system, evolved from more primitive recognition systems common to all living things.

Receptors. That such recognition systems do in fact exist is apparent from such phenomena as mating reactions in protozoans, virus-specific immunity in bacteria, and species recognition in sponges. In each case, components of the cell surface have been implicated in that recognition, and the particular recognition points have been designated as surface receptors. The "germinal theory" of immunity, which invokes inductive tissue interactions and cell-cell interaction and information transfer, extends this concept to immune phenomena. Studies of antigen recognition suggest that the receptors are, in fact, antigen-specific; studies with antisera suggest that the receptors are modified immunoglobulins; and investigations of tolerance suggest that the receptors can be bound or inactivated by antiimmunoglobulins or modified antigens. Techniques for isolation of receptors are being developed. The application of these techniques for isolating receptors obtained from cells at different stages of development, from the B and T lineages, and from cells actively engaged in the process of antibody production offers hope for solution of many of the developmental questions concerning the differentiation of antibody-forming cells.

For background information *see* EMBRYOLOGY; IMMUNOLOGICAL TOLERANCE, ACQUIRED in the McGraw-Hill Encyclopedia of Science and Technology. [ROBERT AUERBACH]

Bibliography: O. Makela and A. Cross (eds.), *Cell Interactions and Receptor Antibodies in Immune Responses*, 1971; J. Sterzl and I. Riha (eds.), *Developmental Aspects of Antibody Formation and Structure*, vols. 1 and 2, 1971.

Intermetallic compound

Intermetallic compounds have functioned as critical components of useful alloys for over 1000 years, yet they were recognized for the first time as chemical and structural entities only in the past century. Today understanding of their formation, structure, and properties is good and increasing. As a consequence, the number and diversity of the technological applications of intermetallics have been proliferating rapidly, ranging from the aircraft jet engine to the razor blade. This article reviews the nature of these unique materials and the recent advances made in their understanding and applications in the past few years.

Structure. An intermetallic compound is a particular alloy in which, as the name implies, a compound is formed between two or more metals. In general its structure and properties are not predictable from the structure and properties of the component metals. Usually, but not always, the different atomic species are arranged in ordered fashion in an intermetallic crystal. In certain compounds the degree of order will change, gradually or abruptly, with increasing temperature, thus offering another property affecting parameter not available with most ordinary compounds. Some of the structure types common among intermetallics are shown in Fig. 1. Which of these, or of other vastly more intricate ordering schemes, a given intermetallic compound adopts depends upon a complex interaction of geometrical, electrochemical, and bonding factors—the main features of which are now beginning to be understood. In any case, it is these special coordinations of dissimilar atoms which give rise to the unique physicochemical and mechanical properties of intermetallic compounds.

Other characteristic features of intermetallics as a class can be visualized with reference to Fig. 2, a typical binary phase diagram. In such a figure the temperature-composition regions of stability of the various phases are represented, as well as the nature of their interactions. It may be seen that when metals form compounds with one another, they rarely obey the ordinary rules of chemical valence and even a given pair may exhibit several different combining ratios, for example, Ni_3Al,

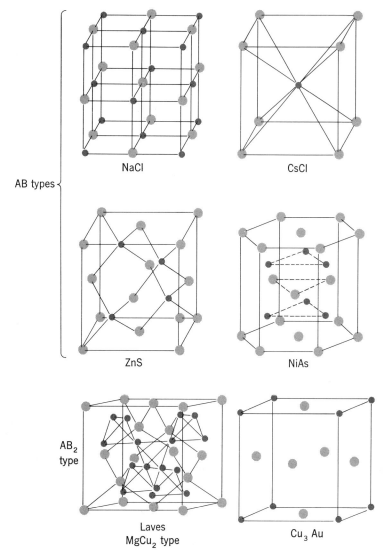

Fig. 1. Crystal structure types common among intermetallic compounds. NaCl structure (AsLa, CaTe); CsCl structure (NiAl, β'CuZn); ZnS structure (InSb, Ga_2HgTe_4); NiAs structure (AuSn, BiRh); Laves $MgCu_2$ structure (Cu_4MnSn, Mo_2Zr); and Cu_3Au structure (Ni_3Al, Ni_3Fe).

With respect to interatomic bonding, intermetallics again exhibit a wonderful diversity. Examples can be found of almost ideal ionic, covalent, and metallic bonds, but the usual case is a hybrid bonding possessing aspects of two or three of the ideal bond types. Indeed sometimes different types of bonding obtain within different structural parts of one and the same compound.

New discoveries. Although dozens of new compounds are discovered each year, few of these individually lead directly to new applications or understanding. There are some exceptional instances, however, where a new method of synthesis or a new class of compounds exerts more than ordinary impact. Such has been the case with the application of "splat cooling" to generate new, metastable, compounds by solidifying melts with ultrahigh cooling speeds (10^5 to $10^{9°}$C/sec) in many binary and higher order systems. Knowledge of the existence and structure of such phases which do not exist under equilibrium conditions is contributing to the general understanding of the structural chemistry of intermetallics.

Scientists at the Philips Research Laboratories in the Netherlands discovered that the Ni_5R and Co_5R compounds, where R stands for a rare-earth metal, have an astonishing ability to absorb and desorb hydrogen at room temperature (up to 7 H atoms per formula unit of Ni_5R, which is nearly twice the density of hydrogen in liquid hydrogen). So-called natural composites have been made by directional solidification of eutectic systems. It has recently been found that intermetallics sometimes form pseudobinary eutectic systems which are amenable to directional solidification to produce filamentary reinforced composites. The Ni_3Al-Ni_3Cb eutectic has shown promising mechanical properties for high-temperature structural applications.

New understanding. Understanding of the crystal chemistry of intermetallics has progressed

NiAl, Ni_2Al_3, and $NiAl_3$. Much more complex ratios are also known, for example, $FeZn_7$, Na_3Si_{136}, and $Mo_{31}Cr_{18}{}^1Co_{51}$. Figure 2 also shows that while some compounds such as $NiAl_3$, sometimes known as "line" compounds, exist only at a discrete stoichiometric composition, others can exist over a range of composition either limited (Ni_2Al_3 and Ni_3Al) or extensive (NiAl) while retaining their structure, homogeneity, and characteristic properties. (Frequently variation of composition within the single phase region can profoundly affect certain properties of the compound.)

It may also be observed from Fig. 2 that some intermetallic compounds are extremely stable, for example, NiAl, having melting points greater than those of their component metals. Some, for example, NiAl, melt congruently, that is, pass directly from the solid to the liquid state, while others form by solid-liquid ($NiAl_3$, Ni_2Al_3, and Ni_3Al) or solid-solid reactions (not shown in Fig. 2). Such circumstances can add to the difficulties of synthesizing intermetallics in the phase-pure condition.

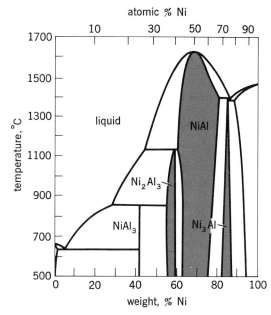

Fig. 2. Typical binary phase diagram, the Ni-Al system, showing formation and composition range of several intermetallic compounds.

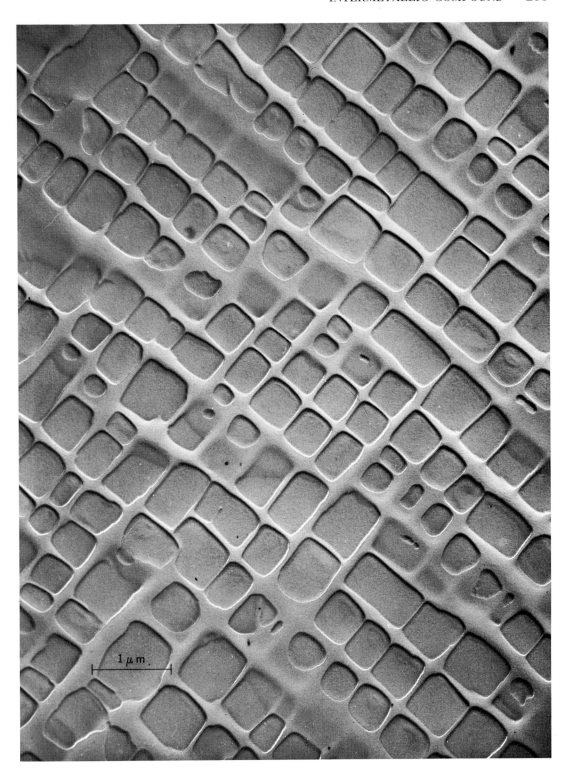

Fig. 3. Electron micrograph of a typical nickel-base high-temperature alloy (IN-100) showing cube-shaped particles of the strengthening phase, Ni_3Al or γ'. Such compounds are used to reinforce jet engine alloys. (INCO)

from the crude empirical groupings of compounds by geometrical and electrochemical factors to more sophisticated analyses derived from the concentration and distribution of electrons. Experiments that describe the Fermi surface directly and resonance techniques which present other information on electron distribution are placing this whole subject on a more funda- mental basis. S. Samson has extended the early concepts of J. Kasper and F. C. Frank of highly coordinated, quasi-close-packed polyhedra of metal atoms to describe and classify the crystal structures of a wide variety of intermetallics, both simple and highly complex (> 1100 atoms/unit cell). Previously unsuspected relationships between structural families have thus been revealed and new compounds have been predicted. B. C. Giessen's book summarizes the present status.

Extensive application has been made of computer calculations quite apart from structure determination. E. Savitskii has calculated the possible formation of new AB_3-type compounds. Computer programs have been designed to tailor the complex compositions of high-temperature alloys in such a way as to maintain electron vacancies at levels which prevent the formation of the embrittling sigma phase. A. Head has used a computer to plot the appearance of dislocation and stacking fault configurations in ordered β-brass and matched these successfully with transmission electron micrographs. *See* EQUILIBRIUM, PHASE.

New applications. The compound γ' (Ni_3Al) continues to be the workhorse reinforcement in the latest jet engine alloys as it has been since the Nimonic alloys were applied in the 1940s; a typical microstructure is shown in Fig. 3. Higher operating temperatures have forced the provision of oxidation-resistant coatings for these alloys usually based on NiAl and CoAl. NiTi is sprayed on steel tubing to reduce sulfur attack in paper mill recovery boilers, and $CbZn_3$ coatings protect columbium alloys from oxidation.

The compound Cb_3Sn has been utilized as the superconductor in the new ultrahigh-field solenoids (> 150,000 gauss-oersteds) and a related compound, $Cb_{0.79}(Al_{0.75}Ge_{0.25})_{0.21}$, has the record superconducting transition temperature, 20.98K, a value which significantly is (0.6K) above the boiling point of hydrogen. A whole new family of hard magnetic materials has been discovered based on the Co_5R compounds, where R is a rare-earth element; energy products of 23×10^6 gauss-oersteds have been attained. A number of ferromagnetic intermetallic compounds have been found which contain no magnetic elements, for example, Au_4V, $ZrZn_2$, and Sc_3In.

The compound NiTi has been extensively studied at the United States Naval Ordnance Laboratory because of its unique property of mechanical memory. A piece of this material deformed at room temperature, upon annealing above a (composition-dependent) critical temperature, will revert instantaneously to its original geometrical configuration. Hosts of applications ranging from simple thermal switches to internal clamps to self-deploying gear for space satellites are currently being tested. Cr_3Pt has been found to have such a favorable combination of hardness and corrosion resistance as to give phenomenal service when applied to edges of steel razor blades. UAl_4 and U_3Si are being explored as possible nuclear fuel materials.

Cr_3Si has become widely used as a thick film resistor for integrated circuit electronics. Such circuits usually use gold and aluminum as conductor materials and hence a problem has arisen in preventing the formation, in processing or in service, of certain Au-Al intermetallics (the "purple plague") which lead to weak connections and poor electronic performance.

Many interesting new applications for intermetallics derive from the possibility of dispersing them in a compatible matrix of vastly different mechanical, magnetic, or electrical properties. Examples of this sort include novel bearings, clutches, brakes, and current collectors. A particularly interesting instance is the aligned InSb-NiSb eutectic whose anisotropic magnetoresistance (InSb is a semiconductor and NiSb a metallic conductor) is utilized in a device known as a field plate. This device is one component of the control mechanism of an electric locomotive brake.

In addition to the well-known group III–group V semiconducting intermetallics, other intermetallic compounds have been found to have unique semiconducting properties for certain applications. For example, new air pollution monitors take advantage of the fact that solid-state lasers of $Pb_{1-x}Sn_xTe$ can be chemically tailored by composition adjustment to emit in the wavelength range from 6.5 to 32 μ, thus matching the strong infrared absorption lines of the most common pollutant gases. *See* LASER, SEMICONDUCTOR INJECTION.

For background information *see* ALLOY STRUCTURES; CRYSTAL STRUCTURES; EQUILIBRIUM, PHASE; INTERMETALLIC COMPOUNDS; NONSTOICHIOMETRIC COMPOUNDS in the McGraw-Hill Encyclopedia of Science and Technology.

[JACK H. WESTBROOK]

Bibliography: B. C. Giessen (ed.), *Developments in the Structural Chemistry of Alloy Phases*, 1969; B. Kear et al. (eds.), *Ordered Alloys: Their Structural Applications and Physical Metallurgy*, 1970; I. I. Kornilov, *Metallides and Reactions Among Them*, 1970; J. H. Westbrook (ed.), *Intermetallic Compounds*, 1967.

Interstellar matter

In the last few years radio astronomical observations have shown that the interstellar medium contains a large variety of molecules. Studies of these molecules have yielded important new information about the temperature, density, and composition of the interstellar medium and simultaneously posed some interesting new problems concerning the formation and excitation of molecules in the unique environment of interstellar space.

Interstellar space is a much better vacuum than can be attained in any laboratory on Earth, and although this space contains but a small fraction of the mass of the Galaxy, its study is vital to astronomy. New stars are continually being formed from clouds of interstellar gas, and stars during the course of their lives return matter to interstellar space either slowly, as do red giants (10^{-8} solar masses per year), or rapidly, as in nature's greatest catastrophy, the supernova explosion, in which a significant fraction of a star's mass is thrown off in one blast. Thus the study of interstellar matter yields information about conditions under which stars are born and about the energy and matter returned to space during a star's death.

Detectable molecules. Recently one of the most fruitful methods of studying interstellar matter has been by use of the newly discovered molecules. These studies have proceeded at such a rapid rate that it is not possible to discuss each molecule in detail. Instead the molecules discovered are summarized in the table, which gives the chemical formula, common name, quantum mechanical designation of energy level transition initially discovered, frequency, discoverers and the date of discovery, and a summary of the other transitions of the molecule so far found. Nearly half of the discoveries were made in 1971. Consequently, for many molecules little more is known than that they

Interstellar molecules observed at radio frequencies as of August, 1971, and some of their characteristics

Formula	Name	States	Frequency, MHz	Discoverers	Other transitions observed
OH	Hydroxyl	$^2\Pi_{3/2}, J=3/2$ (Λ-doubling)	1,667	A. H. Barrett, S. Weinreb, M. L. Meeks, and J. Henry (Lincoln Lab.) 1963	Λ-doubling in: $^2\Pi_{3/2}, J=5/2$ (6035 MHz); $^2\Pi_{3/2}, J=7/2$ (13441 MHz); $^2\Pi_{1/2}, J=1/2$ (4765 MHz); and $O^{18}H$; $^2\Pi_{3/2}, J=3/2$ (1639 MHz)
NH_3	Ammonia	1_1 (inversion)	23,694	A. C. Cheung, D. M. Rank, C. H. Townes, D. D. Thornton, and W. J. Welch (Berkeley) 1968	Inversion in: 2_2 (23723 MHz); 3_3 (23870 MHz); 4_4 (24139 MHz); 6_6 (25056 MHz); 2_1 (23099 MHz); and 3_2 (22834 MHz)
H_2O	Water vapor	$6_{16}-5_{23}$	22,235	A. C. Cheung, D. M. Rank, C. H. Townes, D. D. Thornton, and W. J. Welch (Berkeley) 1969	
H_2CO	Formaldehyde	$1_{11}-1_{10}$	4,830	L. Snyder and D. Buhl (NRAO), and B. Zuckerman (U. of Md.), P. Palmer (U. of Chicago) 1969	$2_{11}-2_{12}$ (14489 MHz); $3_{13}-3_{12}$ (28975 MHz); $2_{11}-1_{10}$ (150498 MHz); $2_{12}-1_{11}$ (140839 MHz); $2_{02}-1_{01}$ (145603 MHz); and $1_{11}-1_{10}$ of $H_2C^{13}O$ (4593 MHz)
CO	Carbon monoxide	0–1	115,267	R. W. Wilson, K. B. Jefferts, and A. A. Penzias (Bell Lab.) 1970	0–1 of $C^{13}O$ (110201 MHz) and 0–1 of CO^{18} (109782 MHz)
CN	Cyanide radical	0–1	113,492	K. B. Jefferts, A. A. Penzias, and R. W. Wilson (Bell Lab.) 1970	Optical lines identified by McKellar in 1940
HCN	Hydrogen cyanide	0–1	88,632	L. Snyder (U. of Va.) and D. Buhl (NRAO) 1970	0–1 of $HC^{13}N$ (86340 MHz)
HC_3N	Cyano-acetylene	0–1	9,098	B. E. Turner (NRAO) 1970	
HCOOH	Formic acid	$1_{11}-1_{10}$	1,639	B. Zuckerman (U. of Md.) and J. A. Ball and C. A. Gottlieb (Harvard) 1970	
CH_3OH	Methyl alcohol	$1_{11}-1_{10}$ (A)	834	J. A. Ball, C. A. Gottlieb, and A. E. Lilley (Harvard) and H. Radford (Smithsonian) 1970	4_2-4_1 (E_1) (24933 MHz); 5_2-5_1 (E_1) (24959 MHz); 6_2-6_1 (E_1) (25125 MHz); 7_2-7_1 (E_1) (25125 MHz), 8_2-8_1 (E_1) (25294 MHz); and 5_1-4_0 (E_2) (84521 MHz).
NH_2CHO	Formamide	$2_{11}-2_{12}$	4,616	R. H. Rubin, G. W. Swenson, R. C. Benson, H. L. Tigelaar, and W. H. Flygare (U. of Ill.) 1971	$1_{11}-1_{10}$ (1540 MHz)
CS	Carbon monosulfide	3–2	146,969	A. A. Penzias, R. W. Wilson, K. W. Jefferts (Bell Lab.) and P. Solomon (Columbia) 1971	3–2 of $C^{13}S$ (138738 MHz)
OCS	Carbonyl sulfide	9–8	109,463	A. A. Penzias, W. W. Wilson, K. B. Jefferts (Bell Lab.) and P. Solomon (Columbia) 1971	
CH_3CN	Methyl cyanide (acetonitrile)	6_0-5_0	110,383	A. A. Penzias, R. W. Wilson, K. B. Jefferts (Bell Lab.) and P. Solomon (Columbia) 1971	6_1-5_1 (110381 MHz); 6_3-5_3 (110364 MHz); 6_4-5_4 (110350 MHz); and 6_5-5_5 (110331 MHz)
SiO	Silicon monoxide	3–2	130,272	A. A. Penzias, R. W. Wilson, K. B. Jefferts (Bell Lab.) and P. Solomon (Columbia) 1971	
CH_3C_2H	Methyl acetylene (propyne)	5_0-4_0	85,457	D. Buhl (NRAO), L. Snyder (U. of Va.) 1971	
HNCO	Isocyanic acid	$4_{04}-3_{03}$	87,925	D. Buhl (NRAO), L. Snyder (U. of Va.) 1971	
CH_3CHO	Acetaldehyde	$1_{11}-1_{10}$	1,065	J. A. Ball, C. A. Gottlieb, A. E. Lilley (Harvard) and H. Radford (Smithsonian) 1971	
?	X-ogen	—	89,190	D. Buhl (NRAO), L. Snyder (U. of Va.) 1970	Suggested identifications: HCO^+, C_2H
?	—	—	90,665	D. Buhl (NRAO), L. Snyder (U. of Va.) 1971	Suggested identification: HNC

exist in detectable quantities. For that reason, this discussion, after some general comments, must center on a few molecules that have been studied in more detail.

Of the molecules found so far, half have been found at wavelengths shorter than 1 cm. Sensitive receivers for those wavelengths have been built only very recently and few exist. Thus much less time has been spent searching for molecules at these wavelengths and much less sensitive limits have been reached. The detection of these molecules surprised many astronomers, not only because such molecules were not expected to be present but because physical conditions are different than previously expected so that they can form detectable lines. The transitions are intrinsically stronger at higher frequencies but, as a consequence, their interaction with their environment is greater. Molecules in thermodynamic equilibrium, on the average, emit as much radiation as they absorb, so that on the whole they neither add to or subtract from the ambient radiation field and are invisible. If they are hotter than the background they are seen against, they are seen as emission lines; if cooler, they are seen as absorption lines.

Since interstellar conditions are so extreme, the investigator cannot assume equilibrium but must analyze in detail the processes taking place. The molecule can absorb photons from the microwave background, or decay back to its ground state by induced or spontaneous emission. The average time for these processes for HCN, for example, is a few days. If these were the only processes in operation, the molecules would come into equilibrium with the microwave background and be undetectable against it. However, the molecule also collides with particles in the gas at a rate proportional to the number of particles. For collisions to be rapid enough to offset the effect of the background photons, for HCN the density must be greater than about 10^4 particles (probably H_2 molecules) per cubic centimeter. This is 10^3 times as great as that previously believed typical of interstellar clouds. For the other molecules observed at short wavelengths, the conclusion is similar; lower limits for the density ranging from 10^3 to 10^7 are obtained. Thus clouds much more dense and massive than any previously found are shown to exist. These massive, dense clouds may be the ones in which stars will soon form. *See* ASTRONOMICAL INSTRUMENTS.

So far the chemically stable molecules are found to predominate; that is, rather than the simple hydrides, such as CH or OH, the chemically stable CO is the most abundant molecule. Similarly HCN is found in much greater abundance than the highly reactive but less complex radical CN. This may mean that the molecules were formed under sufficiently dense conditions that chemical equilibrium was approached. As yet there is no detailed theory for the formation of the more complex molecules. It should be noted that most of the molecules are combinations of hydrogen with carbon, nitrogen, and oxygen, the next most abundant elements (excluding the chemically inert element helium). However, two molecules containing sulfur (about 1/20 as abundant as carbon) and one containing silicon (about 1/16 as abundant as carbon) were found in 1971. Perhaps with increased sensitivity other such molecules will be found, but the rule that simple organic molecules predominate will probably not be changed.

Carbon monoxide. This molecule, CO, is an example that has a fairly simple energy level scheme and for which the astronomical data have a fairly straightforward interpretation. Through the work of R. W. Wilson, K. B. Jefferts, and A. A. Penzias of Bell Telephone Laboratories and P. Solomon of Columbia University, it has been studied more extensively than the other molecules found at very high frequencies. They found intense CO lines from the direction of the Orion Nebula and other bright nebulae, the center region of the Galaxy, dark dust clouds, and several infrared stars—stars that have a great excess of infrared emission compared to normal stars of the same visual brightness. Thus CO is found in many types of sources, and indeed seems to be the most widely distributed interstellar molecule.

Although most of the other molecules appear to come from relatively small sources, the CO emission in Orion extends over as large a region as the visible emission—nearly a degree of arc. Densities of order 10^3 particles per cubic centimeter are required to excite the CO. In Orion and many other of these sources the lines of $C^{12}O^{16}$ are nearly as strong as those of $C^{13}O^{16}$, and the $C^{12}O^{18}$ line is also detectable in some regions. (On Earth C^{13} is only 1/90 as abundant as C^{12}, and O^{18} only about 1/500 as abundant as O^{16}.) The fact that the C^{12} and C^{13} lines are of comparable strength means that they are heavily saturated: The optical depth is so large that the emission is nearly proportional to the excitation temperature of the molecule and nearly independent of the projected density. Projected densities as high as 3×10^{19} cm^{-2} have been reported, which requires that a large fraction of the carbon and oxygen in the region are in the form of CO molecules. Thus, contrary to earlier views, molecules are not trace constituents, but may contain a large fraction of the elements other than hydrogen and helium.

Because the CO lines have such great optical depths, they can be used to estimate the kinetic temperature of the gas directly in most cases. Typical values are of order 20K, with temperatures as low as 10K in the dark dust clouds. These values are to be compared with the often-quoted "typical" value of 100K for interstellar clouds, and show that the large dense clouds in which CO is found are indeed very cold.

In contrast to the CO emission discussed above, the CO emission from the infrared stars is believed to come from shells of material ejected by the star. These are then relatively hot—a few hundred kelvin—and much smaller than the other sources. CS and HCN are also seen in some of these stars, and combination of this data with the CO data will undoubtedly provide important information about the physical conditions and chemical composition in circumstellar shells.

Methyl alcohol. Methyl alcohol, CH_3OH, is an example of the more complex molecules with more complex energy level diagrams. Significantly the observational results are more difficult to interpret. First, CH_3OH is an asymmetric rotor and, second, its spectrum is further complicated by hindered torsional rotation of the OH group with respect to the CH_3 group. Because of the three equivalent positions of the CH_3 group, the molecule has three classes of energy levels, called A, E_1, and E_2.

The first CH_3OH transition found was the 1_{11}–1_{10} in the A group. This line at 834 MHz was found in emission in the direction of the Galaxy center sources Sgr A and Sgr B2 by J. A. Ball, C. A. Gottlieb, and A. E. Lilley of Harvard University and H. Radford of the Smithsonian Astrophysical Observatory. At this low frequency the background radiation in this direction corresponds to a high temperature. Since the line is seen in emission, either the region is very hot—which seems unlikely—or the energy level population must be inverted so that amplification—as in a maser—takes place. Similar arguments for the other complex molecules ($HCOOH$, CH_3CHO, and NH_2CHO) observed at low frequencies suggest that they may be weak masers also.

Recently A. H. Barrett, P. Schwartz, and J. Waters found five CH_3OH emission lines in the direction of the Orion Nebula. The transitions they found all occur near 25 GHz and are of the E_1 type. The excitation for these lines corresponds to the relatively high temperature of 90K, but is probably not maser emission. The 1_{11}–1_{10} transition has not yet been found in Orion and the 25-GHz lines were not found in the Galaxy center sources. Thus differences in excitation conditions are reflected by the appearance of different lines of the same molecule.

Recently a line found fortuitously at 84 GHz by B. Zuckerman of the University of Maryland, B. E. Turner of the National Radio Astronomy Observatory (NRAO), and M. Morris and P. Palmer of the University of Chicago was identified by excellent laboratory "detective work" by D. Johnson of the National Bureau of Standards to be a transition of the E_2 type of CH_3OH. This line was found in the center region of the Galaxy and possibly in other sources. Thus all three types of lines of CH_3OH have been found, covering a frequency range of a factor of 100. Undoubtedly many more CH_3OH lines will be found.

With such rapid progress in the discovery and study of interstellar molecules in the 1970s, even more rapid developments can be expected as increasingly more researchers become interested in this new approach to studying the interstellar medium.

For background information see INTERSTELLAR MATTER; MOLECULAR STRUCTURE AND SPECTRA; RADIO ASTRONOMY in the McGraw-Hill Encyclopedia of Science and Technology.

[PATRICK PALMER]

Bibliography: J. A. Ball et al., *Astrophys. J.*, 162: L203, 1970; A. H. Barrett et al., *Astrophys. J.*, in press; D. Buhl and L. Snyder, *Sky Telesc.*, 40:267, 345, 1970; A. A. Penzias et al., *Astrophys. J.*, 165: 229, 1971.

Ionosphere

Significant new studies of the ionosphere have occurred very recently. One line of research has involved precise measurements of the arctic and antarctic ionosphere (in the high latitudes). As a result, new models of the ionosphere based on changes in time, magnetic activity, and the seasons have been developed. Another investigation involved temporarily modifying the ionosphere with strong radio waves. The measurements obtained in this study elucidate the nature of the ionosphere's electron density and electron temperature.

High-latitude ionosphere. Until recently the arctic (and antarctic) ionosphere had been known mainly in statistical terms because the data were scarce and highly complex. In fact, the data were grossly averaged and consequently many features were "smeared out." As the result of magnetospheric studies, new concepts have emerged and have led to a number of ordering principles for old and new ionospheric observations, including those from aircraft and satellite.

Data-ordering principles. The most suitable coordinate system is that of (corrected) geomagnetic latitude and geomagnetic local time (CGL/CGT). The use of "invariant latitude" appears also sufficiently accurate at this time. For discussion and transformation into geographical coordinates, several aids are available, such as the nomograph by J. A. Whalen. The high-latitude ionosphere varies with a number of parameters, of which universal time (UT), magnetic activity (Kp, Ap, Q), and season are the most dominant. Other parameters less explored are sunspot number and substorm time.

Sporadic E condition. Although highly variable the arctic ionosphere at any instance of time shows

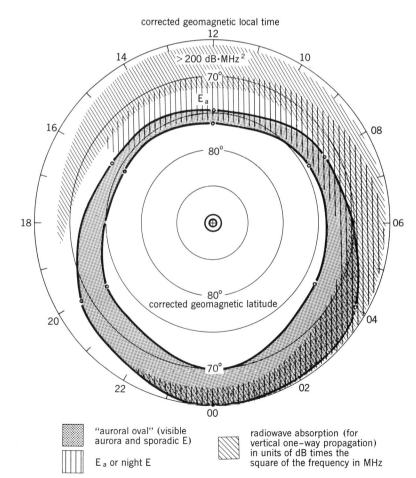

Fig. 1. Schematic example of the arctic ionosphere looking down on the Arctic displayed in CGT/CGL coordinates. The enhanced arctic D layer is at 75–100 km.

distinct features for a given set of parameters. The most conspicuous feature is that of auroral sporadic E. This layer occurs always under or within ± 1.5° latitude of visible auroral arcs. Although most of the time transparent, this region is a strong scatterer of radio waves in heights between 110 and 130 km and supports sky wave propagation for frequencies up to 9 MHz in vertical incidence and up to 50 MHz in oblique circuits. It has been suggested that this region consists of irregularities which are associated with the location of the auroral electrojet, an E-region electric current observable on arctic magnetograms. Its geographic location in statistical terms can best be described by referring to the statistical "auroral ovals" by Y. I. Feldstein. Each of those ovals maps the probability of occurrence of visual arcs for a particular magnetic activity and constitutes a suitable reference for the ionosphere. The auroral oval for moderate magnetic activity is shown in Fig. 1. In the ringlike shaded area in the diagram, sporadic E occurs 75% of the time. The geomagnetic pole is in the center. The direction toward the Sun is marked by 12 at the top of the figure. The Earth rotates underneath this picture in an eccentric way in that the geographic north pole is rolling off along 81.5° CGL. Not shown is the area of strong but short-living absorption centered at 22 CGT and 70° CGL. The whole picture expands and contracts with increasing and decreasing magnetic activity.

A detailed study by E. W. Pittenger has shown that the occurrence of very strong and often blanketing sporadic E, however short-living, is restricted to the evening sector from 20:00 to midnight CGT. This area is where auroral substorms appear to commence, according to S. I. Akasofu.

Night E. Other arctic ionospheric features also appear to have a defined geographical relation to the position of the visual aurora (auroral oval). A genuine E layer with a more or less well defined critical frequency, sometimes spread, often called night E (E_a in Fig. 1), appears to be present all the time in the indicated area during the arctic night and extends possibly fully around the oval. During the arctic summer this region is superimposed on the regular (solar-produced) E region. This region has a northern boundary at or near the latitude where the airglow at 630 nm has a maximum and where visual arcs may occur. From this poleward boundary the region extends by about 5° latitude toward lower magnetic latitudes, while the layer height changes from 140 to 120 km and the critical frequency from 2 to 3 MHz approximately.

Radio-wave absorption. Partly overlapping with the E_a region, but generally extending to lower magnetic latitudes, is an area of radiowave absorption, mild in effect but high in occurrence probability (see area marked >200 dB·MHz^2 in Fig. 1). Canadian studies revealed that the absorption in this area occurs in periods stretched from tens to hundreds of minutes after the onset of individual auroral substorms in the midnight sector. The observed time sequence suggests an eastward movement and gradual tapering off along constant magnetic latitude. In contrast to this fairly smooth and moderate absorption, an area of severe but short blackouts occurs near 22 CGT within the oval (not shown in Fig. 1). However, due to the short duration, the time-averaged probability of occurrence is small. The absorption is clearly due to enhanced D-region ionization superimposed on the regular (cosmic-ray and solar-produced) ionization. It is approximately inversely proportional to the frequency squared. Additional attenuation processes in the arctic ionosphere due to scattering affect radio and radar signal strength. A blackout of F-region propagation in oblique incidence is often caused by the presence of sporadic E along the ray path. Another attenuation occurs when the ray path traverses regions of irregularities. The shape of the irregularities is controlled by the direction of the geomagnetic field lines; consequently the scatter patterns are aspect-sensitive, resulting in various degrees of attenuation that depend on ray path

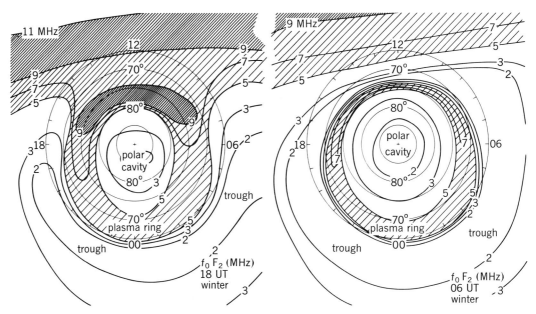

Fig. 2. Arctic F layer. Same coordinates as in Fig. 1. Shown are two situations 12 hr apart in winter.

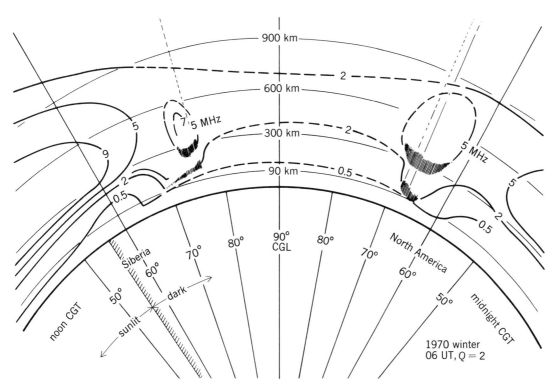

Fig. 3. Schematic vertical profile of the arctic ionosphere across the magnetic pole for 06 UT in winter, for low magnetic activity. Contour lines of plasma frequencies are plotted. Vertical height scale is three times horizontal height scale.

geometry. In addition, there are those very rare polar cap absorption (PCA) events which are caused by solar protons and which may last in the entire polar cap for days.

F region. More so than in the cases of the D and E regions, the polar F region can be described routinely neither in real time nor as hindsight. This is so because of lack of sufficient routine observations. However, individual snapshots can be assembled from existing data. Figure 2 shows a typical situation for the arctic winter at 06:00 and 18:00 UT for moderate magnetic activity. The Sun is beyond the top of the figure. Contour lines of the median critical frequency f_0F_2 are plotted, averaged over 30 min. They have a fixed spatial relation to the auroral oval (not shown). The picture expands and contracts with increasing and decreasing magnetic activity. Superimposed on this picture are small-, medium-, and large-scale irregularities of unidentified origin causing the instantaneous values to fluctuate up to ±50% about the values shown. There is a ringlike area of enhanced ionization, the outer edge of which coincides approximately with the outer edge of the auroral oval. The inner edge is in the polar cap and its magnetic latitude moves systematically with UT; it is closest to the magnetic pole at 18:00 UT. This ring coincides with the "plasma ring," in which satellite observations have noticed an enhanced precipitation of soft electrons. The F region in the entire Arctic, but in particular in this area, produces strong "spread F" in vertical ionograms as recorded both from the ground and in satellites. That the F region is interspersed with irregularities has also been observed from ground recordings of satellite beacons (scintillations). These field-aligned irregularities can cause aspect-sensitive strong backscatter echoes (clutter) on radars of all frequencies.

The area inside the plasma ring has a lower electron density and is marked "polar cavity" in Fig. 2. In the Northern Hemisphere the polar cavity reaches its largest extent at 06:00 UT.

The F region outside both the auroral oval and the plasma ring is characterized by depletion of electrons ("trough" in Fig. 2). The trough extends toward lower magnetic latitudes to about 55° CGL (where the moderate latitude ionosphere begins) and coincides roughly with the magnetic latitude of the plasma pause as determined by magnetospheric observations.

Vertical cross section. Figure 3 summarizes the known phenomenology. It represents a quasi-instantaneous snapshot of the arctic ionosphere's profile across the geomagnetic pole for winter, at 06:00 UT, for moderate magnetic activity ($Q=2$) and for a period when no substorm is in progress. This picture might be regarded as the result of a 30-min averaging. In order to realize the great variability of the high-latitude ionosphere, one must consider that the magnetic activity usually changes within 1 hr. The distinct features in Fig. 3 therefore move to other latitudes in accordance with Feldstein's auroral oval. Considering also that the diurnal and seasonal changes in solar illumination impose an additional variation, it is clear why in the past most averaging processes have produced a smooth and unrealistic arctic ionosphere.

Monitoring. Since the features described follow

Fig. 4. Facility used for ionospheric modification by the U.S. Department of Commerce near Boulder, Colo. View shows ring-array antenna and coaxial transmission lines connecting to transmitter in the building.

certain large-scale patterns, it seems feasible to monitor the high-latitude ionosphere in real time from a limited number of stations. Observations are required of the latitudinal positions of the plasma ring, trough, auroral arcs, absorption, and night E region in the night sector. The entire pattern would follow by inference. Such an approach, however, would not allow the specification of details occurring during auroral substorms.

Arctic-Antarctic conjugacy. Since the geomagnetic field lines dominate in the control of both the ionization (precipitation of charged particles) and deionization (loss by upward diffusion, evaporation, "polar wind") of the polar ionospheres, one might expect conjugacy of the phenomena located at the same magnetic field line. There are indeed certain symmetries in the general description of the phenomena in both the Northern and Southern hemispheres. However, strict magnetic conjugacy is neither observed nor expected. It is obvious that during the seasons not only the solar illumination can become very asymmetric but also the shape of closed magnetic field lines. In addition, magnetic field lines arising at the polar caps are never closed. Even during the equinox season, the tilt of the Earth's magnetic axis against the solar wind changes diurnally and prevents simultaneous symmetry. However, there are two (almost) identical configurations for the Northern and Southern hemispheres occurring 12 hr apart. This diurnal symmetry is probably related to the observed fact that the arctic and antarctic F-region plasma rings attain their largest extent at 18:00 and 06:00 UT, respectively. Seasonal symmetry is best for pairs of days 6 months apart. Considering those seasonal and diurnal symmetries, the descriptions given apply to both hemispheres accordingly. There are unconfirmed fragments of observed differences without identifiable symmetry. *See* SOLAR WIND.

[G. J. GASSMANN]

Ionospheric modification. Various radio-wave, photometric, and launched-probe techniques have been used to observe the region of the atmosphere between about 150 and 350 km, where the constituent gases are partly ionized by solar radiations and have the properties of a plasma. Powerful ground-based radio transmissions are now being used to temporarily modify this F region of the ionosphere so that in some aspects it can be experimented with as though it were a laboratory plasma, but without the complications of container boundaries. Suitable measurements made during excitation and recovery from modification permit delving further into the complexities of the ionospheric plasma and should provide new information on such parameters as electron-ion collision frequency, ambipolar diffusion coefficient, ionic reaction rates, and plasma instabilities which are also generated by some natural phenomena.

Modification theory. Several theoretical studies have shown that ionospheric modification can occur through deviative absorption of a radio wave if its frequency is close to a natural resonance of the plasma. G. Meltz, R. LeLevier, and A. Thompson have made the most recent and refined calculations of ionospheric heating. They showed that a high-power transmitter and a large antenna are required to produce significant changes in electron temperature and electron density in the upper ionosphere. Two high-frequency transmitting facilities which have a transmitter-antenna power-aperture product of about 10^4 MW/m^2 have begun modification experiments. One is located near Boulder, Colo. and is operated by the U.S. Department of Commerce. The second is at the Arecibo Observatory in Puerto Rico. Some of the early results are described below, after a brief description of the heating process and the Colorado facility.

At F-region heights there are about 10^9 particles per cubic centimeter. Of these, about 1 in 1000, or $10^6/cm^3$, are negatively charged electrons; there is an equal number of positive, mainly atomic oxygen, ions. The ambient energy equivalent temperature of the ions and electrons is 1000–1200K. Local plasma frequency ranges from a few to more than 10 MHz and depends upon electron density, which itself is a function of altitude and varies diurnally, seasonally, and with phase of the solar cycle.

Strong absorption of an incident radio wave occurs in a region where the plasma frequency is near the radio frequency, because the wave is slowed near this natural resonance and the electrons moving under the influence of the wave have greater opportunity to collide with heavy particles, usually ions. Thus energy is extracted from the wave and added to the random thermal energy of the electrons. As the electrons become hotter, they transfer only a small part of their excess energy to ions, because the electron mass is so much smaller than the ion mass. Since the collision frequency between ions and electrons in the F region is only about 10^3 collisions per second, because of the low particle density, the time constant for the energy loss is of the order of 10 sec. This slow loss rate makes appreciable heating of electrons possible. As electron temperature rises, the plasma pressure increases, and since the charged particles are constrained by the geomagnetic field to spiral about field lines, the plasma expands along these lines until the pressure decreases to that of the surrounding medium. While the electron temperature reaches its new steady-state value within about 10 sec, the plasma expansion process, which would reduce the local electron density in a field-aligned column, requires some tens of minutes of continuous heating. This is because of the drag placed on electron motion as a result of the relatively low mobility of the heavy positive ions. Electrons cannot move independently of the positive ions, otherwise strong electric fields would develop because of the charge separation. The charged particles move in a way that neutralizes this field, keeping the medium intrinsically electrically neutral and causing the ionized constituents of the gas to move together.

Experimental configuration. The Colorado facility was designed so that experiments could be performed using frequencies between 5 and 10 MHz while transmitting either a right- or left-circularly polarized wave. These polarizations correspond to the ordinary and extraordinary waves which characterize propagation in a birefringent, magnetoionic medium such as the ionosphere. The antenna array is constructed with nine elements of broad-band crossed dipoles which are spaced uniformly about a tenth element in the center of a ring 110 m in diameter. This produces a circular main beam about 20° wide. A view of the antenna array and the coaxial transmission lines connecting it to the transmitter building is shown in Fig. 4. Each element in the array is connected to one of 10 identical amplifier channels capable of producing 200 kW of average power, or 2 MW into the array. This, combined with the gain produced by the array, results in an effective power directed upward into the ionosphere of about 100 MW. Because this power is distributed over a circular area in the

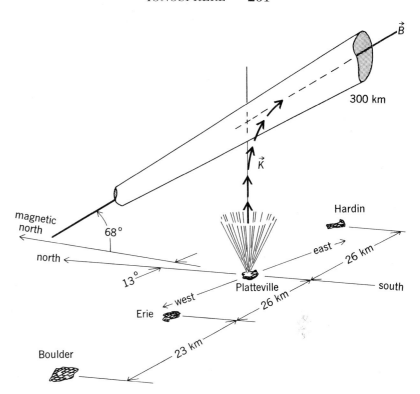

Fig. 5. Schematic view of ionospheric modification experiment. High-power transmitter facility is at Platteville, and radio and optical diagnostic instruments are located at Boulder, Erie, and Hardin sites. Magnetic field orientation is indicated, and a typical extraordinary mode (x-mode) ray path is illustrated.

ionosphere about 100 km in diameter, the power flux density near 300 km in the F region is less than 100 $\mu W/m^2$.

Figure 5 is a schematic view of the ionospheric heating experiment. The high-power transmitter is located at Platteville and radio and photometric diagnostic instruments are located at Boulder, Erie, and Hardin sites. The figure also shows the orientation of the geomagnetic field B (dip angle of 68°, declination 13°) and illustrates a typical flux tube into which energy is directed by the extraordinary magnetoionic component of the heating radio wave. As the extraordinary magnetoionic component progresses into the ionosphere, it is refracted southward and ultimately propagates parallel to the geomagnetic field, as is illustrated by a series of arrows. The ordinary component, however, is deviated northward and eventually propagates in a direction perpendicular to the magnetic field and, for a given transmitted frequency, penetrates to a greater height than does the extraordinary component. Thus different regions of the ionosphere would normally be affected when heating with the ordinary or extraordinary modes, and because of dissimilarities in their propagation characteristics, some different effects were anticipated and observed. Only a brief summary of some salient modification effects is possible here. Some of the results were about as predicted by theory, but others were unanticipated.

Experimental results. M. Biondi was responsible for photometric observations of 630-nm emissions arising from atomic oxygen. From these observations he calculated that the electron temperature in the F region was raised by about 35% when the

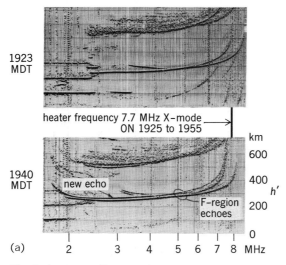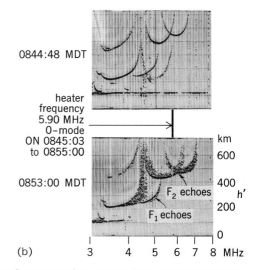

Fig. 6. Ionograms illustrate change in radio echoes reflected from the ionosphere as a result of ionospheric modification by the high-power transmitter. Upper and lower ionograms show, respectively, conditions prior to and following transmitter turnon. (a) Apr. 30, 1970. (b) July 24, 1970.

extraordinary component was used for heating. The method of detecting electron heating is through the attendant effect on the rate of dissociative recombination of electrons and molecular oxygen ions, in which process 630-nm radiation from atomic oxygen occurs. In this process the reaction rate is inversely proportional to electron temperature, so that the emission intensity decreases during heating and increases after the heater is off. Both the rise and fall of electron temperature occurred in tens of seconds, as theory had predicted. Further, heating with the ordinary mode produced an unexpected, and nearly opposite, result; there was an increase in the intensity of emission following turnon and a decrease after turnoff. Although uncertainty exists as to the reactions producing the emission in this case, it is possible that electron temperature was raised to a value in excess of 2200K. If this is true it may provide evidence of the development of plasma instabilities. Airglow measurements of infrared radiation at 1.27 μm were made on a few occasions when heating with the extraordinary mode. It was found that these emissions originated at an altitude of about 200 km, in a region located on the geomagnetic field lines which pass through the higher volume that was initially heated by the radio wave. The process causing these emissions is unclear, as is the case for similar emissions in aurora, but they are believed to originate from molecular oxygen excited by electrons having enhanced energy which was transmitted down the field lines through thermal conduction.

Radio-wave techniques were also used to explore the modified ionosphere. Ionograms obtained with a swept-frequency radar (ionosonde) were the principal diagnostic measurements, but phase and amplitude measurements of probing waves were also measured. Ionograms, which display the virtual range (height) of ionospheric echo returns versus exploring frequency, provide a measure of electron density in the ionosphere. The ionograms in Fig. 6 show some of the changes in echoes from the ionosphere resulting from heating. The upper ionograms in each case were made just before heating began. The echo of particular interest in Fig. 6a is the one identified as the new echo. This echo gradually increased in range and slowly waned from the high frequency end downward. The time history of this echo gives evidence that a major redistribution of electrons occurred and suggests that a heated core of the ionosphere was displaced upward along the magnetic field. Such perturbations disappear shortly after terminating the heating.

Generation of spread F was one of the unexpected results of the modification experiments. Natural spread F occurs occasionally, but usually during darkness and mostly after midnight. It has been found at the Colorado location, however, that it can be generated at any time of day with either ordinary or extraordinary excitation. The lower ionogram of Fig. 6b shows the diffuse echoes resulting from spread F during a daytime observation. The spread F echoes begin to appear promptly at frequencies near that of the heating transmitter and, within a few minutes, extend to frequencies which reflect over a height range of 200 km or more. It is thought that the spread echoes arise from irregularities which grow during heating. They persist after the heater is off for a few tens of minutes during the day and early evening hours and for longer periods later at night.

Heating with the ordinary mode has also produced an unexpected attenuation of 10 dB, or more, of the ordinary component of probing waves. This occurs within seconds and on all frequencies higher than the heating frequency. This anomalous absorption may be caused by plasma instabilities induced by the heating.

The variety of effects which have been observed, more numerous than described here, present a challenge to interpretation. The unanticipated events not only are exciting but, when the phenomena are properly understood, should add significantly to better understanding of this remote region of the upper atmosphere.

For background information see GEOMAGNETISM; IONOSPHERE; MAGNETOSPHERE; PLASMA PHYSICS; RADIO-WAVE PROPAGATION; SOLAR

WIND in the McGraw-Hill Encyclopedia of Science and Technology. [W. F. UTLAUT]

Bibliography: S. I. Akasofu, *Polar and Magnetospheric Substorms*, 1968; D. T. Farley, Jr., *J. Geophys. Res.*, 68(2):401, 1963; Y. I. Feldstein and G. V. Starkov, *Planet. Space Sci.*, 15:1008–1015, 1967; A. V. Gurevich, *Geomagn. Aeron.*, 7(2):291, 1967; D. H. Jelly and L. E. Petrie, *Proc. IEEE*, 57(6):1005–1012, 1969; G. Meltz and R. E. LeLevier, *J. Geophys. Res.*, 75(31):6406, 1970; W. F. Utlaut, *J. Geophys. Res.*, 75(31):6402, 1970; J. A. Whalen, J. Buchau, and R. A. Wagner, *J. Atmos. Terr. Phys.*, 33(4):527–547, 1971.

Jahn-Teller effect

The Jahn-Teller effect was predicted theoretically in 1937 and was first observed experimentally in 1952, but only since 1965 have its most commonly observable consequences come to be understood. Whereas in 1965 it was regarded as a "mystical effect" to be invoked when all other explanations of anomalous data failed, now it is as well understood as most other phenomena in solid-state physics.

The Jahn-Teller effect is most commonly observed in the optical and microwave spectra of point defects and localized impurity centers in solids. Such a center has a certain point symmetry, that is, there is a group of coordinate transformations which leave the center and its surroundings unaltered. If this symmetry is high enough, some of the electronic states of the center will be orbitally degenerate. The Jahn-Teller theorem states that any such degenerate system is unstable against small distortions of the lattice framework which remove the degeneracy.

Because of the great complexity of the interatomic forces in a real solid, it is customary to think of the center as a molecule consisting of the central impurity or defect and its nearest neighbor atoms. The rest of the crystal is regarded as a featureless heat bath. This "cluster" model preserves the symmetry of the center, which is its most essential attribute for purposes of this discussion, and is appropriate when the electronic wave function is sufficiently localized not to be greatly influenced by motion of atoms outside the nearest neighbor shell. In practice, this is also the condition for electron-lattice coupling to be strong enough for the Jahn-Teller effect to be important.

Simple model for Jahn-Teller distortion. As an illustrative example, consider the center shown in Fig. 1, which is a greatly simplified model of the first excited level of the F-center in an alkali halide. (The F-center is an electron trapped at a negative ion vacancy.) The electron is in a p state and is surrounded by a regular octahedron of positive ions. Because of the cubic symmetry, the three possible p states, p_x, p_y, and p_z (of which only p_z is shown), are obviously degenerate. Now suppose that the positive ions move a small distance in the directions indicated by the arrows. If their mean distance from the center is fixed, the mean energy of the p level is unaltered in first order. The p_z state, in which the electron approaches the positive ions more closely, is lowered in energy by the distortion, while the p_x and p_y states are raised, as shown in Fig. 2. The splitting is initially linear in the displacement Q. The surrounding crystal resists the distortion, and by Hooke's law the additional energy due to this resistance is initially quadratic in Q. Thus the total energy of the center in the p_z state goes through a minimum, and equilibrium is reached at some finite value of Q. It is easy to see that this equilibrium value of Q is A/k, where A is the initial (downward) slope and k the opposing force constant, and that the stabilization energy is $E_{JT} = A^2/2k$.

Strong electron-lattice coupling, combined with weak interatomic forces, favors a strong Jahn-Teller effect. E_{JT} can range from several electron volts (eV) for deep states in diamond and silicon to less than 10^{-3} eV for rare-earth ions in ionic crystals. (A Jahn-Teller effect this small is unlikely to be observable.) In transition-metal ions, for which most of the data have been obtained, E_{JT} ranges from 0.01 to 1 eV.

The distortion shown in Fig. 1 could equally well have been along the x or y axes, and the lowest state would have been p_x or p_y, respectively. Thus each electronic state is associated with its own distortion, and if one considers the lowest vibrational state only, the threefold electronic degeneracy is replaced by threefold "vibronic" degeneracy. If no transitions were possible between the different directions of distortion, any one center would remain indefinitely in one state and its corresponding distorted configuration. This distortion might be observable, for instance, in a spin resonance experiment. Such an observable reduction in symmetry is called the "static Jahn-Teller effect." The more common case, however, is the "dynamic Jahn-Teller effect" in which transitions between different distorted configurations are rapid compared with the characteristic measurement time. Such transitions can occur through thermal activation, or through quantum mechanical tunneling due to the zero point vibrational motion about the different equilibrium configurations. The observed spectrum now has the same symmetry as if there were no Jahn-Teller effect, and it might be thought that the effect has "disappeared."

In 1965, F. S. Ham pointed out that this is not the case; there are in fact pronounced observable consequences of the dynamic Jahn-Teller effect. These show up in the effects of off-diagonal electronic operators, that is, operators which connect electronic states associated with different distortions. For instance, consider the angular momentum operator $\hbar L$. The p level in the undistorted octahedron of Fig. 1 has $L=1$. When a magnetic field H is applied parallel to the z axis, the threefold degenerate level is split into three states with wave functions $\frac{1}{2}\sqrt{2}(p_x + ip_y)$, $\frac{1}{2}\sqrt{2}(p_x - ip_y)$, p_z, corresponding respectively to the $+1, -1,$ and 0 eigenvalues of L_z. If one ignores electron spin, the energies are given by $-e\hbar H L_z/2mc$. The $(p_x + ip_y)$ state corresponds to clockwise rotation about the field, and the $(p_x - ip_y)$ state to counterclockwise rotation. In the Jahn-Teller distorted case the electron has to "drag" its distortion with it and can no longer rotate freely. Its mass is thus effectively increased and the orbital contribution to the magnetic splitting correspondingly decreased. At 0K, where rotation is only possible at all because of zero point motion, the reduction factor can be shown to be approximately $e^{-3E_{JT}/2\hbar\omega}$, where ω is an effective frequency of vibration for the cluster.

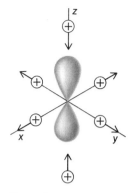

Fig. 1. A p-electron in an octahedral site. The arrows show one possible tetragonal mode of distortion. Only a p_z wave function is shown.

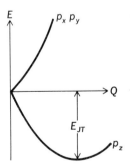

Fig. 2. Energy of the center shown in Fig. 1 as a function of the tetragonal distortion. The lower curve, corresponding to the p_z state, is $E = -AQ + 1/2 kQ^2$; the upper (p_x, p_y) curve is $E = 1/2 AQ + 1/2 kQ^2$.

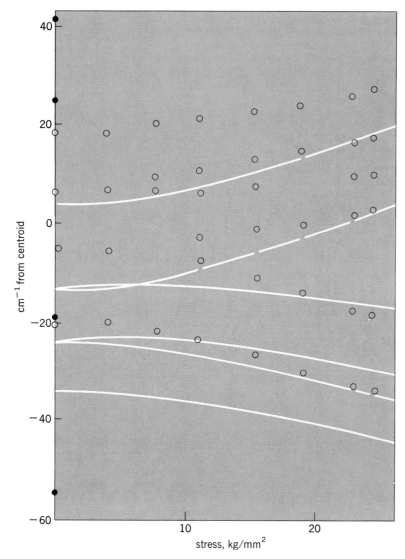

Fig. 3. Splitting of the $^4T_{2g}$ excited state of V^{2+} substituted for Mg^{2+} in cubic $KMgF_3$ as a function of stress parallel to $\langle 001 \rangle$. Solid circles represent calculated spin-orbit levels in the absence of the Jahn-Teller effect. Open circles represent observed energy levels. The curves represent calculated energy levels with $E_{JT}/\hbar\omega = 0.9$. (After M. D. Sturge, Phys. Rev., B1:1005, 1970)

ter in the theory consists of the ratio $E_{JT}/\hbar\omega$.

If the electronic state is doubly degenerate, it has no orbital momentum or spin-orbit coupling in first order. If the Jahn-Teller interaction is strong, the small local strains which are inevitable in a real crystal are sufficient to stabilize the center in one or other distorted configuration at low temperature. In a crystal containing many centers, three uniaxial spin resonance spectra are seen, corresponding to the three possible directions of distortion. As the temperature is raised, transitions from one distorted configuration to another become possible, and the three anisotropic resonances collapse into one isotropic resonance. This process is exactly analogous to motional narrowing due to diffusion in nuclear magnetic resonance. If the Jahn-Teller effect is sufficiently weak, tunneling will average out the spectrum, even at 0K. The spectrum in this case still retains a cubic anisotropy, qualitatively similar to that expected in the absence of Jahn-Teller interaction. Even if tunneling is too slow to produce this averaging, it can still profoundly affect relaxation processes. For instance, it accelerates the decay of spin echoes, and it produces strong damping of acoustic waves whose period is of the order of the tunneling time. *See* SEMICONDUCTOR.

The Jahn-Teller effect can also manifest itself in a pure crystal containing ions with orbitally degenerate ground states if the bandwidth due to overlap between these states is less than E_{JT}. A phase transition is often observed in such crystals, between a high-temperature symmetric structure (usually cubic) and a low-temperature distorted structure. Examples are spinels containing Mn^{3+} as a major constituent. The mechanism of such transitions is not fully understood; in many cases it is not even certain that the Jahn-Teller effect is primarily responsible. When it is, the transition may consist of the lining up of local distortions, which exist in a random arrangement above the transition temperature. Alternatively there may be no local distortion above the transition, while below it the strain due to one distortion stabilizes the next. The first case is analogous to ordinary ferro- or antiferromagnetism, and the second to the ferromagnetism of ions with spin singlet ground states.

Vibrational spectra of molecules. Finally it should be stated that the intimate coupling between electronic and nuclear motion which is a consequence of the Jahn-Teller interaction should have pronounced effects on the vibrational spectra of molecules. However, it has proved quite difficult to pin down these effects, chiefly because of the difficulty of establishing what the vibrational spectrum would be in the absence of the Jahn-Teller interaction.

For background information *see* DEGENERACY (QUANTUM MECHANICS); ELECTRON PARAMAGNETIC RESONANCE SPECTROSCOPY (EPR); QUANTUM MECHANICS in the McGraw-Hill Encyclopedia of Science and Technology.   [M. D. STURGE]

Bibliography: R. Englman, *The Jahn-Teller Effect*, in press; F. S. Ham, The Jahn-Teller effect in EPR spectra, in S. Geschwind (ed.), *Electron Paramagnetic Resonance*, in press; G. Herzberg, *Electronic Spectra of Polyatomic Molecules*, 1966; M. D. Sturge, *Advan. Solid State Phys.*, 20:91, 1967.

Note that the spin operator S is not "quenched" as L is, since the spin direction has no distortion associated with it (except insofar as spin-orbit coupling causes L to follow S).

Crystals with 3d ions. This reduction in the orbital contribution to magnetic splittings has been observed in the spin resonance and Zeeman spectra of many transition-metal ions in crystals. Other off-diagonal operators, such as the spin-orbit coupling (which can often be written in the form $-\lambda L \cdot S$), are quenched in the same way. A detailed comparison of the Ham theory with experiment has been possible in optical spectra of 3d ions in some cubic crystals; such a comparison is shown in Fig. 3. This shows the spin-orbit splitting of an orbital triplet, spin quartet level, and the effect on it of applying a uniaxial stress parallel to one of the directions of Jahn-Teller distortion. The matrix elements of stress, which are diagonal and therefore unquenched, are determined in a separate experiment. Therefore the only free parame-

Jurassic

Some recent research on Jurassic fossils has been concerned with the evolution of *Gryphaea* from the flat oyster (*Liostrea irregularis*).

Among fossil invertebrates, the standard example of an evolutionary sequence has long been the supposed transition from *L. irregularis* to a highly coiled form (*Gryphaea arcuata incurva*) during very early Jurassic times in England. A. E. Trueman, the chief advocate of this sequence, maintained that it represented a case of nonadaptive (and therefore non-Darwinian) evolution. *Gryphaea*, Trueman claimed, was not able to arrest its coiling trend once it had started; the shell coiled tighter and tighter, past the point of utility (for raising the animal above a muddy substrate) to a stage at which the bottom coiled valve pressed against the upper cap-shaped valve (Fig. 1), thus preventing it from opening and sealing the animal within. During the past 10 years, however, three studies have shown not only that Trueman's theory is false but also that a more adequate understanding of the evolution of *Gryphaea* provides an interesting story fully consistent with modern evolutionary theory.

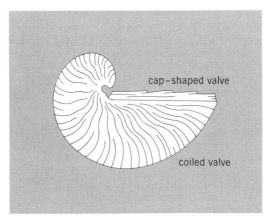

Fig. 1. *Gryphaea arcuata incurva.*

First study. The basis for the argument that *Gryphaea* did not evolve gradually from *Liostrea* in England was established by A. Hallam. In 1959 he showed that no gradual transition could be traced from *Liostrea* in the lowermost Jurassic beds in England to *Gryphaea* in slightly younger rocks. The first *Gryphaea*, he wrote, appear just below the top of the *Angulata* Zone (of the basal Jurassic), sharply replacing the earlier *Liostrea*. Hallam believed that *Gryphaea* either migrated into England at that time from elsewhere or developed rapidly by speciation in an isolated local population. It then replaced *Liostrea* by competition. In any case, *Gryphaea* did not evolve from *Liostrea* by gradual, in-place evolution in England.

Second study. *Gryphaea* did not increase its coiling during lowermost Jurassic times. In 1971, S. J. Gould showed that the data used by Trueman and all later workers to claim an increase in coiling had been misinterpreted. *Gryphaea* did increase in size during its evolution; this trend is accepted by all. Since *Gryphaea* begins its life as a flat shell attached to some hard object, coiling constantly increases during the growth of each individual. True increase in tightness of coiling must be carefully distinguished from a spurious increase due only to growth to larger sizes. This Trueman failed to do, and his data, indicating increased coiling through time, may only reflect the increase of size of *Gryphaea*. G. M. Philip and Hallam attempted to avoid this problem by comparing the growth of ancestral and descendant *Gryphaea* throughout their size range. But they used as their measure for tightness of coiling a parameter that records, rather, the increase in rate of coiling during growth. This is inappropriate. One shell can increase its coiling faster than another during growth but still remain less tightly coiled at any comparable point because it was much flatter to begin with. Both the initial state and the subsequent rate of coiling (corresponding to the *y*-intercept and slope of a bivariate plot) must be considered.

T. P. Burnaby recognized this error and studied both slopes and *y*-intercepts for bivariate plots of coiling versus size. He made the startling claim that descendants were less tightly coiled than ancestors. However, Burnaby based this conclusion on the inappropriate comparison of ancestors with descendants of the same size. Since size increased during evolution, Burnaby compared adult ancestors with juvenile descendants; since coiling increases during growth, the juvenile descendant was less coiled than the adult ancestor of the same size. The comparison should be made at comparable stages of development, for example, between average adults of the two samples. In fact, ancestors and descendants, when correctly compared, display exactly the same tightness of coiling. Differences in plots of coiling versus size between ancestors and descendants are so arranged that descendants have, at their comparably larger sizes, the same shape as their ancestors. In other words, evolution of the coiling versus size relationship assured that shape would remain invariant while size increased.

Third study. The descendants of *G. arcuata incurva* reversed the supposed trend and decreased in coiling. Trueman's original study treated only the basal few zones of the English Lias (Lower Jurassic). The subsequent debate dealt only with these *Gryphaea*. But the *G. arcuata incurva* lineage continues into younger rocks. In 1968, Hallam published an ironic result: Apart from whatever happened to the coiling of *Gryphaea* in Trueman's sequence (and, according to the last section, nothing did), coiling in the descendants of *G. arcuata incurva* steadily decreased through time even though size continued to increase as before (Fig. 2). Moreover Hallam used this information to provide a functional interpretation for the entire sequence. He believed that the original transition from a flat oyster to *Gryphaea* occurred rapidly. This change, which may have had a simple genetic basis, was clearly adaptive because it allowed *Gryphaea* to colonize a new and favorable environment, namely, waters in or near estuaries. Yet, rich as these waters are in food, they also present problems of soft bottoms, high turbidity, and high rates of sedimentation. In such conditions a flat oyster would have no hard substrate for attachment and, even if it could cement, would soon be overwhelmed by the influx of soft sediment.

But tight coiling confers two advantages that

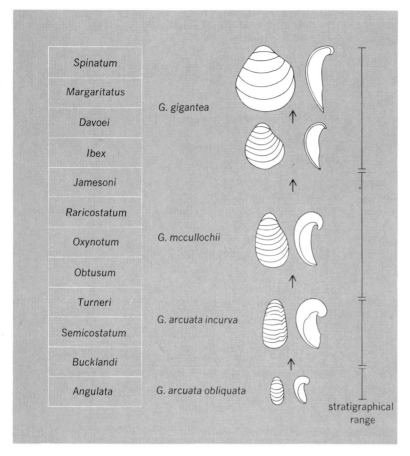

Fig. 2. The evolution of the *Gryphaea arcuata incurva* lineage to larger, thinner, and more loosely coiled shells. Names on the left are successive zones of the English Lias. (From A. Hallam, Morphology, palaeoecology and evolution of the genus Gryphaea in the British Lias, Phil. Trans. Roy. Soc. London, 254:91–128, 1968)

open these waters to habitation by oysters: It raises the mantle margins higher above the sea bed, reducing the chance of suffocation; and it produces a free-living adult no longer restricted to clear areas with large objects for cementation. However, at the same time, the tightly coiled spiral imposes a great disadvantage in that it reduces the stability of the shell, as Hallam demonstrated in flow-channel experiments with actual specimens. Hallam claimed that the subsequent trend to looser coiling in the descendants of *G. arcuata incurva* occurred to improve stability as the shell continued to increase in size. As stability increased, the need for a massive shell diminished and the thick shell of *G. arcuata incurva* gradually became thinner. "The end product," wrote Hallam, "was a thin-shelled, saucer-shaped *Gryphaea* which expressed a good balance between stability and the need to keep the mantle margin above the muddy bottom."

Conclusion. At first sight, the results of this work may seem negative. The most famous example of evolution in fossil invertebrates has been refuted in each of its contentions and the subsequent evolution of the stock has been analyzed as proceeding in a direction exactly opposite to that claimed for the ancestors. Yet the conclusions reported here have two positive aspects. First, "debunking" is often an essential ingredient in scientific progress, especially when the notion under attack had achieved the status of textbook dogma. Debunking is even more essential in this case than in most, because Trueman's notion did not stand simply as an isolated fact; rather, it supported a theory of orthogenesis opposed to modern evolutionary views and provided one of its supposed empirical supports. Although modern Darwinians had rejected Trueman's interpretation, they had not countered his "facts" and the story of *Gryphaea* had remained as an uncomfortable anomaly within modern evolutionary theory. Also, the revised version of the evolution of *Gryphaea* has a good deal of intrinsic interest. The early evolution of *Gryphaea* presents yet one more case of the influence that absolute size plays in prescribing the limits of shape; in this case, the disadvantages of tighter coiling at larger sizes were avoided by maintaining a constant shape during evolution to the larger size of descendants. The later evolution of *Gryphaea* provides a fine example of how functional morphology and the mechanics of form can explain the adaptive significance of evolutionary changes, the "why" of evolutionary events.

For background information *see* JURASSIC; MAGNOLIOPHYTA; PALEOBOTANY in the McGraw-Hill Encyclopedia of Science and Technology.

[STEPHEN JAY GOULD]

Bibliography: S. J. Gould, *Amer. Natur.*, vol. 105, 1971; S. J. Gould, in W. C. Steere, T. Dobzhansky, and M. K. Hecht (eds.), *Evolutionary Biology*, vol. 5, 1971; A. Hallam, *Geol. Mag.*, vol. 96, 1959; A. Hallam, *Phil. Trans. Roy. Soc. London*, 254:91–128, 1968.

Laser, chemical

The recent development of several continuously operating chemical lasers has resulted in a renewed interest in chemical laser research, development, and application. The first purely chemical lasers requiring no external source of energy to initiate or sustain laser excitation have been operated successfully. New techniques for rapid initiation of chemical reactions in premixed gases have been investigated which provide improved peak powers and pulse energies.

Continuous-wave chemical lasers. In 1969 the first continuous-wave (CW) chemical lasers were reported simultaneously by research groups at the Aerospace Corp., the Avco Corp., and Cornell University. Each group had worked independently on different chemical systems. By the end of 1969, D. J. Spencer, H. Mirels, T. A. Jacobs, and R. W. F. Gross had achieved CW power outputs near the kilowatt level from the HF chemical laser with chemical efficiencies of 12%, and extensive research efforts on the new CW chemical lasers were launched at a number of laboratories.

The first chemical laser was developed by Jerome Kasper and George Pimentel late in 1964. This laser, based upon the HCl molecule, and similar lasers, operated between 1964 and 1969 and based upon other hydrogen and deuterium halide molecules and the CO molecule, were pulsed devices capable of operation for only a few microseconds. These devices required either flashlamps or electrical discharges for initial production of radicals, and were capable of operation up to a few kilowatts peak power with energy conversion efficiencies (ratio of laser output energy to external energy input) of less than 0.1%.

The status of the chemical laser was significantly changed, however, with the developments of 1969. The high power output capability, the high chemical efficiency, and the laser wavelengths associated with many of the present CW devices make chemical lasers an attractive alternative to the N_2-CO_2 laser for many applications.

The feature that distinguishes chemical lasers from other types of lasers is a requisite nonequilibrium energy release among the internal degrees of freedom of newly formed product molecules. This can result from certain types of chemical reactions. Present CW chemical lasers are based upon several threebody atom-exchange reactions as shown by reaction (1), where $n = 0, 1, 2, \ldots$ In this type of reaction a large fraction of the energy release is present in the vibrational excitation newly formed AB bond. Recent progress in CW chemical laser development is based upon knowledge concerning the relative probabilities that this reaction generates the AB species in the various possible energy levels identified by the vibrational quantum number v. Such information has been gained through the infrared chemiluminescence techniques developed by John Polanyi and coworkers and the pulsed chemical laser experiments of Pimentel's group, among others.

$$A + BC \rightarrow AB(v=n) + C \quad (1)$$

Figure 1 shows the selective nature of the energy release among preferred vibrational levels for the HF chain reactions (2a) and (2b). Reaction (2a) has a maximum rate constant for reaction into the vibrational level $v=2$ which can directly provide population inversions between the $v=2$ and $v=1$ or $v=1$ and $v=0$ levels. In a similar way reaction (2b) can lead to population inversion on the first few vibrational levels.

$$F + H_2 \rightarrow HF(v=n) + H \quad (2a)$$
$$H + F_2 \rightarrow HF(v=m) + F \quad (2b)$$

The table summarizes the chemical reactions of the type of reaction (1) that have been employed by several investigators for chemical laser operation.

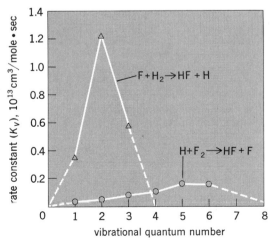

Fig. 1. Detailed rate constants for vibrational excitation for HF by the chain reactions of reaction (2) in the text. (Data from J. C. Polanyi and D. C. Tardy, J. Chem. Phys., 51: 5717, 1969; N. Jonathan, C. M. Melliar-Smith, and D. H. Slater, J. Chem. Phys., 53:4396, 1970)

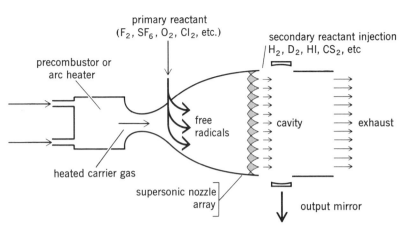

Fig. 2. Schematic diagram of a supersonic-flow chemical laser with atom production by thermal dissociation.

It should be pointed out that several reactions can lead to laser action at 10.6 μm in the HF-CO_2, DF-CO_2, HCl-CO_2, and HBr-CO_2 chemical lasers. The HBr-CO_2 laser has been operated CW; the HBr laser has not.

In addition to operation at various wavelengths from 2.7 to 5.5 μm for partially or totally inverted product molecules AB, it has been possible to employ the nonequilibrium energy release of reactions of the type of reaction (1) to produce laser output from totally inverted CO_2 at 10.6 μm by vibrational energy transfer processes, as exemplified by reaction (3).

$$DF(v=n) + CO_2(000) \rightarrow$$
$$DF(v=n-1) + CO_2(001) \quad (3)$$

The operating conditions of present CW chemical lasers are severely constrained by the relationships between the characteristic rates for the processes of gas mixing, chemical reaction, vibrational energy transfer, and vibrational deexcitation. In order to satisfy the conflicting demands imposed by these processes, it has been necessary to employ rapid gas mixing techniques in high-speed flows. In many cases the gas dynamic cooling provided by supersonic expansion is essential to provide the proper kinetic environment for chemical reaction and population inversion.

Figure 2 illustrates a generalized concept for chemical laser operation currently being investigated by a number of research groups. Atom production in this type of chemical laser is produced by thermal dissociation of one of several species as indicated. The requisite dissociation energy may

CW chemical laser reactions

Reaction				$-\Delta H_f^0$, kcal/mole	Wavelengths, μm
Cl + HI	$\rightarrow$	HCl*	+ I	32	3.5–4.1; 10.6
H + Cl_2	$\rightarrow$	HCl*	+ Cl	45	3.5–4.1; 10.6
F + HI	$\rightarrow$	HF*	+ I	64	2.5–3.0; 10.6
F + H_2	$\rightarrow$	HF*	+ H	32	2.5–3.0; 10.6
F + HCl	$\rightarrow$	HF*	+ Cl	33	2.5–3.0; 10.6
H + F_2	$\rightarrow$	HF*	+ F	98	2.5–3.0; 10.6
F + DI	$\rightarrow$	DF*	+ I	64	3.5–4.1; 10.6
F + D_2	$\rightarrow$	DF*	+ D	31	3.5–4.1; 10.6
D + F_2	$\rightarrow$	DF*	+ F	99	3.5–4.1; 10.6
H + Br_2	$\rightarrow$	HBr*	+ Br	36	(4.0–4.7); 10.6
O + CS	$\rightarrow$	CO*	+ S	75	5.0–5.8

be supplied by either an electric arc heater or by a secondary chemical reaction to provide an all-chemical laser. The supersonic expansion acts to abruptly reduce the translational temperature of the flow and freeze the atom concentrations at the initially high values produced in the dissociator. The use of an optical axis transverse to the flow direction facilitates the scaling of the laser power output to the kilowatt range. Devices of this type have been operated at energy conversion efficiencies of about 5%.

A chemical laser has been developed by T. A. Cool and R. R. Stephens which operates solely by the simple mixing of gases in a subsonic flow. Reaction (4) was used to produce the F atoms necessary to initiate the chain reactions of reaction (2) for HF and DF, and thus provided the basis for the first demonstrations of purely chemical laser operation requiring no external energy sources to initiate or sustain laser excitation. This laser operates at 10.6 μm when CO_2 is added to utilize the processes of reaction (3); Cool and Stephens also found laser output from HF and DF in the absence of CO_2 at 2.7 and 3.8 μm, respectively.

$$F_2 + NO \rightarrow ONF + F \qquad (4)$$

Pulsed chemical lasers. Progress in the development of CW devices has spurred renewed interest in pulsed chemical lasers as potentially attractive sources of the very-high-energy pulses of short duration needed for plasma heating experiments in research directed toward controlled nuclear fusion. To illustrate the potential energy involved, consider that if only 500 g of HF were reacted to convert 15% of the energy release of reaction (2) into laser output, then over 1 megajoule of energy would be released as coherent radiation.

Pulsed operation also provides a convenient means for the screening of candidate reactions for potential CW application. All of the present CW chemical lasers are based on reactions initially successfully used in pulsed chemical laser operation. Many reactions used in pulsed chemical lasers have not yet been found to give CW emission. An interesting recent example is the pulsed OH chemical laser reported by A. B. Callear and H. E. Van Den Bergh.

Much of the current effort has been concerned with the development of new techniques for rapid initiation of chemical reaction to supplement the flash photolysis and low-pressure discharge methods used prior to 1969. A few of the many efforts in this area should be mentioned to illustrate the surprisingly large variety of successful initiation schemes under study, though it is not possible to properly summarize all the current work in this rapidly developing field.

Megavolt electron beams have been used for pulsed initiation of HF laser operation by David Gregg and coworkers in a variety of hydrogen- and fluorine-containing compounds.

The type of high-pressure transverse electrical discharge initially developed by A. J. Beaulieu and coworkers for CO_2 lasers has been widely used for pulsed HF and CO chemical laser initiation with numerous reactions and various compounds. An example is the demonstration by several groups of very high gain (superradiant) and high power output (>50 kW) for HF chemical laser emission from discharges in premixed H_2, SF_6, and He.

Pulsed HF laser emission was observed from a detonation wave propagating through a gas mixture containing F_2O and H_2 by Gross, R. R. Giedt, and Jacobs.

R. J. Jensen and W. W. Rice have reported HF laser action following a chemically initiated explosion containing compounds in the presence of H_2.

Scientific applications. Chemical lasers have scientific importance beyond potential technological applications such as welding, cutting, drilling, power transmission, laser surgery, and controlled nuclear fusion. Noteworthy examples of scientific applications include the use of HF lasers for isotope separation by S. W. Mayer and coworkers, and the use of pulsed chemical lasers for fluorescence measurements of energy transfer rates by J. R. Airey and by C. B. Moore and their respective associates.

For background information *see* KINETICS, CHEMICAL; LASER; OPTICAL PUMPING in the McGraw-Hill Encyclopedia of Science and Technology. [TERRILL A. COOL]

Bibliography: T. A. Cool, J. A. Shirley, and R. R. Stephens, *Appl. Phys. Lett.*, 17:278, 1970; J. A. Glaze, J. Finzi, and W. F. Krupke, *Appl. Phys. Lett.*, 18:173, 1971; T. V. Jacobson and G. H. Kimbell, *Chem. Phys. Lett.*, 8:309, 1971; M. C. Lin and W. H. Green, *J. Chem. Phys.*, 53:3383, 1970; D. S. Spencer, H. Mirels, and T. A. Jacobs, *Appl. Phys. Lett.*, 16:384, 1970; O. R. Wood et al., *Appl. Phys. Lett.*, 18:112, 1971.

Laser, dye

Since their discovery in 1966, dye lasers have been increasingly studied in various laboratories throughout the world. This is largely due to the fact that these lasers possess an unusual degree of tunability. Because the emission bands of fluorescent dyes are very broad, the lasing wavelength of a given dye may be tuned to any chosen value within a wide, continuous spectral region. Furthermore the number of fluorescent dyes is very large, and various compounds may be selected to emit coherent light in any desired portion of the optical spectrum lying between ~ 1.17 μm in the infrared and ~ 0.34 μm in the ultraviolet. The presently observed infrared limit will most likely remain the same in the future since no organic molecules are known to fluoresce beyond this wavelength. The upper frequency limit, on the other hand, will probably be extended at least to the region $\lambda \sim 0.25$ μm once suitable pumping sources are found.

Aside from their tunability, dye lasers are attractive subjects for scientific investigation because of the relative ease with which they permit experimentation to be carried out. The use of a liquid active medium simplifies the problem of obtaining high optical quality and makes possible cooling by means of flow techniques. Moreover the same dyes display lasing characteristics when dissolved in various plastics and even in thin gelatin films. Dye lasers are readily excited by flashlamps or by light beams from other lasers. The nitrogen laser, one of the dozen or so types of lasers that are now commercially marketed, has been shown by various groups over the past year to be a convenient pump for a wide variety of dye lasers. The short wavelength, 3371 A (337.1 nm), of the N_2 laser, coupled with the short, intense nature of the light pulses it emits (~ 10^{-8} sec duration, ~ 100 kW intensity), gives it the capability of efficiently exciting dye laser emission over almost the whole spectral range delineated above. The relatively high repetition rate with which the N_2 laser performs (~ 100 pulses per second) should make its use in conjunction with tunable dye lasers important for such applications as the monitoring of pollutants. *See* LASER, CHEMICAL.

Continuous-wave operation. The biggest advance in the field of dye lasers in the past year occurred when O. G. Peterson, S. A. Tuccio, and B. B. Snavely succeeded in operating the first continuous-wave (CW) dye laser. This event was the successful culmination of a series of advances made by these and other investigators who noted first of all that the problems associated with the accumulation of a large number of dye molecules in the metastable triplet state could be largely eliminated by the introduction of special quenching molecules into the dye solution. These quenching molecules interact with the dye molecules and reduce the average number of them occupying the triplet state in the presence of steady-state pumping to levels such that the associated absorption no longer constitutes an overriding loss. Without the addition of quenching agents, the operation of dye lasers would be restricted to the pulsed regime.

O. G. Peterson and coworkers found a special detergent (Triton X100) to be an effective triplet quencher for the molecule Rhodamine 6G, the dye with which the lowest lasing thresholds had previously been achieved. The detergent, moreover, was found to serve a dual purpose in that it also prevented unwanted dimerization of Rhodamine 6G molecules in water. Water is the liquid for which minimum optical distortions are produced by given temperature gradients, and for this reason it was the obvious choice of solvent for a CW dye laser.

The pumping source employed was a 1-W continuous argon ion gas laser which lases at 5145 A (514.5 nm), a wavelength that falls within the pumping band of Rhodamine 6G. The beam from the gas laser was focussed in an end-on manner onto a thin cell containing the circulating dye solution. Dye laser output powers as high as 100 mW were almost immediately obtained. Subsequently CW dye lasers of this type were made tunable by the introduction of various dispersing elements into the laser cavity. Figure 1 shows one arrangement used. The tuning range of a commercially available unit is given in Fig. 2. The manufacturer of this particular laser asserts that it operates in a single TEM_{00} transverse mode with a long-term stability of ~ 0.004 A (300 MHz) and a short-term stability of ~ 0.004 A (30 MHz). Thus it appears that an entirely new standard of resolution now exists for ordinary visible spectroscopy. It is highly probable that CW gas lasers emitting at wavelengths other than 5145 A (514.4 nm) will be used in the near future to extend the spectral range of CW dye laser emission.

Image amplification. It has long been recognized that laser amplifiers should be able to transmit and intensify optical images. Some experimental work, in fact, was done several years ago with

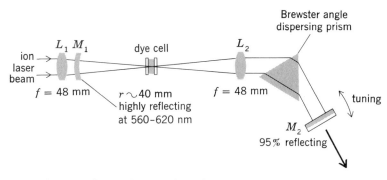

Fig. 1. A tunable CW dye laser configuration.

ruby laser amplifiers, although the gain and resolution obtained were only moderate. Recently a group from Stanford University reported on experiments with wide-angle dye laser image amplifiers in which single-pass gains of 23 dB/mm and diffraction-limited spatial resolution were achieved.

Two cells containing similar organic dye solutions were used, both pumped by the same nitrogen laser. One of these cells was designed to be a relatively weak oscillator and served as a source of illumination. The other, with antireflection-coated windows sealed to the ends under a wedge angle of 10° to avoid multiple reflections, served as the amplifier. A pair of lenses was employed to gather the weak light radiated from the oscillator and focus it onto a small but heavily pumped region of the amplifier cell. Here the light was amplified in a single pass and emerged as a uniform, bright cone from the other end. When an object, such as a fine mesh or a photographic transparency, was placed in the object plane of the projecting lens, it produced a bright image on a screen placed after the amplifier cell, in spite of the fact that it was only weakly illuminated by the light from the oscillator cell. (Sometimes a light diffuser was placed between the oscillator cell and illuminated object. This produced no essentially different results.)

To test the image quality, a slide with a standard TV test pattern was inserted in the object plane and a resolution of up to 200 line pairs per 0.4 rad was obtained, corresponding to the diffraction limit of a circular aperture the size of the pumped spot on the amplifier cell. The measured background

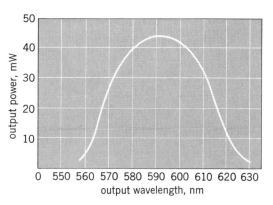

Fig. 2. Dye laser output power versus wavelength for 1-W pump power at 514.5 nm.

emission produced by the amplifier when the light from the oscillator cell was completely blocked corresponded to the proper theoretical value obtained by multiplying the channel capacity of the laser amplifier (number of distinguishable diffraction-limited spatial directions times the number of distinguishable light frequencies of $\sim 10^{-8}$ sec pulses contained within the bandwidth of the amplifier) by the gain of the amplifier. The conclusion was made, therefore, that the dye laser image amplifier behaves in an almost perfectly calculable fashion and that it offers promise for use in TV and color movie projection—the latter with the possibility of greatly reduced frame size—and in high-speed optical information processing. It will be interesting to see if this work does, indeed, stimulate the actual development of any new important devices.

New feedback mechanism. An experiment conducted within the past year utilized organic dye laser action in thin gelatin films in demonstrating a new method for providing optical feedback in laser oscillators. While the essential results of the experiment are quite general and reflect a current intense interest in integrated optical devices, the fact that the experiment was carried out using an organic dye laser medium attests once more that these lasers have an inherent simplicity about them which has drawn, and is likely to continue to draw, the interest of various investigators.

In essence, the experiment of H. Kogelnik and C. V. Shank succeeded in doing away with the usual optical resonator formed by two end mirrors terminating a laser medium, replacing it with a feedback mechanism distributed throughout, and integrated with, the active laser medium. The actual feedback mechanism effective in their experiment was Bragg scattering from a periodic spatial variation of the refractive index of the medium. This scattering was induced by exposing dichromated gelatin to the interference pattern produced by two coherent beams from a gaseous laser and then developing the gelatin with techniques normally employed in holography. The developed gelatin film was next soaked in a solution of Rhodamine 6G to allow the dye to penetrate and then mounted on a glass slide. The beam from a nitrogen laser was used to excite transversely the thin-film dye laser. The latter was observed to produce an output beam comprised entirely of sharp lines, whereas normally a dye laser with broadband mirrors emits radiation that is rather diffuse spectrally. It is apparent that the distributed feedback in a structure of this sort behaves like a grating in that it causes spectral filtering.

As in the case of the image-intensifying experiment described previously, it is too early to predict whether this result will be viewed ultimately as just an interesting essay or whether it will be credited as being one of the harbingers of a new explosion in optoelectronics. The answer partly depends upon the severity of the materials problem. Such unknowns as the stability of dyes for long-term use will have to become better understood.

For background information *see* FLUORESCENCE; LASER; OPTICAL PUMPING in the McGraw-Hill Encyclopedia of Science and Technology.

[PETER P. SOROKIN]

Bibliography: T. W. Hänsch, F. Varsanyi, and A. L. Schawlow, *Appl. Phys. Lett.*, 18:108, 1971; H. Kogelnik and C. V. Shank, *Appl. Phys. Lett.*, 18:152, 1971; O. G. Peterson, S. A. Tuccio, and B. B. Snavely, *Appl. Phys. Lett.*, 17:245, 1970; P. P. Sorokin, *Sci. Amer.*, February, 1969.

Laser, metal-vapor

Recent developments in the control and use of metal vapors have resulted in several new types of visible, ultraviolet, and near-infrared lasers. These metal-vapor lasers are a special class of gas lasers in which the solid or liquid form of the metal is vaporized in the gain region of the laser to form a gaseous medium. In most cases the population inversion (which is necessary in order to have gain) between two excited energy states of the metal atoms is provided by means of an electrical discharge. The most important lasers of this type are the helium-cadmium laser and the helium-selenium laser. *See* LASER, CHEMICAL.

Metal vapors have not been studied as extensively as a source for lasers as have other gases because of the difficulties in working with and controlling the vapors in an electrical discharge. Also, many of the metals require very high temperatures in order to obtain the necessary vapor pressures, and many of the vapors react with the discharge-tube walls, forming unwanted compounds. The advantages, however, in using metal atoms are the existence of low ionization potentials and many low-lying energy levels in both the neutral and ionized states of the gas. Many of these levels are suitable for laser action, and the fact that relatively low energies are necessary to excite them means that the laser process can be relatively efficient.

As a result of overcoming some of the problems in working with metal vapors, there now exists several metal-vapor lasers with high powers, gains, and efficiencies that have advantages over other types of lasers. These lasers can be divided into two categories: the continuously operating or continuous-wave (CW) metal-ion lasers and the pulsed, high-gain, self-terminating lasers. They will be discussed separately in the following paragraphs.

CW metal-ion lasers. The CW lasers that so far have been developed use the vapors of cadmium, selenium, zinc, lead, and tin. Laser action occurs in the singly ionized states of these materials, and in most cases a buffer gas of helium is used. The energy supplied to excite the upper laser level in most cases is provided by transferring energy from long-lived excited states of the helium atoms and ions in the discharge. Thus laser action usually occurs at relatively high helium pressures (3–10 torr or 400–1300 N/m²), where there are large numbers of these excited states, and at relatively low partial pressures of the metal vapor (0.001–0.01 torr or 0.13–1.3 N/m²) so that the electron energies in the discharge can remain large enough to excite efficiently the helium levels. The transfer of energy from the excited helium levels is an effective way of creating a population inversion by preferentially exciting the upper laser level of the metal ions.

The most important laser of this class is the helium-cadmium laser, which provides a power output of up to 250 mW in the blue region of the spectrum at 441.6 nm (4416 A) and up to 50 mW at 325.0 nm (3250 A) in the ultraviolet (this is the shortest wavelength yet attained for a CW laser). The gain at 441.6 nm is approximately 15–20% per meter in 2- to 4-mm-bore discharge tubes at optimum currents of from 100 to 200 mA and typical voltage drops at 2500 volts/meter. The highest gain for the laser is achieved by using a single, even isotope of cadmium which has been separated from the naturally occurring distribution of isotopes. The naturally occurring cadmium can also be used in the laser, but in many cases the increased laser power resulting from the single isotope warrants the additional expense. The maximum efficiency achieved with this laser is 0.05% which is comparable to the best values from other visible gas lasers.

The most convenient way to distribute the cadmium in the laser tube is to use a technique known as cataphoresis pumping. In using this technique, a single source of cadmium metal is located near the anode end of the discharge tube (Fig. 1). The cadmium metal is heated until a sufficient amount of cadmium vapor is flowing into the discharge region. The metal atoms would normally tend to diffuse through the helium gas to the cooler regions at both the anode and cathode ends of the laser tube, where they would condense again as a solid metal. However, when an electrical discharge is established between the electrodes, there are many positively charged cadmium ions created in the discharge region. These ions are pulled toward the cathode by the electric field, thus creating an effective flow of metal atoms from anode to cathode. This flow tends to distribute the metal atoms uniformly in the bore region where the gain is produced. With this technique, the cadmium is used up at a rate of approximately 1–2 mg/hr in a typical laser.

The helium metastable atoms populate the upper laser levels in a manner similar to the way they excite the well-known helium-neon gas laser. In the case of the metal atoms, however, the ionization energy is low enough so that the atom is ionized and an electron is emitted (Fig. 2). This process is known as a Penning ionization. The emitted electron takes up the energy difference between the helium metastable level and the cadmium ion level and thus the collision process is not restricted to levels lying close to the helium metastable levels, as in the case of the helium-neon laser (where the atom is not ionized). The cross sections for this ionization process have been found to be relatively large and, no doubt, are a significant factor in the efficiency of the helium-cadmium laser.

The wavelengths of the helium-cadmium laser are particularly useful for many applications. They occur near the maximum quantum efficiency for most photomultipliers. They are particularly suitable for experiments in photobiology, photochemistry, holography, and Raman spectroscopy. In addition, the simplicity of construction and operation provides a relatively low cost factor for potential applications. See SPECTROSCOPY.

More recently the helium-selenium laser has been developed and promises to be as important as the helium-cadmium laser. This laser produces

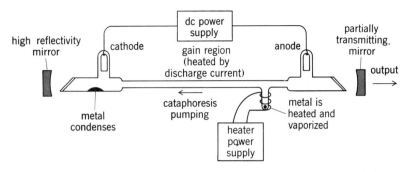

Fig. 1. Diagram of CW metal-vapor laser showing cataphoresis pumping technique.

CW laser action on 46 wavelengths, covering a very broad visible and near-infrared spectrum ranging from 446.7 nm (4467 A) to 1260 μm. The strongest laser wavelengths are in the blue and green regions of the spectrum, with output powers of up to 50 mW per line and total combined powers of 250 mW. As many as 19 of these lines have been observed to lase simultaneously over a spectral range of from 460.4 to 644.4 nm (4604 to 6444 A) using broadband high-reflectivity mirrors. The optimum current for the helium-selenium laser is approximately 400 mA in a 3-mm-bore discharge, and thus no additional cooling requirements are

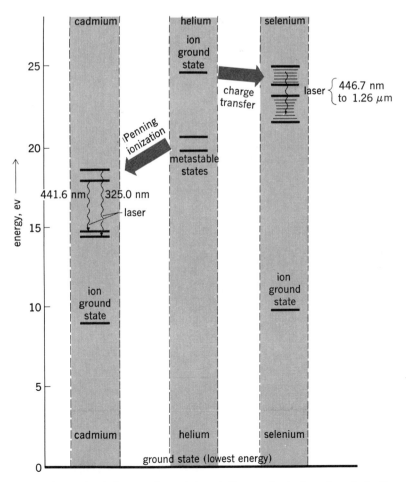

Fig. 2. Energy level diagrams for cadmium, helium, and selenium atoms indicating the energy transfer processes from excited helium states to the laser levels.

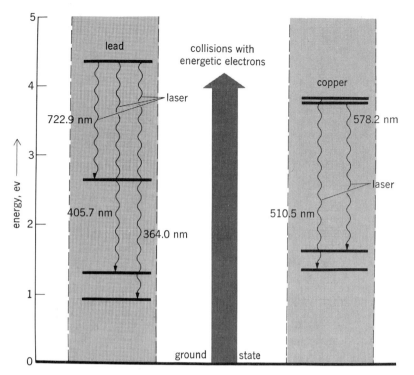

Fig. 3. Energy level diagrams for lead and copper atoms indicating the pulsed electron excitation process to the upper laser levels.

The lead laser was the first laser of this type and has the highest measured gain for a gas laser. The laser output at 722.9 nm (7229 A) has a pulse duration of 10–20 nsec, and the measured single pass gain is 6 dB/cm in a 1-cm-bore, 10-cm-long discharge. Other laser transitions at 405.7 and 364.0 nm have also been observed.

Another laser of this type is the copper-vapor laser, which has produced up to 40 kW at 510.5 nm (5105 A) although the maximum gain is lower than that of the lead laser. The lead and copper lasers require operating temperatures of 1000 and 1500°C, respectively, and thus are not as practical as some other laser systems. Other metals in which this type of laser has been successfully demonstrated include calcium, manganese, and strontium.

For background information see LASER in the McGraw-Hill Encyclopedia of Science and Technology. [WILLIAM T. SILFVAST]

Bibliography: W. T. Silfvast, Appl. Phys. Lett., 15:23, 1969; W. T. Silfvast and J. S. Deech, Appl. Phys. Lett., 11:97, 1967; W. T. Silfvast and M. B. Klein, Appl. Phys. Lett., 17:400, 1970; W. T. Walter, Bull. Amer. Phys. Soc., 12:90, 1967.

Laser, semiconductor injection

The use of heterojunctions in several new types of semiconductor lasers has resulted in a drastic reduction in the room-temperature current required per unit area through the device to initiate lasing, that is, the current threshold. This threshold for lasing is now down to about 1000 A/cm² (100 nm/cm²), compared to ~ 30,000 A/cm² (3000 nm/cm²) several years ago, and the first continuous operation of a pn junction laser at room temperature and above has been achieved as a result.

Homostructure laser. The high current thresholds of the best available junction lasers which did not contain heterojunctions were due to failure of both optical and carrier confinement. Such lasers, now called homostructure lasers, are made of a single semiconductor material, such as gallium arsenide (part a of the figure).

Although contacts are not shown, the four types of laser diodes are illustrated as though current were flowing. The distance, injected electron concentration, and light scales are arbitrary.

Electrons are injected across the pn junction under forward bias. These electrons diffuse some distance into the p-region before recombining with a hole. This distance is a function of the doping of the p-region and it increases with temperature. Thus the injected electrons recombine in a rather large unconfined volume which cannot be varied independently of doping or sharply defined. The light produced by the recombination in homostructure lasers tends to leak out of the lasing region because in that structure it is not confined to a well-defined optical waveguide. A weak and leaky waveguide is formed by the small refractive index step at the pn junction and a second small step presumably resulting from the electron population inversion in the p-region adjacent to the pn junction. At room temperature the homostructure lasers must always be operated pulsed. Pulsed power outputs of as much as 100 W have been achieved with very low duty cycles and pulse lengths of about 100 ns.

needed. The laser operates very well using the cataphoresis pumping scheme to provide a uniform vapor pressure in the active bore region. In addition, no special isotope mixtures are necessary for optimum laser power.

The helium-selenium laser is believed to be excited by charge-exchange collisions between helium ions and neutral selenium atoms, leaving the selenium ions preferentially in the upper laser level, as shown in Fig. 2. As in the case of the helium-cadmium laser, this process appears to be relatively efficient due in part to the moderate excitation energies necessary to excite helium ions.

The simplicity of construction and operation are also important features of this laser. Potential uses include holography, Raman spectroscopy, color displays, and broadband absorption spectroscopy.

The zinc, lead, and tin lasers have CW wavelengths ranging from green to infrared, but most of these lasers are of relatively low power and are thus not presently competitive with other available lasers.

Pulsed metal-vapor lasers. The high-gain, pulsed metal-vapor lasers are inherently self-terminating lasers. The laser energy levels involved lie very close to the ground state or lowest energy level of the atom (Fig. 3) and are populated by very short pulses of current in the discharge tube. With a suitable level scheme such as in lead or copper vapor, the upper laser level is very much more favored by electron excitation than the lower laser level. Thus an initial population inversion is created and a sharp pulse of laser light is generated. During the lasing process, all of the upper-laser-level atoms are stimulated to the lower laser level until the gain is extinguished. At this point in time, the laser action terminates and the system then waits for another current pulse.

Single-heterostructure laser. A great improvement in junction laser thresholds resulted from the discovery that the heterojunction between gallium arsenide and aluminum gallium arsenide does not have a large density of charge carrier traps associated with it. This is presumably a result of an almost perfect lattice match between these two semiconductors. Furthermore aluminum gallium arsenide has a wider bandgap than gallium arsenide, its width being a function of the amount of aluminum present in the crystal. At a heterojunction between p-type gallium arsenide and p-type aluminum gallium arsenide (a pp heterojunction), there is a step in the conduction band at which electrons may be reflected. The single heterostructure laser (part b of the figure) has, in addition to the planar pn junction in gallium arsenide which was present in the homostructure laser, a pp heterojunction (between p-gallium arsenide and p-aluminum gallium arsenide) which is coplanar to the pn junction and a short distance, usually less than 2 μm away. The region between the two junctions is the active region.

When the resulting diode is forward-biased, electrons are injected into the active region. They cannot penetrate their full diffusion length into the p-region because they encounter the pp heterojunction, where they are reflected. The electrons are thus confined to the active region, and when they recombine to produce photons, the recombination occurs in that restricted region. Since a smaller volume must now be brought to the electron density required for lasing than in the homostructure laser, the threshold current for lasing is correspondingly reduced. Furthermore the index of refraction of aluminum gallium arsenide is lower than that of gallium arsenide. As a result there is a refractive index step at the pp heterojunction which prevents leakage of the optical field at the heterojunction and permits further reduction in the threshold. Although some light is lost from the active region by penetration of the optical field into the n-type gallium arsenide and although, for small active region widths, holes may be injected into that region with a resulting threshold increase, single heterostructure lasers with room temperature thresholds as low as 8000 amp/cm² have been achieved. These lasers have not been operated continuously at room temperature, but the threshold reduction has permitted high-power pulsed operation. The peak power output is limited by catastrophic damage to the mirror surfaces when optical power densities reach about 10^6 to 10^7 W/cm² of emitting area. (Typical diode laser emission areas are of the order of 10^{-5} to 10^{-7} cm².)

Double-heterostructure laser. A further considerable reduction in current thresholds results if an additional heterojunction is located at the pn junction, that is, the formation of a pn heterojunction (part c of the figure).

The important part of this structure, which is called a double heterostructure, is an aluminum gallium arsenide – gallium arsenide – aluminum gallium arsenide sandwich comprising three layers on the substrate. There is a pn heterojunction between the first two layers, and a pp heterojunction between the second two. When forward-biased enough for electrons to be injected into the second layer (the active region), electrons are reflected at

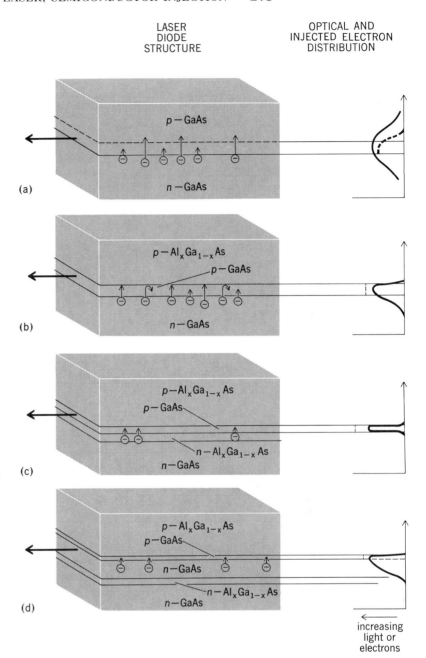

Schematic illustration of four diode laser structures and the injected electron and optical power distribution across each structure. (a) Homostructure. (b) Single heterostructure. (c) Double heterostructure. (d) Large optical cavity. Solid lines represent optical distribution, and broken lines injected electron distribution.

the pp heterojunction and holes at the pn heterojunction. Thus at any voltage near that needed to start lasing there is no hole injection. There are now steps in the refractive index at both the pp and pn heterojunctions, and the light penetration out of the active region is restricted on both sides. In fact, the active region is now an efficient waveguide with well-defined boundaries composed of the two heterojunctions. This structure has, for widths of the active region as small as several tenths of a micrometer, almost complete carrier and optical confinement.

Lasing current threshold densities as low as about 1000 amp/cm² have been obtained at room temperature with the double heterostructure. With an adequate heat sink, such lasers have been operated continuously at and above room temperature with continuous power levels in the tens of milliwatts range. Moreover, with further development, such continuous-wave lasers are expected to be useful as generators of wide-band signal carriers for optical communications. The very narrow active region width used for the lowest threshold operation may be disadvantageous for high-power pulsed operation because the limiting optical power density at the mirror is reached at lower total power levels with these lasers than with lasers with wider active regions.

Large optical cavity laser. Since there are some applications for which high-power pulsed operation is desirable, but for which lower thresholds than those achieved with homostructure lasers are needed, a heterostructure laser which separately confines the carriers and light has been made.

This laser (part d of the figure) is essentially a combination of the single and double heterostructure lasers described above, and consists essentially of a single heterostructure with an nn heterojunction located as illustrated, a short distance from the pn junction on the n side of the structure. In it the electrons are confined between the pn homojunction and the pp heterojunction. The optical field of the recombination light extends past the pn junction as it did in the single heterostructure. The n-type material does not seriously absorb the radiation and it is confined at the nn heterojunction. The result is that separate confinement of electrons and light is achieved, the optical confinement region being rather wide. Such devices show promise for high-power pulsed operation with much reduced catastrophic damage to the mirrors as a result of the much larger mirror surface through which the radiation emerges.

For background information see BAND THEORY OF SOLIDS; LASER; SEMICONDUCTOR in the McGraw-Hill Encyclopedia of Science and Technology.

[MORTON B. PANISH]

Bibliography: Z. I. Alferov et al., Investigations of the influence of the AlAs-GaAs heterostructure parameters on laser thresholds and currents and the realization of cw emission at room temperature, *Physica E. Tech. Poluprov.*, 4:1826, 1970; I. Hayashi et al., Junction lasers which operate continuously at room temperature, *Appl. Phys. Lett.*, 17:109, 1970; I. Hayashi, M. B. Panish, and F. K. Reinhart, GaAs-Al$_x$Ga$_{1-x}$As double heterostructure injection lasers, *J. Appl. Phys.*, 42:1929, 1971; H. Kressel, H. F. Lockwood, and F. Z. Hawrylo, Low threshold LOC GaAs injection lasers, *Appl. Phys. Lett.*, 18:43, 1971.

Leaf

Recent studies have been carried out to further explain the poorly understood symbiotic relationship between bacteria and leaves of certain woody tropical and subtropical members of the dicotyledonous families Rubiaceae and Myrsinaceae. Even though this association has been recognized for about 90 years, the greater volume of work on bacteria-plant symbioses has been concerned with the many economically or ecologically important root-nodulated plants, particularly the legumes. Early studies on leaf-nodulated plants have left unsolved the problems of the role and identification of the bacteria and exactly how the bacteria are passed from one generation to the next. Some workers have suggested that the bacteria are within the plant throughout its complete life cycle and are carried into the next generation by way of the seed. The bacteria appear to serve a dual purpose: They fix atmospheric nitrogen like the root-nodule bacteria, but unlike root-nodule bacteria the leaf-nodule bacteria supply the plant with one or more substances which are necessary for normal plant growth. Most of the anatomical and bacteriological studies related to leaf-nodule symbiosis have been restricted to only a few species of a few genera, namely, *Psychotria*, *Pavetta*, and *Ardisia*.

Location of bacteria within plant. Through the use of serology and light and electron microscopy, the bacteria have been conclusively demonstrated within the terminal and lateral vegetative buds, flower buds (including the ovule—the future seed), and young, expanding, and mature leaves. The bacteria invariably are found either between plant parts in a mucilaginous secretion or in intercellular spaces but never within normal cells. Concentrations of the bacteria (within closed cavities) appear in leaves as visible, spherical, or oblong black spots called leaf nodules. There are numerous related species (for example, *Coprosma*) that have open cavities (domatia) containing bacteria. However, it is not known whether these bacteria and their relationship to the plant are similar to the leaf-nodule bacteria.

Bud relationship. Both terminal and lateral vegetative buds act as bacterial reservoirs. This closed bud system, due to the compactness of young overlapping bud parts (leaf primordia), is filled with a mucilaginous substance (produced by multicellular secretory cells) in which the bacteria reside. The bacteria are rod-shaped, capsulated, and very similar to those found in the floral buds. The young leaves in the bud develop precocious stomates (two guard cells and pore) through which some bacteria pass into small cavities, and establish potential future nodules. It is not known whether each site of bacterial entrance develops into a nodule. When the buds change from the vegetative to reproductive state, the bacteria are carried into the floral parts and apparently are incorporated into the seeds (Fig. 1).

Leaf relationship. The young leaves expand out of the bud, carrying bacteria in closed cavities along with mucilage and bacteria adhering to their surfaces. It is not known whether surface bacteria can later enter the leaf to produce nodules or whether all the potential nodules are initiated while the young leaves are still in the bud. As each leaf increases in surface area, however, more nodules appear, until, depending on the species, several to a few hundred nodules are formed either at random or in a specific arrangement; for example, along the midrib and other veins or on the leaf margin (Fig. 2).

The substomatal cavities (Fig. 3a), where the bacteria initially enter the leaf enlarge while the surrounding mesophyll cells divide to form a network of cells with numerous intercellular spaces (Fig. 3b). A compact layer of mesophyll cells forms at the periphery of the network and apparently acts as a physical barrier to the spread of bacteria throughout the leaf (Fig. 3c). Each nodule is usually in close contact with a portion of a major or minor vein.

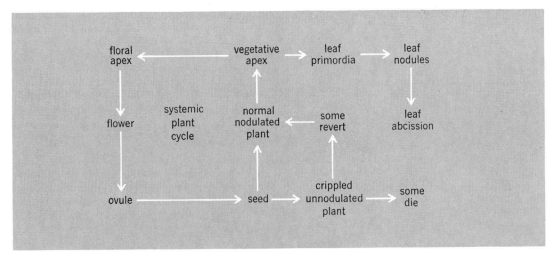

Fig. 1. Proposed cycle of bacterial transmission in leaf-nodulated plants.

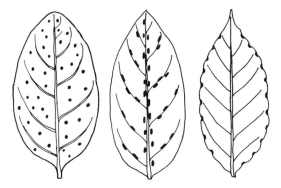

Fig. 2. General arrangements of leaf nodules in nodulated members of the Rubiaceae and Myrsinaceae.

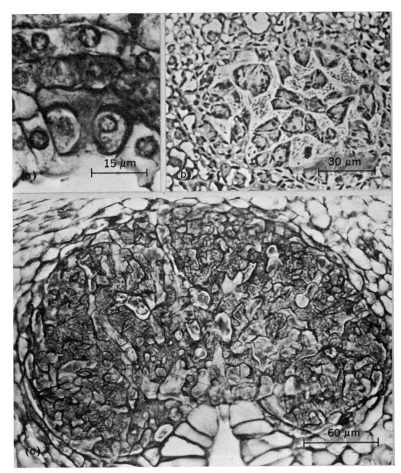

Fig. 3. Light micrographs demonstrating three stages of leaf-nodule development. (a) Stomate and nodule initial (cavity) in leaf primordium. (b) Cross section of young nodule with visible bacteria. (c) Mature nodule.

The bacterial population, at first small (Fig. 4a), continues to increase in size and eventually fills the intercellular spaces of the nodule network (Fig. 4b). The bacteria become more irregular in shape than observed in the buds and no longer are surrounded by capsules. The increased surface area afforded by the cellular network probably facilitates the interaction between the bacteria and their host plant. Each network cell secretes large quantities of carbohydrate material which is utilized by the bacteria. Once the nodule stops increasing in volume, the bacterial population is limited in growth and begins to show signs of degeneration (such as in increasing numbers of bacterial ghosts and masses of bacterial membranes), presumably because of crowding and waste accumulation. Degeneration continues until near the time the leaves fall off the plant (abcission); by this time there are no recognizable bacteria within any of the nodules (Fig. 4c). These observations have been confirmed by the use of a fluorescent microscopic technique localizing ribonucleic acid (RNA) in young, mature, and senescing leaves. Bacterial RNA fluorescence is highest in young and expanding leaves, decreases (beginning in each nodule center) in mature leaves, and finally is absent in leaf nodules near the time the leaves drop from the plant. It is assumed, therefore, that the greater contribution to the symbiosis occurs in the buds and young leaves when the bacterial population is active and growing.

Identification of bacteria. Early workers characterized a number of different bacteria inhabiting the various leaf-nodulated species. The more recent and specific application of the antigen-antibody (serological) technique, using bacterial isolates from germinating seeds of *Psychotria* and *Ardisia*, has shown that the same bacterium is

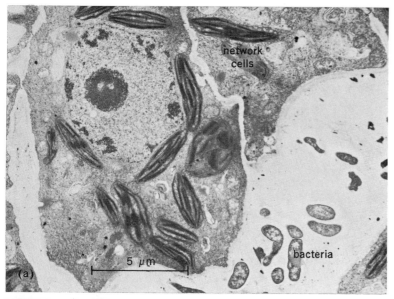
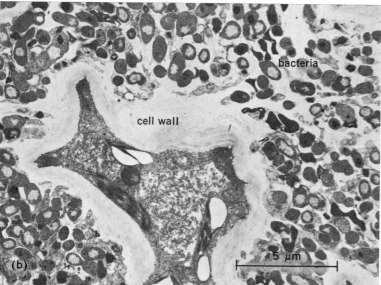
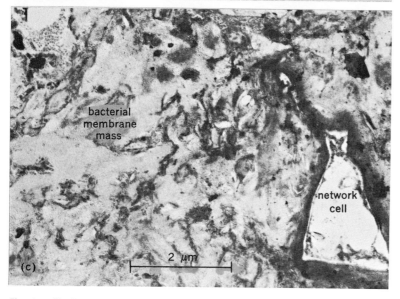

Fig. 4. Electron micrographs of portions of the interior of leaf nodules at three stages of development. (a) Young nodule with few bacteria. (b) Mature nodule filled with bacteria. (c) Old nodule with degenerated bacterial membrane mass.

present in all leaf-nodulated species tested from the genera *Psychotria*, *Pavetta*, and *Ardisia*. However, bacteria have not been successfully isolated and cultured from the leaf nodules. Numerous bacteriological and culture techniques were used to place the bacterium in the *Chromobacterium lividum* group (Rhizobiaceae), but the bacterium is not identical to those normally found in root-nodulated species.

So far it has been impossible to satisfy Koch's postulates which require that the cultured bacteria be reintroduced into a plant free of nodules (bacteria) to produce nodules from which the bacteria can be isolated and characterized. The difficulty lies in obtaining normal plants free of bacteria for reinfection. A certain percentage of seeds from normally nodulated plants do produce so-called "cripples" which are unnodulated and grow poorly (Fig. 1). These cripples either die or revert to normal nodulated plants after some time. The problem with using the crippled plants, then, is that it is impossible to ascertain whether they will revert to the nodulated condition on their own. The production of unnodulated cripples that die, however, has led a few workers to speculate that the bacteria may, in fact, contribute some substance or substances that are necessary for normal plant growth.

Callus cultures. Because of the continued difficulty in obtaining normally nodulated, bacteria-free plants to test Koch's postulates, current work is underway to develop bacteria-free plants from callus cultures. Successful calli, on defined media, have been established from stem internodes of *Psychotria punctata* and they have been carried through a number of subcultures. Roots have also been produced from these calli. However, chemical modifications of the medium have not induced bud and whole plant formation as described for other tissue culture systems. The *Psychotria* callus shows unusual responses to extreme modifications of the medium, including a high requirement for cytokinin and a lack of dependence on exogenous sources of auxin (both control growth in normal plants). Further study of these results may be useful in interpreting the bacterial contribution necessary for normal plant growth. For example, if plants can be formed from the callus by finding the correct proportion of defined chemical ingredients, the next step would be to remove the plants from the medium and inoculate some plants, but not others, with the cultured bacteria. If the symbiosis is obligate, two states would be produced experimentally: normal growth with the bacteria, and death without the bacteria.

For background information *see* LEAF (BOTANY); RHIZOBIACEAE in the McGraw-Hill Encyclopedia of Science and Technology.

[HARRY T. HORNER, JR.]

Bibliography: K. A. Bettelheim, J. F. Gordon, and J. Taylor, *J. Gen. Microbiol.*, 54:177–184, 1968; H. T. Horner, Jr., and N. R. Lersten, *Amer. J. Bot.*, 55:1089–1099, 1968; C. E. LaMotte and N. R. Lersten, *Amer. J. Bot.*, in press; N. R. Lersten and H. T. Horner, Jr., *J. Bacteriol.*, 94:2027–2036, 1967; F. B. Sampson and J. McLean, *N.Z. J. Bot.*, 3:104–112, 1965; R. E. Whitmoyer and H. T. Horner, Jr., *Bot. Gaz.*, 131:193–200, 1970.

Marine influence on weather and climate

In recent years interrelationships between the atmosphere and the ocean (air-sea interaction) have absorbed the interest of an increasing number of meteorologists and oceanographers. Their studies are aimed at achieving the understanding required to predict the marine and atmospheric environment on time scales ranging from hours to decades or longer, and on space scales from a few kilometers to thousands of kilometers. An immense effort is being expended to learn more about the physics of small-scale processes by which heat, moisture, and momentum are transferred through the sea-air interface and within the atmosphere. An example of such an effort is the Barbados Oceanographic and Meteorological Experiment (BOMEX) conducted in the summer of 1969 with the help of ships, buoys, aircraft, balloons, and satellites. It is believed that the transfer of heat and water vapor from any portion of the sea can ultimately be inferred through physical models from widely scattered measurements of wind, barometric pressure, temperature, and humidity.

Another direction in air-sea interaction research aims at describing and understanding the macroscale processes which cause changes in the thermal character of the upper layer of the ocean down to a few hundred meters, and the related variations in the overlying atmospheric wind systems. By "macroscale" one refers to areas roughly a third of the size of the North Pacific Ocean. This macroscale attack is largely concerned with the interactions of "statistical aggregates," for example, the variations in position and intensity of the subtropical anticyclones (Bermuda and North Pacific

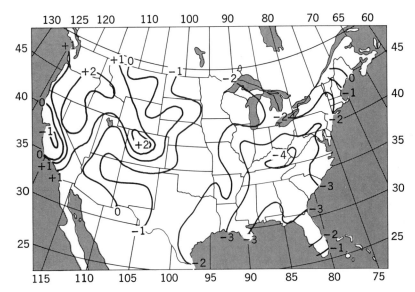

Fig. 2. Average surface temperature departure in degrees Fahrenheit from the 1931–1960 mean of the winters of the 1960s. December through February are defined as the winter months. (*From J. Namias, Climatic anomaly over the United States during the 1960's, Science, 170:741–743, Nov. 13, 1970*)

highs) and the underlying warming and cooling of the sea. Ultimately the small- and large-scale approaches described above should converge, providing scientists with the understanding to predict weather for periods of weeks and general climatic fluctuations for months, seasons, or decades in advance.

Anomalous patterns. If daily sea level pressure maps are averaged over monthly periods, certain large-scale features such as the Aleutian and Ice-

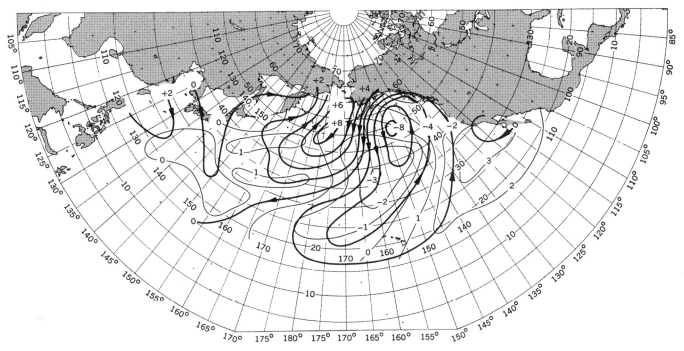

Fig. 1. January, 1959, anomalies of sea surface temperature (light lines, in degrees Fahrenheit) and anomalies of sea level pressure (heavy lines, in millibars), both from a 20-year mean beginning 1947. Arrows indicate anomalous component of resultant wind flow at sea level; 1 mb = 100 N/m². Note tendency of warm water masses to lie under anomalously southerly wind components, and cold water under northerly.

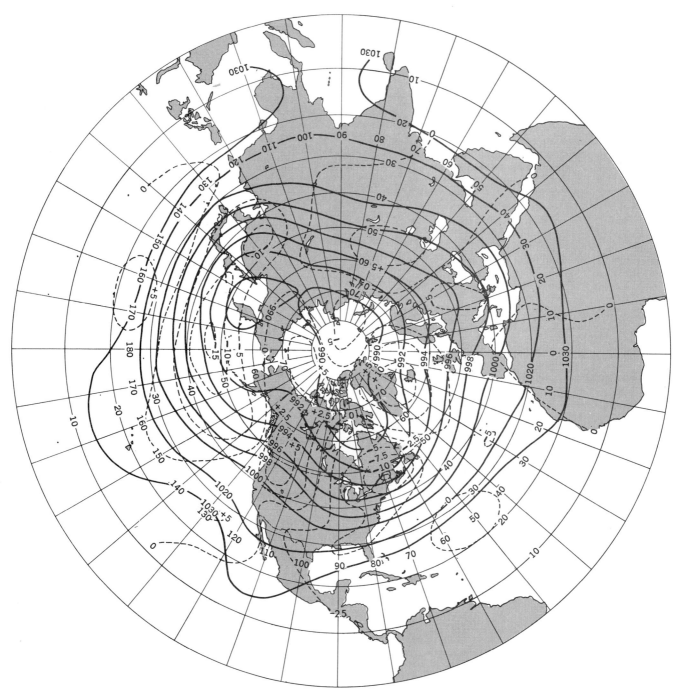

Fig. 3. The 700-mb height contours (solid lines) and isopleths of departure from the 1931–1960 mean (broken lines) for the winters of the 1960s. Contours and isopleths labeled in tens of feet; 1 ft = 0.3048 m. Winds blow along contours with lower height to left of flow. (From J. Namias, Climatic anomaly over the United States during the 1960's, Science, 170:741–743, Nov. 13, 1970)

landic lows come into sharp focus. These time-averaged pressure systems are highly variable in intensity, extent, and position in the same month or season of different years. By subtracting the long-period "normal" pressure field from that of a particular month, one obtains a field of isopleths of pressure anomalies (heavy lines in Fig. 1). The anomaly patterns may be used as indices of the strength and direction of wind flow relative to normal and therefore imply changes in the source and characteristics of the prevailing air masses during that month. Recent work with the help of computers has made this procedure quite objective.

The temperatures at the sea surface are largely controlled by the amount of solar heating and by the amount of heat transferred to the atmosphere—by conduction of sensible heat but more importantly (almost three times greater) by the latent heat of evaporation. The temperatures, humidities, and wind velocities of the air masses all affect this heat exchange. Also important is the transport of water from normally warm areas into cold areas or vice versa. Variations in the rate and direction of transport are due to variations in wind-driven surface flow (the Ekman transport), the speed of the great ocean current gyres, vertical currents (upwelling or downwelling), and turbulent mixing processes in the ocean. Appreciable evidence has recently come to light that all these factors depend on the complex coupling between air and sea. Figure 1 shows an example of coupling between air and sea.

The thermal state of the sea surface, once it is altered by the atmosphere, will usually maintain its character for several months and often years. In this manner, reservoirs of heat or lack of heat are created. These reservoirs appear to have strong influence on the tracks and intensities of cyclones (low-pressure areas) and anticyclones (high-pressure areas). Therefore the sea surface temperature variations lead in their statistical aggregate to short-period climatic fluctuations. Recent studies indicate that areas of abnormal sea surface temperature and areas of strong contrast (gradient) are zones where atmospheric fronts and cyclones intensify. Heat and increased moisture are quickly transferred from warmer water to the overlying air. Prevailing storm tracks tend to be determined by oceanic temperature contrasts (baroclinicity). High-pressure areas seem to grow over abnormally cold water areas, particularly in fall, probably due to diminished frictional leakage of air from the surface layers of the anticyclones.

Meteorological impact. For reasons not yet entirely clear, the temperature pattern in the upper layers of the sea may remain abnormal in the same way (warm or cold) not only for a month or season, but reappear in the same season of successive years. Thus, during winters of the 1960s, North Pacific surface waters averaged from 1 to 3°F (1 to 2°C) warmer than during the first half of the century, and in addition a stronger than normal sea surface temperature gradient existed from California westward to Hawaii (with warmer anomalies off California than north of Hawaii). The associated meteorological impact of these two factors seems to have been (1) increased storminess (lower pressure) in the Central Pacific, which in turn has led to frequent winters with weak (or absent) trade winds in the Hawaiian Islands, (2) more frequent snows in the Atlantic states, and (3) abnormally cold winters over the eastern two-thirds of the coterminous United States but warmer than normal winters in the West (Fig. 2). This pattern of the 1960s still held sway during the 1970–1971 winter. How might this continental temperature pattern be generated by North Pacific atmospheric and oceanic interactions?

To answer this question, it must be realized that the prevailing air currents in most of the troposphere and lower stratosphere are in the form of roughly sinusoidal sweeping undulations with a general west to east drift. The northward and southward bulges (crests and troughs) have dimensions of the order of 5000 km from crest to crest. The jet stream, the high-speed core of these long waves, is usually found just below the stratosphere. If the amplitude of one long wave in the hemisphere is increased, other downstream waves will also be similarly affected. Thus, if storms in the central North Pacific are more intense, the cyclonic vorticity (spin) will spread aloft to intensify the upper level trough. This intensification will lead to amplification of the wave which normally has its crest over the western United States and its trough over the East, as has been the case in the 1960s (Fig. 3). The amplification over the United States causes an increased frequency of outbreaks of Arctic air associated with the stronger than normal prevailing northerly wind component. The coldness in turn leads to more frequent snows than rains over the northeast—providing a further refrigerating effect.

Also in the 1960s a turn to colder than normal weather has set in over much of Europe. By examining the broken lines of Fig. 3, it should be easy to see why this has occurred. The Icelandic Low is weaker than normal, mild maritime (Atlantic) air mass flow into Europe is weakened, and the European climate has become more continental in the 1960s. While the hypothesis is far from proved, it is possible that a complex chain of events hinging upon the North Pacific–North American interaction as described above could have influences as far away as Europe.

Finally some recent studies by J. Bjerknes have indicated that the overall strength of the prevailing west winds in temperate latitudes may be influenced by sea temperatures in equatorial waters, another facet in the complex chain of events causing short-period climatic fluctuation.

For background information see AIR PRESSURE; AIR TEMPERATURE; ATMOSPHERIC GENERAL CIRCULATION; MARINE INFLUENCE ON WEATHER AND CLIMATE in the McGraw-Hill Encyclopedia of Science and Technology.   [JEROME NAMIAS]

Bibliography: J. Bjerknes, *Mon. Weather Rev.*, March, 1969; J. Z. Holland, *Bull. Amer. Meteorol. Soc.*, September, 1970; J. Namias, *J. Phys. Oceanogr.*, April, 1971; R. A. S. Ratcliffe and R. Murray, *Quart. J. Roy. Meteorol. Soc.*, April, 1970.

Marine navigation

Recent advances in marine navigation include the development of a microwave system for precise position measurement of surface ships and a system of radio-navigation coverage for coastal fishing boats (talking beacon system). These two new developments are described in this article.

Precise position measurement. A large number of oceanographic and marine navigational special situations require precise positioning of surface ships relative to shore-based or buoy-marked reference points. For those situations which require all-weather accuracies in the order of 1 or 2 ft (0.3 to 0.6 m) but have available line-of-sight to reference points or towers, high-resolution trilateration radar techniques offer considerable advantages.

Basic theory. A basic feature of pulsed radar-derived systems is an inherent high degree of range-measurement resolution and accuracy. Using nanosecond-pulse techniques and corrections for local atmospheric refractivity, range-measurement accuracies and resolutions on the order of a few inches may be achieved out to distances of several miles. The radar system measures the time taken for propagation of the pulse to the target and back to the receiver. This time is measured to produce an accurate range. Range measurements made to fixed positions permit determination of target location through trilateration computations.

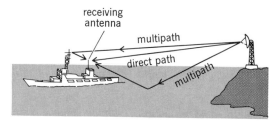

Fig. 1. Multipath propagation between a ship and the shore-based antenna. (*P. N. Migdal and A. J. Hannum, Precision position measurement for surface ships utilizing short pulse microwave transmission, J. Inst. Navig., vol. 17, no. 1, Spring, 1970*)

An interrogator-responsor unit is located aboard the ship the position of which is to be determined. Range measurements are made to two shore-based locations. Each location is equipped with an active signal transponder which transmits a short pulse in response to the short pulse received from the ship-borne equipment. The range measurements are applied to a computer which determines the position of the ship relative to the shore-based transponder antenna locations.

Figure 1 shows the propagation paths for the energy transmitted between the ship and the shore-based antenna. For highest accuracy it is necessary to take into account multipath propagation. Two common sources of multipath transmission are shown in this figure. The direct ray between the two antennas follows essentially an optical path. It is the length of this path which must be measured. However, multipath signals resulting either from ocean surface reflection or from super-structure or other interfering objects on the ship travel a longer path than the direct. Therefore, if the range measurement is influenced by the signals propagated over these longer paths, the result will be in error.

Figure 2 shows the effect of the multipath propagation upon received signals. Continuous-wave (CW) ranging systems are based upon the phase measurement of ranging tones or the radio-frequency (rf) carrier, and the received signal contains both the direct wave and all multipath components. The direct wave and one multipath wave are shown. These waves combine to give a net phase which produces an error in the measurement. The error magnitude is dependent upon the range difference between the direct and multipath signals and upon the relative strength of these signals. A short-pulse system can resolve the longer multipath components as separate pulses. With suitably designed range-measuring circuits, the leading edge of the first arriving pulse may be employed to the exclusion of later-arriving pulses. Thus, while the CW ranging systems may employ lower-power transmitters, a short-pulse system is less susceptible to multipath propagation effects.

Equipment. Figure 3 is a block diagram of the interrogator-responsor unit employed on the ship, and that of one of the two transponder units located at points ashore.

Range-measurement sequences are initiated by the shipboard computer, which sets the identification (ID) code for the first transponder the range of which is to be measured (R_1). The ID code consists of microsecond pulse code modulations transmitted at microwave frequencies. These are followed by nanosecond ranging pulses. Traveling-wave tube (TWT) amplifiers are utilized as transmitters and receivers in all of the equipments. These TWT amplifiers provide an octave of bandwidth, and therefore they accommodate the short-pulse gigahertz bandwidths used by the interrogation and response frequencies. Receipt of proper ID code at a transponder allows the short-ranging pulses to be frequency-translated and amplified for transmission back to the interrogator-responsor unit. A range-measurement circuit determines the time delay between a short pulse applied at the start input and a short pulse applied at the stop input. The start pulse is obtained by detection of the rf interrogation pulse transmitted by the interrogator-responsor. The stop pulse is derived from the short pulse received from the transponder retransmission. The range-measurement circuit produces, in digital form, a number equal to the range between the interrogator-responsor antenna and the transponder antenna, plus fixed internal and cable delays.

The technique used for deriving digital range measurements is as follows: In coincidence with the start pulse, an oscillator is caused to operate which has a period corresponding to 32 ft (9.8 m) of time delay. This mechanism is the coarse count portion of the measuring system. When the stop pulse is received, a second oscillator is initiated. It has a period corresponding to 31.5 ft (9.6 m) per cycle. The time intervals between the two oscillators thus form a vernier measurement system in which the fine increments are determined by the number of cycles required for coincidence between pulses representing the two time intervals. Thus, while the least increment of measure is 0.5 ft (0.15 m), the smallest time increment which must be resolved by the circuitry is 32 ft (9.8 m). This distance corresponds to approximately the 15-MHz counting rates. By this technique accurate digital range measurements can be obtained with fractional foot resolution and without requiring

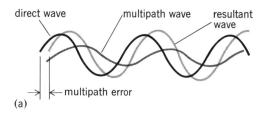

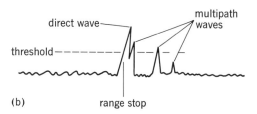

Fig. 2. Effect of multipath propagation upon employed signals. (*a*) CW ranging. (*b*) Short-pulse ranging. (*P. N. Migdal and A. J. Hannum, Precision position measurement for surface ships utilizing short pulse microwave transmission, J. Inst. Navig., vol. 17, no. 1, Spring, 1970*)

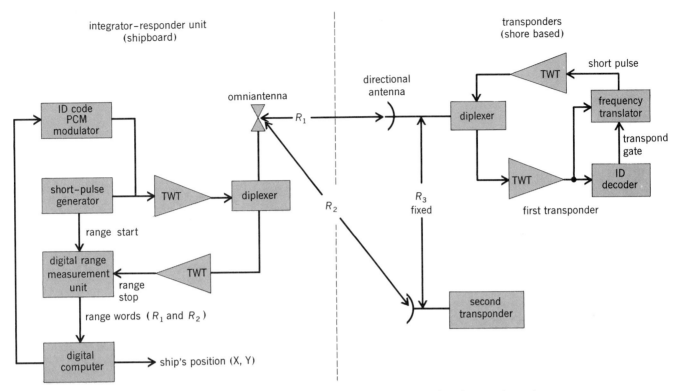

Fig. 3. Block diagram of interrogator-responsor unit on ship and one of two transponder units ashore. (P. N. Migdal and A. J. Hannum, Precision position measurement for surface ships utilizing short pulse microwave transmission, J. Inst. Navig., vol. 17, no. 1, Spring, 1970)

extremely high-speed electronic circuits.

To convert the range measurements to surface coordinates, a trilateration computation is required. This computation readily available by a general-purpose digital computer. The computer employs three destructive 1024-word readout modules of memory, and seven nondestructive readout modules for program and constant storage. Word length is 20 bits, including sign.

The maximum error estimated for a 5×5 mi (8×8 km) operations zone is 2.75 ft (0.8 m) rms. Generally the errors do not exceed 1.7 ft (0.5 m) rms over most of the operating zone.

[PHILIP N. MIGDAL]

Talking beacon system. A talking beacon system has been developed in Japan with the primary intent of providing radio-navigation coverage for important fishing grounds along the coast where fishing boats of less than 10 tons (9070 kg) operate. This system does not require that its users have technical operational knowledge or skill in its utilization. It offers, however, high accuracy with the advantage of low-cost receivers.

System characteristics. Basically the system requires two or more transmitting stations with a mutual connection through a radio network. A combination of three transmitting and monitor stations has been operating in the Tsushima Island area, north of Kyushu, Japan. They are connected over a 400-MHz radio network for operational monitoring and synchronization of the rotation of the transmitting antennas of each beacon. Three talking beacon stations are controlled as one group. They are unmanned, automatically operated, and arranged to provide the users with sufficient voice information for a cross fix. The transmitting antennas have the same specifications as those of marine radars. The transmission of a sharp beam, the power of which is pulse-width-modulated with the voice signals, is swept over an area by the rotation of the transmitting antenna. Since any increase in the rotating speed of an antenna causes difficulty in interpreting the information, each station is equipped with three transmitting antennas. These are structurally arranged with a difference of 120°. This arrangement increases the bearing information. The three antennas are in synchronization with the antennas of the two other stations. The sweep of the transmitting antenna with a sharp beam is in synchronism with those of the other two stations so that the users of this service do not simultaneously receive the signals of the other stations. Strict control of the rotation of the antennas is an essential factor in this system. If the rotation rate of each transmitting antenna were independent, the users could be confused by receiving voice signals simultaneously from one or two stations.

The three stations are given the respective color codes of red (*aka* in Japanese), black (*kuro*), and blue (*ao*) for identification. The transmission of the identification color codes of each station is followed by a voice transmission of the true bearing. These transmissions are made at intervals of every 2°. The verbal information is a series of the numbers which are arranged in order from zero in the true north or south. The bearing information is pulse-width-modulated with voice signals and transmitted in a sharp beam at 9310 MHz. The information arrangement, for example, is as follows: "... *aka* 34, *aka* 35, ...," "... *kuro* 55, ...," or "... *ao* 85, ..." and so forth.

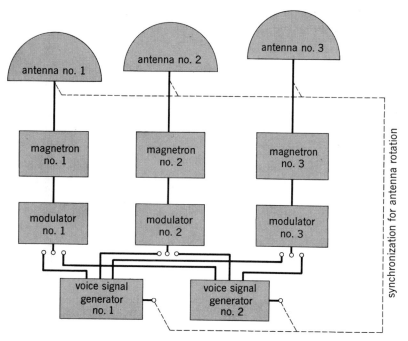

Fig. 4. Schematic diagram of the talking beacon transmitter.

receipt of the incoming synchronizing signal. The bearing information, which has an angle difference of 120°, is given to each of three modulators. The output of the modulators passes to each transmitting magnetron (9M63 type). The pulse-width-modulated signal of 9310 MHz is transmitted from each antenna. Specifications of transmitter and transmitting antenna are as follows:

Frequency	9310 MHz ±2%
Output	1.5 kW/peak
Modulation frequency	0.3–3 kHz
Type of modulation	Pulse width modulation
Antenna type	Slot array
Antenna rotating speed	1/3 rpm
Accuracy of antenna rotation	Less than 0.5°
Antenna directivity	Horizontal 2° ±10% Vertical approx. 15°

Utilization. To use this system, a talking beacon receiver and lattice charts are necessary for obtaining a fix. The radial lines of position (LOPs) are printed on the talking beacon lattice charts in colors corresponding to the identification codes of each station. With the identification code and bearing information, the users on board the ship can fix their position. Since the bearing information is verbal and directly available, it is not necessary for its users to have specific technical abilities. A talking beacon receiver measures 6.7 × 11.4 × 4.1 in. (17.0 × 28.9 × 10.4 cm) and is powered by four dry cells (UM-1 type).

Equipment. Identical electronic equipment is installed at each station. The schematic diagram of the talking beacon equipment is shown in Fig. 4.

Prerecording of verbal information with a female voice is made on a drum-type magnetic recorder. After reproduction and amplification by the voice signal generator, the amplified voice signals go to three modulators. The rotation of the magnetic drums synchronizes with the rotation of the transmitting antenna. The rotation of the magnetic drum of each station is also synchronized with each of the other two stations through the 400-MHz network as described above. To maintain the synchronization of rotation of three antennas of each station, the synchronization signal which is generated in one of the transmitting stations is transmitted to other stations to compare with the gate signal which is generated at each station. In synchronized conditions the antennas are in normal rotation because the received synchronization signals coincide with the gate signals generated at their own stations. In cases where an out-of-synchronization condition occurs between the synchronization and gate signals, the antenna rotation is temporarily stopped and then restarted upon receipt of the incoming synchronizing signal.

Error. When a ship is near a station, the possibility exists that the information received does not give single bearing information but gives more than two or three. Several lines of position, therefore, might be received. To solve this problem, the center value of the information received is made a reference as the position of cross bearing (Fig. 5a); that is, when ambiguity arises in distinguishing the most accurate LOP, the user would have difficulty in distinguishing B from A, B, and C. Center code B (in this example) is made a reference as the probable LOP (Fig. 5b). The maximum error involved is within 1° because a user receives the bearing information every 2°. Another minor error is introduced between the angle of antenna rotation and the transmitted bearing information. This error is held to within a negligible figure.

For background information *see* MARINE NAVIGATION; NAVIGATION; NAVIGATION SYSTEMS, ELECTRONIC in the McGraw-Hill Encyclopedia of Science and Technology. [TOHRU TADANO]

Bibliography: P. N. Migdal and A. J. Hannum, Precision position measurement for surface ships utilizing short pulse microwave transmission, *J. Inst. Navig.*, vol. 17, no. 1, Spring, 1970.

Morphactin

Among the synthetic compounds that affect plant growth, derivatives of fluorene have attracted special attention in recent years, mainly because they have an unusually powerful effect on the morphogenesis of higher plants over a particularly wide range of concentrations (about 10^{-2} to 10^{-7} M). The most active compounds, used as herbicides, are 9-hydroxyfluorene-(9)-carboxylic acid fluorenol and especially the 2-chloro derivative chlorfluorenol (Fig. 1). The collective name of morphactins

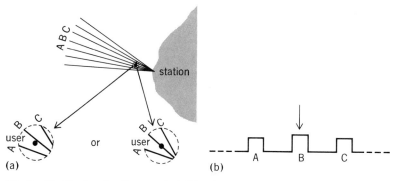

Fig. 5. Clarification for fix error in talking beacon system. (a) Expanded lines of position. (b) Received information.

(morphogenetically active substances) was suggested by G. Schneider for the whole group of fluorene–carboxylic acid derivatives.

Morphogenetic effects. Morphactins affect the adult organs of plants very little, although leaf abscission or shedding of the stem apex may be caused. Much more frequently it is the new growth that is affected: Stem internodes are shortened, leaf laminae are reduced, and apical dominance (suppression of lateral shoots by the apical bud) is deranged, so that branching is increased. Plants that have developed under the influence of morphactin show, therefore, a dwarf bushy habit (Fig. 2) with numerous morphological anomalies; but, protected from competition with other plants, they normally survive and may bloom (though sometimes with deformed flowers) and even form fruit. The effect of a single treatment with morphactin usually lasts only for a limited time, and generally the plants fairly soon resume normal growth.

Other noticeable effects are abnormal development of embryos within the seeds, inhibition of germination, stopping the spreading of rosette plants, change in the number and placing of flower buds, and a strong inhibition of lateral root formation.

These striking changes are produced by morphactins in multicellular plants with either apical cells or meristems. The division of unicellular organisms is not affected, or at least only at high dosages. In multicellular plants the rate of cell division in the meristems may be reduced greatly. Particularly noticeable and morphogenetically significant is a change, by no means rare, in the orientation of the spindle axis of the dividing cells. This may affect the growth form of the organ considerably and, together with the reduction of cell division, is probably responsible for the formation of abnormal dwarfs, which has already been described as typical of plants treated with morphactins.

The development of the cross wall which even-

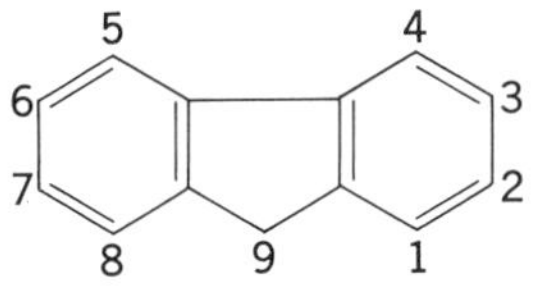

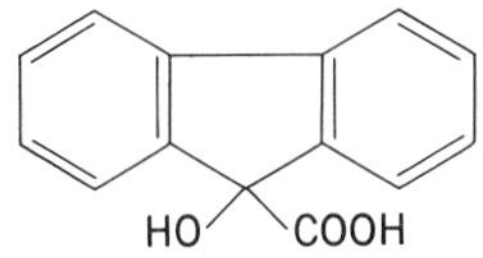

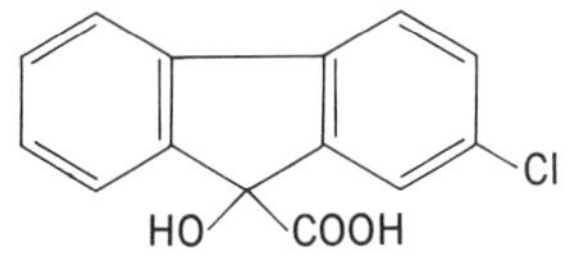

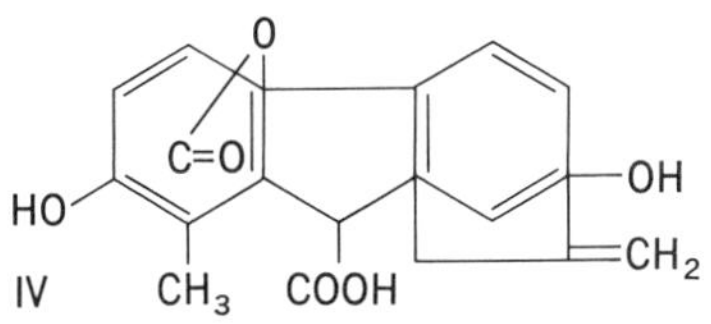

Fig. 1. Structural formulas of fluorene (I), fluorenol (II), chlorfluorenol (III), and gibberellic acid (IV).

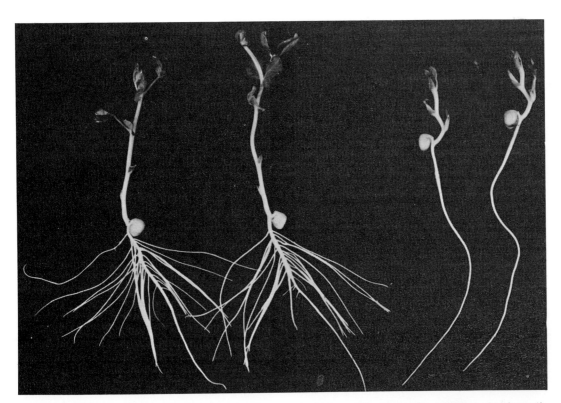

Fig. 2. Ten-day-old seedlings of *Pisum sativum* (cultivar Kleine Rheinländerin). (*Left*) Two control plants grown in water. (*Right*) Two plants from seeds that had been soaked for 12 hr in 5×10^{-5} M chlorfluorenol showing shortening of the internodes, inhibition of leaf growth, and suppression of side roots. There is no inhibition of the elongation growth of the main root.

tually divides the two daughter cells resulting from a cell division is also interfered with: Many mitoses are found in root tips treated with morphactins in which the vesicles in the equatorial plane have not fused, even when the daughter nuclei have reached interphase. Because the material in the vesicles consists mainly of pectic substances, the effect may indicate that morphactins interfere with the metabolism of these substances. Other data support this assumption. These derangements of cell division die away after a while and normal growth and development may be resumed.

Low, but morphogenetically fully effective, concentrations of morphactins cause no structural changes in the adult cells. The new growth formed after treatment, on the other hand, shows striking cytological aberrations, among which the most conspicuous are a special growth of individual mitochondria which become much elongated and form semicrystalline inclusions, a change in the fine structure of the nuclei, and an accumulation of starch in the amyloplasts of the guard cells of the stomata of *Pisum*.

Metabolic effects. Unlike some other herbicides, chlorfluorenol usually has no effect on the gas exchanges of respiration and photosynthesis, so long as there is no visible effect on growth of the organ concerned. Photophosphorylation and $C^{14}O_2$ fixation by isolated spinach chloroplasts are also not affected by chlorfluorenol. Nevertheless roots arising in the dark from peas treated with chlorfluorenol showed a considerable increase of respiration, and green tissues of pea and *Lemna* grown up under the influence of the morphactin showed a marked depression in photosynthesis as compared with the controls.

Interrelations with growth hormones. Various evidence pointed to an antagonism between morphactins and natural growth hormones, such as the gibberellins. Actual synthesis of gibberellin in *Fusarium moniliforme* was not inhibited by morphactins, and thus a competitive interaction between gibberellins and morphactins was considered likely. The apparent similarity between the structure of the fluorenes and the gibbane nucleus of gibberellins (Fig. 1) also seemed to point in this false direction. The three-dimensional molecular structures of morphactins and gibberellins are really quite different, and it is therefore not surprising that further analysis revealed no direct interaction between the two classes of growth regulators. Chlorcholine chloride (CCC), which also causes dwarfing, does antagonize gibberellin formation and produce its effect in this way. Chlorfluorenol also differs from CCC in that it inhibits the elongation of pea shoots much more in the dark than in the light, whereas CCC works the other way round. Growth in darkness thus appears to depend predominantly on a system sensitive to chlorfluorenol, and growth in light on one sensitive to CCC. Since CCC depresses the level of gibberellin, its inhibiting action in light can be reversed by gibberellic acid, and this effect can be almost wholly prevented by chlorfluorenol. Gibberellin, indeed, is in this way fully effective only in the absence of chlorfluorenol, not in its presence. The process inhibited by chlorfluorenol that evidently mainly determines growth in the dark—when the system regulated by gibberellin is out of action—must therefore be coupled in the light with the one affected by gibberellin and CCC, even though it may be slower under these conditions than in the dark. Since the growth of pea shoots, without addition of external growth substances, is about the same in light and dark, the two processes are presumed to act in a compensatory manner (Fig. 3).

As gibberellin is a prerequisite for the action of auxin, the latter may play a part in the process sensitive to chlorfluorenol. How do morphactins affect the auxin (β-indoleacetic acid, IAA) economy of the higher plants? Determinations of the IAA content of pea roots treated with chlorfluorenol indicated a definite reduction in the amount of IAA. It was also found that pea roots germinated in chlorfluorenol showed a greater activity of IAA oxidase (an IAA-destroying enzyme), and it is possible that this is the cause of the reduction of IAA concentration. Morphactins affect the IAA economy in yet another way: They inhibit its transport downward from the apical bud. It is likely that the inhibition of apical dominance described above is due to this cause. The activities of the fluorenol growth regulators and of IAA thus have a number of connections; but the morphactins are not true IAA-antagonists. Most of their effects cannot, for example, be reversed by supplying additional IAA.

Finally chlorfluorenol behaves similarly to kinetin in promoting regeneration in leaf discs of *Begonia*; but here also the effect is only analogous.

Thus one comes to the conclusion that the morphactins can indeed be influenced by the natural growth hormones in their effects and, conversely, can modify the actions of the latter, but that the primary effect of the fluorene derivatives is not an interaction with one of the well-known natural hormones.

Influence on plant movement. One of the most striking effects of morphactins on plant growth is the inhibition of the geotropism and phototropism (curvatures due to gravity and light, respectively). It seems plausible at first sight that the reduction of IAA concentration in tissues might be responsible for the disturbance of root geotropism by chlorfluorenol (Fig. 4). Negative geotropism, that is, an upward growth of the root, which is what usually happens under the influence of morphactins, would be expected if the IAA content of the root were no longer above optimal, but suboptimal as in the shoot. A gravitational enrichment of growth

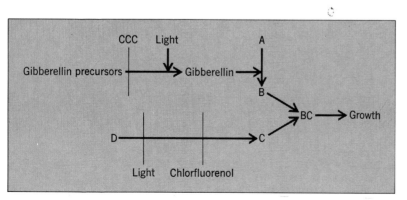

Fig. 3. Diagram of the possible points of operation of CCC and chlorfluorenol. (From D. Köhler, Effect of 2-chloro-9-fluorenolo-9-carbonic acid on stem elongation of peas in light and darkness, Planta, 79:50–57, 1968)

Fig. 4. Four-day-old *Avena sativa* seedlings whose grains had been soaked in 10^{-5} M chlorfluorenol. Roots (with hairs) are growing upward out of ground (negatively geotropic), while coleoptiles (without hairs) are behaving normally (also negatively geotropic).

hormone on the physically lower side of the root would then lead to this side becoming convex and thus to a negative geotropism. This explanation proves to be incorrect because (1) an effect on geotropism is also observed in shoots; (2) the effect cannot be reversed by supplying chlorfluorenol-treated plants with IAA; and (3) the phototropic responses of higher plants are also blocked by chlorfluorenol.

It is probable that the effect of morphactins on tropism is caused by suppression of the lateral transport of IAA in the organs. Such an effect has been demonstrated directly for 2,3,5-triiodobenzoic acid (TIBA), an herbicide, which is also antitropistic and very like the morphactins in its effects on growth. It is rendered probable for the morphactins by the observation that the curvature of cucumber hypocotyls, due to applying IAA pastes of different concentrations on opposite sides, was considerably increased by pretreating the seedlings with morphactin. Presumably diminution of the concentration difference between the two sides was prevented by the growth regulator, that is, the lateral transport of IAA was inhibited. Since fluorenols affect phototropism similarly to geotropism, it would be natural to think that the same primary reaction occurs in both inhibitions. This assumes, however, that a lateral transport of auxin under the influence of unilateral illumination causes phototropism, and not a photocatalyzed destruction of auxin as is occasionally supposed.

It is noteworthy that roots growing upward under the influence of chlorfluorenol show a normal displacement of the statoliths in the root cap. They always move toward the physically lower side. The direction of curvature of a root is therefore not always determined by the position of the statolith amyloplasts alone.

The lack of geotropic sensitivity caused by morphactin is particularly striking when the normal placing of an organ depends on an interaction of geotropism and a nastic movement (a movement in which the direction is dictated by the structure of the organ and becomes induced only by the stimulus). Nastic movements are definitely not affected by the fluorene derivatives. One can, so to speak, use the morphactins as "chemical clinostats" to exclude geotropism. It is possible that the antitropistic substances may become important aids to plant physiologists in the causal analysis of plant movements.

Transport and metabolism in plants. The movement of morphactins in plants was studied by means of C^{14}-labeled compounds. Acropetal and basipetal movements of morphactin through the parenchyma of coleoptiles were at the same rates. Over longer distances in the plant, the compound is shown by autoradiography to be carried predominantly in the phloem. It is thus transmitted from the point of application particularly into the youngest, still-growing regions, for example, into the apical meristem and young leaves. That the labeled material actually travels in the sieve tubes is shown by the fact that it is demonstrable in the honeydew of aphids which tap them.

An important point in the practical utilization of morphactins is that they are relatively quickly converted—partly by formation of glycosides and partly by removal of the OH group at position 9—and deposited in forms that cannot be further translocated. Wheat plants which had been supplied with labeled morphactin still contained mainly the unaltered substance 1 day later, while after 3 days conversion products appeared, and after 9 days the initial substance was no longer detectable. Meanwhile an ever-increasing amount of the radioactive material was converted into insoluble form. The fact that in wheat, 30 days after application of the growth substance, the newly grown parts no longer receive any translocated radioactivity can be attributed to this. If the morphactin is given at the right time, there is no danger that it, or its metabolic products, will get into the grains of cereals or possibly even into the flour.

Practical applications. At present there are two possible practical applications of this group of growth substances. One consists of chemically damping down growth, and here combinations with maleic hydrazide are outstanding as regards breadth of action; chlorfluorenol attacks the broadleafed plants, and maleic hydrazide the grasses. The second application is the combination of fluorenols with phenoxyacetic acid derivatives to give compound herbicides which are used to combat weeds in corn and grassland. One cannot yet foresee whether further applications will become of practical importance. The future may hold more promise for the use of morphactins in combination with other growth regulators rather than alone.

For background information see PLANT GROWTH; PLANT METABOLISM in the McGraw-Hill Encyclopedia of Science and Technology.

[HUBERT ZIEGLER]

Bibliography: G. Mohr and H. Ziegler, *Vortr. Ges. Geb. Bot.*, N.F. 3, 1969; G. Schneider, *Annu. Rev. Plant Physiol.*, 21:499–536, 1970; G. Schneider, *Naturwissenschaften*, 51:416, 1964; H. Ziegler, *Endeavour*, 29:112–116, 1970.

Navigation systems, electronic

Yachting has recently grown in popularity throughout the world to the extent that several firms are now engaged solely in the development and manufacture of electronic aids to yacht navigation. These aids extend from the measurement of such basic quantities as speed through the water, distance run, depth of water, and wind velocity to position fixing by high-accuracy radio systems such as Loran C, Decca, and Omega. Much of the development work has been a scaling down of the systems already in use in commercial and naval ships, and this has been facilitated by the availability of an ever-widening range of solid-state electronic devices. Recent developments of particular interest in this field relate to echo sounding, the computation of the boat's position at sea by automatic dead reckoning, and the optimization of the performance of a sailboat when proceeding to windward.

Echo sounding. The trend toward miniaturization has resulted in the development of electronic sounders which use a meter or digital lamp readout for indicating the depth of water. These indicators replace the conventional motor-driven recorders. In both meter and lamp systems the time delay between the transmission of an ultrasonic pulse into the water and its return from the seabed is used as a measure of depth. Sounding rates are in the region of 20 pulses/sec and frequencies used are generally in the 100–200 kHz range. In the meter-type instrumentation the transmission pulse and the echo signal are used to close and open an electronic bistable relay. This relay switches a constant current to the meter. The mean current, and hence the pointer deflection, is proportional to the echo delay period. In the digital display system the transmission pulse opens a gate at the output of a stable oscillator of a binary counting chain. The gate is closed by the echo signal. A binary count proportional to the echo delay period is decoded in decimal form and causes the indicator lamps to glow.

The main problem to be overcome in using these simple displays is the elimination of spurious readings due to echoes from fish, plankton, and other mid-water echoing sources and to pulses of interference from seaborne noise. These present little or no difficulty in the case of the conventional display systems since the operator can usually distinguish between the seabed echo and the spurious signals.

In the Hecta meter-display instrument, by Brookes & Gatehouse Ltd, a high degree of immunity has been achieved by the combined use of a time-variant receiver gain control and a computer. The computer samples the amplitude of the echoes in each cycle of transmission, selects the largest, and on subsequent cycles causes all echo signals preceding the largest (in order of time of arrival) to be eliminated from the range-measuring circuit. The success of the computer depends on the premise that at the output of the receiving amplifier the seabed echo signal is invariably of greater amplitude than any of the unwanted echoes.

Dead-reckoning computer. Dead reckoning is the process of advancing one's geographical position to give a position at a later time by the addition of one or more vectors representing course and speed. By the classical method, speed and heading are observed frequently to achieve the desired accuracy. The calculated changes of latitude and longitude occurring over each interval are recorded. The modern computer accepts continuous signals from a transmitting compass and log and displays the present latitude and longitude, provided that these parameters have been set correctly at the time of the last "fix."

A yacht computer named Hadrian has been designed in which, in the interest of economy, the readout is in the form of the distance traveled along, and perpendicularly to, a preset course line. The navigator simply sets the prescribed course to steer on the dial of an electronic compass. The computer accepts heading error signals from the compass and a speed signal from the log (Fig. 1). Its output is a direct-current voltage which actuates an integrating motor. The rotation of this motor is representative in both magnitude and sine of the speed of the vessel resolved in a direction perpendicular to the set course. A pointer attached to the shaft of the motor thus shows the accumulated error of position perpendicular to the set course. The distance moved along the course line

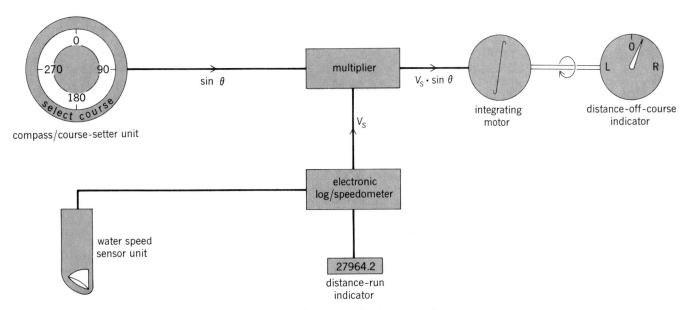

Fig. 1. Block diagram for Hadrian dead-reckoning computer. V_s is speed of ship; θ is heading error.

is given by the electronic log. The two outputs thus establish the dead reckoning position. The compass unit is perhaps remarkable in that it has no magnet system. The horizontal component of the Earth's magnetic field is sensed directly by means of a flux gate, which is referred to the horizon plane by means of an oil-damped pendulum. The gate is of the saturable reactor type and delivers an alternating-current output signal proportional to the sine of the angle between the axis of the flux gate and the Earth's magnetic meridian.

Sailing computer. Yacht races are largely won and lost in the windward phases of the course, for the degree of skill required in achieving high performance is greater in windward sailing than in any other mode. A compromise has to be achieved between the conflicting requirements of obtaining the highest speed through the water and the smallest angle between the yacht's course and the direction of the wind, since it is the component of speed measured in the direction of the wind vector that determines how rapidly the yacht is approaching the upwind mark (Fig. 2). This resolved velocity is generally called the "speed made good" or V_{MG}, and the relationship is shown in the equation below, where V_S is the boat's speed through the water, and γ is the angle between the course and the wind direction.

$$V_{MG} = V_S \cdot \cos \gamma$$

Up to the present time V_{MG} has been maximized intuitively by the helmsman and sail trimmers, but the advent of accurate electronic speedometers and wind gages has made it possible to calculate V_{MG} and hence to maximize it by trial and error. Slide-rule calculators have recently been developed for this purpose. The mathematical problem is complicated by the fact that γ cannot be measured, since the yacht's own velocity introduces a headwind component which compounds with the true wind vector. The resultant vector, which is the one measured by the gage, lies closer to the fore and aft axis of the yacht and has a greater magnitude than the true wind vector. It is called the "apparent wind." Gage correction factors have to be applied to compensate for errors introduced by the heeling of the yacht. However, γ can be calculated when V_S and the speed and relative direction of the apparent wind are known, but the complexity of the final expression for V_{MG} precludes the development of an exact-solution electrical computer that would meet market acceptance.

Brookes & Gatehouse Ltd, in collaboration with the University of Southampton, have carried out investigations utilizing a digital computer of the effect on accuracy of introducing various simplifying assumptions. Working prototypes of a prac-

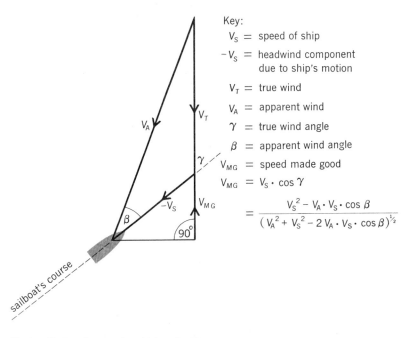

Fig. 2. Vector diagram for windward sailing.

tical on-board analog computer solving a simplified expression have been used with success, bringing much nearer the day, which many sailors will regret, when yachts will be sailed "hands off" by computer and autopilot.

For background information see COMPUTER; NAVIGATION SYSTEMS, ELECTRONIC in the McGraw-Hill Encyclopedia of Science and Technology. [R. N. B. GATEHOUSE]

Bibliography: R. N. B. Gatehouse, *J. Inst. Navig.*, 23(1):60–69, January, 1970.

Neutron star

During the last several years the discoveries of pulsars and the uniform 3° blackbody radiation have revived much interest in the study of stellar structure at nuclear density and beyond, as well as in the study of the physics of the early universe. Considering pulsars as spinning objects with extreme densities, what can be said about them from the present knowledge of nuclear physics and high-energy physics? Is there a maximum mass? Should the star be superconducting? Are they formed during stellar explosion? What is the radiation mechanism that produces pulses?

Early work and recent developments. Before the discovery of neutrons in the laboratory in 1932, S. Chandrasekhar and L. D. Landau had already constructed stellar models composed of neutrons. In 1939 J. R. Oppenheimer and G. M. Volkoff did some realistic calculations on the stellar model with a neutron core and noninteracting neutron gas. For the last 10 years or so many attempts have been made to improve on the treatment of the neutron gas in the model by including nuclear interactions. Recent success in applying the Brueckner-Bethe-Goldstone calculation of the nuclear many-body problem to the realistic nuclei assures one's confidence in calculating the properties of neutron matter, a matter composed of pure neutrons near the nuclear density. On the other hand, the richness of information one deduces from the dual-resonance model in recent high-energy physics encourages one to construct some realistic models of hadron matter, a matter composed of particles with strong interactions. See ELEMENTARY PARTICLE.

Physics of very dense matter. For very dense matter the degenerate Fermi energy of the matter is of the order of tens of millions of electron volts, which is large in comparison with the estimated maximum stellar temperature of a neutron star, of the order of 10^8K. As long as the thermal energy is less than a few percent of the Fermi energy, one can consider the matter to be totally cold, that is, the approximation of 0° temperature should be very accurate. Now consider the qualitative features of the composition of matter as an increasing function of density, that is, viewed from the stellar surface of a neutron star and going inward (Fig. 1).

To begin with, solid iron is not very compressible up to a pressure of 10^6 newtons/cm² (10^5 atm.), where the iron atom's outer electrons begin to have a rather deformed orbit (~1 eV). Further increase of pressure ionizes the atom, the ionization increases as pressure increases, and the pressure of the matter is generated by the Fermi energy of the nonrelativistic electron gas. At a density of 10^6 g/cm³, the electrons begin to be relativistic and are virtually free. When the density reaches 10^9 g/cm³, the electrons are energetic enough to enter nuclei and induce inverse β-decay, which makes neutron-rich nuclei energetically favorable. When the symmetry energy from neutron excess is large enough, the neutrons drip out from nuclei at a density of 4×10^{11} g/cm³ and begin to contribute pressure as a neutron gas. The neutron-rich nuclei grow in size and their boundaries begin to touch so that the matter becomes basically one giant nucleus. There are still a few percent of protons clustering together with neutrons, and neutrons themselves may condense into superfluiding pairs. At a density of 2×10^{14} g/cm³, almost all clustering disappears and the proton number density decreases to less than 1%; the matter is now composed of almost pure neutrons. The binding energy of pure neutrons from the attractive nucleon interactions is not nearly enough to overcome the repulsive neutron Fermi energy, and therefore the gravitational force is required to bind the neutron

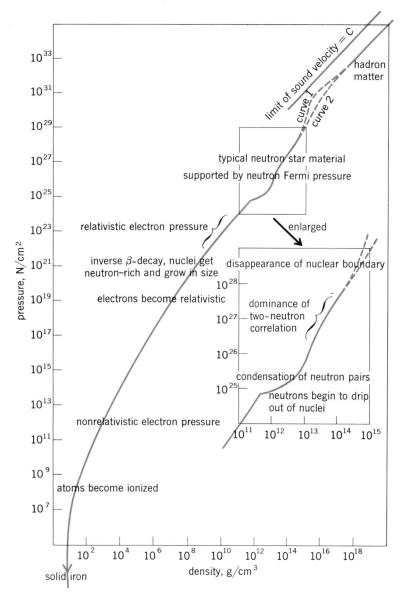

Fig. 1. Equation of state in neutron stars. Curve 1: Nuclear potential with strong repulsive core. Curve 2: Repulsive and attractive interactions are balanced in the core of nuclear potential.

matter together. Up to a few times the central density of a heavy nucleus, the matter is basically pure neutrons and the equation of state can be calculated very accurately by applying the Brueckner-Bethe-Goldstone approximation of nuclear matter to some well-considered nuclear potentials.

The Brueckner approximation is a self-consistent two-body calculation, and the usual phenomenological nuclear potentials are derived from nonrelativistic two-body scattering phase shifts with low angular momentum states (with partial waves $l = 0,1,2$). From the two-body wave function, one can estimate the three-body contribution, which is at most a few percent and can be neglected. At higher densities, the two-body contribution becomes less dominant, the nuclear potentials become poorly known as the internucleon distance approaches the repulsive nuclear core, and the phenomenologically (obscure) higher angular momentum states become more and more important. Unfortunately the equation of state at this uncertain region is most sensitive to the maximum mass of a neutron star. Various estimates of the limiting stable mass now range from 0.4 to about 2 $M_\odot$ (2 × solar mass). Any mass beyond this limiting mass would presumably collapse to a singularity (black hole) due to the enormous gravitational attraction. See GRAVITATIONAL COLLAPSE.

At the densities of 10^{16} g/cm^3 and beyond, the energy per particle is so high that the nucleons are themselves excited, and knowledge of high-energy physics is required. Unlike the study of neutron matter, in which one can rely upon the knowledge of heavy nuclei, here one has to be rather speculative and draw only qualitative conclusions. When nucleons get excited, the number of degeneracies increases exponentially as a function of energy.

Rolf Hagedorn has proposed an asymptotic bootstrap model and, within the framework of statistical mechanics, found that the exponential increase of degeneracy exactly cancels out the exponential Boltzmann dependence on temperature and thus gives rise to an ultimate temperature of 2×10^{12}K, beyond which the partition function would not converge. The same conclusion has now been derived from the dual-resonance model.

Recently G. Veneziano has further shown that the interaction among hadrons can be considered effectively at all orders of virial coefficients by assuming the nonzero but narrow resonant width in the dual-resonance model, which proves that the single-body, noninteracting formulation of statistical mechanics is indeed a good approximation for hadron matter. Qualitatively hadron matter is very compressible; any increment of energy to the system would be readily converted into rest mass and is therefore the most effective factory for creating heavy hadrons. However, a stellar structure composed of hadron matter is very fragile and unstable against gravitational collapse, and the maximum stable mass of a neutron star is the one which just begins to contain some of the fragile hadron matter. A graph of gravitational mass versus central density is shown in Fig. 2.

Spinars. By the process of elimination, pulsars are now believed to be spinning neutron stars with enormous magnetic field, as first suggested independently by Franco Pacini and by Thomas Gold.

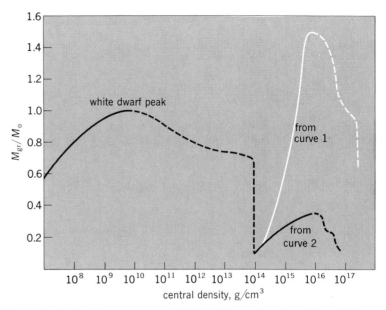

Fig. 2. Gravitational mass versus central density in neutron stars. The solid line represents a stable star; the broken line represents an unstable star.

The dynamics of a strong and rotating magnetic field has not yet been fully understood, but the estimated energy output from pulsars agrees rather well with the kinetic energy loss from the spindown of a typical neutron star. In 1969 Philip Morrison argued from dimensional analysis that the quasi-stellar radio source (quasar) 3C345 is in many ways analogous to the pulsar NP0532 (the Crab Pulsar). Their masses, rotational period, and radius differ by a scale of a few billion; they both have contracted by a factor of 10^5 in dimension and increased by a factor of 10^{10} in magnetic field strength. They both are coupled by the magnetic field to an outer surrounding, have very effective mechanisms of coherent radiation, and have a flare up (or pulse) of about one-twentieth of their respective period. The neutron stars give off energy while slowing down, but the quasars, a contracted kind of galaxy, are still in the process of contraction and are trading off their enormous gravitational energy to their rotational energy, which in turn supplies the energy for coherent electromagnetic radiation. Unlike neutron stars, which do not change size, quasars spin faster as their dimensions are reduced and eventually their outer part may break away. Due to the evolutionary change of forces and dimensions, there can be many forms of mass loss through gravitational instabilities. See GALAXY CLUSTERS.

Quasars are also extremely distant objects, and therefore as scientists see them, they were evolving in their early stages of life, while by comparison the pulsars are objects very nearby and are already at their final stage of evolution. Very little is known about the various mechanisms of coherent radiation from quasars. At a late stage in the evolution of a quasar, the magnetic fields may be "rubbed" smooth, the electromagnetic radiation nearly stop, and the spinning object become an undetectable giant top. One of these compact giants may exist in the center of the Milky Way galaxy.

Pulsars, the spinning quasars, and other contracted spinning objects form a new class of celestial objects now called spinars. They may very well be the main source of all the cosmic rays observed.

For background information see ELEMENTARY PARTICLE; GRAVITATIONAL COLLAPSE; PULSAR; QUASARS in the McGraw-Hill Encyclopedia of Science and Technology. [C. G. WANG]

Bibliography: G. Baym, H. A. Bethe, and C. J. Pethick, *Nuclear Physics*, in press; A. G. W. Cameron, *Annu. Rev. Astronaut. Astrophys.*, 8:179, 1970; H. Lee, Y. C. Leung, and C. G. Wang, *Astrophys. J.*, 166:387, 1971; P. Morrison, *Astrophys. J.*, 157:L73, 1969.

Nitrogen fixation

During the past 5 years considerable progress in understanding nitrogen fixation or, alternatively, dinitrogen fixation, the term favored now, has been made. The enzyme system nitrogenase has been purified and many of its properties are now known. The discovery that other compounds similar to dinitrogen (N_2) are reduced by nitrogenase has led to an insight into the way in which N_2 is reduced. The discovery that nitrogenase reduces acetylene to ethylene has led to the development of a rapid, sensitive, and accurate technique for measuring N_2 fixation in cell-free extracts and whole cells. With this technique one may now directly measure the extent of N_2 fixation in lakes, streams, and other natural habitats. Scientists are now in a position to determine in chemical and physical terms how N_2 is reduced.

Most, and probably all, of the useable nitrogen on Earth is present because of N_2 fixation and yet, until recently, research revealed little about the process. Dinitrogen fixation implies the conversion of N_2 to a fixed (nongaseous) form considered directly useable by man or living agents. In the biological system this fixed form is ammonia. In the chemical systems the reaction of N_2 to form ammonia and stable N_2 metal complexes can be considered dinitrogen fixation. The major chemical process (Haber process) for obtaining ammonia from N_2 and H_2 requires high temperatures and pressures, in addition to metal catalysts, whereas biological N_2 fixation is an enzymatic reaction that takes place at normal temperatures and atmospheric pressure. Recently chemists have synthesized many metal coordination complexes in which N_2 is a ligand (complexed to the metal). Two chemical processes for fixing N_2 at normal temperatures and pressures also have been developed recently, although the yield of product (ammonia) is very low.

One of the problems in fixing N_2 is its resistance to chemical reaction, in particular, chemical reduction. This is the reason for the requirement for high temperature and pressure in the chemical processes and the reason that the biological process requires a unique enzyme system, in addition to a source of "high-energy" electrons and adenosinetriphosphate (ATP), a biological energy storage compound.

Requirements. The breakthrough in research on biological N_2 fixation came in 1960, when conditions were developed for removing the enzyme nitrogenase from bacterial cells and demonstrating its activity. The requirements were found to be (1) the complete absence of oxygen, (2) the supporting substrate (pyruvic acid), and (3) an appropriately prepared extract of N_2-fixing bacteria. Pyruvate was needed because during its breakdown in the cell extract two components were produced that were absolute requirements for N_2 fixation. These were ATP and reduced ferredoxin (Fd). Magnesium ion (Mg^{++}) is required for ATP utilization. Fd was discovered during research on N_2 fixation when researchers looked for and found the acceptor (electron carrier) of electrons released during pyruvate oxidation. This reduced electron carrier, Fd, was required for N_2 reduction. The reactions of pyruvate necessary to produce these required reactants are summarized in Eqs. (1–3), where CoA is coenzyme A, ADP is adenosinediphosphate, and Pi is inorganic phosphate.

The overall reaction for N_2 fixation can be written as Eq. (4).

More ATP is consumed than specified in reaction (4), but of the total ATP consumed only six moles appear to be coupled to each mole of N_2 reduced. The utilization of ATP when coupled to N_2 fixation requires reduced Fd or a chemical reductant, sodium dithionite, that will substitute for reduced Fd. ATP utilization coupled to the reduction process of N_2 fixation is termed "electron activation" or reductant-dependent ATPase and can be compared with reverse electron flow that can occur in mitochondria when the oxidative phosphorylation system is intact.

In the last few years several unexpected substrates for the dinitrogen-fixing system (nitrogenase) were found. Compounds such as acetylene, cyanide, azide, and N_2O that are isoelectronic with dinitrogen were found to be reduced to ethylene, methane and ammonia, ammonia and N_2, and N_2 and water, respectively. These compounds, previously shown to be "competitive" inhibitors of N_2 fixation, inhibit by competing for the reductant or "activated electrons" and themselves become reduced.

Another substrate found to be reduced by nitrogenase was hydrogen ions or protons. In the absence of N_2 and to some extent even in its presence, protons, always present in the reaction mixture, were reduced and released as hydrogen gas. The reduction required ATP and reduced Fd (or dithionite) and was termed ATP-dependent hydrogen evolution. One unique feature of the reduction of protons over the reduction of all the other sub-

$$CH_3COCOOH + Fd + CoA \cdot SH + \text{Pyruvate dehydrogenase} \rightarrow$$
$$CH_3CO \cdot SCoA + \boxed{\text{reduced Fd}} + CO_2 \qquad (1)$$

$$CH_3CO \cdot SCoA + Pi + \text{Phosphotransacetylase} \rightarrow CH_3CO \cdot PO_3H_2 + CoA \cdot SH \qquad (2)$$

$$CH_3CO \cdot PO_3H_2 + ADP + H_2PO_4^- + \text{Acetate kinase} \rightarrow \boxed{\text{ATP}} + CH_3COO^- \qquad (3)$$

$$N_2 + 6Mg\text{-ATP} + 6H^+ + 3 \text{ reduced Fd} \rightarrow 2NH_3 + 3Fd + 6ADP + 6H_2PO_4^- \qquad (4)$$

strates is that it is not inhibited by carbon monoxide. Carbon monoxide inhibits all the other reductions at concentrations as low as 0.01 atm but, although similar electronically to N_2, is not reduced by nitrogenase. Since there are no free intermediates formed during N_2 reduction, the differential effect of CO inhibition is the best evidence that there is more than one catalytic site involved in the overall process.

Nitrogenase composition. Nitrogenase from all organisms examined is composed of two easily separable metalloproteins: an oxygen-sensitive protein containing iron (called Fe protein or azoferredoxin) and a less-oxygen-sensitive protein containing molybdenum and iron (called MoFe protein or molybdoferredoxin). Azoferredoxin (AzoFd) from *Clostridium pasteurianum* is isolated as a dimer with a molecular weight of 55,000. Based on end group analyses the two monomers, with a molecular weight of 27,500 each, appear to be identical. AzoFd is brown, and the chromophore of each monomer seems to consist of a complex of two iron atoms, two "acid-labile" sulfide groups, and six protein sulfhydryl groups. Fe protein from other organisms has not been characterized yet.

MoFd has been purified from both *C. pasteurianum* and *Azotobacter vinelandii*. The clostridial protein has a molecular weight of 168,000 and appears to be composed of three subunits, one of molecular weight of 50,700 and two of molecular weights of 59,500. It is brown and contains per 168,000 molecular weight: 1 molybdenum atom, 11–12 iron atoms, 15–17 "acid-labile" sulfide groups, and 22–24 protein sulfhydryl groups. The distribution of these constitutents among the subunits has yet to be determined. MoFe protein from *A. vinelandii* is isolated as a dimer of molecular weight 270,000. It also is brown and contains per 270,000 molecular weight: 34–37 iron atoms, 2 molybdenum atoms, and 26–28 "acid-labile" sulfide groups. Its subunit composition has not been reported.

Neither MoFd nor AzoFd has activity alone, although AzoFd does bind magnesium-ATP to a much greater extent than MoFd. In addition, AzoFd binds ADP, an inhibitor of nitrogenase. The active unit, nitrogenase, contains at a minimum two AzoFd for each MoFd. In fact, preliminary data indicate that three "27,500" AzoFd subunits are required for each "168,000" MoFd, because when activity is plotted against AzoFd concentration with MoFd constant, a sigmoidal curve is obtained. When the latter data are plotted by the method of R. Hill, $\log v/V - v$ against $\log$ AzoFd, a straight line with a slope of 2.9 is obtained. Thus the assay for AzoFd must take into account the fact that with MoFd in excess, the active form [(AzoFd)$_{2\ or\ 3}$·MoFd] is the minor species present and that most AzoFd will be in the inactive form, AzoFd·MoFd.

Control of nitrogenase. There appear to be at least two cellular controls over nitrogenase, one over the enzyme activity and another over enzyme biosynthesis. The activity of the enzyme is under control of the ATP/ADP ratio in the cell. ADP, a product of the action of nitrogenase, at a ratio of 1:1 with ATP inhibits activity 50–60%, whereas at a ratio of 2:1 activity is inhibited completely. Thus under conditions where ATP in the cell is in short supply nitrogenase will not needlessly consume ATP.

Ammonia, the product of N_2 fixation, does not inhibit the enzyme at concentrations as high as 20 mM, whereas when the cell is growing in a batch culture on ammonia, nitrogenase biosynthesis is completely repressed. Cells of *C. pasteurianum* and other N_2-fixing organisms grown in a chemostat under ammonia limiting conditions in the absence of N_2 synthesize nitrogenase to a concentration about three times that present when such cells are grown on N_2 in the absence of ammonia. Cells grown on N_2 accumulate enough of an ammonia pool to partially inhibit nitrogenase biosynthesis. N_2 is not required as an inducer; that is, cells that exhaust ammonia in the absence of N_2 synthesize high levels of nitrogenase. Mutants of N_2-fixing bacteria have been selected that lack the ability to synthesize active MoFd (MoFe protein) and ammonia still inhibits the synthesis of AzoFd (Fe protein). Other mutants have also been obtained but are not as well characterized. Presumably nitrogenase requires at least three structural genes, as well as one for producing a repressor and possibly one or more operator and regulator genes.

Mechanism. No precise mechanism for N_2 fixation can be delineated at this time. The principle function of nitrogenase appears to be production of activated electrons, which then readily reduce chemisorbed dinitrogen (and other substrates). Data now available suggest that the Fe protein first accepts electrons from reduced Fd and "activates" them when in combination with the MoFe protein. The reaction with ATP probably is required to change the structure of the site or sites where the accepted electrons are localized. This seems to facilitate N_2 reduction by changing the environment of the electrons, which in turn makes the complex more electronegative. Metals are involved in the electron activation, as well as in the reduction of the substrates, because when they are removed from the Fe and MoFe proteins, activity is lost. Since all activities do not decrease to the same extent as metals are removed, some of the metals seem to play a role in "electron activation," whereas others are involved in substrate reduction.

For background information *see* NITROGEN COMPLEXES; NITROGEN CYCLE; NITROGEN FIXATION in the McGraw-Hill Encyclopedia of Science and Technology.   [LEONARD E. MORTENSON]

Bibliography: R. H. Burris, *Proc. Roy. Soc. London Ser. B*, 172:339, 1969; J. Chatt and G. J. Leigh, in E. J. Hewitt and C. V. Cutting (eds.), *Recent Aspects of Nitrogen Metabolism in Plants*, 1968; R. F. W. Hardy and R. C. Burns, *Annu. Rev. Biochem.*, 37:331, 1968; L. E. Mortenson, *Surv. Progr. Chem.*, 4:127, 1968.

Nobel prizes

The Nobel prizes for 1971 were awarded to an American scientist and an economist, a Canadian scientist, a Chilean poet, a British scientist, and a West German politician.

Physiology or medicine. Earl W. Sutherland, Jr., of Vanderbilt University, won this prize "for his discoveries concerning the mechanisms of the action of hormones." Sutherland's research uncovered the intermediary role of cyclic adenosinemonophosphate in the mechanism by which many hormones exert their control over various metabolic activities throughout the human body.

Chemistry. Spectroscopist Gerhard Herzberg, of the National Research Council of Canada, was chosen "for his contributions to the knowledge of electronic structure and geometry of molecules, particularly free radicals."

Physics. Dennis Gabor was awarded this prize for his invention and development of the holographic method of three-dimensional imagery. He divides his time between the Imperial College of Science and Technology, London, England, and Columbia Broadcasting System Laboratories, Stamford, Conn.

Literature. Pablo Neruda was honored "for poetry that, with the action of an elemental force, brings alive a continent's destiny and dreams." His major work is *Canto General* (1950). Neruda had been appointed Chilean ambassador to France shortly before the award.

Peace. West German Chancellor Willy Brandt won this prize because he had "stretched out his hand to reconciliation between countries that have long been enemies."

Economic science. Simon Kuznets, of Harvard University, was honored for his "empirically founded interpretation of economic growth which has led to new and deepened insight into the economic and social structure and process of development."

Noise, acoustic

Noise can cause progressive loss of hearing in those employees exposed to the relatively high but common levels experienced in industry. In cities noise seldom reaches the high levels found in industry except for a few locations in which most city residents find themselves only for a short period each day, that is, subways. However, it is apparent that noise in the city is an unwelcome component of the environment, at least at the levels commonly experienced within residential buildings and in nonindustrial workplaces and in recreational areas both indoors and out. One major contributor to noise in the contemporary city, as well as in the suburban-rural environments, is jet aircraft overflight. Late in 1969 the Federal Aviation Administration, authorized by Public Law 90-411, issued rules under an amendment to the Federal Aviation Act that regulated the noise levels of future subsonic aircraft that would be certificated for commercial flight. This was the beginning step in a long-range program to reduce aircraft noise exposure in the vicinity of existing and planned airports.

The effects of noise on persons not engaged in industrial tasks, whether at work or in leisure activities are not clearly understood, although it is well known that distraction plays an important part in human response to noise when carrying out creative tasks. In any case some specific ranges of noise levels can be defined. The figure shows how to interpret noise levels in terms of voice level and distance between a speaker and a listener. For example, jet aircraft cabin noise is roughly 80 +2 dBA. At 80 dBA in their expected (raised) voice level, seatmates can converse at 2 ft and by moving a little closer can lower their voices to normal level and converse at 1 ft.

Federal regulation. Federal regulation in the industrial noise area started in 1969, with revisions to the Walsh-Healey Public Contracts Act. The act requires that contractors engaged in government contracts in any amount exceeding $10,000 must protect their employees under conditions where the noise exposure is in excess of the permissible values (see table). The values in the table are only for exposure to one level of noise for one period of time or day. When the daily noise exposure is composed of two or more periods of noise exposure of different levels, their combined effect should be considered rather than the individual effect of each. If the sum of the fractions $C_1/T_1 + C_2/T_2 \ldots C_n/T_n$ exceeds unity, then the mixed exposure should be considered to exceed the limit value. C_n indicates the total time of exposure at a specified noise level, and T_n indicates the total time of exposure permitted at that level. Furthermore administrative controls or engineering measures (noise abatement procedures) or both shall be used to limit the exposure to values less than those stated. In addition, when the exposures are above those in the table, "a continuing, effective hearing conservation program shall be administered." In essence, this last requirement means that all current employees in exposed areas, and all new employees scheduled for these areas, must have their hearing measured and their hearing must be remeasured at appropriate intervals (from 12 to 18 months). The equipment used is an audiometer, which produces pure tones through headphones at known sound intensities to each ear in turn. Either the trained audiologist or the audiometer automatically records the threshold level at which the employee being tested hears the test signal. Where the test signal presented by the audiometer has to be raised considerably above the normal level at which people can just barely hear each tone (the threshold of hearing), the employee is said to have a hearing loss. The amount of impairment is determined by different formulas, depending on the exact purpose of the measurement.

During 1970 new legislation was promulgated which will extend the Walsh-Healey provisions to essentially all businesses engaged in interstate commerce.

Noise levels in industry. Surveys in many industries show that noise levels at many work stations are in excess of the allowable exposure, although the provisions for higher level exposure for shorter durations does cover probably one-third of the otherwise excessive level situations. Efforts by industries of all types are based on current technology in the noise control field using small and large mufflers on airflow system fans and blowers, engines, and pneumatic control valves. Some mufflers cause energy losses that are not readily made up by the system and thus require successive changes in order to provide the quieting without reducing machine efficiency, causing excessive temperatures to exist, or reducing the combustion air available to a process.

Protection for employed. Many work stations must be protected by the use of baffles between the noise source and the employee. These baffles reduce the noise by acting as acoustical barriers, and they also form visual barriers even when made of clear plastic. In some situations employees can be enclosed and thereby protected from a high-

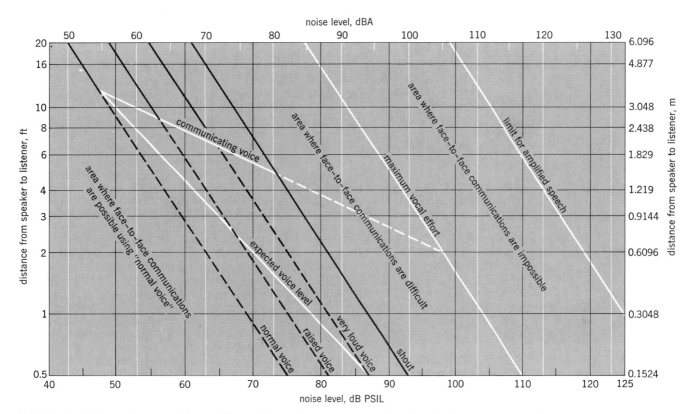

Voice level and distance between talker and listener for satisfactory face-to-face speech communications as limited by ambient noise level. Along the abscissa are two generally equivalent objective measures of noise level: the average octave-band level in the octaves centered at 500, 1000, and 2000 Hz, called the three-band preferred octave speech-interference level (PSIL), and the A-weighted sound level meter reading (dBA).

level-noise environment. Typical of this method are the operators' locations in power generating stations and refrigeration plants. In some instances a single machine or group of machines may be fully enclosed, even if they must then be force-ventilated either for cooling or fume removal using small blowers which in turn should be equipped with mufflers. The enclosures are frequently fabricated from 2–4 in. thick "acoustical" panels having a solid sheet-metal exterior surface and a perforated sheet-metal interior surface with a rockwool or glass fiber packing, varying in density from 0.5 to 6 lb/ft^3 (8 to 96 kg/m^3), depending on the particular application. In essence, one can reduce noise at the source by machinery modification, by operating on the path to either contain the noise or the employee, by making the path longer, which in turn reduces the noise (except in ducts and pipes), and by operating on the receiver or employee. This, of course, involves the use of either muff-type or insert-type ear protectors.

New York City noise control report. The report of New York City's Mayor's Task Force on Noise Control concluded that noise is certainly an undesirable environmental contaminant and that it can be brought under adequate control by several different concurrent methods. These include administrative actions, such as having a city department specify maximum noise level limits in the purchase specifications for garbage trucks and other mechanical devices and vehicles purchased by the city. Other conclusions of the Mayor's Task Force were that regulations controlling noise from construction, transportation systems, and air-conditioning systems within the city were needed. The report also pointed out that a first step had been taken in Section 1208 of the city's new building code *Noise Control in Multiple Dwellings* enacted into law Dec. 6, 1968. The code sets limits for the noise that a building's environmental control system may make at its own and neighboring occupied spaces and set limits, probably too low, for the sound isolation provided by partitions within the building. As a result of the activity of the Mayor's Task Force, the city entered into a contract with the U.S. Department of Housing and Urban Development to define the problems of establishing the present noise environment throughout the city and to develop long-term noise abatement procedures.

Interference with creative activity. Noise interferes with creative activity through distraction. Although it may not readily interfere with sleep, it can prevent onset of sleep, especially in sensitized

Permissible noise exposures for workers on Federal contracts

Duration per day, hr	Sound level, dBA
8	90
6	92
4	95
3	97
2	100
1½	102
1	105
½	110
¼ or less	115

SOURCE: *Federal Register*, vol. 34, no. 96, May 20, 1969.

individuals. At moderate noise levels, noise can interfere with speech communication, the audition of the sound portion of television programs, music listening, and audio-visual instruction. Thus, even at relatively low sound levels, below 50 dBA, noise may be undesirable. It was also indicated in several applied research efforts that noise may at low levels, in the neighborhood of 35–40 dB, screen people from less desirable sounds. Thus a quiet fountain or nearby ocean "roar" can shield the auditor from the sounds of traffic, children, and distant aircraft that otherwise might be distracting. The character of this "masking" sound must be bland and practically nonvarying or, if varying, must vary slowly and contain no threatening or unpleasant information.

Current activity in the area of defining the acceptable noise environment for man away from the workplace is going on in both the United States and abroad. Aspects being considered include the intrusiveness of noise, or how much louder specific single occurrences are than the background or ambient noise; the spread in the statistical parameters known as the 10th and 90th percentile noise levels and the standard deviation of the noise level as a function of time during the three socially important periods of the day: day, evening, and night; the effects of social and political factors; and the views on the part of the auditor as to whether the particular noise is necessary or not and whether the auditor believes that it may or may not be readily abated.

For background information see NOISE, ACOUSTIC in the McGraw-Hill Encyclopedia of Science and Technology. [LEWIS S. GOODFRIEND]

Bibliography: *Airport Environs: Land Use Controls, An Environmental Planning Paper*, Department of Housing and Urban Development, 1970; J. D. Chalupnik (ed.), *Transportation Noises: A Symposium on Acceptability Criteria*, 1970; Mayor's Task Force on Noise Control, *Toward a Quieter City*, New York City, January, 1970; J. C. Webster, Effects of noise on speech intelligibility, in W. D. Ward and J. E. Fricke (eds.), *Noise as a Public Health Hazard*, American Speech and Hearing Association, 1970.

Nuclear explosion engineering

An important aspect of nuclear explosions since 1957 has been a government-industry program to develop a spectrum of scientific, civil, and industrial uses. Much interest has been focused on engineering applications of nuclear explosions for such uses as public works construction and natural resources development. This new engineering technology has advanced during the period since 1957, but only two industrial application experiments have been achieved thus far in the United States. The second experiment involved a court fight to prevent it on the grounds of environmental hazard. Although several civil construction works have been proposed over the years, none has been achieved under the United States program, and as of 1971 no excavation experiments are presently proposed.

In contrast to the decreasing activity of the United States program during 1969–1971, a remarkable international interest in peaceful nuclear explosions has grown, including exchanges of technical information in meetings between the Soviet Union and the United States. These meetings resulted in the first panel meeting organized by the International Atomic Energy Agency (IAEA) in Vienna in March, 1970. Technical papers were presented by British, French, Soviet, and American workers. During the latter part of 1970, additional information was released by the Soviet Union describing their achieved projects in both civil construction and stimulation of natural resource production. A second IAEA panel was convened in January, 1971, at which results of application in underground resource development were presented. A comparison of the United States and Soviet programs is given in the table. Efforts are underway in the four "nuclear" nations for developing the technology and safety assurances for industrial utilization of nuclear explosives.

Explosion technology. The state of the art of nuclear explosions for engineering application is described in the proceedings of two recent international conferences. Further studies are underway to develop the capability of achieving desired configurations from single or multiple nuclear explosions. The configuration desired varies with the intended application. It also varies with the size and depth of the explosion and the properties of the geologic mediums involved. Fully contained explosions are generally designed to create a column (chimney) of rubble in hard-rock mediums to increase permeability for fluid flow or to create volume for fluid storage. Excavation explosions are generally designed to create a mound of rubble rock to serve as rock-fill for a dam, a hole in the ground to serve as a reservoir or a harbor, or a channel to serve as a waterway or a mountain cut. The engineering design for each of these configurations requires a considerable degree of information about the mechanical effects of nuclear explosions and the properties of the geologic mediums. Experimental explosions at the Nevada Test Site have been used for developing empirical scaling laws for contained and excavation explosions. Computer programs have been developed to predict explosion-produced configurations from basic information about the effects of nuclear explosions and measured properties of the host geologic mediums. These programs have been tested for Schoo-

Summary of United States and Soviet developments in nuclear explosion engineering

Application	United States projects		Soviet projects	
	Actual	Proposed	Actual	Proposed
Excavation				
Experiments	Many	0	~3	?
Highway cuts	0	1*	0	0
Harbors	0	2*	0	0
Canals	0	1*	0	1
Water resources	0	0	1	1
Contained				
Experiments	Many	1	Several	?
Gas stimulation	2	Several	1	1
Gas-fire blowout	0	0	2	0
Oil stimulation	0	0	1	2
Oil shale	0	1	0	0
Storage	0	1	2	2
Geothermal heat	0	?	0	0
Mining	0	2	0	2

*Cancelled.

ner, the last excavation experiment at the Nevada Test Site, executed in 1968, and the first industrial experiment, Gasbuggy, near Farmington, N. Mex., executed in December, 1967.

During the past few years, increased efforts have been made to design a nuclear explosive expressly for industrial use. The desired characteristics for such an explosive include a small diameter to fit in conventional sizes of wells drilled by oil companies, low production of radioactive materials, and low cost. The first test of such a specific device was executed successfully on July 8, 1971.

Safety assurance. A difficult problem in nuclear explosion engineering is the demonstration that applications involving nuclear explosions can be achieved without hazard to populations, private property, and the environment. Among the potential hazards involved, production of radioactive materials is considered far more important by the public than mechanical damage from airblast and seismic shock. During the last few years, engineering methods have been devised for estimating the extent and cost of settlement for structural damage. Airblast is not considered a significant hazard for fully contained explosions and is dependent upon meteorological conditions in excavation explosions. Structural damage due to ground motion can be an important hazard for large-yield explosions near inhabited areas. Methods to estimate structural damage by analysis of the properties of the structures and their response to ground motion have been developed. Estimates of costs for complaints due to structural damage can be made for specific applications and added to the project cost.

Fig. 1. A crater lip dam that was constructed in the Soviet Union. (a) The outer reservoir. (b) The inner reservoir constructed with a nuclear explosive of over 100 kt. (*Lawrence Radiation Laboratory*)

These are generally small compared to the project benefits.

The problems due to radioactivity contamination and radiation exposure are more difficult to resolve. In addition to the technical problems of determining the contamination and exposure levels as a function of location and time, social problems exist of relocating people from exclusion areas, if necessary, and of setting maximum permissible exposures. The exposures must be evaluated in relation to existing radiation standards for all nuclear energy programs, in addition to exposures

Fig. 2. Stimulation of natural gas reservoirs by multiple nuclear explosions spaced horizontally and vertically to maximize the recovery from an entire formation. (*Lawrence Radiation Laboratory*)

from medical treatment and natural background.

The types of radioactivity problems which must be considered are the amounts of radioactivity produced, the initial distribution, and the long-term transport of the radioactive materials. Recently much has been done to decrease the production of radioactivity in nuclear explosions. One of the important announced remedies is the shielding of the explosive with neutron-absorbing elements, such as boron. The problem of venting and fallout had already received much attention during the period of nuclear weapons testing in the atmosphere prior to 1963 and is of importance primarily for excavation applications. Underground nuclear explosions for commercial use have two potential radioactivity problems: the contamination of groundwater and surface water supplies, and the contamination of the commercial resources (for example, minerals, oil, and gas) removed from the explosion environment. The public health aspects of these problems are continually examined.

The basis for setting standards for allowable concentrations of radionuclides in commercial or environmental materials is still unresolved. The natural gas removed from the two experimental test wells has not been used but was flared at the well head. Conservationist groups in Colorado attempted in 1970 to prevent production tests from the Rulison well on the basis of hazard due to environmental pollution by the tritium in the natural gas. However, the court ruled that the environmental radiation would be well within limits prescribed by law. Research is currently underway for establishing a standard suitable for nuclear-explosion-stimulated natural gas. See RADIOACTIVE WASTE DISPOSAL.

Civil construction. Early efforts in the United States program involved primarily a search for potential applications of excavation with nuclear explosives. Consideration was given to a variety of construction projects, including harbors, canals, reservoirs, mountain cuts, dams, mineral deposit uncovering, and quarrying. Several specific projects were proposed, such as the Chariot harbor in Alaska, the Cape Keraudren harbor in Australia, the Carryall cut through the Bristol Mountains in California, and a quarry for a rock-fill dam in Idaho. The most ambitious project was a trans-isthmian sea-level canal along one of several possible routes considered by the Atlantic-Pacific Interoceanic Canal Study Commission appointed by President Johnson in 1964. For one reason or another, each of these proposed projects has been abandoned. In December, 1970, after 6 years of evaluation, the Canal Study Commission reported to President Nixon that the technology of nuclear explosion excavation had not yet reached a satisfactory level for recommendation as a means to construct such a canal. Many problems, such as national budget restrictions, public concern about environmental radioactivity, and the provisions of the 1963 Limited Test Ban Treaty which prohibit nuclear explosions that result in fallout beyond the national borders have caused the deletion of excavation experiments from the United States program since 1969.

However, during this same period, the Soviet Union announced a series of excavation experiments, culminating in the release of a movie in 1970 showing the construction of a crater lip dam with a nuclear explosion estimated to be over 100 kt in magnitude (USSR project "1004"). This water reservoir system is shown in Fig. 1. The crater for the inner reservoir has lip heights of 20–35 m, a lip diameter of 520 m, and an area of 0.14 km², contains about 7×10^6 m³ water, and backs up about another 10×10^6 m³ in the 3.5 km² outer reservoir during the runoff season. The Soviet program also plans the creation of a larger water reservoir in central Asia with 2×150 kt crater lip dams. To stem the continuous drop in level of the Caspian Sea, an important water resource, the Soviet Union plans to divert surface waters southward to the Caspian Sea. The diversion will be accomplished by a series of dams and a canal connecting the Pechora and Volga rivers, to be constructed partly with nuclear explosives. Although the development of excavation construction with nuclear explosions is essentially dormant in the United States, the efforts of the Soviet Union may demonstrate the long-range feasibility of such public works.

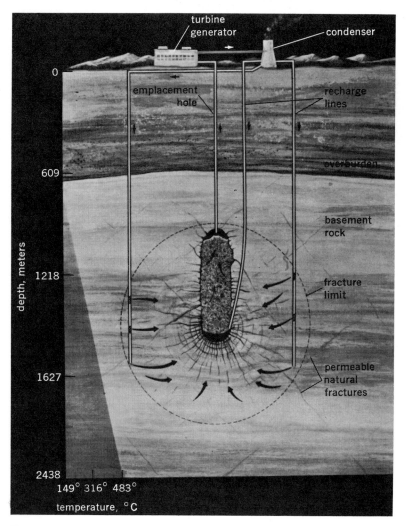

Fig. 3. Production of electricity from steam from geothermal deposits stimulated by nuclear explosions. This process may be an important source of power in the future. (*Lawrence Radiation Laboratory*)

Resource development. Most of the effort in the United States program and much of the Soviet announced program is involved with the development, stimulation, or storage of natural resources in deep geological formations. Of the major projects for contained nuclear explosions being considered, one has been reduced to practice in the United States, while three have been achieved in the Soviet Union. The shortage of "clean-burning" natural gas in the United States has led to a greater interest by industry in the stimulation of natural gas production by nuclear fracturing of deep, low-permeability formations uneconomical to develop by conventional drilling techniques. The first industry-government experiment in the United States was the 29-kt Gasbuggy experiment accomplished on Dec. 10, 1967, in the San Juan Basin of New Mexico. Production tests have indicated that the production from the Gasbuggy well over an 18-month period already exceeds the 10-year production from the nearest conventional well and that the projected 20-year production will be 5–8 times greater. Production tests and measurements of the gas chemical and radiochemical quality are continuing.

The second industry-government experiment in natural gas stimulation was the 40-kt Rulison experiment detonated Sept. 10, 1969, in the Mesaverde formation near Grand Junction, Colo. Preliminary production results which followed a lengthy shut-in period have indicated a successful nuclear stimulation with both higher rates of production and lower levels of radioactivity than anticipated. Current investigations for economic recovery of natural gas by nuclear explosion stimulation consider the effects of multiple explosions in large natural gas–bearing formations. The concept of such natural gas reservoir stimulation is shown in Fig. 2.

Stimulation of oil production has been examined in the Soviet Union. A pilot experiment was conducted several years ago in the middle of a depleted deposit. The nuclear stimulation consisted of two stages of explosions, the first was two 2.3-kt charges and the second was a single 8-kt charge. Production results indicated that the stimulated production was 30–60% greater than the yield expected without stimulation. The oil is being used commercially. Two additional experiments in oil stimulation are planned, one involving three explosives of 40 kt each and the other involving three explosives of 20–30 kt each. An interesting application achieved by the Soviet Union has been the use of nuclear explosives to extinguish fires in runaway oil wells. Two such blowouts have been reported.

Two types of projects which are being evaluated both in the United States and the Soviet Union are the creation of void space by the explosion for storage of gas and oil products and the crushing or removal of rocks for mining of minerals. Underground storage has also been considered by the United Kingdom, especially in the form of offshore storage of petroleum for loading and unloading large oil tankers.

A newly proposed project in the United States is the concept of steam production from geothermal heat deposits stimulated by nuclear explosion (Fig. 3). Electrical power production from geothermal steam may be an important contribution to the ever-increasing demand for power as the technology of the world continues to grow.

For background information see NUCLEAR EXPLOSION; RADIATION SHIELDING; RADIOACTIVE FALLOUT in the McGraw-Hill Encyclopedia of Science and Technology. [PAUL KRUGER]

Bibliography: *Engineering with Nuclear Explosives*, CONF-700101, 2 vols., U. S. Atomic Energy Commission, 1970; O. L. Kedrovsky, *Prospective Applications of Underground Nuclear Explosions in the National Economy of the USSR*, UCRL-Trans-10477, July, 1970; *Peaceful Nuclear Explosions*, International Atomic Energy Agency, Vienna, 1970; E. Teller et al., *The Constructive Uses of Nuclear Explosives*, 1968.

Nuclear materials safeguards

A problem that is assuming more importance as the use of nuclear materials increases relates to the safeguards of the nuclear materials. The term safeguards refers to a system of controls designed to assure that nuclear materials and the means of their production are devoted to peaceful ends in keeping with national policy and international commitments. In this context, nuclear materials consist of both source and special nuclear materials. Source materials are the natural elements uranium and thorium; special nuclear materials are uranium enriched in the isotopes U^{233} and U^{235} and plutonium, derived from the source materials through irradiation in nuclear reactors or through the separation of the rare U^{235} isotope from the predominant U^{238}.

Safeguard coverage. The control system must cover the complete cycle of refining, purification, isotope separation, reactor use, chemical recovery, and recycling and the flow of materials between these phases. The cycle begins with the processing of uranium and thorium ores and proceeds through the production of pure metals, alloys, or compounds. A gaseous uranium compound, UF_6, is the form used for U^{235} isotopic enrichment and subsequent conversion to other compounds or to metal. Other compounds or alloys, including mixtures with thorium, are used as fuels for nuclear reactors to produce heat and neutrons, which in turn produce the fissionable plutonium or U^{233}. These fissionable materials can remain in the fuel to contribute to the chain reaction, or they may be separated chemically from the fuel materials and used for enhancing new reactor fuel, or in essentially pure form they may be used for building nuclear explosives. Worldwide production in power reactors will soon approach quantities of plutonium sufficient for tens of nuclear weapons per day. See REACTOR LICENSING, NUCLEAR.

A special difficulty arises in the need to estimate the content and quality of fissionable material in finely machined products and in nonhomogeneous scrap resulting from mechanical and chemical processes. This estimation often can be met only by the use of reliable methods for measuring the radioactivity emitted by nuclear material isotopes. Some methods may be passive, measuring selected radiations continuously emitted by the material. Another course is to bombard the material with radiations, x-rays or neutrons, and measure selected reaction emissions induced therefrom.

Whenever feasible, of course, more precise measurements are made by sampling and chemical and mass spectrometric analytical techniques. These definitive techniques are also used to validate the nondestructive techniques mentioned above.

Accounting system. The measurement activities form the basis for recording in an accounting system the location and responsibility for all nuclear materials. For the United States, the Atomic Energy Commission (AEC) maintains such a system in the form of an automated Nuclear Materials Information System (NMIS).

NMIS receives reports of all production, transfers, and other significant inventory changes from all nuclear material custodians, regardless of ownership of the material. The significant inventory changes include fabrication discards, consumption, and growth of fissionable isotopes in nuclear reactors.

Custodians are required to establish and maintain administrative and physical controls over their facilities and operations to protect nuclear materials from theft or misuse.

Special controls that include prescheduling, monitoring, tracing, and reporting govern shipments of materials, since this is the phase in which material is most vulnerable to theft or loss.

The effectiveness of the government-prescribed accounting and physical controls is determined by a program of surveillance by government inspectors and auditors. These staffs review custodian systems for applying the controls and verify that the systems are effective, are adhered to, and are modified to meet changing conditions. These reviews are planned also to include examination of construction design of facilities and equipment to assure that location and quality of materials in process can be verified.

Improving measurements. The program outlined above is supported by continuing research, development evaluation, and testing activities. Primarily these activities are needed to improve measurement capabilities in terms of accuracy and promptness.

One approach to active interrogation is the use of accelerators as neutron generators which provide a copious source of neutrons of well-defined energy that can be varied. Californium-252, a spontaneous fission source, also offers a practical maintenance-free neutron source in extremely compact form. Mobile nondestructive assay laboratories, embodying such systems, are being evaluated in field use of nondestructive assay techniques to verify nuclear materials inventories.

Passive gamma or neutron assay techniques are useful for items of a size and shape that allow natural radiation to emerge from the item. The development of gamma detectors with high energy resolution makes it feasible to apply this technique in many situations. Significant developments in fission neutron-counting assemblies permit wider applications in the assay of plutonium which undergoes spontaneous fission.

Calorimetry is another nondestructive method which provides improved accuracy when the isotopic composition of heat-emitting isotopes is known. By using sensitive calorimeters to measure the total power output of a sample, the mass of each isotope can be calculated. Work with PuO_2-UO_2 mixtures (heat output 0.08 to 2.5 watts) attains an accuracy of a calorimetric measurement of ± 0.1 to 0.3%.

Systems studies. A somewhat broader research and development effort is devoted to safeguards systems studies. Systems studies encompass the evaluation of material balances and the limits of error relating to material balances and the measurements on which they are based. They involve the development, testing, and implementation of statistical techniques to determine and analyze measurement uncertainties and error propagation. These studies are also designed to improve understanding of where losses occur in the nuclear fuel cycle, their causes, and the practical limits for reducing them. In addition, the studies should help identify those portions of the nuclear fuel cycle where diversions are most likely to be difficult to detect because of the inherent loss mechanisms at work.

For background information see REACTOR, NUCLEAR; REACTOR, NUCLEAR (CLASSIFICATION) in the McGraw-Hill Encyclopedia of Science and Technology. [DELMAR L. CROWSON]

Bibliography: W. C. Bartels and W. A. Higinbotham, in *Safeguards Techniques*, vol. 1, International Atomic Energy Agency, Vienna, 1970; D. L. Crowson, in *Safeguards Techniques*, vol. 1, International Atomic Energy Agency, Vienna, 1970; W. A. Higinbotham, *Phys. Today*, 22(11):33–37, 1969; *Safeguards Glossary*, WASH-1162, U.S. Atomic Energy Commission, 1970; B. W. Sharpe, *Phys. Today*, 22(11):40–44, 1969.

Oil and gas well completion

New liquid or slurry explosives are available to introduce totally random fracture patterns within the formation. Basically there are two methods for stimulating a well: to concentrate the detonation in the borehole, or to inject the explosive into an existing fracture pattern. Both systems are new techniques and extend explosive fracturing to deeper and hotter zones. Results range from the spectacular to the mediocre, but the mediocre results stem from tests in wells with poor production potential. Materials are available which can be used safely in such hostile environments as 300°F and 10,000 psi (422K and 6.9×10^7 N/m²).

The renewed interest in explosives as a well completion and stimulation technique is due to the application of several new families of explosives. While the "shooting" of wells with explosives was among the earliest of the methods of stimulating petroleum production, it declined rapidly with the advent of hydraulic fracturing some 25 years ago. This decline was precipitated by safety problems encountered as wells were drilled deeper to hotter formations, post-shot clean-out problems, the increased use of fully cased holes, and, of course, the successes experienced with hydraulic fracturing and acid treatments. Now only a few relatively shallow wells are shot with the nitroglycerin-base materials and the shooters and torpedo companies are a vanishing breed.

Recently the newer liquidlike slurry explosives have been applied to well completion problems.

This is due to a combination of factors. First, safer, higher energy explosives became available. Second, the mechanics of how rock breaks up under high impulse loads was applied to the downhole situation. Third, there are many rock strata which do not respond to conventional hydraulic fracturing treatments.

Hydraulic fracturing exploits planes of weakness in the formation. When the stresses in the pressurized fluid system exceed the strength of the rock, the rock yields and the fluids enter the newly formed cracks. This volume of fluid pumped continues to tear along the rock's plane of weakness, thus extending the fracture from the initial fissure created at the wellbore. In many deep petroleum reservoirs, these planes are vertically oriented and parallel. The new area opened for drainage taps only a small portion of the oil reservoir. It would be more desirable to introduce many randomly oriented fractures. These omnidirectional fractures can be introduced by detonating a high-energy explosive in intimate contact with the formation. This is what makes explosive fracturing attractive, for the permeability is increased over a wide area, largely independent of any existing fracture pattern or formation planes of weakness.

System reflection dynamics. The way in which random fractures are created from explosive detonation is a major subject of rock mechanics. However, a summary of system reflection dynamics reveals the more important and relevant features.

Detonation in a borehole creates a large quantity of gas at high temperature and pressure. Gas pressure acting against the containing rock generates compressive stress and strain pulses that travel radially into the formation faster than the speed of sound in rock. This initial compressive pulse of several million psi steadily declines with time and distance. A tensile strain pulse is associated with the radial compressive strain pulse due to the rarefaction wave traveling behind the high-pressure pulse. High gas pressure and the strain wave generate hoop stresses around the borehole. Radial cracks are caused by these stress interactions.

If no free boundaries were near the explosion, the only rock damage would be the crushed zone around the shothole and some associated radial (hoop stress) cracking. However, many oil and gas reservoirs have changes in rock heterogeneity and stratigraphy which act as free surfaces or boundaries. At these free surfaces, additional rock breakage occurs due to reflected tensile and shear pulses which are generated when the compressive pulse strikes these free surfaces. The amount of energy in each reflected pulse depends on the angle of incidence of the impinging pulse. This complex dynamic situation produces a variety of shocks going back and forth in the rock. Since the strength of rock in compression is 50 to 100 times greater than in tension, the reflected tensile pulses are able to break the rock in tension far out in the rock.

The total strain in the formation is a function of phase interaction of several types of pulses. The combination of these stresses—compressive, shear, tensile, radial, hoop—result in a totally random fracture pattern independent of existing fractures or weakness planes.

To achieve multiple, large, disoriented fractures requires both an energetic explosive and efficient energy transfer. The most efficient coupling of explosive energy to rock is obtained with the explosive in direct contact with the formation face, as in open holes. In a cased well, detonation energy is lost in tearing apart and heating the casing and cement.

This is one of the major drawbacks of the nitro system, for if containers, such as nitro torpedoes (cans), are used to place the explosive, an annular barrier of well fluids must first be loaded with energy before it is transferred to the formation. This results in an energy and momentum transfer mismatch which reduces fracture propagation.

Another attribute affecting fracturing is the ability of rock to accept energy and momentum and to transfer both back into the formation. This energy transfer process is associated with the explosive's detonation characteristics. As detonation velocity increases, the efficiency of transfer decreases.

Based on this fact, the new slurry explosives developed for oil wells have lower detonation velocities than typical oil field gelatin dynamites or liquid nitroglycerin. This means the shock wave acts more like a "shove," resulting in less energy waste in rubblizing rock at the borehole and making more energy available for fracture extension.

Chemical explosives produce large amounts of high-pressure gases which act as energy reservoirs. It is logical to expect fractures produced by chemical explosive to be longer than those generated solely by heat and shock, as in a nuclear blast. In fact, comparing on an equal-energy basis, chemical explosives create fractures about twice as long as do nuclear explosives. As yet, no suitable mathematical model exists to accurately describe fracturing stress waves in rock. *See* NUCLEAR EXPLOSION ENGINEERING.

Borehole blasting. At present, two major methods of explosive stimulation are being proposed and evaluated in the industry. The first method is the borehole blasting concept in which the borehole is filled in intimate contact with a slurry explosive. This mass is then detonated, creating the gas energy and shock waves necessary to fracture the formation. This process can be repeated, creating a nuclear-type cavity after multiple shots. For example, a 7 in. (0.18 m) borehole in a 100 ft (30.48 m) thick formation could be shot with 2800 lb (1270 kg) of explosive. After cleanout, the new cavity will hold 2 to 4 times the original quantity of explosive. After a third shot, the hole would contain up to 250,000 lb (135,200 kg) of explosive and fractures would extend out 450 ft (15.75 m), depending on rock type.

The explosive may be placed in several ways: pumped down tubing between plugs, lowered on a line in polyethylene bags, or dropped from the surface in special deformable but impact-resistant bags. Typically a sealed clock-actuated detonator is lowered into the hole just prior to the final 5 to 10 ft (1.5 to 3.1 m) of explosive fill-up. Then pea gravel is dropped down the casing to form a barrier of 10 to 50 ft (3.1 to 15.5 m) on which a fast-setting cement mixture can set. The total length of this stemming column is 30 to 90 ft (9.45 to 28.4 m) and effectively contains the explosive charge. Upon detonation, all of the energy released does useful work in that the hot gaseous products are not re-

Physical properties of typical oil field explosives

Property	Slurry explosives	Nitroparaffins	Liquid nitroglycerin
Specific gravity*			
(lb/gal)	13.4	10.8	12.3
(kg/m^3)	1606	1294	1474
Viscosity*			
(centipose)	200 to semisolid plastic	30,000 to 50,000	7 to 35
$\left(\dfrac{\text{newton sec}}{\text{m}^2}\right)$	0.2 to semisolid plastic	30 to 50	0.007 to 0.035
Detonation velocity*			
(ft/sec)	19,700	23,200	24,500
(m/sec)	6000	7070	7470
Critical diameter*			
(inch)	2	1/64	<1/32
(meter)	0.05	0.004	0.008
Heat of explosion*			
(kcal/gm)	1.50	1.38	1.40
(joules/kg)	6.3	5.8	5.9
Deepest well treated			
(feet)	7600	5900	550
(meter)	2320	2000	170
Detonate with no. 8 blasting cap	No	Yes	Yes
Detonate with rifle bullet impacting steel pipe	No	Yes	Yes
BuMines 2-kg weight drop test, cm to detonate	>100	>100	15

*At 70–80°F or 294–300K.

leased to vent the hole.

Fracture detonation. The second method is the use of the pourable liquid slurry which was displaced back into the existing fracture pattern of the formation as well as in the borehole. The explosive is pumped down the tubing or the casing, depending on the bottom hole situation. When a static fluid level cannot be maintained, it is necessary to use the tubing with a modified cement retainer to act as a check valve and prevent slurry movement after the borehole and formation fractures have been filled with the desired quantity of slurry.

If the well is less than 4000 ft (1220 m) deep, it is possible to use the simpler technique of pumping down the casing. The special detonator assembly in a modified drillable packer is spotted with a wire line at the desired depth. The slurry is pumped down the casing followed by a flexible cementing wiper plug assembly. The volumes of water or other pusher fluid pumped behind the slurry is closely monitored to determine whether wiper plug seats on the packer. The wells are stemmed with 100 to 200 ft (30.5 to 71 m) of sand and detonated. In some cases, the sand tamp is blown out of the well.

The technique developed for loading the desensitized liquid nitroglycerin utilized a dump bailer. The well was killed with salt water and a static fluid level was maintained. The denser nitroglycerin was spotted on the bottom with the dump bailer and displaced into the formation by the hydrostatic head of fluid.

The explosives used in these techniques differ. The borehole blast fracturing technique has used explosives with established safety records adapted from the mining industry. These explosives are basically ammonium nitrate–aluminum slurries with high-explosive sensitizers. Some types have hollow glass spheres as sensitizers so that the explosive will detonate under pressure. The explosive is not cap-sensitive and has a relatively large critical diameter, about 2 in. (50.8 mm). The critical diameter is defined as the minimum diameter of an explosive column necessary for the detonation wave to continue and not die out.

The fracture detonating materials are usually nitroparaffins, such as nitromethane, sensitized by the addition of a small quantity of an amine. The critical diameter has to be very small in order to have the detonation occur in the narrow confines of the fracture system. Desensitized liquid nitroglycerin has been used by the U.S. Bureau of Mines scientists in a test program, but it is not available as a commercial service.

The table shows many of the properties of these three broad classes of liquid slurry explosives. But it cannot show end results. The two methods, fracture detonation and borehole shots, have yet to be used in identical reservoirs so that the effective comparisons and evaluations of the differing techniques and materials can be made. The denser borehole slurry packs more energy into every downhole region. The fracture detonation process admittedly spreads the energy thinly over a large area, but the borehole also contains a sizable portion of the energy. There are environmental restrictions, primarily temperature and chemical compatibility, for these materials. At present, the liquid nitroglycerin has been used only in relatively shallow wells. 500 ft (172.5 m) deep, where the severe temperature limitation—autodecomposition begins at 150°F (339K)—has not been important. Some 40% improvement in production was achieved. The nitroparaffin system has been used in wells as deep as 5900 ft (1799.5 m) with production improvements as great as 940%. The slurry explosive has been placed in wells as deep as 7800 ft (2979 m) and production increases of over 1100% have occurred.

These two liquid explosives are probably limited to the 180–210°F (341–379K) range before autodecomposition begins to take place. However, materials are available which have not yet been used in wells which will endure 300°F (420K) at 10,000 psi (6.9×10^7 N/m^2) for more than 72 hr. It is anticipated that high-thermal-stability slurries will be useful in stimulating the deeper, hotter wells.

For background information see EXPLOSION AND EXPLOSIVE; OIL AND GAS WELL COMPLETION; PETROLEUM RESERVOIR ENGINEERING in the McGraw-Hill Encyclopedia of Science and Technology.   [ARTHUR M. SPENCER]

Bibliography: M. A. Cook, *Industrial and Engineering Chemistry*, July, 1968; C. G. Laspe and W. H. Weigelt, *World Oil*, November, 1970; A. M. Spencer, A. L. Anderson, and G. R. Dysart, *World Oil*, November, 1970.

Optics, nonlinear

When matter is illuminated with the high-intensity electromagnetic fields available from lasers, nonlinear responses of this medium to the incoming radiation can occur. The dependence of the index of refraction of a medium on the intensity of a beam of light is the basis for the various nonlinear optical phenomena that have been studied recently. Recent advances have been made in understanding the microscopic origin of the nonlinear optical effect, which is responsible for changing the frequency of a light beam; ultrashort pulses ($\sim 10^{-11}$ sec in duration) have been used to produce an ultrafast camera and study fast relaxation times; continuously tunable Raman oscillators operating in the infrared wavelength ($\lambda \sim 10.9-13~\mu$) have been developed; and the desire to study nonlinearities at even shorter wavelengths has led to the investigation of nonlinearities at x-ray wavelengths ($\lambda \sim 10^{-8}$ cm).

Nonlinear susceptibility. When an electric field E (for example, the electromagnetic field of a light wave) is applied to matter, a polarization P is produced. This polarization can be expressed by a power series in the applied field, as in Eq. (1). The first term, which is linear in E, relates to the usual linear susceptibility χ, which is directly related to the index of refraction. The second term describes the lowest order nonlinear optical susceptibility d and vanishes if the medium posesses a center of inversion. If the applied field has components oscillating at frequencies ω_1 and ω_2, then E^2 will have frequency components at both the sum (ω_3) and difference frequencies. Eq. (2) exemplifies this.

$$P = \chi E + dE^2 \qquad (1)$$

$$\omega_1 + \omega_2 = \omega_3 \qquad (2)$$

Thus the nonlinearity d makes the generation of new light frequencies possible and is the basis of important devices such as optical mixers and parametric oscillators.

A good understanding of the microscopic origin of the nonlinear susceptibility d has been achieved through the use of a model which attributes the susceptibilities to the motion of a bonding charge located between the atomic sites. This model relates the nonlinear susceptibility to the microscopic parameters describing the bond, such as the bond polarizability, ionicity, and atomic sizes.

This bond-charge model shows that there are two important contributions to the acentricity (and hence the nonlinearity), namely the difference in electronegativity and the difference in atomic sizes of the two atoms composing the bond. Since these two contributions can be either positive or negative, cancellation effects leading to small coefficients can occur. (This must be avoided in device applications.) The agreement with experiment is excellent (Fig. 1), explaining both the magnitude and sign of the second harmonic coefficient. The absolute sign is of fundamental importance and has been measured recently for a large number of materials.

Ultrashort pulses. Pulses of the order of 10 picoseconds (1 psec = 10^{-11} sec) have been obtained by mode locking neodymium glass lasers. These very short pulses can be used to study a wide variety of ultrafast processes, such as relaxation times in gases, liquids, or solids. *See* SEMICONDUCTOR.

An extremely fast camera having a shutter capable of opening and closing in 10 psec has been demonstrated. This new shutter employs a Kerr cell between crossed polarizers and is activated by the intense electric field of the laser pulses, instead of the usual electrically operated Kerr cell. This camera has been used to photograph the enormously fast-moving light pulses themselves.

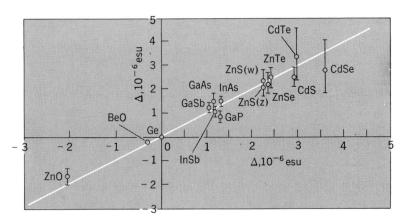

Fig. 1 The straight line shows the good agreement between the calculated nonlinear susceptibility (using the bond-charge model) plotted horizontally and experiment plotted vertically. The quantity plotted is the normalized nonlinear susceptibility (Miller's delta), defined as $\Delta = d/\chi^3$; w is wurtzite and z is zincblend; esu is electrostatic units.

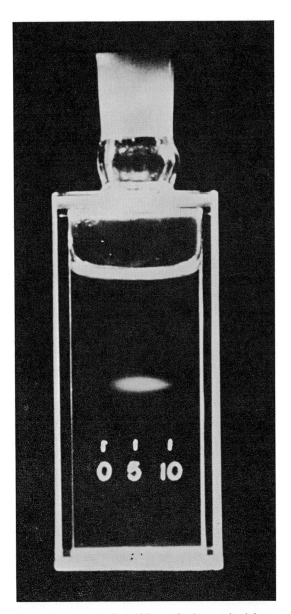

Fig. 2 Photograph of a rapidly moving laser pulse taken with the ultrafast camera.

Figure 2 shows a light pulse traveling from right to left; because of the great speed of the shutter, it has been photographically "stopped" and appears to be standing still inside the liquid cell. This is analogous to the "stopping" of a bullet in flight by illuminating it with a short burst of light.

The speed of the shutter is determined both by the duration of the optical driving pulse as well as by the molecular relaxation time of the Kerr liquid used. Hence, by measuring the shutter response, this relaxation time can be determined.

Raman effect. When a beam of light interacts with matter, some of the energy in an incident light photon may be lost by creating an excitation in the medium. If this occurs, the scattered (outgoing) photon will have less energy, that is, a lower frequency (ω_{out}), than the incident photon (frequency ω_{in}), as in Eq. (3), where ω_{ex} is the energy of the

$$\omega_{out} = \omega_{in} - \omega_{ex} \qquad (3)$$

excitation in the medium. If the incident light beam (the pump) is intense enough and the scattering efficiency is high enough, then one can make an oscillator (called a Raman oscillator) with an output frequency ω_{out} given by Eq. (3). Thus intense light sources at new discretely shifted frequencies can be obtained.

Recently a continuously tunable Raman oscillator has been demonstrated by scattering off of electron spin excitations. The energy of the spin in a magnetic field H is proportional to this field ($\omega_{ex} \propto H$); by varying the magnetic field, the excitation energy ω_{ex} can be changed. Hence Eq. (3) shows that the output frequency of the oscillator can be continuously tuned. Pumping a crystal of InSb (which has a large spin interaction with the magnetic field) with a 10.6-μm laser resulted in a highly monochromatic oscillator having a line width of less than 0.4 nm. By varying H between 15 and 100 kG, this oscillator could be tuned from 10.9 to 13 μm. These features of monochromaticity and tunability can be used to advantage in spectroscopy, replacing the conventional spectrometer in this frequency region. This source has also been used as part of an ultralow-level gas detector, capable of measuring concentrations as low as 0.01 parts per million. See SPECTROSCOPY.

X-ray nonlinearities. There is a natural tendency to push the study of nonlinear phenomena to ever higher frequencies. Recently a jump in frequency by over three orders of magnitudes (into the x-ray region) has been accomplished. This has proved possible due to theoretical and experimental work at Bell Laboratories and has resulted in the prediction and observation of parametric down-conversion of x-rays. This parametric process is the inverse of the mixing of two frequencies described by Eq. (2); that is, a high-energy photon at frequency ω_3 decays into two lower energy photons of frequency ω_1 and ω_2.

Quantum-mechanical calculations have related this new nonlinear phenomena to the well-known linear atomic scattering factor, enabling highly accurate predictions of the optimum crystals, orientations, and expected conversion efficiency. The experiment involved shining a 0.07-nm x-ray beam onto a properly oriented crystal. The fraction of pump photons at 0.07 nm which were converted into two x-ray photons having twice the wavelength (0.14 nm) was in accord with the theoretical predictions.

The wavelengths of the two photons resulting from the decay of a single x-ray pump photon (having a frequency ω_x) need not both be in the x-ray region (Fig. 3). In fact one wavelength could occur with an optical frequency (ω_o), while the other photon remains in the x-ray region; energy conservation, as expressed by Eq. (2), requires this photon to have a frequency ($\omega_x - \omega_o$). This x-ray–optical nonlinearity leads to one of the most interesting discoveries in this new field, namely, that it should now be possible to study in detail the valence electrons which form the chemical bonds. Ordinary x-ray experiments, while having the necessary resolution (wavelengths comparable to atomic dimensions), are not sensitive to the valence (that is, optical) electrons which determine a material's most important physical and chemical properties. The new nonlinear x-ray optical technique is very sensitive to these electrons, since it involves photons at optical frequencies. Thus, when such experiments are performed, they should lead to new insights in the fields of physics, chemistry, and biology.

For background information see ELECTRONEGATIVITY; LIGHT; OPTICS; OPTICS, NONLINEAR in the McGraw-Hill Encyclopedia of Science and Technology. [BARRY F. LEVINE]

Bibliography: M. A. Dugay, *IEEE J. Quant. Elec.*, QE-7:37, 1971; P. Eisenberger and S. L. McCall, *Phys. Rev. Lett.*, 26:684, 1971; Isaac Freund and B. F. Levine, *Phys. Rev. Lett.*, 25:1241, 1970; B. F. Levine, *Phys. Rev. Lett.*, 25:440, 1970; C. K. N. Patel and E. D. Shaw, *Phys. Rev. Lett.*, 24:451, 1970.

Orogeny

Orogeny, a term proposed by C. K. Gilbert in 1890 and also known as orogenesis, is the process by which mountainous tracts such as the Alpine-Himalayan, Appalachian, and Cordilleran orogenic belts are formed. Characteristically, orogenic belts are long and linear or arcuate, with distinctive zones of sedimentary, deformational, and thermal patterns that are, in general, parallel but asymmetric to the belts. Orogenic belts have complex internal geometry, involving extensive mass transport of very dissimilar rock sequences of dominantly marine sediments and continental crust (sial)

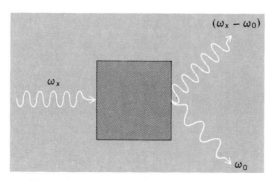

Fig. 3 Parametric decay of an input x-ray photon at frequency ω_x generating an optical photon at frequency ω_o and a slightly shifted x-ray photon at the difference frequency.

OROGENY 303

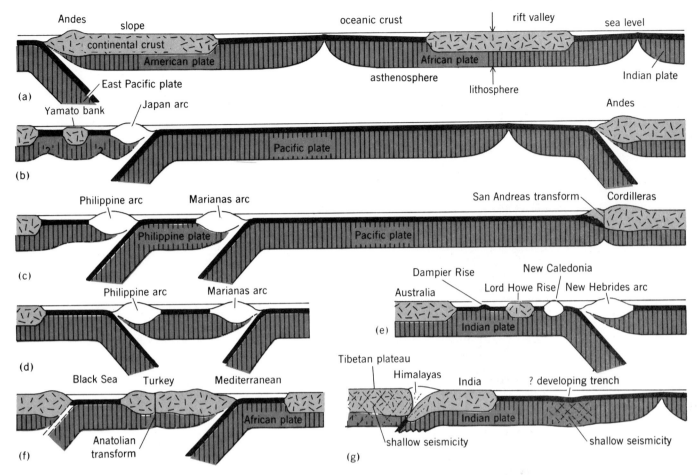

Fig. 1(a–g). Schematic sections showing world lithosphere plate, ocean, continent, and island arc relationships. (*From J. F. Dewey and J. M. Bird, Mountain belts and the new global tectonics, J. Geophys. Res., 75:2625–2647, 1970*)

and oceanic crust and mantle (ophiolite suite). Intensive deformation and metamorphism of a particular orogenic belt is relatively short-lived when compared to the time during which much of the sedimentary rock of the belt was deposited. Typically, orogenic belts are sites of abnormally thick accumulations of sedimentary and volcanic rock and severe deformation and thermal alterations.

Mountain belts are objects of intense study, argument, and mystery for geologists. Their observable features, entirely land-based, are exceedingly complex, primarily because of severe deformation; their internal features for the most part have been interpreted from surface data, although over the past several decades seismological studies have provided much new information

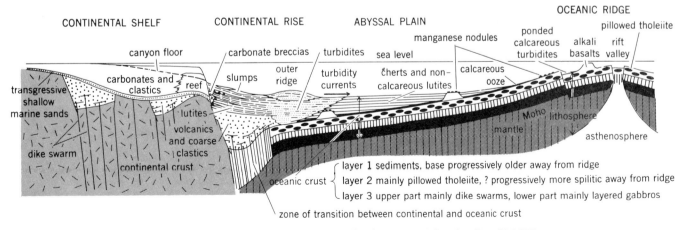

Fig. 2. Schematic section of Atlantic-type half-ocean, relationships of continental and oceanic crust, and sediments. (*From J. F. Dewey and J. M. Bird, Mountain Belts and the new global tectonics, J. Geophys. Res., 75:2625–2647, 1970*)

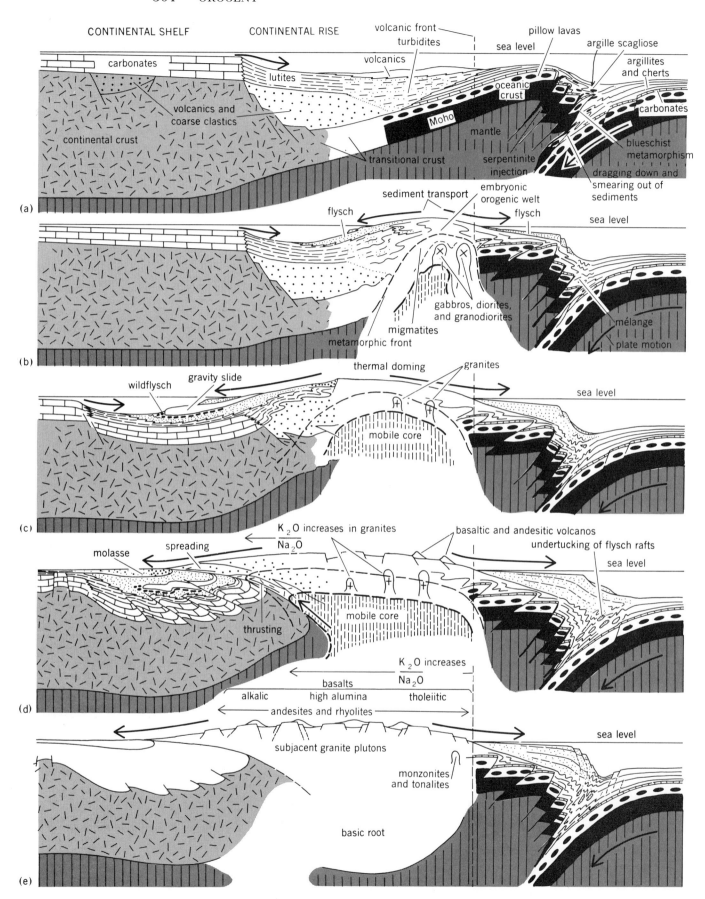

Fig. 3(a–e). Schematic sections showing conversion of continental margin to cordilleran-type mountain belt by the progressive subduction of lithosphere. (From J. F. Dewey and J. M. Bird, Mountain belts and the new global tectonics, J. Geophys. Res., 75:2625–2647, 1970)

about the interior of orogenic belts. However, it is now apparent that results of ocean-based study of the Earth provide fundamental keys to understanding orogeny.

Seismological and oceanographic studies during the past several decades have revealed a globe-encircling, seismically active belt of mountains, called oceanic ridges, and trenches with associated volcanic island arcs; this oceanic ridge and trench system is fundamentally different from orogenic belts as perceived by Gilbert. Recent models propose that orogeny is a consequence of the evolution of oceanic ridges and trenches and continental drift. Although oceanic ridges and trench–island arc systems are mountainous, they are not orogenic belts in the classical sense which restricts orogenic belts to continental masses. Orogeny results from interactions of this global, continuously evolving system of oceanic ridges and trenches, according to concepts of sea-floor spreading or, currently, lithosphere plate tectonics.

Lithosphere plate tectonics. Lithosphere plates are spherical segments of upper mantle and crust, varying in thickness from approximately 5 km at ridges to 150 km under central areas of continents, that are generated by growth of crust and mantle at oceanic ridges (accreting plate margins) and consumed in trenches (consuming plate margins or subduction zones). Plates are generated at accreting plate margins such as the Mid-Atlantic Ridge between the American and African plates (Fig. 1a), and consumed in subduction zones such as the Peru-Chile Trench, just west of the Andes, and the Japan Trench (Fig. 1b). Island arcs (unpatterned in Fig. 1) develop above subduction zones within entirely oceanic portions of plates. Various combinations of subduction zones, island arcs, and microcontinents, such as the Lord Howe Rise, are shown in Fig. 1c–f. The collision of the Indian continent with the Tibetan continental mass, with the resulting termination of subduction and development of the Himalayan collisional mountain belt, is shown in Fig. 1g.

Plates move symmetrically away from ridges, over the low-velocity channel of the mantle, with rates ranging from about 1 to 10 cm/year. They behave as more or less rigid masses; their motions are described by poles of rotation with respect to one another. Continental masses such as Africa and Australia are merely passive passengers on evolving plates; continental drift is prescribed by plate evolution. Vectors of motion (magnitude and direction) of plates are determined from F. J. Vine and D. H. Matthews's interpretation of oceanic magnetic anomaly patterns about oceanic ridges. Ocean evolution, such as Atlantic opening and Mediterranean (Tethyan) closing of the past approximately 200,000,000 years, results from plate evolution.

Fundamental concepts of orogeny based on plate tectonics are that continental margins of the Atlantic type, which develop during ocean opening, are converted to island arc–cordilleran-type (Andean) orogenic belts as lithosphere is consumed in subduction zones, below trenches, along the continental margin during closing of an ocean. A collision-type (Himalayan) orogenic belt is superimposed on a cordilleran belt upon final ocean closing, as opposing continental margins suture.

Ocean sediment accumulation. Fundamental to this mechanism of orogeny is the evolution of the whole assemblage of continental margin–deep oceanic sediments and new oceanic crust of an opening ocean. The Atlantic is taken as typical and is shown schematically in Fig. 2. Here lithosphere is being generated at the oceanic ridge and moves symmetrically away from the ridge axis. Upon opening of an ocean, overlying continental crust separates and, as it is carried away from the ridge, its trailing edges become continental margins of the continuously opening ocean and the sites of the bulk of sedimentation in the ocean.

Essentially, the sediment accumulation in the ocean is synchronous with plate accretion at the ridge. After initial rupture and distension of continental crust over an evolving accreting plate margin (the Red Sea is an example), the continental margins are stabilized as the ocean widens. Vast amounts of sediment may accumulate along the continent-ocean interface as the ocean becomes large. If a consuming plate margin (subduction zone) develops along the continental prism of sediment (as has happened along the west coast of the Americas), lithosphere plate is consumed, the ocean commences to close, vulcanism develops above the subduction zone, and the continental margin is converted into a cordilleran-type orogenic belt. This is shown schematically in Fig. 3; lithosphere is consumed in a subduction zone as the ocean closes. If the subduction zone develops near the continental margin, as shown in Fig. 3a, the margin is progressively converted to an orogenic belt, dominantly by thermally driven processes originating in the subduction zone. Schematic sections of this progressive conversion that is synchronous with lithosphere consumption

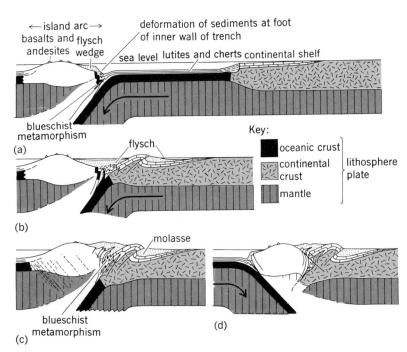

Fig. 4(a–d). Schematic sections showing collision of Atlantic-type continental margin and island arc, followed by change in direction of lithosphere plate consumption. (From J. F. Dewey and J. M. Bird, Mountain belts and the new global tectonics, J. Geophys. Res., 75:2625–2647, 1970)

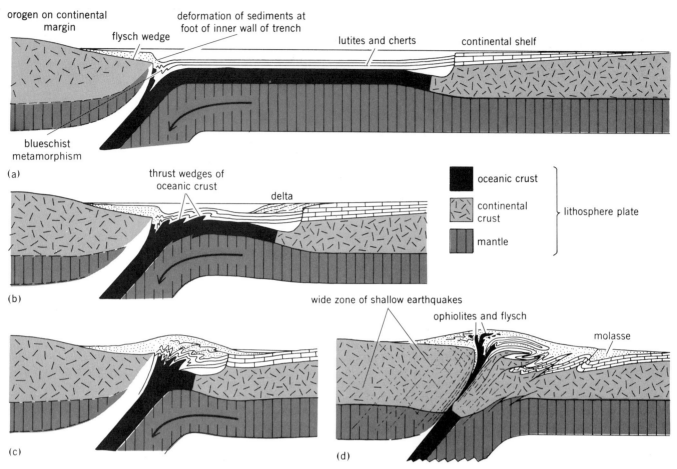

Fig. 5 (a–d). Schematic sections showing collision of two continental margins. (From J. F. Dewey and J. M. Bird, *Mountain belts and the new global tectonics*, J. Geophys. Res., 75:2625–2647, 1970)

follow in Fig. 3b through e. The trench, above the subduction zone, is the site of sedimentation of debris from the orogenic belt; high pressure–low temperature deformation occurs in the upper few kilometers of the subduction zone. Sedimentation and synchronous deformation occur on the continent-ward side of the orogenic belt.

Marginal basins such as the Philippine Sea may develop behind the trench by several mechanisms of diffuse, inter-arc sea-floor spreading, as proposed by D. E. Karig, which complicate time-space relationships of sedimentation, orogeny, and island arc vulcanism. Furthermore, depending on the direction of dip of the subduction zone, island arcs may collide with Atlantic-type margins. Several schematic relationships are shown in Fig. 4. If ocean closing commences with the lithosphere plate consumption direction as shown in Fig. 4a, in contrast to that shown in Fig. 3a, the resulting island arc and associated trench sediments and the continent, with its associated continental margin assemblage of sediments, approach one another by the consumption of the intervening ocean. Sediments fill the small ocean remaining just before collision and then are incorporated into the deformed belt resulting from final suture (Fig. 4c). However, because the main ocean continues to close, a "flip" in plate descent direction occurs (Fig. 4d), and continued cordilleran-type orogeny is superimposed on the preexisting belt.

Ultimately, as lithosphere is consumed in a closing ocean, the opposing continental margin will arrive at the island arc–cordilleran belt, resulting in a collision (Fig. 5). This results in a Himalayan-type orogeny, and because continental crust is less dense than mantle and therefore not significantly consumed in the subduction zone, plate consumption ceases (Fig. 1g). If plate consumption occurs on only one side of a closing ocean (Fig. 3), the trench-bearing margin may be associated with an existing or developing cordilleran-type orogen or a margin resulting from an island arc–continent collision (Fig. 4). The opposite continental margin will arrive at this margin as the ocean finally closes. The resulting collision, predominantly mechanically driven, superimposes an additional orogenic "event" on the margin. Ultimately the ocean disappears, the margins are sutured, and because of buoyancy restraints of the continental crust, plate consumption ceases.

Fundamental mechanism. The plate tectonics model indicates that mountain building results from two fundamental mechanisms. Island arc–cordilleran orogeny, for the most part thermally driven, occurs along leading lithosphere plate edges, above subduction zones. Continent–island arc or continent-continent collision orogeny is for the most part mechanically driven and occurs subsequently to island arc development or during final closing of an ocean or both. Actual orogenic belts are usually the result of complex combinations of these basic mechanisms. Widely differing rates of consumption, vulcanism, and deformation along consuming plate margins result because of varying distances from the poles of rotations of the involved plates. Further complexities result from expanding and contracting transform offsets of consuming plate margins. Diachronous events, more or less normal to the plate margin, may migrate along the orogenic belt as the result of oblique intersection of triple junctions (for example, the southward migration of the East Pacific Rise intersection with the Americas Trench off California).

The foregoing concerns active orogenic belts such as the Alpine system. Older, extinct orogenic belts such as the Appalachians and Urals, by analysis of the time-space relations of their rock assemblages, also resulted from ocean evolution. Although they are presently within continental masses, they mark the site of preexisting oceans that resulted from plate evolution.

As yet, no satisfactory mechanism has been found to account for the driving forces of plates. However, the existence of the world's present system of lithosphere plates cannot be denied. The Joint Oceanographic Institutions for Deep Earth Sampling (JOIDES) results indicate that practically all of the oceanic crust of the Earth underlying the oceans has been generated over the past 200,000,000 years (about 6% of geologic time) at the globe-encircling oceanic ridge system. All of the world's active orogenic belts and island arcs are along consuming plate margins, where oceanic crust is being consumed in subduction zones at rates commensurate with rates of accretion.

Therefore, either plates now in existence evolved from about Permian times concurrently with about a 1.8/1.0 global expansion and no significant consumption, or plates existed in Paleozoic and even during some Precambrian time and that plate evolution has been the mechanism by which even ancient (Precambrian) linear orogenic belts were formed. Analysis of pre-Permian orogenic belts such as the Appalachians and Urals strongly indicate they were the consequence of plate evolution. Plate tectonics models have not yet been applied to Precambrian (pre-600,000,000 years before present) orogenic belts. However, initial study suggests that plates have existed on the Earth at least to about 2,000,000,000 years before present and that their evolution is independent of any factor of global expansion or contraction.

For background information see CONTINENT FORMATION; OROGENY in the McGraw-Hill Encyclopedia of Science and Technology.

[JOHN M. BIRD; JOHN F. DEWEY]

Bibliography: J. M. Bird and J. F. Dewey, *Bull. Geol. Soc. Amer.*, 81:1031–1060, 1970; J. F. Dewey and J. M. Bird, *J. Geophys. Res.*, 75:2625–2647, 1970; J. F. Dewey and J. M. Bird, *J. Geophys. Res.*, 76:3179–3206, 1971; W. R. Dickinson, *Rev. Geophys. Space Phys.*, 8:813–851, 1970; W. Hamilton, *Bull. Geol. Soc. Amer.*, 81:2553–2576, 1970; B. J. Isacks, J. Oliver, and L. R. Sykes, *J. Geophys. Res.*, 73:5855–5900, 1968.

Parkinson's disease

Results of recent research in the most common type of idiopathic Parkinsonism—a disorder of the central nervous system—appear to indicate that the disease is related to degenerative changes in the corpus striatum, substantia nigra, and related dopaminergic neurons; the sympathetic nervous system also seems to be involved because there is a decreased production of noradrenaline, the neurohormone of the sympathetic nerves. The disease reflects itself as an aberration in the biosynthesis of dopamine (3,4-dihydroxyphenylethylamine) and noradrenaline (norepinephrine). The biosynthetic pathway for the formation of these two compounds is shown in the illustration.

Clinical picture. Parkinson's disease (paralysis agitans) is a chronic disorder of the central nervous system first described by James Parkinson in 1817. It is characterized by slowness of movement, muscular weakness, rigidity, and tremor. The patient commonly has a characteristic fixed facial expression (masked facies) and walks with a slow short, shuffling step, often demonstrating a festinating gait. There is slowness of all voluntary movements. Muscular rigidity is frequently present, and when an attempt is made to passively bend the arm or leg, it responds in a jerking fashion described as "cogwheel rigidity." The involuntary tremor commonly seen in this disorder usually begins in the upper extremity but may spread to other extremities and even the jaw and neck. When the tremor involves the fingers and thumbs, it takes on a characteristic "pill-rolling" motion. The tremor is most severe at rest, intensifies with excitement and fatigue, and disappears during sleep and with voluntary motion. Writing is difficult (micrographia). The speech is usually impaired but there are no mental changes. Sensation and reflexes are normal.

Discoveries. Whereas most of the precursors involved in the synthesis of dopamine and noradrenaline were discovered many years ago, it was not until 1951 that McC. Goodall showed that L-dopa (3,4-dihydroxyphenylalanine) and dopamine occur naturally in mammalian tissue. Since then dopamine has been demonstrated in multiple tissues, but of particular significance is its relative high concentration in the corpus striatum, substantia nigra, and dopaminergic neurons. In 1960, H. Ehringer and O. Hornykiewicz showed that in Parkinsonism there is a subnormal amount of dopamine and noradrenaline in the corpus striatum. Later Goodall demonstrated that Parkinsonism patients could not adequately synthesize noradrenaline. G. Cotzias was the first to show that the oral administration of L-dopa (the precursor of dopamine) in small amounts produced varied results but in large daily doses produced a sustained improvement.

Mode of action of L-dopa. The theoretical basis for the use of L-dopa in patients with Parkinsonism is related to a depletion of dopamine in the corpus striatum, substantia nigra, and related dopaminergic neurons. Presumably this depletion is secondary to degenerative changes in the corpus striatum and related dopaminergic neurons. Whereas the introduction of dopamine might seem to be the logical approach to treating the depletion of dopamine in idiopathic Parkinsonism, nevertheless this has not proven successful since the corpus striatum, substantia nigra, and dopaminergic neurons are considerably less permeable to circulating dopamine than to circulating L-dopa. For this reason L-dopa was used. L-Dopa enters the cells of the corpus striatum and dopaminergic neurons; there is evidence that the intraneural L-dopa is converted (decarboxylated, catalyzed by a pyridoxal phosphate linked enzyme) to dopamine, which is hydroxylated to noradrenaline by way of the same pathways as it is converted by the sympathetic nerves and ganglia (see illustration); but in the corpus striatum and in the dopaminergic neurons the conversion is more restricted to the formation of dopamine. In the sympathetic nerves the synthesis proceeds to the formation of noradrenaline.

When idiopathic Parkinsonism patients are infused with dopamine-2-C^{14}, they do not synthesize noradrenaline to the same degree as normal subjects. When these same patients are infused with L-dopa-3-C^{14}, they do not appear to utilize the L-dopa to the same degree as normal subjects, and this is reflected by a decrease in the formation of both dopamine and noradrenaline and a shift toward the formation of L-dopa metabolites. The implication of these findings is that the pathogenesis of Parkinsonism is in some way related to the synthesis of dopamine and noradrenaline.

If the therapeutic value of L-dopa were simply to replace the deficit in dopamine and noradrenaline, then one would expect that patients with severe degenerative changes to respond well. Such is not the case; the clinical results are inconsistent, since good results have been obtained with the severe as well as the mild cases. Further, the fact that not all patients improve with the use of L-dopa and often demonstrate a different response to the same dose of L-dopa emphasizes the complex nature of this syndrome. In elucidating the pathogenesis of Parkinsonism, one must also consider possible enzymatic aberrations, formation of aberrant metabolites, and/or a defect in synthesis further back than dopamine, such as the conversion of tyrosine to dopa.

L-Dopa is the current treatment of choice in Parkinsonism. By administering dopa decarboxylase inhibitors in conjunction with L-dopa, it is possible to reduce the daily intake of L-dopa by as much as one-half.

For background information see CENTRAL NERVOUS SYSTEM (VERTEBRATE); PARKINSON'S DISEASE in the McGraw-Hill Encyclopedia of Science and Technology. [MC CHESNEY GOODALL]

Bibliography: A. Barbeau and F. H. McDowell, L-*DOPA and Parkinsonism,* 1970; G. C. Cotzias et al., in *Proceedings of the 3d Symposium on Parkinson's Disease,* 1969; H. Ehringer and O. Hornykiewicz, *Klin. Wochenschr.,* 38:1236, 1960; McC. Goodall and H. Alton, *J. Clin. Invest.,* 48:2300, 1969.

Particle accelerator

Advances in the design and construction of different types of particle accelerators have been made recently. These developments are covered in this article.

Superconducting accelerator. Of the many applications of low-temperature technology to the field of nuclear physics, the one that has the greatest potential impact is the development of the superconducting linear accelerator. This device is a powerful new tool capable of providing particle beams with characteristics far superior to those available from present accelerators and at reduced cost.

Studies of the feasibility of high-power radiofrequency (rf) superconducting devices began at Stanford University in 1961, and the success of this work led to the planning for a superconducting electron accelerator and to the development of the large-scale cryogenic and superconducting materials technology necessary for this undertaking. These new techniques can be applied to the acceleration of protons, heavy ions, and mesons, as well as of electrons; and they are also applicable to the problem of secondary particle separation and to other problems in the fields of chemistry, medicine, and atomic physics. More recently a number of other laboratories have become active in the development of microwave superconductivity.

General characteristics. Superconductors are perfect conductors only for direct current (dc). For alternating current (ac), there are inherent losses which increase with the frequency and decrease exponentially with decreasing temperature for temperatures well below the superconducting transition temperature. These losses are low, however, even for frequencies as high as 10 GHz. For suitably prepared niobium below 2K, the measured surface resistance is a factor of 10^6-10^7 less than that of copper at room temperature, and as a result, the Q factor of a niobium microwave resonator is in the $10^{10}-10^{11}$ range.

Experiment shows that at some value of the rf magnetic field this high Q decreases abruptly. In niobium cavities this value has been observed to be as high as 1100 oersteds. For type I superconductors (lead, tin, and so on) this ac critical field is identical to the dc critical field at which the superconductor returns to the normal state, and for type II materials the ac critical field can be associated with the lower dc critical field, that is, the point at which magnetic flux penetrates the superconducting surface.

One should notice in what has been said two important differences between the technology of microwave superconductors and that of superconducting magnets. The first of these is the contrasting requirements for the superconducting

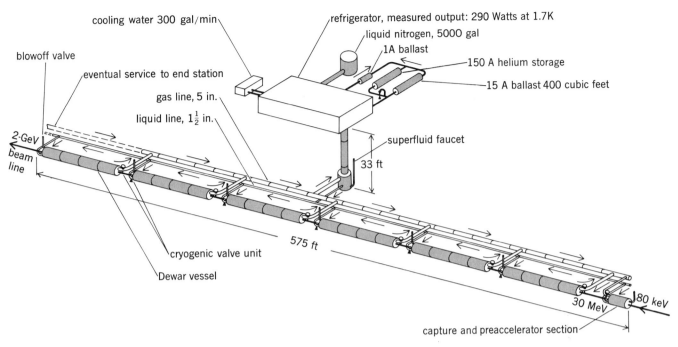

Fig. 1. Diagram of the physical layout of the superconducting linear accelerator.

materials in the two cases. In the microwave superconductor all the current flows in the penetration layer (10^{-5} cm thick) at the surface of the metal; thus careful surface preparation is essential in achieving good results. In the case of niobium, ultrahigh vacuum heat treatment and chemical or electropolish techniques are employed. For the high-field magnet, on the other hand, the alloy is chosen for high upper critical field, the current flows through the material, and surface preparation is irrelevant.

The second difference to be noticed is in the cryogenic requirements of the two applications. The microwave devices are dissipative and must operate at a very low temperature, so that the cryogenic problem centers in the areas of refrigeration, heat transport, and thermal stability. For the superconducting magnet, however, the problem is simply insulation and support of the device.

Operating techniques. Conventional copper linear accelerators require several megawatts of microwave power per foot of waveguide in order to maintain a reasonably high accelerating gradient; and because of this enormous power requirement, the machine can be turned on for only a few microseconds every few milliseconds. This poor duty cycle and the control and stability problems that go along with it limit the usefulness of the device for many types of nuclear measurements of great interest, notably high-resolution, coincidence, and low-cross-section studies.

A superconducting accelerator, with its factor of a million reduction in losses, needs only a few watts per foot, in addition to the beam power, to keep it on essentially continuously. This high duty cycle permits feedback stabilization of field gradient, phase, and current in the accelerator, resulting in an energy resolution more than two orders of magnitude better than can be achieved with conventional methods.

In addition, it is only a matter of time before the high energy gradients that have been obtained in single cavity tests (greater than 9 MeV per foot) can be realized in actual accelerating waveguides. This high gradient capability is important in accelerators for heavy particles.

The accelerator under construction at Stanford is an electron linear accelerator designed to produce a beam of 100 µA at 2 GeV energy with an energy resolution ($\triangle E/E$) of 10^{-4}. The physical layout of the machine is shown in Fig. 1. The 80-kV injected beam is first accelerated to 30 MeV by two short superconducting sections, one 2.5 ft long and the other 7.5 ft long, both in one cryogenic vessel. Following these are 24 identical sections 18.5 ft long and each in its Dewar tank module. Each of these 26 accelerator sections has its 20-kW klystron amplifier and feedback control electronics on amplitude and phase of the accelerating field; all are tuned to a single frequency within a part in 10^8 (1300 MHz); and all are locked to a reference phase line so that the beam will pass through each at the appropriate phase. There is in total about 450 ft of superconducting waveguide in the system.

The Dewar modules are assembled as shown in Fig. 1 in six groups of four, making cryogenic units about 80 ft long. Between these the electron beam passes through room-temperature pipes. The 80-ft units are manifolded onto vacuum-insulated pipelines which supply liquid helium to each from a central refrigeration plant. Each of these units contain valves which when closed allow it to be emptied of helium and warmed up independently of the rest of the system. The superfluid faucet device is a liquid helium take-off to supply experimental apparatus elsewhere in the laboratory.

The total dissipation of the accelerator operating at 2 GeV energy and the expected Q of 3×10^{10} will be about 1 kW at 1.8K. The refrigeration plant is capable of 250 W net, so that the duty cycle of the machine will be one in four. However, the system contains 20,000 l of superfluid helium, which provides an adequate thermal reservoir for accelerator on-times of several minutes at 1 kW. Thus,

Fig. 2. Photograph of sections of the superconducting accelerator.

from the users point of view, this 25% duty cycle represents no serious disadvantage.

Current status. The present state of construction of the superconducting accelerator is represented by the pieces of accelerator structure shown in Fig. 2. These are the 2.5- and the 7.5-ft accelerators mentioned above. The short section is on the right. These structures consist of a series of niobium resonators about 8 in. in diameter and 4 in. long coupled together. The vertical pipes are microwave power inputs; the mechanical devices on the side are frequency tuners; and the cables connected at various places couple out small amounts of power for monitoring and for input to the feedback systems.

These pieces of accelerator were in operation at 10 MeV output energy in July, 1971, together with the cryogenic system and units of all the other components of the accelerator system. It was expected that improvements in the niobium surface treatment techniques would permit operation of these sections at the full 30 MeV design energy before the end of 1971 and that construction would proceed rapidly enough to permit 300-MeV operations and the beginning of an experimental physics program by the end of 1972. [M. S. MC ASHAN]

Intersecting storage rings. During the past decade, experiments on high-energy proton accelerators (the CERN and Brookhaven 30-GeV proton synchrotrons in Europe and the United States and more recently the Serpukov 70-GeV accelerator in the Soviet Union) have revealed to physicists the great complexity of the nucleon. Although no breaking up of the proton is observed in high-energy collisions, many new particles (hadrons) are produced. Hadrons are of two types: baryons, which eventually decay into protons, and mesons. Models have been proposed to describe some of the observed data, but physicists are far from a complete explanation of the nucleon's complexity. It has become clear that further experiments are needed and that the most important new factor required is higher energy. Indeed, this higher energy would not only enable the production of hypothetical particles of very high mass but would also give, by means of scattering experiments, a more precise picture of the internal structure of the nucleon.

The most straightforward approach to high-energy experiments is the construction of multi-hundred-BeV accelerators, one of them being very near completion at Batavia, Ill., and another just started at CERN. A less conventional way of reaching very high energy is to observe head-on collisions of protons. In this way the available energy is considerably increased, since the collision of two protons of 30 GeV each colliding head on is identical to the collision of a 2000-GeV proton on a proton at rest. In spite of the technical difficulties involved in the construction of such an instrument and the limitations due to the low intensities, the construction of the intersecting storage ring (ISR) was started at CERN at the beginning of 1966. The first experiments were started in spring, 1971.

ISR program in context of CERN. CERN was created in 1953 in order to provide common research tools in high-energy physics for Western European scientists. Figure 3 shows the three major steps taken in the development of this center.

Phase I started in 1954. For the past 10 years research has been done using the 600-MeV synchrocyclotron and the 25-BeV proton synchrotron (PS), which were the essential elements of the initial construction program.

Phase II began in 1965, when the CERN council approved a large improvement program, the major part of which was the construction of the intersecting storage rings which started operation at the beginning of 1971.

In February, 1971, phase III began. The CERN council approved the construction of a multi-hundred-BeV proton synchrotron which is being built in such a way that many of the existing large

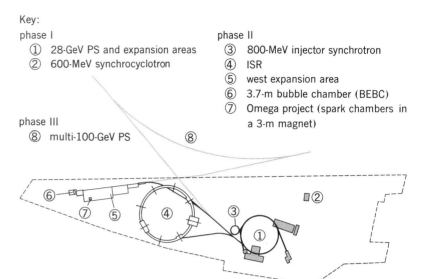

Fig. 3. Place of ISR program in CERN's general development. (*CERN*)

pieces of equipment can be easily transferred to the new machine. Thus in the coming decade the CERN laboratory should offer a complete variety of the modern pieces of equipment necessary for the study of elementary particles.

Synchrotron. As can be seen in Fig. 3, protons accelerated in the proton synchrotron (PS) are extracted and then injected into one of two concentric rings of magnets. These are interlaced and slightly distorted in order to intersect at eight points where head-on collisions may be observed.

Fig. 4. Photo showing array of counters placed downstream of each beam. (*CERN*)

The center-of-mass energy is clearly the sum of the energy of the two protons. A simple calculation shows that the same center-of-mass energy would be reached only with a multithousand-GeV accelerator (for instance, two beams of 30 BeV each provide the same center-of-mass energy as a 2000-BeV proton colliding with a proton at rest).

However, a considerable problem concerns the intensity of the beam. The interaction rate would be one per second if only one pulse from the present PS were circulating in each of the two rings. The solution to the problem is to stack many successive pulses, a delicate operation which requires picking up each pulse with an rf system and depositing it side by side with others in the vacuum chamber. The design intensity is 20 A (or 4×10^{14} particles); at the present time an intensity of 5 A has been reached. For the full 20 A the expected rate of interaction is a few 10^5 interactions per second, still many orders of magnitude lower than the ones that are obtained at conventional accelerators.

Finally it is essential to achieve an extremely good vacuum for two reasons. Stacking a beam of this intensity requires about 1/2 hr. For experiments to be possible, the lifetime of the beams must be at least of the order of hours. Furthermore during that time multiple scattering on the residual gas must not unduly blow them up. The average vacuum required around the ring is 10^{-9} torr (1 torr = 133.32 N/m²) for a lifetime of 1 day. Moreover it is essential to keep the rate of beam-gas interactions in the intersecting region significantly below the rate of beam-beam interactions: the two rates would be similar for a vacuum of 10^{-9} torr. Soon after the ISR began to operate, the average vacuum in the ring was about 10^{-10} torr. Special pumps kept the vacuum in the intersecting regions about 10^{-11} torr.

Early in 1971, although full intensities had not yet been reached in the ISR, an exciting research program on ultrahigh-energy proton-proton collisions was started. Figure 4 shows the simple array of counters that were placed downstream of each beam to detect the first collisions between protons in an energy range nearly two orders of magnitude above the present CERN PS energy.

Experimental program. Some 12 experiments were approved a year before the ISR was commissioned. Only very scarce data are available from cosmic-ray experiments at these energies, and knowledge of strong interactions is certainly not sufficient to enable physicists to predict, with any degree of certainty, the detailed features of the collisions. Thus many of these experiments are exploratory and deal with problems such as finding the production spectrum of known hadrons. Another question is whether new stable particles will be produced. Since the existence of possible heavy objects which may decay into leptons has been predicted, a search will be made for the production of muons or electrons, either singly or in pairs. *See* COSMIC RAY.

Other experiments aim at measuring the total proton-proton cross section and the elastic scattering cross section. Experiments at presently known energies show a shrinkage of the diffraction peak. Scientists ask if this continues and according to which law. Very preliminary answers to these questions were reported at a July, 1971, Amsterdam conference. *See* ELEMENTARY PARTICLE.

In parallel with these early experiments, it was thought that a powerful analyzing instrument would be needed for further investigations. A deci-

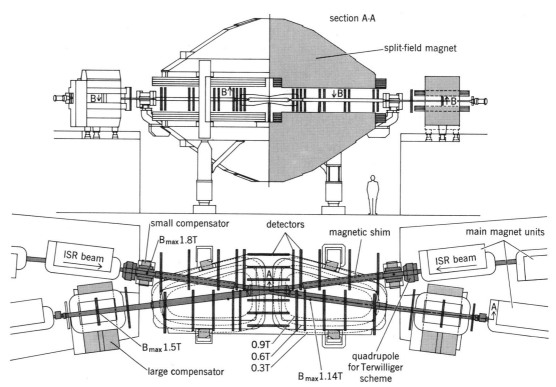

Fig. 5. Layout of the split-field magnet facility. (*CERN*)

sion was therefore made, early in 1969, to provide at one of the intersecting regions a very large magnet, shown in Fig. 5. Momentum measurement is an essential feature for the analysis of particles, and this magnet will provide the possibility of observing secondary particles in a nearly 4π solid-angle configuration. The two beams must, of course, be bent back into the ISR vacuum tube, this being done by sets of auxiliary magnets placed before and after the main one. In Fig. 5 also shown is a set of wire chambers which is being built as a standard set of detectors for the experiments still in the design stage. The complete facility will be installed at the end of 1972, and experiments will start at the beginning of 1973.

Prospects. The ISR instrument, a new method of investigating ultrahigh-energy proton-proton collisions, was proven to be feasible in 1971. Not only will the present ISR be the only instrument for many years to come to enable research to be done at energies beyond 1000 GeV, but it opens up the possibility of reaching energies that could not be obtained by other means. Intersecting rings at 300 GeV would provide collisions which, if attempted by the construction of a conventional proton synchrotron, would require a ring 1000 mi in diameter.

For background information *see* PARTICLE ACCELERATOR; Q (ELECTRICITY); SUPERCONDUCTIVITY in the McGraw-Hill Encyclopedia of Science and Technology. [B. P. GREGORY]

Bibliography: E. E. Chambers, *Nuclear Instruments and Methods*, 87:73, 1970; L. R. Suelzle, *IEEE Trans. Nucl. Sci.*, NS-18(3):146, March, 1971; J. Turneaure and V. Nguyen, *Appl. Phys. Lett.*, 16(9):333, May, 1970.

Petroleum geology

It is estimated in a recent report by the National Petroleum Council, which was based on studies by 141 geologists, that possibly 436×10^9 bbl of oil and 1227×10^{12} ft^3 of gas remain to be discovered in the many sedimentary basins of the United States. Despite favorable prospects the exploration effort has been declining, and the results have been disappointing. It is generally recognized that methods different from those of the past must be used in searching for the remaining hard-to-find accumulations.

Stratigraphic studies, which make it possible for geologists to determine the history of sedimentary basins, are being emphasized with the expectation that more successful exploration will result. Refinements in geophysical surveys and in the analysis of the data are being made, and the results are integrated with geological data in evaluating basins and individual prospects. Theories of continental drift, sea-floor spreading, and plate tectonics are studied by petroleum geologists who are pondering how the global movements may have influenced the formation of oil and gas and the present location of the accumulations. Thermal gradients in the sedimentary basins are receiving increased attention and are believed by some geologists to be the controlling factor whether oil, gas, or no hydrocarbons are present. *See* PROSPECTING, PETROLEUM.

Environmental control. It has been determined empirically from stratigraphic studies of many sedimentary basins that oil and gas occur only where the depositional environments were favorable for the development and preservation of abundant organic material. These conditions obtained in large deltas which were constructed in rapidly subsiding coastal areas and in the margins of carbonate platforms which had rapid subsidence and periodic influx of clay and silt. Most of the known petroleum accumulations are in deltaic sands and porous shelf-edge limestone reefs. Porous sand which was deposited in stream valleys or in desert dunes has no oil or gas even where the potential reservoir is sealed by nonporous sediment and is uplifted in domes or anticlines. Porous limestone which was deposited in shallow broad parts of the seas where ocean waves and currents were constantly active also is barren of hydrocarbons, though the limestone may be sealed and uplifted over structures. The slowly deposited deep-ocean sediments have little or no oil and gas potential.

The close relationship between hydrocarbon occurrence and the conditions under which the sediments containing the accumulations were deposited forces the conclusion that the depositional environment controlled the occurrence of oil and gas. It must also be concluded that the hydrocarbons were formed near where they now occur.

It appears probable that the organic material (remains of plants and animals) which was deposited with the sediment was converted to petroleum soon after burial. Gas bubbles and oil globules resulting from decay of the organic matter were able to push apart the sand and carbonate grains before the sediment was solidified. This created or increased porosity and permeability of the rock; however, continued rapid sedimentation in the deltas and the margins of carbonate platforms prevented loss of the newly formed hydrocarbons to the air.

As the hydrocarbons formed from the organic debris in the sediments, the oil and gas accumulated in porous sand or limestone at their highest structural position. The most favorable deltaic and carbonate platform environments coincided with structurally active areas of rapid subsidence; however, local segments of the sedimentary basin went down at a slower rate than the general region. Geologists refer to the local features which were relatively high during deposition as growing structures. Most oil and gas fields are in anticlines, domes, and fault traps; therefore petroleum geologists have concluded that structure controlled the accumulation. The truth is that oil and gas are not found in structural or stratigraphic traps where the depositional environment was unfavorable for the generation and preservation of abundant organic matter. Therefore the anticlinal theory of petroleum occurrence is valid only for those parts of sedimentary basins which had rapid subsidence and deposition.

Possible source beds of petroleum have been studied by geochemists who conclude that dark-colored marine shale and limestone are the "mother rocks" of the oil and gas which are concentrated in porous sandstone and limestone. It is probable that these dense petroliferous rocks, which were deposited slowly and lithified early, have retained most of their hydrocarbons. The closely associated porous rocks with petroleum accumulations may have had the source material.

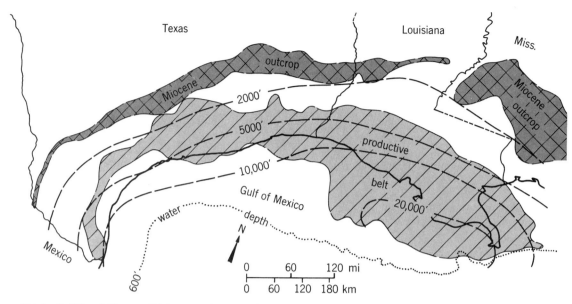

Fig. 1. Gulf Coast Miocene thickness, outcrop, and productive belt.

The probability that the occurrence of oil and gas is largely controlled by conditions which existed in the sedimentary basins during deposition makes it imperative that emphasis be placed on stratigraphic study. Instead of looking exclusively for structures in any part of the sedimentary basins, the exploration effort should be directed to areas where the depositional and structural history was favorable for the development of petroleum source material, porous reservoir rocks, and local uplifts.

The geologic history of sedimentary rocks cannot be determined without an understanding of the factors which control sedimentation. Therefore geologists are studying Recent sediments, faunas and floras in deltas, coastal interdeltaic regions, and areas where limestone and dolomite are forming. The knowledge gained is making it possible to determine the conditions under which geologically old sedimentary rocks were deposited (Fisher and others in 1969). The physical criteria are applicable to rocks of all ages, but the biological data are most useful in interpreting the depositional environments of young sediments whose fossils are similar to living plants and animals.

Gulf Coast Miocene. The very thick wedge of Miocene sedimentary rocks in coastal and offshore Louisiana and Texas clearly demonstrates the close relationship between depositional conditions and the occurrence of oil and gas. Figure 1 shows the belt in which more than 600 fields are productive from sands of Miocene age (deposited during the interval 25,000,000 to 13,000,000 years before present). It also shows how the section thickens from less than 2000 ft at the outcrop to more than 20,000 ft in southeastern Louisiana. Since several thousand wells have penetrated part or all of the Miocene in this area, much is known about the depositional history of this group of sediments and the geologic occurrence of oil and gas in the Miocene sands. This was reported by E. H. Rainwater in 1964 and 1970. This knowledge is being used in the search for Miocene accumulations within and outside the presently productive belt, and also for accumulations in formations of all geological ages in many sedimentary basins.

More than 80% of the oil and gas production is from fields in Louisiana, and all of the many large Miocene fields are in that state. The Miocene Mississippi River, like the modern one, was large and it drained an extensive area. It built deltas from Texas to the mouth of the modern river. During the early part of the period the river discharged in southwestern Louisiana; by the end of the Miocene the deltas were constructed in the general area of the modern Mississippi River delta. The numerous oil and gas accumulations are in sands which were deposited in deltas where sedimentation was rapid and organic production was prolific

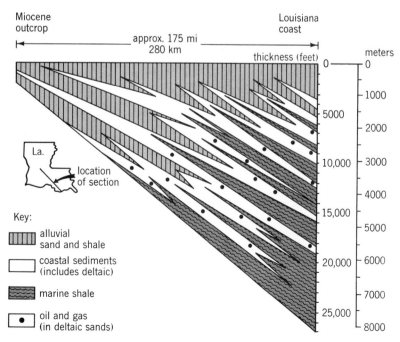

Fig. 2. Schematic section of Miocene sediments in southern Louisiana showing oil and gas accumulations in deltaic sands.

(Figs. 2 and 3). Countless sands which are equally good as reservoirs and which are sealed by impervious clay and silt have no hydrocarbon accumulations because they and the contiguous sediments were deposited where organic material was destroyed by oxidation, as in river valleys, or where the slow rate of sedimentation allowed the organic material in the marine and brackish clay and silt to be sealed in the fine-grained rock by early compaction.

The many small Miocene oil and gas fields in Texas are located where rivers, much smaller than the Mississippi River, constructed small deltas in embayed parts of the coast during early Miocene time. The growth of salt domes and ridges, which caused the flank areas to subside because salt was squeezed out, influenced the location of the deltaic depocenters and created local conditions favorable for the generation and early entrapment of small amounts of hydrocarbons.

Nearly all of the production is from sands which occur over or around salt domes and ridges or in anticlines associated with faults. These structures were forming during deposition of the sediment, and they coincide with areas which had the fastest sedimentation and the greatest production and preservation of organic material.

Probably most of the shallow and deep-seated salt domes and the prominent faults in the Miocene productive belt have been found, mainly by geophysical surveys. Numerous accumulations of oil and gas remain to be discovered in lenticular deltaic sands which are not associated with the local uplifts or which "pinch out" on flanks of the domes and anticlines. Exploration for these stratigraphic accumulations is in the early stages. Detailed studies of samples recovered from the numerous wells will make it possible to determine more accurately the depositional history of the Miocene in the Gulf Coast and to predict the location of these hidden accumulations.

For background information see PETROLEUM, ORIGIN OF; PETROLEUM GEOLOGY; PROSPECTING, PETROLEUM in the McGraw-Hill Encyclopedia of Science and Technology. [E. H. RAINWATER]

Bibliography: W. L. Fisher et al., Delta systems in the exploration for oil and gas, *Tex. Bur. Econ. Geol.*, 1969; National Petroleum Council, *Future Petroleum Provinces of the United States*, 1970; E. H. Rainwater, Geological occurrence of oil and gas in the Miocene of the Gulf Coast of the United States. *Bull. Amer. Assoc. Petrol. Geol.*, 54:5, 1970; E. H. Rainwater, Regional stratigraphy of the Gulf Coast Miocene, *Trans. Gulf Coast Assoc. Geol. Soc.*, 14:81–124, 1964.

Petroleum reservoir engineering

The purpose of petroleum reservoir engineering is accurate prediction of oil and gas reservoir performance under a variety of alternative well configurations or operating schemes or both. The need for these predictions stems largely from well-drilling and completion costs, which in offshore or remote areas currently range up to several million dollars per well. In the face of these rising costs and an increasing demand for petroleum, producers must maximize deliverability (flow rate) from existing wells and maximize the additional deliverability and oil or gas recovery contributed by newly drilled wells. It is the reservoir engineer's task to determine means of maximizing this deliverability and recovery for the lowest possible cost. See PROSPECTING, PETROLEUM.

More specifically, in the case of a given reservoir, the engineer must typically determine (1) whether to inject gas or water to maintain pressure and hence oil deliverability from existing wells; (2) whether gas or water injection will increase or decrease ultimate oil recovery; (3) optimal locations of injection wells if pressure maintenance is desired; (4) optimal drilling locations to minimize the number of additional wells required to meet a given deliverability requirement; (5) the effect of producing rate on recovery; and (6) the effects of various surface equipment designs on reservoir deliverability.

The tools for studying these questions have evolved dramatically in the past decade from the engineer's intuition and judgment to complex mathematical models which require large-scale digital computers. Recent advances in numerical computation techniques and computer size and speed, combined with engineering judgment and knowledge pertinent to the reservoir fluid mechanics, have resulted in computerized reservoir simulation models which in many cases outstrip one's ability to describe the reservoir. Thus a problem of increasing concern in reservoir engineering is the determination of the rock formation and fluid prop-

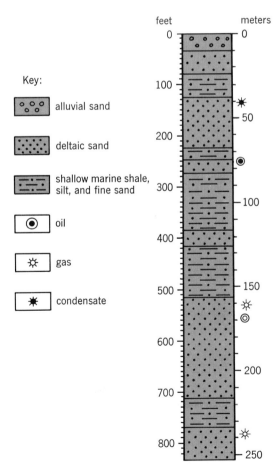

Fig. 3. Part of middle Miocene in a field in south-central Louisiana which produces oil and gas from deltaic sands.

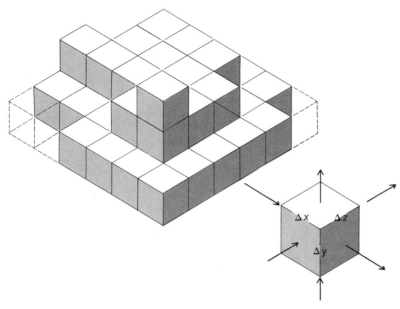

Fig. 1. Drawing of a mathematical model of a petroleum reservoir showing its finite block components or elements and the flow in and out of a typical reservoir element. Cumulative flow in minus cumulative flow out equals rate of accumulation.

relating fluid flowrate to pressure gradient, of Fourier relating heat flow to temperature gradient, and of Fick relating solute transport rate to concentration gradient. Finally, various assumptions may be invoked, such as those of one- or two-dimensional flow and single- or two-phase flow, and negligible gravity or capillary effects. The resulting set of equations is generally so complex that numerical solution on a high-speed digital computer is required. Use of computational techniques recently developed in the area of numerical analysis results in efficient computer programs capable of predicting years of reservoir performance in minutes of computer time.

Reservoir description. Reservoir description data such as permeability, porosity, structure, and boundary and fault locations constitute the bulk of the required input data for reservoir simulation models. Thus the accuracy of these data largely determines the validity of simulator predictions of reservoir performance. Until recent years, reservoir descriptions were obtained from laboratory analyses of rock samples recovered during drilling, interpretation of well-pressure buildup tests, and geological interpretations. The shortcomings of reservoir descriptions based solely on laboratory rock analyses become obvious when one considers the fraction of the total reservoir thus sampled. A rock sample from a well drilled on a typical 80 acre (323,750 m²) spacing represents a volume fraction of the 80 acres of about 6×10^{-8}. Information derived from well tests is generally more useful but frequently fails to define reservoir properties in large areas between wells sufficiently for use as input data to the simulator.

A more basic question in reservoir description is simply that of how much and what type of data are erties which are necessary input data to the predictive simulation model. Recent successful treatments of this reservoir description problem have involved trial-and-error or automatic history matching of observed reservoir performance. The computerized simulation model is run many times with the user varying the input reservoir description or fluid property data between runs or both in order to obtain a match between calculated and observed reservoir behavior. The description resulting in this agreement is then used in subsequent predictive simulator runs to study the questions listed above. See PETROLEUM GEOLOGY.

Reservoir simulation. Reservoir simulation refers to the construction and operation of a model whose behavior assumes the appearance of actual reservoir behavior. The "model" itself is either physical, such as a scaled laboratory sandpack, or mathematical. The mathematical model consists of a set of equations which, subject to certain assumptions, describes the physical processes active in the reservoir. The model obviously lacks the reality of the oil or gas field, but the behavior of a valid model assumes that of the field itself.

The mathematical model of the reservoir breaks the actual porous body into small finite blocks. Each block, although interconnected with all other blocks, assumes its own identity; that is, the properties of each block can vary, thus describing reservoir heterogeneity. Figure 1 depicts this building-block concept of the reservoir formation. This figure also shows connection of the blocks through the maintenance of a material balance around each block. Gas, oil, or water or all three flow between blocks, and equations are written expressing conservation of mass for each block and for each fluid. These equations include various phenomenological "laws" describing the rate processes active in the transport of mass or energy or both between the grid blocks representing the reservoir. Examples of these laws are those of Darcy

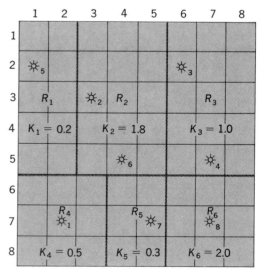

Key:

K_1 to K_6 = reservoir description parameters of permeability

R_1 to R_6 = regions into which the reservoir is divided

☼ = oil well

Fig. 2. Drawing of a single-phase gas field with a closed exterior boundary.

necessary for use in predicting future reservoir behavior under various proposed operating conditions. Data gathering, whether by rock analysis or well testing, can involve considerable expense. Thus the engineer would like to determine in advance which particular description data have the greatest effect on the answers to the questions he is facing regarding future reservoir performance. The reservoir simulator itself is employed in the context of this problem as described next.

History matching. Reservoir simulator predictions are frequently the basis of multimillion dollar decisions by management regarding field development and operation. Thus the question of validity or accuracy of the simulator is extremely important. A partial but necessary indication of this validity is the ability of the model to duplicate or nearly duplicate past reservoir performance (history). Nearly all reservoir studies using simulators involve a history matching phase to determine a reservoir description resulting in a good match of past performance. Ideally this phase is preceded by a sensitivity study which consists of a number of computer runs to determine which portion of the reservoir description data has a significant effect as simulator input data on the simulator predictions. This avoids the senseless task of attempting to define through history matching many reservoir description parameters which have minimal effects on predicted reservoir performance. In addition, the sensitivity study shows which reservoir parameters must be uniquely defined through history matching in order to place reliance on subsequent predictions. Most sensitivity studies show that a relatively small set of parameters is important.

The purpose of the history matching phase is to uniquely determine those reservoir parameters indicated by the sensitivity study to be important. Field history or performance data are presumed available for some period of time designated as the match period. The available data may reflect single- or multiphase, multidimensional flow, and the performance data to be matched may be any mix of observed pressures, producing rates, and gas-oil or water-oil producing ratios or both. The observed field performance may correspond to a period of depletion or injection or both. The engineer adjusts input data values of the important reservoir description parameters between successive runs of the simulator until he obtains an acceptable match of observed performance. A rule of thumb for an acceptable match is a deviation between calculated and observed values of no more than 5% of pressure drawdown; that is, 5% of the difference between initial pressure and current pressure. For example, if the initial reservoir (or well) pressure were 1000 psi (690 N/cm²) and the current pressure level is 700 psi (483 N/cm²), the rule of thumb indicates that the engineer should strive for an average absolute deviation between observed and calculated pressure points of 0.05 times 300 or 6 psi (207 or 10.35 N/cm²).

In some cases, depending upon the amount and quality of available reservoir performance data, the engineer finds that he can obtain equally acceptable matches of history with significantly different values of the important reservoir description parameters. This then indicates what type of additional data or well tests or both are necessary in order to refine the reservoir description so that it is sufficient for prediction purposes. The simulator can be used to help design the additional indicated field or well tests to obtain maximum information at minimal expense.

Automatic history matching. Recent developments in the area of determining reservoir description from performance data include techniques designed to minimize the engineering time required for the history matching phase of a reservoir study. One such approach for determining a viable reservoir description requires a number of runs using a reservoir simulator, each run using a reservoir description that is random within limits specified by the engineer. Then a second program which utilizes least squares and linear programming techniques processes this computed data to determine a reservoir description. A simulator run using this description is then made in order to determine the need for "fine tuning" or local adjustments to any of the reservoir description parameters. A possible need for subsequent local (well) adjustments results from the fact that the reservoir is divided into zones or regions (Fig. 2). A set of description parameters is computed for each region. Since these regions frequently include more than one well, the local parameters such as wellbore skin and turbulence must be adjusted manually. However, these local parameters may be included in the list of reservoir description parameters used in the automatic history matching process. Thus, at the expense of additional com-

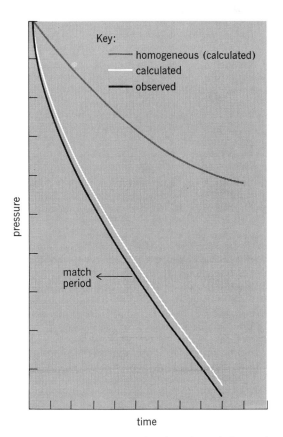

Fig. 3. Comparison of calculated results and observed results from 105–110 days for automatic history matching technique.

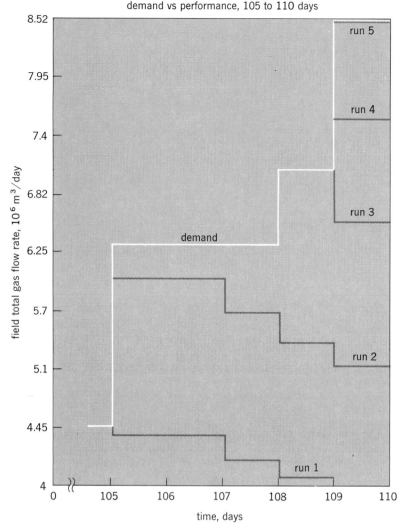

Fig. 4. Graph showing five predictions of demand versus performance in a dry gas storage field.

neous reservoir description which generated this curve is not representative of the actual reservoir. The curve labeled calculated represents the calculated performance of the reservoir using the description generated by the automatic history matching program. This good agreement indicates an excellent reservoir description, and in fact the values of permeability generated by automatic history matching agreed within 1% of the actual values given in Fig. 2.

Once a reservoir description is obtained by history matching, the simulation model becomes an excellent predictive tool. An example of this is illustrated in a published study by J. R. Dempsey and J. H. Henderson. This study was performed to maximize deliverability at minimum cost from a dry gas storage field. The sensitivity study showed the permeability-thickness product distribution throughout the field to be the important description data. This distribution was obtained using the automatic history matching program and back-pressure data (deliverability tests) as performance data. The objective of the study was to design the reservoir development and operation to meet the demand schedule shown by the broken line in Fig. 4. The highest demand rate is required at the end of the demand (withdrawal) schedule when the reservoir energy (ability to produce) is at its lowest level. For storage reservoirs, this normally occurs in late February or early March.

A number of predictive runs were made with this model to determine the most economic design that would accomplish the given deliverability schedule. Run 1 in Fig. 4 shows the result of a case where a large number of new wells were clustered in the highest deliverability area of the field. Notice that this clustering of the wells is very ineffective. The next strategy tried involved locating the same number of wells as for run 1 at the extremities of the field in order to drain the reservoir more uniformly. Run 2 shows the result of this case. Further strategies of well locations and operating procedures yielded results as shown in runs 3, 4, and 5. This application of the reservoir description as determined from matching performance demonstrates the value of this technique. The company was able to meet its required demand schedule with several fewer wells than originally predicted at a savings of more than $100,000 per well. See OIL AND GAS WELL COMPLETION.

For background information see NUMERICAL ANALYSIS; PETROLEUM RESERVOIR ENGINEERING; PROSPECTING, PETROLEUM in the McGraw-Hill Encyclopedia of Science and Technology.

[J. R. DEMPSEY; H. S. PRICE; K. H. COATS]

Bibliography: K. H. Coats, *J. Petrol. Technol.*, 21:1391–1398, 1969; K. H. Coats, J. R. Dempsey, and J. H. Henderson, *Soc. Petrol. Eng. J.*, 10(1): 66–74, 1970; J. H. Henderson, J. R. Dempsey, and J. C. Tyler, *J. Petrol. Technol.*, 22(11):1239–1246, 1968.

puting time, this manual "fine tuning" is eliminated.

Examples. The example shown in Fig. 2 represents a single-phase gas field with a closed exterior boundary. The reservoir is overlain with an 8×8 grid and this grid is divided into six regions, R_1, ..., R_6. The reservoir contains eight wells. Permeability is chosen as the reservoir description parameters, $K_1, \ldots, K_6$, to be "backed out" by the technique described above. Figure 2 also shows the true values of regional permeability.

A performance curve of pressure versus time is available for each well. Thus there are eight curves for history matching, each curve exhibiting its own characteristics and anomalies with respect to its location in the reservoir and the manner in which it had been produced. An example of the performance curve for one of these wells is the curve marked observed in Fig. 3.

Figure 3 also shows the "first picture," that is, the curve labeled homogeneous. This is usually the result of the existing initial information and the experiences and judgment of the engineer and geologist. Notice that this curve is not at all close to the curve labeled observed. Therefore the homoge-

Petroleum secondary recovery

As new oil sources become more difficult to find, the petroleum industry has increasingly applied methods which improve recovery from known oil reservoirs. A substantial amount of the potentially recoverable oil remaining in these reservoirs has a relatively high viscosity, a characteristic that has resulted in low primary recovery. As an extreme, example, the Athabasca tar sands in Canada contain over 300×10^9 bbl of high-viscosity oil with no primary recovery. This amount is almost equal to all the oil that has been discovered in the United States to date. Thermal methods of reducing oil viscosity include the injection of steam, the injection of hot water, and the application of the in situ combustion process. Of these, steam injection is by far the most successful. In California alone, approximately 200,000 bbl/day of additional oil was produced by thermal operations in 1970. Most of this was a result of steam injection. Two processes are currently being used, the steam soak and steam drive.

Steam soak. Steam soaking is a thermal stimulation method in which (1) steam is injected into a well for a number of days, (2) the well is shut in for a short period, and (3) the well is put back on production. Response from a successful soak is evidenced by a dramatically increased oil production rate along with recovery of some of the injected water. One explanation for the improvement in oil production for viscous oils is that the radius of the well bore is effectively increased to the radial distance heated around the well bore. The oil viscosity is greatly reduced with heat so that the resistance to flow is small in this area compared to the unheated region. For large reductions in oil viscosity, the well-bore radius is effectively increased to the radius of the heat front. The oil rate after steaming, Q_{hot}, can be expressed by the equation below, where Q_{cold} is the rate before

$$Q_{hot} = Q_{cold} \frac{\log(r_e/r_w)}{\log(r_e/r_h)}$$

steaming, r_e is the effective radius of drainage, r_w is the well-bore radius, and r_h is the radius of the heated zone. In addition to the beneficial effects of reducing viscosity, steam injection cleans the well bore, perforations, and producing sand face to further reduce flow resistance into the well. Over a period of months, however, the oil production rate declines and eventually another steam soak is required. Most wells respond favorably to several steam soak cycles.

Casing failures, a problem for steam stimulation at depths greater than about 3000 ft, have largely been overcome through new design techniques which now permit application at depths of $4500 \pm$ ft ($1370 \pm$ m) with 2200 psi (1.57×10^8 N/m^2) steam pressure. High-strength casing is cemented from the pay zone to the surface to prevent buckling. In some instances, the casing is also set in tension from the top of the pay zone to further prevent buckling as a result of thermal expansion when the well is heated. By using an accelerated cement at the bottom and light-weight retarded cement to the surface, the casing is pulled in tension after the accelerated cement has set and before the retarded cement hardens.

Because of its simplicity, steam soaking is the most widely used thermal process. In 1968 about 14,000 wells were being steam soaked in California.

Steam drive. The steam-drive process involves the continuous injection of steam into selected wells in an oil field and production from other, nearby wells (see illustration). Very high oil recoveries can be obtained by steam displacement. Residual oil saturations of 15% are common for relatively nondistillable oils. In reservoirs containing very volatile oils, nearly 100% of the oil can be recovered in the steam-swept region by a combination of displacement and steam distillation. If only a fraction of the oil is steam distillable, these light components are transported to the steam front, where they recondense and mix with the oil bank to form a solvent slug. As the steam zone advances, this solvent slug is displaced and redistilled to further increase recovery. Thermal expan-

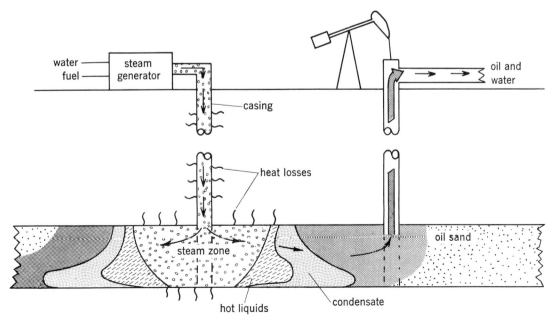

Schematic diagram of the steam-drive process.

sion of the oil and gas drive from the steam vapor phase are other mechanisms responsible for additional oil relative to that recoverable by waterflood.

The steam drive is a relatively stable displacement process. This stability is enhanced both by fluid convection and thermal conduction effects. The first stabilizing influence results because of higher volumetric vapor flow rates in the steam zone compared to the condensate flow rates ahead of the steam front. This increases the pressure gradient behind the oil bank. The second stabilizing effect causes irregularities in the steam zone to collapse as a result of heat losses normal to the direction of fluid flow.

Research, using both physical models and sophisticated analytical and numerical mathematical models, has provided a better understanding of the steam-drive process. Physical models have shown that oil displacement is generally a direct function of the steam-zone size. Therefore accurate methods of calculating steam-zone volume becomes very important for simplified prediction models. G. Mandl has shown analytically that heat flow across the steam front depends upon the injection rate, temperature, and quality of the steam. Before a certain "critical time," the heat flow across the condensation front is purely conductive, and after the critical time, heat flow is predominately convective. Steam-zone growth rate, and thus oil displacement, is reduced after the onset of convective heat transport across the condensation front. N. D. Shutler has developed a thermal numerical simulator model which couples the mass and energy equations to describe the simultaneous flow of three phases—oil, water, and gas. This type of model uses less restrictive assumptions than analytical models and should lead to improvements in the design of steam-injection projects. *See* PETROLEUM RESERVOIR ENGINEERING.

Steam driving is becoming more popular. A survey, published by the *Oil and Gas Journal* in 1966, indicated that out of 144 steam projects in operation, 44, or nearly one-third, were steam drives. The number of steam drives has increased substantially since then. One example is the Kern River, Calif., steam drive project reported by C. G. Bursell. This project has 85 injection wells with a total injection rate of 30,000 bbl/day. The oil production from 157 producing wells is 6700 bbl/day. The increasing application of steam drive compared to the steam-soak process is a result of significantly higher oil recoveries from the steam-drive process.

For background information *see* PETROLEUM RESERVOIR ENGINEERING; PETROLEUM SECONDARY RECOVERY in the McGraw-Hill Encyclopedia of Science and Technology.

[CHARLES W. VOLEK]

Bibliography: C. G. Bursell, Steam displacement—Kern River field, *J. Petrol. Technol.*, 22: 1225–1231, October, 1970; G. Mandl and C. W. Volek, Heat and mass transport in steam-drive process, *Soc. Petrol. Eng. J.*, 9(1):59–79, March, 1969: N. D. Shutler, Numerical three-phase simulation of the linear steam flood process, *Soc. Petrol. Eng. J.*, 9(2):232–246, June, 1969; B. T. Willman et al., Laboratory studies of oil recovery by steam injection, *J. Petrol. Technol.*, 13(7):681–690, July, 1961.

Photorespiration

The respiration of green plants in sunlight is quite different from "dark" respiration, although for many years most scientists tacitly assumed that they were identical. In the late 1950s the pioneering investigator of photorespiration J. P. Decker questioned this assumption and performed experiments based on the rationale that an illuminated leaf, when suddenly placed in darkness, might exhibit a momentarily perturbed dark respiration if light and dark respiration were different. The rationale proved to be correct for certain plants, as shown in Fig. 1 with *Panicum bisculatum* leaves. As the light is extinguished, photosynthesis ceases almost immediately but the leaf rapidly releases a burst of CO_2 for about a minute before gradually returning to the normal steady dark rate of respiration. The complex shape of this postillumination CO_2 curve is not yet fully explainable, but clearly light has a large momentary effect on the subsequent dark respiration, and it is assumed that light respiration also is affected.

As these types of measurements were extended to other plants, it was discovered recently that, with certain species, their postillumination CO_2 metabolism simply returns to the normal dark respiration rate, as in Fig. 1 with crabgrass. Thus, in certain species, photorespiration is not easily detectable. It was quickly recognized that most major crop plants of the world had photorespiration, whereas many other plants apparently did not. Photorespiration apparently is a completely wasteful process energetically in that no useful form of energy, such as adenosinetriphosphate (ATP), is derived from the oxidation reactions and CO_2 is lost. Hence much current research is aimed at discovering methods for controlling photorespiration either chemically, genetically, or environmentally.

Photorespiration is defined as all respiratory activity in the light in which CO_2 is released and O_2 is taken up. Immediately upon giving this definition, the problem of measuring photorespiration comes into focus because the most dominant biochemical process occurring in green plant tissues during illumination is photosynthesis. Photosynthesis in full sunlight results in a net uptake of CO_2 and evolution of O_2 which may exceed the rate of normal "dark" respiration some 10–30 times, depending upon the plant species. Obviously photorespiration and photosynthesis both involve the same gases, so that there occurs simultaneously opposite processes of gas uptake and release in an illuminated leaf. Seemingly all gas measurements to determine absolute rates of photosynthesis or photorespiration in intact tissues are somewhat in error due to this intractable problem of CO_2 and O_2 cycling inside the tissue.

Due to the inherent problems of accurately measuring photorespiration, the early research was not readily accepted by the scientific community. Recently, however, other types of information have been discovered and a great deal of older literature has been reevaluated, such that photorespiration can now be described in biochemical terms with reasonable confidence. The major substrate for photorespiration is glycolic acid. The enzymes involved in glycolic acid metabolism in leaves have been studied and a new leaf organelle, the peroxisome, which contains some of the enzymes of the glycolic acid pathway, has been isolated and partially characterized. The contribution of normal mitochondrial respiration to photorespiration is unknown, so that although peroxisomal respiration will be considered primarily, the entire process of photorespiration involves the mitochondrion.

Methods of estimating photorespiration. All measurements of photorespiration with intact tissues employ indirect methods, each of which includes at least one limiting assumption. The following methods have been employed to estimate photorespiration. The postillumination CO_2 burst discussed above and shown in Fig. 1 is a method in which it is assumed that a remnant of light respiration can be measured in the dark. A second method is to measure the release of CO_2 by an illuminated leaf into CO_2-free air. Unfortunately this method may be confounded because in photosynthesis CO_2 is required and glycolate is produced. A third method is to measure the rate of net photosynthesis in low O_2 (1%, approximately) and in 21% or some higher concentration of O_2, since photorespiration is dependent upon O_2 concentration, whereas photosynthesis generally is not. The decrease in rate of net photosynthesis in 21% O_2 compared to the lower O_2 levels is a measure of photorespiration in higher plants. A common method used to detect photorespiration is to place a plant in an illuminated sealed chamber and to allow the plant to reduce the CO_2 concentration to a point where CO_2 output equals CO_2 uptake, which is defined as the CO_2 compensation concentration. These are the most popular methods employed in current research for measuring photorespiration.

Physiological factors. Photorespiration is favored by high-intensity illumination, high temperature, high O_2 concentration, and atmospheric CO_2 concentration.

Light. The magnitude of photorespiration is correlated with light intensity in that photorespiration increases as the light intensity increases in a fashion similar to a light intensity response curve for photosynthesis. The action spectrum for photorespiration indicates that chlorophyll is the light-absorbing pigment. Photosynthetic electron transport inhibitors also inhibit photorespiration in intact tissues.

Temperature. Temperature has a pronounced effect on photorespiration. Although a reliable Q_{10} has not been reported, photorespiration increases severalfold as the temperature rises from 15 to 35°C. The temperature optimum for photorespiration is near 35°C.

Oxygen. Photorespiration in leaves increases as the concentration of oxygen is varied from 0 to 100%. Dark respiration usually is saturated near 1 to 2% oxygen, which clearly distinguishes photorespiration from dark respiration.

Carbon dioxide. Changing the CO_2 concentration between 0 to 300 ppm probably affects photorespiration indirectly in that photosynthesis is proportional to these concentrations; thus the production of glycolate is affected. At CO_2 concentrations much higher (2000- to 5000 ppm CO_2) than physiological (300 ppm CO_2), the synthesis of glycolate is markedly inhibited. Therefore at high CO_2 concentrations the metabolic contribution of photorespiration probably is minimal.

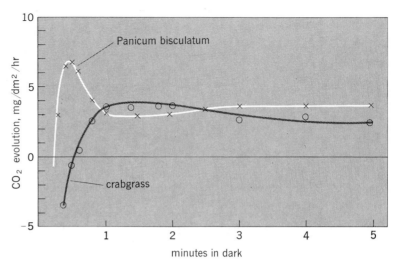

Fig. 1. Leaf postillumination CO_2 metabolism of *Panicum bisculatum*, a pentose-cycle plant, and of crabgrass, a C_4-cycle plant. The respective rates of leaf photosynthesis prior to extinguishing the lights were 34 and 45 mg of CO_2 taken up per square decimeter per hour. (*Data courtesy of R. H. Brown*)

Substrate. Several years ago it was discovered that, during photosynthesis in certain algae and in leaves, a sizeable amount of carbon can pass through glycolate, glyoxylate, glycine, and serine. Indeed in some instances as much as 50% of the photosynthetic carbon has been estimated to pass through the glycolic acid pathway. In Fig. 2 some likely pathways of glycolic acid metabolism are outlined as a guide for the path or paths of photorespiration carbon flow, and the metabolic sites of O_2 uptake and CO_2 evolution during photorespiration are indicated.

Primarily through the work of I. Zelitch it has been established that glycolic acid is the major source of carbon for the CO_2 evolved during photorespiration. For example, through the use of glycolate labeled in carbon positions 1 or 2, it has been shown that CO_2 evolved in the light can arise from the carboxyl carbon of glycolate. In Fig. 2 the asterisk denotes a labeled carbon atom; in this figure the CO_2 arises from the carboxyl carbon of glycolate. In another type of experiment glycolate oxidase, which catalyzes the oxidative conversion of glycolic to glyoxylic acid, can be inhibited fairly specifically by adding an α-hydroxysulfonate. With tobacco leaf discs at 35°C, the addition of the inhibitor stimulated photosynthetic CO_2 uptake an average of about 3.8-fold. Furthermore the inhibitor caused glycolate to accumulate in illuminated tobacco leaves, since glycolate oxidation was specifically blocked.

Synthesis of glycolic acid. Exactly how glycolate is synthesized is uncertain. It was discovered over two decades ago that glycolate is a very early labeled product of photosynthesis in $C^{14}O_2$, and in all plants studied labeled glycolate can be detected during photosynthesis. The site of glycolate production is the chloroplast, and it most likely is derived from carbon atoms 1 and 2 of a sugar phosphate of the photosynthetic reductive pentose phosphate cycle (pentose cycle). Probably a transketolase addition complex, dihydroxyethylthiamine pyrophosphate, is oxidized by a photochemically produced reductant to glycolate or 2-phosphoglycolate, depending upon whether a mono- or diphosphate sugar is utilized. Plant chloroplasts contain an enzyme, phosphoglycolate phosphatase, which irreversibly removes the phosphate.

Glycolic acid oxidation enzymes. Two enzymes in higher plants may oxidize glycolic to the corresponding keto acid, glyoxylic. Glycolic acid oxidase irreversibly catalyzes the oxidation, using oxygen as the electron acceptor, with the production of hydrogen peroxide. Glycolic acid oxidase is localized in the peroxisome. Glyoxylate reductase reversibly catalyzes the oxidation, using a pyridine nucleotide as an electron acceptor. The equilibrium for this reaction, however, strongly favors glycolic acid formation. Glyoxylate reductase is localized in the chloroplast and may act in a shuttle system with glycolic acid oxidase in the peroxisome, having a net action of effectively oxidizing excess reduced pyridine nucleotides in the chloroplast.

Catalase also is present in the peroxisome in large quantities and efficiently removes the hydrogen peroxide produced by glycolic acid oxidase action. There are reports suggesting that the glyoxylic acid is nonenzymatically decarboxylated by H_2O_2 to produce formic acid and the CO_2 of photorespiration (Fig. 2). However, this appears very unlikely, since catalase is over a 1000 times more active than any other glycolate pathway enzyme in the peroxisome. In addition, the isolated peroxisome does not produce CO_2 when fed glycolate.

Glyoxylic acid is converted to glycine through the action of an aminotransferase. Both glutamate-glyoxylate aminotransferase and serine-glyoxylate aminotransferase are present in the peroxisome and catalyze this essentially irreversible production of glycine.

Two molecules of glycine then are converted to serine, with the release of NH_3 and the CO_2 of photorespiration. Serine hydroxymethyl transferase is the enzyme which catalyzes this conversion. The peroxisome, however, does not contain this transferase; rather it is localized in the mitochondrion. Hence the CO_2-releasing reaction of photorespiration is in the same organelle which also releases the CO_2 of dark mitochondrial respiration! Again further proof of the site of CO_2 release is the discovery that isolated peroxisomes when fed glycolate do not evolve CO_2.

The fate of the serine is uncertain since it could be used in many processes, such as protein synthesis. However, by feeding labeled serine to illuminated leaves, it is known that serine is converted primarily to carbohydrates. Serine-pyruvate aminotransferase converts the serine to hydroxypyruvic acid in the peroxisome, and hydroxypyruvate reductase in the peroxisome converts the hydroxypyruvic acid to glyceric acid by utilizing reduced pyridine nucleotides. Chloroplasts contain a glyceric acid kinase which phosphorylates glyceric acid with ATP to form 3-phosphoglyceric acid, the well-known intermediate of the photosynthetic pentose cycle.

Leaf organelle for photorespiration. Plant electron microscopists a decade ago described a leaf particle of unknown function bounded by a single membrane. Recently, through the efforts of N. E. Tolbert, this leaf particle was isolated and partially characterized. The leaf particle has been termed a peroxisome since it somewhat resembles the peroxisome found in animal tissues. Morphologically peroxisomes are characterized by a single limiting membrane and a granular matrix and frequently exhibit a dense or crystalline inclusion. Peroxisomes have a buoyant density of 1.24–1.26 gm/cm³ in sucrose, which is utilized in isolating peroxisomes by gradient centrifugation. Leaf peroxisomes tend to be circular and to vary in diameter from about 0.2 to 1.5 μm.

Figure 3 is an electron micrograph of a leaf cross section demonstrating the general appearance of a leaf peroxisome in comparison to a mitochondrion and a chloroplast. Frequently in electron microscopy studies with leaves these three organelles are found in close proximity. The peroxisome frequency in leaves relative to mitochondria and chloroplasts varies with species and even in adjacent cells within leaves of plants with the C_4-dicarboxylic acid cycle (C_4 cycle) of photosynthesis, such as crabgrass. In spinach and sunflower, peroxisomes are quite frequent and represent 1 to 1.5% of the total soluble leaf protein.

Isolated leaf peroxisomes are characterized by a specific enzyme content: glycolic acid oxidase, catalase, glutamate-glyoxylate aminotransferase, serine-glyoxylate aminotransferase, serine-pyruvate aminotransferase, and hydroxypyruvic reductase. Several other enzymes are present but their role in photorespiration is uncertain. Therefore the glycolate pathway given in Fig. 2 is only partially localized in the peroxisome.

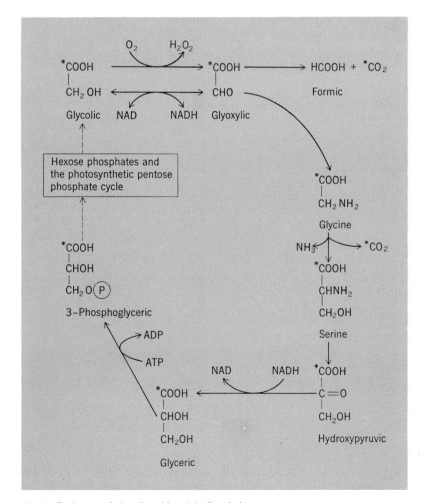

Fig. 2. Pathways of glycolic acid metabolism in leaves.

Peroxisomal respiration does not involve phosphate or phosphate esters. Thus, in contrast to mitochondrial respiration, the oxidation reactions are not linked to energy-conserving reactions such as ATP synthesis. Rather the peroxisome is characterized by the production of hydrogen peroxide and the loss of energy as heat through the destruction of hydrogen peroxide by catalase.

The complete process of photorespiration involves three cellular organelles. As depicted in Fig. 3, glycolate is synthesized in the chloroplast and moves to the peroxisome, where it is converted to glycine. The glycine moves to the mitochondrion, where CO_2 is released and serine is formed. The serine moves back to the peroxisome, where glycerate is formed, which then moves to the chloroplast and enters the pentose cycle through 3-phosphoglyceric acid.

As the cyclic pathway is given in Fig. 2, the ratio of O_2 uptake to CO_2 evolution is 1:1, since the hydrogen peroxide is split to form $H_2O + \frac{1}{2}O_2$ and two molecules of glycine are required to produce a molecule of CO_2 and serine.

Apparent absence of photorespiration. Most methods of measuring photorespiration fail to detect CO_2 from photorespiration in some plant species, such as crabgrass, corn, sugarcane, and pigweed, although an oxygen uptake with illuminated corn leaves has been observed which may be attributable to photorespiration. It should be noted that all these plants fix CO_2 photosynthetically by way of the C_4 cycle. However, leaves of these plants contain peroxisomes and enzymes of glycolate metabolism and produce glycolate. It has been speculated from the first experiments that the lack of detectable photorespiration in certain plants is due to an internal leaf recycling of the CO_2 released during photorespiration.

In relation to internal leaf CO_2 recycling, it is noteworthy that in fully developed leaves of C_4-cycle plants two distinct types of cells are present in intact tissues. Recently these distinct cell types, bundle sheath and mesophyll, were isolated from leaves of crabgrass. The glycolate pathway enzymes are present in both cell types, but the specific activities of most glycolate pathway enzymes are about fourfold higher in bundle sheath cell extracts than in mesophyll cell extracts, and in electron micrographs the number of peroxisome profiles are about threefold higher. Each layer of bundle sheath cells is surrounded by at least one layer of mesophyll cells which are low in peroxisome content and enzyme activity; thus most of the CO_2 released would have to pass through or between the mesophyll cell layer before escaping to the atmosphere. The refixation of CO_2 by the active C_4 cycle in the mesophyll cells before it can escape would greatly decrease the amount of CO_2 which eventually reaches the outside of the leaf. This situation is quite different from that in pentose plants, such as spinach, which have loosely packed mesophyll cells containing many peroxisomes and very poorly defined bundle sheath cells. Thus, because C_4 plant leaves have an unusual leaf anatomy combined with a spatial distribution of enzymes and organelles, they do not exhibit photorespiration, although the process is present in the intact leaf.

In unicellular algae the photorespiration process is somewhat different from higher plants in that

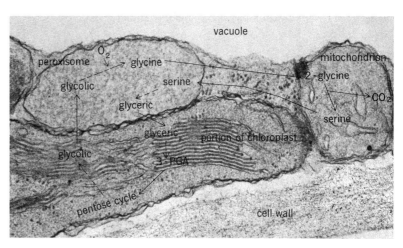

Fig. 3. Electron micrograph of a section from a *Sedum* leaf mesophyll cell. The glycolic acid pathway is inserted in outline, where broken lines indicate that other components of the pathway have been omitted. (*Courtesy of H. H. Mollenhauer*)

spatial specialization may not occur. However, the enzymes of glycolate metabolism are present and glycolate is produced. Indeed, in culture, some algae excrete large quantities of glycolate, but until additional data are available few conclusions can be reached about photorespiration in algae.

Function of photorespiration. Photorespiration is a firmly established physiological process in plant leaf metabolism. Certainly many unanswered questions remain, but the foundation for future research has been laid in that a substrate (glycolate) has been identified, the enzymes to operate the glycolate pathway are present in leaves, and a spatial separation of function has been demonstrated. Clear data are not available on the rate-limiting step or steps in photorespiration, although one appears to be the synthesis of glycolate. The ratio of CO_2 release to O_2 uptake in the most controlled experiments is about 0.25, which is much lower than the theoretical ratio of 1.0. This raises the question of oxidation of substrates external to the glycolate pathway. Clearly the control of photorespiration is a complex process, since many intermediates of other metabolic pathways are involved, plus three cellular organelles and the cytoplasm. Future research holds the possibility of uncovering means for controlling photorespiration, which could result in phenomenal increases in the production of man's food.

For background information *see* PLANT RESPIRATION in the McGraw-Hill Encyclopedia of Science and Technology. [CLANTON C. BLACK]

Bibliography: G. E. Edwards and C. C. Black, *Photosynthesis and Photorespiration*, 1971; S. E. Frederick and E. H. Newcomb, *Planta*, 96:152–174, 1971; M. Gibbs, *Ann. N.Y. Acad. Sci.*, 168:356–368, 1969; W. A. Jackson and R. J. Volk, *Annu. Rev. Plant Physiol.*, 21:385–432, 1970; N. E. Tolbert, *Annu. Rev. Plant Physiol.*, 22:45–74, 1971.

Plant growth

Chilling temperatures in the range of 0 to 15°C greatly affect plant growth. Recent research has added to the knowledge of negative and positive effects of low temperature on growing plants.

Positive effects of chilling. Cool temperatures within the biokinetic range are required by many

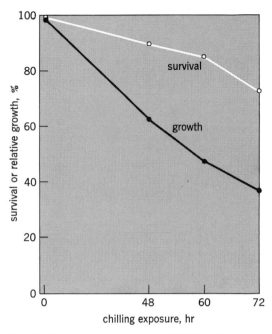

Fig. 1. Survival and relative growth (48 hr after return to 21°C) of Rainbow Flint corn seedlings subjected to 0.3°C and four exposure intervals.

plants to successfully complete their life cycles. Some plants need cold temperature to initiate bud development. Others need a period of cool temperatures before they will germinate. Night temperatures that are cooler than day temperatures are required for flowering in tomatoes and peas. Cool temperatures, along with other environmental factors, are required for development of cold-hardiness in cold-resistant perennials.

Negative effects of chilling. Many plants, especially those native to tropical and subtropical regions, are subject to chilling injuries that may prove fatal at temperatures 10°C above their freezing point. Chilling sensitivity may vary as a function of temperature and exposure time, species, growth stage, and other environmental conditions, but chilling temperatures that cause injury can be as high as 10.5°C, which is the case for bananas.

Germination. Chilling injury may occur when seeds are imbibing water at the initiation of the germination process. Lowering the temperature decreases the imbibition rates because of the increased viscosity of water and decreased membrane permeability. The extent of chilling injury during imbibition depends both on temperature and exposure time. Viability of cacao seeds, a tropical plant, is reduced to less than 1% at 6°C exposure, but the chilling injury may be reversed if the seeds are returned to warmer temperatures. However, when the seeds are exposed for 20 min to 4°C, the reaction is not reversible.

While wheat may germinate at slightly greater than 0°C, corn requires 5 to 10°C, and cotton requires greater than 10°C to initiate germination. Chilling cotton seeds during imbibition causes "nub root," resulting in poor stands. Dry lima bean and soybean seeds exhibit injury at the beginning of the imbibition period at 15 and 5°C, respectively. Injury is not observed if the seed has initially imbibed a portion of its water at room temperatures.

Growth. Young plants generally suffer more severely than do older plants exposed to similar chilling conditions. For example, 28-day-old velvet bean seedlings suffered 10% injury when exposed to temperatures of 0.5 to 5.0°C for 24 hr, whereas 14-day-old seedlings suffered twice as much injury when exposed to these same conditions. Exposing 10-day-old cotton seedlings to temperatures of 2 to 4°C proved fatal after 24 hr, but exposure for 48 hr was required to kill older seedlings.

While survival, or death, is frequently a measure of harsh injury, visual symptoms may also be seen. These include a general chlorotic appearance in cow peas, brown spots in cotton, white spots in soy beans, and white bands across the leaves in grasses, including corn.

Chilling reduces water and nutrient uptake and likewise translocation of photoassimilates. These processes frequently return to normal at warmer temperatures. R. P. Creencia and W. H. Bramlage reported that survival of Rainbow Flint corn seedlings decreased with increased exposure time to 0.3°C (Fig. 1). Corn seedlings that survived the exposure to 0.3°C continued to grow at increasing rates which, 96 hr after chilling, were similar to those of nonchilled plants.

Chilling of seedlings at certain growth stages may cause injury that is easily confused with herbicide damage. This type of chilling injury has been observed in potatoes and beans (Fig. 2).

Polycellulosic deposits called callose plugs form in phloem sieve tubes of beans exposed to low temperatures, but plugs have not been found in tomatoes. Plugs also occur in peach trees, but at lower temperatures than required for tender plants. These callose plugs undoubtedly interfere with assimilate translocation in the plant.

Chilling injury reduces marketability of certain fresh fruits and vegetables held at temperatures between freezing and 10°C. Crops subject to chilling injury are usually native to tropical or subtropical areas and include such diverse species as bananas and cucumbers. The storage life of cucumber fruits is reduced by a factor related to the product of exposure time and the degree of chilling below 10°C.

Physiological responses. Chilling produces physiological changes within tissue that may be responsible for later development of visual symptoms. A characteristic change is increased respiration rate, possibly resulting from oxidative phosphorylation uncoupling. Another effect is increased ion leakage from plant tissue, possibly resulting from increased membrane permeability. This membrane damage may allow entry of disease organisms. Cold-hardened plants generally resist these forms of injury.

Chlorophyll synthesis is sensitive to low temperatures and tomato chloroplasts are known to degenerate at temperatures below 8°C. Resulting visual symptoms are chlorosis or striping of leaves, as mentioned earlier. *See* PHOTORESPIRATION.

Inflexibility of mitochondria membranes, one site of adenosinetriphosphate (ATP) synthesis, may result from low temperature. Cotton seedlings exposed to 5°C showed continual decrease in ATP concentration with time. Plants chilled for 1 day

Fig. 2. Trifoliate bean leaf showing injury caused by 2-hr exposure to −2°C at time when this trifoliate was in the growing tip (no ice present).

and then returned to optimum temperatures restored their initial ATP concentration, but those chilled for 2 days did not (Fig. 3). Oxidative and photophosphorylation must be more sensitive to low-temperature inhibition than systems that use ATP. The decrease in ATP with chilling is avoided when seedlings are hardened to the cooler temperatures.

Enzyme responses. Dry-matter production at unfavorably low temperatures can sometimes be sustained by supplying specific metabolites. The substances may not be the same for all plants. At least part of the growth response following exposure to low temperatures is caused by a temperature-induced shortage of one or more essential metabolites. Thiamin, for example, compensates for a low-temperature-caused growth reduction in cosmos.

Ribonuclease development in lima bean seedlings normally parallels growth; however, under low-temperature stress, enzyme formation is reduced to a greater degree than is growth. The decreased enzyme formation may be due to an initial oxygen-dependent reaction, membrane damage, or blocking of intercellular air spaces during imbibition thus retarding oxygen diffusion.

Causes of low-temperature injury are still uncertain, but a theoretical basis is indicated by studies of enzyme behavior at low temperatures. At chilling temperatures well below the optimum, certain enzyme reaction rates no longer fit the Arrhenius formulation but possess an apparent activation energy (E) higher than that expected. There may be a rather sudden shift in E at a particular temperature, or a gradual transition. Thus a single enzyme reaction may become limiting to growth below a critical temperature where the activation energy changes to a high E value. This inhibition may then be expressed as a low-temperature injury reparable with a single substance.

J. K. Raison and R. M. Lyons have shown that the activation energy for succinate oxidation by mitochondria from chilling sensitive plant tissue increases from approximately 5 kcal/mole, within the temperature range of 11 to 25°C, to 35 kcal/mole between 1 to 10°C, indicating that a phase change occurs at about 10°C. No change in the activation energy was observed with mitochondria from chilling resistant tissues. Enzyme inactivation at chilling temperature may be attributable to an increase in intramolecular H-bonding, so that active centers lose their specific or essential configuration or are no longer exposed to the substrate.

For background information *see* PLANT GROWTH in the McGraw-Hill Encyclopedia of Science and Technology. [H. F. MAYLAND]

Bibliography: R. P. Creencia and W. H. Bramlage, *Plant Physiol.*, 47:389, 1971; I. L. Eaks and L. L. Morris, *Proc. Amer. Soc. Hort. Sci.*, 69:388, 1957; H. F. Mayland and J. W. Cary, *Advan. Agron.*, 22:203, 1970; J. K. Raison and J. M. Lyons, *Plant Physiol.*, 46(suppl.):38, 1970; E. E. Roos and B. M. Pollock, *Crop Sci.*, 11:78, 1971; J. McD. Stewart and G. Guinn, *Plant Physiol.*, 44:605, 1969.

Polymerase

Although the discovery of the "reverse transcriptase" (a nucleic acid–polymerizing enzyme that is present in certain RNA viruses and that synthesizes DNA molecules with a base sequence complementary to the viral RNA) was perhaps the most dramatic development in the field of nucleic acid polymerases during the past year, some important discoveries were also made concerning the nature and biological function of polymerases which use a DNA template to synthesize a complementary DNA product. In particular, it was shown that the well-studied enzyme DNA polymerase (DNA polymerase I), which was first isolated by A. Kornberg and collaborators in 1957, is not the only bacterial enzyme capable of catalyzing this reaction. A new polymerase (DNA polymerase II) has been isolated by several groups of researchers working independently. The biological role of the new enzyme is at present unclear. It may be the enzyme actually responsible for the replication of DNA, or it may be (as DNA polymerase I is now thought to be) merely involved in the repair of certain types of radiation or chemically induced damage to DNA. *See* DEOXYRIBONUCLEIC ACID (DNA).

DNA polymerase I. The discovery of DNA polymerase II was really an outgrowth of much that has been learned about DNA polymerase I since its discovery. DNA polymerase I catalyzes the reaction shown below. In the reaction dATP, dGTP, dTTP, dCTP represent the nucleotide triphos-

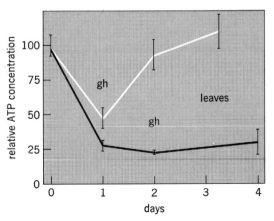

Fig. 3. Effect of chilling on subsequent adenosinetriphosphate (ATP) levels in cotton seedling leaves. Plants were chilled 1 or 2 days at 5°C and then returned to the greenhouse for 2 days.

$$\begin{matrix} n_1\,\text{dATP} \\ + \\ n_2\,\text{dGTP} \\ + \\ n_1\,\text{dTTP} \\ + \\ n_2\,\text{dCTP} \end{matrix} \xrightarrow[\text{DNA polymerase I}]{\text{Mg}^{++}\;\;\text{DNA template}} \begin{bmatrix} \text{dAMP} \\ \text{dGMP} \\ \text{dTMP} \\ \text{dCMP} \end{bmatrix}_{2n_1 + 2n_2} + 2(n_1+n_2)\text{PP}_i$$

phates derived from the bases adenine (A), guanine (G), thymine (T), and cytosin (C). PP$_i$ represents the pyrophosphate liberated by polymerization. The reaction requires a divalent cation (Mg^{++} is best), a template DNA that provides the base sequence which the enzyme copies, and the four triphosphates shown. (Certain exceptions to the above statement are not of concern here.) The initiation of polymerization requires, in addition, a primer DNA containing 3′-OH groups on the deoxyribose sugar. Replication proceeds only in the $5' \to 3'$ direction. The DNA product appears to be (except as noted below) a faithful replica of the DNA template that produced it; for example, it has the same base ratios of A, G, C, and T as the template, and the "nearest neighbor frequencies," that is, the frequency with which each base is followed by one of the four bases, is the same in template and product. These data offer good evidence that the enzyme is accurately copying the base sequence of the template DNA. Even more impressive evidence of the fidelity of replication was offered in 1967 when Kornberg, M. Goulian, and R. Sinsheimer showed that DNA polymerase, in conjuction with a second enzyme called DNA ligase, was capable of producing in the test tube a fully infective viral DNA molecule using as a template the covalently closed single-stranded circular DNA of bacteriophage ϕX174.

In spite of these successes in duplicating with DNA polymerase I in the test tube the general sort of DNA replication known to be carried out in the cell, many workers have for several years expressed doubts that DNA polymerase I was the enzyme responsible for DNA synthesis in the cell. There were reasons for such skepticism — none completely compelling by itself, but taken together, serious enough to cast doubt on the idea that polymerase I was the enzyme responsible for DNA replication.

First, polymerase I was incapable of producing biologically active DNA from a double-stranded template. Only the experiments with the rather unusual single-stranded ϕX174 DNA template were successful. The product synthesized from double-stranded templates showed a branched structure under the electron microscope — very different from the linear structure invariably seen in DNA molecules extracted from living cells.

The second piece of evidence that polymerase I was not a replication enzyme was that a large number of bacterial mutants were known with a conditional block in DNA synthesis in the cell. Unlike wild-type cells the mutants fail to replicate DNA at high temperatures, although replication proceeds normally at lower temperatures. If polymerase I were a replication enzyme, one might have expected some of these mutants to produce a defective polymerase I activity in the test tube at high temperature, but this could not be demonstrated for the bacterial polymerase. A fairly similar enzyme, which Kornberg had shown was synthesized by bacteriophage T4 when it infected *Escherichia coli*, was shown to be temperature-sensitive in the test tube when the enzyme was purified from cells infected with T4 phage carrying a mutation which led to temperature-sensitive DNA synthesis in the cell. The correlation of temperature-sensitive DNA synthesis in the cell and in the test tube for the phage enzyme made the failure to find such a correlation for the bacterial enzyme even more puzzling.

The third piece of evidence against the role of polymerase I as a replication enzyme was that the rate of DNA synthesis by polymerase I in the test tube was slower by about two orders of magnitude than the rate of DNA synthesis seen in the cell. Although one might argue that the enzyme failed to work as efficiently when removed from its normal cellular environment, this argument was not very satisfying.

Fourth, evidence from experiments with cells labeled for very brief periods with radioactive thymine suggested that recently synthesized DNA molecules were associated with the cell membrane — very likely via some of the enzymes and structural proteins which constituted the replication complex. The fact that the bulk of polymerase I was found free in the cytoplasm and not associated with the membrane argued that the enzyme was not an essential part of the replication complex.

Fifth, replication with polymerase I proceeds only in the $5' \to 3'$ direction, that is, the $5'$-triphosphate of the incoming nucleotide reacts with the $3'$-OH group of the previously polymerized nucleotide to extend the chain by one unit. Replication in the $3' \to 5'$ direction is not observed. However, several experiments with cells suggest that both strands of the parental double helix are replicated simultaneously, and since the two strands have opposite polarity (one strand $5' \to 3'$, the other $3' \to 5'$), the simplest interpretation of the data is that polymerization should occur in both the $5' \to 3'$ and $3' \to 5'$ directions. Although theoretical schemes were developed to allow polymerase I to copy both strands, they had a distinctly contrived character.

Sixth, certain antibiotics, such as nalidixic acid, which are effective inhibitors of DNA synthesis in the cell do not affect the activity of polymerase I in the test tube, suggesting either that the site of antibiotic action is not the DNA polymerase or that polymerase I is not responsible for polymerization in the cell.

Seventh, the discovery of the enzymatic repair of DNA and the realization that highly purified polymerase I contained several nuclease activities that might play a role in the removal of damaged DNA prior to repair suggested that the polymerizing activity of DNA polymerase I might be used solely for the repair of damaged DNA.

DNA polymerase II. All these data suggested that polymerase I was not the enzyme responsible for DNA replication in the cell, but they did not prove it. Clearly if one could isolate a mutant completely devoid of polymerase I activity, one could prove that the enzyme was not essential for DNA synthesis. The discovery of such a mutant (designated as strain P3478) after a long search was announced by P. De Lucia and J. Cairns in 1969. Test-tube assays showed less than 0.5% the polymerase I activity found in wild-type cells. (Cells possessing normal polymerase I activities are said to be pol A$^+$, while strains lacking the enzyme, such as P3478, are said to be pol A$^-$.) Although P3478 grew normally, it was more radiation-sensitive than the parent strain which possessed normal polymerase I activity, suggesting that polymerase I may indeed play an important role in repairing certain kinds of damage to DNA.

It could be argued that the mutation in P3478 merely affected the ability to assay polymerase I in the test tube and did not prevent the functioning of the enzyme in the cell. However, further experiments suggested that this explanation was unlikely and that the simplest interpretation, that P3478 contains little or no polymerase I activity in the cell as well as in the test tube, is correct.

But if polymerase I is not the enzyme responsible for DNA replication, what is? A possible alternative enzyme was discovered when D. Smith, H. Schaller, and F. Bonhoeffer demonstrated that a very gentle lysis procedure could be used to show the presence of what appeared to be a somewhat different polymerizing activity in E. coli. Like polymerase I, this activity required all four deoxyribonucleotide triphosphates, Mg^{++} ion, and template DNA for activity. Unlike polymerase I, this activity was sensitive to reagents, such as p-chloromercury benzoate (p-CMB), that react with the sulfhydral groups of many proteins and inactivate the enzymatic activity. In addition, unlike polymerase I, the activity was stimulated by adenosinetriphosphate (ATP). This ATP stimulation was of interest in that several recent experiments suggest that ATP is needed for DNA synthesis in the cell. This ATP-stimulated synthesis is sensitive to nalidixic acid. The DNA product of this new activity was fully denaturable on heating, like natural DNA and unlike the branched DNA product of polymerase I which fails to separate into separate strands when heated. The activity appeared to be associated with the cell membrane and replication proceeded at rates close to that found in the cell—at least during the initial phases of the reaction. Most importantly, the activity was present in undiminished amount in extracts of P3478, and it was largely insensitive to specific antibody prepared against purified polymerase I.

Recently several research groups have managed to purify a new DNA polymerase to apparent homogeneity as judged by polyacrylamide gel electrophoresis. In the course of purification, polymerase activity is clearly separated from polymerase I activity by any of several chromatographic techniques. The new enzyme is called DNA polymerase II.

The purification of polymerase II, however, has raised as many questions as it has answered. First, the purified enzyme—unlike the activity present in crude extracts—is no longer stimulated by ATP. Second, the rate of DNA synthesis seen for the purified enzyme is nearly as slow as is seen for polymerase I. Third, no mutant possessing temperature-sensitive DNA synthesis in the cell has yet been shown to produce a purified temperature-sensitive polymerase II, even though crude extracts of such mutants do show a temperature-sensitive polymerizing activity. Fourth, the purified polymerase II seems to show maximum activity when its DNA template has been partially damaged by sonication or enzymatic treatment. Thus purified DNA polymerase II differs substantially in several respects from the activity seen in crude extracts and appears to have properties very similar to polymerase I. Yet it is clearly a different enzyme based on its sensitivity to p-CMB, its resistance to antipolymerase I antiserum, and the fact that it can be isolated from cells lacking polymerase I. The difference seen between the activity found in crude extracts and purified polymerase II are at present unexplained.

The biological role of this new enzyme is currently a topic of great interest. It may ultimately turn out to be another repair enzyme, similar to polymerase I, or it may really be the long-sought enzyme responsible for DNA replication. Already there are hints of still other new polymerizing activities in the extracts, and it seems inevitable that the present concept of the replication of DNA must become a good deal more complicated before it can ever become simpler.

For background information see DEOXYRIBONUCLEIC ACID (DNA); MOLECULAR BIOLOGY; RIBONUCLEIC ACID (RNA) in the McGraw-Hill Encyclopedia of Science and Technology.

[PHILIP L. CARL]

Bibliography: R. Knippers and W. Stratling, *Nature*, 226:713, 1970; T. Kornberg and M. Gefter, *Proc. Nat. Acad. Sci. U.S.*, 68:761, 1971; R. Moses and C. Richardson, *Biochem. Biophys. Res. Comm.*, 41:1557, 1971; D. Smith, H. Schaller, and F. Bonhoeffer, *Nature*, 226:711, 1970.

Polywater

Until recently the structure of water, except for minor details, was thought to be well understood. However, during the past few years there have been several reports claiming that a new phase of water, called polywater, had been discovered. Only very small amounts of polywater are produced at one time. Consequently, important experiments which would quantitatively characterize the structure and properties of polywater have been difficult, and in some cases impossible, to perform. The absence of definitive experiments has resulted in a vigorous controversy between those who believe that the phenomenon results from artifacts and those who believe that it results from a new form of water. However, the evidence now available rather convincingly refutes the proposal that a new phase of water has been discovered.

Polywater is made by exposing very fine glass capillaries to a saturated atmosphere of water vapor. After a few days the capillaries contain a condensate which, compared to normal water, is very viscous and dense, melts at a high temperature, does not freeze, and has a refractive index similar to that of glass. Initial analyses of the infrared absorption and Raman scattering spectra of this viscous material suggested that it was a new form of water characterized by a symmetrical hydrogen bond and having a repetitive structure like a polymer. It was therefore called polywater. (Its discoverers called it "anomalous water," "modified water," or "water II.") Although early chemical analyses of polywater revealed only insignificant amounts of impurities, more recent detailed analyses have indicated very high impurity concentrations which may explain many of the reported properties. A reevaluation of the spectroscopic data has offered alternative interpretations to explain the spectra without invoking the presence of polymerized H_2O.

Historical development. In 1962, N. N. Fedyakin of Kostrama Technical Institute in the Soviet Union was studying the effect capillaries had on the physical properties of water. In these experiments Fedyakin suspended 1–10 μm diameter capillaries over a supply of water to which a small amount of sulfuric acid had been added, thereby reducing the relative vapor pressure to about 0.98 of the saturation pressure. Although condensation

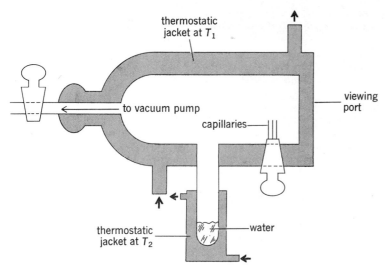

Fig. 1. Chamber for making polywater. The apparatus is attached to a vacuum system to remove the air. Temperature T_1 is slightly greater than T_2 so that no normal condensation can occur in the capillaries.

would not be expected under these conditions of unsaturation, columns of liquid were observed growing in the capillaries. Fedyakin found that the thermal expansion of this condensate differed from that of normal water and he concluded that this capillary water did not have the same structure as water in the bulk liquid.

Later in 1962, Fedyakin went to Moscow and began working with B. V. Deryagin, a member of the Soviet Academy of Sciences and an eminent surface chemist. Working together and with several collaborators, they prepared numerous samples of polywater and determined its physical properties. Their standard apparatus for making polywater is shown in Fig. 1. A source of high-purity water is placed in a chamber which is surrounded by a water bath to very accurately control the temperature. A set of quartz or Pyrex capillaries is placed in a different part of the chamber which is surrounded by a water bath at a slightly higher temperature. This temperature differential keeps the vapor pressure of the water surrounding the capillaries slightly below saturation and thereby ensures that normal condensation will not occur in the capillaries. With partial saturation pressures as low as 0.93, Deryagin and colleagues obtained condensation in the capillaries within a few days. Typically an individual capillary contained about 1 μg of material.

The condensate that is first formed is a mixture of normal water and polywater. By placing the capillary in a dry atmosphere the normal water may be removed and the polywater is concentrated. The properties of this resulting viscous residue, listed in the table, are very different from normal water. Rather than freezing at 0°C with expansion, polywater changes from a viscous liquid to a glass at −50°C with no significant volume change. Due to the very small quantities of material available, the boiling point is difficult to measure but a thermal stability determination has been made. Deryagin and coworkers found that they could distill polywater through a 500°C hot zone without modifying its properties. However, by increasing the temperature of the hot zone to 700°C, they reported that they were able to decompose the polywater into normal H_2O. Polywater is very viscous, with a consistency much like that of petroleum jelly. Although the viscosity depends on how thoroughly the normal water has been removed from the sample, in one measurement Deryagin found that it was 15 times greater than that of normal water. The refractive index is between 1.48 and 1.50, compared with 1.33 for H_2O, and the density of polywater is 1.4 (see table).

In 1969, E. R. Lippincott and G. L. Cessac from the University of Maryland and R. R. Stromberg and W. H. Grant from the National Bureau of Standards repeated some of the physical property measurements and also obtained infrared absorption and Raman scattering data on samples of polywater that they had prepared. Although they reported later that their Raman spectrum was not repeatable, their mid-infrared absorption spectrum (Fig. 2) is very reproducible and consists of strong lines at 1600, 1420, and 1365 cm^{-1} and two weaker lines near 1100 cm^{-1}. The remarkable feature of the spectrum is the complete absence of any absorptions near 3400 cm^{-1}, the band which is so characteristic of normal H_2O. On the basis of these spectroscopic results it was concluded that the sample contained no normal H_2O and that this capillary water was a polymer composed of H_2O units. Furthermore the scientists from the University of Maryland and the National Bureau of Standards claimed that their data indicated that polywater contained a new type of hydrogen bond in which the hydrogen atom was placed symmetrically between two oxygen atoms. The distance between these oxygen atoms was calculated to be about 0.23 nm, in contrast to an O . . . O distance in normal water of about 0.285 nm. Similarly they predicted that the bond energies of each of the two

Comparison of some physical properties of water and polywater

Property	Water	Polywater
Freezing point	0°C	−50°C
Boiling point	100°C	~300°C
Density	1	1.4
Refractive index	1.33	1.48–1.50
Viscosity	1	~15
Surface tension	72 dynes/cm (72 × 10^{-5} N/cm)	75 dynes/cm (75 × 10^{-5} N/cm)

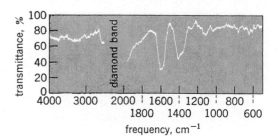

Fig. 2. Infrared spectrum of polywater. The absorption near 2200 cm^{-1} results from the diamond substrate on which the polywater was placed.

bonds would be about 50 kcal (209.3 kjoules), in contrast to normal liquid water in which the hydrogen atom is asymmetrically bound to two oxygen atoms by 120 and 4 kcal (502.4 and 16.7 kjoule) bonds.

To accommodate this new type of hydrogen bond, a model consisting of a network of hexagonal units (Fig. 3) was proposed by Lippincott and coworkers. Several additional structures have been proposed, including rhombic dodecahedra, diborane-like arrangements, and ice II–like structures.

However, the common feature of all of the proposed structures is a hydrogen atom equally spaced between two oxygen atoms separated by about 0.23 nm.

Dissent. In the early studies of polywater the emphasis was placed on evaluating the physical and spectroscopic properties and very little attention was directed toward chemical analysis. There were very many skeptics, but most of them attributed the polywater properties to a silica gel formed by the action of water leaching the walls of the capillary tubing. Deryagin argued against this possibility, stating that it would be impossible to dissolve the needed amount of silica to reproduce the polywater properties.

To resolve this question, several laboratories began microanalytical experiments on these very small samples.

In early 1970 two groups independently reported on the first of these analytical experiments. D. L. Rousseau from Bell Telephone Laboratories and S. P. S. Porto from the University of Southern California grew many polywater samples by the standard techniques, characterized their material by the infrared absorption spectrum, and then analyzed it by x-ray fluorescence, neutron activation, electron microprobe analysis, and spark-source mass spectrometry. They found that all of their samples contained large quantities of sodium, potassium, carbon, and chlorine, in addition to trace amounts of calcium, boron, silicon, nitrogen, and sulfur. Independently S. L. Kurtin and C. A. Mead from the California Institute of Technology, in collaboration with W. A. Mueller and B. C. Kurtin from Stanford Research Institute and E. E. Wolf from Hughes Research Laboratories, made electron microprobe analyses on samples they had prepared and which were characterized by the infrared spectrum. They found that their polywater contained substantial quantities of potassium and chlorine and smaller amounts of sodium and sulfur. Neither of these groups found significant amounts of silicon and therefore concluded that the properties of polywater could not be attributed to a silica gel. However, they argued that many of the reported properties probably resulted from a complicated mixture of salts.

Many other laboratories have now done analytical experiments and impurities are always found, although the relative concentrations vary considerably. Sodium is most consistent and seems to be present in nearly every sample studied. Often carbon and chlorine are also seen. In addition to these contaminants, one group found high concentrations of boron in their samples and another group found large amounts of silicon.

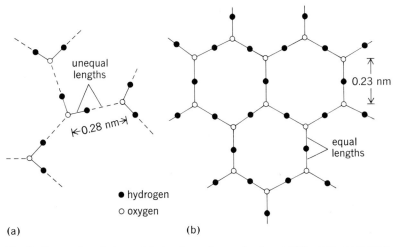

Fig. 3. Comparison between (a) structure of normal water and (b) a proposed structure for polywater. In normal water the hydrogen is asymmetrically placed between two oxygen atoms. In this model for polywater the hydrogen atom is equally spaced between two oxygen atoms.

An extremely important test for the origin of the polywater infrared spectrum is the examination of the spectrum of polywater grown from D_2O. If the polywater infrared spectrum resulted from only H_2O units, then the major lines in the spectrum involve motions of the hydrogen atoms. In polywater prepared from D_2O rather than H_2O, all of the hydrogen motions should be shifted in frequency due to the large mass change. These shifts would be readily observable in the infrared spectrum. However, if the spectrum results from impurities and the absorption bands do not result from proton motions, the basic spectral frequencies should be unchanged. In experiments performed independently in three different laboratories, polywater has been prepared by using high-purity D_2O as a source. In each case infrared analysis of the material condensed in the capillaries yielded spectra identical to those of polywater made from H_2O. No frequency shifts were seen, conclusively demonstrating that the infrared spectrum of polywater does not result from a polymer of H_2O units.

Interpretation. Polywater samples have been made in numerous laboratories and chemical analysis consistently shows that all samples are highly contaminated, although the analyses do not agree quantitatively. The specific amounts of various impurities found in each sample apparently depends on minor details in the preparative procedure. However, the presence of impurities is in itself considered a strong argument against the existence of a polymer of water. Many of the physical properties, such as the high density, refractive index, and viscosity, may be accounted for by the presence of these ionic salts.

The infrared spectrum, demonstrated to have a non-H_2O origin by the deuteration experiments, is not as readily rationalized as are the physical properties. Therefore considerable emphasis has been placed on trying to find specific materials in the chemical analysis that may give the characteristic absorption spectrum. Several possibilities have been suggested. One group of researchers have pointed out that nitrates may be formed by corona discharge in the moist air due to a charge buildup in the freshly drawn capillaries. Since nitrates have a strong absorption band near 1400 cm^{-1}, it was argued that they may be partially responsible for the infrared spectrum. Because boron was found in one set of analyses, another group has demonstrated that borates also have infrared absorption bands in the 1400-cm^{-1} region and may influence the spectrum. The presence of silicon in some analyses has led to suggestions that basic solutions of these salts absorb CO_2 from the atmosphere to form bicarbonates which have absorption bands in the 1600- and 1400-cm^{-1} regions. Simple carboxylic acid salts such as formates and acetates are consistent with most of the analyses and qualitatively have spectra very similar to the polywater spectrum. Although each of the above-mentioned contaminants may very likely influence the infrared spectrum, they all suffer from serious drawbacks when trying to quantitatively fit all of the spectroscopic data.

In a recent mass spectral analysis made on some of Deryagin's samples it was learned that they were highly contaminated with biological impurities—lipids and phospholipids. Furthermore it was pointed out that this type of material is released along with human perspiration. W. W. Mansfield of CSIRO in Melbourne, Australia, and Rousseau have independently demonstrated that residues from perspiration can very accurately reproduce the infrared spectrum. It is argued that these materials reach the capillaries either by accidental handling or by vapor condensation. Materials released with perspiration not only give a very close fit to the infrared spectrum but they are also consistent with nearly all the reported chemical analyses. Furthermore the similarities between the properties of polywater samples made in different laboratories, even in different countries, may result from the pervasiveness of these biological contaminants.

Conclusion. The large variety of impurities found in polywater can explain nearly all of its physical and spectroscopic properties. Those few that remain difficult to explain, such as the high-temperature distillation and decomposition behavior, are currently being studied by several laboratories in greater detail. Also the mechanism by which the impurities enter the capillaries has yet to be fully understood. However, a complete evaluation of all the available evidence indicates that it is extremely unlikely that a polymer of water has been discovered, but that instead a rather specialized set of conditions has led to the accidental concentration of a complicated mixture of impurities in very small capillaries.

For background information see HYDROGEN BOND; INFRARED SPECTROSCOPY; WATER in the McGraw-Hill Encyclopedia of Science and Technology.

[DENIS L. ROUSSEAU]

Bibliography: B. V. Deryagin, *Sci. Amer.*, 223: 52, 1970; J. B. Hasted, *Contemp. Phys.*, 12:133, 1971; S. L. Kurtin et al., *Science*, 167:1720, 1970; E. R. Lippincott et al., *Science*, 164:1482, 1969; W. W. Mansfield, *Search*, 1:332, 1970; D. L. Rousseau, *Science*, 171:170, 1971; D. L. Rousseau and S. P. S. Porto, *Science*, 167:1715, 1970.

Powder metallurgy

Powder metallurgy is emerging from its traditional position of being largely a technique for cold pressing and sintering of pure metal powders or blends of elemental powders to one involving the predominant use of alloyed powders. Powders of high alloy content are increasingly employed utilizing a variety of consolidation techniques which lead to fully dense, high-strength, high-integrity materials. Improved atomization techniques, under protective atmospheres, offer high yields of useful powders, and are rapidly opening up the full potential of the powder metallurgy process. A number of related studies (for example, rapid solidification of liquid metals) have shown the major advantages which result from utilizing the fine-grained, fine-structured, homogeneous powders produced by liquid atomization.

Background. Powder metallurgy until fairly recently was based on the availability of fine powders (100 mesh) (150 μm) produced by atomization, mechanical attrition, reduction of fine oxides, and various small-scale specialty techniques. Production was based largely on small batches of material and led to variability of product, high costs, and contamination. The powders are usually in elemental form and require blending of elemental powders, followed by sintering, to produce alloy compositions. Metals (and their alloys) of prime interest are iron, copper, aluminum, tungsten, and molybdenum, with smaller amounts of nickel and cobalt. The products are generally mechanically cold pressed and sintered, or are cold rolled into fragile sheet and strip, followed in each case by sintering and reworking. Full density was seldom achieved unless the product was rerolled or coined and resintered. Accordingly, properties depended on the degree of density, alloy content, and amount of cold work.

In the past several years the use of iron and steel powders, in particular, has outpaced the growth of other metals, growing from 112,000 tons in 1968, to 127,000 tons in 1969, to a predicted 135,000–150,000 tons in 1970. In the meantime, iron and steel powder production has increased well beyond 200,000 tons per year, a production capability well in excess of demand.

Powder production. Important developments have occurred in powder production capability and utilization for the metals aluminum, copper, iron and steel, molybdenum, and tungsten, which more than satisfy current requirements for press and sinter products, and cold-rolled, sintered, and rerolled sheet and strip. New uses are expected, particularly for steel powders in the area of powder metallurgy preforms for forging. This growth is estimated to be about 20% per year and will utilize some of the excess powder production capacity. The automotive industry, a major consumer of powder parts, is active in this field.

Critical to the expansion of steel powder utilization is the need for larger tonnages of low- and intermediate-alloy steel powders at low cost. High-pressure water atomization appears to be the most promising of current techniques, permitting single atomization runs of 10–20 tons, in contrast to the more usual 1000 lb (500 kg) common in recent practice. This increased production rate will result in important price decreases for the raw powder.

Of importance in the emerging powder production technology are the stainless steels, titanium, tool steels, high-speed-tool steels, and nickel-base and cobalt-base superalloys. Here powder production is more difficult because of alloy reactivity. Advantages are greater, however, because of the high cost of the alloys, machinability problems, forming difficulties, and so forth. The high-chromium alloys (stainless steels, but especially the superalloys), produced as fine powders (usually less than 100 mesh), oxidize readily and are very difficult to reduce to clean powder; accordingly, inert gas atomization, usually employing argon at high pressures is the atomization medium. Tool steels contain readily oxidizable elements such as chromium and vanadium but also are subject to carbon oxidation (decarburization).

Titanium is especially reactive with atmospheric elements and is difficult to melt in the large volumes desired for economical atomization. Therefore a number of specialty methods of powder production are utilized. In the simplest case, sponge (pure) titanium is comminuted to the desired size. To produce alloys, elemental additions or master alloy powders are blended with the titanium powder, cold pressed, and then sintered, during which time alloying takes place. Alloyed powders can be produced by comminution of alloyed scrap (which is relatively cheap) but involves problems of contamination and cost. Alloyed scrap titanium is also hydrided, becoming embrittled, is then crushed to powder, and is later dehydrided in vacuum at elevated temperatures. This tends to produce a lower level of contamination. Titanium is also produced by novel techniques, such as by atomization of a rotating electrode in an inert gas atmosphere. Because of problems in the casting of titanium parts, powder metallurgy offers one of the more attractive alternative manufacturing routes.

A new development of considerable interest, because of significantly lower costs, is coarse powder production of these same specialty materials (except titanium). These coarse powders have a usual screen analysis between 200 and 4 mesh (74 and 500 μm). While inert gas atomization would give the cleanest powders, the coarse powders are atomized instead by water, steam, air, or nitrogen and are quenched into water. The resultant powders are fine grained and are bulky or spheroidal in shape to minimize surface area and therefore contamination. Cobalt-base superalloys are produced

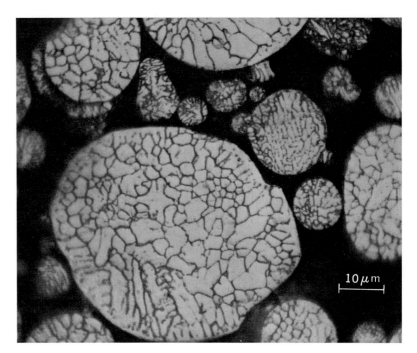

Fig. 1. Microstructure of Cu–1% Zr alloy powders.

successfully by water or steam atomization; however, nickel-base superalloys, with their very high aluminum and titanium contents, generally form flakes and are of high oxygen content, which is removed with extreme difficulty. Figure 1 shows the fine grain structure of rapidly cooled copper–1% zirconium alloy powders.

Consolidation. Alloyed powders do not cold press nearly as readily as do elemental powders. Very high pressures, die lubricants, and binders must be used. Cold isostatic pressing is a common alternate technique to die pressing since high pressures (up to 100,000 psi) (7070 kg/cm²) can be achieved, and die friction and die wear are eliminated. Better density distribution is thus achieved. However, the increase in density from cold pressing of alloyed powders is small; the prime purpose is to obtain a compact or powder body which can be handled in subsequent sintering operations or in hot forging, rolling, or extrusion. The availability in the past several years of cold isostatic presses capable of pressing compacts ranging in size from 3 to 60 lb (1.3 to 27 kg) or more has increased the size potential of powder metallurgy parts by more

Mechanical property comparisons for hot isostatically pressed powders versus conventional forms

Alloy	Condition	Yield strength, ksi (kg/cm²)	Ultimate tensile strength, ksi (kg/cm²)	Elongation, %
X-45*	As cast†	66 (4,686)	119 (8,449)	17
	Hot isostatically pressed	84 (5,964)	164 (11,644)	28
M-509*	As cast†	85 (6,035)	97 (6,887)	5
	Hot isostatically pressed	145 (10,295)	200 (14,200)	10
IN-100‡	As cast†	123 (8,733)	145 (10,295)	7
	Hot isostatically pressed	160 (11,360)	171 (12,141)	9

*A cobalt-base superalloy. †Usual condition for the alloy. ‡A nickel-base superalloy.

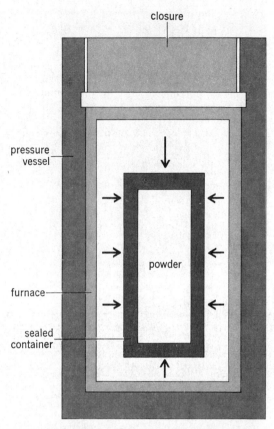

Fig. 2. Hot isostatic press showing component parts.

than an order of magnitude. Further progress will depend strongly on the availability of still larger isostatic and mechanical presses.

Of particular importance is the recent availability of hot isostatic pressing equipment capable of achieving pressures up to 30,000 psi at 2600°F (2120 kg/cm² at 1425°C). Three major industrial groups are producers of such equipment. Full density is being achieved with alloyed metallic, ceramic, and intermetallic powders.

Whereas cold isostatic pressing resulted in increases in density of perhaps 5–10% above that of gravity packing of properly sized powder charges, hot isostatic pressing achieves 100% density for exposure times of 1–2 hr at high temperatures at maximum pressure. Hot isostatic pressing is illustrated in Fig. 2. The powder, whether gravity packed, cold isostatically pressed, or mechanically compacted, is enclosed in a plastically or viscously deformable envelope, usually of thin sheet metal. The envelope is then evacuated and sealed. The sealed container is placed in the working zone of a heated furnace inside the pressure vessel. The system is purged and then pressurized with argon gas. The furnace is then heated, resulting in a pressure build up, currently from 15,000 to 30,000 psi (1060 to 2120 kg/cm²). The relatively short holding time of 1–2 hr results usually in full densification.

Spark sintering is a new technique of hot pressing which offers considerable promise. Utilized for smaller parts typically less than about 1 lb (0.45 kg), further developments await equipment development.

The hot pressed compacts can be further densified if necessary, or converted to some intermediate shape which can be hot worked by rolling, forging, or extrusion. If one starts with fine-grained, fine-structured powders, these desirable features can be retained during consolidating. Excellent work to date has shown that hot isostatic pressing can be utilized to achieve a variety of shapes and preforms. The table shows a comparison of some of the properties achieved by hot isostatically pressed alloys compared to those of cast products of the same composition.

Future. Current hot isostatic pressing equipment is largely restricted to a hot volume of perhaps 6–8 in. (15–19 cm) in diameter by about 12–20 in. (30–50 cm) high. One unit has been sold which has a work zone of 18 in. in diameter by 24 in. high (45 by 60 cm). As soon as working zones 24 in. in diameter by 36 in. high are available, which will occur shortly, costs per pound of material processed will fall to perhaps 20 cents per pound. This is an attractive, competitive price for high-priced alloys which are subject to difficult hot working with subsequent poor yield of useful product (25–50% of a cast ingot). Further increases in the hot working zone will decrease the price to about 10 cents per pound, a highly competitive price.

Parallel improvements in atomization, with yields of 85–95% of the ladle weight, will decrease atomization costs to perhaps 10 cents per pound over ladle costs for fine powders and 5 cents per pound for coarse powders. The indicated low costs should enhance the growth of the high-alloy powder materials and will stimulate cost reductions and process improvements among the pure metals and low-alloy materials for the press and sinter powder industry.

For background information see ALLOY; METAL FORMING; POWDER METALLURGY in the McGraw-Hill Encyclopedia of Science and Technology.

[NICHOLAS J. GRANT]

Bibliography: G. W. Cull, *Powder Met.*, 13:156–164, 1970; W. M. Long and P. Snowden, *Powder Met.*, 12:209–218, 1969; G. R. Sellors, *Powder Met.*, 13:85–99, 1970; Where powder metallurgy is growing, *Metal Progr.*, April, 1971.

Power plant

The siting, or location, of power plants for the generation of electricity has become the concern of many people. There is an increasing demand for more electric energy for an industrialized civilization (Table 1). How can this demand be satisfactorily met without impairment of the environment, without sacrifice of natural resources, without curtailed reliability of supply, without prohibitive increase in dollar costs to the consumer, and without

Table 1. Annual electric energy generation in the United States

| | | Percentage of total generation | |
Year	Total generation, 10⁹ kwhr	Hydro	Thermal
1930	91	34	66
1940	142	33	67
1950	329	29	71
1960	753	19	81
1968	1326	17	83

Table 2. Rankine cycle work

Initial pressure, psi (bars)	Initial steam temperature, °F (°C)	Exhaust pressure and temperature, in. Hg abs. (mm Hg), °F (°C)	Initial enthalpy, Btu/lb (J/g)	Exhaust enthalpy, Btu/lb (J/g)	Rankine cycle work, Btu/lb (J/g)
100 (6.89)	Saturated 330 (166)	29.92 (760), 212 (100)	1187 (2760)	1047 (2440)	140 (320)
100 (6.89)	Saturated 330 (166)	2.0 (50.8), 101 (39)	1187 (2760)	894 (2080)	293 (680)
100 (6.89)	Saturated 330 (166)	1.0 (25.4), 79 (26)	1187 (2760)	861 (2000)	326 (760)
148 (10.1)	Superheat 400 (204)	2.0 (50.8), 101 (39)	1220 (2840)	894 (2080)	326 (760)
800 (55)	Saturated 520 (271)	29.92 (760), 212 (100)	1200 (2790)	922 (2040)	278 (650)
800 (55)	Saturated 520 (271)	2.0 (50.8), 101 (39)	1200 (2790)	790 (1840)	410 (950)
800 (55)	Saturated 520 (271)	1.0 (25.4), 79 (26)	1200 (2790)	761 (1770)	439 (1020)
1200 (83)	Superheat 595 (313)	2.0 (50.8), 101 (39)	1229 (2860)	790 (1840)	439 (1020)

paralysis of the industrial economy? The answer is a complex problem with many facets, many of them not fully identified or resolved. *See* ENVIRONMENTAL ENGINEERING.

The industrial economy of the United States is predicated on an abundant, reliable supply of electric energy at the lowest cost. Raw energy exists in nature in many forms, but few of those sources are utilized today for the basic operation of electric power plants. Actually, the choice is (1) an elevated water supply, (2) heat from the burning of a fossil fuel, or (3) heat from the fission of a nuclear fuel. All other sources (wind, tides, sun, waves) are of no significance for the generation of electric power. *See* ELECTRICAL UTILITY INDUSTRY.

Hydropower is of decreasing importance, since most of the better sites have already been harnessed (Table 1). The new source is atomic. The prime source today is fossil fuel. Atomic and fossil-fuel systems use the heat developed in a reactor and in a furnace, respectively, for the production of steam at high pressures. Steam plants will increasingly dominate the scene in the years to come. Direct energy conversion systems are far from reality at this time.

The location of hydroplants is simple. They must be located at the site of an elevated water supply, or its equivalent, a good hydraulic gradient. Steam plants offer greater, but not unlimited, choice of location.

Rankine cycle. The underlying principles of thermodynamics dictate the selection of the proper steam cycle for the conversion of heat to electric energy. The efficiency of conversion is reflected in Table 2 for the ideal Rankine cycle. High pressures and high temperatures improve cycle efficiency, lower fuel consumption per kilowatt-hour, and conserve irreplaceable fuel resources. However, the exhaust pressure on the steam prime mover is even more influential than the live steam condition. The earliest steam engines, James Watt's, recognized the value of low exhaust pressure. The data in Table 2 demonstrate the magnitude of the economy which results from the use of exhaust pressures below the atmosphere (29.92 in. Hg at sea level, or 101,300 N/m^2). In Table 2 the Rankine cycle work with a steam pressure of 100 psi (6.89 × 10^5 N/m^2) equals 140 Btu/lb (320 J/g) of steam when exhausted at atmospheric pressure. If the exhaust pressure is lowered to 2 in. Hg abs. (6772.8 N/m^2), this work more than doubles to 293 Btu/lb (680 J/g). By further improvement in the vacuum to 1 in. Hg abs. (3386.4 N/m^2), the work is increased another 33–326 Btu/lb (80–760 J/g). Instead of lowering the back pressure 1 in. Hg abs., the same work output, 326 Btu/lb (760 J/g), would require raising the steam pressure to 148 psi.

The same reasoning is applied in the lower part of Table 2, where the live steam pressure would have to be increased from 800 psi to 1200 psi to get the same gain (29 Btu/lb or 70 J/g) in work by lowering the exhaust pressure from 2 to 1 in. Hg abs.

These data are striking evidence of the fundamental value of vacuum in the interest of minimum fuel use for the production of a kilowatt-hour of electricity. The value of vacuum cannot be gainsaid. It is the most effective part of the steam power plant cycle for the economic generation of electric power.

This desirable vacuum condition on the exhaust of the prime mover calls for the maintenance of an appropriate low temperature in the steam condenser of the plant. In practice, every effort must be made to use the lowest available ambient temperature—the atmospheric or a natural water supply. Water is preferred since it gives a much more practical construction for the condensing part of the plant. The high heat capacity of water and its excellent heat transfer coefficients lead to the overwhelming preference for water rather than air.

The engineer is thus guided by the fundamental requirements of thermodynamics to seek the lowest cost solution in a practical, reliable power plant which will operate without damage or hazard to the environment and deliver power at the lowest cost to the consumer and with maximum conservation of fuel resources. This is the basic pattern of good technology. It can be expressed in equation form, namely, sound engineering equals sound science plus sound economics and sound sociology.

Cooling water. The power engineer will consequently seek the lowest available water temperature for the cooling medium. A natural flowing river is usually the most attractive when all factors are considered. With every pound of steam to be condensed it is necessary to provide 100 lb (45 kg) of circulating water, with an allowable temperature rise of 10°F (5.6°C) on the water. On a modern coal-fired steam plant, this translates into the need for 700 lb (315 kg) of circulating water for 1 lb

(0.45 kg) of coal burned; or 500 lb (225 kg) of circulating water for 1 kwhr send out of electric energy. The magnitude of the circulating water requirement is such that the water cannot be pumped very far. It is much more economic to haul the fuel than to pump the water. Therefore sites are sought where full advantage can be taken of an abundant, lowest temperature water supply.

The substitution of nuclear fuel is of no help in this problem of circulating water, even though the fuel is essentially weightless. In fact, with present-day electric power station technology, the thermal efficiency for the conversion of heat from a nuclear reactor into electric energy is much lower than with fossil fuels. Reactor plants presently cannot be built to operate at the highest steam pressures and temperatures, 3500 psi and 1100°F (593°C), that prevail in conventional fuel–fired plants. The heat required to produce a kilowatt-hour is roughly 11,000 versus 9000 Btu (2800 versus 2300 kcal). The resultant heat exhausted to the circulating water of a nuclear power plant is some 50% more than that for a coal-, oil-, or gas-fired plant. This makes the thermal load on the flowing stream much worse than with the conventional fuel–fired plant. *See* REACTOR LICENSING, NUCLEAR.

The engineering profession tries to find alternative ways of providing the necessary condensing water capacity. Evaporative cooling towers, spray ponds, and air-cooled condensers all offer possibilities. Each, however, has its own disadvantages. These may range from poorer plant thermal efficiency and more extravagant use of natural fuel resources to fog, ice, and snow nuisances.

Other problems. These problems must be supplemented by consideration of many other elements, such as power transmission and distribution. All such problems are essential considerations in the siting of electricity generating plants. Continuity and reliability of electric supply, at the point of use, is the prime requirement for the effectiveness and well being of an industrialized civilization. There is no other service which is so requisite. Existence in an urban area today without electric energy, for a month, a week, or even a day, is tantamount to disaster. *See* ELECTRIC POWER SYSTEM, INTERCONNECTING.

The conservation of fuel resources alone prompts maximum efficiency in plants for the conversion of raw energy to the electric form. Discharge of waste products in solid, liquid, or gaseous forms ranges from coals containing 10–20% ash, to particulate matter, to unburned combustibles, to sulfur and nitrogen compounds, and to radioactive wastes and emanations from nuclear fuels. *See* AIR POLLUTION; NUCLEAR MATERIALS SAFEGUARDS.

When such problems must be combined with changing and improving technology and with the demands for the lowest cost to the electrical consumer, all without damage to the environment (including even the aspects of its beauty), it is rational to expect that solutions will be difficult, diverse, and costly. There is no easy road to the proper siting of electric power stations. There must be a rational, practical balance of a wide assortment of many component problems if the growing community is to be assured an adequate, reliable, low-cost supply of electric power.

For background information *see* CONDENSER, VAPOR; POWER PLANT; PRIME MOVER; RANKINE CYCLE; STEAM TURBINE; THERMODYNAMIC CYCLE in the McGraw-Hill Encyclopedia of Science and Technology.   [THEODORE BAUMEISTER]

Bibliography: T. Baumeister (ed.), *Standard Handbook for Mechanical Engineers*, 7th ed., 1967; *Elec. World*, 173(3):51, 1970; R. W. Holcomb, *Science*, 167(3915):159–160, 1970; L. H. Roddis, Jr, *The Future of Nuclear Power*, Science and Technology Advisory Council to the Mayor, New York City, Jan. 15, 1970; P. M. Stern, *Elec. World*, 173(4):22–27, Jan. 26, 1970.

Printing, color

Reports of recent developments in color printing have been concerned with the development of dry photomechanical processes and electronic scanners for color separations.

Dry photomechanical processes. Color brings new dimension to graphic communications. It adds attention, interest, and impact. Increased demand for color has encouraged development of new technologies to simplify placement of color in all forms of graphic communications, including package design, visual aids, printing, motion picture titling, television animation, and medical education.

The 3M Company attached the term Colorgraphics to this increased color emphasis and uses the term to cover a wide variety of dry photomechanical and technically related processes.

Color transparency film. Commercialization of the 3M Company's Thermo-Fax line of copiers paved the way to development of color transparency films that use the infrared "thermography" process and produce color images on clear or color backgrounds.

Dramatic, colorful visual transparencies now can be produced in home or office by desk-top transparency makers in as little as 4 sec (Fig. 1). Imaging originals may be taken from newspapers, magazines, or trade periodicals to supplement the user's own artistic abilities.

Transparency films are also available for use with process copiers. This process uses visible light exposure and heat development. The system produces a positive black image on a light red, blue, yellow, or green background. Average copy time from exposure through finished $8^{1}/_{2} \times 11^{1}/_{4}$ in. (21.59×28.57 cm) transparency visual is 20 sec.

Direct color reproduction. This has become a reality with the introduction of a dry positive-image process. This system involves special machine design, optics, and existing technology in dry dyes, coatings, and films. The process produces completely dry color visual transparencies, color separations, and color copies in 30 sec in colors as permanent as most color photographs.

An original is placed on an image platen, where it is scanned optically and reproduced automatically. In the system, intermediate color separations are made that transfer yellow, magenta, or cyan dyes in register to a special receptor paper or to a matrix for image transfer to other surfaces.

Although the system is automatic, it has manual controls to increase its flexibility as a color graphics tool. A print can be held in the system for a second cycle so that lettering or other information can be overprinted. The reproduction cycle can also be halted after each color step. The original can be moved to produce animation, or information can be added or blocked out for additional color effect.

Print paper locks in the dry dyes for permanency. For image transfer, the paper is replaced with a matrix sheet that holds the dry dyes in suspension. By placing the matrix in contact with other materials at 280°F (137.8°C) for 2 min, the image is transferred to almost any material that can withstand the heat.

Areas of application include textile, fashion, and packaging design, advertising, audiovisual education and training, graphics services, and reproduction of color-coded materials such as diagrams or charts.

Signs, posters, contemporary art. The use of polyester-based camera plates processed by a special camera plate system has opened additional avenues for color graphics. The processed plates are easily hand-colored for use as signs, posters, or contemporary art.

Original copy is placed on the camera's copy board and the processor activated by a foot switch. Inside the machine, the polyester-base plate is imaged by the camera, automatically cut to a predetermined size, and developed. The processed plate is delivered completely dry for immediate use.

Standard lithographic printing inks, dyes, or paints may be applied with cotton or a soft cloth over the plate's image areas to produce the desired color combinations. Three-dimensional objects may be integrated into the composition, increasing the creative potential.

Slide preparation and TV animation. Film sheets coated with ultraviolet-sensitive ink pigment are being used to simplify motion picture titling, slide preparation, television animation, and the addition of color to medical graphics.

Producing both positive and negative images, sheets are coated with an ultraviolet-sensitive ink pigment and may be imaged with a standard photo floodlight in from 1 to 3 min, but is best imaged by stronger sources of ultraviolet light such as carbon arcs or mercury-vapor light equipment.

Any translucent or transparent original containing an opaque image can be used to selectively filter out light to image a sheet of the film. Once exposed, material is chemically developed in one step and rinsed with water. Once blotted dry, it is ready for immediate use.

Posterization techniques—selective reproduction of tonal values of a continuous-tone image—have received wide public viewing in many national magazines. Television stations also use this procedure to convert black-and-white graphic stills to color slides. The addition of color helps achieve smooth transition between color shows and lends originality to color TV presentations.

Posterized x-rays are being used by medical educators to add highlighting color to visual transparencies. Exposure of each Color-Key sheet is carefully adjusted to highlight the study area. Several separate sheets are produced and are combined as overlays into one visual which may be projected as is or photographed for filing.

Animated graphics. Computer animated graphics—use of electronic equipment to create and

Fig. 1. A colorful visual transparency being made with a desk-top transparency maker and color transparency film. (*3M Company*)

display dimensional scenes—has advanced from simple line drawings and diagrams to sophisticated synthesis of halftone images. Though the mechanics of the various available systems differ greatly, the visual outputs are quite similar.

The animation artist can now sit at an electronic console and create movement and graphic effect by manipulating knobs and switches.

Designed initially for "special effects" and animation, the systems employ computers to produce a dimensional picture from the three-dimensional representation of objects. Color is added electronically or photographically, depending on the system. Lifelike movement and voice synchronization are programmed into the system by translating movement and voice into electronic signals.

Original inputs may be mathematical or visual. One system uses a line negative which is displayed in front of a TV camera. Signals generated by the camera are fed into a computer, and the computer in turn generates visual images from the input signal and animates and displays them in real time on a special cathode-ray tube. Input originals for animation generally are hand-drawn on acetate film. However, when dimensional stability is required, as in industrial and technical drawings and displays, originals may be produced on 3M's Color-Key material. The dimension stability of the polyester-based film ensures perfect continuity of motion, and the color pigment consistency ensures continuity of color. [J. M. FOLEY]

Color scanners. A color scanner is a device which electronically produces color separations from a colored transparency or reflection copy. In addition it may perform some or all of the following functions: (1) color correction to compensate for ink deficiencies; (2) tone correction to compensate for distortions introduced in the various stages of the reproduction process; (3) enlargement or reduction from the size of the original to the size at which it is to be reproduced; and (4) screening to produce halftone positives or negatives. Only

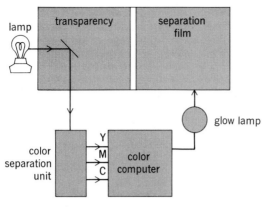

Fig. 2. Block diagram of Diascan 3000 color scanner.

green, yellow, magenta, and cyan areas. The black printer and undercolor removal are scientifically computed from the three color channels.

The computer can be preprogrammed on two plug-in boards to produce separations suitable for a given process. By interchanging these boards with alternative sets, the program is easily changed to suit different types of printing, inks, and paper.

A simple block diagram of the Diascan 3000 is shown in Fig. 2. A narrow beam of light passes through the transparency into the color-separation unit, where it is split into red, blue, and green components which are picked up by photomultipliers. The photomultipliers feed electrical signals representing the uncorrected yellow (Y), magenta (M), and cyan (C) separations into the computer, where all the necessary corrections are applied. The output of the computer is then fed to a glow lamp which is focused onto the separation film. As the cylinder rotates, it is driven axially by a special servo-controlled motor to produce a picture at 2000, 1000, or 500 lines per inch.

those scanners introduced since 1969 will be described here.

Diascan 3000. This scanner is produced by Crosfield Electronics Ltd., England, and is marketed in the United States by Rutherford Machinery Division of Sun Chemical Corporation. The Diascan 3000 was introduced in 1969. It produces color and tone-corrected, positive or negative, continuous-tone separations, the same size as the original, which can be either a transparency or flexible reflection copy up to 14×18 in. (35.56×45.72 cm).

The color correction system is extremely flexible and uses 18 different controls. Each of the three color channels—yellow, magenta, and cyan—has six color controls which permit the amount of each ink to be independently controlled in red, blue,

Magnascan 450. The Magnascan is produced and marketed by the same companies as the Diascan 3000 and was first marketed in 1971. The two scanners use identical color computers, but the Magnascan has the added features of enlarging, reducing, and screening. It accepts copy up to 10×12 in. (25.4×30.48 cm) and produces separations to a maximum size of 20×24 in. (50.8×60.96 cm). The separations can be continuous-tone or halftone positives or negatives, and the enlargement range is 0.5 to 16.5.

A block diagram of the Magnascan 450 is shown in Fig. 3. The colored original is mounted on an interchangeable transparent input drum and the separation film is mounted on an opaque output drum, with a contact screen superimposed when required. The screen and film are held in tight contact by a vacuum system built into the drum.

A single, constant-intensity xenon lamp is used for analyzing, viewing, and exposing, and the computer controls the brightness of the exposing light spot by means of an electrooptical crystal modulator. A digital system working in conjunction with an electronic memory facilitates the enlarging/reducing function.

Each line of information coming from the computer is digitized and then temporarily stored in the memory. After a short period the information is removed from the memory, converted back to analog form, and fed to the crystal modulator in the exposing head. By reading information out of the memory at a different speed to that at which it was written, a circumferential size change is obtained. The difference in read and write rates is governed by the digital control unit, into which the desired enlargement is dialled. The same control unit governs the speeds of the special servo-controlled motors that drive the analyzing and exposing heads along their respective lead screws. In this way the two heads are driven at different speeds, producing an axial size change equal to the circumferential one.

Other scanners. Very recently other companies have announced scanners which are in various stages of development.

Venture Research in Princeton, N.J., is working

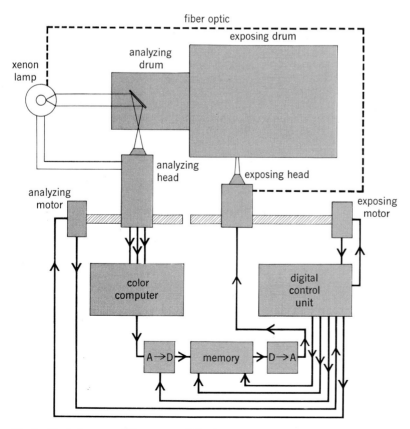

Fig. 3. Block diagram of Magnascan 450 color scanner.

on a system which uses digital techniques throughout and produces electronically generated halftones with triangular dots. This machine does not enlarge.

Other machines of the same size announced recently are the Linoscan 202 and 204. The 202 produces two separations at a time, and the 204 produces four at a time. These scanners are said to be capable of screening. They are manufactured by Linotype-Paul Ltd., England.

Based on a similar principle to the Magnascan is the DC 300, which has been announced by H. Hell of Kiel, West Germany. Like the Magnascan, the DC 300 enlarges and reduces by a digital electronic technique. It is also said to screen by way of a contact screen using a glow lamp as the exposing light source.

At the present time color scanners function as isolated units within a large system. However, with the introduction of digital techniques, it is quite feasible that in a few years time the scanner will be interfaced with other equipment as part of an integrated system.

For background information *see* PRINTING, COLOR in the McGraw-Hill Encyclopedia of Science and Technology.   [CLAUDE KERSH]

Bibliography: E. Chambers, *Litho Printer*, April, 1969; M. Grayson, *Web Printer 5*, vol. 1, no. 4, 1968; T. H. Nelson *Comput. Decis. Mag.*, May, 1971; Posterization, *Graphic Arts Buyer*, February, 1971; M. J. Schultz, *The Teacher and Overhead Projection*, 1965.

Printing plate

Among the recent developments in this area have been advances in gravure platemaking and a new printing method, driography, that utilizes a novel flat-surfaced plate.

Gravure printing. Most printers acknowledge that gravure printing gives the closest reproduction to the original art or copy in terms of continuous-tone effect and richness of color. Gravure also has an economical press operation and utilizes low-cost paper stock and fast-drying solvent-type inks. In spite of these advantages, the growth of gravure has been hampered by the lack of control and the high cost of gravure platemaking. This is particularly evident when gravure is compared to offset lithographic platemaking. However, significant advances have been made recently in the gravure preparatory operations directed toward overcoming these disadvantages. These advances have been in preproofing, color duplicating, etching resists and etching control, and electronic engraving.

Preproofing. The gravure preparatory operations consist of the photographic steps leading up to platemaking and the actual platemaking, or more properly, gravure cylinder making, since most gravure printing is done from engraved copper-plated steel cylinders.

The photographic operations of gravure cylinder making are similar to those of the other processes. The operations consist of color separation, color correction by color scanner, masking or hand retouching, layout or stripping, and preproofing of the gravure separation positives which are then used for exposing the etching resist. A major shortcoming in gravure has been the lack of a suitable continuous-tone color preproofing system for checking gravure positives prior to cylinder engraving. *See* PRINTING, COLOR.

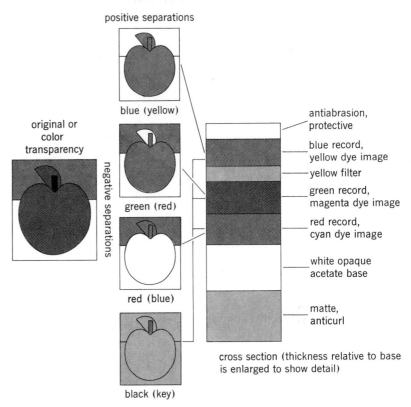

Fig. 1. Cibachrome Graphic process as used in gravure proofing. Terms shown in parentheses are those often used in the printing industry to identify the separations.

Most preproofing systems currently available (3M Color Key, Staley Colex, Agfa-Gevaert Gevaproof, and others) are halftone processes suitable only for the offset and letterpress processes which use full-scale halftones for platemaking. To fill this need for a continuous-tone positive proofing system, Ciba-Geigy has successfully introduced the Cibachrome Graphic process which yields positive color proofs for gravure from continuous-tone separation positives.

Cibachrome Graphic material is a silver halide–type integral tripack photographic product coated on a white opaque triacetate support. It consists of three light-sensitive emulsion layers, a yellow filter layer, and a thin antiabrasion top layer on one side and a matte anticurl gelatine layer on the other side (Fig. 1).

A gravure proof is prepared by exposing the color separation positives in sequence with matching color filters (Wratten 47B, 58, and 29) placed in the light path. Since Cibachrome Graphic is a three-color emulsion, the black printer record is obtained by giving a partial exposure of the black or key separation positive to each of the color filter exposures.

Processing consists of black-and-white development, stop-fixing, washing, dye bleaching in which the yellow, magenta, and cyan azo dyes are bleached in proportion to the negative silver image in each color emulsion layer, washing, silver bleaching, washing, fixing, and stabilizing. Processing is carried out in vertical tanks with nitrogen burst agitation and precision temperature control to ±0.5° F and careful control of chemical baths with preexposed quality control strips and replenishment as required. Processing may be also carried out in drum-type processors in which the chemical solutions are used once and discarded.

Cibachrome Graphic preproofing enables the gravure printer to check positives before engraving gravure cylinders, after which it is too late to make corrections except by laborious reetching and hand finishing in copper or by remaking the cylinder entirely, which is expensive. While the Cibachrome dyes do not exactly match the hues of the gravure inks used in magazine, catalog, and newspaper supplement printing, adequate correlation can be obtained to predict the printing characteristics of a set of gravure positives.

For scanned separation positives, it is possible to preproof before any retouching is done so that the amount of retouching can be controlled to the minimum. Cibachrome Graphic also has applications in color duplicating or preparation of "second-generation" color copy.

Color duplication. There is an increasing trend toward the use of second-generation color copy in the gravure field. Color duplicates made to size and photocomposed to page layout can be retouched with color dyes and electronically scanned in the new high-speed digital scanners of Crosfield Electronics Ltd. and Rudolf Hell to produce positive color separations directly in 5 min or less. Color duplication materials currently being used are furnished by Kodak, Agfa-Gevaert, and Ansco, in addition to a Cibachrome duplicating film. Kodak has announced the introduction of a new Ektachrome duplicating film with a contrast or gamma of 1.0, which, like Agfa Duplichrome, requires no contrast reduction masking in making duplicates.

Chemical etching. In the area of gravure platemaking or cylinder engraving, significant developments have occurred in both of the engraving methods currently in use; namely, chemical etching and electromechanical scanner engraving.

Chemical etching of copper-plated gravure cylinders is accomplished by using ferric chloride as the etchant and either a diffusion-type resist, such as carbon tissue, or a direct engraving stencil resist, such as Kodak Process Resist (KPR). Du Pont, whose silver halide gravure resist film, Rotofilm, has been available for more than a decade, has been successful in recent years in capturing a major portion of the market for diffusion resists in the publication field. Carbon tissue, a pigmented gelatin-coated paper sensitized by the user by immersion in a 3.5% potassium bichromate solution, is still the most widely used gravure diffusion resist because of its low cost and simplicity of use, even though lacking in control as compared to Rotofilm. Very recently Du Pont announced the development of a new gravure resist film, Cronavure, which falls in price between carbon tissue and Rotofilm and is expected ultimately to replace both of these diffusion-type resists.

The new Du Pont gravure resist film is a presensitized photopolymer coating on a dimensionally stable polyester film base (Fig. 2). It is exposed like carbon tissue and transferred to the gravure cylinder in a wet laydown machine. After laydown the film base is peeled off and the resist is ready for staging and etching immediately, without the necessity for alcohol washes, hot water development, cooling, and drying as with the other diffusion resists. Exposure controls the opening time or rate of penetration of ferric chloride through the resist layer, with the rate of etching into the copper being independent of exposure and etchant concentration. Etching progresses at a reproducible rate of 1 μm etched depth per minute of etching. The 35-μm cell depth required in the shadow tones of a typical magazine page would thus require 35 min etching time plus the opening time, which might be 20 min, resulting in a total etch time of 55 min. The new gravure resist film from Du Pont is being successfully used by a number of gravure printers, primarily on line work, but with a continuous-tone product to be available shortly.

Since chemical etching is the most widely used gravure cylinder production method, attention has been given to the etching machines as well as etching resists. Automatic machine etching is the rule rather than the exception today. Many ingenious etching machines have been developed, particularly in Europe, such as the Crosfield Etchomatic in England, which uses the continuous dilution principle; the Gravomaster in Switzerland, which uses roller application of the etchant to the cylinder and magnetic tape cassette programmed control of the etchant baths; and in West Germany the Rotary single bath spray etching machine and the Burda-Ottava system.

Etching control. What has been lacking in all of these etching systems for truly automatic control of the termination of etch by other than visual observation by the machine operator or by a predetermined time cycle has been an instrument for measuring the depth of etch during the etching process. A system recently developed in West Germany by Manfred Pauly, in cooperation with the Darmstadt gravure printer Habra, appears to close the loop in gravure cylinder etching control. In the Habra-Pauly system a 10-step grayscale and an adjacent blank area of the cylinder are scanned during the etch by inductive measuring and reference heads. The change in magnetic flux in the measuring head is proportional to the amount of copper removed. Cell depth is calculated, by an electronic computer, as a function of copper removed. As different steps reach programmed depths, the etching depth of the shadow step is used to determine if corrections should be made. If the depth is smaller than a preprogrammed value, the cylinder is rotated faster. If the depth is greater than the preprogrammed value, the rotational speed of the cylinder is reduced. The Habra-Pauly etching control is to be marketed by K. Walter of Munich.

Direct etching of gravure cylinders through KPR or similar stencil-type resists has been used for line cylinders and some tone work in the packaging and specialty gravure printing fields for many years. However, high-quality color process work in both of these fields and in publication printing has been restricted to diffusion resist etching because of the lack of tonal gradation with direct etching.

In etching through a stencil resist, the etchant attacks bare copper directly, and thus little or no graduation in cell depth from highlights to shadows is achieved, with tonal gradation being ob-

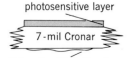

Fig. 2. Cross-sectional diagram of the Du Pont gravure resist film.

tained by variation in cell diameter only. Since gravure must use cell diameters, or halftones, which range up to only about 50% of an offset or letterpress halftone to preserve the cell walls, the effect on tonal quality is apparent. In a diffusion resist, both the cell diameter and cell depth can vary with the rate of penetration through the diffusion resist. Direct engraving has advantages in control and low cost of cylinder production if these shortcomings in tonal reproduction can be overcome.

The Gravure Research Institute (GRI) has successfully adapted the Platemakers Educational and Research Institute (PERI) powderless etch process for letterpress plate etching to direct engraving of gravure cylinders. The thiourea-based banking agent of the PERI etchant additive, used to protect the side walls in letterpress etching, provides in the GRI etching process total gradation by gradually stopping the gravure cells from etching, progressively from the smaller highlight cells through the midtone to the shadow cells (Fig. 3). A cell depth range of from 5 to 35 μm can be obtained in the GRI powderless etching process, comparable to that obtained with a diffusion resist. A major packaging gravure printer is currently introducing this process into commercial production.

Electronic engraving. The first electronic gravure cylinder engraving machine was developed by the firm of Rudolf Hell of Kiel, West Germany, more than 10 years ago. While at least two other companies have in the meantime developed electronic gravure cylinder engravers, the Hell Helio-Klischograph is the only one to make a significant impact on gravure printing. There are now approximately 50 pairs of Helio-Klischographs in use in 10 countries. A machine pair consists of a scanner which scans black-and-white reflective copy, either negative or positive, and an engraver which simultaneously engraves pyramidal-shaped gravure cells with an electromechanically driven diamond stylus cutting at a rate of some 4000 cells per second per engraving head. A cylinder is engraved in about 1 hr using multiple scanning and engraving heads, depending on the number of ribbons or pages or width of the cylinder. The scanner engraver pairs may be two separate machines linked together electronically or a single machine linked mechanically. The separate machine configuration is used for larger publication cylinders, while the single type is used for packaging and specialty engraving, for example, wood grains. The majority of the Helio-Klischograph installations are in West Germany, with several German gravure printers committed exclusively to electronic engraving. Two of these printers, Gruner & Jahr and Axel Springer and Sohn, each have seven pairs of machines in operation. The new plant of Diversified Printing Corp., in Atglen, Pa., is also committed exclusively to electronic engraving. There are a total of six Helio-Klischograph installations in the United States, including publication, packaging, and specialty printing applications. Hundreds of magazines are produced monthly by electronic engraving throughout the world, and there is no longer any question that it is both technically and economically feasible. It is interesting to note also that the Soviet Union and Yugoslavia have five

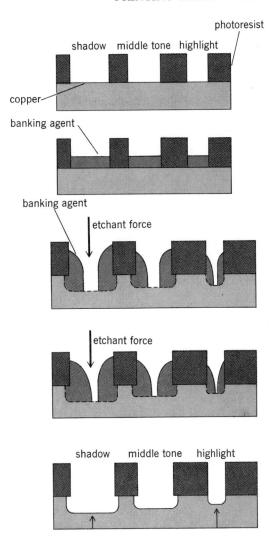

Fig. 3. Powderless etching principle for gravure printing. (*Gravure Research Institute*)

pairs of Helio-Klischograph machines installed; there is also an installation in Leipzig, East Germany.

A demonstration was made in West Germany in March, 1971, to indicate the technical feasibility of linking a color scanner to a Helio-Klischograph so that gravure cylinders can be engraved directly from color copy, by-passing the making of black-and-white color separation positives. This is not a new concept but something that is already done in the Hell Vario-Klischograph letterpress plate engraver, of which Hell has delivered over 1700 machines since 1957. For this test, color copy was supplied by participating printers, including a magazine page, a mail-order catalog page, and a package design. Cibachrome Graphic prints of all copy enlarged to page size were produced by Ciba-Geigy in Switzerland. The test was run at Gruner & Jahr in Itzehoe, where the Hell engineers mounted the color optic head of a Chromograph scanner on a Helio-Klischograph scanning machine. The electronic signals were transmitted via the Chromograph color computer to the Helio-Klischograph engraving machine. By changing filters the four colors were engraved one at a time. The four cylinders thus produced, after some trials

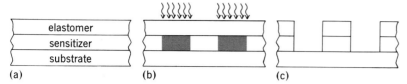

Fig. 4. Driographic platemaking process. (a) Cross section of the Dry Plate. (b) Exposure through a screened negative. (c) Cross section after removal of exposed areas. (3M Co.)

and adjustment, were proved without further reetching or finishing. The results, while not perfect, were considered commercial quality or close to it. Commercial development of such a system is probable for 1973 or 1974.

The introduction by Crosfield Electronics Ltd. in London and Hell of digital enlarging scanners opens up further possibilities for computer-controlled electronic gravure cylinder engraving. Since these color scanners convert the analog signals generated by the scanning head into digital data, picture originals could be stored on magnetic tape without major new machine development. Manipulation of copy after scanning and storage in digital form could be accomplished through computer input-output equipment in the form of a color TV display console with a light pen of input. The light pen could be used to effect minor color changes, shading, smoothing, and so on. When an acceptable image has been obtained, the cathode-ray-tube display unit could be used to expose a preproof similar to Cibachrome. At the point when the preproof is exposed, the computer would have made the corrections necessary for color and tonal reproduction characteristics of the press-paper-ink system. When all pages are ready, the data would be fed from the computer to the cylinder engraving equipment by way of magnetic tape. The engraver could use either the electromechanical system of the Helio-Klischograph or, if available by that time, a laser engraving head to replace the electromechanically driven diamond stylus of the Helio-Klischograph. Such a system is probable for the late 1970s.

Laser engraving. As far as laser engraving is concerned, the present state of the art indicates that lasers of sufficient average power are available to excavate single gravure cells by evaporation of the copper metal. The required developments are in the switching and modulation of the laser beam so as to obtain a sufficiently high rate of cell cutting. In mid-1971 there was no known work being done on laser engraving of gravure cells. A few years ago RCA discontinued the work it was doing in this area. The Quantronix Corp., Farmingdale, N.Y., had proposed a laser engraving system for gravure cylinders but had not found sufficient commercial interest to undertake this development. [HARVEY F. GEORGE]

Driography. A new system of printing, driography, was announced recently by 3M Co. after several years of laboratory development. The novel part of this printing method is the Dry Plate, a planographic or flat-surfaced plate which can be used on conventional offset presses without dampening fluids.

Conceptually the Dry Plate relies on a background or nonprinting area which is highly "abhesive" in nature and thus tends to reject printing ink on a printing press. In a composite sense the Dry Plate consists of three elements: a low-surface-energy polymeric layer, a light-sensitive diazo intermediate coating, and a supporting thin, flexible metallic substrate.

The platemaking process is outlined in Fig. 4. Exposure of the plate is made by contacting it to a halftone or line negative in conventional ultraviolet light. Liquid developer poured over the entire plate surface penetrates the polymeric coating and reacts with those areas of the plate which have been exposed to the ultraviolet light. Easy scrubbing of the plate surface removes the polymeric and diazo layers, leaving metallic substrate in the areas corresponding to the clear regions on the negative. The highly ink-rejecting background surface remains.

Plate characteristics. If one examines the plate surface after development, it can be seen that the highlight areas contain dots representative of those contained in the negative surrounded by the preserved background coating. By contrast, the shadow areas of the plate are mainly devoid of elastomeric coating and retain coatings only in the halftone dot areas. This is just the reverse of a litho plate with respect to the amount of background metal left exposed after development.

Dot quality on the plate is a high fidelity reproduction of that contained in the negative. There is no need to adjust negatives to correct for sharpness or dot growth when contacted to the plate.

Plate resolution characteristics are comparable to that obtained in high-quality litho plates.

Exposure times for the Dry Plate provide relatively wider latitude than do litho plates. This comes about by the fact that litho is "boxed in" by the necessary light reaction in the image areas to develop hardness sufficient to yield press life and, on the other hand, to not overexpose so that halation becomes a problem.

Since the Dry Plate removes the coating from the image areas resulting in printing from the base metal, the lower exposure limitation is essentially removed. Thus, an open 5 on the grayscale is recommended primarily for development ease rather than for any other press requirement.

Any conventional offset press may be used to print with the Dry Plate. In the case of a litho press, a simple disengaging of the dampening system is all that is required to accommodate this system. Other factors such as form roll pressure, type of blanket, and back cylinder squeeze do not require change on presses that are routinely maintained.

Press life of the Dry Plate in its present state of development may be characterized as medium run. That is to say, maximum plate life on an offset press is controlled by press, paper, and ink-related factors.

Ink. A necessary adjunct of the driographic system is that the printing ink be optimized for use with the Dry Plate. These inks have been compounded from the same raw materials, resins, solvents, pigments, and driers as are lithographic inks, the principal difference being the physical properties that are built into the driographic ink.

Since the Dry Plate does not rely on any secondary fluid to preserve cleanliness of the printed

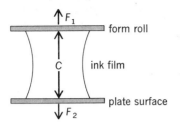

Fig. 5. Ink splitting on press involves cohesive forces (C) and adhesive forces (F) between form roll and plate image in the background regions. (*3M Co.*)

background, as does lithography, the cohesive forces (C) involved in splitting the ink at the press form roll–plate interface must be sufficient to resist toning or tinting of the ink in unwanted areas. Figure 5 is a diagram of the relating forces in the ink splitting. As indicated, if F_1 is greater than C and F_1 is greater than F_2, then for a clean background F_2 will have to be much less than C.

Ink must transfer efficiently to the image areas and resist splitting in the background regions. Thus, to optimize this aspect of the system, the energy differential between background and image areas is maximized.

Two of the more important ink properties explored with respect to the driographic system have been tack values and ink viscosities.

Tack may be considered a measurement of the forces required to split an ink film under controlled conditions. However, it is not sufficient to relate tack as measured on an Inkometer to describe an ink's total physical properties.

Under the dynamic running conditions on a press, loss of ink solvent due to operating press temperatures, the imbibing of solvent into the rubber rolls of the ink train, and the speed of the press all must be considered when the tack of an ink is discussed. It is safe to say, however, that ink with lower tack values will generally produce better print qualities on paper than will inks of higher tack values.

Control of ink viscosity also must be considered as a desirable property in efficiently transferring a uniformly controlled film thickness through the press and deposited at the proper density on the printed sheet. In general, the requirement here is to minimize viscosity index change with respect to press operating temperature.

Because of the recent introduction of the driographic printing system, many of its potential advantages have yet to be confirmed. Included in such a list would be immediate rollup on the press, better ink density control, reduction in paper waste, no ink emulsification problems, and cost savings by the elimination of sleeves, fountain solution, gums, and reduced press maintenance.

For background information *see* PRINTING; PRINTING PLATE; PRINTING PRESS in the McGraw-Hill Encyclopedia of Science and Technology.

[JOHN L. CURTIN]

Bibliography: J. L. Curtin, in H. Spencer (ed.), *The Penrose Annual 64*, 1971; H. F. George, *GTA Bull.*, vol. 22, no. 3, 1971; *Graphic Arts Buyer*, vol. 3, no. 8, Mar. 14, 1971; W. N. Welch and H. F. George, *TAGA Proc.*, 1969; R. W. Woodruff, *GTA Bull.*, vol. 21, no. 4, 1970.

Process radiation

Few of the many areas which come under the category of radioactivity applications have experienced the exciting development and growth of the area of process radiation, or radiation chemical processing. In the following are presented concepts and definitions of terms used in this field, the growth and present status of commercial development, and recent key developments and new products made with radiation.

Process radiation may be defined as the use of large isotopic sources, electron accelerator machines, or ultraviolet light for irradiation of materials to bring about chemical, physical, or biological change on a relatively large scale. Isotopic sources, such as cobalt-60 or cesium-137, emit gamma rays (x-ray like) and are used in amounts measured in kilocuries or more usually in megacuries. Electron accelerator machines are large devices wherein electrons are accelerated to kinetic energies up to 10 MeV. These devices are commercially available in various energy ranges and designs. Sources of ultraviolet light used commercially are high-pressure mercury arc lamps which emit actinic radiation down to about 3200 Å (320 nm) if the tubes are made of glass or below that (to the far ultraviolet) if quartz or Vycor tubes are used. A comparison of important properties of gamma rays, electrons, and other means for reaction initiation is given in Table 1. The function of the radiation (whether it be gamma rays, electrons,

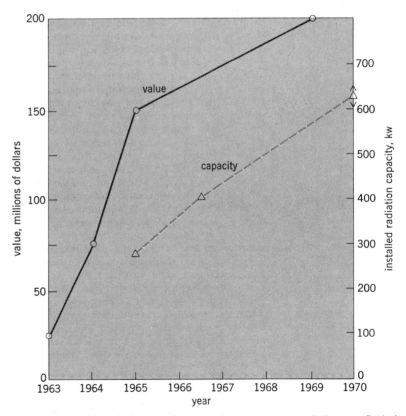

Solid line of graph shows the increase in commercial process radiation as reflected in the increase in value of irradiated products in the United States, exclusive of food products. Broken line shows increase in installed radiation capacity in the United States. (*Adapted from AEC estimates, and from A. D. Little, Process radiation, Service to Inventors, Nov. 25, 1966*)

Table 1. Comparison of radiation with other methods in chemical reaction initiation

Factor	Radiation		Thermal catalytic	Ultraviolet light
	Gamma cobalt-60	Electron machines		
Can initiate reaction at low temperature	Yes	Yes	No	Yes
Can be used at any temperature (i.e., temperature independence)	Yes	Yes	No	Yes
Can be used external to the system (i.e., leaves no chemical residual behind)	Yes	Limited; need thin metal for penetration	No	Yes, if quartz windows are used
Can be used in thick solid material (thickness more than a few millimeters)	Yes	No	Limited; get temp. buildup	No
Can achieve better chemical reaction control	Yes	Yes	No	Perhaps
Physical shielding needed	Yes	Yes	No	No
Can be turned "on-off"	No	Yes	Limited	Yes
Need AEC or state license to operate	Yes	Yes	No	No
Relative cost of installation	High	High	Low	Low

or ultraviolet rays) is to cause a temporary, localized molecular breakdown of the material into active chemical fragments. These recombine to form modified or new molecules and related end products. In radiation processing the objective is to make a new or improved commercially valuable material in situations where the use of radiation offers advantage over competing nonradiation (chemical) techniques.

Growth and status. Before the 1960s the field of process radiation as known today was virtually nonexistent; now it is a fast-growing $200,000,000 per year commercial activity. The figure indicates the growth in commercial process radiation in terms of two parameters: dollar value of irradiated products (exclusive of food products) and installed industrial capacity in kilowatts. A U.S. Atomic Energy Commission (AEC) study in 1968 revealed a 20–25% growth in value of irradiated products per year.

This impressive growth has resulted from a combined effort on the part of both industry and government, beginning in the 1950s and continuing to the present time. Although the electron accelerator was invented in 1931, it was not applied to commercial chemical irradiation until 1955–1959. Similarly the increasing availability of cobalt-60 (made from ordinary cobalt metal by neutron irradiation in reactors) during the 1950s led the AEC to institute a long-range process radiation program.

The ability of radiation to induce chemical change in substances, whether it be the sterilization of bacteria or the polymerization, cross-linking, or grafting of plastics, has led to the present known list of commercial process radiation activities given in Table 2. Several aspects of this table bear comment. First, the use of electron machine accelerators dominate the picture, since at present some 90% of installed industrial radiation capacity are accelerator machines. Second, seven of the total 12 processes listed in the table were announced in 1969–1970. However, it must be mentioned that the announcement of a process does not always mean that the process becomes an immediate commercial reality since several years may elapse between date of announcement and date of commercial use.

In terms of annual products sales, the area of radiosterilization of medical supplies has been, and continues to be, the most important single commercial activity. The increasing use of disposable medical supplies, such as surgical gloves, blades, and catheters, has resulted in a huge growth in this field in the last 6 years. It should be pointed out that irradiation of medical supplies is not limited to organizations shown in Table 2 but is also carried out in service irradiation facilities available for general use on a cost per hour basis.

Recent key developments. The most important recent developments occurring in the field of process radiation include the development of low-voltage (high-current), reliable machine irradiators; the commercialization and coming on-stream of the Ford Motor Co. Electrocure process; the use of ultraviolet-ray curing of paints; and the development of a new product, concrete-polymer material.

Electron accelerators (that is, machine irradiators) have been in commercial use since about 1955. However, these were of the higher voltage type (0.75 to 3 MeV) and lacked high-percentage continual operation capability. Since about 1964 electron machine operation reliability has been increased significantly. In addition, in recent years, low-voltage (0.25 to 0.50 MeV), high-current electron accelerators have been developed by several industrial organizations and have become commercially available. Low-voltage accelerators are excellent for surface treatment (that is, curing of paints and so on), while the higher voltage machines are more useful for conditions where electron penetration is needed. The availability of these low-voltage machine irradiators, coupled with high reliability of operation, will be an impor-

Table 2. Commercial chemical process radiation facilities in the United States

Process or product	Company	Year announced	Radiation source
Cross-linking of wire and cable insulation	Raychem Corp., Menlo Park, Calif.	1959	Accelerators
	ITT, New York, N.Y.	1964	Accelerators
Heat-shrinkable film and tubing	Raychem Corp., Menlo Park, Calif.	1959	Accelerator
	W. R. Grace Co., Cryovac Div.	—	Accelerator
	Electronized Chemicals, Subsidiary of High Voltage Engr., Burlington, Mass.	—	Accelerator
Specialty copolymers	RAI Research Corp., L.I. City, N.Y.	1962	Accelerator
Sterilization of medical products	Ethicon, Inc. (Div. of Johnson & Johnson), Somerville, N.J.	1963	Cobalt-60
	Becton, Dickinson, Rutherford, N.J.	1964	Accelerator
Ethyl bromide synthesis	Dow Chemical Co., Midland, Mich.	1963	Cobalt-60
Curing of surface coatings	Ford Motor Co., Dearborn, Mich.	1966	Accelerator
Soil release and permanent-press fabrics (cross-link cotton of cotton-polyester blends)	Deering Milliken Co., Blacksburg, S.C.	1966	Accelerators
	Cone Mills, Greensboro, N.C.	1969	Accelerator
Wood-plastic composites	American Novawood Corp., Lynchburg, Va.	1967	Cobalt-60
	ARCO Chemical Co. (NUMEC Div.), Karthaus, Pa.	1968	Cobalt-60
	Radiation Machinery, Parsippany, N.J.	1969	Cobalt-60
	Radiation Technology, Inc., Rockaway, N.J.	1969	Cobalt-60
Controlled degradation of polyethylene oxide	Union Carbide Corp., Charleston, W.Va.	1969	Cobalt-60
Concrete-polymer wall tile (trade name, RADCRETE)	Radiation Technology, Inc, Rockaway, N.J.	1969	Cobalt-60
Synthesis of flocculant (very high molecular weight polyelectrolyte; trade name, ATLASEP)	Atlas Industries, Inc., Wilmington, Del.	1970	Cobalt-60
Polyethylene foam	Voltek, Inc. (joint venture of High Voltage Engr. Co. and the Sekisui Chemical Co., Japan), Lawrence, Mass.	1970	Accelerator

tant factor in the future growth of process radiation.

The first commercial process for accelerator curing of surface coatings became a reality in 1970. The Ford Motor Co. is now using its patented Electrocure process to mass-produce painted plastic instrument panels. The accelerator electrons are used to fast-cure the special paint developed for this process. These paint formulations, developed both by Ford and by several paint companies for this application, have interesting properties: They can be cured in a few seconds, contain no solvents and thus are 100% curable, and exhibit improved adhesive properties. Another advantage of this process is that since there is no solvent in the paint, there is no need for solvent reclamation in a plant and thereby a decreased danger of possible toxic effects to workers. The paint formulations are custom made for the end-use application but generally consist of a prepolymer possessing varying degrees of chemical unsaturation plus a reactive vinyl monomer. The radiation causes the vinyl monomer to copolymerize with the unsaturated prepolymer to yield a cross-linked network of molecules in the final cured paint product. Electron accelerator machines useful for this application are made by several companies. This technique has potential in several areas, such as in the curing of adhesives and drying of printing inks. Generally one obtains a stronger bond of the irradiated layer to its substrate.

Recently there has been increased activity in the use of ultraviolet lamps for rapid curing of certain coatings and printing inks. These photochemical lamps emit ultraviolet light and have been manufactured with increased reliability of operation. Curing speeds of approximately 2000 ft/min (610 m/min) are believed to be practical with this technique. Active in this area are Sherwin-Williams, Glidden-Durkee, and the Sun Chemical Co.

The development of concrete-polymer material and the present governmental program aimed at its applications are worthy of comment because of their importance and significance. Concrete polymers are new materials which are much stronger and more corrosion-resistant than ordinary concrete.

If one heats and evacuates air and moisture out of ordinary, cured concrete, then impregnates it with a liquid chemical monomer such as methyl methacrylate, and finally hardens (polymerizes) the monomer with gamma-rays (or by thermal-catalytic means), one obtains a composite product called "concrete polymer." This product shows such significant improvements in its physical and chemical properties that it has been termed the most important single development in the field of concrete in 50 years. These improvements include up to fourfold increase in both compressive and tensile strengths; increased resistance to freeze-thaw; increased abrasion resistance; and increased chemical resistance to sulfate soils, acids, and distilled water. The potential applications are very promising and seven other government groups (Office of Saline Water, Bureau of Mines, Federal Highway Administration, Departments of the Navy and Army, Bureau of Reclamation, and U.S. Department of Agriculture) plus one industrial association (American Concrete Pipe Association) have joined the AEC in cooperative programs to develop concrete-polymer materials and products to meet their needs. Interest in concrete-polymer materials is now international in scope, with such countries as Japan, Denmark, France, West Germany, and South Africa actively pursuing potential applications.

For background information *see* ISOTOPE; PARTICLE ACCELERATOR; RADIATION BIOLOGY in the McGraw-Hill Encyclopedia of Science and Technology.

[GEORGE J. ROTARIU]

Bibliography: D. S. Ballantine, *Isotopes Radiat. Technol.*, 6(2):164, Winter, 1968–1969; D. E. Harmer and D. S. Ballantine, Radiation processing for chemical and polymer systems, *Chem. Eng.*, 79:9–10, April–May, 1971; G. J. Rotariu, *Nucl. News*, 13(9):45–47, 1970; G. J. Rotariu, The status of process radiation in the United States, *Proceedings of the Conference on Utilization of Nuclear Technology in Industrial Operations, Charleston, W.Va.*, June 8–9, 1970; M. Steinberg et al., *Concrete-Polymer Materials*., 3d Topical Rept. no. BNL-50275(T-602), January, 1971.

Propeller, marine

Marine propellers exist today in essentially the same form used successfully by John Ericsson 100 years ago, the primary difference being in the refinement of details and the adaptation to specific ship designs. In recent years propeller developments mostly have been directed at more exact propeller theory and the solution of technical problems in various propulsion devices that have limited the general application of these devices.

The theoretical basis for screw propeller design was developed by F. W. Lancaster in 1910 and subsequently modified and improved by L. Prandtl, S. Goldstein, K. E. Schoenherr, L. C. Burrill, and J. G. Hill.

A significant contribution to propeller theory was made by H. W. Lerbs in the development of the induction factor principle, in which the effect of the shed vorticity was related in detail to the induced flow at the propeller.

These theories were based on the replacement of the actual propeller blades by "lifting lines," or imaginary line vortices having the same distribution of vorticity as the actual blades. The effect of the relatively close spacing of the blades and the finite blade thickness were taken into account by various correction factors developed with varying degrees of success in correcting the theoretical predictions with actual propeller performance.

With the advent of high-speed, large-capacity computers, propeller lifting surface theory has been developed which, by considering small elements of the propeller blade geometry, has considered the overall effects of the forces generated, as well as the interference of one blade on the others and the effects of finite blade thickness. In addition, extension of the theory has permitted the calculation of the fluctuating forces on the adjacent ship structure due to the passage of the propeller blades.

All these theories fail in the practical case of a propeller operating in the varying-flow field in the wake of a ship and hence in the practical application of propeller theory to a specific ship problem. The earlier and simpler lifting line theories yield just as reliable results as the more complicated theories requiring the use of a larger computer.

Controllable-pitch propeller. In recent years controllable-pitch propellers have been developed in increasing powers, and designs for 40,000 shaft horsepower are now under development for specific ship applications. In these propellers the blades are mounted in the propeller hub so that they can be rotated from ahead to astern pitch, thus providing a degree of flexibility in the propeller to accommodate the power output characteristics of the prime mover; for example, a diesel engine which is limited in torque output can be coupled to a controllable-pitch propeller and the pitch adjusted to load the engine to its torque limit over a range of operating conditions. This is particulary advantageous where the output load varies to an extreme extent, as when a tugboat is running by itself at high speed or towing a string of barges at low speed or as when a ship is heavily loaded on one passage and light on the return trip.

Another attractive application is the case of a prime mover consisting of a number of diesel engines coupled to a single shaft where the maximum number of revolutions per minute (rpm) is limited by the engine but the shaft horsepower can be varied by using any combination of one or more engines. Similarly the increasing application of gas turbines to ship propulsion requires the use of one or more turbine units on the same shaft.

Contrarotating propeller. Propellers sustain a loss of efficiency due to the energy transferred to the water in the form of the axial velocity imparted to the propeller slipstream and also the rotational velocity behind the propeller. The axial acceleration cannot be recovered since it is equivalent to the thrust developed by the propeller. However, the rotational velocities usually can be canceled by the contrarotating propeller, in which two coaxial propellers rotate in opposite directions, and where the rear propeller is designed to cancel out the rotational velocities from the forward propellers.

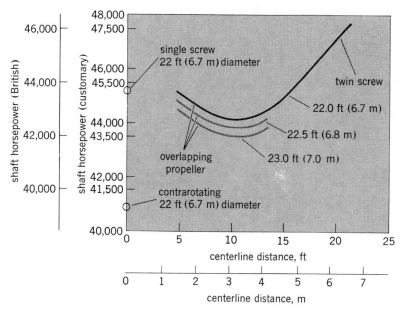

Fig. 1. Relative performance of a single propeller, twin propellers at various spacings, and contrarotating propellers on a fast cargo ship model.

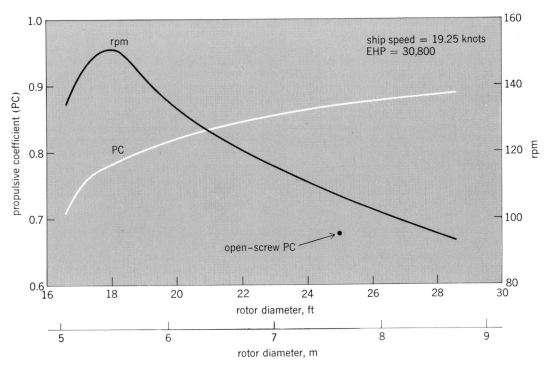

Fig. 2. Relative performance of an open propeller and a ducted propeller on a tanker hull.

While torpedoes have used contrarotating propellers for many years, the complexity of the power plant required has limited the application in ship propulsion. There are only two known ship installations of these propellers, even though the gain in propulsion efficiency has been demonstrated to be as high as 15%, depending on the particular application. Although numerous model studies have been made of contrarotating propellers, the optimum power distribution between the propellers is not adequately defined and in any specific application must be verified by model experiments. Figure 1 shows the relative performance of a single propeller, twin propellers at various spacings, and contrarotating propellers on a fast cargo ship model. The overlapping propellers noted in Fig. 1 are twin propellers where the centerline distance is less than half the propeller diameter.

Ducted propeller. The theory of ducted propulsion systems, in which the propeller is inside a coaxial duct or cowling, has also been greatly advanced in recent years, and installations of this system are now being made in tankers as large as 230,000 tons. Earlier applications of ducted propellers in the form of Kort nozzles have found widespread application in towboats, where the high ratio of power to size handicaps the performance of open propellers. The earlier designs were based on systematic model tests, while the recent developments have been aimed at theoretical solutions which consider the distorted flow field behind a ship. The advantages of the ducted propeller lie in the smaller propeller diameter and higher rpm compared to the open propeller at the same efficiency and the homogenizing effect of the duct which improves the cavitation performance. In cases where the propeller loading is high, such as on towboats, a substantial improvement in efficiency is obtained. In addition, the transverse vibrations normally experienced with open propellers are partially canceled by counteracting forces on the duct. Figure 2 shows the relative performance of an open propeller and a ducted propeller on a tanker hull.

For background information see PROPELLER, MARINE in the McGraw-Hill Encyclopedia of Science and Technology. [JOHN G. HILL]

Prospecting, petroleum

The use of numerous geophysical methods in prospecting for petroleum has resulted in the worldwide discovery of 750×10^9 bbl of oil, of which 525×10^9 bbl remain to serve the future needs of the world. These remaining reserves constitute only a 20-year supply. To maintain even this minimum level and in order to meet the swiftly rising demand, the petroleum industry must find an additional 900×10^9 bbl during the next 15 years. This is more oil than has been discovered during the past 100 years' history of the industry. In order to accelerate the finding rate, exploration for oil will not only become more widespread but will require the refinement of present techniques, particularly the multidisciplinary approach to the interpretation of physical, geological, and geophysical data.

Exploration sequence. In the past the sequence of developing an oil field has consisted of the following five stages.

1. Field geology. Examination of surface outcrops defines sedimentary basins which are potential scenes of hydrocarbon entrapment.

2. Magnetic surveys. The magnetometer has been used on land and from aircraft to determine the thickness and extent of sedimentary rocks. This instrument measures the intensity of the Earth's magnetic field, small variations of which can be interpreted to calculate the depth to the

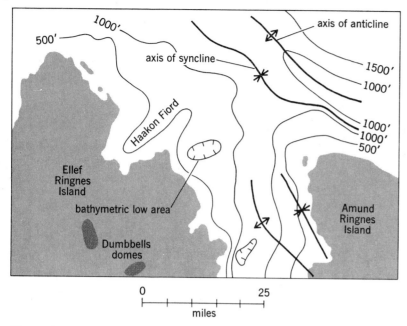

Fig. 1. Contour map derived from the Haakon Fiord geobathymetric survey. The arrows indicate the direction of dip of beds into and out of the syncline and anticline, respectively (5280 ft = 1 mi; 0.621371 mi = 1 km).

magnetic "basement" which underlies the sedimentary cover.

3. Gravity surveys. The gravimeter measures local variations of the Earth's gravitational force. These changes are introduced by varying rock densities within the sedimentary column. Gravity observations can be interpreted to yield not only the thickness of the sedimentary section but to indicate the presence of types of lithology such as salt masses, reefs, and shale-carbonate boundaries.

4. Seismic surveys. Such surveys are of two types: refraction and reflection.

In refraction surveys the horizontal travel time of sound waves measured in line of profile by the seismograph establishes the velocities of subsurface layers. These velocities can be correlated to individual formations. When these velocities are interpreted in conjunction with surface geology, one can postulate the depth and structure of the exposed rocks into their subsurface position.

In reflection surveys the measurement of reflection times from subsurface boundaries with the seismograph provides a clear picture of subsurface structure.

The seismograph provides a "direct" method of mapping the subsurface in comparison to the "field potential" methods of gravity and magnetics, which are of much lower resolution since the measurements made at the Earth's surface are affected by surrounding areas and thus subject to options of interpretation.

5. Drilling. After establishing an area of presence of sediments by geological, gravity, and magnetic surveys and after defining a suitable oil entrapping structure by seismograph survey, a hole is drilled. Even if oil is not found from this well (which is likely the case in new basins), the information derived from the hole is vital to future exploration in the area. Records kept of the material taken from the borehole help to determine various physical characteristics of the subsurface formations. Of particular interest is the calculation of rock porosity and permeability to define the potential reservoir beds within the sedimentary section. Rock velocity and density can be determined throughout the length of the hole. This information is of infinite value to the geologist and geophysicist in refining the interpretation of geological and geophysical measurements previously made. This newly acquired information is then used to define areas where further geophysical surveys should be conducted, which in turn will determine the next drill site. Historically this sequence has been cycled and recycled, with the eventual discovery of an oilfield.

Relative costs of prospecting methods. Assuming a reconnaissance but meaningful survey spacing, the relative costs per square mile of the prospecting methods are as follows.

Geological	$10
Magnetic	$10
Gravity	$20
Seismic	$1500
Drilling	$25,000–250,000

From this comparison can be seen why seismic and drilling programs must be very selective by comparison to the others. See TERRAIN SENSING, REMOTE.

Trend to offshore exploration. Statistically, on the average, one can expect to find 50,000 bbl of oil in every cubic mile (4.16×10^9 m^3) of sediments. The major source for new oil discoveries is thus areas of thick sediments where little drilling exploration has taken place to date. The shelves surrounding the continents provide the area with

The multidiscipline exploration program

Physical disciplines	Geological disciplines	Geophysical disciplines
Bibliography (compilation and analysis of all published information of the various disciplines)	Bottom sampling	Gravity
Ecology	Geobathymetry	Magnetics
Environment	Geologic field	Magnetotellurics
Logistics	Photogeology	Remote sensing
Meteorology	Stratigraphic mapping	Seismics (reflection and refraction)
Oceanography	Stratigraphic tests	
Offshore drilling feasibility		

greatest potential meeting these criteria. Drilling technology has adapted to offshore areas with wells initially being drilled in water depths up to 200 ft (60.96 m). Now drilling and production technology is moving forward rapidly to enable drilling on the high seas in water depths not only to the edge of the continental shelf (generally 600 ft or 182.88 m water depth) but beyond in progressively deeper waters.

Offshore exploration is largely an extension of the onshore methods but with some significant modifications. Because the cost of drilling a hole located offshore is in many cases 10 times or more higher than an onshore location, the dependence on a precise geophysical interpretation is increased. Many enigmas in interpretation of geophysical data were solved from stratigraphic information gained during the subsequent drilling of geophysical anomalies. These solutions must now be provided from geophysical surveys to a much larger degree because of the relatively fewer wells which will be drilled as a result of their higher cost. Also, in offshore exploration, surface geological information is much less available because of the ocean cover. Need for physical information in addition to geophysical data is introduced by the nature of the offshore, and it is necessary to carry out surveys encompassing oceanography, bathymetry, environment, ecology, and logistics preparatory to entering the drilling stage. See ENVIRONMENTAL ENGINEERING.

Multidiscipline approach. In order to provide the most sophisticated interpretation of physical, geological, and geophysical data, a multidiscipline approach has become essential.

Rather than continuing the sequential, recycling system of the past, in which each stage was considered an individual entity, the trend is toward the gathering of all geological and geophysical data concurrently and interpreting them in an integrated fashion. The disciplines of geology, gravity, magnetics, and seismics are first interpreted in their own right to provide a best approximation. Invariably in areas where little is known of the subsurface geology since few or no wells have been drilled, the interpreter is faced with a choice of conclusions from the data of each discipline. Each of the individual methods provides influential information toward the interpretation of the geology of the subsurface. These individual interpretations can be greatly enhanced by injecting the conclusions derived from surveys of other disciplines.

The final stage of mapping and evaluation of the potential of an area is the reinterpretation of each of the surveys after having concluded the unidiscipline interpretation. It is a progressive system in which one must first provide one's best interpretation uniquely from each method, followed by an integrated interpretation influenced by each other method. A "cross-fertilized" interpretation is the result.

The table summarizes the factors in a multidiscipline exploration program.

An example of a multidiscipline interpretation is provided by this survey in the Sverdrup Islands of the Canadian Arctic. On Ellef Ringnes Island piercement structures (Fig. 1) were delineated by surface geology. These structures resulted from

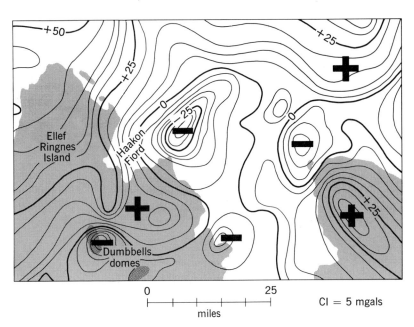

Fig. 2. Contour map derived from the Haakon Fiord gravity survey. The minus and plus signs indicate contoured areas of minimum and maximum gravity readings, respectively. Gravity readings are in milligal units (1/1000 of a gal; 1 gal = acceleration of 1 cm/sec^2).

the movement of deep rocks upward through the overlying sediments. Salt beds under increased load from overlying sedimentation sought density equilibrium and flowed upward as piercements. The question is whether similar structures could exist offshore. Anomalous variations in water depth can be seen east of Haakon Fiord.

Gravity measurements indicate a gravity minimum (Fig. 2) over the Dumbbells piercement and a similar anomaly offshore in the area of a bathymetric high. The negative gravity is likely an indication of evaporitic materials of lower density having replaced the normal sequence of sediments.

An area of high magnetic intensity (Fig. 3) is

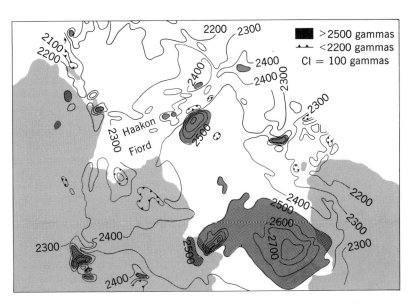

Fig. 3. Contour map derived from the Haakon Fiord aeromagnetic survey (a gamma unit is the equivalent of 10^{-5} gauss; 1 weber/m^2 = 10^4 gauss).

associated with the Dumbbells piercement, indicating that magnetic materials normally very deep in the area are close to the surface in the vicinity of the structure, conforming the intrusive or piercement nature of the structure. A similar magnetic anomaly is observed offshore.

From this evidence it can be concluded that a structure similar to the Dumbbells dome exists offshore and defines an area where oil accumulation can be expected.

As well as the technicalities of the sciences themselves, equally important to the process of finding oil is the establishment of scientific interdiscipline communication. In the continued development of each technology the practicing scientist has by necessity become progressively a restricted specialist of an individual discipline as a result of its tremendously increased complexity. The use of digital recording techniques and computer processing, for instance, has provided a means of enhancing and displaying information, resulting in the ability of the derivation of much more information from processed data. However, this has required a high degree of specialization in the various component processes, from the electronics of data recording to the computer processing of the results. The challenge in the future is to provide the most reliable conclusion from combinations of data.

For background information see OIL AND GAS, OFFSHORE; PETROLEUM GEOLOGY; PROSPECTING, PETROLEUM in the McGraw-Hill Encyclopedia of Science and Technology.   [A. E. PALLISTER]

Bibliography: A. E. Pallister, *Oil and Gas J.*, Nov. 9, 1970; A. E. Pallister, *Proceedings of the 2d International Symposium on Arctic Geology*, American Association of Canadian Geologists, 1970.

Prosthesis

A form of sight for the blind can be provided by impressing images onto the skin by means of hundreds of vibratory stimulators or points of electrical stimulation. A small television camera is utilized as the artificial "eye." This can either be held in the hand or can be head-mounted, typically on a pair of glasses to be worn by the blind. The video output of the television camera is used to control the array of mechanical or electrical stimulators in contact with the skin. Visible images are thus converted point for point to a form of energy to which the skin can respond. Matrices of up to 400 stimulators have been employed permitting a 20-line television image to be projected onto the skin. The skin then relays this image information to the brain.

Such devices are still in the experimental stage but may in the future be useful in the areas of education, employment, and mobility. A reading machine has been developed which projects printed words onto the fingertips, a letter at a time, permitting the blind to read at rates up to 80 words per minute. The brain mechanisms involved in the success of visual prostheses have been discussed elsewhere.

Recognition. In response to images projected onto the skin, subjects first perceive movement. Direction and rate of movement are quickly learned. Subsequently recognition of simple, basic shapes is gradually learned. With repeated presentations, the time to recognition of objects falls markedly; in the process, students discover visual concepts, such as perspective, shadows, and shape distortion, as a function of viewpoint and apparent change in size as a function of distance. When more than one object is presented at the same time, subjects learn to discriminate overlapping objects and to describe the positional relationship of three and four objects in one field. The visual analysis techniques and concepts thus developed are then used in letter recognition and in the exploration of other persons standing before the camera. Subjects learn to discriminate between individuals, to decide where they are in the room, and then to describe their posture, movements, and individual characteristics, such as height, hair length, and presence or absence of glasses. Figure 1 illustrates the appearance of a woman's face as digitally presented to the skin through a 400-point matrix of stimulators. The tactile presentation in Fig. 1 can be recognized as a woman's head by blind individuals experienced in this technique. Visual perception of this type of display is often enhanced by squinting or otherwise further blurring the image.

As blind individuals become more familiar with objects, they learn to recognize them from minimal or partial cues. This skill permits them to describe with accuracy the layout of objects on a table, in depth and in correct relationship, even though the objects may be overlapping and only partially visible. Blind subjects have developed spatial concepts that they had not previously attained, such as the determination of distance as a function of size change. After approximately 10 hr of training, the blind subjects no longer subjectively locate the stimulus on the skin; they subjectively project it out into the three-dimensional space in front of them, similar to visual experience.

Resolution. Twenty-line tactile television equipment has been extensively tested by blind sub-

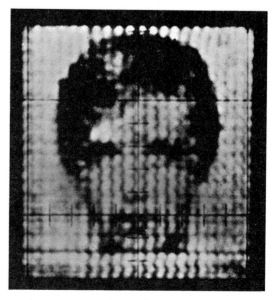

Fig. 1. Digitized representation of a woman's face as seen on the monitor oscilloscope of a tactile television prosthesis.

jects. Higher resolution systems are being built, making use of some of the latest advances in electronic technology. The ultimate resolution attainable will be limited by factors such as the two-point and pattern resolving power of the skin and its finite area. It is encouraging to note that the fovea of the eye (subserving the central 2° of clear, detailed vision) is made up of a matrix of approximately 34,000 light receptor cone cells, an array with only about 200-line resolution.

Stimulus. Electrical stimulation of the skin of the trunk has been shown to require less power, weight, and cost than mechanical stimulation and is consequently well suited to portable operation. Concentric silver electrodes 1 or 2 mm in diameter with 0.5 to 1 mm annular insulated spacing from an outer grounded silver ring appear to present the optimum stimulus geometry. The small annular spacing tends to limit current spread to the immediate surface of the skin, thus tending to restrict stimulation to touch rather than to deeper (dull) pain receptors. Direct current from a stimulus electrode produces polarization skin damage and consequently the stimulus must be coupled to the skin through a small capacitor preventing dc current flow.

Pain often associated with electrical stimulation has been eliminated by limiting both the magnitude of stimulating current and the area to which it is applied. The sensation evoked by such constant current–limited area stimulation is typically described as mechanical vibration. Stimulus currents up to 5 mA and pulse widths from 10 to a few hundred milliseconds have been found to be optimum for transmitting image information through the skin. It appears that the skin responds to delivered charge, that is, the product of pulse width times pulse amplitude. The threshold of sensation is approximately 40 nanocoulombs. Thus either pulse width or pulse amplitude may be modulated to convey intensity information in a projected picture. The range of stimulus magnitudes between the threshold of feeling and the threshold of pain is more than 10 to 1. With a single electrode, subjects have distinguished over 50 just-noticeable differences of intensity within an 8-to-1 pulse width variation.

Working system. An example of a portable electronic seeing aid is shown in Fig. 2. This prosthesis weighs 4 lb (1.8 kg) and is battery-operated. A 10-g miniature television camera electronic package is mounted with an optical system on a pair of glasses weighing 4 oz (113 g) total. The video signal from the camera is distributed to the array of stimulating electrodes mounted in a flexible matrix held elastically in contact with the skin of the abdomen. The system's rechargeable batteries give comfortable operation for 8 hr. The stimulus pattern is applied to the skin in point-for-point correspondence with the television image and thus produces an electronic image on the skin, which the blind have quickly learned to recognize and associate with an object in front of them. Objects are scanned by moving the head. Shades of gray are discernible from the stimulus which is pulse duration modulated in correspondence with the brightness at each point of the visible object. Brightness differences of 6% are detectable, resulting in some 16 just-noticeable differences of intensity being

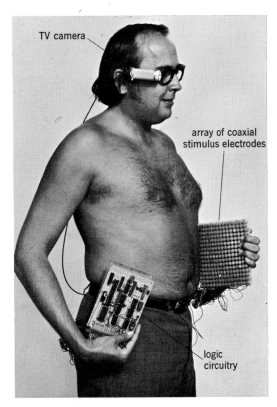

Fig. 2. Photograph of an experimental portable visual prosthesis showing television camera, array of coaxial stimulus electrodes, and logic circuitry.

detectable, which has proved more than adequate for transmitting shades of gray in closed-circuit tactile television systems.

Blind users of visual prostheses have been able to search their environment, select a named object, and to talk to and pick up the object, such as a telephone, in 10–12 sec. Portable electronic seeing aids have proved quite comfortable, being worn for periods of up to 8 hr without discomfort. For the first few hours, objects appear as projected on the skin. With sufficient experience, these haptic images are perceived as originating from the object in front of the subject rather than from the skin. Thus the tactile television system appears to become an electronic extension of the visual apparatus.

For background information see PROSTHESIS in the McGraw-Hill Encyclopedia of Science and Technology.

[CARTER C. COLLINS; P. BACH-Y-RITA]

Bibliography: P. Bach-y-Rita et al., *Nature*, 221: 963, 1969; J. C. Bliss, *IEEE Trans. Man-Mach. Syst.*, MMS-10:1, 1969; C. C. Collins, *IEEE Trans. Man-Mach. Syst.*, MMS-11:65, 1970; C. C. Collins, *J. Biomed. Syst.*, 5:3, 1971; C. C. Collins and P. Bach-y-Rita, in J. H. Lawrence and J. W. Gofman (eds.), *Advan. Biol. Med. Phys.*, in press.

Quantum solid

A quantum solid is one in which the zero-point motion of the atoms about the equilibrium lattice sites $\sqrt{\langle u^2 \rangle}$ is a large fraction of the near neighbor distance Δ. In a typical nonquantum solid, for example, solid argon, $\sqrt{\langle u^2 \rangle}/\Delta \ll 1/10$; in a typical quantum solid, for example, solid helium, $\sqrt{<u^2>}u/\Delta \approx 0.2$ at $T = 0K$. The quantum solids which have received the attention of most researchers in the past few years are the helium solids He^3 and He^4, the hydrogen solids H_2, D_2, and HD, and most recently the Coulomb crystal of nuclei that constitute the crust of neutron stars.

Microscopic properties. The interesting properties of the quantum solids are a consequence of their large zero-point motion. For example, the interaction between helium atoms is adequately represented by a Lennard-Jones potential, shown below. The potential is graphically shown in Fig. 1a.

$$v(r) = 4\epsilon \left[\left(\frac{\sigma}{r}\right)^{12} - \left(\frac{\sigma}{r}\right)^{6} \right]$$

where $\epsilon = 10K$ and $\sigma = 2.56$ angstroms (0.256 nm). This interaction is attractive for $r > \sqrt[6]{2}\sigma$ and strongly repulsive at $r < \sqrt[6]{2}\sigma$; σ is called the "hard-core" radius. An atom on its equilibrium lattice site at distance Δ from a near neighbor is described by a wave function (Fig. 1b) that overlaps that of its neighbor.

At the microscopic level the three important consequences of this zero-point motion are: (1) Neighboring atoms in the lattice encounter one another away from respective lattice sites at distances comparable with the radius at which there is a hard-core repulsion between the atoms. (2) The small parameter of conventional lattice dynamics, $(\sqrt{<u^2>}/\Delta)$, is not small. (3) There is a finite overlap between the wave function of an atom localized on lattice site 1 and an atom localized on lattice site 2. Thus the atoms in the solid can change place.

Macroscopic properties. Although these microscopic properties are unique to the quantum solids, there is very little experimental evidence to suggest that these quantum solid properties lead to behavior fundamentally different from that of the nonquantum solids. The most apparent departure from conventional behavior is seen in the phase diagram of solid He^3 shown in Fig. 2. Two points may be mentioned.

First, even at an absolute temperature of zero kelvin the solid does not form unless an external pressure in excess of 30 atm is applied (1 atm = 101,325 newtons/m²). A similar statement is true of only one other solid, solid He^4. The solid heliums have no "triple point."

The second point is that the thermodynamic and transport properties of the solid are related to its elastic properties, for example, the Debye temperature or the compressibility. These parameters can be changed by a factor of two by raising the pressure from 30 to 100 atm. The solid is soft and often regarded as a liquid in its behavior.

Except for these gross macroscopic properties, there is little else to distinguish solid He^3 from other solids. Experiments on the thermodynamic properties and transport properties of solid helium reveal behavior qualitatively similar to that of other dielectric solids. The most dramatic departures from conventional solid behavior are seen in nuclear magnetic resonance experiments on the Fermi solid, solid He^3. Solid He^4 is referred to as the Bose solid.

Theoretical description. While the thermodynamic and transport properties of the quantum solids lead to few surprises, it is not easy to understand how to quantify the behavior of these solids. For example, solid He^4 exhibits a low-temperature phonon specific heat like that of solid argon. But to construct a theory of this phonon specific heat is difficult. The phonons of conventional lattice dynamics which exist in argon and which explain its specific heat do not exist in solid helium because of the second point mentioned above.

The ground-state energy of solid He^4 cannot be calculated in the usual Hartree approximation that is so successful for other solids such as solid argon. Because of the first point, the relative motion of the atoms in the region of space midway between neighboring lattice sites is much more highly correlated than the Hartree approximation permits. Thus the theoretical description of a quantum solid must account for (1) the short-range correlations in the relative motion of a pair of neighboring particles that approach one another at hard-core distances, and (2) the motion of the particles over a large region of space in the vicinity of their lattice site where conventional lattice dynamics fails.

The short-range correlation problem has been dealt with successfully by L. Nosanow and coworkers and others using the Jastrow wave function within the framework of a variational calculation of the ground-state energy. The outcome of these calculations is that a pair of atoms in a quantum crystal interact with one another through an effective interaction which is the product of the bare interaction and a correlation function for the

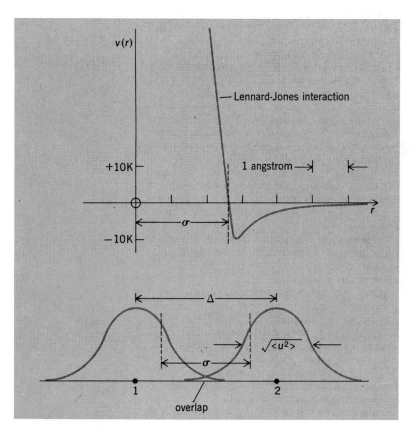

Fig. 1 Representation of the interaction between two helium atoms. (*Top*) Plot of the Lennard-Jones potential for the atoms. (*Bottom*) Representation of each atom's wave function.

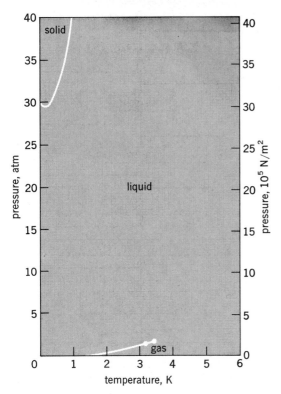

Fig. 2. Phase diagram of He³. The solid does not form until a pressure of at least 30 atm is applied. It melts at temperatures of only a few kelvins.

pair, that is, approximately the t-matrix result. This effective interaction has a softened hard core for which short-range correlations are relatively unimportant.

The long-range correlation problem (phonons) has been dealt with successfully by many workers. The outcome of these calculations is that, for the purpose of finding the phonons, the spring constant of the interaction between a pair of particles in a quantum solid is given by the second derivative of the bare interaction (that is, the Lennard-Jones interaction of Fig. 1) between the pair averaged over their relative motion. The phonons in the solid are the collective modes for particles coupled by these springs. Finally, in detailed numerical calculations for the phonons in a quantum solid, the bare interaction called for in the phonon theories is replaced by the effective interaction: The phonons are taken to be the collective modes for pairs of particles coupled by springs whose spring constant is the second derivative of the effective interaction averaged over the relative motion of the pair.

A large body of computational results (principally on solid helium) have been generated by Nosanow and coworkers and others for the ground-state thermostatic properties of solid helium (energy, E; pressure, P; compressibility, β; phonon spectrum, $w(q)$; and so on). These results are in reasonable qualitative agreement with experiment.

Recently F. Iwamoto and H. Namaizawa and R. A. Guyer and coworkers have developed a theory of quantum solids using what are substantially the techniques of the theory of nuclear matter. This approach has the advantage of yielding the t-matrix and phonons within the same computational framework. Detailed calculations of the ground-state properties of solid helium within the framework of this theory yield both a lowering of the ground state from 3K/particle to 1K/particle, and a pressure dependence of E and β in excellent agreement with experiment. Aside from conceptual problems which are clarified by this approach, its most important contribution is to show that a simple but careful treatment of the short-range correlation part of the problem yields substantial improvements in the quantitative features of the theory.

There is a third important consequence of the large zero-point motion of the atoms in a quantum solid. There is a finite overlap between the wave function of an atom localized near lattice site 1 and the wave function for an atom localized near lattice site 2. Thus the atoms can change place. In solid He³ the atoms are fermions (there is one unpaired nuclear spin) so that, because of the finite overlap, there is a nuclear exchange interaction. The energies associated with the exchange process are on the order of 1 mK. Thus the exchange process is unimportant to the ground-state properties of the solid. It is the tunneling of particles from one lattice site to another that leads to the properties of the Fermi solid that are observable in nuclear magnetic resonance (NMR) experiments. In NMR experiments one sees evidence for the existence of systems of excitations, the particle motion excitations that are not present in conventional solids, for example, the vacancy waves and mass fluctuation waves.

Further the evidence for the particle motion excitations that is available from the NMR experiments suggests that there is the possibility of mobility of the single particle density in the Bose solid, solid He⁴, that might lead to superfluidity in this solid system. It is tempting to suggest that a superfluidity is also possible in the solid crust of the neutron stars and may be responsible for some of the interesting properties of these systems. See NEUTRON STAR.

For background information see BOSE-EINSTEIN STATISTICS; HELIUM, LIQUID; INTERMOLECULAR FORCES; QUANTUM THEORY, NONRELATIVISTIC in the McGraw-Hill Encyclopedia of Science and Technology. [R. A. GUYER]

Bibliography: R. A. Guyer, *Sci. Amer.*, August, 1967; R. A. Guyer, *Solid State Phys.*, 23:402, 1969; W. E. Kellers, *Helium 3 and Helium 4*, 1969; J. Wilks, *The Properties of Liquid and Solid Helium*, 1967.

Radiocarbon dating

The usefulness of the radiocarbon dating technique strongly depends on the limits that can be placed on the variability of specific atmospheric carbon-14 (C^{14}) content during the past. Recent studies of tree rings, varves, and historically dated samples have greatly increased knowledge of these limits. The inherent limitations of radiocarbon dating are important because the use of C^{14} as a dating tool and isotope tracer is still increasing in a variety of disciplines, such as archeology, atmospheric sciences, geology, oceanography, and palynology. Somewhat paradoxically the greater use of

C^{14} dating is accompanied by the spread of an uncomfortable feeling that the dates are less reliable than originally assumed. A sense of frustration is particularly evident in archeology, where radiocarbon years are used in the A.D./B.C. framework. For many it is extremely difficult to realize that this is an unjustified marriage of two chronologies.

Computation method. A radiocarbon age, given in radiocarbon years since the origin of the sample, is based on C^{14} decay and is calculated from the ratio of the present C^{14} activity of the sample to the C^{14} activity of the standard (95% of the C^{14} activity of the National Bureau of Standards' oxalic acid). The standard C^{14} activity is assumed to be identical to the atmospheric C^{14} activity for the time interval over which C^{14} dating is used. Thus the conventional radiocarbon age does not allow for past deviations in atmospheric C^{14} content. As a result, radiocarbon years need not always be equal to solar years.

Tree rings. Variations in atmospheric C^{14} content can be studied in many ways. Tree rings are the most suitable for such studies. A tree-ring chronology is obtained through dendrochronological studies where each consecutive annual growth layer is assigned to the calendar year in which it was formed. For a living tree, the outermost ring has a precisely known date, and successive annual growth layers are assigned to sequentially earlier years by counting inward from the bark layer. Cross dating among different specimens is obtained by matching identical patterns of wide and narrow rings between specimens. Through this technique, fossil trees that partly overlap each other timewise can be used for extending the chronology. In addition, information is obtained on missing and "double" rings. The oldest living bristlecone pine is about 4500 years old but, through cross dating of fossil specimens, C. W. Fergusson at the University of Arizona has been able to establish a 7500-year bristlecone chronology.

C^{14} concentration in wood. By measuring the present C^{14} concentration in wood samples whose ages are dendrochronologically determined, it is possible to calculate the initial C^{14} concentration in the sample at the time of formation. The atmospheric C^{14} activity is calculated from the initial wood C^{14} activity by applying a correction for the isotope fractionation between wood and atmospheric CO_2. Basic to this method is that each ring is formed during one year only and afterward ceases to accumulate or exchange carbon.

C^{14} changes in atmosphere. A large number of tree-ring samples has been analyzed by several laboratories, and both long- and short-term variations in the C^{14} concentration of wood samples have been proven. Short-term variations involve changes for single consecutive years, as well as oscillations lasting several hundred years. Within one solar cycle, large changes in the C^{14} production rate occur in the upper atmosphere through modulation of the cosmic-ray flux by the magnetic field associated with the solar wind. This would, theoretically, result in only small C^{14} changes in the large reservoirs (atmosphere, oceans, and biosphere) on Earth. However, it has recently been suggested that the C^{14} production of several years is stored in the stratosphere and released to the

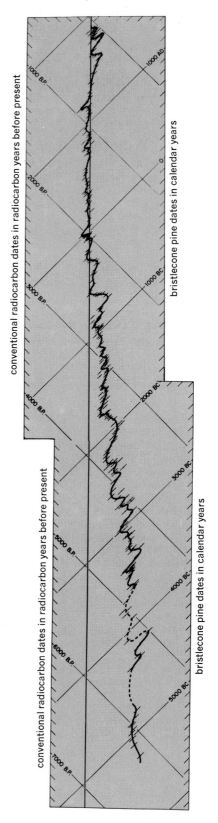

Fig. 1. Conventional radiocarbon ages of tree-ring-dated wood versus tree-ring dates for the last 7400 years. Conventional radiocarbon ages are calculated with a 5568-year half-life; zero age is A.D. 1950. The coordinate is tilted 45° in order to save space. (From H. E. Suess, Bristlecone pine calibration of the radiocarbon time scale 5200 B.C. to the present, in I. U. Olsson, ed., *Radiocarbon Variation and Absolute Chronology: Nobel Symposium 12*, Wiley, 1970)

troposphere at a certain stage of the 11-year solar cycle. This would explain year-to-year changes of a few percent in atmospheric C^{14} content experienced in the 20th century.

Many data are available for the short-term oscillations of a few hundred years. These oscillations appear to correlate significantly with solar cycle modulation of the cosmic-ray flux. Sunspot maxima often form patterns; for instance, a series of several 11-year cycles may have low sunspot maxima and be followed by a series with high maxima. This results in a few percent variation in atmospheric C^{14} content over intervals of the order of 100 years. An additional 400-year cycle in atmospheric C^{14} content, probably also associated with solar modulation of the cosmic-ray flux, has recently been suggested.

A long-term change in atmospheric C^{14} concentration is also evident. For the past 2500 years the average C^{14} concentration has been close to the baseline (standard) activity, but it was nearly 10% higher for the interval 7500–6000 before present (B.P.). Between 6000 and 2500 years B.P., C^{14} content was gradually reduced from this high level.

Geomagnetic effect. Earth magnetic field changes are the most likely cause of a major portion of the C^{14} change in the atmosphere. Measurements of paleomagnetic field intensities show changes over the last 8000 years amounting to 50% of present-day values. The less energetic component of the cosmic radiation is subject to greater deflection than the more energetic, and when the Earth magnetic field is stronger, a smaller portion reaches the upper atmosphere where C^{14} production takes place.

Climatic effect. In addition to this shielding effect, there is the possibility that climatic factors are responsible for part of the long-term change in C^{14} concentration. Sea-level changes, pH changes of ocean water associated with temperature changes, and changes in exchange rates between the surface layer and deep ocean can all be associated with climate changes and can influence the C^{14} concentration in the atmosphere. At present, it is not entirely clear what portion of the long-term trend should be assigned to geomagnetic or climatic effects.

Dating variability. For short time intervals, the exponential decay of C^{14} can be approximated by a linear decay of 1% per 80 years. An increase in original C^{14} level of 1% results in a radiocarbon age 80 years too young, because it takes this time interval before the C^{14} is back to the standard baseline from which the conventional C^{14} dates are determined. The relationship between calendar years and radiocarbon years, determined through tree-ring C^{14} analysis by Hans Suess, is given in Fig. 1 for the last 7400 years. Superimposed on this major trend are the short-term oscillations of a few percent that correspond with age errors of 100–200 years. Thus, even with this calibration curve, the basic inaccuracy of a date has to be estimated in centuries.

A portion of the calibration curve obtained from tree-ring studies has been confirmed by analysis of historically dated samples. C^{14} analysis of materials from the various dynasties of Egypt going back to about 5000 B.P. result in the proper historical age when corrected for the known C^{14} variations

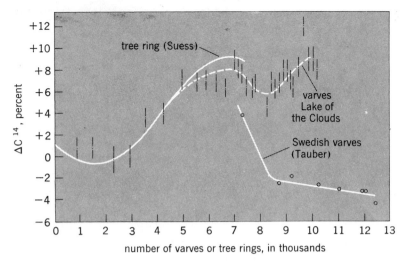

Fig. 2. Percent deviation of atmospheric carbon-14 content (ΔC^{14}) plotted against varve years and tree rings. Lines are approximate. (*From M. Stuiver, Tree ring, varve and carbon-14 chronologies, Nature, 228:454–455, 1970*)

from tree-ring studies.

Varves. Varves also are useful for the study of long-term C^{14} variations. Varves are produced in sedimentary environments by annual deposition of layers of different composition or texture. The most extensively studied varve series consists of silt and clay laminae deposited as couplets in proglacial lakes. Prominent in these investigations is the Swedish varve chronology of G. DeGeer. Although the Swedish varves contain insufficient organic material for C^{14} measurements, it is possible to correlate C^{14}-dated climatic episodes with varve-dated rates of recession and geomorphological features in southern Sweden. In addition, pollen-zone boundaries that are varve-dated can be correlated with C^{14} pollen-zone boundaries in nearby peat bogs. The bottom curve in Fig. 2 gives the calculated C^{14} deviations resulting from the differences in C^{14} age and the Swedish varve age.

Another varve series involves fresh-water sediments in Lake of the Clouds in northern Minnesota. These varves contain sufficient organic material for a direct C^{14} analysis, permitting a comparison of the C^{14} ages with the number of varves deposited (Fig. 2). Agreement between the tree-ring and Lake of the Clouds varve series is excellent, but is lacking between the two varve series. This discrepancy may have been caused by errors in either the Swedish or the Lake of the Clouds varve chronology, or both. However, both varve sequences show that deviations in atmospheric C^{14} content are maximally 10%. Thus conventional radiocarbon ages are expected to deviate less than 800 years from the true age over the 7500–12,000 year B.P. interval for which tree-ring data are lacking.

For background information *see* LOW-LEVEL COUNTING; RADIOCARBON DATING in the McGraw-Hill Encyclopedia of Science and Technology.

[MINZE STUIVER]

Bibliography: I. U. Olsson (ed.), *Radiocarbon Variations and Absolute Chronology: Nobel Symposium 12*, 1970; M. Stuiver, Tree ring, varve and carbon-14 chronologies, *Nature*, 228:454–455, 1970.

Reactor licensing, nuclear

Commercial nuclear power reactors are licensed by the Atomic Energy Commission (AEC). Figure 1 shows a model of a typical nuclear power plant. The rapid growth of the nuclear power industry and growing public awareness of environmental questions have increased public interest in the AEC regulatory program. Recently new laws on environmental quality were enacted that enlarged the AEC responsibility concerning nonradiological effects of nuclear power reactors.

AEC regulatory program. The principal objective of the AEC's regulatory program of licensing nuclear reactors is to protect public health and safety. This program is carried out independently of the agency's operational and developmental activities. The two stages in the AEC licensing process are: (1) the construction permit stage, in which the AEC evaluates the adequacy of the proposed site for a reactor of the power level and type planned and determines whether there is reasonable assurance that the proposed reactor can be constructed and operated at the proposed location without undue risk to the health and safety of the public; and (2) the operating license stage, in which the AEC determines through a technical review that the construction of the plant conforms with the permit issued earlier and that the reactor can and will be operated safely.

Applicants for construction permits are required to submit a safety analysis report which is reviewed in detail by the AEC regulatory staff. The AEC obtains advice and recommendations from other Federal agencies and specialized consultants, particularly in areas such as hydrology, geology, and seismology and on environmental and ecological factors. An independent technical review is also made by a statutory Advisory Committee on Reactor Safeguards (ACRS). When these reviews are completed, an Atomic Safety and Licensing Board (ASLB), drawn from a qualified panel of legal and technical experts, conducts a public hearing in the vicinity of the proposed plant location where all parties interested may be heard. The AEC regulatory staff and the ACRS also conduct intensive technical reviews at the operating license stage of the review process. A public hearing is not required at this stage, unless requested by affected persons. The AEC on its own initiative may schedule such a hearing.

Two Federal laws enacted during 1970 have enlarged the AEC's responsibility concerning environmental matters: the National Environmental Policy Act of 1969 (NEPA) (Public Law 91-190, Jan. 1, 1971) and the Water Quality Improvement Act of 1970 (Public Law 91-224, Apr. 3, 1970). This legislation has increased the scope of licensing activities both during the regulatory review process and at the public hearings. The AEC has issued a policy statement implementing its new responsibilities under one of these laws in a manner consistent with the need for providing adequate electric power and for protecting environmental quality.

Civilian nuclear power program growth. The growth of the nuclear power industry since 1965 reflects the utility industry's acceptance of nuclear power as an energy source. The expansion of the industry is also a measure of the confidence placed on the capability of nuclear power plants to operate economically, reliably, and safely.

As shown in the table, central station nuclear power plants in operation, under construction, or on order totaled 114 units, approximately 92,000,000 kW of electrical generating capacity. The growth of nuclear power and the projected status in the future are shown in Fig. 2. This chart includes those units for which an application has been filed with the AEC and a projection of applications expected in the future.

The nuclear units under construction during the spring of 1971 range in capacity from 330,000 to 1,130,000 kW and are located in 26 states. Nearly all of these units are scheduled by the utilities for operation by 1975. New applications for construction permits continued the recent trend as shown in Fig. 2. The AEC projected that by the end of fiscal year 1971 applications from 22 utilities for permits to construct 32 nuclear reactors would be under active consideration by the agency.

AEC licensing process. The potential hazards associated with the operation of nuclear power reactors and considered by the AEC fall into two main categories. The first relates to the remote possibility of a reactor accident that conceivably could release large quantities of radioactive material from the reactor. The second relates to the risk associated with normal reactor operation, primarily the effects of releasing small quantities of radioactive material under controlled conditions. In each category the characteristics of the reactor site must be considered.

Reactor siting. The evaluation of the proposed site for a power reactor is an important aspect of the AEC's licensing review. Those factors considered by the AEC in its safety evaluation of proposed sites include: (1) the population distribution in the area surrounding the proposed site, the dis-

Fig. 1. Artist's conception of Consumers Power Company's Midland Nuclear Power Station. The two units shown will generate 1,310,000 kW of electricity.

REACTOR LICENSING, NUCLEAR 355

Nuclear power plants in the United States*

State	Operable	Being built	Planned (reactors ordered)	State	Operable	Being built	Planned (reactors ordered)
Alabama	—	3	2	Nebraska	—	2	—
Arkansas	—	1	1	New Jersey	1	2	3
California	2	3	4	New York	3	3	3
Colorado	—	1	—	North Carolina	—	2	3
Connecticut	2	1	—	Ohio	—	1	1
Florida	—	4	1	Oregon	—	1	—
Georgia	—	1	1	Pennsylvania	2	5	4
Illinois	3	4	2	(Puerto Rico)	—	—	1
Indiana	—	—	1	South Carolina	1	3	1
Iowa	—	1	—	Tennessee	—	2	4
Louisiana	—	—	1	Vermont	—	1	—
Maine	—	1	—	Virginia	—	4	—
Maryland	—	2	—	Washington	1	—	1
Massachusetts	1	1	—	Wisconsin	2	2	—
Michigan	2	3	3				
Minnesota	1	2	—	Total	21	56	37

*Nuclear plant capacity: operable, 8,306,800 kW; being built, 47,102,000 kW; planned (reactors ordered), 36,727,000 kW—for a total of 92,135,800 kW. Total electric utility capacity as of Mar. 31, 1971: 340,718,926 kW.

tance to the nearest boundary of a densely populated center, and the uses made of this area, such as industrial, farming, or residential; (2) activities within the area which could have an effect on the safe operation of this plant, such as airports, industrial complexes, and railroads; and (3) the physical characteristics of the site, including the seismology, meteorology, geology, and hydrology of the area. See POWER PLANT.

Reactor safety. Licensed nuclear facilities are designed, constructed, and operated so as to prevent accidents from occurring and to mitigate their consequences if they should occur. Concentrations and quantities of radioactive material re-

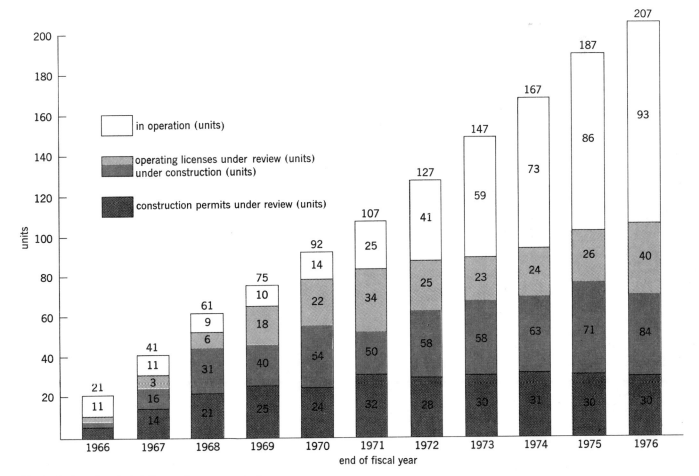

Fig. 2. Status of central station nuclear power plants.

leased during normal plant operation are also maintained as low as practicable. To assure these objectives are satisfied, the AEC performs an extensive safety review of every licensed nuclear facility.

The safety reviews performed by the AEC are based on the important principle of defense-in-depth. All structures, systems, and components important to safety are required to be designed so that the probability of an accident occurring is very small. Effective quality assurance programs must be applied at all nuclear power plants to assure that they are designed, built, and operated in a manner that is consistent with this objective. Experience with operating plants has encouraged confidence in this primary line of defense.

In spite of the measures required to be taken to prevent accidents and in accordance with the defense-in-depth concept, the conservative hypothesis is nevertheless made that highly improbable accidents could occur. Engineered safety features are required to be provided to mitigate the consequences of these postulated accidents. The design of engineered safety features is reviewed by the AEC to assure that they will function properly under accident conditions. These stringent design requirements are considered to be appropriate, even though the probability of such postulated accidents occurring is very small and even though there is a high degree of redundancy in engineered safety systems. Each line of defense is reviewed carefully to assure that the defense-in-depth concept is effectively implemented.

Radioactive waste treatment systems are required by the AEC for all nuclear power reactors to maintain control over radioactive effluents. The design and proposed operation of these systems are reviewed by the AEC prior to licensing, and periodic AEC review continues over the operating life of the plant to assure that radioactivity releases are maintained as low as practicable. *See* NUCLEAR MATERIALS SAFEGUARDS.

The AEC standards and criteria related to the normal releases of radioactive material from nuclear plants were developed to control the quantity and concentration of radioactivity reaching the environment. These limits are consistent with the radiation protection guides set by the Federal Radiation Council (FRC). The radioactive releases from operating power reactors have generally been less than a few percent of the limits in AEC regulations, and resultant exposures to the public in the vicinity of such plants have been small fractions of the FRC guides.

AEC regulations require licensees to maintain a program to monitor the offsite environment surrounding the plant and to monitor the releases of radioactive material from the plant. The results of these monitoring programs are reviewed by the AEC, which in some cases performs independent monitoring to verify the licensee's results. The U.S. Public Health Service has performed detailed environmental monitoring around selected nuclear plants. The results from these monitoring programs show that the radiological impact of these plants on the environment is negligible.

Reactor inspection program. Field compliance inspections of reactor facilities are conducted as an important part of the AEC's regulatory program during construction, test, and startup and during operation phases to verify that AEC regulatory requirements are being met. Over 600 inspections were conducted by the AEC's inspection force during 1970, indicating the level of effort in this area.

Environmental impact. Consideration of the principal environmental effects of nuclear power facilities, particularly radiological and thermal effects, is accomplished in conjunction with the AEC licensing process and through application of water quality legislation which has established a system of Federally approved state standards. To meet these water quality standards it has been necessary for some utilities to take additional steps in the design of power plants, such as cooling ponds and cooling towers. Figure 3 shows cooling towers at Metropolitan Edison Company's Three Mile Island Nuclear Station on the Susquehanna River.

Each applicant for a construction permit or operating license for a nuclear power reactor is required to submit to the AEC an environmental report on his facility. A detailed environmental statement is then prepared by the AEC based on this environmental report, on a technical review of the facility and its site, and on recommendations made by other Federal and state agencies. In general, the environmental statements discuss (1) site and reactor characteristics, (2) power needs, (3) the environmental impact, including both radiological and nonradiological effects, (4) any provision for enhancement of environmental amenities, such as recreational or ecological facilities, (5) alternatives to the proposed action, (6) any adverse environmental effects that cannot be avoided, (7) relationship between local short-term uses and maintenance and enhancement of long-term productivity, and (8) any irreversible and irretrievable commitments of resources.

The AEC's general policy statement on environmental issues under the National Environmental Policy Act authorizes atomic safety and licensing boards to consider whether the issuance of a license or permit is likely to result in a significant adverse effect on the environment. Each nuclear power reactor license contains a condition requiring the licensee to observe all validly imposed Federal and state environmental quality standards and requirements that are determined by the AEC to be applicable to the facility involved. Under the Water Quality Improvement Act of 1970, applicants must also provide certification that construction or operation of the facility will not violate applicable water quality standards (including thermal standards). This certification is to come from the state or interstate water pollution control agency or the administrator of the Environmental Protection Agency, as appropriate.

Antitrust considerations. A Dec. 19, 1970, amendment of the Atomic Energy Act eliminated the requirement that the AEC make a finding as to whether nuclear power plants are sufficiently developed to be of practical value for industrial purposes and therefore should be licensed under the commercial licensing section of the act (Section 103). Previously all nuclear power reactor licenses

Fig. 3. Photograph of Metropolitan Edison Company's Three Mile Island Nuclear Station on the Susquehanna River. Visible in this view are the two 372 ft (113.4 m) natural draft cooling towers.

had been issued under the research and development licensing section of the Act. Most nuclear power reactors will now be licensed under Section 103. One of the principal effects of licensing under this section is that applications for nuclear power plant licenses are now subject to antitrust review by the attorney general and the AEC. Recent changes in AEC regulations provide for hearings on antitrust matters, if appropriate. These hearings would generally be held separately from hearings on radiological and environmental matters.

Public hearings. AEC procedures for public hearings provide for participation by individuals or organizations whose interests may be affected. Hearings are playing an increasingly important role in the licensing of nuclear power plants. This increased emphasis on public hearings is due to the increased public concern and awareness of environmental matters and to the broadening of AEC responsibilities concerning the environment. These factors have resulted in an increased number of contested hearings and an increase in the complexity of issues under consideration.

Safety criteria and standards. The establishment of a comprehensive framework of reactor safety criteria, codes, and standards is important both to safety and to the efficiency of the licensing process. The AEC and the nuclear industry have intensified efforts in the past several years to develop comprehensive standards programs for light-water reactors. Substantial progress has been made in this area; some examples follow.

Quality assurance criteria. Quality assurance criteria that cover the design, construction, and operation of the safety-related aspects of reactor plants from design throughout the operating life of the plant were added to the AEC's regulations pertaining to reactor licensing, effective July 27, 1970. These requirements for quality assurance of nuclear power plants are expected to enhance the overall safety of nuclear reactor operation.

General design criteria. The AEC's regulations require an applicant for a permit to construct a nuclear reactor to provide assurance that his principal design criteria encompass all facility design features required in the interest of public health and safety. To provide guidance in establishing the principal design criteria for nuclear power plants, general design criteria have been developed. These criteria were first issued in proposed form for interim guidance in 1965, and a revised version was issued in 1967. The recent publication of these criteria as a regulation effective on May 21, 1971, marked an important step in the standards development program.

Emergency plans. The AEC published an amendment effective Jan. 24, 1970, establishing the minimum requirements for emergency plans for nuclear facilities, including nuclear reactors. All nuclear reactor licensees are required to develop plans for coping with the potential consequences of a severe accident. A guide to aid prospective licensees in the development of emergency plans has been published.

Nuclear power plant safety guides. While general guidance on safety requirements in the design of nuclear facilities is provided in AEC regulations, detailed guidance has not been established in a number of areas important to safety. Regulatory decisions in these areas have been made on an individual case basis in licensing actions. The AEC has recently begun making available a series of safety guides to assist the nuclear power industry in determining the acceptability of specific safety-related features of light-water power reactors. These new guides, while not containing mandatory requirements, indicate positions developed by the AEC regulatory staff and the ACRS on specific safety problems and describe principles and specifications that will represent acceptable solutions.

Release of radioactivity. Technological progress has demonstrated increasingly that modern nuclear power reactors are capable of normal operation with radioactive releases at levels far below the limits in AEC regulations. Because of this and the desire by the AEC to provide additional assurance that total radiation exposure levels from licensed activities remain low, amendments to the Federal regulations were recently developed. These amendments, which require that reasonable efforts continue to be made to keep releases of radioactive material in effluents from nuclear reactors as low as practicable, became effective on Jan. 2, 1971.

For background information *see* REACTOR, NUCLEAR in the McGraw-Hill Encyclopedia of Science and Technology. [HAROLD L. PRICE]

Bibliography: Licensing of Production and Utilization Facilities, Code of Federal Regulations, Title 10—Atomic Energy, Chapter 1—Atomic Energy Commission, Part 50; Amendments to 10 CFR Parts 20 and 50, *Fed. Regist.*, Dec. 3, 1970; Appendix A to 10 CFR Part 50, *Fed. Regist.*, Feb. 20, 1971; Appendix B to 10 CFR Part 50, *Fed. Regist.*, June 27, 1970; Appendix D to 10 CFR Part 50, Statement of General Policy and Procedure: Implementation of the National Environmental Policy Act of 1969, Public Law 91-190; Appendix E to 10 CFR Part 50, *Fed. Regist.*, Dec. 24, 1970; Safety guides for nuclear power plants, *Fed. Regist.*, Apr. 16, 1971.

Root (botany)

Theoretical studies of the effects of root geometry on nutrient uptake from the soil show that the length of roots in unit volume of the soil tends to have an overriding influence on the rate of uptake. The length of roots within the topsoil is commonly so great that the resistance to transfer to the root network is small, except for the relatively immobile nutrients. Large geometrical effects of root hairs on the uptake of the less mobile nutrients are predicted from theory, and experimental evidence has been obtained that root hairs appreciably increase the rate of uptake of phosphate from clays. In contrast, the presence of hairs appears to have no effect on the uptake of phosphate from a stirred solution.

Although it has become widely accepted that the resistance offered by the soil to nutrient transfer exerts a strong influence on the rate of uptake of nutrients, most of the arguments in support of this view relate to experiments or models describing the transfer of ions to single roots. The question arises as to whether this resistance continues to be important in relation to the uptake of nutrients by root systems or "sets" of roots.

Soil property and root relationship. The model described in Fig. 1 shows the relationship between the various properties of the soil and of the root system that are concerned in uptake. Given information on the soil, a knowledge of the following plant properties is generally sufficient to enable a useful estimate to be made of the likely rate of uptake of a nutrient in scarce supply: L_v (cm^{-2}), the length of roots per unit volume of the soil, termed the rooting density; A (cm^2 sec^{-1}), an uptake coefficient defined for a particular nutrient ion by $Q = 2\pi AC$, where Q is the rate of uptake per unit length of root from a stirred solution having a concentration C equal to the initial concentration of the nutrient in the soil solution; and W (cm^2 sec^{-1}), the rate of water uptake per unit length of root.

Distance nutrient travels. The distance through which nutrients move to reach the root network can be calculated easily if the network is assumed to form a regular lattice. A more realistic assump-

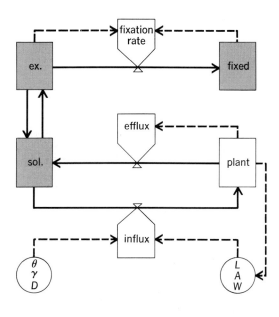

Key:
- θ = water content
- γ = slope of exchange isotherm
- D = diffusion coefficient
- ex. = amount of nutrient present in soil in exchangeable form
- sol. = amount of nutrient present in soil solution
- L = rooting density
- A = ion-uptake coefficient
- W = rate of water uptake per unit length of root
- □ = quantity absorbed per unit volume of soil
- ▨ = quantities of a specified nutrient ion per unit volume of soil
- ○ = auxiliary values
- ⋈ = rates
- --→ = transfer of information
- ⟶ = transfer of nutrient

Fig. 1. Relational model of nutrient supply and uptake.

tion is that the straight lines representing the individual segments of sinuous roots form a random distribution. The volume fraction of the soil lying within any given distance Δ from points of nearest contact with the axial parts of roots in the network can then be derived as a function of L_v from A. G. Ogston's theory for random lines. Values of the volume fraction are shown in Fig. 2. Cross reference to the general values of L_v given below Fig. 3 shows that under established crops and pastures nearly all the topsoil is within 5 mm of the axial surface. Root hairs reduce the distances involved in nutrient transfer to even smaller values.

Influence of L_v. The influence of L_v on the rate of uptake of nutrients from a particular compartment of the root zone may be estimated through the use of "hollow cylinder" theory, applied originally in studies of water uptake by J. R. Philip and subsequently in studies of nutrient transfer to sets of roots by J. B. Passioura and M. H. Frere. The rooting density exerts its strongest influence on nutrient uptake when transference is least efficient, that is, when the initial concentration of nutrients in the soil solution is low and transpiration is slow. Figure 3 shows the depletion likely to be achieved in these circumstances by sets of roots operating at various densities over a period of several days. L_v will be less influential in soils that are well supplied with nutrients, or at high values of W.

Nutrient mobility. For a set of roots, as opposed to a single root, the resistance to nutrient transfer offered by the soil is important only when the rooting density is such that the path lengths operating in transfer to the set are sufficiently large. Figure 3 shows that the store of mobile nutrients in scarce supply can be depleted rapidly in topsoils. For the most mobile nutrients, variation of L_v within the usual range of values found in topsoils has little influence on depletion. This implies that the path lengths are so small that there is little resistance to transfer to the root network. (It should be noted that a negligible resistance to transfer to the roots

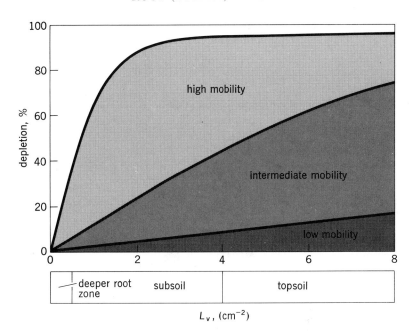

Fig. 3. Predicted depletion of the isotopically exchangeable store of scarce nutrients of low, intermediate, and high mobility by roots in a set of density L_v. $A = 5.10^{-7}$ cm^2 sec^{-1}, $W = 0$, $t = 4$ days. The usual range of values of L_v under established crops or pastures is indicated in the bar below the graph.

does not imply an adequate or unlimited rate of uptake.) For the less mobile nutrients and in the deeper parts of the root zone, the degree of depletion can be seen to be highly dependent on L_v.

In relation to the uptake of less mobile nutrients, the effect of root hairs on the resistance to nutrient transfer needs to be considered in addition to that of L_v, as the hairs increase the volume fraction of the soil lying within very short (< 100 μm) distances of the epidermal wall. (In mycorrhizal roots outgrowths of mycelium have similar but even more pronounced effects on the geometry.)

Root hair presence. Theoretical estimates made by M. C. Drew and P. H. Nye suggest that the presence of root hairs might be expected to almost double the rate of uptake of potassium from moderately buffered potassium-deficient soils. Experimental evidence on the effects of root hairs on the uptake of phosphate from a highly buffered clay soil has been obtained recently by K. P. Barley and A. D. Rovira. Roots were grown in the clay with or without root hairs, the growth of the hairs being controlled by varying the mechanical strength of the clay. The roots with hairs absorbed phosphate nearly twice as rapidly as the hairless roots. For reasons given by Barley and Rovira, their measure of the effectiveness of root hairs in increasing the uptake of phosphate is likely to prove conservative.

For background information see ROOT (BOTANY) in the McGraw-Hill Encyclopedia of Science and Technology.

[K. P. BARLEY]

Bibliography: K. P. Barley and A. D. Rovira, *Soil Sci. Plant Anal.*, 1:287–292, 1970; M. C. Drew and P. H. Nye, *Plant Soil*, 31:407–424, 1969; A. G. Ogston, *Trans. Faraday Soc.*, 54:1754–1757, 1958; J. B. Passioura and M. H. Frere, *Aust. J. Soil Res.*, 5:149–159, 1967.

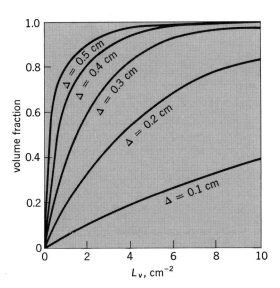

Fig. 2. The volume fraction of the soil occurring within distance Δ (cm) from points of nearest contact with the axial parts of roots in a set of randomly distributed roots of density L_v (cm^{-2}). Root radius = 0.02 cm.

Seed (botany)

Considerable attention has recently been given to the process of nutrient mobilization in cereal grains whereby the developing embryo receives nourishment from the endosperm. In certain cereals this process has been shown to be directed by the plant growth hormone gibberellic acid (GA_3). In barley (Hordeum vulgare) seed, gibberellic acid has been shown to enhance the synthesis and release of at least six hydrolytic enzymes in the aleurone layer. Of these enzymes, α-amylase and a protease have been shown to be synthesized anew in this tissue.

Structure. In barley the aleurone layer consists of 3–4 cell layers surrounding the starchy part of the endosperm. This tissue is characterized by the presence of prominent thick cell walls and nonvacuolated cells which are interconnected by plasmodesmata. Histochemical evidence indicates that the cell wall is composed primarily of a carbohydrate polymer containing both β-1,3 and β-1,4 linkages. In the imbibed state, aleurone cells are completely filled with various organelles and devoid of the typical plant vacuole (Fig. 1). The largest (2–4 μm) and most prominent organelle is the aleurone grain. This organelle is spherical, bounded by a single unit membrane, and contains protein and salts of phytic acid. The aleurone cell also contains numerous lipid bodies or spherosomes. Quantitation of organelles by centrifugation of whole cells indicates that the aleurone grains and spherosomes occupy at least 80% of the volume of these cells. The imbibed aleurone cell also contains a nucleus, mitochondria, dictyosomes, rough endoplasmic reticulum (RER), glyoxysomes, and leucoplasts.

Cellular cytological changes. The time course of hydrolytic enzyme release from aleurone layers after GA_3 treatment has been well documented. β-1,3-Glucanase release begins after 4 hr of GA_3 treatment, α-amylase and protease after 10–12 hr, and ribonuclease after 24 hr. Ultrastructural changes, however, are evident in barley aleurone cells within 1–2 hr of GA_3 treatment. The aleurone grains increase in volume and lose their spherical appearance. This increase in aleurone grain volume reaches a maximum after 8–10 hr of GA_3 treatment. Changes in volume exhibited by the aleurone grain result from an increase primarily in the region of the grain occupied by protein. It can be inferred that since the aleurone grain is the only region for the storage of large amounts of protein, the changes in this region are a prerequisite for the subsequent new synthesis of enzymes.

Associated with this change in aleurone grain volume is a pronounced increase in the amount of RER. This proliferation of RER continues up to 24 hr of GA_3 treatment. During this period the width of the cisternae of the RER increases markedly. It would seem reasonable to conclude that this developmental stage of the aleurone cell represents the period when maximum hydrolytic enzyme synthesis is occurring. Following the period of RER distention is a phase of vesicle proliferation from the ER. Initially the RER can be recognized as a true reticulum throughout the cell but afterward (14–20 hr of GA_3) becomes fragmented into successively smaller units, having diameters of 0.2–0.4 μm.

During the period of RER distention and fragmentation, the aleurone grains become gradually reduced in size, and at 20–22 hr they cannot be recognized as protein containing organelles (Fig. 2). Evidence indicates that the depleted aleurone grains contribute to the formation of vacuoles which appear in the ground cytoplasm of these cells after 24 hr of GA_3. That aleurone grains could give rise to vacuoles is consistent with information concerning their origins, that is, as deposits of protein within preexisting vacuoles. The process of vacuole formation continues in GA_3-treated aleurone cells until a stage is reached, at about 36 hr, when the cell is seen to contain one large central vacuole (Fig. 3).

Structural changes in wall. Although cytological changes in the aleurone layer have focused on

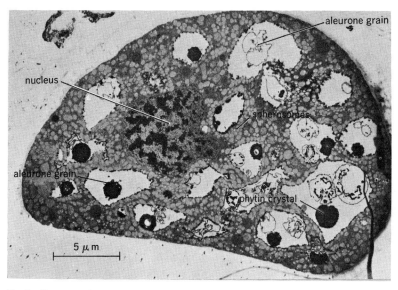

Fig. 1. Electron micrograph of an imbibed aleurone cell.

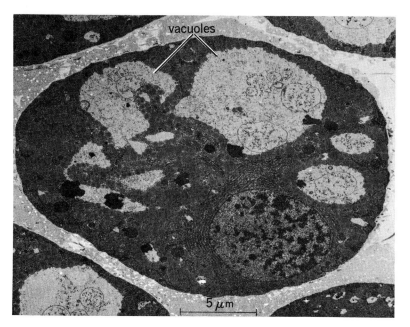

Fig. 2. Electron micrograph of an aleurone cell treated with GA_3 for 24 hr showing appearance of larger vacuoles in the ground cytoplasm.

those occurring within the cell, dramatic changes occur in the cell wall. Within 12 hr of GA_3 addition, the aleurone cell walls become digested, as indicated by histochemical staining at the light and electron microscope levels (Fig. 4). This digestion proceeds with duration of GA_3 treatment until as much as 80% of the cell wall is removed. Since this wall contains β-1,3 linkages, cell wall digestion is presumably caused by the β-1,3-glucanase released from this cell after GA_3 treatment. There are two features of interest concerning the way in which cell wall digestion occurs. First, digestion of the cell wall occurs in a polarized manner and predominates in the region of the wall which lies closest to the starchy endosperm. Polarized digestion occurs in this region of the wall irrespective of the position of the cell in the aleurone layer. It has been speculated that hydrolysis of the cell wall must occur to allow the rapid diffusion of enzymes from the cell. This speculation is supported by the observation that β-1,3-glucanase release begins after 4 hr of GA_3 treatment and at least 4 hr before the release of α-amylase and protease and 20 hr before ribonuclease.

The second feature of interest concerning cell wall digestion lies in the observation that, during the early phase of this process, digestion of the cell wall can be observed not only in the region of the wall immediately adjacent to the plasmalemma, but also in areas of the wall several micrometers removed from the plasmalemma. However, this more remote pattern of digestion takes place only in that region of the wall surrounding the plasmodesmata.

Enzymes release. Within a period of 36 hr of GA_3 treatment, aleurone cells synthesize and release several hydrolytic enzymes and are transformed from cells characterized by their protein reserves to cells possessing one large central vacuole. Although correlations can be made between the synthesis of hydrolytic enzymes and the observed cytological changes, little cytological evidence has been obtained to elucidate the way in which the enzymes are released from these cells. Thus, although the RER fragments to smaller units, there is no evidence to indicate that these fragments are involved in the export of these en-

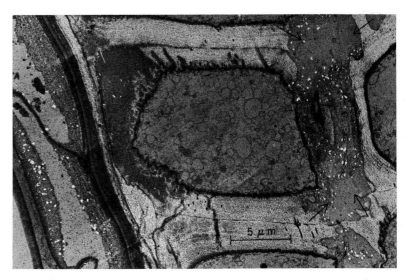

Fig. 4. Electron micrograph of an aleurone cell treated with GA_3 for 18 hr and stained to emphasize the cell wall. Note the digestion of the cell wall at arrows only.

zymes. Further, there is no evidence that in this cell the release of hydrolytic enzymes occurs in a manner analogous to that seen in animal cells (for example, the pancreas). In the pancreas secretion occurs by way of a specific Golgi-derived organelle. Attempts to isolate a subcellular organelle containing the secreted hydrolases have not succeeded, in spite of the fact that other enzyme-containing organelles (for example, glyoxysomes) can be isolated and fractionated on sucrose density gradients. Indeed, the observation that the cell-wall-digesting enzymes can migrate out of the cell through the plasmodesmatal canal (diameter 0.1 μm) indicates that distinct secretory organelles are not involved in the process of hydrolytic enzyme release from aleurone cells.

For background information see SEED (BOTANY) in the McGraw-Hill Encyclopedia of Science and Technology.   [RUSSELL L. JONES]

Bibliography: R. L. Jones, *Planta*, 85:359–375, 1969; R. L. Jones, *Planta*, 87:119–133, 1969; R. L. Jones and J. M. Price, *Planta*, 94:91–102, 1970; L. Taiz and R. L. Jones, *Planta*, 92:73–84, 1970.

Semiconductor

A new semiconductor regime is realized when the dielectric relaxation time τ_D, which is the product of the resistivity ρ and dielectric constant ϵ, exceeds the carrier diffusion-length lifetime τ_0. This condition of $\tau_D > \tau_0$ defines the "relaxation case." The conventional semiconductor behavior encountered in pn junction rectifiers and transistors is the "lifetime case" of $\tau_0 \gg \tau_D$. For semiconductors, ϵ is typically 10^{-12} farads/cm which gives $\tau_D = \rho \times 10^{-12}$ sec with ρ in ohm-cm. In crystalline or amorphous semiconductors doped with deep traps, which shorten the lifetime to less than 10^{-8} sec, the relaxation case may be expected for $\rho > 10^4$ ohm-cm, while the lifetime case usually occurs for $\rho < 10^4$ ohm-cm. Because the transport of carriers is drastically different from that of the familiar lifetime case, new and novel relaxation-case devices may be anticipated. At the present time, only the fundamental theory and the results of a few basic experi-

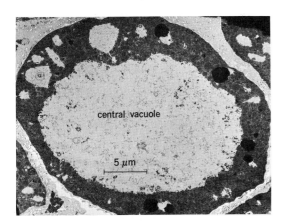

Fig. 3. Electron micrograph of an aleurone cell treated with GA_3 for 36 hr. Cytoplasm possesses one large central vacuole.

ments are available. However, several potential areas of application are obvious. Recognition that amorphous semiconductors are relaxation-case materials should facilitate understanding of their electronic behavior. Many wide-energy-gap semiconductors also readily satisfy the condition

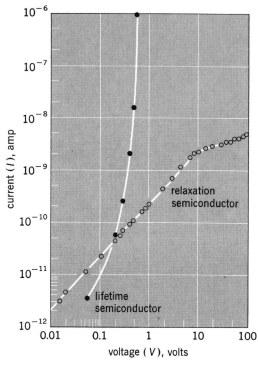

Fig. 1. Forward-bias current-voltage plot of GaAs lifetime and relaxation semiconductor *pn* junctions. Temperature is 22°C.

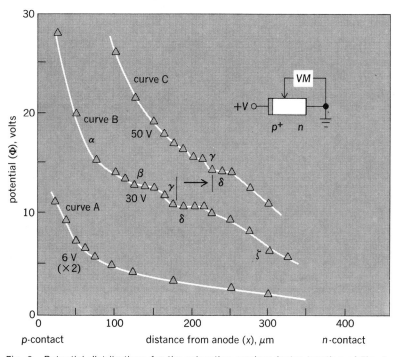

Fig. 2. Potential distributions for the relaxation semiconductor junction of Fig. 1. The voltmeter is represented in the insert by *VM*, and the applied voltage is given on each curve.

$\tau_D > \tau_o$. The physics of semiconductors has become a mature field, and new concepts such as the relaxation case will hopefully permit extension of the many uses and applications of semiconductors.

Theory. Theoretical analysis based on the solution of Poisson's equation, the continuity equations, and the total current density has demonstrated the fundamental differences between the lifetime and relaxation semiconductors. These differences are summarized in the table. The first property given in the table is simply the condition that defines each case.

In the solution of the differential equations for carrier transport, the local space-charge neutrality ($\Delta q = 0$) assumption is a good approximation for the lifetime semiconductor, while departures from local neutrality ($\Delta q \neq 0$) and pronounced space charge are predicted for the relaxation semiconductor. The net local recombination rate R is nonzero at steady state for the lifetime semiconductor, but for the relaxation semiconductor, R approaches zero after an initial rapid recombination. This condition of $R = 0$ requires that the product of the steady-state hole and electron concentration n and p is constant: $np = n_i^2$, where n_i is the intrinsic carrier concentration. The enhancement of the carrier densities in a lifetime-case junction as $np = n_i^2 \exp(qV/kT)$, where V is the applied forward bias, is essential to operation of devices such as the transistor. Minority carrier injection reduces resistivity in the lifetime semiconductor. In the relaxation semiconductor, the injected minority carriers can both reduce the majority carrier density and fill ionized traps. Injected minority carriers thus increase resistivity because of this recombinative depletion.

Experimental verification. Experimental measurements of a relaxation semiconductor both illustrate and verify the theoretical predictions. A room-temperature current-voltage [$I(V)$] characteristic for a GaAs relaxation semiconductor *pn* junction is shown in Fig. 1 together with the $I(V)$ characteristic for a GaAs lifetime-semiconductor *pn* junction. It may readily be seen that the forward characteristic of the relaxation-semiconductor junction does not exhibit the exponentially increasing forward current of the lifetime-semiconductor junction. The relaxation semiconductor was *n*-type GaAs with $n = 3 \times 10^7$ electrons/cm^3 and $\rho = 1.5 \times 10^8$ ohm-cm at 22°C, while the lifetime semiconductor had $n = 3 \times 10^{17}$ electrons/cm^3. In both devices the Zn diffusion created a 3×10^{-4} cm deep *p*-layer with a surface concentration of 10^{20} holes/cm^3. The diode areas were 4×10^{-3} cm^2 and the relaxation semiconductor was 4.5×10^{-2} cm long.

The forward branch of the relaxation semiconductor $I(V)$ characteristic extends linearly to about 10 V, and the device resistance is dominated by the junction space-charge region (SCR). Holes and electrons are swept into the SCR where rapid recombination ensures $np = n_i^2$. The sublinear region with $d[\ln(I)]/d[\ln(V)] \approx 0.4$ begins when injected holes traverse the entire SCR width. These holes deplete majority electrons and widen the SCR, increasing resistance. Space charge is built up through hole capture by the Coulomb-attractive centers, which become neutral and no longer compensate the positive donors. The de-

Comparison of lifetime and relaxation semiconductors

Property	Lifetime case	Relaxation case
Definition	$\tau_D < \tau_o$	$\tau_D > \tau_o$
Space charge (Δq)	$\Delta q = 0$	$\Delta q \neq 0$
Net local recombination (R)	$R \neq 0$	$R = 0$
Forward-bias junction electron-hole density product (np)	$np = n_i^2 \exp(qV/kT)$	$np = n_i^2$
Effect of carrier injection	reduces ρ	increases ρ

tailed theory is complicated, but this SCR widening may be verified by considering the potential distributions shown in Fig. 2.

The potential distributions for the junction of Fig. 1 were obtained by mechanically dragging a tungsten tip across the diode. Curve A is typical for the linear $I(V)$ region and shows that the largest voltage drop is across the 7×10^{-3} cm junction SCR. The average resistivity in the junction SCR is 2×10^9 ohm-cm and is an order of magnitude greater than the resistivity of the semiconductor bulk. The maximum resistivity ρ_{max} may be specified by $\mu_n n_m = \mu_p p_m$, where μ_n and μ_p are the electron and hole mobility and n_m and p_m are n and p for ρ_{max}. When the net local recombination rate approaches zero, the condition of $np = n_i^2$ holds, and $n(\rho_{max}) = n_m = \sqrt{\mu_p/\mu_n}\, n_i$ and $p(\rho_{max}) = p_m = \sqrt{\mu_n/\mu_p}\, n_i$. The maximum resistivity is $\sqrt{\mu_n/\mu_p}/2qn_i\mu_n$. At 22°C, $n_i = 9 \times 10^5$ carriers/cm³, $\mu_n/\mu_p = 20$, and thus $\rho_{max} \approx 3 \times 10^9$ ohm-cm. In addition to the numerical agreement between the average resistivity of the junction SCR and ρ_{max}, the resistivity has been found to have the temperature dependence of n_i as predicted.

Curve B is typical of the sublinear region. Region α is the junction SCR, region ζ is the semiconductor bulk, and the regions β, γ, and δ represent the transition between these two regions. The γ region is the high-field "depletion drift region" in which carrier transport is by drift, while β is a predicted low-field region in which the current is by hole diffusion. The boundary between γ and δ is the "recombination front" where holes transported through the junction SCR recombine with electrons that flow by drift in the semiconductor bulk. The low field in region δ is due to an enhancement of majority carrier electrons near the recombination front. As the forward bias is further increased in curve C, the boundary of the junction SCR, as shown by the arrow, moves further toward the negative contact. This broadening of the junction SCR by minority carrier injection increases the device resistance and results in the observed sublinear $I(V)$ characteristic. There are many other properties predicted for the relaxation semiconductor, but this example illustrates and verifies one of the more fundamental predictions of the theory.

For background information see BAND THEORY OF SOLIDS; RELAXATION TIME OF ELECTRONS; SEMICONDUCTOR in the McGraw-Hill Encyclopedia of Science and Technology.   [H. C. CASEY, JR.]

Bibliography: H. J. Queisser, H. C. Casey, Jr., and W. van Roosbroeck, *Phys. Rev. Lett.*, 26:551, 1971; W. van Roosbroeck and H. C. Casey, Jr., *Phys. Rev.*, in press; W. van Roosbroeck and H. C. Casey, Jr., in S. P. Keller, J. C. Hensel, and F. Stern (eds.), *Proceedings of the 10th International Conference on the Physics of Semiconductors*, p. 832, 1970.

Ship, merchant

The past several years have witnessed the advent of two new types of merchant ships and tremendous technological changes in certain types already in service. The two new ship types are generally classed as bulk carriers. One of these is the ore/bulk/oil (O/B/O) ship. The other is the slurry carrier, now classified usually as a mineral tanker. The O/B/O is, as its name implies, a ship capable of carrying its full deadweight in bulk cargo such as coal and grain, high-density cargoes such as iron ore, as well as crude petroleum products. The mineral tanker is capable of loading, transporting, and discharging dry bulk commodities in slurry form, as well as carrying crude petroleum products.

The mineral tanker design was developed by the Marcona Corp. primarily as a means of improving the economics of transporting raw materials from the mine source to the ultimate receiver through the use of fluid pumping techniques, thus eliminating the necessity of costly railroad and port loading facilities at the mining end, together with the requirement for equally costly shore-based discharging facilities at the receiving end. The mineral tanker is, in fact, the ultimate configuration of the basic ore/oil (O/O) carrier. The original of this type of carrier was, as its name implies, capable of carrying cargoes of dry iron ore as well as crude petroleum products.

O/S/O. In 1969 Marcona converted a 50,000 ton O/O carrier to an ore/slurry/oil (O/S/O) carrier. In 1971 the world's largest O/S/O was delivered, a

Fig. 1. *San Juan Exporter*, a 142,000 dwt combination ore/slurry/oil (O/S/O) carrier capable of loading, transporting, and discharging bulk commodities as slurry or, alternatively, handling dry iron ore products.

ship of 142,000 deadweight tons (dwt) capable of transporting finely ground iron ore in slurry form, dry ore, and crude petroleum products (Fig. 1). A 200,000 dwt mineral tanker is now being designed.

The Marconaflo system, reduced to its simplest arrangements, requires sizing or grinding or both of any bulk product which may be thus transported. The product is slurried through the introduction of water or other compatible medium, and the material is pumped aboard ship in varying densities. Once aboard ship the fluid carrier is rapidly decanted prior to the vessel's departure from the loading port. On the way to the discharging port additional decantation takes place, so that when the vessel arrives at the receiving port the cargo is essentially a compacted dry mass. For discharge the material is reconstituted by the introduction of high-pressure water, or other suitable medium, through revolving jets operating in sumps regularly spaced within the tank top (Fig. 2). The cargo flows out through the sumps into special collecting tanks, from where slurry pumps discharge it to relatively simple shore receiving facilities consisting of either tanks or ponds. As a specific example of this process, iron ore finely ground to pellet feed form is pumped aboard the mineral tanker at approximately 70–75% density. Decantation prior to sailing increases the density to approximately 90%, with additional decantation at sea increasing the density to approximately 93%. During discharge at the receiving port, pulp densities average approximately 70%. Discharge rates of up to 6,000 dry long tons per hour are being achieved.

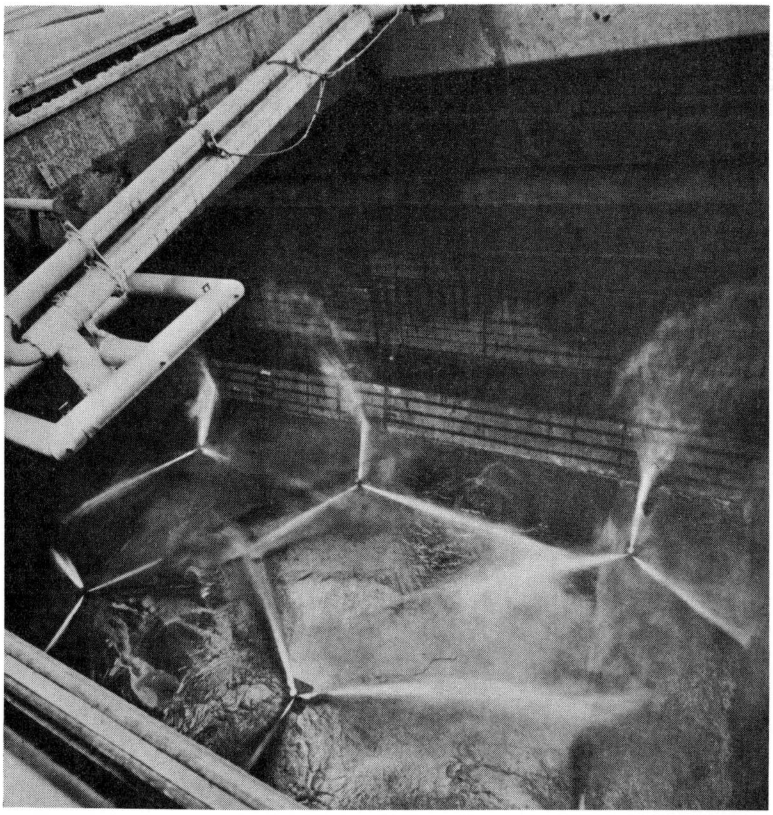

Ships designed for the transport of dry bulk commodities have previously been limited in size by draft restrictions at loading and receiving ports and, additionally, by the higher construction costs of such ships as compared to costs of the more simple tanker configuration. The mineral tanker, incorporating the simplicity of tanker design and construction and capable of discharging through offshore single-point mooring systems, thus makes it feasible to transport dry bulk commodities in the largest ships possible to design and build.

O/B/O. The O/B/O carrier (Fig. 3), although not introducing a completely new type of transportation system, nevertheless represents a major advance in the carriage of bulk commodities. The introduction of this triple-purpose ship, capable of moving several types of commodities on separate legs of a single voyage, resulted in major reductions in transportation costs. Prior to the introduction of O/O and O/B/O carriers, tankers and iron ore carriers normally operated in ballast 50% of the time. The modern O/B/O, on the other hand, frequently operates in cargo as high as 85% of steaming time, with only 15% steaming time in ballast, that is carrying no cargo. On a typical voyage from an iron ore loading port in Peru to a Japanese receiving port with backhaul of oil or coal, this utilization of cargo capacity can result in reducing the transportation costs of iron ore from approximately $4.90 per ton to approximately $2.33 per ton. The large O/B/O is, because of critical hull design problems resulting from the absence of longitudinal bulkheads and the exceptionally large hatch openings required, coupled with other hull structural requirements, more costly per deadweight ton to construct than are simple tankers, ore carriers, or O/O carriers. It follows, therefore, that the efficient use of the triple-threat characteristics of the O/B/O is mandatory to offset the higher capital costs of construction.

Specialized container ships. A third major development during the past few years has been the changeover from general-purpose cargo liners to specialized container ships. The container ship, as such, was developed many years ago but in its initial form was generally a conversion of then existing ship types such as C-4s, C-3s, C-2s, Mariner class vessels, and T-2 tankers. Such converted ships rarely exceeded 17,000 dwt, with speeds not in excess of 17 knots and container capacities of 500–600, depending on size. Container ships being built today are typified by the series under construction at Howaldtswerk, Germany. These specialized vessels have a deadweight of 49,000 tons, a length between perpendiculars (LBP) of 900 ft (274.3 m), a beam of 106 ft (32.3 m), with container capacities of 1400–1600. With 80,000 shaft horsepower, they are capable of speeds exceeding 27 knots. Highly automated both as regards the propulsion and navigational equipment as well as the container handling facilities, they have revolutionized the efficient transportation of general cargo.

Although the designers of the lighter aboard ship (LASH) type of vessel and of the SEABEE type would not necessarily agree with this statement, it can generally be indicated that these two

Fig. 2. Overhead view of one of four holds of 52,000 dwt *Marconaflo Merchant*, world's first combination ore/slurry/oil (O/S/O) carrier showing rotating, high-pressure water jets sweeping out remains of 40,000 tons of iron ore pellet feed concentrates discharged by ship. Educator mechanism (device at top left) is used to decant water from hold after slurry is loaded.

ship types are modifications or refinements or both of the straight container ship. The SEABEE vessels can carry loaded and empty barges which are hoisted to the appropriate deck level by a high-capacity elevator system. Additionally they can carry containers or can act as roll-on/roll-off ships. The LASH vessels handle barges which are hoisted aboard and stowed by means of a gantry crane operating over the stern of the vessel. They also can handle containers. The SEABEE barges are larger than the LASH barges. In principle, these two types operate as do container ships, taking aboard and discharging loaded and empty barges rather than containers. Where the container ship has to go alongside shore container terminals, the LASH and SEABEE vessels can anchor in a harbor or any other convenient place, discharging their barges and towing them to shore distribution centers and receiving barges in the same manner. Both of these ship types have followed the trend toward large size and high speed indicated for the Howaldtswerk container ship.

Other technological improvements. The fourth major merchant ship development of the past few years has been in the field of technological improvement. This applies primarily to size of ships as well as to propulsion systems, automation, and other areas.

Size increase. Typical tanker size has increased from 50,000 dwt approximately 10 years ago to 250,000 dwt in 1971. Additionally, six tankers of approximately 326,000 dwt are in operation, with three tankers of approximately 500,000 dwt on order. One of the major classification societies, working closely with the shipbuilding industry, is already developing plans for a tanker of 1,000,000 dwt.

The typical O/O carrier has increased in size from approximately 70,000 dwt in 1963 to approximately 260,000 dwt on order for delivery in 1973. The O/B/O types have increased in size from the 64,000 dwt *San Juan Trader*, which at the time of its delivery in 1966 was the largest O/B/O capable of carrying its full deadweight in any cargo, up to approximately 165,000 dwt for O/B/Os on order for delivery in 1972.

Use of new alloy and steels. To a considerable degree the construction of increasingly larger ships of all categories has been made possible by the development and use of special alloy and high-tensile steels, resulting in substantial weight reduction for the same strength requirements. Of major importance in the design of such large ships, however, has been the increasing use of computer programs to optimize the hull structural design characteristics. These factors have combined to decrease the steel weight/deadweight tonnage ratio. For example, this ratio for a 70,000 dwt O/O carrier built in 1962 is approximately 0.198, for a 132,000 dwt O/O carrier built in 1970 approximately 0.180, and for a 260,000 dwt ton oil carrier for delivery in 1973 approximately 0.140. Continuing research by the world's major ship classification societies, in cooperation with major shipyards throughout the world, together with various owners, has been the key to this continuing progress in increased ship size.

Automation. The extensive use of automation has resulted in more efficient vessel operation and

Fig. 3. *San Juan Trader*, an approximately 64,000 dwt ore/bulk/oil (O/B/O) carrier. One of the first of its type, on delivery it was the largest such ship capable of carrying its full deadweight in any commodity.

in the reduction of crew requirements. On a typical O/O, for example, the complement has been reduced from 60–65 officers and men in 1963 to an average of approximately 35–40 officers and men in 1971. Some ships have operating complements as low as 27 officers and men. The unmanned engine room has been introduced as a result of automation. Typically the engine room is unmanned only during the night hours, with workers present for normal maintenance and repair during the daylight hours.

Propulsion plants. Improvements and modifications have continued in all types of propulsion plants. The large slow-speed diesel has been redesigned and improved so that such a plant can now be built in horsepowers of up to 40,000 in a single engine, using as fuel the same type of Bunker C residuals that normally fire a steam generator. Steam turbine plants have generally been improved in design and this, together with the introduction of the reheat cycle, makes it possible to achieve, on a steam turbine ship, fuel consumption rates comparable to those of the large slow-speed diesel plants. In a typical reheat cycle, steam is generated at 1450 psig (102 kg cm^{-2}g) and 950°F (510°C). Following the passage of steam at these initial characteristics through the high-pressure turbine, the exhaust at 310 psig (22 kg cm^{-2}g) and 620°F (327°C) is routed through a reheater in the steam generator, changing its characteristics to 280 psig (20 kg cm^{-2}g) and 950°F (510°C). From there the exhaust is routed to the low-pressure turbine and thence to the condenser. The typical fuel rate for such a plant is approximately 0.41 lb (0.19 kg) per shaft horsepower per hour for all purposes, as compared to 0.37 lb (0.17 kg) per shaft horsepower per hour for propulsion only for a slow-speed diesel plant.

A major improvement in the propulsion of large vessels was introduced on the San Juan Vanguard class of ships delivered in 1970. These O/O carriers of approximately 132,000 dwt are the first commercial vessels to couple a controllable-pitch propeller to a nonreversing steam turbine plant. The main propulsion unit generates 25,000 hp, of which 23,500 hp is delivered to the propeller shaft, with the remaining 1,500 hp being used at sea to drive the main ship service generator, the feed pump, and the lube oil pressure pumps. Reversing is accomplished simply by changing the pitch of the propeller blades from ahead to astern position. Finite maneuvering speeds are achieved by varying the pitch of the propeller blades. Operation is completely remotely controlled by the master on the bridge, with the master having finite control of ahead and astern motion. This type of plant results in increased propulsion efficiency inasmuch as the pitch can be changed to meet varying conditions of weather, wind, and loading. Maneuverability is greatly improved, with stopping distances amounting to less than half of those required for the conventionally propelled ship. The turbine operates as an essentially constant-speed unit similar to those employed in shore power plants.

The past few years have witnessed increasing experimentation with the gas turbine plant as a propulsion medium for large ships. This type of plant has obvious advantages, among these being simplicity, ease of maintenance, reduced space requirements, and somewhat lower initial costs. Up to the present, however, to a considerable degree, these advantages have been largely nullified by the requirement of using highly refined fuels, distillates, and so on. The development of a heavy duty industrial-type gas turbine capable of burning residuals which have been properly treated aboard ship has eliminated the major drawback to the use of this type of main propulsion plant. A 200,000 dwt mineral tanker for delivery in 1974–1975 is now in the design stage. It is anticipated that this vessel will incorporate a propulsion plant consisting of a heavy duty industrial-type gas turbine power plant coupled to a controllable-pitch propeller. Additionally the gas turbine unit will be operated in port to develop the power required for discharge of the slurry and oil cargoes.

For background information *see* SHIP, MERCHANT in the McGraw-Hill Encyclopedia of Science and Technology. [HUGH C. DOWNER]

Bibliography: H. C. Downer, Revolution in bulk ocean transportation, *Amer. Soc. Nav. Eng. J.*, 82 (6):49, December, 1970; H. C. Downer et al., *The Triple Purpose Ore/Bulk/Oil Carrier*, Soc. Nav. Architects Mar. Eng. Tech. Pap., October, 1968; G. P. Lutjen, Marconaflo: The system and the concept, *Eng. Mining J.*, 171(5):67, May, 1970; G. M. McManus, The sea changes mineral logistics, *Iron Age*, 204(5):79, July 31, 1969; T. B. Thomas et al., *Marine Transportation of Mineral Slurries*, Soc. Nav. Architects Mar. Eng. Tech. Pap., October, 1970.

Ship design

Ship design, an ancient art and a modern engineering science, is going through a process of adaptation to the rapid advancements of computer technology. Computer-aided ship design has become a new specialty, supporting the evolution of a more rational and systematic style in ship design.

Design objectives and tasks. Ship design aims at an optimal technical solution to an owner's given functional requirements. For example, a merchant ship owner may specify the amount of cargo to be transported from A to B. He will judge the success of a design by the expected return on his investment. The design at the same time has to meet numerous technical and regulatory requirements related to the safety and seaworthiness of the ship, especially its strength, floatability, and stability (intact and damaged), and other factors concerning the safety of human life at sea. Current efforts by many governments and the Intergovernmental Maritime Consultative Organization (IMCO) toward reducing the probability of ocean pollution resulting from ship collisions are adding a new dimension to the requirements for liquid bulk cargo ships.

In short, ship design may be regarded as an economic optimization problem with numerous constraining side conditions. This viewpoint is not new, and economic studies have accompanied many developments of new ship types, for exam-

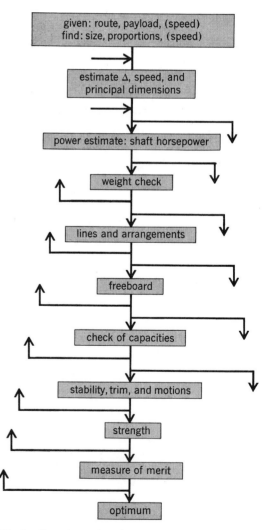

Fig. 1. Flow chart of preliminary ship design. The sequence of steps is flexible and varies with circumstances, as implied in the arrows not being connected to the nodal points. (*From H. Nowacki, Ship design and the computer, Mar. Technol., Society of Naval Architects and Marine Engineers, July, 1970*)

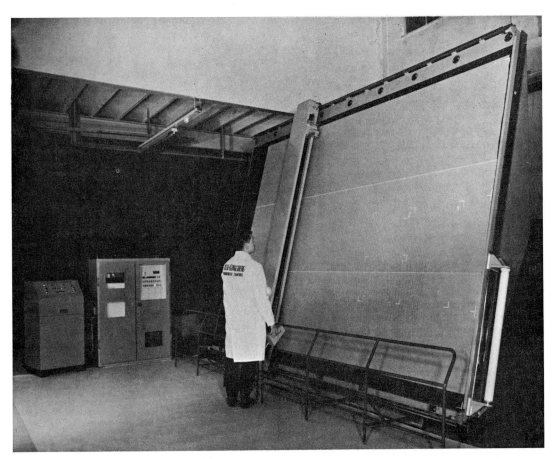

Fig. 2. Computer-faired ship body plan plotted on 1:10 scale by automatic drafting machine. (*Kongsberg Co.*)

ple, the famous Mariner class ships of the U.S. Maritime Administration. However, recent computer studies have recognized the pertinence to ship design of the modern tools of operations research and mathematical optimization (random search, direct search, nonlinear programming). The most economic size, speed, and principal dimensions of a ship are thus obtained rather automatically by computer.

For this purpose, the individual tasks of preliminary ship design (Fig. 1) have to be converted into computer algorithms. This is not difficult where it is merely a matter of encoding existing analytical techniques. But ship design also involves the synthesis of ideas and geometrical concepts not easily expressed in analytical terms. Much thought has been devoted to this problem recently.

Lines fairing and creation. Ship surfaces are usually empirical, not mathematical, functions obtained by manual drafting techniques assisted by a fairing tool, the batten. Many mathematical techniques for ship surface definition have therefore simply simulated the behavior of an elastic batten. The Theilheimer spline, a polynomial of discontinuous curvature, is widely used. Spline methods have been found adequate for interpolating a surface through a given set of ship offset points, or for fairing a smooth approximation through such a set of points. As a measure of fairness for a ship surface, the criterion of minimum strain energy of flexure in the batten is generally favored in recent work. The quality of computer-faired ship lines is equivalent to the results of manual lofting and adequate for ship production purposes (Fig. 2).

Computers are also beginning to be used in the earliest design stage to generate a hull form, not from given offsets, but from given specifications as to underwater hull volume, centroids, slopes, and other geometric parameters. G. Kuiper in Holland, K. H. Kwik in Germany, A. Reed and H. Nowacki in the United States, and the U.S. Navy have explored methods of computer-assisted lines creation. Light-pen-controlled computer graphics seems to hold good promise for this purpose.

Systems approach. A ship is a complex system of many interacting components or subsystems. Moreover the ship is only part of a larger transportation system, a fact underscored by the recent rapid expansion of intermodal transport of unitized cargo. Planning the transport system and designing the ship have to go hand in hand.

For these reasons, ship design studies follow the systems approach with increasing frequency and tend to become more elaborate. Figure 3 shows the possible scope of a ship systems study. Many uncertain influences are involved (market, weather and seaway, delays) that have to be treated statistically.

The systems approach also provides the methods for dealing with such additional complexities and for decomposing the problem into simpler, more tractable ones. It reduces the design task to a mathematical optimization problem.

Optimization. Ship design presents us with an abundance of feasible technical solutions and calls

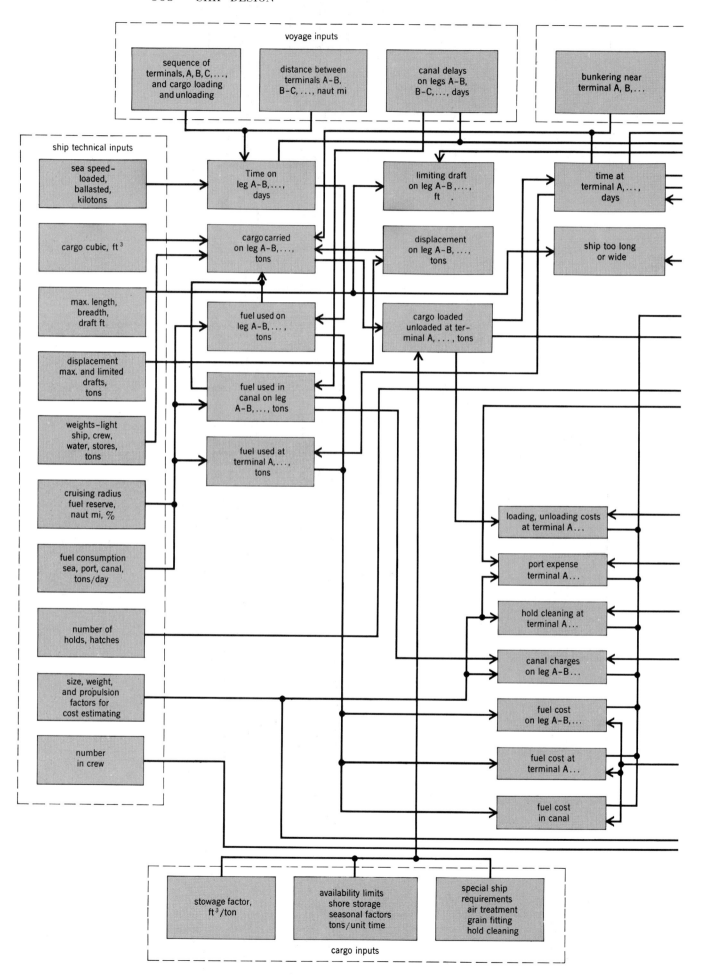

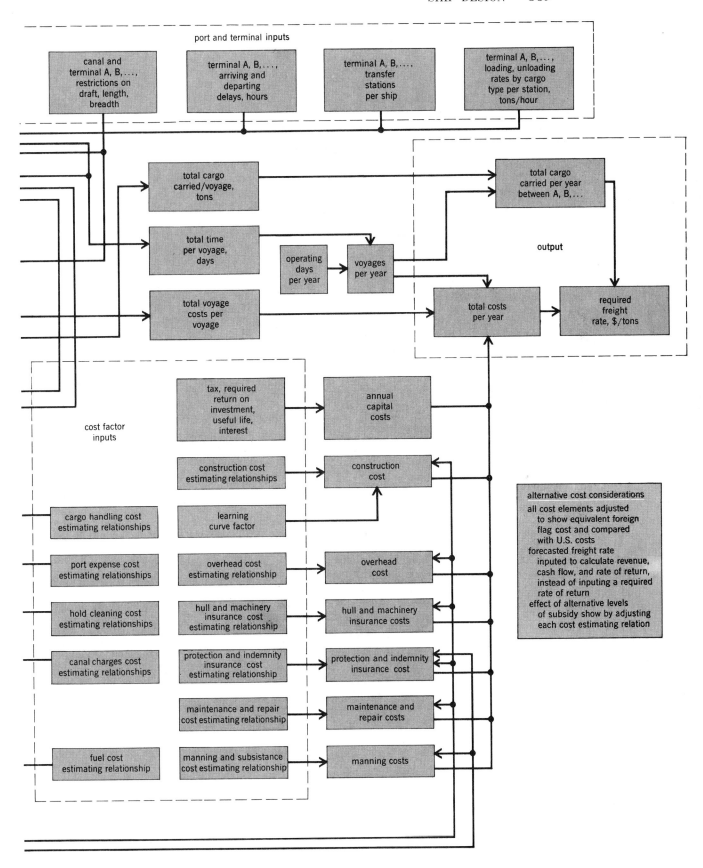

Fig. 3. Format of systems study in ship design shown by a bulk carrier evaluation model developed by Booz, Allen Applied Research, Inc., under the Maritime Administration's Competitive Merchant Ship Project. (*Booz, Allen Applied Research, Inc.*)

for rational choices and, in fact, optimization. Economic, hydrodynamic, and structural optimization studies are providing new perspectives to modern ship design.

In the structural optimization of ships, the work done by J. Moe and coworkers in Trondheim, Norway, has been most notable. Moe formulates the problem of finding the least weight or least cost ship structure for a given set of structural load assumptions. Corresponding load assumptions also underlie the rules of the classification societies and form the basis of conventional structural design of ships. Moe then solves for the optimal scantlings of the ship's plating and stiffening by a nonlinear programming computer algorithm called the sequential unconstrained minimization technique (SUMT). It is noteworthy that, in this connection, the computer serves as more than merely a tool of structural analysis. It selects the best combination of scantlings for a given structural configuration, and in this sense it actually "designs."

Structural design of ships has in the past several years also benefitted greatly from advancements in finite element stress analysis. In the DAISY design program, which was developed at the University of Arizona under sponsorship of the American Bureau of Shipping and Chevron Shipping Co., an automated approach to ship structural analysis is taken. The user provides a description of the hull geometry and structural configuration, loading condition data, and some general rules for the arrangement of nodal points. From this the computer program generates a grid of finite elements approximating the ship structure and calculates the stress distribution in the hull. A coarse grid analysis of the complete ship is the first step, followed by a fine-mesh examination of critical local areas if required.

The problem of minimizing the hydrodynamic resistance of a ship has intrigued many generations of naval architects, but its complete solution still is elusive for lack of a full understanding of the ship's boundary layer, especially near the stern, and of its waves. There has nonetheless been some encouraging success in obtaining low-resistance hull forms by optimization methods. Outstanding examples in recent years are studies by T. Inui, H. Maruo, and M. Bessho on "waveless" hull forms and bulbous bows of optimal size and location; by J. V. Wehausen on the minimum total resistance problem; and by P. C. Pien on low-wave-making forebodies. The insights gained from these studies are visible in many modern hull forms, although it must be noted that experimental research also contributed a fair share.

In ship operations the prospects of optimum weather routing have been investigated in several recent studies. For a known ship and a given weather map, it is not too difficult to determine the minimum time or minimum fuel track of a ship, avoiding at the same time areas where ship or cargo might get damaged in severe sea states. This type of service is now commercially available. Unfortunately, long-term weather forecasts still are not fully reliable, so that the ship's recommended route should be updated a few times during the voyage. The time savings due to optimal routing may be in the order of 6–12 hr in a single North Atlantic crossing. Reducing the probabilities of cargo damage may be of equal economic importance.

Computer graphics. The processing of geometric data is extremely intensive in ship design and construction. A design idea is often born as a visual concept and must be displayed graphically to be communicated easily and uniquely.

Many computer systems today are strictly analytically oriented, so that there exists a mismatch between the computer and a user who prefers pictures as input and output. Digital plotters or cathode-ray-tube displays with light-pen control (active graphics) remove this deficiency.

Plotter output is an essential ingredient in lines fairing applications and as a check medium for numerically controlled flame-cutting equipment. Most computer graphics applications in the shipbuilding industry are directed to ship production. But following the example of the United States aircraft and automotive industries, there is some experimentation with active graphics in early ship concept formulation, especially in ship arrangement plan design, engine room and piping layout, and ship lines creation. It appears that computer graphics is beginning to play an eminent role in all phases of ship design.

Design and production. Design and production form a continuum in data processing. Consequently wherever ship production technology motivated the development of a major software system, a good climate exists for its symbiosis with design-oriented computer software. Numerous program packages have emerged in the last few years, all by and large with capabilities ranging from earlier stages of design to the generation of numerical control tape for cutting steel. A common data base unites the components of the program system.

Some of the better known integrated ship design and production systems are: AUTOKON with PRELIKON (developed in Norway), CASDOS (U.S. Navy), BRITSHIPS (Britain), FORAN (Spain), and VIKING and the Kockum's Shipyard System (Sweden).

These systems differ significantly in scope and implementation, but in a generic sense display the following capabilities. It is incorrect to say that each system named here has all six capabilities listed. Some do, some do not, and some are still developing. However, as a group they do.

1. Each system has a hull-definition program which, from a given set of offsets, can produce a fair ship surface that is kept in the master file for reference throughout the design.
2. Each system has a set of calculation routines for hull-form analysis, hydrostatic and stability analyses, and other computations in the early design stage.
3. Each system has a structural detailing capability to perform the engineering calculations for the design of the ship structure and for specification of its parts and scantlings. This feature necessitates a sophisticated bookkeeping system for the structural members and their connections (list processor in CASDOS).
4. Each system has part definition which is performed in a parts-programming language in conventional ship design terminology.
5. Each system has nesting, that is, a step with

manual intervention in which the parts are arranged on the large plate from which they will be cut so as to minimize scrap. A nesting table is often used to distribute the part patterns optimally.

6. Each system has numerical control tape generation for flame cutting.

The style of design work in such an environment is one of man-computer dialog. A fully automated design process is not favored because the creative and critically selective role of the human is crucial in all stages of design. However, ship design, especially in the detailing phase, involves a vast amount of tedious routine work which can be performed by the computer. Because of the repetitious character of many details in the ship's structure, it is feasible to develop computer languages that express the desired properties of the structural elements in very compact form in the ship designer's terminology. For example, a single one-line statement in such a ship design language is sufficient to define a ship floor plate in enough detail for drawing and cutting.

For background information see SHIP, MERCHANT; SHIP DESIGN; SHIPBUILDING in the McGraw-Hill Encyclopedia of Science and Technology. [HORST NOWACKI]

Bibliography: H. A. Kamel et al., *Trans. SNAME*, vol. 77, 1969; W. Marks et al., *Trans. SNAME*, vol. 76, 1968; J. Moe et al., *Trans. RINA*, vol. 110, 1968; H. Nowacki, *Mar. Technol.*, July, 1970.

Silica, biology of

Knowledge of the importance of silica, the dioxide of silicon, in the nutrition of plants and animals has continued to expand in recent years. Silicon is the most abundant element after oxygen, comprising 27.6% of the Earth's crust. This abundance has overshadowed the possible importance of small amounts of the element in the nutrition of plants and animals. Traditionally thought to be biologically relatively inert, this element now appears to play a more important role than was previously thought.

Plants. Members of the plant kingdom may be divided into two classes: those species that actively cumulate silica and those that do not. Important examples of the former include diatoms, *Equisetum* (horsetail), and rice and many other graminaceous plants. Most broad-leafed plants, including vegetables, are noncumulators and contain low amounts. Typical levels found in various tissues are shown in the table. The function of silica is partly structural, in that the element forms the skeleton of diatoms and parts of the cell-wall structure of grasses and other higher plants, where it occurs as opaline silica. Grasses have been shown on the basis of solubility-phase data to contain at least two major silicon fractions. *Equisetum* and grasses contain significant amounts of a soluble silicon fraction that may contain organosilicon compounds, although these have not been characterized. Methyl silicones have been isolated from tobacco smoke and are considered to be the degradation products of silicon structures in the original tobacco.

The soluble silicon fractions of grasses and *Equisetum* protect the plants from attack by cellulolytic microorganisms and detract from the feed value when such plants are consumed by ruminants and other herbivores. Considerable work on sugarcane has shown that silicon moderates manganese toxicity in leaves and that orthosilicic acid is an effective inhibitor of invertase. Exploitation includes spraying of sugarcane fields prior to harvest with solutions of sodium silicate to improve sugar yield. Fertilization of sugarcane fields with calcium silicate or cement slags has also increased sugar tonnage, an effect that is partly attributable to silicon and partly to the liming effect. Calcium silicate will precipitate aluminum and manganese, which may become toxic in tropical acid soils. This diminishes the phosphate-binding capacity, thus releasing phosphorus for plant use. At the same time silicon fertilization produces a decrease in protein content and an increase in sugar content of the plant. The same changes are also produced when solutions of sodium silicate are applied to the leaves.

Silicon is regarded as an essential element for diatoms and is so claimed for rice and *Equisetum*. The question is debated in the cases of sugarcane and other agricultural crops. Grasses are tolerant to a wide range of silicon levels—a characteristic that may cause the composition of the plant tissue to vary greatly, as shown in the table.

Animals. Animal tissues contain lower amounts of silica, which nevertheless forms an important proportion of the ash. Care must be taken in interpreting literature values because of the problem of contamination by soil or dust. This is particularly a problem in the analyses of skin, hair, and feathers. Earlier studies with grazing ruminants noted the problem of siliceous urinary calculi in animals grazing grasses of high silica content. Attempts to reproduce calculi by dosing ruminants with sodium silicate were unsuccessful. However, when a silicon-containing ester, tetraethylorthosilicate, was given, calculi have been produced. No reports of such calculi exist for nonruminants. Other effects of high silicon intake are diminished reproductive activity (for example, as a result of

Reported silica content of various biological materials

Plants and animals	Material	Silica, SiO$_2$	
		Dry tissue, mg/100 gm	Total ash, %
Equisetum	Aerial growth	6,000–13,000	32–56
Crimson clover	Aerial growth	120	1.4
Peas	Aerial growth	250	3.2
Alfalfa	Aerial growth	200–500	2.5–7
Reed canary grass	Aerial growth	600–8,760	8–56
Tall fescue	Aerial growth	900–4,800	17–51
Coastal Bermuda grass	Aerial growth	700–6,700	10–53
Oats	Leaf	5,300	40
	Culm	1,030	—
	Inflorescence	7,720	—
Rice	Polished grain	50	9
	Bran	4,600	37
	Hulls	20,000	99
	Straw	13,600	73
Human	Brain	22	0.4
	Skin	11	0.4
	Hair	9	1.9
	Kidney	11–22	1.3–5.0
	Heart	72–124	1.2–7.2
Bovine	Muscle	24–37	1.0
	Liver	4–10	0.1–0.2
Chicken	Feathers	34	9
Sheep	Wool	24	6

dosing with sodium silicate) and a lowered digestion coefficient in the case of forages high in silicon. The latter effect is a result of the protective action of natural silicon compounds that inhibit the cellulolytic action of rumen bacteria. Such inhibition is not possessed by silicic acids either in solution or as deposited upon cellulose fibers, whereas the reaction of sodium silicate or tetraethylorthosilicate with cellulose to form a silicate bond does produce inhibition. The latter products show a marked solubility drop on treatment with traces of acid. The drop in solubility as large as two orders of magnitude and detectable at pH 4 is characteristic of all natural plant silicas and of no inorganic sources so far tested.

Studies on the possible essentiality of silicon in animals have recently been renewed by Edith Carlisle, who has fed diets very low in silicon to rats. Evidence that silicon plays a role in calcium metabolism in bone mineral deposition has been presented. Silicon-deficient rats are less tolerant to a marginal calcium level than are rats receiving normal levels of silicon.

An unresolved problem in regard to the nutritional biochemistry of silicon resides in the failure thus far to characterize organic linkages or mechanisms by which the element is transported. Active transport dependent on a high-energy phosphate source appears to occur in diatoms and possibly in rice. Since silicic acid is essentially unionized at physiologic pH, transport may be dependent upon development of covalent organic linkages. The impediment in characterizing such linkages appears to reside in their fragility and the difficulty of fractionating plant cell-wall tissue under mild conditions.

For background information *see* BIOSPHERE, GEOCHEMISTRY OF; SILICON in the McGraw-Hill Encyclopedia of Science and Technology.

[PETER J. VAN SOEST]

Bibliography: E. Carlisle, *Science*, 167:279, 1970; R. L. Fox et al., *Plant Soil*, 30:81, 1969; G. Samuels and A. G. Alexander, *J. Agr. Univ. Puerto Rico*, 53:14, 1969; P. J. van Soest, *Proc. Cornell Nutr. Conf.*, p. 103, 1970.

Soil

Some of the reports of research in soil have been concerned with (1) the net loss of nutrient elements that occur during soil formation; (2) agriculture's role in pollution; and (3) analysis of soil aeration with the platinum microelectrode method.

Nutrient element losses. The soil mantle is an open system to which substances are added and from which they are lost. This has been long recognized, but interest in losses of nutrient elements has increased sharply in recent years because of growing concern about pollution. During the formation of soils, taking the world or the United States as a whole, there seems to have been a net loss of nutrient elements.

Estimates of losses must be based on indirect evidence because continuing measurements are not feasible over the long time spans of soil formation. Estimates rest on the chemical and mineralogical composition of soils and rocks, the measurement of short-term leaching losses from soils in lysimeters, and the amounts of nutrient elements carried by streams. Roy W. Simonson has recently discussed magnitude of losses for two broad groups of soils, Ultisols and Mollisols.

Ultisols. These soils of warm, humid regions have been formed in strongly weathered regoliths from a variety of rocks. Mostly the soils occupy old land surfaces. Major areas are in southeastern portions of the United States and Asia (Fig. 1).

Much of the calcium, magnesium, potassium, and phosphorus originally present in rocks is lost during weathering and the accumulation of parent materials for Ultisols. Chemical and mineralogical data for specimen soils and rocks indicate that as much as 90% of the original amounts disappear near the weathering front. Calcium is lost more readily than magnesium, potassium, and phosphorus. Phosphorus losses are smallest but still substantial.

Losses of nutrient elements continue from the soils until remaining quantities are very low. Thus Ultisols have extremely low levels of exchangeable calcium, magnesium, and potassium, the form readily available to plants. Amounts of phosphorus are also low and much is not available to plants.

Quantities of nitrogen are much greater in soils than in source rocks or parent materials because of biological fixation of the element. Part of the nitrogen so fixed is later lost through leaching and volatilization. Amounts in the soil of 1 acre (4047 m^2) to a depth of 40 in. (1 m) are commonly near 2 tons (2032 kg) for Ultisols.

Some indication of losses through leaching during formation of Ultisols is provided by the composition of stream waters. The average amounts per year of five elements carried by the Neuse River in east-central North Carolina and the Hiwassee River in east Tennessee, expressed as pounds per acre, are given in the table. Quantities of calcium and magnesium are small, whereas those of potassium plus sodium and of nitrogen are moderate. Similar amounts are carried by other streams draining Ultisols.

Mollisols. These soils of cool-temperate grasslands have been formed in weakly weathered regoliths on young land surfaces. Major areas are in the north-central United States and adjacent Canada, the Ukraine and Siberia, and the pampas of Argentina (Fig. 1).

Some of the calcium, magnesium, potassium, and phosphorus in source rocks is lost during formation of Mollisols. Roughly one-third of the calcium and potassium, a larger share of the magnesium, and a very small part of the phosphorus originally present disappear.

Mollisols normally have high levels of exchangeable calcium, magnesium, and potassium. Expressed as milliequivalents per 100 g of soil, rough

Amounts of five nutrient elements carried by four rivers in 1 year, expressed as pounds per acre of watershed*

River	Ca	Mg	K	K + Na	N
Neuse	20	7	—	32	1.0
Hiwassee	33	11	0	—	—
Cedar	86	32	—	22	1.4
Iowa	78	29	—	22	1.0

*One lb/acre = 0.1255 g/m^2.

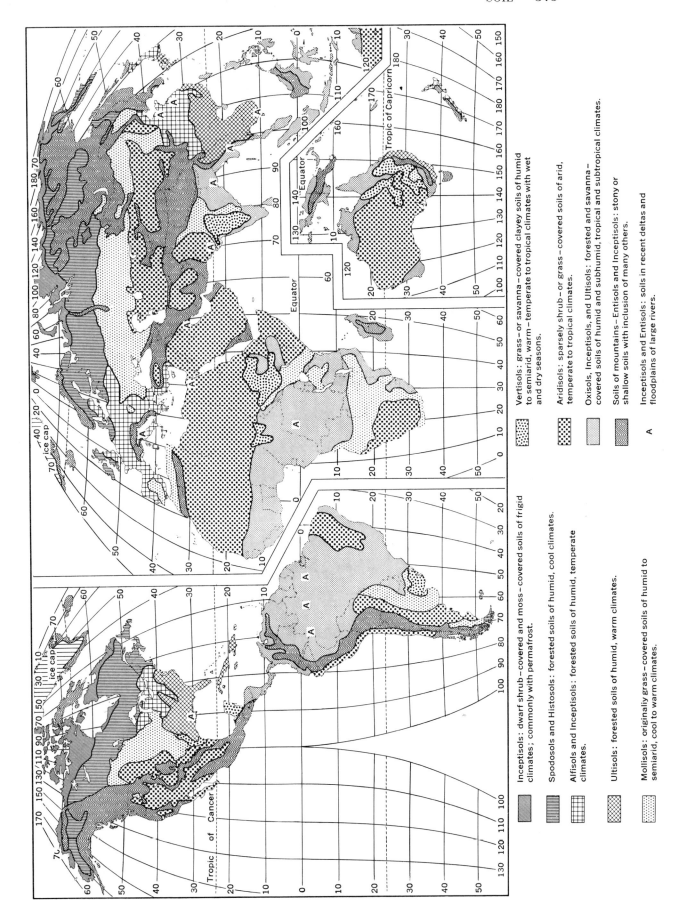

Fig. 1. Map of the world showing regions in which certain broad soil groups are important.

average figures are 15 for calcium, 6 for magnesium, and 0.8 for potassium. Amounts of phosphorus are also high and much is in forms accessible to plants.

Quantities of nitrogen are commonly five times as large in the soils as in parent materials for Mollisols. Amounts in the soil of 1 acre (4047 m²) to a depth of 40 in. (1 m) are commonly near 8 tons (8128 kg).

The high levels of nutrient elements in Mollisols are reflected in composition of streams. Average amounts per year carried by the Cedar and Iowa rivers in Iowa, expressed as pounds per acre, are given in the table. Quantities of calcium and magnesium are high, whereas those of potassium plus sodium and of nitrogen are moderate. Similar quantities are carried by other streams draining Mollisols.

Comparison of Mollisols and Ultisols. The two broad soil groups illustrate extremes in the spectrum of nutrient element losses during soil formation.

If chemical compositions of soils to a depth of 5 ft (1.5 m) are compared, rough average ratios in favor of Mollisols are 10:1 for calcium and magnesium, 3:1 for potassium, 2:1 for phosphorus, and 5:1 for nitrogen. Ratios are larger for exchangeable calcium, magnesium, and potassium.

Losses of nutrient elements in streams are related in part to soil composition. Ratios in favor of the two Iowa rivers listed in the table are approximately 3:1 for calcium and magnesium, 2:3 for potassium, and 1:1 for nitrogen. Only part of the calcium and magnesium comes from soils in Iowa; much is from the remainder of the regolith. The larger amounts of potassium in streams draining Ultisols come from the underlying rock, because the soils are very low in that element. Amounts of nitrogen are far higher in Mollisols, though losses in streams are about the same.

Ultisols occur on older land surfaces than Mollisols and under climates that effect greater weathering and leaching. Past losses of nutrient elements have therefore been greater. Present fertility levels are much lower. Mollisols are among the most fertile soils of the world, whereas Ultisols are near the other end of the scale. The low fertility levels have caused many people to believe that Ultisols have been "worn out" by long-continued cropping. Rather, such soils were "worn out" as they were being formed.

Losses from other broad soil groups. Similar to Mollisols in losses of nutrient elements during formation are the Aridisols, Inceptisols of cold or dry regions, and Vertisols. If anything, losses from these soils are smaller than from Mollisols. Collectively these broad groups and the Mollisols occupy 40% of the land surface of the Earth.

The Oxisols and Inceptisols of tropical and subtropical regions are similar to Ultisols in losses. Collectively these and the Ultisols occupy about 25% of the land surface.

The remaining broad groups of soils, the Alfisols, Entisols, Spodosols, and many Inceptisols, fall between the two extremes, both in nutrient losses during formation and in fertility levels. Mountainous regions, with their great variety of soils, also belong to this intermediate group. Collectively these occupy about 35% of the land surface.

Appreciable losses of nutrient elements are normal to the formation of most soils of the world. Losses are small, perhaps even negligible, for a few kinds, are very large for others, and are intermediate for still others. The magnitude of such losses is reflected in present fertility levels of all soils, which bear directly on their usefulness for food and fiber production under any level of technology.   [ROY W. SIMONSON]

Agriculture's role in pollution. Because of growing concern for the environment and of need for a fuller understanding of agriculture's role in pollution, more attention is now being given to soil than at any other time since the dust bowl days of the 1930s and the advent of the U.S. Soil Conservation Service.

Nutrient mobility. Perhaps the most significant characteristics of the Earth's land surfaces are that the chemical elements at or near the surface are subject to movement and that the outgo of inorganic plant nutrients from agricultural land to water greatly exceeds the input over most of the Earth.

Fortunately the same degree of negative balance does not apply to the nonmineral nutrients (carbon, nitrogen, oxygen, and hydrogen). With them there is continuous turnover through the atmosphere, land, organic matter, and water, with outgo from land almost being balanced by inputs, except for soils devoted to farming, where there annually is a net deficit due to crop and livestock removals, and for shorelines, swamps, and bogs, where plant remains are accumulating to give a huge net addition.

Along with the nutrients contained in plant remains, one cannot overlook the vast amounts deposited annually in silt and clay sediments. In the United States alone such sediments amount to 4×10^9 tons (4064×10^9 kg) a year. Certainly it would appear that the supply of available (and potentially available) nutrients in and underlying most waters of the Earth will be more than ample to support marine plant and animal life for ages to come even though no new inputs are received from land.

Nitrogen. Because nitrogen is so abundant and such vast quantities of it are constantly moving about, a few facts should be kept in mind. Even though the atmosphere contains about 35,000 tons of nitrogen over each acre of the Earth's surface (8787 kg/m²), this vast quantity represents only about 2% of the Earth's supply of the element, the major portion being in minerals, rocks, and soils.

According to W. D. P. Stewart of the University of London, 9×10^7 tons (8165×10^7 kg) of nitrogen are fixed biologically on this globe every year. According to Samuel Aldrich of the University of Illinois, the United States (exclusive of Hawaii and Alaska) receives 10^7 of these tons (9×10^9 kg). Add to this amount the 5×10^6 tons (4536×10^6 kg) reaching United States soils and waters in rainfall, 7×10^6 tons (6350×10^6 kg) from fertilizers, and some 5×10^6 to 6×10^6 tons (4536×10^6 to 5443×10^6 kg) from combustion of fuels in factories, autos, and homes, and it is obvious that the amount of nitrogen transient in the environment is indeed very great. Beyond these sources one must not overlook the 10^7 tons (9×10^9 kg) in livestock wastes plus the 8×10^6 tons (7257×10^6 kg) which go down the sewer every year from the food man eats. Still another major source of transient nitrogen is the net deficit of 2×10^7 tons (1814×10^7 kg) arising from decay of organic matter in United States soils, according to George Stanford of the U.S. Department of Agriculture.

Perhaps it is because nitrogen is present everywhere and because it is such a mobile nutrient in water, land, and air that it has received so much of the recent attention with respect to excessive growth of unwanted aquatic plants, especially algae. However, if one considers that rainfall contains from 0.5 to 1.0 ppm of nitrogen and that 40 some species of algae get their own nitrogen from the air as needed, it does not appear likely that nitrogen from fertilizer is generally increasing this growth. It must be remembered that only 0.3 ppm of nitrogen in a water body will support growth of algae.

Phosphorus. Phosphorus, unlike nitrogen, is highly immobile in soil. Nevertheless it is receiving a great deal of attention in connection with pollution. Like nitrogen, however, it is an essential constituent of every plant and animal cell and thus a major component of the human diet as well as that of livestock, pets, and wildlife. Because phosphorus is ever-present in both soil and water, and always will be, there is little cause for concern over it, especially when less than 10 ppb of it in a lake is enough to support massive algal blooms. Underwater deposits of organic matter and soil, silt, and clay will provide enough available phosphorus for decades to come to grow many aquatic species of vegetation. It must be remembered, too, that certain rooted plant species obtain their supply of mineral nutrients through their roots. Upon death these nutrients are released back to the medium, where they are subject to being recycled over and over again by countless forms of aquatic life.

Thus it is difficult to envision a water environment in nature, let alone find one, where growth of plant life could be greatly limited by supply of phosphorus.

Soil as resource. Research programs, although meager, are seeking more effective means of conserving soil and keeping it from blowing and washing away. When land is farmed and otherwise handled as best as one knows how, practically no sediment gets into streams and lakes. Unfortunately there is yet a long way to go until all land is properly maintained and adequate protection provided to areas undergoing transformation into highways and housing projects, not to mention the most difficult and costly task of preventing streambank erosion.

The public needs to be aware that both intensive research and extensive action programs are going to be required, and although these will appear to be costly, in reality they will return untold dividends because only fertile soils will support man in the future and because it is twentyfold cheaper to keep soil in place than it is to dredge it out of reservoirs and lakes. [WILLARD H. GARMAN]

Soil aeration. Recent electrochemical analysis with the platinum microelectrode method by D. S. McIntyre of electrode processes occurring in three-phase media, such as unsaturated soil, indicates that activation and concentration polarization occur simultaneously at different points on the electrode surface. Results in the literature are consistent with this postulate. Existence of activation polarization is indicated by the O_2 reduction current being continuously dependent on voltage and exponentially related to it. It follows that the electrical resistance of the soil must be taken into account when measuring O_2 flux with the microelectrode at constant applied voltage. A low pH may also cause errors, particularly at low O_2 concentrations and high applied voltages. The mechanism of operation of the method under different soil conditions is probably now understood, but certain experiments are necessary to quantitatively evaluate the method. Except for K. J. Kristensen and associates, most workers in recent years have still made measurements with the model which assumes complete diffusion control of current (concentration polarization) as the basis of the method applied. As a result, the problem of markedly different results between soils in the O_2 flux values critical to plant response has not yet been resolved.

O_2 flux and plant growth. M. W. Gradwell tested the response of mature white clover plants to different levels of O_2, controlling the O_2 flux by permeation of gases through unsaturated soil. Depression of shoot growth attributable to O_2 deficiency occurred only when the mean flux was less than 10×10^{-8} g cm^{-2} min^{-1}. Growth did not differ significantly from that at high levels of O_2 when O_2 flux was 15×10^{-8} g cm^{-2} min^{-1}. These values are very close to Gradwell's previous critical values for rye grass, using the same techniques and presumably the same soils. They are lower by a factor of 2 to 5 than many previous measurements of O_2 flux critical to shoot growth. For example, R. W. Rickman and coworkers in 1966 found that an O_2 flux below 40×10^{-8} g cm^{-2} min^{-1} could detrimentally

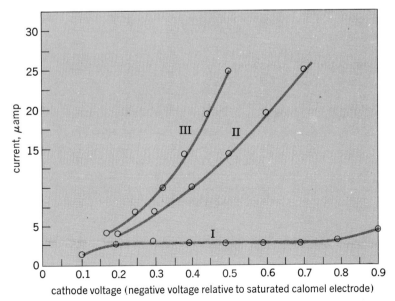

Fig. 2. Current as a function of applied voltage in a saturated soil (curve I), and unsaturated soil (curve II), and of effective voltage in the same soil (curve III).

affect shoot growth of tomatoes, and that the effect was large at a flux value of 20×10^{-8} g cm^{-2} min^{-1}. Similar figures have been obtained for other plants.

C. F. Shaykewich and B. P. Warkentin found no adverse effect on tomato plants of O$_2$ flux values as low as 10×10^{-8} g cm^{-2} min^{-1} and concluded that critical values were even less. They, as Gradwell, used mature plants, but imposed poor aeration conditions by flooding, and changes in O$_2$ flux resulted from changes in water content as the water was used. Daily measurements were made of O$_2$ flux and shoot growth. Gradwell kept moisture constant over a long period. The approximate agreement between these workers is probably fortuitous but may reflect the observation that critical values found in areas of leached soils with low total soluble salts (and hence high electrical resistance) are in general lower than those found in soils of higher total soluble salts.

S. Dasberg and J. W. Bakker grew beans in soils having varying bulk densities and imposed different aeration conditions by different irrigation frequencies and the maintenance of different O$_2$ concentrations in the gas at the soil surface. Differences were imposed after an initial growing period of 3 weeks. The level of O$_2$ flux in the pots most frequently watered was mostly below 20×10^{-8} g cm^{-2} min^{-1}, but the effect of low O$_2$ flux was quite small.

Mechanism of operation. It was originally assumed that, for three-phase media, current in the electrode circuit was controlled by the diffusion rate of O$_2$ (as it is in two-phase media) through films of water surrounding the electrode. The current voltage relations for such a system are shown in curve I of Fig. 2. It is now recognized that for three-phase media (and perhaps for two-phase media if the water is under suction) the O$_2$ reduction current is continuously dependent on voltage in the manner shown by curve II of Fig. 2. This means that the electrical resistance of the soil will affect current for a constant applied voltage, by ohmic loss of some of this voltage in the soil, and not by electroreduction of O$_2$. Analysis and experiment have also shown that in some circumstances—very low O$_2$ concentration or an acid pH or both—H$^+$ ions may reach the electrode and, on being reduced to H$_2$ give rise to current which is interpreted as O$_2$ reduction current. If a high negative voltage is applied (for example, -0.6 to -0.8 V versus Ag-AgCl), H$^+$ decomposition current may result without the pH being very acidic. J. D. F. Black and D. W. West have shown this effect and recommend application of a lower negative voltage than previously (-0.45 as against -0.65 V).

With this knowledge the disparities occurring in published critical O$_2$ values can be explained if not reconciled. Of the possible effects, that of electrical resistance is likely to cause the greatest differences. Kristensen and McIntyre have shown independently that electrical resistance of three-phase media can be quite large. Properties which affect it are the volumetric content of electrolyte (water content) and its tortuosity, electrolyte concentration, and type. The resistance is also dependent on the dimensions of the microelectrode and its distance from the reference electrode if that distance is small.

The ohmic drop in the soil caused by its electrical resistance makes the effective voltage of microelectrode less than that voltage applied between it and the reference electrode and kept constant. The effective voltage is that between the electrode and the adjacent solution. Hence different resistance values, unknown in all experiments relating to plant growth except those of Kristensen, will make for different effective voltages. With the current-voltage relations of the type shown in curve II of Fig. 2, the O$_2$ flux value calculated from the current at constant applied voltage for an individual three-phase system will in most cases differ from, and will not be comparable with, that from any other system. In fact, a change in moisture content in one soil will make the correspondingly determined O$_2$ flux incomparable with the previous value. Published O$_2$ flux–water content curves, obtained at constant applied voltage, are consistent with an increase in resistance with decreasing moisture and a subsequent decrease in effective voltage. In three-phase systems comparison of results can be made only at constant effective voltage. This involves sufficient measurements

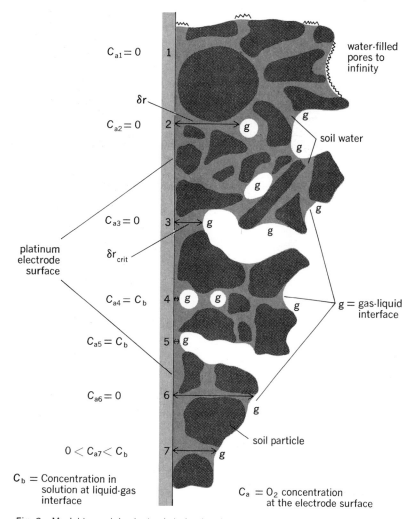

Fig. 3. Model to explain electrode behavior showing phase geometry at the electrode surface. Electrode activation polarization takes place at points 4 and 5 on electrode surface where water film is less than a critical value (δr_{crit}). This results in the rate-limiting process becoming the electrochemical reduction of O$_2$ which is dependent on the effective electrode voltage.

of current and resistance at different applied voltages, and construction of the relation shown by curve III in Fig. 2 between current and effective voltage.

Operative model. McIntyre has proposed a new model to describe operation of the electrode (Fig. 3) which differs from the previous (diffusion) model. Instead of current being limited completely by the diffusion rate of O_2 in solution as water content decreases until supposed rupture of the film, it is proposed that in places where film thickness (δr) is less than a certain critical value (δr_{crit}) that part of the electrode is undergoing activation polarization. This means that the rate-limiting process is the electrochemical reduction of O_2, which is dependent on the effective electrode voltage. At other parts of the electrode, current may be still diffusion-controlled. Where $\delta r < \delta r_{crit}$, C_a (the O_2 concentration at the electrode) will be greater than zero, and when δr becomes very small, C_a may equal C_b (the equilibrium concentration of O_2 in solution at a liquid-gas interface). Changing the moisture content, or the soil, will alter the relative proportions of the electrode undergoing activation and concentration polarization and hence may invalidate use of the microelectrode method for any but rough correlation purposes. Published current-effective voltage relationships are consistent with the occurrence of activation polarization, for most show the exponential relation between current and voltage, characteristic of that type of polarization. The relations between current and (1) O_2 concentration, (2) water content, and (3) time are also consistent with the model proposed.

For background information *see* FERTILIZING; SOIL; SOIL, SUBORDERS OF; WATER POLLUTION in the McGraw-Hill Encyclopedia of Science and Technology. [D. S. MC INTYRE]

Bibliography: S. R. Aldrich, *12th Sanitary Engineering Conference, University of Illinois*, Feb. 11–12, 1970; S. Dasberg and J. W. Bakker, *Agron. J.*, 62:689, 1970; R. C. Ellis, *Aust. J. Soil Res.*, 7: 317–323, 1969; G. E. Likens et al., *Ecology,* 48: 772–785, 1967; D. S. McIntyre, *Advan. Agron.*, 22: 235, 1970; R. W. Rickman, J. Letey, and L. H. Stolzy, *Agron. J.*, 30:304, 1966; C. F. Shaykewich and B. P. Warkentin, *Can. J. Soil Sci.*, 50:205, 1970; R. W. Simonson, *SSA Spec. Publ. No. 4*, 1970; G. Stanford, *Plant Food Rev.*, vol. 15, no. 1, 1969; W. D. P. Stewart, *Science*, 158:1426, 1967; S. J. Toth and A. N. Ott, *N.J. Acad. Sci. Bull.*, 14:29–36, 1969; A. W. Thomas, J. R. Carreker, and R. L. Carter, *Ga. Agr. Exp. Sta. Res. Bull.*, no. 64, 1969.

Soil chemistry

Clay minerals interact with organic compounds to form a variety of complexes. Such complexes are formed in nature through interaction of clays with organic materials resulting from biological processes in soils, or they may be created by industry for various uses. Adsorption of organic materials by clays in soils is known to protect such compounds from biological degradation. In addition, some compounds may bridge between neighboring clay particles, creating relatively stable aggregates, which in turn influence the water and aeration properties of the soil. Pesticides added to the soil may be adsorbed by clays which may render them nontoxic. The pesticides then may be released at a later time to the soil solution, where they may exert their toxic properties in the biological community. In some cases certain pesticides have been observed to be catalytically degraded at clay surfaces. Industry utilizes clay-organic complexes for such disparate purposes as paper, paint, cosmetics, and lubricants. The importance of clay-organic interaction in nature and in industry is thus obvious. *See* CLAY.

The nature of naturally occurring organic materials in soils and sediments is still not completely understood. They are extremely complex and difficult to study. In consequence, studies on their complexation with clays have not yielded much fundamental information. On the other hand, studies of clay interactions with organic compounds of known structures and properties have resulted in considerable basic information on the mechanisms of such adsorption. The recent application of infrared spectroscopy to such systems particularly has provided a rigorous approach to these problems.

The fact that organic cations may be adsorbed at clay-mineral surfaces by ion exchange with metal ions has been recognized for many years. The mechanisms of adsorption of nonionic but polar molecules has been less well understood until recently. With the advent of high-quality infrared spectrophotometers and breakthroughs in sample preparation and spectrophotometer cell construction, important new information on these interactions has been obtained. Extremely thin self-supporting clay films can be made that may be placed directly in the beam of the spectrophotometer. Upon adsorption of organic molecules on the films, spectra of the adsorbed species are obtained. Shifts in frequency of diagnostic bands may then give information on the mode of interaction between the organic molecule and the clay.

Protonation. While organic cations are adsorbed to clay surfaces by way of ion exchange, many compounds become cationic after adsorption at the clay surface through protonation. The sources of the protons may be exchangeable H^+ occupying cation-exchange sites, water associated with metal cations at exchange sites, or proton transfer from another cationic species already at the mineral surface. A number of workers have shown that the proton-donating properties of clay-mineral surfaces are considerably greater than would be expected from pH measurements of the clays in water suspension. These excessive acidic effects have been attributed to increased dissociation of water associated with exchangeable metal cations above and beyond that predictable from their hydrolysis constants. The nature of the exchangeable cation present and the water content have been shown to be the dominant factors in determining the proton-donation capabilities of clay-mineral surfaces. With regard to water content, it has been shown that as water content decreases, proton-donating capabilities increase.

Coordination. Where polar organic molecules are not basic enough to accept a proton, they may bond to clay surfaces through coordination or ion-dipole interactions with exchangeable cations. The greater the affinity of exchangeable cations for electrons, the greater will be the energy of interaction with polar groups of organic molecules capable of donating electrons. Transition-metal cations on the exchange complex will interact strongly with electron-supplying groups because of their unfilled d orbitals. The primary, and therefore most energetic, mechanism of adsorption of water by clay minerals is solvation or coordination of exchangeable cations. For the 2:1 type minerals, where there are no exchangeable cations in the interlamellar regions to solvate, no expansion occurs (for example, talc and pyrophyllite). In nature, polar organic molecules thus compete with water for ligand positions around the cations. Their ability to compete successfully is dependent upon their relative electron-donating properties as well as concentration, molecular size, and so forth. Once coordinated, such organic ligands can often be displaced by additions of water or other polar molecules capable of competition for ligand positions.

Hydrogen bonding. When exchangeable cations have high solvation energy, they may retain their primary hydration shell in spite of importunities of neighboring polar organic molecules. The polar groups on the organic compound may then interact by hydrogen bonding through a linking water molecule of the primary hydration shell of the cation. An example of such an interaction would be a carbonyl group interacting with a hydrated exchangeable cation M^{+n}, as shown in linkage (1).

$$M^{+n}\!-\!\overset{H}{\underset{}{O}}\!-\!H\cdots O\!=\!\overset{R}{\underset{R}{C}} \qquad (1)$$

This type of interaction has recently been established through infrared absorption studies. Other kinds of possible hydrogen bonding include interaction with surface oxygens or hydroxyls of clay minerals, although these have been shown by infrared absorption to be considerably weaker than would be assumed from a reading of the older literature on clay-organic interactions. In fact, it has been shown that for H_2O, intermolecular hydrogen bonding is more energetic than that with surface oxygens of clay minerals. Organic-organic interaction through hydrogen bonding may take place at clay surfaces when the exchange site is occupied by an organic cation, as in linkage (2).

$$R\!-\!\overset{H^+}{\underset{H}{N}}\!-\!H\cdots O\!=\!\overset{R}{\underset{R}{C}} \qquad (2)$$

Other mechanisms. Physical forces and entropy effects may also be important factors in organic adsorption by clays, particularly for large molecules and polymers. Recently it has been shown that organic molecules that do not have polar functional groups but that do have π-electrons available for donation may complex with clays saturated with transition-metal cations. In particular, this has been demonstrated by H. E. Doner and M. M. Mortland and by Mortland and T. Pinnavaia for benzene and some of its alkyl derivatives adsorbed on Cu(II) montmorillonite. Infrared absorption showed without ambiguity that the arenes occupied ligand positions around the Cu(II) ions. In order to create these complexes it was necessary to remove H_2O from the coordinating position by dehydration over P_2O_5 or by evacuation and low heat. Since Cu(II) arene complexes have not previously been reported in the chemical literature, their appearance at the mineral surface is strongly suggestive of a unique environment existing there which is not duplicated in solution chemistry.

An increased understanding of the nature of clay-organic complexes and of reactions that take place at these interfaces will lead to still more important utilization of these systems in industry and a better comprehension of clay-organic complexes in soils and sediments.

For background information see SOIL CHEMISTRY in the McGraw-Hill Encyclopedia of Science and Technology.   [M. M. MORTLAND]

Bibliography: H. E. Doner and M. M. Mortland, *Science*, 166:1406–1407, 1969; M. M. Mortland, *Advan. Agron.*, 22:75–114, 1970; M. M. Mortland and T. Pinnavaia, *Nature*, 229 (3):75–77, 1971.

Soil management

Lowland rice is grown on soils which are flooded all or part of the growing season, unlike other important crops which are produced on drained soils. Physical, chemical, and biological processes in the soil are affected by flooding. Some plant nutrients are increased in availability by flooding, whereas nitrogen is subject to serious loss. The rice farmer must know how to manage the soil and floodwater so that plant nutrients are conserved and high yields of rice are obtained. Recent studies have been aimed at understanding the biological and chemical reactions taking place in the flooded soils and the way in which management practices can be modified to improve the utilization of nutrients and increase the yield of rice.

Flooded soil environment. A flooded soil constitutes a unique environment which provides both advantages and disadvantages to the rice crop. The lack of competition from many weeds due to their inability to grow under flooded conditions and the constant availability of an adequate water supply for the nutritional needs of the crop are among the most important advantages resulting from flooding. Disadvantages of flooding are the extra expenses of preparing the land for irrigation and obtaining an adequate supply of water.

Soil physical properties. Flooding changes the physical properties of a soil. The bonds that hold soil particles together into clods and stable aggregates are weakened by excess water. In countries where machine power is not readily available farmers take advantage of this condition and flood the field before preparing the seed bed. The soil is softer and easier to manipulate with animal-drawn implements if excess water is present. This puddling operation breaks down soil aggregates which are important in maintaining the structure and tilth of nonflooded soils but which do not appear to be important for rice growth.

Atmospheric oxygen. Flooding also causes marked changes in some of the chemical and biological processes taking place in the soil. These changes are caused by the curtailment of atmospheric oxygen entering the soil. In drained soils there are enough air-filled pores to permit adequate diffusion of atmospheric oxygen into the soil. When these soil pores are filled with water and the soil surface is covered with several inches of water, the amount of oxygen reaching the soil surface is reduced by a factor of over 10,000. Even though the supply of oxygen reaching the soil is greatly curtailed by flooding, microbial respiration is not diminished. The imbalance between the oxygen requirement of the soil and the oxygen supply reaching the soil surface results in a rapid depletion of soil oxygen within hours after flooding. The only part of a flooded soil that is oxygenated is a thin layer at the soil surface (Fig. 1). This occurs because oxygen diffusing through the floodwater penetrates only a short distance into the soil before it is consumed. An important property of a flooded soil is the thickness of this surface oxygenated layer. In soils that have a low microbial requirement for oxygen this layer is relatively thick, 2–5 cm, while in soils that have high oxygen requirement the layer may be no thicker than 1–2 mm. The oxygen requirement is usually governed by the supply of decomposable organic matter.

Aerobic soil layer. In the surface oxygenated or aerobic soil layer, microbial and chemical conditions are very much like those in drained soil. In the underlying oxygen-free soil layer, pronounced biological and chemical changes are set in motion when oxygen disappears. As long as oxygen is present, oxidized components of the soil (such as nitrate, manganic compounds, and ferric compounds) are not biologically or chemically reduced. After oxygen disappears following flooding, many soil microorganisms can substitute one or more of these oxidized chemical components for the oxygen required in respiration. Nitrate, the higher oxides of manganese, and hydrated ferric oxide are reduced if oxygen becomes absent or limiting and if an energy source (organic matter) is available to the microorganisms. Nitrate and manganic compounds are readily reduced since the energy required for their reduction is low, and a number of species of microorganisms can carry out this process. Ferric compounds are more difficult to reduce, but the large amount of reducible ferric iron in most soils make ferric compounds an important oxidation-reduction component of the soil. These reduction reactions are carried out by facultative anaerobes, microorganisms which can substitute other reducible compounds for respiratory oxygen. Sulfate can also be reduced to sulfide by anaerobic microorganisms, but this reaction is carried out only under strictly anaerobic conditions by a few species of microorganisms.

The difference in ease of reduction of the inorganic oxidation-reduction systems in the soil results in a more or less sequential reduction of the various components following flooding. Free oxygen in the soil is reduced first and is at least partially depleted before nitrate and manganese compounds begin to be reduced. Ferric compounds are then reduced, and if no oxygen enters the soil, sulfate will be reduced to sulfide. There is some overlap between the end of reduction of one system and the beginning of reduction of another, depending on the physical availability of the compounds to soil microorganisms. For example, sulfate is more difficult to reduce than ferric compounds, but both processes may occur simultaneously due to the relative inaccessibility of iron compounds. Almost always, however, all oxygen and nitrate

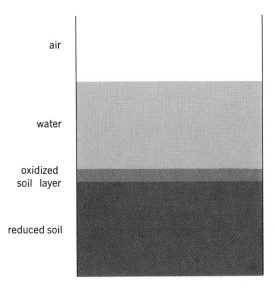

Fig. 1. Differentiation of a flooded soil into a surface oxygenated or oxidized layer and an underlying reduced layer as a result of an imbalance between oxygen requirement and oxygen supply.

have disappeared from the soil before iron reduction begins. The availability of an energy source and the absence of oxygen are the only necessary requirements for these reduction reactions to occur, outside of favorable environmental conditions, since the microorganisms that carry out these reactions occur in almost all soils, even those where flooding has not been known to take place.

Plant nutrient availability. The biological and chemical changes in the soil that accompany flooding have a marked effect on the availability to rice of several of the major plant nutrients. Phosphorus, iron, and manganese are increased in availability by flooding, while nitrogen is more readily lost from a flooded than from a drained soil.

Nitrogen. Nitrogen is the most important plant nutrient for rice, as well as for other cereal crops, but the rice plant is notably less effective in utilizing nitrogen than are the other cereals. The poorer efficiency with which rice uses nitrogen is due to the relative instability of nitrogen in flooded soil. Nitrate nitrogen is readily converted to gaseous nitrogen under reducing conditions and is lost from the soil. The key to nitrogen management of flooded soils is prevention of ammonium nitrogen from being oxidized to nitrate. The only part of a flooded soil where microbial ammonium oxidation can take place is in the surface oxygenated layer, since the microorganisms responsible for this process require oxygen. Nitrate formed from ammonium is not reduced in the oxygenated layer but it can readily diffuse downward into the reduced layer, where the conversion to nitrogen gas is carried out rapidly (Fig. 2). Three precautions can be taken by the rice farmer to minimize the loss of nitrogen: (1) Only ammonium-type nitrogen fertilizers should be applied to lowland rice. (2) The nitrogen should be applied 6–8 cm deep in the soil so that it will not be located in the oxygenated layer after the soil is flooded. (3) The floodwater should be maintained continuously on the field during the growing season, since the entry of oxygen into the soil as a result of draining the field will

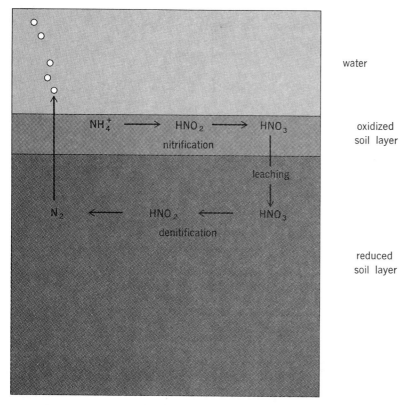

Fig. 2. Diagram of the process by which ammonium nitrogen is lost from a flooded soil as a result of nitrification in the oxygenated (oxidized) layer and denitrification in the underlying reduced layer.

allow much of the ammonium nitrogen to be oxidized to nitrate. If the soil is flooded again this nitrate will be rapidly lost. By observing these precautions for minimizing denitrification, the rice farmer can increase the efficiency of nitrogen fertilization to a level almost comparable with that of other cereal crops.

Phosphorus. The availability of phosphorus to lowland rice is increased by flooding. Soil phosphorus is almost always more available to rice grown under flooded conditions than to upland crops grown on the same soil under drained conditions. In many cases the farmer can apply phosphorus fertilizer to the upland crop grown in rotation with rice and depend on residual phosphorus for the rice crop's needs. The reduction processes caused by flooding release forms of phosphorus that are ordinarily not available to the crop under drained conditions. Two of the major sources of the released phosphorus are ferric phosphate and other phosphate compounds (ferric phosphate, aluminum phosphate, calcium phosphate) that are coprecipitated with hydrated ferric oxide. The reduction of these insoluble ferric compounds to the more soluble ferrous forms releases phosphate which can be utilized by the rice crop. Because of the absence of reducing conditions in drained soils, these forms of phosphorus are not available to upland crops grown on the same soil. As a result of the greater availability of phosphorus under flooded conditions and also because of the lower likelihood of fixation of added phosphorus, fertilizer phosphorus applied to rice should be placed deep enough in the soil so that the phosphorus will be in the reduced layer after the soil is flooded. This usually requires a placement depth of 6–8 cm. Although proper placement will improve the efficiency of phosphorus fertilizers added to rice, the improvement obtained is not nearly as great as can be obtained by proper placement and management of nitrogen fertilizer.

For background information *see* RICE in the McGraw-Hill Encyclopedia of Science and Technology. [WILLIAM H. PATRICK, JR.]

Bibliography: F. E. Broadbent and D. S. Mikkelsen, *Agron. J.*, 60:674–677, 1968; W. H. Patrick and D. S. Mikkelsen, in *Fertilizer Technology and Use*, 1971; F. N. Ponnamperuma, in *The Mineral Nutrition of Rice*, 1965.

Soil microorganisms

Recent studies of the microorganisms of grassland emphasize the extent to which the very small forms of life participate in the flow of energy in nature. Soil microorganisms, despite their minute size, metabolize the lion's share of that portion of the Sun's energy trapped yearly by photosynthesis. According to A. Macfadyen, the green plants in terrestrial ecosystems usually accomplish a net dry matter production of about 1 kg/m²/year, equivalent to about 5000 kcal (21,000 kJ) of energy. Such production in a grassland commonly supports cattle or other large mammals, as well as diverse populations of mice, rabbits, birds, grasshoppers and other insects, and a host of smaller invertebrates, all of which feed on the primary vegetation. These herbivores in turn support secondary food chains involving such predators as coyotes, hawks, owls, snakes, wasps, and spiders. To the observer of such a landscape, it may seem incredible that the yearly consumption of energy by the soil microorganisms exceeds the combined yearly consumption of energy by all the easily seen inhabitants of the grassland, both herbivorous and carnivorous.

Of the total herbage utilization in grassland, no more than one-third of the total, and quite frequently much less, is consumed by herbivorous insects and vertebrates. The remainder, barring such events as fire, goes largely to the soil microorganisms. Those who may question this amount of microbial consumption need only to be reminded that in grassland the belowground biomass in the form of plant roots is several times greater than that of the aboveground herbage. Because of the mechanical difficulties and inefficiencies involved in belowground feeding for animals too large to move through the soil pores, nearly all the larger herbivores feed on aerial plant parts, leaving the soil microorganisms as the principal feeders on plant roots. Furthermore both the invertebrate and vertebrate grazers of the aerial plant parts usually metabolize only about half of the herbage consumed; the remainder passes through the gut and is returned to the soil as excreta. Even that portion of the food intake of an animal that has been converted to, and exists as, body tissue at the time of the animal's death is principally or in part returned, sometimes by circuitous routes, to the soil microorganisms.

Microbial utilization of standing crop. Green plants support a leaf-surface, or phyllosphere, population of microorganisms that feed on leaf exudates. Of the total energy flow in an ecosystem, the leaf surface microorganisms metabolize a relatively minor fraction. Estimates made by F. E. Clark and E. A. Paul range from as high as 1% in a tropical rainforest to as low as 0.01% in semiarid grassland. Microbial utilization of standing dead vegetation, although it may be relatively intense during intervals of high atmospheric humidity, likewise is relatively insignificant when viewed on a yearly basis. Standing vegetation, however, does contribute food material to microbes over and beyond that which is metabolized on the leaf surfaces. During rainstorms, soluble organic substances, as well as inorganic nutrients, are leached from the vegetation. The actual microbial utilization of the leachate takes place at the soil surface. The organic content of the leaf drip can be measured during the course of either natural or simulated rain. Such measurements indicate that although microbial utilization of leachate at the soil surface may exceed the microbial use of exudates on the leaf surfaces, it nevertheless constitutes only a very minor percentage of the total energy flow.

Standing vegetation pathways to litter. There are two major pathways by which standing vegetation reaches the soil microorganisms. One of these is by way of herbivorous animals, both vertebrate and invertebrate. The other is by way of direct transfer of standing vegetation to the surface litter.

In grassland reasonably stocked with large mammals, estimates of the standing crop intake by such animals range from as low as 10% to as high as 70%. On the assumption that for some given grassland the value is 30% and that 50% of the food intake passes through the gut and is returned to the soil surface as feces, the fraction of the standing crop reaching the soil microorganisms by way of large mammals would be of the order of 15%.

It is not as easy to estimate the percentages of the standing crop that reach the soil microorganisms by way of the digestive tracts of small mammals, insects, and other small to very small invertebrates. This is partly because of the factor of diet sharing; that is, as one group of herbivores eats more or less of the standing crop, the amount left for the other groups is correspondingly decreased or increased. A further complication lies in the fact that the populations of the small to very small herbivores may change drastically from year to year, or even within the same year, thereby adding greatly to the variability in the amount of the standing crop that is consumed by any one subgroup.

Under almost any combination of primary consumers, however, one can be quite certain that all of the standing crop does not pass through herbivores so that only the digestive residues reach the soil surface. Large mammals trample some of the standing crop into the surface litter. Rabbits in the course of feeding may let fall to the ground as much of the standing crop as they swallow, and grasshoppers feeding on green leaves may let 10 times as much vegetation fall to the ground as is actually consumed. Accordingly, it is probably safe to assume that grazing animals waste, in the sense of adding to the surface litter, as much of the standing crop as they consume. That portion of the standing crop that is not disturbed by herbivores goes directly to surface litter by gravity, aided by such events as wind, rain, snow, and hail.

Decomposition of surface litter. Following the death of plants or of plant parts and their accretion as litter, the rate of decomposition of the residues is controlled by many factors—moisture, temperature, reaction, degree of fragmentation, degree of soil contact, and plant chemical composition, to name only a few. Especial emphasis is usually placed on the need for favorable conditions of moisture and temperature for achieving optimal rates of litter decay. However true this may be, it must be remembered that the ambient environment also affects plant growth. Viewed on an annual basis, interventions of drought or of cold do decrease microbial activity, but they also decrease plant growth and the amount of plant material that will become available for decomposition. For most vegetation, the amount of organic material in the ecosystem remains approximately constant from year to year. A. Burges has suggested that a rough estimate of the total amount of material being decomposed annually above- and belowground can be obtained by estimating annual litter fall and annual death of plant roots. Although total yearly decomposition approximates annual productivity, this does not imply that each annual crop is fully decomposed during the current or following year. The major portion of decomposition during a given year is either of the current or the preceding year's growth or a combination of them. This is accompanied by successively smaller amounts of decomposition of progressively smaller fractions of earlier crops. This is because plant material contains chemical components and structures that differ in their susceptibility to decomposition.

Decay rates of plant constituents. At the risk of oversimplification, plants can be characterized as containing readily decomposable, slowly decomposable, and resistant components. In the readily decomposable category are sugars, proteins, fatty acids, hemicellulose, and cellulose. Inasmuch as hemicellulose and cellulose account for roughly 50% of the total carbon in many plants, a corresponding percentage of the plant tissue can be expected to be readily decomposable. Under environmental conditions that are favorable for microbial activity, sugars and cellulose decompose within a matter of days or weeks. In field environments characterized by drought, freezing temperature, or other unfavorable interludes, decomposition of the readily decomposable components may require several months to several years.

The intermediate or slowly decomposable plant material includes principally lignins, waxes, and lignin-occluded cellulose. Under conditions favorable for microbial activity, this category can be expected to have a half-life in the soil of several months to a year, but under less favorable conditions, the half-life may be tenfold longer.

In the third and final category are plant constituents that form a continuum from the above-mentioned lignins and waxes to complex polyphenols and long-chain aliphatics. The half-life of materials in this category is several times that noted above for the slowly decomposable category.

To illustrate the variable rates of decomposition

of plant constituents, the following data from a recent study by G. Minderman on woodland litter are cited. The percentage losses in weight during the first year of decay in the field for six identifiable plant constituents were as follows: sugars 99%, hemicellulose 90%, cellulose 75%, lignin 50%, waxes 25%, and phenols 10%.

Agents of litter decomposition. It appears indisputable that the soil microflora, principally the bacteria and fungi, are the agents largely responsible for litter decomposition. What is controversial is just how much of the decay process is to be attributed to them and how much to the soil invertebrate fauna. The soil invertebrate fauna includes one subgroup collectively called the decomposer reducers; these are known to consume litter, carrion, and dung and embrace such forms as the earthworms, isopods, diplopods, dipterans, and coleopterans. A second subgroup of the soil invertebrate fauna contains such forms as the nematodes, enchytraeids, collembolans, and mites, collectively called the microbial grazers because of their extensive feeding on bacteria and fungi.

With the advent of the nylon net litter bag, it became possible to expose given quantities of litter under field conditions and to measure rates of disappearance of the litter. By varying the size of the mesh openings or by treating the litter with repellants, in order to exclude or not to exclude members of the soil fauna, it appeared possible to measure the importance of the invertebrate fauna in the processes of decomposition. Some rather striking results were published. Several investigators noted that litter exposed in bags to which the soil fauna had access decomposed 3–6 times faster than did litter from which the fauna was excluded. Consequently they were willing to ascribe rather large annual energy flows to the soil invertebrate fauna.

Other workers have been much more conservative. Some have suggested that much of the weight loss from nylon-bagged litter represents not decomposition as such but animal transport of litter from the bags as well as acceleration of gravitational litter losses because of animal movements in the litter. In a recent field study C. R. Malone and D. E. Reichle observed that litter confined in mesh bags disappeared most rapidly in enclosures treated to eliminate all invertebrate fauna. The investigators suggested that the invertebrate fauna sufficiently disturbed the microflora to interfere with their optimum potential in decomposing litter. Although such an experiment nicely demonstrates that litter decay can at times proceed at a comparable rate regardless of the presence or absence of the soil fauna, it does not mean that such fauna in the natural ecosystem do not play a significant role in the comminution and decomposition of litter.

Decay of plant parts belowground. In nearly all grassland, more than half the net primary production goes into the root system. In turn, nearly all the root material, either through exudates and exfoliations of the live root or upon senescence and death of roots, goes to the soil microorganisms. Bacteria and fungi are easily the principal agents of decay belowground. The invertebrate fauna, which may account for one-fifth of the energy flow in surface litter, are comparatively less active belowground. Because the quantity of belowground plant material exceeds that in the surface litter, the total energy turnover accomplished by the belowground fauna may still exceed that of the fauna in surface litter. Of the soil fauna belowground, the nematodes, either as root parasites or as saprophages, are easily the most active group. Current evidence is that they accomplish more of the belowground faunal energy flow than do all the other soil fauna groups combined.

Accepting that nearly all the belowground decomposer energy flow is accomplished by bacteria, fungi, protozoans, and nematodes, the question that may be asked is on what substrates or in what sites are these soil microorganisms active.

Several studies involving organic carbon measurements in substrates in which plants have been grown axenically, that is, in the absence of microorganisms, have shown that young, vigorously growing roots can lose roughly 2% of their carbon as exudates and sloughed cells. That figure may be an underestimate. Recently S. Shamoot, I. McDonald, and W. V. Bartholomew grew plants within closed chambers containing radioactive carbon dioxide. At harvest there was meticulous removal of all recognizable root material and all top growth. Total and tagged contents of organic carbon remaining in the soil were then determined. It was found that root-derived organic debris accumulating in the soil during plant growth amounted to slightly over 10% of the combined top and root growth harvested.

Whether the carbon loss value be taken as 2 or 10%, it represents an appreciable pathway of energy flow. The root-derived material supports an active microbial population on the root surfaces. Comparisons of the root respiration of plants grown axenically in sand plus nutrients with that of plants in similar substrate but in the presence of microorganisms have shown that the respiration rate of the axenic roots is only about 50% of that of roots with microbial contaminants. Stated otherwise, 50% of the carbon dioxide being produced by plant roots in field soil is probably being produced by the microorganisms on the root surfaces.

Senescing and dead roots. Assessing microbial activity on senescing and dead roots that are still structural components of the gross root system poses a difficult problem. Workers concerned with the aboveground foliage of plants can quite readily divide that foliage into the standing green and the standing dead vegetation. Workers concerned with what loosely may be called the standing crop belowground, however, have no easy way of determining the amounts of living, senescing, and dead roots in the total root biomass that is recoverable by washing techniques. Consequently measurements on the respiratory activity of gramineous root systems have commonly been made on roots carrying their normal complement of root surface microorganisms, and the total recoverable root system has been employed, without regard to the actual viability of the individual roots therein.

Studies of this sort yield wide-ranging values, so much so that one hesitates to cite any specific experiment. As a general approximation, however, it appears that of the belowground energy flow in a grassland during the growing season, no more than 15% of the total flow is due to root respiration; a like amount is due to the microorganisms associated with the roots; and perhaps 70% is due to decomposer activity on the soil organic matter.

Microbial biomass in soil. It may be stated that just as the soil microorganisms are responsible for more of the total energy flow in a grassland system than are all the other components combined (excepting of course the primary vegetation), so too do the microorganisms, again with the same exception, constitute the major part of the ecosystem biomass. In grassland reasonably stocked with large mammals, the biomass of such animals is of the order of 5–10 g/m². Small-mammal biomass is likely to be in the range of fractions of a gram to 1–2 g/m², as is also that of birds, reptiles, and many species of insects. In contrast, the microflora in the top 30 cm of grassland soil has been estimated by Clark and Paul to be as large as 100–200 g/m². On the basis of this very great amount of biomass and of its potential metabolic activity, the soil microorganisms could be accomplishing an annual energy turnover far greater than that available in the net primary production. Stated briefly, there are many microorganisms in the soil, and they are nearly always hungry.

For background information *see* SOIL MICROORGANISMS; TERRESTRIAL ECOSYSTEM in the McGraw-Hill Encyclopedia of Science and Technology. [FRANCIS E. CLARK]

Bibliography: A. Burges, in A. Burges and F. Raw (eds.), *Soil Biology*, 1967; F. E. Clark and E. A. Paul, *Advan. Agron.*, 22:375–435, 1970; A. Macfadyen, in J. Phillipson (ed.), *Methods of Study in Soil Ecology*, 1970; C. R. Malone and D. E. Reichle, *Health Physics Division Annual Report*, Oak Ridge Nat. Lab. Publ. no. 320, 1970; G. Minderman, *J. Ecol.*, 56:355–362, 1968; S. Shamoot, I. McDonald, and W. V. Bartholomew, *Soil Sci. Soc. Amer. Proc.*, 32:817–820, 1968.

Solar wind

The concept of the solar wind represents a revolutionary change in scientific thought regarding the nature of the solar system. Astronomers traditionally viewed interplanetary space as a vacuum, of little scientific interest. A theory of the continuous expansion of the outermost visible layer of the Sun, the solar corona, was developed by E. N. Parker in the 1950s. This theory led to the new concept of interplanetary space as pervaded by an ionized gas (or plasma) and a magnetic field of solar origin. The rapid outward flow of solar material led to the name "solar wind." Until the advent, in the 1960s, of spacecraft observations beyond the influence of the Earth's magnetic field, Parker's theory remained unproven, even subject to great skepticism. The past decade of spacecraft exploration has not only established the existence of the solar wind but has provided a detailed and remarkably complete description of its characteristics. Some of these characteristics can now be used to infer properties of the Sun.

Quiet solar wind. Spacecraft observations have revealed that all physical properties of the solar wind vary widely. The theories of Parker and other investigators usually assumed, for the sake of mathematical simplicity, a steady, structureless coronal expansion. This ideal "quiet" state of the solar wind is most nearly approached when the flow speed is about 325 km sec^{-1}, near the lower end of its total range of variation (from 300 to 750 km sec^{-1}). The table summarizes some of the physical properties typical of the quiet solar wind. The number densities of the protons and electrons (the main constituents of the ionized gas) are nearly equal, as they must be to maintain electrical neutrality. The temperatures of the protons and electrons are not equal. This is a direct consequence of the extremely low density; collisions between the particles occur too rarely to ensure an equilibrium state in which all components of the gas would have a common temperature. The weak interplanetary magnetic field is a remnant of the solar magnetic field, "dragged" into interplanetary space by the expanding plasma. The magnetic field assumes a spiral configuration (Fig. 1) under the combined effects of this expansion and the rotation of the Sun.

The properties of the quiet solar wind given in the table are in general agreement with the predictions of theoretical models. However, precise comparisons reveal some significant discrepancies. The solar wind speed is generally higher than predicted; the proton and electron temperatures

Typical properties of the quiet solar wind

Density	8 protons (or electrons) cm^{-3}
Flow speed	325 km sec^{-1}
Proton temperature	5×10^4K
Electron temperature	1.5×10^5K
Magnetic field strength	5×10^{-5} gauss

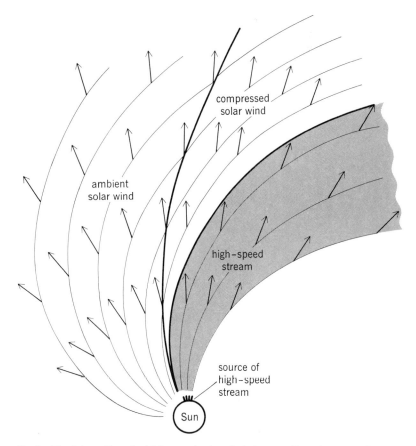

Fig. 1. The interaction of a high-speed solar wind stream with a slower moving ambient solar wind. The arrows indicate the flow of plasma, and the light lines show the spiral magnetic field configuration.

are respectively higher and lower than predicted. These discrepancies are the breeding ground of much current theoretical work, which attempts to resolve them by postulating the influence of new physical processes upon the coronal expansion. For example, it has been suggested that waves generated deep within the solar atmosphere pass unabated through the corona to interact with the solar wind, dissipating their energy well out in interplanetary space. Plasma instabilities, generated internally by the nonequilibrium state of the solar wind, have been suggested as a means of exchanging energy between the electrons and protons. In exploring such ideas, solar wind research overlaps the realm of laboratory plasma research. Present observations, largely performed in a narrow range of heliocentric distance near the Earth's orbit, seem incapable of determining the validity of these suggestions. Future observations performed on spacecraft missions to Mercury and Jupiter may help to determine the importance of such mechanisms.

Disturbed solar wind. The solar wind is rarely observed in the quiet state described above but is usually found to be in a "disturbed" state, characterized by fluctuations in physical properties on a temporal scale of about a day and by flow speeds higher than the 325 km sec^{-1} value representative of the quiet state. Several distinct patterns of disturbance have been identified. "High-speed streams" lasting for a few days are observed several times during a typical month. It has been suggested that these streams of fast-moving plasma emanate from localized centers of solar activity, but no precise source identifications have been established. In propagating through interplanetary space, these streams are distorted into a spiral configuration similar to that of the magnetic field, as shown in Fig. 1. The "ambient," slower moving solar wind just ahead of the high-speed stream is overtaken and compressed, producing unusually high densities and proton temperatures. Some high-speed streams appear to recur at intervals near 27 days, the rotation period of the Sun. This recurrence implies that the streams (and their localized solar sources) persist for several months and are swept past an interplanetary observer during each rotation of the Sun. See IONOSPHERE.

A pattern of disturbance more clearly associated with a precise form of solar activity is the "shock wave" observed near the Earth's orbit several days after some solar flares. Flares are violent explosions occuring above the Sun's photosphere and have been observed optically for over a century. Some flares appear to eject a cloud of fast-moving gas. As this cloud propagates into interplanetary space, it sweeps up and compresses the ambient solar wind in its path, as shown in Fig. 2. A "shock front," at which an abrupt compression and heating of the swept-up gas takes place, forms at the leading edge of the compressed shell. To an interplanetary observer, this type of disturbance begins suddenly with the arrival of the shock front. Densities as high as 50 protons cm^{-3} and solar wind speeds as high as 750 km sec^{-1} may occur. Spacecraft observations of these shock waves reveal an energy release of $\sim 10^{25}$ joules within a period of hours by solar flares (compared with the worldwide electrical energy consumption of about 10^{19} joules per year). The mass in the wave is $\sim 3 \times 10^{16}$ g, equal to that of all the material in the solar atmosphere above the flare site. These energy and mass determinations demonstrate the cataclysmic nature of a solar flare and pose a difficult task for any theory attempting to explain them.

Chemical composition. The solar origin of the interplanetary plasma lends special interest to studies of its chemical composition. The ionic component of the plasma is mainly H$^+$, ionized hydrogen, as would be expected of solar material. The next most abundant ion is found to be He^{2+}, fully ionized helium (He$^+$, singly ionized helium, is much less abundant). Several spacecraft have provided long-term observations of the relative solar wind abundance of helium and hydrogen; all give an average abundance ratio of 4–5% by number. The solar helium abundance can be determined only by indirect means and is believed to be 5–10%. It is unclear whether the difference between the helium abundances of the Sun and the solar wind reflects observational uncertainties (probably in the solar value) or a real effect. The latter could be produced by a separation of these elements in the coronal expansion, with the lighter H$^+$ ions escaping solar gravity more readily than the heavier He^{2+} ions. This separation process would lead to a low helium abundance in the solar wind and a high helium abundance in the lower corona. In fact, the solar wind helium abundance is found to fluctuate widely, ranging from 1 to 25%. The extremely high abundances are observed during the passage of flare-produced shock waves. It has been suggested that this helium-rich solar wind is the material originally ejected from the lower corona by the flare.

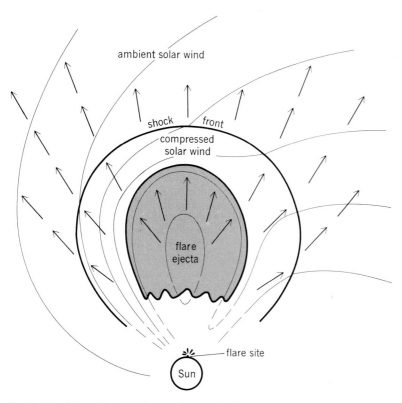

Fig. 2. The interaction of a fast-moving cloud of gas, ejected by a solar flare, with the ambient solar wind.

Solar wind ions other than H^+ and He^{2+} can be identified by present-day spacecraft instruments only under favorable circumstances. A few observations of the rare helium-3 isotope mark its first detection in solar material. The ion O^{6+} (oxygen ionized by the loss of six electrons) is found to be the third most abundant solar wind ion. Its observation is of special interest because the predominance of this ion over the neighboring oxygen charge states (O^{5+} and O^{7+}) implies an origin at a temperature of 1×10^6 to 2×10^6K. This is not the local temperature of these ions in the solar wind, but corresponds to the coronal temperature derived from traditional spectroscopic techniques. The ionization state of solar wind oxygen has retained coronal characteristics in traveling from the Sun to the orbit of the Earth because interplanetary densities are so low that atomic processes occur too slowly to modify the ionization state. Thus observations of the solar wind ionization state, although performed far from the Sun, permit a direct determination of the coronal temperature! Very recently the silicon ions Si^{7+}, Si^{8+}, and Si^{9+} and the iron ions Fe^{7+} to Fe^{12+} have been identified in the interplanetary plasma. These observations permit a more precise temperature determination for the coronal region from which these ions came, leading to a value near 1.5×10^6K. Routine determination of the silicon and iron ionization states by future spacecraft instruments would permit not only the monitoring of coronal temperatures, but would provide a "tracer" for aid in identifying the coronal sources of such solar wind features as the high-speed streams.

For background information *see* INTERSTELLAR MATTER; SOLAR WIND; SUN in the McGraw-Hill Encyclopedia of Science and Technology.

[A. J. HUNDHAUSEN]

Bibliography: J. C. Brandt, *Introduction to the Solar Wind*, 1970; T. E. Holzer and W. I. Axford, *Annu. Rev. Astron. Astrophys.*, 8:31, 1970; A. J. Hundhausen, *Rev. Geophys. Space Sci.*, 8:729, 1970.

Space flight

Both the United States and Soviet Union space exploration programs reached a faster tempo in 1971. In the wake of the Soviet success with its unmanned lunar surface vehicle, Lunokhod, the United States successfully used its manned vehicle, the Lunar Rover. In this same year the Soviets were the first to put into space an orbiting manned space laboratory, the Soyuz-Salyut Program. The Soviet space laboratory is smaller than the United States Skylab planned for 1973. Also in 1971 the Soviet government revealed plans for establishment of a permanent space station on the Moon and ultimately a manned landing on Mars, while the United States continued planning for unmanned exploration of deep space and establishment of a large Earth-orbiting laboratory with a space-shuttle commuting service.

Lurking behind the apparent race for scientific competence in space are military objectives which, in a sense, render the broad range of unmanned flights (see table) more important than the manned missions. Despite the enormous long-term political significance of this trend in military strategy, attention of the public has focused on the Soviet Soyuz-Salyut and the United States Apollo programs.

Fig. 1. Photograph of Alan B. Shepard using handtools from the mobile equipment carrier to take samples of the Moon's surface. (*NASA photo via UPI*)

Apollo 14. In 1971 the *Apollo 14* and *15* missions succeeded in attaining nearly all objectives with only minor difficulties. The emphasis of the Apollo Program shifted with *Apollo 14* from testing space systems and equipment to gathering of scientific data and samples.

At 6:31 P.M. EST on Jan. 31, after a 40-min weather delay, *Apollo 14* was successfully launched with Capt. Alan B. Shepard, Jr., as mission commander, Maj. Stuart A. Roosa as command module pilot, and Cmdr. Edgar D. Mitchell as lunar module pilot. Hopes for the continuance of the Apollo Program depended to some extent on the success of this mission because of the near disaster of the *Apollo 13* flight in April, 1970.

The mission met all but a few goals. At 10:38 P.M. EST on Feb. 1, a midcourse correction aimed *Apollo 14* into a perfect translunar coast. At 1:59 A.M. EST on Feb. 4, a second burn inserted *Apollo 14* into lunar orbit. After a day of photographic assignments in lunar orbit, the lunar module, called Antares, separated from the command module, the Kitty Hawk. Antares made lunar touchdown at 4:18 A.M. EST on Feb. 5.

Shepard and Mitchell made two extravehicular excursions to the lunar surface. The lunar module had touched down exactly on the target site, Fra Maura. During the 33½ hr Antares rested on the lunar surface, the two astronauts spent about 9 hr outside the craft.

The astronauts explored the lunar surface utilizing a two-wheel cart, handtools, sample bags, cameras, and a magnetometer (Fig. 1). They collected raindrop-shaped pebbles, took corings from 2 ft below the surface, and photographed a young

Principal space-flight events in 1971

Launch date	Vehicle	Source	Mission
Jan. 26	*Intelsat 4F2*	United States	Fourth-generation, spin-stabilized commercial communications satellite; last in series; carried 5000 circuits
Jan. 31	*Apollo 14*	United States	Manned extravehicular exploration of Fra Mauro highlands of the Moon
Feb. 3	*Nato 2*	NATO	Communications
Feb. 9	*Cosmos 394*	Soviet Union	Rendezvous with *Cosmos 397*
Feb. 16	*Tansei 1*	Japan	Atmospheric studies
Feb. 17	*Cosmos 395*	Soviet Union	Mission undisclosed
Feb. 18	*Cosmos 396*	Soviet Union	Mission undisclosed
Feb. 25	*Cosmos 397*	Soviet Union	Rendezvous with *Cosmos 394*; precursor of satellite intercept experiment
Feb. 26	*Cosmos 398*	Soviet Union	Experimental studies in orbital plane of *Salyut 1* and *Soyuz 10* and *11*
Mar. 3	Undesignated craft	Peoples' Republic of China	Second in series designed for surveillance of Japan
Mar. 13	International Monitoring Platform (*Explorer 43*)	United States	Studies of interplanetary fields and particles
Mar. 18	*Cosmos 400*	Soviet Union	Rendezvous with *Cosmos 404*; precursor of satellite intercept experiment
Apr. 1	International Satellite for Ionospheric Studies (*ISIS 2*)	Canada	Fixed- and swept-frequency sounders for electron and ion density measurement
Apr. 1	*Cosmos 402*	Soviet Union	Orbital plane maneuver experiments
Apr. 4	*Cosmos 404*	Soviet Union	Orbital plane maneuvers, followed by rendezvous with *Cosmos 400* and return to Earth
Apr. 7	*Cosmos 405*	Soviet Union	Mission undisclosed
Apr. 15	*Tournesol*	France	Series of eight simultaneous launches
Apr. 17	*Meteor 8*	Soviet Union	Meteorological observations
Apr. 19	*Salyut 1*	Soviet Union	Large orbiting space station
Apr. 23	*Soyuz 10*	Soviet Union	Rendezvous and docking with *Salyut 1*
Apr. 23	*Cosmos 407*	Soviet Union	Mission undisclosed
Apr. 24	*San Marco 3*	Italy	Atmospheric studies; last in series
Apr. 28	*Cosmos 409*	Soviet Union	Mission undisclosed
May 1	*Oscar*	United States	Amateur radio communications
May 7	*Cosmos 411–418*	Soviet Union	Series of eight simultaneous launches
May 8	*Mariner 8*	United States	Martian atmosphere studies; mission aborted
May 19	*Cosmos 421*	Soviet Union	Mission undisclosed; high-inclination, low-altitude orbital characteristics
May 19 and May 28	*Mars 2* and *3*	Soviet Union	Scientific investigation of Martian atmosphere
May 29	*Cosmos 425*	Soviet Union	Mission undisclosed; high-inclination, low-altitude orbital characteristics
May 30	*Mariner 9*	United States	Scientific investigation of Martian atmosphere
June 4	*Cosmos 426*	Soviet Union	Mission undisclosed; high-inclination, highly eliptical orbital characteristics
July 8	SOLRAD (*Explorer 44*)	United States	Mapping of solar radiation sources in the universe
July 16	*Meteor 9*	Soviet Union	Meteorological observation
July 20	*Cosmos 429*	Soviet Union	Surveillance
July 26	*Apollo 15*	United States	Manned investigation of Hadley Rille area of Moon
July 28	*Molynia 18*	Soviet Union	Communication
July 30	*Cosmos 431*	Soviet Union	Surveillance
Sept. 3	*Luna 18*	Soviet Union	Unmanned exploration of lunar surface using Lunokhod rover vehicle
Fall	*Statsionar 1*	Soviet Union	Synchronous orbit communication satellite

crater within a larger older one. They took magnetometer readings and began to climb the 18° slope toward their goal, the rim of Cone Crater. However, Mission Control measured their heartbeat and ordered them to turn back. The biomedical information gained was considered as valuable as the findings the astronauts may have made had they traveled the remaining distance to the crater rim.

The Apollo Lunar Science Experiment Package (ALSEP) deployed by the astronauts included seismic sensors and vibration-inducing devices such as high-explosive, rocket-launched grenades, which were activated by Mission Control on Earth after the lunar lift-off. Another device was a reflector to enable Earth-based lasers to measure wobbling of the Earth on its axis and attempt to correlate such effects with major meteorological and geological phenomena. Biomedical experiments during the subsequent voyage back to Earth included the study of effects of weightlessness and other factors on several physiological functions.

After lift-off, Antares rendezvoused with Kitty Hawk at 3:36 P.M. EST on Feb. 6. Subsequently the Antares crew entered Kitty Hawk, jettisoned Antares, and prepared for the trans-Earth insertion some 5 hr later. Only one mid-course correction was required in the coast back to Earth, at 1:37 P.M. EST on Feb. 7. Splashdown was 900 mi south of Samoa in the South Pacific Ocean at 4:04 P.M. EST on Feb. 9.

Apollo 15. The largely successful Apollo 14 mission restored hopes for the continuance of the Apollo Program. At a later date the complete success of Apollo 15 confirmed the program's continuance. Apollo 15 was launched without a flaw on July 26 at 9:34 A.M. EST. On board were David R. Scott as mission commander, Maj. Alfred M. Worden as pilot of the command module, Endeavor, and Lt. Col. James B. Irwin, Jr., as pilot of the lunar module, Falcon.

The target was Hadley Rille, a 1000-ft-deep canyon in the Apennine Mountains, 465 mi north of the lunar equator. The objective was to gather information on the mountains and rille pertinent to the theory that the mountains had been heaved up by the meteor impact that formed the nearby Sea of Rains.

An innovation on Apollo 15 was an advanced ALSEP left by the Falcon crew to continue a lunar heat flow experiment. Also new was the first use of the two-man vehicle, Lunar Rover, to enable the Falcon crew to explore terrain several miles away from the touchdown site (Fig. 2).

The chronology of major events in the Apollo 15 mission does not resemble those of previous Apollo missions. The Falcon rested on the lunar surface nearly 3 days. The crew completed three excursions outside the Falcon, lasting up to 7 hr each. These excursions covered a much greater area than was possible in any previous flight.

The first excursion began at 9:39 A.M. EST on July 31. Scott and Irwin lowered the Lunar Rover from the Falcon to the lunar surface, made the contingency collection of rocks, and set up the television camera and transmitted the first video coverage of Hadley Base. At 11:20 A.M. EST they began driving south in the Lunar Rover. They picked up rocks with specially designed tongs and took corings from depths of about 3 ft. After the first 5-mi drive, they returned to Hadley Base and deployed ALSEP.

In the 2 days to follow, Scott was able to describe color bands in the Apennine Front to scientists at Mission Control. A gamma-ray spectrometer was used to detect a low differentiation between radiation levels at highland and lowland regions. The use of an x-ray spectrometer showed that highland areas are richer than lowlands in silicon and aluminum. Information obtained with a mass spectrometer indicated the presence of carbon dioxide being released from the lunar interior. The Fairchild mapping camera and the RCA laser

Fig. 2. Photograph of James Irwin walking away from the Lunar Rover, which is parked near the edge of Hadley Rille. (*NASA*)

altimeter increased the accuracy of the data taken for maps. Subsequent analysis by scientists will be done on much of the data obtained.

The lift-off and return to Earth were routine when compared with the 3 days of lunar excursion. At 1:11 P.M. EST on Aug. 2, Falcon lifted off the lunar surface, leaving the Lunar Rover and ALSEP. At 3:09 P.M. EST, Falcon docked with Endeavor and was jettisoned at 6:55 P.M. EST. A 2-hr delay was required to recheck suspect hatches before jettisoning. ALSEP seismometers from Apollo 12, 14, and 15 recorded the impact of Falcon when it hit the Moon at 9:05 P.M. EST.

Endeavor left lunar orbit the next day at 5:18 P.M. EST, and the $445,000,000 mission neared its completion. About 120,000 mi on route from the Moon, Scott photographed a lunar eclipse. The only incident in the entire mission which gave cause for major concern was the failure of one of three parachutes to open at splashdown, which occurred 320 mi south of Hawaii on Aug. 7.

Mars missions. Left largely unnoticed in 1971 were three unmanned missions to the planet Mars, one by the United States and two by the Soviet Union. The opportune alignment of Mars and Earth only occurs once every 25 months, and both the United States and the Soviet Union almost always undertake missions at this time.

The first Mars mission in 1971 was the United

States *Mariner 8*, which was launched unsuccessfully on May 8. The second and third missions were the Soviet *Mars 2* and *3*. The Soviet Union had made at least two previous attempts to reach Mars. One was the *Mars 1* mission in 1962 and the other was *Zond 2* in 1964. Both incurred loss of radio contact. *Mars 2* and *3* were very large craft, weighing about 5 tons each or about five times the weight of *Mariner 8*.

United States scientists think that the very large craft launched by the Soviet Union are themselves, or include, Mars landers with equipment for biological and chemical detection of life on the planet. The United States does not plan such a mission until 1975 (Viking mission) because of budget cutbacks and priority assigned to the Skylab Space Station and Space Shuttle programs. There was also some comment by critics of the Mars exploration program that the 1975 Viking Mars lander might balloon into a gigantic manned Mars landing program with great budgetary demands.

The United States plans for a 1973 Mariner flight past Venus and Mercury are still in effect, as are plans for the 1977–1980 Jupiter, Saturn, Pluto and 1978–1980 Jupiter, Uranus, Neptune missions. These deep-space probes will take advantage of highly favorable planetary alignments which occur only rarely. United States scientists speculate that the Soviet Union will not wait for this 1977–1980 launch window and that they will launch several vehicles into heliocentric orbit past the planets before that time.

Cosmos and Soyuz. Precedent for ambitious multiple flights by the Soviet Union was set by the Cosmos Program, which includes some 450 recorded missions since 1967. Cosmos missions are unmanned Earth-orbiting flights; some are associated with scientific investigation, some with military objectives, and most of them with both.

The numerous Cosmos flights listed in the table represent only the numbered missions. With each numbered Cosmos flight are as many as seven back-up launches. Military systems experts in the United States associated multiple-craft launches by the Soviet Union with the attempt to develop a fractional orbital bomb system. Another military objective attributed to Cosmos missions is the development of a satellite intercept capability. Rendezvous and orbital plane shifting capabilities are required for intercept. Another purpose attributable to the Cosmos craft rendezvous maneuver is the testing of docking systems for subsequent use in Salyut manned space stations.

Rendezvous capability and orbital maneuverability are basic requirements for Salyut's effectiveness as a platform for scientific investigation and for military purposes. The first Salyut was launched on Apr. 19 from the Baikonur Cosmodrome at Kazakaska. The 25-ton station, when linked with a Soyuz craft, is 66 ft long and 12 ft in diameter.

Soyuz 10 docked with Salyut on Apr. 23 but did not transfer cosmonauts. The next spacecraft, *Soyuz 11*, docked with Salyut on June 7, only 26 hr and 50 min after launch from Baikonur. The three cosmonauts transferred and then broke the record for the longest mission in space, lasting over 28 days. Incidentally, test engineer Viktor Patsayev had a birthday party on Salyut with flight engineer Vladimir Volkov and flight commander Col. Georgy Dobrovolsky.

Reports from Salyut indicated that the crew conducted life support, navigation, and geographic mapping experiments from very-low-altitude orbits. The mission was considered a success until 25 min before splashdown. It appears that the *Soyuz 11* hatch was not tightly closed after undocking with Salyut and that abrupt decompression upon entering the Earth's atmosphere caused aeroembolism and the death of the three cosmonauts. If the cosmonauts had been wearing space suits, as United States astronauts do on reentry, they might not have been killed. This tragedy took place on June 29.

For background information *see* MOON; SPACE FLIGHT in the McGraw-Hill Encyclopedia of Science and Technology. [JOHN KING]

Spectroscopy

Raman spectroscopy, the study of inelastic scattering of light by molecules, is a branch of spectroscopy that is rapidly coming into prominence because of the development of continuous-wave lasers in the visible part of the spectrum. A few specialists have used the Raman effect for molecular structure investigations and evaluation of molecular potential functions, but the techniques with mercury arc sources have seemed forbidding to the majority of chemists and to almost all biologists. Use of the Raman effect for analysis has generally been bypassed in favor of infrared absorption spectroscopy, for which the experimental techniques are well developed. Along with the recent development of stable lasers, with 0.1–1.0 W power in one spectral line, have come interferometrically ruled gratings practically free of "ghosts" and improved double monochromators and low-noise detectors. Sampling techniques for solid, liquid, or gaseous compounds, to be studied with the new instrumentation, are generally much more convenient and containers or cells much less expensive than for infrared absorption spectroscopy. Hence it is safe to predict that use of Raman spectroscopy will rapidly increase among chemists and biologists.

The Raman effect reveals vibrational energy levels of molecules by addition to, or subtraction from, the energies of the incident light photons. When a photon of visible light of frequency f and energy hf, where h is Planck's constant, collides with a molecule, a scattered photon results. In the vast majority of scattering cases, no change in frequency occurs, but in something like one in a million times the scattered photon has a different energy. This may be $h(f+\nu_i)$ or $h(f-\nu_i)$, where ν_i is a vibrational frequency of the molecule. The molecular frequencies ν_i are in the infrared spectral range, but in the Raman effect one observes not ν_i but usually $f-\nu_i$, and hence one can choose the part of the spectrum that is convenient by selecting f appropriately. One chooses visible light because sensitive detectors are available and because convenient materials such as glass may be used to hold samples.

Lasers are more effective light sources than the mercury arcs used previously for the Raman effect for several reasons. First, they provide a single frequency of light rather than a whole spectrum of lines. Second, the laser beam is precisely parallel and therefore may be focused to a small diameter with a very high density of photons in the sample region. This means that a slender cylinder (~0.1 mm in diameter) of molecules is receiving intense illumination, and the light scattered at right angles from several millimeters' length of this cylinder can efficiently be collected by a fast lens and focused on the entrance slit of the monochromator. Third, the laser light is plane-polarized, and this is needed to distinguish between totally symmetric vibrational levels and nontotally symmetric ones. Fourth, many intense lines from neon, argon, or krypton lasers are available from the blue Ar^+ 488.0 nm to the red Kr^+ 647.1 nm. Thus the blue or green region may be used for greatest scattering and detection efficiency, the red may be used for samples that fluoresce or decompose in blue light, and other lines may be used to avoid absorption bands in colored samples.

Liquid samples. Figure 1 shows a Raman spectrum of liquid CCl_4. It illustrates the flexibility of sampling techniques, since the liquid was left in a laboratory reagent bottle. The focused laser beam entered through the bottom of the bottle and the scattered light emerged through the side of the bottle toward the collection lens and monochromator slit. The figure shows the frequency range from 16,250 cm^{-1} (601.5 nm) at the right edge to 14,610 cm^{-1} (684.5 nm) at the left. Near 15,449 cm^{-1}, the frequency of the incident 647.1 nm Kr^+ laser light, the shutter was closed, since the peak intensity would have corresponded to thousands of times full-scale deflection. All the photons elastically scattered (Rayleigh scattering) would contribute to this intensity. In order to reject this extremely intense single frequency and still record the weak inelastically scattered light (Raman spectrum), it is necessary to have a well-designed double monochromator with nearly perfect gratings.

The scale shown on the bottom of Fig. 1 gives frequency shifts from the exciting frequency, 15,499 cm^{-1}. They are marked negative for the anti-Stokes region and positive for the Stokes region. For the anti-Stokes region the scattered photons had more energy than the incident ones, that is, the molecules added a quantum of vibrational energy to the energy of the colliding photon. Except for a frequency factor that is only slightly different from 1, the probabilities for Stokes and anti-Stokes shifts are equal. The anti-Stokes bands are weaker because there are more molecules in the ground state than in any excited state. For the molecular energy level corresponding to the 218 cm^{-1} shift, the Boltzmann factor is relatively high and the anti-Stokes band is over half as high as the Stokes band. For the 459 cm^{-1} vibration the anti-Stokes band is about one-third as high as the Stokes band, and for the 762–791 cm^{-1} doublet the ratio is less than one-tenth. In fact, the anti-Stokes doublet is just barely perceptible at the extreme right of Fig. 1.

In the experimental arrangement for Fig. 1, the polarization of the incident light was perpendicular to the plane containing the incident light beam and the slit length. If, by insertion of a suitable crystal, the polarization had been rotated by 90°, the 459 cm^{-1} band would have been missing and the other bands would have been approximately 75% as

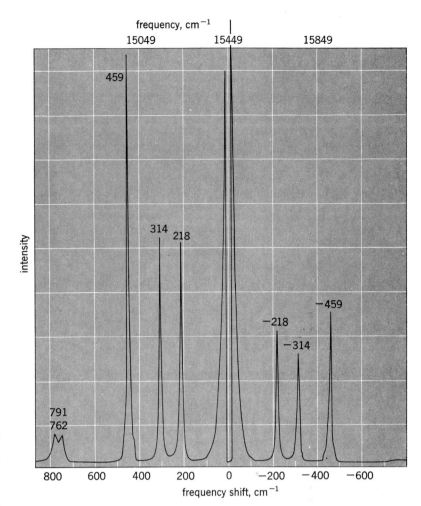

Fig. 1. Raman spectrum of liquid CCl_4.

high. This would clearly show the 459 cm^{-1} band to be due to the totally symmetric vibration of the tetrahedral molecules. In general, Raman bands representing totally symmetric molecular vibrations are polarized, that is, they are attenuated more by rotation of the polarization than are bands arising from nontotally symmetric vibrations. Polarization measurements are therefore an essential part of Raman data to be used for identifying symmetries of vibrations.

Solid samples. Figure 2 is an example of a Raman spectrum of a solid. A Pyrex tube was pulled to a funnel shape and some Na_2SO_4 powder poured in. It was then placed horizontally so that the laser beam from below struck the tube where its outside diameter was about 2 mm and the spectrum of the scattered light was recorded. As is the usual practice, in Fig. 2 only the Stokes spectrum is shown and only the frequency shifts are given. What is observed is the vibrational spectrum of the tetrahedral sulfate ion. In this case, crystal forces cause a splitting of the degeneracies, and one can at a glance assign the multiplets as due to a nondegenerate vibration, a doubly degenerate vibration, and two triply degenerate vibrations.

Gaseous samples. Figure 3 shows the spectrum of a gas. The sample, 2 atm (202.6 kN/m^2) of MoF_6 at about 60°C, was contained in a short length of Pyrex tubing of 25 mm outside diameter. The highly symmetrical molecule MoF_6 has only six vibrational frequencies, three of which are Raman-active, two of which are infrared-active, and one of which is inactive in both spectra. The three intense bands at 741.5, 651.6, and 318 cm^{-1} represent the three Raman-active fundamentals. In addition, three weak Raman bands are observed. The most intense of these, at 233 cm^{-1}, is the overtone of the inactive frequency and thus locates it at about 116 cm^{-1}. This is an unusually low frequency of vibration for a fluoride, but the low value had already been predicted from heat capacity measurements before this spectrum was obtained.

The use of Raman spectroscopy for qualitative and quantitative analysis is rapidly gaining in popularity. Several instrument companies are marketing complete units that are convenient for the nonspecialist to use. There are a number of advantages over infrared spectroscopy that are often important: (1) Glass may be used as a sample cell instead of fragile and hygroscopic crystalline materials; (2) solid samples may be of any shape or thickness; (3) the complete vibrational spectrum from 20 to 4000 cm^{-1} may be observed in one scan; (4) water may be used as a solvent with only very weak bands of its own; (5) Raman bands tend to be narrower than infrared bands and hence there is less overlapping of bands in multicomponent solutions; and (6) the intensity of Raman bands are often quite accurately proportional to concentrations. One disadvantage of the Raman technique is that occasional samples fluoresce and the fluoresence tends to mask the weak Raman effect.

For background information *see* INFRARED SPECTROSCOPY; LASER; MOLECULAR STRUCTURE AND SPECTRA; RAMAN EFFECT; SPECTROSCOPY in the McGraw-Hill Encyclopedia of Science and Technology.   [HOWARD H. CLAASSEN]

Bibliography: H. H. Claassen, H. Selig, and J. Shamir, Raman apparatus using laser excitation and polarization measurements, *Appl. Spectrosc.*, 23:8, 1969; J. Loader, *Basic Laser Raman Spectroscopy*, 1970; H. A. Szymanski (ed.), *Raman Spectroscopy: Theory and Practice*, vol. 1, 1967, vol. 2, 1970.

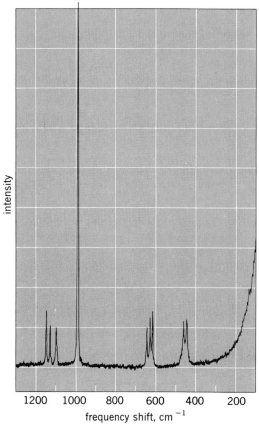

Fig. 2. Raman spectrum of solid Na_2SO_4.

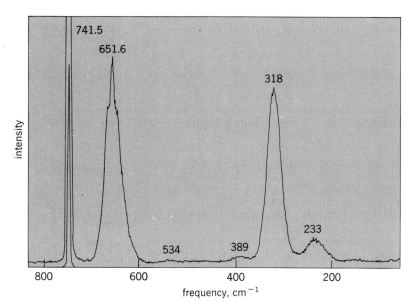

Fig. 3. Raman spectrum of gaseous MoF_6.

Speech

Electroencephalographic (EEG) recording techniques have been applied recently to the identification and localization of brain structures underlying man's language behavior. The tiny electrical currents generated by the brain and recorded through the scalp have been shown to vary systematically with speech output and linguistically significant visual and auditory input.

Substrates of language. During the 19th century, brain scientists first made explicit statements about the areas of the cerebral cortex responsible for particular language functions. Paul Broca in 1861 pinpointed the center for articulated language in the area which now bears his name: the foot of the third frontal convolution in the left hemisphere. Since that time, neurologists have delineated most of the language areas of the brain; these are summarized in Fig. 1, a diagram of the brain's left hemisphere. Evidence from a wide variety of sources has established that language function is carried on in one of the two hemispheres of the brain, usually the left.

Wernicke's area is primarily involved in the auditory processing of language input. The angular gyrus integrates visually presented language input and thus subserves both reading and writing. The supramarginal gyrus, together with anterior portions of the angular gyrus and posterior portions of

Wernicke's area, is part of the central integrating area for language and therefore would be involved in the elaboration of both syntactic and semantic structures. Exner's center is a motor association area for the part of the primary motor cortex which controls the hand and thus it is part of the structures underlying writing. Broca's area is a motor association area for the part of the primary motor cortex which controls the muscles of the head, face, and neck and thus is the structure underlying articulated speech. Numerous fiber paths connect these areas with each other; these are not represented on the diagram. The exact boundaries of these areas are not indicated in the diagram because there are none; the labels and functions indicate only in a very general way what is known today about the language system in man's brain.

All of the earlier knowledge about the brain's language functions has been obtained under abnormal conditions: brain damage (stroke, wound, tumor), surgical or pharmacological intervention, or electrical stimulation of the brain during neurosurgery. This is very difficult evidence from which to draw conclusions since the damage is often diffuse, its borders are usually not well defined, verification of the lesion site is often fragmentary or unobtainable, behaviors other than language are usually adversely affected, and good observations of behavior prior to the brain damage are ordinarily of an anecdotal nature. Beyond this, there are basic questions of how the brain is organized. In other words, the functional interrelations of separate brain areas must be considered, especially when studying an integrated group of behaviors as complex as language.

Electrical concomitants. Only with the recent development of experimental EEG techniques has it been possible to investigate these brain functions in the normal human. The history of experimental EEG has been one of increasing sophistication in separating the electrical signals of interest from the background of ongoing EEG activity. Beginning with the pioneering work of G. D. Dawson in the mid-1940s and continuing through recent advances in computer technology, methods of data collection and analysis have been developed which allow neuropsychologists to investigate brain concomitants of complex behaviors. Of particular salience are the discoveries of W. Grey Walter and subsequent investigators, who have shown that very slow potentials recorded at the scalp are potent predictors or accompaniments of such complex processes as motivation, attention, and memory. Other slow potentials have been shown to be related to intended motor activity; these are the so-called readiness potentials, which have been shown to be topographically distributed across the scalp in a manner consistent with the organization of underlying cortical areas subserving motor functions. For example, clenching the left fist is anticipated by a readiness potential which is maximal over the hand area of the contralateral right motor cortex. Additionally, renewed interest has been focused on sensory averaged evoked responses, and especially upon their relationships to complex behaviors. For example, a visual pattern that is perceived and attended to is followed by a larger averaged evoked response than one which is ig-

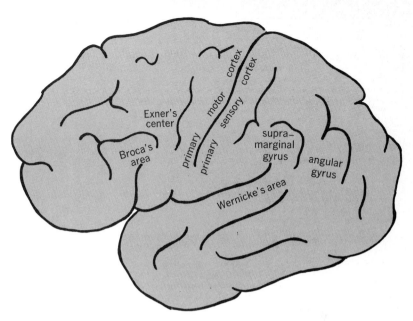

Fig. 1. Simplified diagram of surface of left hemisphere of human brain.

nored. In short, technology and theory have advanced to the point that scientists are now able to investigate, relatively directly, the electrical activity of the mind.

As yet, few reports exist on the brain electrical concomitants of language behavior in themselves. Two studies have looked at brain processing of language input. M. Buchsbaum and P. Fedio in 1969 investigated the averaged evoked responses in the occipital lobe to visually presented words, random-dot patterns, and designs. Their data indicated that the verbal and nonverbal stimuli generated averaged evoked responses of greater difference in the left occipital lobe than in the right and they also found that the average evoked responses for words had shorter latencies than those for nonverbal stimuli. R. Cohn in 1971 demonstrated that the amplitude of the averaged evoked responses is larger over the right hemisphere for click noises and equal or larger over the left hemisphere for words. In studies of speech production, J. Ertl and E. Schafer presented data as early as 1967 which demonstrated systematic variations in EEG activity preceding speech. D. Galin and R. Ornstein in 1971 showed differential left-right distribution of alpha-rhythm activity (a component of the normal EEG) depending on whether the subject was doing a spatial-motor task or a linguistic task.

Readiness potentials and speech. D. W. McAdam and H. A. Whitaker applied techniques for observing readiness potentials to the investigation of localization of speech areas underlying language production in normal humans, and have showed maximal readiness potentials over the left hemisphere and, furthermore, over Broca's area. Recording electrodes were placed symmetrically left and right over the precentral gyri (primary motor areas) and over left and right inferior frontal areas (Broca's area in the left hemisphere and its homolog on the right). Signals from these electrodes were referred to linked electrodes over the mastoids, and were amplified using direct-coupled

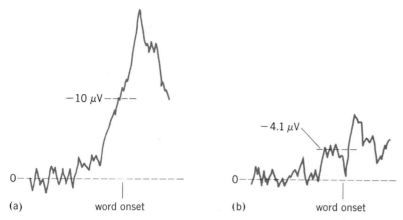

Fig. 2. Readiness potentials from (a) left hemisphere (Broca's area) and (b) right hemisphere. The time of word onset is indicated, and the broken lines show baselines and measured amplitudes of the readiness potentials. These were simultaneously recorded and averaged electronically during the articulation of a sample of 30 /p/-initial words. The time represented by each sample is 2 sec.

recording techniques and averaged 1.5 sec prior to articulation. Activation of a voice trigger by the subject's verbalizations started the sweep of the averager. Samples of 30 of each of the responses were taken for analysis.

Each subject produced four sets of responses: a spitting gesture, a set of words beginning with the phoneme /k/ having at least three syllables, a coughing gesture, and a set of words beginning with the phoneme /p/ also having at least three syllables. The spitting and coughing gestures were chosen as nonspeech controls which involved parts of the vocal musculature in a manner analogous to the /p/-initial and /k/-initial words. For the two different types of words, subjects were told to think of a different word spontaneously each time, to avoid repetition, and to articulate the word rapidly, naturally, and accurately.

Figure 2 shows readiness potentials from one subject which occurred over Broca's area and its homolog on the right prior to the articulation of a series of /p/-initial words. The larger negative potential (upward deflection) over Broca's area is easily seen. When the data from all five subjects were treated statistically, it was found that larger readiness potentials occurred over the left hemisphere, and, furthermore, that the differences between Broca's area and its homolog were larger than the differences between left and right precentral areas.

Further work on potentials related to speech production has supported a hypothesis stating that Broca's area serves as a staging area for speech articulation rather than as a dictionary (storage area for words and their meanings). In a follow-up study, nonsense words were used instead of real words. Nonsense words are ones which could be English but, fortuitously, are not, for example, *polufratz* and *peenotine*. Readiness potential preceding the articulation of nonsense words were also maximal over Broca's area compared with its homolog in the right hemisphere. This indicates that the specific role of Broca's area is to convert the words one wishes to say into articulatory commands to the vocal musculature. Interestingly enough, this is virtually the hypothesis proposed by Broca in 1861.

These studies are examples of work being done to explore the electrical signs of neural activity. When data such as these are coupled with developments in the understanding of the physical and chemical substrates of these electrical signals, scientists will have moved much closer to the goal of full knowledge of the biological bases of mental activity.

For background information see BIOPOTENTIALS AND ELECTROPHYSIOLOGY; ELECTROENCEPHALOGRAPHY; SPEECH in the McGraw-Hill Encyclopedia of Science and Technology.

[DALE W. MC ADAM; HARRY A. WHITAKER]

Bibliography: M. Buchsbaum and P. Fedio, Visual information and evoked responses from the left and right hemispheres, *Electroenceph. Clin. Neurophysiol.*, 26:266, 1969; J. Ertl and E. W. P. Schafer, Cortical activity preceding speech, *Life Sci.*, 6:473, 1967; D. W. McAdam and H. A. Whitaker, Language production: Electroencephalographic localization in the normal human brain, *Science*, 172:499, 1971; H. A. Whitaker, *A Model for Neurolinguistics*, Occas. Pap. no. 10, University of Essex Language Centre, Colchester, England, 1970.

Stereophonic sound, four-channel

Recent advances in the technology of four-channel stereophonic sound have been concerned mainly with techniques for recording and reproducing four-channel signals.

Four-channel stereo, as considered here, refers to a recently devised system for reproducing sound in the home by means of four loudspeakers properly situated in the listening room, with each loudspeaker being fed its own identifiable segment of the total signal comprising the program. In the usual, stylized arrangement, a loudspeaker is placed near each corner of the room, with two speakers in front of the listening area and two speakers to the rear, as depicted in the figure. Modifications of this basic arrangement have been suggested and demonstrated, but all arrangements are intended to achieve the same result, namely, to provide the listener with an enhanced subjective experience by surrounding him with a carefully controlled sound field. The precise nature of the sound field is determined by the manner in which the program signals are distributed among the four channels, and effects ranging from simulation of the ambiance and auditory perspective of a concert hall to the intimacy of chamber music or a jazz combo and on to avant-garde experimentation with moving sound sources are possible. This article is concerned with the technical aspects of recording and reproducing the four-channel signals and not with esthetics, although the latter is strongly influenced by the former.

First of all, assume that a recording director has succeeded in mixing down his multiple microphone inputs into a four-channel master tape whose signals, when fed to loudspeakers, create the sound field desired by the director. The following discussion considers the problems involved in transferring these signals from the master tape in the recording studio onto tape or disks suitable for reproduction in the home. The transfer process of the four-channel signals from a master tape to another tape is straightforward when each channel is allotted its own discrete track on the copy tape. Four-channel stereo tapes have been commercially available for some time on four-track open reels, in eight-track endless-loop cartridges, and more recently in four-track cassettes.

Phonograph records. Considerable effort is being devoted to developing techniques for carrying four discrete channels in the single groove on a phonograph record. One such system under development supplements the conventional two-channel stereo modulation of right and left groove walls with a modulated 30-kHz carrier on each groove wall to record and reproduce the additional information needed for four-channel signals. The sum of the right front (RF) and right rear (RR) signals is recorded directly as an audiofrequency signal on the right-hand groove wall. Similarly, the sum of left front (LF) and left rear (LR) signals is recorded on the left-hand groove wall. The difference signals, (RF)−(RR) and (LF)−(LR), are each recorded as frequency modulation of 30-kHz carriers on the right-hand and left-hand groove walls, respectively. When played back with a wideband two-channel pickup (response to at least 45 kHz), the (RF)+(RR) and (LF)+(LR) signals are immediately available at the pickup output terminals. These two signals can be combined in simple networks with (RF)−(RR) and (LF)−(LR) signals, obtained by demodulating the 30-kHz carriers, to produce the original four signals (RF), (RR), (LF), and (LR) with negligibly small crosstalk. The four-channel record may be played with a conventional two-channel stereo pickup that is insensitive to the high-frequency carriers, in which case the (LF)+(LR) signal is fed to the usual left-hand loudspeaker and the (RF)+(RR) signal goes to the right-hand loudspeaker. A monophonic pickup and system reproduce all four signals together through a single loudspeaker. Conversely, if the four-channel pickup is used to play a conventional two-channel stereo record, only the L and R baseband signals are reproduced, since no high-frequency carriers are present. Thus there is a high degree of compatibility between this four-channel system and existing stereo and mono records and pickups—an important consideration when introducing a new system. With further improvement, particularly in the development of inexpensive pickups capable of wideband response even at the innermost groove diameters, a four-channel disk system of this type will be close to satisfying commercial requirements.

Compression and sorting out. Another major approach to four-channel stereo involves compromising the discreteness of the original four channels in order to compress the total information into two conventional stereo channels, whether on disk or tape, and subsequently sorting out the signals in reproduction into four components resembling, but not duplicating, the original signals. This technique is known as matrixing, a term derived from the algebraic procedures used in analyzing such a system. A very simple example will serve to illustrate the technique and to point out its shortcomings and advantages. If the conventional two channels—left and right—are designated L and R, one way, arbitrarily selected from an infinite choice of ways, matrixing the original four channels into two channels is represented by Eq. (1). Minus signs in matrixing equations indicate a phase reversal of the signal.

$$L = (LF) + 0.25(RF) - 0.25(RR) + (LR)$$
$$R = 0.25(LF) + (RF) + (RR) - 0.25(LR) \quad (1)$$

One sees that if this combination of signals is reproduced by a two-channel system, the left-hand loudspeaker will carry predominantly the original left-hand signals (front plus rear) but with 25% crosstalk from the right-hand channels. A corresponding result occurs at the right-hand loudspeaker. Thus some degree of compatibility exists between the four- and two-channel systems. The composite two-channel signal may be matrixed in playback to synthesize four new signals for four-channel reproduction, the new signals being indicated by $(LF)'$, $(LR)'$, $(RF)'$, and $(RR)'$. Out of an infinite number of choices, Eqs. (2) are selected for illustration. For this combination, the desired signal predominates in each channel, but considerable crosstalk occurs between front and rear on each side and a lesser amount of crosstalk between right and left. Zero crosstalk occurs between diagonally opposite channels.

$$(LF)' = L + 0.25R = 1.1(LF) + 0.5(RF) + 0.9(LR)$$
$$(RF)' = 0.25L + R = 0.5(LF) + 1.1(RF) + 0.9(RR)$$
$$(RR)' = -0.25L + R = 0.9(RF) + 1.1(RR) - 0.5(LR) \quad (2)$$
$$(LR)' = L - 0.25R = 0.9(LF) - 0.5(RR) + 1.1(LR)$$

By choosing other values for the coefficients in Eqs. (1) and (2), one can obtain other compromises in crosstalk distribution. For example, the right/left crosstalk in the front channels can be

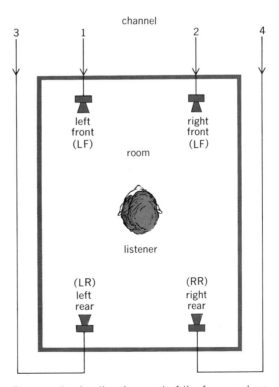

Diagram showing the placement of the four speakers used in four-channel stereo.

reduced if one is willing to accept greater crosstalk in the real channels, and so on. The psychoacoustic effects of some of the crosstalk can be reduced by introducing a 90° rather than a 180° phase shift between the two (RR) components in Eqs. (1), and also between the two (LR) components, before matrixing. A complementary phase shift introduced in the matrixing represented by Eqs. (2) yields crosstalk components that are out of phase in diagonally opposite speakers, leading to a diffuse, rather than a localized, sound field for those components if they are present. A further embellishment is sometimes added, consisting of continuously sampling the program during reproduction and using logic networks to select the predominant channel at each instant. The gain is increased in this channel, while the gain in the other three channels is reduced correspondingly. For many types of program this technique appears to enhance the simulated impression of four-channel sound.

Several versions of the four- into two- into four-channel (4-2-4) matrixing systems are being vigorously studied. However, even at best this technique is a compromise that degrades the original sound field. In exchange for this degradation one gains the advantage of immediate compatibility with all existing two-channel systems—tape, disk, or FM broadcast.

For background information *see* CROSSTALK; PSYCHOACOUSTICS; SOUND-REPRODUCING SYSTEMS; STEREOPHONIC SOUND in the McGraw-Hill Encyclopedia of Science and Technology.

[J. G. WOODWARD]

Bibliography: J. M. Eargle, *J. Audio Eng. Soc.*, 19:552, July/August, 1971; T. Inoue. N. Takahashi, and I. Owaki, *J. Audio Eng. Soc.*, 19:576, July/August, 1971; P. Scheiber, *J. Audio Eng. Soc.*, 19:267, April, 1971.

Strigiformes

Recent studies of strigid birds include morphological descriptions of several peculiar features of the wing, one being a bony spur at the wrist and another a bony arch on the radial bone which is associated with a mass of sensory bodies (Herbst corpuscles). Many of the peculiar characteristics of the nocturnal birds of prey—the owls of the order Strigiformes—are adaptations for life after dark. Some of the most extreme of these specializations are associated with their silent hunting under conditions of low light intensities. Owls possess a soft fringe along the leading edge of their flight feathers which muffles the sound caused by air rushing past a stiff feather; their flight is noiseless, permitting owls to approach their prey without warning. Owls have large, forwardly directed eyes which contain only rods in the retina. Rods, one of the two types of visual cells, permit vision under dim light conditions. Cones, the second type of visual cells, are specialized for color vision. Owls are thus able to see under very low light intensities but lack color vision. Moreover, owls are able to locate and capture their prey in absolute darkness. They possess asymmetrical external ears that are reflected in asymmetry of the auditory bullae of the skull. Asymmetrical ears allow directional hearing, which is highly developed in owls. These birds are able to locate and pounce upon a mouse at distances of over 25 ft (7.6 m) by hearing alone.

Os prominens. In addition to these specializations, owls possess several other unusual or unique features, some well known, such as the ability to twist their head in an almost complete circle, and some only poorly known but which may also be associated with their nocturnal habits. They have a large sesamoid bone in the wrist—the os prominens (Fig. 1)—which redirects the pull of the tensor patagii longus muscle to the hand; this muscle and its tendon are responsible for maintaining the leading edge of the wing. Although other birds possess a sesamoid bone in this location and some birds, for example, some hawks, have an enlarged bone, no other avian group possesses a structure as large or with the shape of the owl os prominens.

Osseous arch of radius. Perhaps the most peculiar of these poorly known features is a low bony arch on the shaft of the radius (Fig. 1). The radius is a wing bone which, together with the ulna, comprises the central part of the wing; both bones articulate with the humerus at their proximal end and with the hand at their distal end. The radius lies at the leading edge of the wing and does not have any flight feathers attaching onto it. The primary flight feathers attach onto the bones of the hand, and the secondaries onto the ulna. The osseous arch of the radius is located at the proximal third of the posterior edge of the shaft, so that it faces the ulna. The nutrient foramen of the radius is located at the distal end of the arch; blood vessels pass into the radius through this foramen.

A bony arch on the shaft of a long bone is a very unusual structure. Spines, ridges, and processes of all shapes are common, but arches similar to the osseous arch of the radius are sufficiently rare that another such example within the class Aves, or indeed within the tetrapods, has not as yet been reported. All owls possess the osseous arch of the radius, but no other bird shows any indication of a like structure on this bone. Hence it is important to know what other structures are associated with it.

Muscles. Two muscles attach onto the bony arch (Fig. 2). One is the pronator profundus muscle, which originates from the distal end of the humerus and inserts on the osseous arch and surrounding shaft of the radius. The second is the extensor indicis longus muscle, which arises from the arch and from the shaft of the radius distal to the arch and inserts on the base of the distal phalanx of

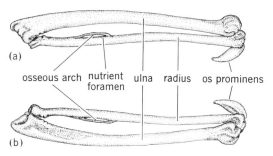

Fig. 1. Forearm of *Ninox strenua* as seen from (a) below and (b) above showing the os prominens, osseous arch, and nutrient foramen. (*From W. J. Bock and A. McEvey, The radius and relationship of owls, Wilson Bull., 81(1): 55–68, 1969*)

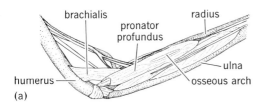

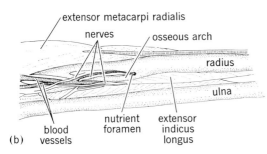

Fig. 2. Forearm of *Asio flammeus* showing (a) the attachment of the pronator profundus muscle and (b) the attachment of the extensor indicus longus muscle to the osseous arch. (*From W. J. Bock and A. McEvey, The radius and relationship of owls, Wilson Bull., 81(1):55–68, 1969*)

digit II, the main digit of the avian hand.

Herbst corpuscles. No structures, for example, nerves or blood vessels, pass through the arch from dorsal to ventral, although blood vessels do extend through the nutrient foramen. However, nerves run into a mass of whitish tissue lying between the arch and the radial shaft; this tissue fills the entire space below the osseous arch. Under high magnifications of a dissecting microscope, cigar-shaped objects about 0.75 mm long and 0.3 mm in diameter were observed. Examination of histological sections of the osseous arch and the tissue beneath it revealed that these oblong bodies are Herbst corpuscles (Fig. 3). These are deep pressure sense organs—the counterpart of mammalian Pacinian corpuscles. Herbst and Pacinian corpuscles are cigar-shaped structures composed of successive fluid-filled shells of thin epithelial cells and a central bare nerve ending. When the corpuscle is deformed by mechanical pressure, the nerve ending is stimulated. The strand of Herbst corpuscles is suspended in connective tissue between the shaft and the osseous arch of the radius with their longitudinal axes parallel to the longitudinal axis of the radius and hence the osseous arch. About 200 corpuscles are present beneath the arch in a great horned owl.

Herbst corpuscles have been reported lying along the shafts of the tibia of the avian leg and of the radius and of the ulna of the wing, in addition to many other locations in the bird's body. Those corpuscles on the radius in other groups are in the same approximate position as those seen below the osseous arch in owls. Thus owls are not unique in the position of Herbst corpuscles along the radial shaft, but they are unique in the possession of a bony arch spanning the mass of sensory corpuscles. It may be reasonably argued that the osseous arch is associated with some special function of the Herbst corpuscles lying along the strigiform radius.

The sensory nerves coming from these corpuscles join the main nerve trunk (the brachialis longus inferior) just distal to the branching of the nerve that innervates the pronator profundus muscle. The nerve innervating the extensor indicus longus muscle arises from a different major nerve (the brachialis longus superior) to the arm. Both nerve trunks arise directly from the brachial plexus. Possibly the close morphological association of the nerves suggests that the sensory function of this collection of Herbst corpuscles is related to the action of the pronator profundus muscle.

Deformation of the concentric shells of epithelial cells stimulates the nerve ending in the center of the corpuscle. However, the exact type of deformation required for stimulation is not known. One could imagine the corpuscle being squeezed or being bent along its longitudinal axis. Pull of the extensor indicus longus muscle would squeeze the corpuscles, while pull of the pronator profundus muscle or bending of the radius would tend to bend the corpuscles along their longitudinal axis. The most reasonable possibility appears to be longitudinal bending resulting from pull of the pronatus profundus, the fibers of which lie at an angle to the longitudinal axis of the Herbst corpuscles. More importantly, part of this muscle inserts on the bony arch and part on the shaft of the radius. Hence the force of different parts of this muscle would place a differential deformation on the sensory corpuscles. The bony arch thus acts much like a strain gage, with the arch serving either to dampen the force of the muscle or to provide a

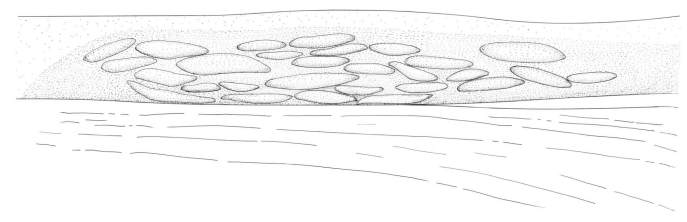

Fig. 3. Enlarged view of Herbst corpuscles lying in position under the osseous arch.

differential in the force of the muscle that can be detected by the Herbst corpuscles or both.

Whatever the exact sensory function of the radial Herbst corpuscles may be in owls, it is reasonable to assume that the osseous arch is associated with this function. Presumably this complex structure is a proprioceptive sense organ correlated with some aspect of owl flight. But here one's knowledge ends and the need for more study begins. Recording of nerve impulses from the Herbst corpuscles can be made and correlated with different patterns of forces placed on the radius and bony arch. Attempts could be made to sever the two ends of the arch and then compare the new impulses between the normal and experimental conditions. Exact observations of owl flight might provide additional clues. Whatever its exact function proves to be, the osseous arch of the radius and strand of Herbst corpuscles lying below it is a highly unusual proprioceptive sense organ and is a unique feature of the owls which characterizes this order of birds as definitely as many of their better known features.

For background information see MECHANORECEPTORS; MUSCLE SYSTEM (VERTEBRATE); SENSE ORGAN; SKELETAL SYSTEM (VERTEBRATE); STRIGIFORMES in the McGraw-Hill Encyclopedia of Science and Technology. [WALTER BOCK]

Bibliography: W. J. Bock and A. McEvey, *Wilson Bull.*, 81(1):55-68, 1969.

Tectonophysics

During the last decade an international program of investigations of the upper 100 km of the Earth's crust and mantle was carried out. This program, the Upper Mantle Project, was instrumental in encouraging and coordinating research activities and in improving knowledge of that part of the Earth's interior which has had the greatest influence on the development of the surface features. This program came to an end in 1971, and was succeeded by the Geodynamics Project, a program initiated by the International Council of Scientific Unions and devoted to understanding of the dynamics and dynamic history of the Earth and the deep-seated foundations of geological phenomena. This project is particularly timely since the efforts during the Upper Mantle Project resulted in a revolutionary change in thinking about the origin of the surface features of the Earth. The change of thinking was not due to evidence for a mechanism which produced the surface features, but was due to new data which could be interpreted as meaning that certain motions had taken place.

New data. The major breakthrough depended on three independent sets of data related to the Earth's magnetic field and its effect upon the rocks at the surface. The first of these involved the history of the magnetic field. Suggestions had been made in the past that the field had reversed periodically with time. This was not universally accepted, and alternative explanations were sought for the frequent occurrence of rocks containing magnetic minerals which were magnetized in a direction opposite from that of the present field. A. Cox, R. R. Doell, and B. G. Dalrymple settled this issue by carefully collecting oriented samples of volcanic rocks from many parts of the world, measuring the direction in which they were magnetized, and measuring their age. From these data they constructed a history of the magnetic field for the last 3,500,000 years which showed as a number of periods of normal and reversed magnetization of the rocks. This time scale has since been extended further back in time.

The second observation dealt with magnetic anomalies found over ocean ridges. When a magnetometer is towed over the ridges, a series of high and low magnetic measurements are found. These tend to form trends paralleling the axes of the ridges. In trying to construct models to explain these anomalies, the usual method was to assume that magnetic blocks of rock were separated by nonmagnetic blocks so that the highs appeared over the former and lows over the latter. This explanation had difficulties since it implied two rock types of very different character, an implication not supported by dredging of rocks from the ridge, and further it required an unusually high degree of magnetization of the rocks. F. J. Vine and D. H. Matthews in 1963 offered an alternative explanation. Accepting that the magnetic field had reversed with time, they proposed that the anomalies were caused by the intrusion or extrusion of basic igneous rocks at times when the Earth's field was normal or reversed. Normal magnetism would result in augmentation of the field, while reversed magnetization would result in a decrease in measured field strength. This explanation removed the need for two rock types and reduced the necessary magnetization by half.

The third set of data resulted from magnetic data taken aboard a U.S. Coast and Geodetic Survey vessel off western North America and reported by A. D. Raff and R. B. Mason. These data indicated that the rocks of the ocean floor were magnetized in such a way that they produced a series of magnetic stripes, linear strips of alternately positive and negative magnetic anomalies. These were not clearly understood until a magnetic survey was flown by the Naval Oceanographic Office over the Reykjanes Ridge south of Iceland, a segment of the mid-Atlantic ridge. This survey, reported by J. Heirtzler, X. LePichon, and J. G. Baron, showed that the magnetic pattern over the ridge was remarkably similar on either side of the axis.

Then, in a series of papers by Vine, T. Wilson, Heirtzler, W. Pitman, LePichon, and others, these three observations were drawn together. It was noted that the relative widths of the magnetic anomalies outward from the axes of the ridges were in the same proportions as the relative intervals of normal and reversed magnetization in the time scale of Cox, Doell, and Dalrymple. Thus the magnetic anomalies could be associated with a time sequence of emplacement of the crustal rocks which implied that the crustal rocks were progressively older away from the axes of the ridges. Mechanisms were advanced by H. Hess, R. Dietz, B. Heezen, and others which suggested that new crust was being created along the mid-ocean ridge system and that the older crust was moving laterally in order to make room for it. This idea was supported by studies of seismicity of the ridges by L. R. Sykes which favored separation of the two sides of the ridge over alternative crust motions.

It was further supported by the data from the JOIDES Deep Sea Drilling Project. In this program a series of holes were drilled through the sediments of the deep-ocean floor into the underlying crustal rocks. The ages of these crustal rocks increased in distance from the ridge axes in a manner very close to that predicted by the magnetic anomalies.

The concept of crustal creation at ridge axes meant either that the Earth must expand to provide room for it or that some mechanism, or mechanisms, must be found to dispose of the excess crust. Investigations of expansion led to the conclusion that it was geologically and physically unlikely, and thus alternatives were sought.

Mechanism. In the Atlantic Ocean the idea of continental drift developed by Alfred Wegener in 1910 was revived and expanded. This concept supposed that the continents around the Atlantic were joined some 200,000,000 years ago and that they fragmented and moved to their present positions in the subsequent period. Reexamination of the data and the addition of further supporting data leant credence to the concept, and the time constants of continental drift as conceived corresponded very well to the rate of crustal production along the mid-Atlantic ridge.

In the Pacific this explanation would not do, since the continents, in moving away from the Atlantic, must be moving toward the Pacific. The earthquake seismologists provided a different mechanism to account for the excess crust in this case. K. Wadati in 1927 had contoured the earthquake hypocenters in the vicinity of Japan and noted that they increased in depth from east to west. H. Benioff, in several papers in the 1950s, showed that the hypocenters in the earthquake belt around the Pacific fell in zones of limited thickness, dipping from the ocean trenches beneath the continents or beneath the island arcs. J. Oliver and coworkers at the Lamont-Doherty Geological Observatory established some temporary seismograph stations in the Tonga-Fiji area where deep-focus earthquakes are most abundant. The added instruments gave greater accuracy to hypocenter determinations and also permitted studies of variations in propagation along different paths from deep-focus earthquakes. The seismicity studies indicated that the earthquakes fell along a very narrow band; first motion studies suggested that this band was the upper surface of a lithospheric plate 50–100 km thick that was being thrust into the underlying asthenosphere; and studies of the propagation of shear waves indicated that the characteristics of this underthrust plate more closely approximated those of the lithosphere where it was not underthrust rather than those of the asthenosphere into which it was being pushed. By making a few assumptions, it was possible to conclude that the rate of underthrusting was compatible with the rates of production of new crust on the ridges.

Dynamic picture. The complete concept, then, pictures the Earth's surface as made up of a small number of large lithospheric plates moving relative to each other—separating along the ocean ridges, colliding in the island arcs and in the young mountain systems, and sliding along each other in areas such as the western Aleutians or southern California. *See* OROGENY.

The Geodynamics Project has been designed to take advantage of this state of affairs. It is an international program of research on the dynamics and dynamic history of the Earth—the causes of deformation and mountain building, the forces which produce earthquakes and volcanoes, and the basic factors behind economic concentrations of minerals or energy sources. It has four major elements: (1) examining phenomena in the seismically active belts where actual movements can be identified with geological features; (2) studying the interior of the Earth to determine the driving mechanism responsible for the surface features; (3) determining the nature of past movements to see if these can be related to the basic mechanism which appears to be causing deformation at present; and (4) probing the nature of the vertical movements which have taken place or are taking place within the plates and which have no obvious connection to the basic mechanism which is moving the plates relative to each other.

So far theoretical and experimental studies of the deformational history of the Earth and of its interior have tended to support the plate tectonics concept. However, many unexplained details persist, and the forces producing the deforming and the sources of energy are at best poorly understood.

For background information *see* CONTINENT FORMATION; MARINE GEOLOGY; TECTONOPHYSICS in the McGraw-Hill Encyclopedia of Science and Technology. [CHARLES L. DRAKE]

Bibliography: Deep Sea Drilling Reports: Initial Reports, vols. 2–6, National Science Foundation, 1970–1971; *EOS*, 51:712–713, 1970; J. Heirtzler, X. LePichon, and J. G. Baron, *Deep Sea Res.*, 13: 427–443, 1966; B. Isacks, J. E. Oliver, and L. R. Sykes, *J. Geophys. Res.*, 73:5855–5899, 1968; L. R. Sykes, *J. Geophys. Res.*, 72:2131–2153, 1967; F. J. Vine, *Science*, 154:1405–1415, 1966; A. Wegener, *The Origin of Continents and Oceans*, 4th ed., 1966.

Terrain sensing, remote

Petroleum companies have become interested in remote-sensing technology as an exploration tool. To date, side-looking radar (SLAR) continues to be the most widely used remote-sensing instrumentation for petroleum exploration. Most other instrumentation is still restricted to research and development testing by the petroleum companies.

Remote sensing for petroleum exploration is defined as the utilization of the electromagnetic spectrum for the detection of potential petroliferous zones; thus it is a branch of geophysics. Broadly speaking, uses of gravity, seismic, and magnetic energy might be included, but in view of the half-century of successful application of these force fields, they are fully operational and proven. As with other geophysical tools, remote sensing does not find petroleum directly but is being used to reduce the number of dry holes and consequently increase the efficiency of the exploration process. *See* PROSPECTING, PETROLEUM.

Ultraviolet signals. Very little interest has been demonstrated in the utilization of the ultraviolet sector of the electromagnetic spectrum. The signals here tend to be weak. They tend to be dispersed owing to their short wavelength, and with the exception of carbonate terranes, few surficial rocks yield strong reflectance in this region.

Visible sector. In the visible sector of the spectrum, two activities are coming into use in petroleum exploration. One is the use of aerial Ektachrome infrared film, and the second is the experimentation with multispectral photography. Some companies have determined that aerial Ektachrome infrared film is assisting their conventional field exploration by attracting attention to anomalous sedimentation patterns in sedimentary rocks. An example would be the earlier recognition and identification of local areas of turbidite facies in a sedimentary rock unit elsewhere interpreted as having a quiet-water depositional environment. These differential sedimentation patterns are usually not detected with other film emulsions. Data of this type can be generated inexpensively by utilizing commercially available 35-mm film and the lightest aircraft. Usually, oblique photographs are taken. Data of this type permit the field teams to go directly to the outcrop of interest. These photographs are also more valuable than other aerial photographs because the longer wavelengths of the infrared allow better penetration of most atmospheric degradation and may yield slightly increased resolution.

Multispectral photography. It is difficult to determine the extent petroleum companies are experimenting with multispectral photography. This technique utilizes photographs collected simultaneously in three or four narrow band-pass zones (approximately 100 nm) ranging from the blue to the infrared sectors (400–900 nm). Reconstitution of the color is possible by superposing projections of the photographs and placing the appropriate color filter in each projection path. Substitutes of randomly selected filters frequently yields "enhanced" images which may very well reveal details not seen in natural colors. Full understanding of this process requires the acquisition of additional basic data on the spectroreflectivity of sedimentary rocks, plus perhaps some psychological insights as to how or why interpreters can see more detail under false color patterns than normal color patterns.

Thermal infrared segment. The use of the thermal infrared segment (3.5–5.5 and 8–14 μm) for petroleum exploration presently appears to be lagging. Demonstration aircraft missions have shown the detection of data which have been interpreted as evidence of subsurface faulting and surface folding. These interpretations have been confirmed by ground studies. Thermal infrared surveys yield a heat map of the Earth's surface. They are best collected in predawn hours, when the previous day's solar radiation has been reradiated back into the atmosphere. Data cannot be collected from above clouds or through precipitation. Thermal infrared imagers are sufficiently reliable to meet commercial needs and the cost for infrared surveys is constantly being lowered, although it is still usually 2–3 times as expensive as aerial photography. Presently there is no record of operational exploration flights within the United States with this technique. Such flights as have been flown are generally classified as experimental and in most companies have been handled by the research and development teams in cooperation with the exploration department. Outside the United States some American companies have used this technique operationally, covering entire concessions, but the operations themselves are generally confidential and seldom acknowledged.

In a nonexploration activity, the companies have utilized infrared in determining the extent of oil dispersal in the widely publicized spills from tankers and offshore drilling sites.

SLAR. Petroleum companies have utilized SLAR operationally more frequently than any other type of remote sensing. They have found the SLAR image is generally clear and has sufficient resolution to be meaningful to their exploration programs. SLAR, because of its low illumination angle, tends to accentuate surficial geologic structures; yet the texture of the returns yields useable information concerning the surficial rock types. SLAR was apparently first used commercially during 1969 in Ecuador by a consortium of petroleum companies, and in 1970 was flown in the Republic of Indonesia for both petroleum and mining companies. The cost of SLAR imagery is currently well above the cost of all other remote-sensing activities, requiring large transport-type aircraft. While the system has the advantage of seeing through all kinds of weather, its present high cost should restrict it to those situations where deleterious weather precludes the use of other remote-sensing spectral bands.

Satellite data. Remote-sensing-oriented petroleum companies are anticipating the potential to be derived from satellite-generated data. A remote-sensing satellite at orbital altitudes would permit them to study sedimentary basins in their entirety. In June, 1970, a consortium of seven companies sought arrangements with NASA for the launching of a privately owned satellite for petroleum exploration. Presumably these negotiations are either terminated or proceeding extremely slowly. NASA's plans include the launching on Mar. 31, 1972, of the Earth Resources Technology Satellite (ERTS), the first with sensors to look at the surface of the earth rather than the atmosphere. *ERTS A* will offer multispectral sensors in the visible and nearinfrared sectors, collected by two techniques. *ERTS B*, a year later, will also offer data in the thermal infrared area. Several petroleum companies are keenly interested in this program and are expected to submit proposals to NASA concerning the interpretation of ERTS data for petroleum exploration if the government will make sensor data available to them. See EARTH-RESOURCE SATELLITES.

For background information see AERIAL PHOTOGRAPH; PHOTOGRAMMETRY; TERRAIN SENSING, REMOTE in the McGraw-Hill Encyclopedia of Science and Technology.

[JOSEPH LINTZ, JR.]

Bibliography: D. J. Barr and R. D. Miles, *Photogramm. Eng.*, 36(11):1155–1171, 1970; J. Lintz, Jr., *Amer. Assoc. Petrol. Geol. Bull.*, 54(5):857, 1970; H. C. MacDonald et al., in *6th International Symposium on Remote Sensing of Environment*, University of Michigan, 1969; E. W. Wolfe, *Photogramm. Eng.*, 37(1):43–48, 1971.

Transplantation biology

It seems clear today that all strong human transplantation antigens belong to two genetic systems, the ABO blood group system and the HL-A system (H=human, L=leukocyte, and A=the first genetic region described). Besides these strong antigens, there exist undoubtedly a substantial number of weaker transplantation antigens belonging to other genetic systems. Nothing is known about these weak systems, except that the immunological reactions induced by them seem to be controllable for years by the immunosuppressive therapy which is always used in human transplantations. An exception to this rule is a transplant performed between identical twins. However, immunosuppressive agents may even be used in this situation, but only in an attempt to "control" the original disease—not the possible rejection of the graft.

Any human graft should be ABO-compatible with the recipient, that is, the recipient must not contain ABO antibodies active against ABO antigens present in the donor. This a generally accepted rule in human transplantations. However, in any transplantation situation where there is a possibility for a graft versus host (GVH) reaction, such as in bone-marrow transplantation, the possibility of the donor organ producing antibodies against the recipient should also be considered.

The other major transplantation system in man is the HL-A system, which is an equivalent to the H2 system in mice and similar systems in other animals. The HL-A system is fairly well explored; however, present knowledge about this system has been achieved only recently and it is already known that it is the most complicated genetic system so far envisaged in man, so that a lot of the details are still left to be explored. This article will discuss the HL-A system.

It has been established that the genetic determinants for the HL-A antigens are localized on a rather narrow region of a single pair of autosomal chromosomes. It is still unknown how many loci the system contains, but only two seem to have major importance for human transplantations, because they code for the most important strong transplantation antigens. These two loci are called the LA (or first) and the FOUR (or second), respectively. There is a substantial number (the final number is still unknown) of genes (or alleles) belonging to each of the two loci. These genes are mutually exclusive, which implies that only one of the genes can be present on one of the chromosomes. This again implies that any individual can have, at most, two different LA and two different FOUR genes—a total of four different genes and corresponding antigens. There will be less than four determinants and antigens if there is homozygosity at one or both of the two loci involved. *See* CHROMOSOME MAPPING, HUMAN.

Antisera problem. The major problem in the determination of the HL-A types is the difficulty in finding reliable antisera reacting only with one antigen (monospecific reagents), since it seems to be a characteristic feature of HL-A antisera that they very often react not only with the antigen with which the antiserum donor was stimulated but also with other HL-A antigens "related" to the stimulating antigen. The most resonable explanation to this seems to be that different "clusters" of HL-A antigens have some antigenic determinants in common (although they are not identical). Sometimes stimulation with a given HL-A antigen may result in appearance of a monospecific reagent (antibody), reactive only with this specific antigen, but on other occasions the antiserum produced may have a much broader reactivity than expected since it may also react with a varying number of the "related" HL-A antigens. This phenomenon is also referred to as cross-reactivity between various HL-A antigens. Typically this may be revealed by an antiserum being nonreactive with a cell suspension having a given HL-A type; nevertheless, if this nonreactive cell suspension has an HL-A antigen which is cross-reactive with the stimulating antigen, it may absorb out all antibody activity from the antiserum, which becomes nonreactive even against cells from the stimulating donor.

This phenomenon has been, and still is, one of the most intriguing problems in HL-A serology, and it has to a great extent hampered the exact specification and characterization of the individual HL-A antigens, irrespective of the methods used for HL-A typing.

Table 1. Genes belonging to the LA (or first) HL-A locus

Previously recognized broad specificities	Narrow specificities	Comments and suspected specificities
	HL-A1	The Torino research group suspects splitting
Ba-8[a]	{HL-A2 Ba*(=W28=Da15)	
ILN	{HL-A3 HL-A11 (=ILN*)	
	HL-A9	The Paris group suspects splitting
HL-A10	{To31 To40	This splitting is supported by the Paris, Amsterdam, and Danish groups (To=Torino)
Li (=>W19)	{Thompson (<W19) LA-W (<W19) Ao28 (<W19)	Li is suspected to be split into two or three more narrow components
	"0"	As yet undetectable gene(s)
Total number of possible genes	11	
Total number of possible genes, including one unknown gene	12	

HL-A typing. The methods used today in HL-A typing are mainly lymphocytotoxicity tests, with isolated living lymphocytes as antigen or complement fixation tests, with platelets as antigen. The previously widely used leukoagglutination reaction is disappearing, mainly because it is too sensitive. A high sensitivity results in an increased risk of reactivity with cross-reactive antigens, which makes it quite difficult to find monospecific reagents while using the sensitive technique.

The principle in cytotoxicity and complement fixation tests (using lymphocytes and platelets, respectively, as antigens) is the same. When an antiserum contains an antibody active against HL-A antigen present on lymphocytes or platelets, the antibody combines with the antigen on the cell surface and simultaneously binds complement. In lymphocytotoxicity the complement lyses and kills the living cells, which may be shown in different ways but most frequently through the use of a dye exclusion test. Living lymphocytes do not take up dyes such as trypan blue and eosin, while these dyes readily penetrate the killed cells. When platelets are used as antigen, the complement consumption is shown by adding sensitized sheep red cells as in any other complement fixation test.

Present knowledge. Through the use of lymphocytotoxicity and complement fixation and through extensive international collaboration organized mainly through four international workshops in histocompatibility testing—at Durham, N.C.; Leiden, Holland; Turin (Torino), Italy; and Los Angeles, Calif.—a rather detailed knowledge about the HL-A system has been achieved, and today one can account for the vast majority of the genes and corresponding antigens belonging to the two loci. The present status appears in Tables 1 and 2. In these tables HL-A and a number is the World Health Organization (WHO) nomenclature. W and a number refer to the terminology used during the Los Angeles workshop. The remaining designations are the locally used identifications and are not officially accepted specificities.

It is apparent from Tables 1 and 2 that a number

Table 2. Genes belonging to the FOUR (or second) HL-A locus

Previously recognized broad specificities	Narrow specificities	Comments and suspected specificities
R (= Rafter = 4c)	HL-A5 R* (= W5)	The Scandinavian group suspects splitting into HL-A5–AJ and HL-A5*; it is suspected to be split into at least two components and may include CM*
7c (= HL-A7 + FJH)	HL-A7 (FJH) HL-A8 HL-A12	A shorter HL-A12 has been claimed by different groups (see TT)
BB (= W10)	HL-A13 BB (= W10) JA	The Danish group suspects splitting into two (or more) components; some BB antisera reacting with an antigen (JA) rather frequent in Eskimos
FJH (= W27)	FJH–AJ FJH*	The Scandinavian group has split antigens FJH, LND, and AA into two components utilizing the antiserum AJ
LND (= W15)	LND–AJ LND*	
AA (= W22 = Bt22)	AA–AJ AA* MaKi (= W14) ET*	
SL (> W17)	SL-ET SL-MaPi (= W17 = Orlina) SL-CM CM* 407	Defined with a single antiserum from Thulstrup, Copenhagen (low-frequency antigen)
> HL-A12	TT	Some HL-A12 antisera react also with the low-frequency TT antigen
	U18 (= Sa533)	A "new" FOUR antigen defined with a number of different Scandinavian antisera
	"0"	As yet undetectable gene(s)
Total number of possible genes	23	26
Total number of possible genes, including one unknown gene	24	27

of HL-A antigens, even some of the WHO specificities, are being split into two or more components, and that if all possible antigens and corresponding alleles belonging to the two series are taken into consideration, one can estimate that there must be at least 12 LA genes and 27 FOUR genes, considering only a single missing gene at each locus. From the mutually exclusive genes belonging to the two loci, it has been calculated that there can be 324 different haplotypes, 52,650 different genotypes, and 25,024 different phenotypes or tissue types.

It is also apparent from Tables 1 and 2 as marked with "0" that unknown genes and antigens belonging to both series do exist, although amorph genes (genes not expressing their product as HL-A antigens on the cell surface) are not supposed to exist. The gene frequencies of these missing genes in the Scandinavian population are 1–2% and 5% for the LA and FOUR series, respectively. It is, of course, pertinent to find antisera defining these "missing" genes.

Genetics. It is known from extensive family studies that the HL-A antigens are inherited in accordance with the two-loci concept, that is, in families the LA and FOUR antigens are inherited from the parents by the children either in coupling or in repulsion. In coupling, the same pairs of maternal or paternal LA and FOUR antigens are found in the children, since the corresponding genes are present on the same chromosome. In repulsion, the genetic determinants for the paternal or maternal LA and FOUR antigens are located on different chromosomes and do not appear together in any of the children (except when recombination occurs). The parental haplotypes (the genetic information present on one chromosome) can be followed from the parents to the children and further on to their children. Thus in any mating there are four HL-A haplotypes (two paternal and two maternal), and this implies that in any sibship there will be a 25% chance for HL-A identity, a 50% chance of two siblings sharing one haplotype, and a 25% chance for two siblings having none of the parental haplotypes in common.

Recombination between the LA and FOUR genes should be expected to occur during meiosis if they belong to separate loci. Extensive Scandinavian family materials involving 1362 parental meiotic divisions showed that recombination had occurred in 11 cases, which gives a recombination fraction of 0.8%. This is quite similar to the recombination frequency within the H2 system in mice.

Linkage between the HL-A system and the phosphoglucomutase (PGM_3) system has recently been demonstrated, and this is the tenth human linkage so far demonstrated with reasonable certainty. Linkage between HL-A and haptoglobin (Hp) has been postulated by the Durham group, but it is has not as yet been confirmed by other groups. A possible linkage between HL-A and haptoglobin is of great interest since it is already known that Hp is located on chromosome 16.

Mixed lymphocyte culture test. The mixed lymphocyte culture (MLC) test is a laboratory measure of the immune response of living lymphocytes when confronted with allogeneic (generally mitomycin-treated) lymphocytes (that is, a one-way culture) assayed by the degree of DNA synthesis, mostly using the cell uptake of radioactive thymidine. As has been clearly demonstrated in family studies there is a close association between the HL-A system and the results of MLC tests. MLC tests between HL-A-identical siblings result in no stimulation, while MLC tests performed between nonidentical siblings always result in stimulation. MLC tests between siblings differing for two haplotypes generally result in a higher stimulation than those differing for only one haplotype.

An association between the MLC test and the HL-A system has been much more difficult to demonstrate when cultures are performed with lymphocytes from unrelated individuals. In contrast to the results with HL-A-identical siblings, MLC tests performed between HL-A-identical unrelated individuals having the same two LA and the same two FOUR antigens ("full-house" donors) result mostly in some, although generally rather low, stimulation and, in a few cases, no stimulation.

This is an intriguing problem, but when it is considered in relation to the universal nonstimulation when MLC tests are performed between HL-A-identical siblings, the most probable explanations may be either that further loci linked with LA and FOUR are of importance for MLC tests, or that some of the HL-A antigens, as defined today, are heterogeneous. Both explanations would explain nonstimulation in the HL-A-identical sibling situation and stimulation when cultures are performed between HL-A-identical unrelated individuals.

Clinical transplantation. The importance of the HL-A system for clinical transplantations has been unequivocally shown by the universally good results obtained in kidney transplants performed between more than 100 HL-A-identical siblings, and recently also by an increasing number of successful transplants performed with HL-A-identical ("full-house" donor) cadaver kidneys. However, with the use of cadaver donors, it has been more difficult than in the related donor situation to show a correlation between the degree of HL-A mismatch (number of antigens mismatched) and the clinical outcome. This is, of course, of pertinent interest when donors are selected for the recipients in the large international kidney exchange programs. An early 1971 analysis of the results obtained in "Scandiatransplant" has shown a highly significant correlation between graft survival and the number of HL-A antigens mismatched. The analysis included 216 consecutive cadaver transplants, where 130 recipients were transplanted with C-matched (one antigen mismatched) and 86 with D-matched (two antigens mismatched) kidneys. The match grade used was "the worst possible" match; that is, when there was a possibility for incompatibility for an as yet undetectable

Table 3. Scandiatransplant correlation between C and D matches (worst possible matches) and graft survival in 216 consecutive cadaver transplants*

Match grade	Number of grafts	Graft survival	Graft rejection
C	130	94	36
D	86	42	44
Total	216	136	80

*$\chi^2 = 11,241$ (Yates' correction); $p < 0.001$.

antigen, this was counted as an incompatibility and added to the incompatibilities actually found by the HL-A typing of recipient and donor. The results appear in Table 3.

Many other results point in the same direction. Some of the difficulties involved in establishing a clearer correlation between HL-A matching and the results of kidney transplants, especially using unrelated donors, may be caused by one or more of the following problems: (1) The polymorphism of the HL-A system is very great, which influences typing more in the unrelated than the related donor situation—especially when typing is performed only for a limited number of antigens. (2) Some of the HL-A antigens, as presently defined, may be heterogeneous (as discussed for the MLC test), and this will also make HL-A matching less reliable in the unrelated than in the related situation. (3) Additional loci other than LA and FOUR may exist, which again will have greatest influence in the unrelated situation. (4) Immunosuppressive therapy is becoming increasingly more effective, and this will make it increasingly difficult to establish the importance of HL-A matching. (5) The "immune response" may vary from one individual to another, and this may also blur the picture.

For background information see TRANSPLANTATION BIOLOGY in the McGraw-Hill Encyclopedia of Science and Technology. [F. KISSMEYER-NIELSEN]

Bibliography: V. C. Joysey, *Int. Rev. Exp. Pathol*, 9:223–285, 1970; F. Kissmeyer-Nielsen et al., *Transplant. Proc.*, 3:1019–1029, 1971; F. Kissmeyer-Nielsen et al., *Tissue Antigens*, 1:74–80, 1971; F. Kissmeyer-Nielsen and E. Thorsby, *Transplant. Rev.*, 4:1–176, 1970; A. Svejgaard et al., *Vox Sang.*, 18:97–133, 1970: E. Thorsby et al., *Tissue Antigens*, 1:32–39, 1971.

Transportation engineering

Since World War II the urban population of the United States has doubled to 149,000,000, and both the nation and the average individual have reached a high level of affluence. The transportation industry has been drastically affected, as shown by statistics on travel in the United States (see table).

Highways. Conscious of the trends in automobile ownership and use, the Federal and state governments initiated a massive program of highway construction and improvement in the 1950s. The most dramatic element was a Federal system of interstate and defense highways composed of 42,500 mi (68,400 km) of grade separated, limited access, divided roads. Projections of travel volumes for the future indicate that major additions to the highway system will be required, particularly in urbanized areas.

These modern highways and those of lower design standards directly benefited the public through greater safety and comfort with reduced travel time and cost and through faster and less costly distribution of goods, particularly food. Nevertheless, adverse reactions appeared in the late 1960s relative to community and environmental impacts from some of the completed and proposed major urban highways. Research and development efforts were directed toward various methods of increasing capacity of existing highways, therefore, such as computer-controlled and coordinated traffic signal systems and metering of traffic entering congested freeways.

Public transportation in cities. Urban public transit services suffered a steady decline in use during 1950–1970 due to population shifts, deterioration of service, and increasing ownership and use of automobiles. Interest grew after 1960 in improving urban public transportation to provide an acceptable alternative to automobiles. Major cities, notably San Francisco and Washington, D.C., are engaged in development of wholly new rapid transit systems, while other cities with rapid transit systems have made or are planning substantial additions. Many other cities are moving toward improvements in bus service.

Intercity travel. The airlines, through constant improvements in aircraft and service, have captured a substantial share of intercity travel since 1945 by offering fast, comfortable, and prestigious service at costs competitive with railroads, intercity buses, and automobiles.

The National Rail Passenger Corporation (AMTRAK) was created by Congress in 1970 in response to popular demands for improvements in rail passenger service between key cities. The AMTRAK system, in operation since May 1, 1971, over selected routes with conventional railroad passenger equipment, will initially serve only those metropolitan areas offering highest patronage potential. It is hoped that the initial AMTRAK service will attract new patronage and revenue that will justify further improvements and possibly expansion of intercity service.

Transportation innovations. Inventors, manufacturers, universities, governmental agencies, and others have proposed many new systems, vehicles, and techniques for improving transportation of people. In general, objectives have been to reduce operating expenses, travel time, traffic congestion, and air pollution. Several of the de-

Statistics for travel in the United States

Parameter	1940	1968	Increase or decrease, %
Vehicle miles by automobiles, $\times 10^9$	249	814 (estimated)	227
Domestic air passenger miles, $\times 10^9$	1.0	87.5	8750
Intercity rail passenger miles, $\times 10^9$	19.7	8.7	−56
Urban transit revenue passengers, $\times 10^9$	10.5	6.5	−38

vices are merely concepts, others are in prototype stage, and still others are far enough advanced to permit their early use. Specific innovations under study or being applied are described in the following.

Urban systems. Development of extensive urban systems of automated and individually routed personal vehicles has been restrained by serious questions related to cost, capacity, operating feasibility, safety, and availability of funds. Interest has been displayed, however, in limited applications to serve heavy concentrations of person movements such as at airports and universities, where trips would otherwise be made on foot. Moving pedestrian walkways are in use at several major air terminals, including Dallas, Montreal, and San Francisco. These all use the end loading technique, and belt speeds are in the range of 1.5 to 2 mph (2.4 to 3.2 km/hr).

Schemes for some belt systems are predicated on higher belt speeds between stations with low-speed belts at stations for passenger boarding. None has yet been installed due to space, cost, and potential accident hazards. Other proposals would use "window shade" belt designs, with the belt contracted at low speed in station areas and expanded at higher speeds between stations. Individual belt-mounted frames or boxes would provide passenger support.

The Goodyear Carveyor concept, exemplified in the Disneyland People Mover (Fig. 1), would provide individual or coupled passive four-passenger car units propelled by powered rollers mounted in a guideway. The passenger would board the slowly moving vehicle from a platform or belt moving at the same speed, after which the vehicle would be accelerated to a speed on the order of 7 mph (11.3 km/hr) by variable-speed rollers, returning to unloading speed at each successive station. Figure 2 shows a proposed use of this concept in a mall in the center of a city.

Tampa, Fla., and Seattle, Wash., airports are installing versions of the automated vehicle system (Skybus) developed by the Westinghouse Electric Co. and demonstrated at South Park, Pittsburgh, Pa. At Tampa a central core area will be connected to each of four plane-side satellite buildings by 1000 ft (304.8 m) double-track shuttles, using 100-passenger vehicles. Up to 840 passengers will be moved in 10 min, it is claimed. At Seattle similar vehicles will interconnect airline terminals on two one-way loops and a shuttle, the system being intended to handle 1200 one-way passengers in 5 min. During peak movements, all stations will be served; in off-peak periods, the vehicles will operate "on call." The vehicles will be electric-powered and rubber-tired, with guidance afforded by horizontal wheels below the vehicle floor engaging a center-mounted guideway rail.

Plans for the Dallas–Fort Worth Regional Airport include provision of an intraairport small-vehicle system. Two systems are being considered, the Dashaveyor and the Varo Monocab. The Dashaveyor vehicle, as evaluated in other studies, would have capacity for 30 passengers and would be bottom-supported on pneumatic tires. The Monocab vehicle, as previously considered, would accommodate six passengers and would be sym-

Fig. 1. Disneyland People Mover application of Goodyear Carveyor principle.

metrically supported from an overhead monorail. The propulsion, guidance, communications, and automatic controls would be above the cab. Service is proposed to be "on call" from station positions.

A small-vehicle system is being planned at the University of West Virginia, Morgantown, using the Alden StaRRcar concept. A fleet of small electrically-powered, rubber-tired vehicles will operate under computer control, each with point-to-point service with no intermediate stops.

In general, small-vehicle systems do not present insuperable technical problems. However, such

Fig. 2. Proposed installation of Goodyear Carveyor in central city mall.

Fig. 3. Model of French Aerotrain tracked air-cushion vehicle.

systems usually contemplate automatic computer-controlled operation with very close vehicle spacing to approximate automobile standards of comfort and availability and to develop required capacities. Application has been delayed by insufficient in-service experience to prove that such devices could be operated safely as public carriers offering service to people of all ages, the partially incapacitated, the illiterate, and those with little knowledge of English. In addition, each system is insufficiently developed to demonstrate that automatic operation of guideway switches with close vehicle spacing would be feasible and reliable in branched systems. The proposed installation, hopefully, will develop solutions of such problems. Other proposed small-vehicle systems include several types of air-cushion-supported passenger cubicles, notably the Uniflow, Transportation Technology, Inc., and PitKanen systems. These had not reached application stage by May, 1971.

High-speed ground transportation. U.S. Government agencies initiated research and development of high-speed ground transport in the mid-1960s to provide a competitive alternative to highways and airlines, already overburdened in some metropolitan corridors. Desired speeds for medium distances were stated to be in the 300 mph (500 km/hr) range.

The Metro rail service between Washington, D.C., and New York City, a government-industry venture, uses vehicles designed for 160 mph (257 km/hr) peak speeds. Due to existing track alignment and grade crossing restrictions, operating speeds do not achieve this maximum.

The Japanese National Railroad operates its Tokaido line at peak speeds in the 125 mph (200 km/hr) range and is designing an extension for maximum speeds of 155 mph (250 km/hr). Studies by the Japanese National Railroad indicate that practical maximum speeds for steel-wheeled passenger vehicle operation on conventional high-type railroad tracks are in the 155–175 mph (250–280 km/hr) range, considering passenger comfort, wheel-rail adhesion, and power collection aspects.

Development work by English and French government agencies and industrial firms has shown potential for high-speed operation with a tracked air-cushion vehicle (TACV), eliminating the wheel and its attendant friction, noise, and vibration problems. The English Hovertrain and French Aerotrain are in test phases. One Aerotrain test vehicle is an 80-passenger gas turbine–powered unit, operating at speeds up to 180 mph (290 km/hr) on an elevated test track at Orleans, France (Fig. 3). The vehicle is supported on a 0.75 in. cushion of air from vehicle-mounted fans, with a uniformly distributed air pressure at guideway surface of less than 1 psi (33.5 N/m^2) (Fig. 4). The guideway is a concrete inverted T, supported on piers at 65 ft (19.8 m) intervals on average, and constructed to precise line and grade for smooth high-speed operation. Vertical air cushions in the vehicle center slot prevent lateral contact with the guideway.

Conventional rotary electric motors have been considered infeasible for high-speed vehicle operation until recently; test vehicles have been powered by aircraft-type jet and propeller systems. Attention has been focused on the linear induction motor (LIM) as a propulsion system for TACV. The LIM resembles an "unrolled" conventional electric motor with one element forming either a primary or secondary winding the full length of the trackway, and the other element fixed in the vehicle. The electromagnetic field set up by the primary winding pulls the TACV along the trackway similar to the way it turns a conventional rotor. Efficiency and power factors are indicated to improve with speed. Operational testing of the LIM is needed to identify and solve problems. These include reliability under adverse climatic conditions and maintenance of a uniform gap between the track and vehicle-mounted LIM elements. The U.S. Department of Transportation is constructing a 6 mi (9.7 km) conventional test track near Pueblo, Colo., for testing the LIM at high speeds. Additional facilities at the site are contemplated to test TACV vehicles.

For background information see TRANSPORTATION ENGINEERING in the McGraw-Hill Encyclopedia of Science and Technology.

[ROBERT B. RICHARDS]

Bibliography: Collection and Distribution Systems: Technical Reviews of Six Baseline Defini-

Fig. 4. French Aerotrain tracked air-cushion vehicle in operation on aerial guideway.

tions, Johns Hopkins University–Applied Physics Laboratory, June, 1970; B. Richards, *New Movements in Cities,* 1966; *Tomorrow's Transportation: New Systems for the Urban Future,* U. S. Department of Housing and Urban Development, 1968; *Transportation Within Airports,* ASCE Committee on Terminal Facilities, April, 1967.

Triassic

The most important recent discoveries in rocks of Triassic age have been in Antarctica (Fig. 1). In 1967, the lower jaw of a labyrinthodont amphibian was found there. In 1969 a large assemblage of fossil reptiles and amphibians was collected, among which one jaw was identified as *Lystrosaurus,* a mammallike reptile also found in the Triassic rocks of South Africa, India, and China. In 1970 complete skeletons of several reptiles were excavated. All these antarctic finds were in the Fremouw Formation of the Lower Triassic in the general vicinity of the Beardmore Glacier. These fossils are the strongest paleontological evidence found so far to prove that Antarctica was, during Triassic time, connected to the southern continents, especially to Africa, and that it drifted to its present position at a later time. Thus this evidence gives strong support to the theory of continental drift. *See* TECTONOPHYSICS.

Graphite Peak discovery. In December, 1967, Peter Barrett, a New Zealander working under the auspices of the Institute of Polar Studies at Ohio State University, found what he thought was a bone in the Fremouw Formation at Graphite Peak at 85°3.3′S latitude, 172°19′E longitude in the Transantarctic Mountains of Antarctica. This fragment was identified by Edwin H. Colbert of the American Museum of Natural History in New York as the back portion of the lower jaw of a labyrinthodont amphibian. This was the first fossil record of a four-legged land-living animal in Antarctica.

Triassic amphibians and reptiles have long been known from the other southern continents, and it had been postulated by Lester King in 1965 that Triassic vertebrate fossils would eventually be found in Antarctica. Therefore this discovery added another facet to the already striking similarity in fossils and rock types of the upper Paleozoic and Mesozoic geology in the southern continents.

Coalsack Bluff find. Owing to the importance of the Graphite Peak amphibian, a team of vertebrate paleontologists led by Colbert, was sent to the Beardmore Glacier area of Antarctica during the austral summer of 1969–1970. In addition to Colbert, this group consisted of James Jensen from Brigham Young University in Utah, William Breed of the Museum of Northern Arizona in Flagstaff, and Jon Powell from the University of Arizona in Tucson. A field camp was set up near Coalsack Bluff by the Institute of Polar Studies, Ohio State University. On the first day of field work, bone fragments were discovered on the southwestern slopes of Coalsack Bluff in the middle bed of the three successive sandstone units. At this locality and at another locality 1/2 mi (0.8 km) to the northwest, 450 specimens were excavated. Although many of these specimens were fragmentary, one jaw fragment was identified by Colbert as the mammallike reptile *Lystrosaurus* (Figs. 2 and 3). This reptile is also found very commonly in South

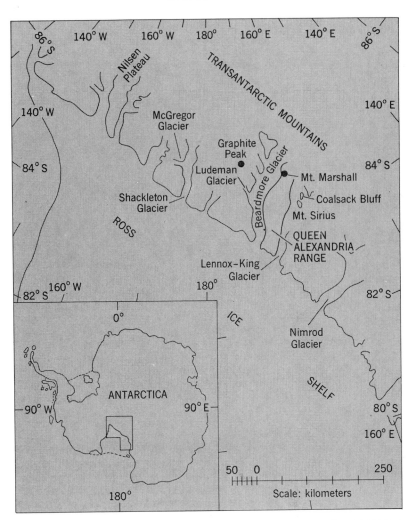

Fig. 1. Map of Antarctica showing location of Triassic discoveries. (*From D. H. Elliot et al., Triassic tetrapods from Antarctica: Evidence for continental drift, Science, 169: 1197–1201, Sept. 18, 1970*)

Fig. 2. Jaw fragment of *Lystrosaurus* from Coalsack Bluff.

Africa, India, and China. Other fossils found at Coalsack Bluff include labyrinthodont amphibians, as well as thecodont and therapsid reptiles.

McGregor Glacier discoveries. Although plans were made for the 1969–1970 season to set up a camp at McGregor Glacier, an area east of Graphite Peak, bad weather and helicopter trouble forced postponement until the 1970–1971 season. Three vertebrate paleontologists went to the Antarctic that season: James W. Kitching of the Bernard Price Institute of Paleontology, University of Witwatersrand, Johannesburg, South Africa, assisted by John Ruben of the University of California at Berkely and, during the early part of the season, by Thomas Rich of Columbia University, New York. Again, on the first day of field work important finds were made. The initial find of the 1970–1971 season was the imprint of a complete skeleton of the carnivorous mammallike reptile *Thrinaxodon* (Figs. 4 and 5). Subsequent field collections included not only more thecodonts and labyrinthodont amphibians such as were known from previous years but also complete skeletons of *Thrinaxodon* and another small reptile, *Procolophon*. Other small lizardlike reptiles known as eosuchians were also found in abundance.

Significance of discoveries. The presence in Antarctica of fossil reptiles and amphibians of Early Triassic age which are related to similar forms in South Africa can only be explained in two ways: (1) There was a land bridge between Antarctica and South Africa, or (2) the continental mass of Antarctica was once directly connected with South Africa. The first of these alternatives was the one most favorably considered by geologists in the past for explaining faunal similarities between continents now separated by oceans. However, there has never been any evidence to support that hypothesis. On the other hand, evidence for the second alternative, which implies continental drift, has been abundant from various other lines of geological and geophysical evidence.

Similarities between the Triassic faunas of South Africa and South America have been known for many years. These similarities could be explained by supposing the reptiles to have spread by way of land connections in the Northern Hemisphere. However, in the case of Antarctica and South Africa there is no "long way around." There had to be land connections for *Lystrosaurus* and *Thrinaxodon* to migrate between Antarctica and South Africa. The best explanation at present is that these land masses were once contiguous and are now separated by 800 mi (1300 km) of ocean as the result of continental drift since Triassic time.

Other developments. One of the first true lizards ever found in the fossil record has been described recently from the Triassic of northern New Jersey. This lizard, *Icarosaurus*, is closely related to *Kuehneosaurus* from the Upper Keuper of Great Britian. Both lizards are unique for the enormous elongation of the ribs on both sides—an obvious adaptation for gliding. In this respect they are sim-

Fig. 3. Reconstruction of *Lystrosaurus*.

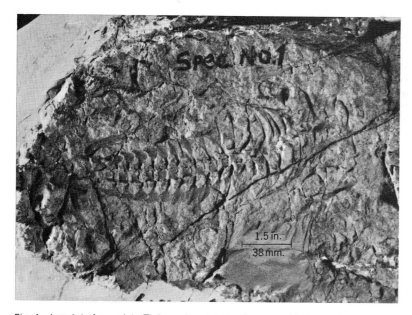

Fig. 4. Imprint of complete *Thrinaxodon* skeleton from near McGregor Glacier.

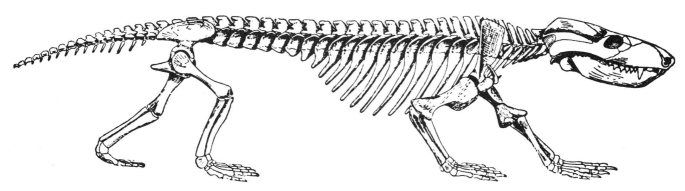

Fig. 5. Reconstruction of *Thrinaxodon*.

ilar to the Asian gliding lizard *Draco*, but there is no close relationship between these two forms. It is surprising to find these Triassic lizards so highly adapted for a specialized mode of life.

A new dinosaur has been described from the Triassic Kayenta Formation of northern Arizona. This reptile is known from two skeletons: one excavated in 1942 and the other in 1964. In preparing the second specimen a highly unusual double crest was found on the skull. Because of this new information, the dinosaur is no longer considered to be *Megalosaurus wetherilli*, but the new genus *Dilophosaurus*. Although S. P. Welles, who discovered and named the dinosaur, considers it to be a Jurassic form, most other geologists and paleontologists, on the basis of much evidence, consider the Kayenta Formation to be Triassic.

Samples of the Lower Triassic Moenkopi Formation of western Colorado taken at intervals of 9–12 in. (228.6–304.8 mm) have been analyzed recently by Charles Helsey from the University of Texas at Dallas. An investigation of the magnetic field of these samples has revealed that there were at least eleven reversals during this interval. This change in the magnetic field from normal (with the north and south magnetic poles in their present position) to reversed (with north and south poles opposite from present) is at present poorly understood, but it is hoped in the future that these reversals might be useful for regional correlations.

For background information *see* AMPHIBIA; ANTARCTICA; CONTINENT FORMATION; REPTILIA; TRIASSIC in the McGraw-Hill Encyclopedia of Science and Technology. [WILLIAM J. BREED]

Bibliography: P. J. Barrett, R. J. Baillie, and E. H. Colbert, *Science*, 161:460–462, 1968; E. H. Colbert, *Bull. Amer. Nat. Hist.*, 143(2):85–142, 1970; D. H. Elliot et al., *Science*, 169:1197–1201, Sept. 18, 1970; C. E. Helsley, *Geol. Soc. Amer. Bull.*, 80(12):2431–2450, 1969; L. C. King, *Geol. Soc. S. Afr. Ann.*, 68:1–32, 1965; S. P. Welles, *J. Paleontol.*, 44(5):989, 1970.

Venus

Definitive evidence that the surface of Venus is seething with hot gases was provided on Dec. 15, 1970, by the epic flight of *Venera 7* into the atmosphere of Earth's nearest planetary neighbor. The unmanned Soviet spacecraft descended through the impenetrable visible clouds of Venus, through its dense lower atmosphere, and landed on the planetary surface, where it continued to transmit temperature measurements for 20 min. This is the first successful acquisition of data from the surface of another planet.

Temperature and pressure data. Even before the first direct measurements of the atmosphere of Venus in October, 1967, by *Venera 4* and *Mariner 5*, it had been possible to infer conditions near the surface of Venus from ground-based radioastronomical data. Carl Sagan and colleagues had shown that there were serious difficulties with all explanations of Venus radiowave emission other than emission from a hot surface, and deduced surface temperatures of about 700K and surface pressures of about 50 bars (50×10^5 N/m²), which is 50 times the surface atmospheric pressure on the Earth. The most recent spacecraft data imply, according to V. S. Avduevsky, M. Y. Marov, and colleagues, a surface pressure of about 90 ± 15 bars ($90 \times 10^5 \pm 15 \times 10^5$ N/m²) and a surface temperature of 750 ± 20K. The data are consistent with the ground-based radar radius of Venus of 6050 ± 5 km. Throughout the lower atmosphere to the surface, the atmosphere is close to convective equilibrium; that is, a rising bubble of hot air and a falling bubble of cold air are not stabilized in their journeys by the atmosphere but continue moving. The temperature is observed to be close to the constant adiabatic lapse rate of about 8.6K/km predicted for a Venus atmosphere in convective equilibrium. There is no detectable isothermal region near the surface. An apparently significant and yet unexplained departure from this linear temperature gradient was observed by *Venera 7* between 27 and 5 km above the surface.

The most recent ground-based radioastronomical results show, contrary to earlier reports, negligibly small temperature differences at the planetary surface, between equator and pole, and between day and night hemispheres. Thus surface winds driven by temperature differences are weak; according to G. S. Golitsyn, they are no stronger than 1 m/sec.

Chemical composition. *Venera 4*, *5*, and *6* all appear to have been crushed by the weight of the overlying Venus atmosphere before reaching the surface. To prevent such a recurrence, the entry capsule of *Venera 7* was constructed about 100 kg heavier in its structural support than were its predecessors. Consequently there was no room for a chemical composition experiment on *Venera 7*, as on previous Venera spacecraft. The chemical composition sensors on *Venera 4*, *5*, and *6* showed the Venus atmosphere to be composed of 97(+3, −4)% carbon dioxide, in agreement with ground-based spectra. The nitrogen abundance is less than 2% and the oxygen abundance less than 0.1% according to these sensors.

The water vapor abundance was determined by A. P. Vinogradov and colleagues on *Venera 4*, *5*, and *6* in two different ways. In one, the dessicant calcium chloride, $CaCl_2$, was exposed to a captured volume of Venus atmosphere, and the resulting decline in pressure was used to determine the water vapor content. In the other, phosphorous pentoxide, P_2O_5, was exposed to the atmosphere of Venus, and the resulting change in electrical resistance of the pentoxide was employed to estimate the water vapor abundance. All measurements were made beneath the visible clouds of Venus, a region in which the water vapor mixing ratio is expected to be roughly independent of altitude. The calcium chloride sensors gave water vapor mixing ratios of about 0.1%; the phosphorous pentoxide sensors gave results somewhat smaller.

These results should be accepted with some caution: Scientists do not know how other known minor constituents of the Venus atmosphere—for example, HCl and HF—might affect the measurements; and they do not yet understand the reported variation of water vapor mixing ratio with altitude. Nevertheless the Venera water results are not in contradiction with the much lower values for the water vapor abundance determined by ground-based infrared spectroscopy. The latter measurements are strongly weighted towards the region near the tops of the visible clouds, where the water vapor abundance is determined by the vapor pressure curve and where the air at low temperature may not even be saturated. Something like 0.1% mixing ratio of water in the lower atmosphere of Venus appears necessary to explain the radio and radar spectra of that planet. In addition, approximately 0.1% of water vapor, added to the known quantity of carbon dioxide in the Venus atmosphere, appears adequate to explain the high surface temperatures as being due to the greenhouse effect.

If there is 0.1% of water vapor below the clouds, the water vapor will condense as ice crystals at higher altitudes—forcing there to be a minimum of one cloud layer composed at least in significant part of ice crystals. Attempts to search for hexagonal ice crystals in the clouds of Venus through the 23° refraction "halo" have led to equivocal results. The interpretation of ground-based infrared spectra now suggests strongly the presence of at least two cloud layers, as does more recent reduction of the *Mariner 5* radio occultation experiments. There is no general agreement on the composition of these clouds, and a wide range of mutually exclusive constituents has been proposed: condensed water, HCl solutions, various oxides and silicates of abundant elements, ammonium chloride, halides and other compounds of the element mercury, carbon suboxide, polywater, and ferrous chloride dihydrate. Further progress on this question is clearly needed.

Observations of the upper atmosphere of the sunlit side of Venus by the ultraviolet photometer aboard *Mariner 5* imply temperatures in the Venus exosphere (the level from which escape of atoms to space can occur) of between 300 and 600K. The corresponding value for the Earth's exosphere is about 1500K or larger. Thus it is extremely unlikely that any gas heavier than helium can be significantly depleted from the Venus atmosphere over geological time, if the present exospheric conditions are typical. Accordingly, if at the time of its formation Venus had a significantly larger amount of water vapor than it does now, the water could not have been lost by ultraviolet photodissociation and the escape of hydrogen to space, unless the resulting oxygen thoroughly reacted with the Venus surface.

Topography. The topography of the Venus surface is beginning to emerge with the use of radar delay Doppler interferometry. Preliminary results are suggestive of large mountain ranges and impact craters. But such studies are in their very early stages. The postlanding signal characteristics from *Venera 7* are inconsistent with the strength of a solid or a liquid surface, but are consistent with the properties of powders. The radar reflectivity of Venus at long wavelengths also agrees with powders, and much of the surface of Venus may be a roasting desert.

Radar studies of surface features on Venus have shown the planet to be rotating in a retrograde sense with a sidereal period of about 243.16 days. Ultraviolet photographs show features in the Venus clouds to be moving around the planet, also in a retrograde sense, but with a period of only about 4 days. Some optical Doppler spectroscopic confirmation of this 4-day figure is now emerging. The nature of the atmospheric dynamics in this rapidly moving region near the upper visible clouds is under intensive theoretical investigation. The relative velocity between this layer and the ground is ~ 100 m/sec.

For background information *see* GREENHOUSE EFFECT, TERRESTRIAL; SPACE PROBE; VENUS in the McGraw-Hill Encyclopedia of Science and Technology. [CARL SAGAN]

Bibliography: V. S. Avduevsky et al., *J. Atmos. Sci.*, 28:263, 1971; P. J. Gierasch, *Icarus*, 13:25, 1970; G. S. Golitsyn, *Icarus*, 13:1, 1970; C. Sagan, *Comments Astrophys. Space Phys.*, 1:94, 1969; C. Sagan, T. C. Owen, and H. J. Smith (eds.), Planetary atmospheres, *Proceedings of International Astronomical Union Symposium #40*, 1971.

Video recording and playback

The 1970s will see the emergence of video record and playback systems for home and other nonprofessional uses. The ability to record, rent, or purchase recordings for playback by way of a home television receiver at the convenience of the viewer will allow television to realize much of its potential for education and information dissemination, as well as entertainment.

A practical video playback system depends on its ability to store and reproduce television images reliably and at low cost. A 1-hr program will contain over 100,000 television image frames. In order to make such a system economically feasible, it has been conjectured that the price to the consumer per television frame must be in the range of 0.01 to 0.05 cents. In the past 2 years the television industries of the world have announced that systems are under development. Because of the great interest of program producers and the consumer, there is much public discussion of such systems in advance of full fruition of the technical developments.

As indicated in the table, there are at present four major types of video playback systems, one of which offers a recording capability. The current popularity of color television has led all playback system developers to promise at least an option for color playback. Of the four systems, two handle the stored video information in the one-dimensional form of the television video signal. The other two handle the information as a two-dimensional image. The four systems are also divided as to the type of storage medium. Two use a medium consisting of a substrate coated with an appropriate active layer. The two other systems use a homogeneous storage medium with the information carried as modulations of the surface contour. Those using the homogeneous medium offer the potential of lower program replication cost. In balance, however, those systems using the complex storage media are currently more highly developed, since the information readout is generally less difficult.

VTR. The video magnetic tape recorder (VTR) is the best known video record and playback system. It has been in use for over 15 years in television broadcast operations. David Sarnoff spoke of the potential for "electronic photography" in the middle 1950s. He envisioned the combination of a compact television camera with a tape recorder as a substitute for silver halide amateur photography. Such systems have been available in reasonably

Major types of video playback systems

Video playback system	Storage medium	Storage format
Video tape recorder (VTR)	Magnetic tape	Video signal
Video disk	Homogeneous plastic	Video signal
Electronic Video Recording (EVR)	Silver halide tape	Image
SelectaVision holographic tape	Homogeneous plastic	Image

compact form for a few years but are not comparable in price or performance with color photography. However, there are many applications where the immediate playback feature of the VTR has made its use indispensable. Figure 1 shows a portable VTR used with a compact vidicon camera.

A large number of firms throughout the world are working on nonprofessional VTR systems to improve the quality and lower the cost. Among these are Ampex Corp. and Avco Corp. in the United States, Matsushita and Sony in Japan, and Philips in Europe. These systems vary widely in the approach to the form of the tape storage cassette and the mechanics of getting the tape from the cassette into the recorder proper. However, the principles of operation can be explained without attention to the details of tape handling.

The principal differences between an audio-frequency tape recorder and a video tape recorder are all traceable to the considerable difference in the frequency ranges of the signal involved. The hundredfold increase in bandwidth requires that the linear speed of the tape in a video recorder be increased from the neighborhood of 3 to 200–500 in./sec (7.6 to 500–1270 cm/sec). In order to avoid having the tape and reels moving at this impractical speed, video tape recorders use a rotating "head wheel" which scans the slow-moving tape at the required high speed. This head wheel contains the record or playback transducers and some means for electrical connections to the rest of the recorder. The wheel turns at a carefully controlled speed related to the television frame rate. The tape is wrapped around the rotating wheel so that contact is maintained between the transducer and the tape for the duration of one television field. The tape is moved forward at a slow speed, in the range 3 to 7 in/sec (7.6 to 17.8 cm/sec). The result is a series of parallel strip recordings, each a television field in duration. By proper choice of the tape speed, wheel rotational speed, and the angle of inclination of the wheel with respect to the tape, the entire width of the tape can be filled. Typically the low-cost VTRs use tape widths of 0.5 or 0.75 in. (12.7 or 19 mm).

The cost of unrecorded magnetic tape and high mechanical accuracy required in the player may limit the market for the VTR as a home video record and playback system. A further problem exists in making low-cost, high-quality program copies for mass distribution of prerecorded material.

Teldec Video Disc. The Teldec Video Disc system has attacked the problem of program replication cost by using a homogeneous plastic storage medium in a format resembling the ordinary phonograph record. The required information density is achieved by using a groove pitch of several thousand grooves per inch and a record rotational rate of 1500 rpm, corresponding to one rotation per frame in the European 25 frames per second television standards.

The recordings are made in the same manner as audio records but with the cutting stylus moving in a vertical direction in response to the video signal. In order to use nearly conventional audio recording equipment, the video signal is generated at a rate substantially below standard television rates and recording is done with the master record turning at a similarly scaled rate.

The finished master record consists of finely spaced grooves containing a replica of the video signal in the form of vertical undulations in the record surface. The record is then converted into a replicating master by plating a heavy metal surface on the record, and this metal "stamper" is

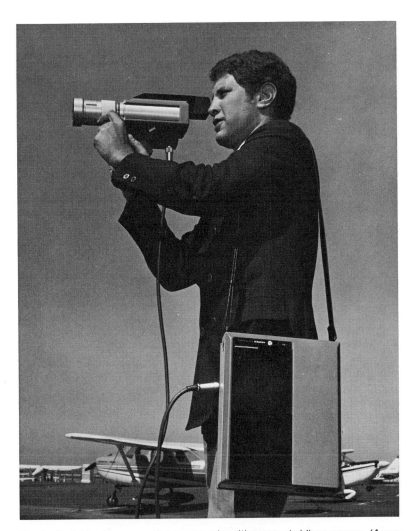

Fig. 1. Portable videomagnetic tape recorder with compact vidicon camera. (*Ampex Corp.*)

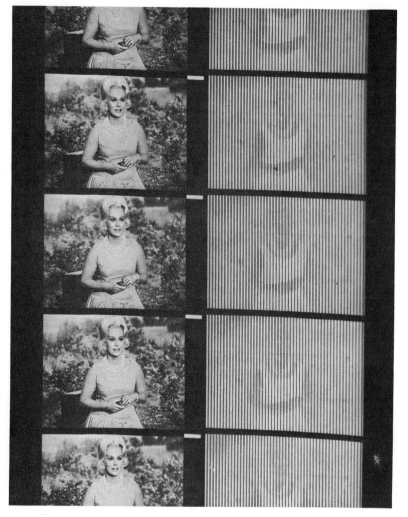

Fig. 2. An enlargement of the Electronic Video Recording system film. At left is luminance image, and at right chrominance image.

monochrome silver halide photographic film to produce full-color television recordings. The recording system converts the video signal into an image by direct electron exposure of a silver halide photographic emulsion. The apparatus, operated in a high vacuum, consists of a film transport and an electron gun with means for modulating and scanning the beam as in a conventional television picture tube. The silver halide emulsion is so highly sensitive to the energetic electron beam that an extremely fine-grain emulsion can be used. The use of direct electron exposure also frees the system from the limitations of light optics and light scattering within the emulsion. It is possible to achieve good television system performance in an EVR image 0.1 in. (2.5 mm) high.

Since the recording system is quite complex, the system is offered to the consumer only as a playback system. The small image size results in a lower film cost than either 16- or 8-mm film systems. The EVR system includes equipment for production of a large number of copies of a master program film. The film format consists of two parallel picture tracks on an unperforated film 8.75 mm wide, with a magnetic sound stripe along each edge. A synchronizing mark is located between each pair of images. This mark is used for image "framing" in playback.

In the case of monochrome program material, the two picture tracks are used independently. To produce a color television output, the two images are used together. One image is used to convey the luminance, or conventional monochrome "brightness," of the objects televised. The other image conveys the hue and color saturation information for the color picture. This image is a spatial representation of the color television chrominance carrier, which, in more usual temporal form, is described as being phase-modulated with hue information and amplitude-modulated with color saturation information. Figure 2 is an enlargement of the EVR film.

The EVR player uses a capstan drive transport, which resembles a conventional audio magnetic tape transport. A scanning raster on a small cathode-ray tube is imaged on the two picture tracks by means of a pair of lenses. The light transmitted by the film is incident on two electron multiplier phototubes. The output currents of these phototubes carry the information of the luminance and chrominance components of a color television video signal. They are reencoded onto carriers to allow introduction into a standard color television receiver.

The performance of EVR is probably the best of the currently available prerecorded video systems. Further reduction in image size would result in an unacceptable increase in image movements and undue disturbance from dirt and scratches on the film surfaces. The principal economic limitation, in spite of masterful achievements in image size reduction, lies in the cost of the film. The film consumption rate of 50 ft²/hr (4.6 m²/hr) puts the purchase price of program material at a level which will probably discourage a mass market.

used to emboss many replicas of the original record in vinyl.

Reproduction of the video signal on the record is accomplished in a way which superficially resembles conventional phonographic reproduction. The transducer for the video disk consists of a piezoelectric element bonded to the stylus. Whereas the stylus in audio reproduction is displaced by the modulating groove, the video disk stylus elastically deforms the undulations in the record. The piezo element responds to the pressure waves created by passage of the stylus over the deformable information contours. The electrical output of the stylus-piezo element is a recreation of the video signal used to drive the cutting stylus.

At present the playing time for each record is shorter than for other systems, and tracking of the grooves at the required high velocity is difficult to achieve without occasional "drop-outs" which cause streaks and other disturbances in the picture. Nevertheless this system has potential for very low cost for both the playback equipment and the records. An attractive application, not limited by the short playing time, is the insertion of video disks in newspapers and magazines as a moving-picture supplement.

EVR. The Electronic Video Recording (EVR) system of CBS Laboratories uses a conventional

SelectaVision holographic tape. The SelectaVision system, under active development at RCA, uses a low-cost material for its playback medium in an attempt to reduce the program tape cost to a level that will allow development of a mass market for a video playback system. In addition to low material cost, the process involves a unique form of optical encoding which frees the system from some of the problems of material uniformity and accuracy of replication. It is hoped that this will further reduce costs.

The SelectaVision process starts with an electron-beam-recorded silver halide film, as in the EVR system. Instead of using conventional silver halide technology to make copies of the original, the RCA system uses the silver halide master film as the object in a holographic system.

In the holographic process the images on the master film are encoded as phase holograms on a special light-sensitive relief medium. This medium yields a hologram in its surface contours in response to the exposure of the medium to the coherent light of a laser modulated by the image on the silver halide master. The holograms are formed in highly redundant arrays so that local disturbances on the medium do not produce local dirt spots or scratches on the reconstructed image. A series of contiguous holograms is made on the relief medium, each from a different frame of the silver halide master frame.

To replicate the phase holograms on the master relief tape, it is necessary to replicate only the surface of the master in some homogeneous material, in this case polyvinyl chloride tape. To prepare the master tape for replication, its surface is first rendered electrically conductive and then metallic nickel is electroplated on it. When the nickel plating is heavy enough to be self-supporting, it is separated from the master tape. The surface contours of the master are now mirrored in the surface of the nickel tape. The nickel tape is placed in rolling contact with blank vinyl tape under heat and pressure. The vinyl surface assumes the contours of the nickel tape and therefore contains the holograms which are replicas of those on the original tape. This process can proceed at high speed and is largely automated. In this way many thousands of copies can be made quickly and at low cost.

To reproduce the image encoded in the hologram, the vinyl tape is passed through the beam from a small gas laser. A vidicon camera tube is used to convert the image to a television video signal for coupling to a color television receiver.

An important property of the Fraunhofer holograms used in this system is that the reconstructed image appears to originate infinitely far from the television camera. As a consequence of this, the image position on the vidicon target is largely independent of the position of the holograms in the laser beam. Therefore the tape can be moved through the laser beam at any arbitrary speed, without regard for synchronization with the television standards. Furthermore the position of the tape, or of the holograms on the tape, does not affect the position of the reproduced image. The use of redundant Fraunhofer holograms therefore removes the image motion and dirt and scratch susceptibility limitations from the system. This is achieved with a very low cost medium and replication process.

For background information *see* HOLOGRAPHY; MAGNETIC RECORDING in the McGraw-Hill Encyclopedia of Science and Technology.

[ROBERT E. FLORY]

Bibliography: G. Dickopp et al., *J. Audio Eng. Soc.*, 18(6):618–623, 1970; *Electronics*, 43(20):89–90, 1970; P. C. Goldmark, *IEEE Spectrum*, 7(9):22–23, 1970; W. J. Hannan, *RCA Eng.*, 16(1):14–18, Repr. no. RE-16-1-11, 1970.

Volcano

Gaseous volcanic emissions contain sulfur dioxide, SO_2, and hydrogen sulfide, H_2S. These enter the atmospheric sulfur cycle, along with emissions from other natural sources and from pollution sources. Hydrogen sulfide in the atmosphere is thought to be rapidly oxidized to sulfur dioxide. Atmospheric sulfur dioxide is generally readily oxidized to form sulfate compounds in particulate form. These particles may then take part in atmospheric processes by serving as nuclei for the formation of cloud droplets and by scattering and absorption of solar radiation. Recently attention has been given to the possible effects of volcanic emissions upon the atmosphere and upon climate. *See* AIR POLLUTION.

Composition of volcanic emissions. Volcanic gases are difficult to sample without contamination by atmospheric air. Furthermore, direct sampling of volcanic plumes is difficult except during the later stages of an eruption when temperatures are tolerably low. Studies of the physical chemistry of magmatic mineral systems indicate that the composition of gaseous effluents changes during the course of an eruptive episode, particularly with respect to the relative amounts of sulfur dioxide present. Studies of gases from Kilauea in Hawaii, Surtsey in Iceland, and some volcanos in the Kamchatka Peninsula in the Soviet Union have shown considerable variation in the composition. J. J. Naughton and coworkers recently used infrared spectrometric methods to study gases in lava formations in Kilauea. This technique of remote measurement avoided the problem of obtaining samples directly. As a result of their measurements, they suggested the following composition for "average" volcanic gaseous emissions:

H_2O 95% by volume
CO_2 4% by volume
SO_2 1% by volume

R. D. Cadle and associates sampled and analyzed particles from volcanic fumes and found that they contain significant amounts of sulfuric acid and sulfate salts, in addition to mineral particles directly associated with the magma.

Effects of volcanic emissions. Volcanic sulfur-bearing gases, H_2S and SO_2, emitted to the atmosphere are rapidly oxidized (over a period of 8–10 days) to sulfate, probably by catalytic reactions in cloud droplets and on moist surfaces. Tropospheric particles are characterized by residence times ranging from a few days to several weeks. During their residence in the atmosphere these sulfate particles (in the form of sulfuric acid droplets, ammonium and calcium sulfate particles, or various admixtures) may act as nucleating agents for the formation of cloud droplets. These in turn may affect rainfall. Because of the generally sporadic nature of volcanic eruptions and the short lifetimes of tropospheric particles, volcanic emissions are not thought to have a significant effect upon global climate by the processes discussed above.

Volcanic eruptions in which large explosions

occur can inject gaseous and particulate material to great heights into the stratosphere. (The stratosphere is the atmospheric layer above the tropopause, the upper 15% of the atmosphere. The height of the tropopause varies over the globe from less than 10 km in polar regions to more than 16 km in the tropics.) Material in the stratosphere, of sufficiently small particle size (radius less than 1 μm), can persist there for 6 to 9 months at high latitudes and 1 to 2 years in tropical regions. Atmospheric circulations and transport by turbulent diffusion gradually carry stratospheric particulate materials down into the troposphere, where they are removed by the processes of precipitation scavenging and dry deposition.

Most volcanic eruptions are not associated with explosions large enough to inject materials up into the stratosphere. But the occurrence of such events is by no means rare. The table lists names, approximate locations, and observed heights of ash, dust, and vapor clouds resulting from explosive eruptions since 1947. These are the maximum heights attained by the clouds during eruptive episodes which individually may last up to several months. The list may not be complete because some volcanic eruptions in remote regions may be unobserved and unreported.

Sulfur dioxide is thought to be quite rapidly oxidized by oxygen atoms in the lower stratosphere to form sulfuric acid and ultimately also ammonium sulfate. These sulfates exist in particulate form and may be mixtures of the acid and the sulfate salt. The so-called stratospheric sulfate aerosol, originally discovered by C. E. Junge, may be formed by such reactions. It may be seen from the foregoing that volcanic sulfur dioxide reaching the stratosphere is rapidly converted to particulate sulfate. These particles, along with any others of submicrometer-size in the volcanic cloud, are globally dispersed in the lower stratosphere.

Particulate material in the stratosphere may interact with solar radiation by both absorption and scattering processes. The ultimate effects of these interactions have been the subject of much discussion in recent years. It is sufficient to say that clear-cut relationships between stratospheric particulate concentrations and climate changes have not been demonstrated. However, it is generally agreed that the immediate effect of increased backscattering due to enhanced stratospheric particulate concentration is to cause cooling tendency at the surface of the Earth. This is so because radiation which normally would reach the Earth's surface and be absorbed there is prevented from doing so. Also, it is agreed that the absorption of radiation by particles in the atmosphere tends to cause warming of the air in that region.

H. H. Lamb and J. M. Mitchell, Jr., separately studied the possible effects of volcanic eruptions on global climate. Lamb introduced the term dust-veil index as a measure of the combined effects of the size and energy of a volcanic eruption and dispersion of its particulates in the stratosphere on the obscuration of solar radiation. An index value of 1000 was assigned to the great eruption of Krakatoa in 1883. Lamb showed that, among other things, hemispheric average temperature decreases of the order of 0.5°C for a 2–3 year period occurred after Krakatoa and other great eruptions. Mitchell suggested that the hemispheric average surface temperature decrease which has taken place since the 1940s may be the result of volcanic material in the stratosphere injected over a period of time starting with the 1947 eruption of Hekla. Mitchell's argument countered suggestions that the latter-day hemispheric temperature decrease was due to widespread pollution of the lower atmosphere by particles from fossil fuel–burning and metal-refining processes.

Gunung Agung eruption. Gunung Agung erupted in two main paroxysmal events in March, 1963. Lamb assigned a dust-veil index value of 800 to the eruption. Optical effects of the resulting particle concentrations in the stratosphere were observed around the globe. Direct sampling of particles in the stratosphere of both Northern and Southern hemispheres by S. C. Mossop and by J. P. Friend showed enhancement in the concentration of sul-

Year and location of eruption and estimated maximum height reached by volcanic cloud during entire eruptive episode

Year	Name	Location	Cloud height, km
1947	Hekla	Iceland	27
1953	Mt. Spurr	Alaska	23
1956	Bezymjannaja	Kamchatka	45
1960	Puntiagudo	Chile	Up to tropopause
1963	Gunung Agung	Bali	15
1963	Mt. Trident	Alaska	15
1963–1965	Surtsey	Iceland	9
1964	Sheveluch	Kamchatka	15
1965	Taal	Luzon	16–20
1966	Awu	Celebes	Uncertain, but probably reached atmosphere
1966	Mt. Redoubt	Alaska	12–16
1967	Deception Is.	63°S 60½°W	10
1968	Fernandina	Galapagos	Uncertain
1967–1968	Mt. Redoubt	Alaska	10
1967–1968	Mt. Trident	Alaska	11
1970	Hekla	Iceland	15
1970	Deception Is.	63°S 60½°W	Well into stratosphere
1970	Beerenberg	Jan Mayen Is. (Greenland Sea)	10

fate particles by a factor of more than 10 over the pre-Agung concentrations. Laser radar studies by G. Fiocco and G. Grams showed that the enhancement in stratospheric aerosol backscattering remained undiminished, except for seasonal variations, until 1966. Studies of global solar radiation attenuation by A. J. Dyer and B. B. Hicks showed a large effect beginning at the time of the Agung eruption and continuing through 1965 with relatively little change in amplitude. Finally, R. E. Newell showed that temperatures in the stratosphere above 13 km altitude over Australia increased markedly following the eruption of Agung. The temperature did not return to the pre-Agung average value by the end of 1966.

Friend recently pointed out that the apparently prolonged and undiminished effects of the Agung eruption are most likely due to stratospheric injections of material, primarily SO_2, by other volcanoes between early 1963 and the end of 1966, six of which are listed in the table.

Volcanic sulfur emissions. Friend estimated that, on the average, volcanic activity releases about 2×10^6 metric tons of sulfur as SO_2 into the atmosphere annually. This emission rate is only 0.5% of the annual emission of sulfur as H_2S, SO_2, and sea spray sulfate by natural and anthropogenic sources. However, the atmospheric sulfur cycle of emission, dispersion, chemical transformation, and removal takes place almost entirely in the troposphere. The total amounts of sulfur in the troposphere and the stratosphere (pre-Agung) are estimated to be 4×10^6 and 3×10^4 metric tons, respectively. The average lifetime of sulfur in the troposphere is only a few days. Little, if any, sulfur of nonvolcanic origin ever reaches the stratosphere. By contrast, most volcanic sulfur is placed in the atmosphere at heights above 5 km. A single volcanic eruption could temporarily double the sulfur content of the entire atmosphere. Only about 1% of total volcanic sulfur emissions, averaged annually, is needed to account for all of the sulfur in the stratosphere (pre-Agung). On this basis, it may be appreciated that because of the large amount of sulfur released in a single event and the high altitude of injection of such releases, volcanoes may be responsible for much, if not all, of the sulfate aerosol usually present in the stratosphere.

For background information see ATMOSPHERIC GENERAL CIRCULATION; ATMOSPHERIC POLLUTION; VOLCANO in the McGraw-Hill Encyclopedia of Science and Technology. [J. P. FRIEND]

Bibliography: *Annual Report of the Center for Short-Lived Phenomena, 1970*, Smithsonian Institution, 1970; H. H. Lamb, *Phil. Trans. Roy. Soc. London*, 266:425–533, 1970; R. E. Newell, *J. Atmos. Sci.*, 27:977–978, 1970; S. F. Singer (ed.), *Global Effects of Environmental Pollution*, 1970.

Wastes, agricultural

The food and fiber industry, including production and processing, produced over 2×10^9 tons (1.81×10^9 metric tons) of waste in 1970, or about 58% of all the solid wastes produced in the country. These wastes include the obvious manure and organic residues from farms and forests, as well as various solid materials and water discharged from processing or manufacturing plants that use any form of an agricultural product as a raw material. Some of these wastes are utilized or recycled, but most of them require disposal. Many of the present disposal techniques simply move the waste to another place rather than solve the problem. Therefore the only true solution to the waste problems of agriculture is to consider the material as a national resource to be recovered by recycling and utilization.

Solid wastes. The following sections discuss the recycling of various types of solid waste.

Animal wastes. Disposal of animal wastes poses a major problem to the agricultural industry. Tremendous amounts of manure are produced (Table 1), but it is the concentration of this manure in small areas that constitutes the real problem. Agricultural science and technology met the challenge to increase animal production by developing procedures to raise large numbers of animals in confinement, as in cattle feedlots. Unfortunately the technology for managing animal waste has not developed as rapidly. Where usual disposal techniques are being curtailed by social pressure or by possible danger to the environment, the animal producer is forced to operate under a narrowing profit margin and higher overhead costs. Recycling and utilization may partly solve these problems.

Many systems for injecting manure into the subsoil have been investigated. Such systems reduce the insect and odor problems, but the operations are not economical and probably will be used only in areas where the social pressures are extreme. At present, land spreading must be considered as a least expensive rather than economical or profit-making method of waste handling.

Many methods have been investigated for the potential utilization of animal wastes, for example, selling as fertilizer, composting, animal feeding, and energy (methane) production. The primary purpose of composting is to eliminate putrescible organic matter while conserving much of the original plant nutrients. Manures may be composted alone but frequently are combined with high-carbon, low-nitrogen wastes such as sawdust, corn cobs, paper, and municipal refuse. The resulting material is suitable for use as a soil conditioner and organic fertilizer. Full-scale operations have been technically successful with poultry, beef, and dairy manure in a unique regional situation, but in general the market for this product is very small. Without a market, essentially all the dry matter remains for further disposal. Manure, either as a fertilizer or soil conditioner, is difficult to sell profitably, regardless of preparation procedures, because of the low cost of commercial fertilizers.

Dried and dehydrated manure is sometimes sold as a soil conditioner, organic fertilizer, or animal feed supplement, but the cost of drying and dehydration is generally greater than the return realized.

Use of animal wastes in the feed of animals has recently received much publicity as a potential method for utilizing agricultural wastes. When nutritional principles are followed, the technique has produced good results, especially if the waste of a single-stomached animal is added to the rations of ruminants or if the waste of the ruminant is chemically treated before use. However, a Food and Drug Administration regulation (1970) prohibits the use of animal wastes as feed supplements

Table 1. Numbers of livestock and their total waste production in the United States in 1970

Livestock	Total population, millions	Solid wastes, millions of metric tons/year	Liquid wastes, millions of metric tons/year	Total wastes, millions of metric tons/year
Cattle	107	1015.0	391.0	1406.0
Hogs	57	62.5	36.5	99.0
Sheep	21	9.6	5.8	14.4
Horses	3	17.5	4.4	21.9
Chickens	2950	54.0	—	54.0
Turkeys	106	21.7	—	21.7
Ducks	11	1.6	—	1.6
Total	—	1181.9	—	1618.6

because of the possible transmission of drugs, feed additives, and pesticides to another animal and to some agricultural products, such as milk and eggs. On-going research should clarify the situation.

Much of the nutritional value of animal wastes might be captured by growing housefly larvae and insects as a source of protein for animal feed supplement. The method has been tested and shows considerable promise. Algae can be commercially grown on a manure substrate and can also be used as a feed supplement. The remaining growth media could be used as a soil conditioner or fertilizer.

Manure can be used directly as a fuel or as the substrate to produce methane anaerobically. Manure must be collected and dried to be used as a fuel, a costly operation. Therefore it is doubtful that any appreciable quantity will ever be utilized in this manner.

Regardless of the disposal procedure or recovery process, the soil will be the ultimate disposal site of most of the bulk of animal wastes. Therefore the challenge is to develop techniques of recycling to incorporate animal wastes in land management programs without damaging the environment or causing a nuisance to the human population. This will most likely be accomplished by convincing the farmer that manure improves the physical condition of a soil as well as supplies plant nutrients.

Crop and orchard residues. The tonnage of plant residues left on the farms exceeds by far the tonnage of the crops taken to market. These wastes consist of straw, stubble, leaves, hulls, vines, tree limbs, and similar trash. Most of the residue is burned to eliminate troublesome plant diseases, pests, and weeds. Unfortunately this pollutes the air with smoke and volatile organic compounds. A small part of these materials is being used for mulch and as a soil builder when the mulch is plowed under, ensilage, bedding for animals, and as bulk material in the manufacture of corrugated cartons, insulated board, and specialty paper. Although the handling of crop residues is not a major problem in terms of amount, these wastes can be a focal point of major infestation by harboring insects and plant diseases. Obviously, new and better methods are needed to handle these wastes without polluting the environment.

Food processing wastes. The meat processing industry produces few wastes, other than water, that require disposal (Table 2). This industry has illustrated that the utilization of wastes can be profitable. In recent years over 35×10^9 lb (15.89×10^9 kg) of meat and similar amounts of other animal parts, or wastes, have been processed annually. These wastes, including those collected during water disposal procedures, are the raw material for the manufacture of soap, leather goods, glue, gelatin, and animal feeds. Glands and organs are processed to produce hormones, vitamins, enzymes, liver products, bile acids, and sterols. Other animal parts are the source of certain fatty acids, oils, grease, and glycerine. Bones can be processed to produce proteins and fats, as well as bone meal. Finally fuels and solvents can be made from the waste products of the above industries. The waste waters leaving the processing plants still contain small amounts of solids, and disposal procedures are necessary. The utilization of these suspended materials must be preceded by a more efficient use of water which will result in concentrating the wastes and improving the economics of recovery.

Although the fruit and vegetable processing industry does not produce large amounts of solid wastes relative to other segments of agriculture, these wastes are difficult to handle because of their varied nature. Peels, skins, pulp, seeds, and fibers are suspended in billions of gallons of water which may be saline, alkaline, or acidic and may contain a wide variety of soluble organic compounds. Of the 12.7×10^6 tons (11.52×10^6 metric tons) of solid waste produced, only about 4×10^6 tons require disposal — but at a cost of $\$25 \times 10^9$. Most of the remaining 8.7×10^6 tons are utilized as animal feed and consist principally of the by-products of the processing of citrus fruits, potatoes, and corn. These by-products are usually given away, resulting in a cost-free disposal procedure which is very valuable to the industry. Increasing the efficiency of the in-plant use of water is a necessity before any new salvage operations will be economical.

About 118×10^9 lb (53.4×10^9 kg) of milk were converted in 1970 to fluid milk, cheese, butter, condensed and powdered milk, and ice cream. The amount of solid wastes coming from this industry is small compared with the amounts of other agricultural wastes, but they require costly handling. Milk solids have very high pollution potentials. Therefore these solids must be removed from millions of gallons of water daily before this water can be released to streams or used for irrigation. The collected milk solids do have value as animal feed, feed supplements, raw material for the production of some chemicals such as alcohol, and as a basic

Table 2. Wastes produced by selected agricultural industries in the United States during 1970

Type of processing industry	Total solid waste discharged, millions of metric tons/year	Solid waste salvaged, millions of metric tons/year	Solid waste requiring disposal, millions metric tons/year	Comment
Meat	17.5	17.5	0.003	Values include all livestock and broilers
Vegetables and fruit	12.7	8.7	4.0	Organic and inorganic wastes in disposed fraction
Dairy	0.2	*	*	As fat-free solids
Forestry	36.0	*	*	Logging residue from lumber mill
	55.0	*	*	As sawdust and edging
	41.5	*	*	From pulp and paper mill

*No estimates are made because the amount of salvaged and disposed by-product depends upon the location and size of the plant or operation.

ingredient in growth medium for microorganisms in the production of pharmaceutical chemicals. Lowering the cost of production and increased utilization are, as before, related to a more efficient use of water.

Forest waste products. A large part of a tree harvested for pulp paper or lumber becomes waste. It has been estimated that 19% of the tree is left in the forest. Approximately 16% of the log is wasted as sawdust and 34% as slabs and edgings during milling. Finally pulp and paper mills discharge effluents containing about 50% of the log. Wastes from the wood industry have great pollution potential and are usually destroyed by burning because other disposal procedures are too costly. Burning of forest debris is deemed to be necessary in the control of forest diseases and insects, but it results in air pollution.

There have been many processes developed by wood science and technology laboratories to recover a large variety of by-products, but these processes have been only sparingly applied, primarily because of economic limitations. Wood is composed primarily of lignin and cellulose. Lignin can be processed to obtain many valuable chemicals, such as artificial vanilla, or used directly as a binder, a dispersant, an emulsion stabilizer, and a sequestrant. Lignin can also be utilized as the raw material for plastic production. Cellulose is readily processed to simple sugars, alcohol, fodder, yeasts, and chemicals such as furfural. Utilization methods that may be economically feasible in the future include the use of wastes for the production of specific chemicals, through fermentation, and in new building products such as board, paper, and blocks.

Recycling of paper does not seem to have a promising future. Now less than 20% is recycled as compared to about 30% 20 years ago. If the cost of removing the adulterations of the paper, such as ink, plastic, and metal clips, is decreased, waste paper recycling probably will be increased, thereby decreasing the overall wastes produced by the wood industry.

Liquid wastes. The agricultural industry uses water in almost every operation of production, processing, and manufacturing. Irrigation and food processing use the largest amounts. These waters must be considered agricultural wastes in the same sense as are solid wastes and must either be recovered or disposed of. In the past the axiom "control pollution by dilution" was readily accepted. Consequently, large volumes of water were used to dilute the agricultural wastes before they were discharged into lagoons or other surface waters, sprinkled on a field, or discharged into a municipal sewage system. The great increase of soluble and solid wastes to be transported, the lack of usable water in some areas, and social pressures are causing a reevaluation of water use for this purpose. Processing plant procedures are being changed to increase recycling of water in the plant by segregating highly contaminated water and using the clean discharged water to irrigate crops. The highly contaminated water is handled by conventional disposal procedures before it is utilized or recycled.

Irrigation of agricultural crops uses vast quantities of water. In the past this was not carefully controlled and streams were contaminated by runoff and drainage waters containing sediment, soluble salts, pesticides, and so forth. However, important changes are taking place, such as better controls in the ditches and fields and reuse of runoff irrigation waters. Irrigation water is reused by collecting the normal runoff water in shallow ponds and then pumping it to other fields. Less than 10% of the liquid water is lost by this method. Also, most of the soluble and solid materials usually lost in runoff are kept on the farm, thereby decreasing the possible spread of a polluting agent, although salt buildup may be a problem in some areas.

Agricultural waters can also be used to recharge groundwater. Uncontaminated water can be discharged directly into the groundwater. Contaminated water can be processed, or in some cases the soil can be used as a filter.

For background information *see* AGRICULTURE; MANURE; WATER POLLUTION in the McGraw-Hill Encyclopedia of Science and Technology.

[WALTER R. HEALD]

Bibliography: R. B. Enghahl, *Public Health Service Publ. No. 1856*, 1969; E. P. Taigamides, *Proceedings of the 24th Purdue Industrial Waste Conference*, pp. 542–549, 1969; C. H. Wadleigh, *USDA Misc. Publ. No. 1065*, 1968; N. H. Wooding, Jr., *Spec. Circ. No. 113*, Pennsylvania State University Extension Service, 1970.

Weather modification

The improvement of visibility in fog has been one of the primary goals of weather modification. The detrimental effects of fog on surface and air transportation is a problem of major proportions. In recent years the U.S. Department of Transportation has found it necessary to limit the number of flights at the nation's busiest airports during periods of low ceiling and visibility. The loss of revenue caused by one fog at a major airport due to aircraft diversions, delays, and cancellations is estimated at $100,000. The cost of one fog occurrence in the era of jumbo jets is expected to rise to $500,000. Costs in excess of $300,000,000 per year are incurred by fog-associated accidents on highways in the United States. Militarily, fog is a serious factor in the movement of personnel, cargo, and combat material at airfields and staging areas.

Considerable progress has been made in the development of fog dispersal technology during the past few years. Limited operational systems to dissipate supercooled fog at airports are being successfully used in the United States, Germany, France, and the Soviet Union. Fog dispersal operations are being conducted at 12 major airports in the United States as part of an airlines industry supported program. During the winter 1969–1970 fog dispersal operations, costing approximately $80,000, resulted in a saving of over $900,000 in operating expenses. During this period the technical feasibility of dissipating warm fog has also been demonstrated. Significant visibility enhancement or clearing of warm fog has been accomplished with several techniques. These techniques, however promising, are still in the developmental stage.

Scientific problem. Fog can be classified into three general types according to its constitution and temperature, namely, ice fog, supercooled fog, and warm fog.

Ice fog is a suspension of ice crystals existing at temperatures well below freezing. Its occurrence is, therefore, restricted to the far northern latitudes in winter. At present there is no known method of dispersing ice fog once it has formed. Ice fog can, however, be prevented by controlling the moisture and nuclei sources which contribute to its formation.

Supercooled fog is composed of water droplets that exist at below freezing temperatures. Ice crystal formation in these cases is inhibited by the lack of suitable ice embryos in the atmosphere. Since supercooled fog is in an unnatural, unstable state, it can be easily upset and thereby dissipated. Ice crystals are introduced into the fog, grow at the expense of the water droplets, and fallout as snow, thereby producing a clearing. The ice crystals required to trigger this process are produced either by seeding the fog with freezing nuclei such as silver iodide particles or by local cooling of the air to below $-40°F$ ($-40°C$) by dry-ice seeding or by the expansion of propane gas.

The most prevalent type of fog is warm fog, which consists of water droplets at above-freezing temperatures. Warm fogs are stable cloud systems in the atmosphere. In contrast to supercooled fog, there is no latent phase instability in warm fog that can be exploited to promote the artificial dissipation process. Warm fog dispersal methods are necessarily "brute force" in character. Whatever energy is required to dissipate the fog must be supplied by the dispersal method. Careful engineering is required to make any warm fog dispersal technique reliable and cost-effective.

Warm fog. The development of warm fog dispersal methodology in the past few years has been based on an interrelated program of computer simulation of the artificial dissipation processes and field experimentation. The acceleration of progress in warm fog dispersal can be attributed to this approach. Three techniques, all designed to promote the evaporation of the water droplets, have been found to be effective in improving the

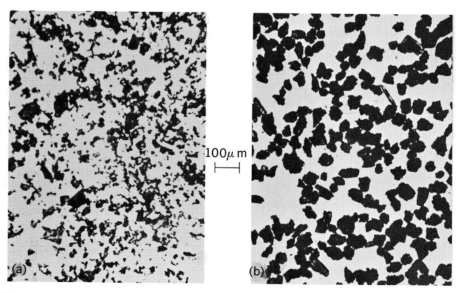

Fig. 1. Effect of microencapsulation on raw urea (a) before microencapsulation and (b) after microencapsulation. Microencapsulation provides for the sizing and stabilization of the urea particles.

visibility in warm fog: (1) drying of the air with hygroscopic chemicals, (2) heating of the air, and (3) mechanical mixing of the fog with drier, warmer air from above.

Hygroscopic particle seeding. When hygroscopic chemicals in the form of dry particles or saturated solution droplets are injected into a fog, they absorb water vapor from the air and, in drying, the air causes the fog droplets to evaporate. Computational and experimental studies have shown that the effectiveness of the hygroscopic treatment is highly dependent on the size and concentration of the seeding particles, the optimum value of the seeding parameters being contingent upon the intensity of turbulence and wind shear in the fog. Urea has been found to be the most practical seeding agent since it is nontoxic and noncorrosive to metals, protected surfaces, and animals and it is highly beneficial to plant life. Raw urea cannot be used, however, because it has a soft, friable crystalline structure which fragments easily during handling, producing large numbers of submicron particles that are ineffective in clearing fog. A technological breakthrough by the Air Force Cambridge Research Laboratories (AFCRL) has solved this problem. Microencapsulation technology, whereby single crystals are chemically packaged inside thin, harmless shells, was exploited to provide for the sizing and stabilization of the urea particles, thereby optimizing the efficiency of urea as a warm fog seeding agent. The microencapsulation process produces a narrow size distribution that is completely devoid of the very small particles (Fig. 1). Microencapsulation techniques are widely used in foodstuffs, pharmacology, and chemical engineering. Perhaps the best known example is the "tiny time pills" advertised by several commercial cold remedies which employ microencapsulation to effect a controlled, sustained release of the pharmaceutical contents.

AFCRL and its contractor, Meteorology Research Inc., have conducted a series of tests at McClellan Air Force Base in California to investigate the practicability of using the airborne microencapsulated urea particle seeding technique to improve the visibility in fog at an airfield to operationally useful levels. Although significant clearing of fog was produced by the seeding (Fig. 2), the lack of constancy of the wind field made targeting of the cleared zone a difficult problem. This technique cannot be employed reliably on an operational basis until satisfactory procedures for positioning the cleared area over the desired location are developed. The seeding of relatively large areas, as in supercooled fog dispersal operations, may be the solution to this problem.

Heating. In this technique the temperature of the fog environment is raised by approximately 5°F, supplying enough heat to evaporate the fog droplets and sustain the additional water vapor. Warm fog dispersal by air heating was used successfully in England during World War II and at Los Angeles International Airport during 1948–1953. The burning of fuel oil provided the required heat in this fog incendiary dispersal operation (FIDO). It was abandoned after 1953, primarily because it was considered too expensive to warrant routine use by commercial aviation. Serious operational problems such as pollution due to in-

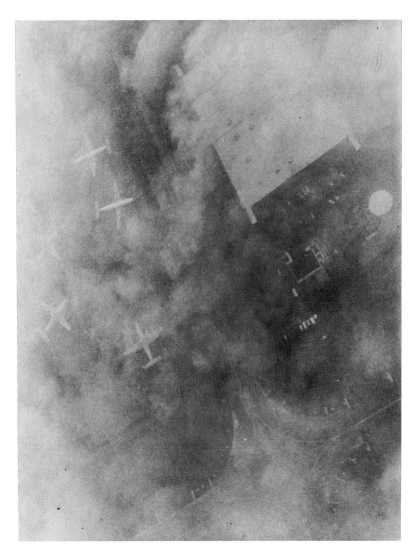

Fig. 2. The result of an AFCRL microencapsulated urea seeding experiment at McClellan Air Force Base in California.

complete combustion of the fuel and the danger of landing aircraft near open flames were also factors.

The increased losses in revenue due to fog with the advent of the jet age has made the heating technique economically attractive for at least those airports having a high volume of air traffic. In recent years the French have been investigating a system which uses the exhaust heat from jet engines to disperse the fog. They have encountered problems with air and noise pollution in tests of this technique at Orly Airport in Paris. They report that the turbulence generated by the high-speed jet exhaust may also be a problem for landing aircraft. In the United States, AFCRL is sponsoring the design of a heating system based on the controlled merging of heat plumes generated by a large number of enclosed liquid-propane burners. It promises to be an efficient, flexible system that is noiseless, smokeless, and safe for landing aircraft.

Helicopter downwash mixing. The helicopter, during the clearing operation, either hovers or moves slowly forward in the clear air above the fog layer. The downwash action of the helicopter

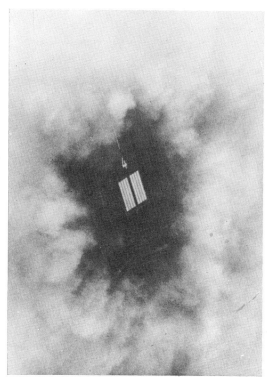

Fig. 3. Typical result of AFCRL-ASL helicopter hover test at Greenbriar Valley Airport, Lewisburg, W. Va. The clearing is 1000 ft in diameter.

forces this relatively dry, clear air downward into the fog. The wake air, on descending, entrains and mixes with the fog. The resulting air mixture becomes subsaturated and the fog droplets are thereby caused to evaporate. The fog clearing capability of the helicopter downwash mixing technique was conclusively demonstrated in a series of field experiments jointly conducted by AFCRL and the Army Atmospheric Sciences Laboratory (ASL) at Lewisburg, W.Va. Cleared zones large enough to permit helicopter landings were routinely created in fog less than 300 ft deep (Fig. 3), and on six occasions helicopter landings were accomplished. On two occasions the helicopter succeeded in clearing the fog over the full 6000-ft length of the airport runway.

Future prospects. Research and development of warm fog dissipation technology is being vigorously pursued by military, airline, and government transportation agencies in the United States, France, and Italy. The increased attention to and support of this effort is expected to result within the next 5 years in the development of several operational systems having limited but extremely beneficial applications. Each fog dispersal system will include specifications as to which meteorological and fog conditions the technique can be applied with reliability, simple operational methods for identifying when such conditions exist, and optimized procedures for implementing the technique. The user will be able to select the system that satisfies the meteorological, logistical, and/or economical requirements of the intended application.

For background information see FOG; WEATHER MODIFICATION in the McGraw-Hill Encyclopedia of Science and Technology.

[BERNARD A. SILVERMAN]

Bibliography: H. Appleman and F. Coons, *J. Appl. Meteorol.*, vol. 9, no. 3, 1970; V. Plank, A. Spatola, and J. Hicks, *Environmental Research Paper no. 335*, AFCRL-70-0593, 1970; B. A. Silverman, *Bull. Amer. Meteorol. Soc.*, vol. 51, no. 5, 1970; B. A. Silverman and B. A. Kunkel, *J. Appl. Meteorol.*, vol. 9, no. 4, 1970.

Work measurement

During 1970 there were several significant developments pertaining to combined or "second generation" methods time measurement (MTM) data. First, the United States–Canada MTM Association for Standards and Research approved the MTM-2 and MTM-3 data systems developed in Europe in the 1960s. Second, Maynard Research Council released its set of combined MTM data known as Maynard Research Data (MRD).

MTM-2, MTM-3, and MRD are based exclusively on MTM data and represent excellent procedures for economically applying basic MTM to the lower quantity and longer cycle operations. The systems can be applied more quickly and without sacrificing too many of the advantages of the basic system. They were developed in an attempt to eliminate the cost and effort of the many groups individually involved in developing data systems. They drew on the experience of some earlier combined data systems that were also developed in the 1960s. The most popular and widely used are probably Universal Standard Data (USD), Master Standard Data (MSD), and MTM General Purpose Data (MTM-GPD). These earlier systems were designed primarily for longer cycle work and tailored to a specific work measurement problem, or tailored to cover specific tasks such as machining, material handling, maintenance, and clerical. Excluding the tailored sets of task data, the most widely used and probably the best known sets of combined MTM data are shown chronologically in Fig. 1. These data are horizontal in nature, that is, they normally can be applied universally on a broad, general-purpose base.

USD. The basic concept of a combined MTM data system was first formulated by a group of consulting management engineers of the Methods Engineering Council who were faced with the task of developing a large number of long-cycle standards in a plant assembling tractors on a common progressive assembly line. They recognized that basic MTM would be very time consuming and expensive to apply and that a new, less-time-consuming approach was needed.

The procedure worked well on this initial application and opened a way for applying this same approach to other types of work. With each new application, the data were refined and expanded and were finally modified to give universal application, that is, they were not limited to any particular operation or process. The data consisted primarily of getting objects, placing objects, walk displacements, and miscellaneous elemental and body motions. They were set up in tables and coded for fast application.

To predict the accuracy of the system, a computer program was developed to handle the variables and their respective probabilities. In the final analysis, it was determined that USD was on parity with MTM when the job cycles were 0.010 hr and longer.

MSD. These data were similar to USD in approach and included data for obtaining, placing, and miscellaneous elemental and body motions. They were developed by a group of engineers from Serge Birn Associates to provide a system for use where extreme accuracy was not important and which would be quicker and easier to apply than basic MTM but would retain many of the advantages of basic MTM.

The accuracy of MSD was proved when it was compared with time studies and with standard data. A cycle length criterion was not established, but on the basis of the computer analyses for USD and MTM-GPD, it is estimated that MSD would also be on parity with MTM when the cycle lengths reach approximately 0.010 hr.

MTM-GPD. USD and MSD are proprietary systems and are widely used by their originating companies on consulting assignments. A number of similar systems with the same given objective were developed in the 1960s. The MTM Association for Standards and Research recognized that it would be highly desirable if these many systems could be combined into a single system which all Association members could use. This was then established as an Association project, and many industrial and professional members donated their data to the Association. The Association made a survey of the needs of industry and then reviewed the data contributed by its members. Finally there evolved a comprehensive set of data, namely, MTM-GPD. Not only did the data contain the GET and PLACE data that were the primary data in USD and MSD, but they also contained a set of the frequently used general-purpose data. The total data package contains 18 data tables, and they employ a seven-position alphamnemonic code.

To determine when MTM-GPD becomes an acceptable tool, the data variables and their respective probabilities were programmed into a computer, and it was determined that MTM-GPD is on parity with basic MTM when the job cycles reach approximately 0.010 hr.

MTM-2. This data system is an international set of combined MTM data. The international situation on combined data systems was similar to that which existed in the United States prior to the acceptance of MTM-GPD. The International MTM Directorate (IMD) recognized the problem and set up the machinery to develop an international set of data. When the data were eventually developed, they were presented to the IMD in Munich, Germany, in 1965 and approved. They were later presented to the United States–Canada MTM Association, which conducted a validation study and approved the data in 1970.

The approved data (Fig. 2) were constructed by combining and averaging basic MTM motion times. MTM analyses from different industries and workshops with different degrees of mechanization were collected and checked. These analyses were then summarized by a computer which was programmed to yield information about the motion sequences and frequencies. The number of motions summarized totaled over 14,000, and the analyses represented 30,000 man-hours of work.

On the data card, the CODE column denotes the distance range in inches. The next three columns contain the GET data, and the following three columns the PUT data. The remaining data on the card are elemental and body motion data. All time values are expressed in time measurement units or TMUs (1 TMU = 0.00001 hr).

The Swedish MTM Association conducted a series of tests to establish the accuracy of the results and the speed of application of MTM-2 compared with basic MTM. These tests were made in actual industrial situations. They determined that MTM-2 is on parity with basic MTM at operations cycles of 1 min or more. They also determined that the speed of analysis during the test was at least twice that of basic MTM.

MRD. This data system was compiled in 1970. The data were developed because extensive research into the aforementioned sets of data, coupled with considerable field experience, indicated that although they were excellent sets of data, several adjustments in the data and the coding could make a significant contribution. MRD is very similar to MTM-GPD. The Basic Series of data includes the GET and PLACE data. The General Series and Tool Series include many of the common elements of work that are characterized by hand tools and the like. The total data package contains 24 data tables in a $4\frac{1}{2} \times 8$ in. booklet. All data are coded with a simple four-position code. The first two code positions denote the data set, and each set is provided with a dual code—alphamnemonic and numeric. The data also contain the latest developments and research in MTM.

The accuracy of the data is comparable to USD and MTM-GPD. MRD is on parity with basic MTM when the job cycle is approximately 0.010 hr.

MTM-3. This data system is an international set of combined MTM data. These data, approved by the United States–Canada MTM Association in 1970, are shown in Fig. 3. They were constructed by a further combination and averaging of the MTM analyses used in the development of MTM-2. In essence, MTM-3 is a condensation of MTM-2

WORK MEASUREMENT
MTM-2

code	GA	GB	GC	PA	PB	PC
2	3	7	14	3	10	21
6	6	10	19	6	15	26
12	9	14	23	11	19	30
18	13	18	27	15	24	36
32	17	23	32	20	30	41

GW 1–2 Lbs.			PW 1–10 Lbs.		
A	R	E	C	S	F	B
14	6	7	15	18	9	61

Fig. 2. MTM-2 data card. (*Maynard Research Council Inc.*)

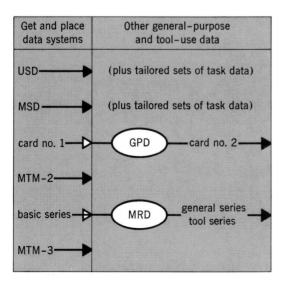

Fig. 1. Combined MTM data systems.

WORK MEASUREMENT
MTM-3

code	HA	HB	TA	TB
6	18	34	7	21
32	34	48	16	29
SF 18		B 61		

Fig. 3. MTM-3 data card. (*Maynard Research Council Inc.*)

and represents a coarser set of GET and PUT data. The CODE column was reduced to two distance ranges. The next two columns contain the HANDLE data and represent a combination of GET and PUT. The last two columns contain the TRANSPORT data and are an expanded but rougher version of PUT.

MTM-3 accuracy is on parity with basic MTM when the job cycle is approximately 4 min or longer. Also, MTM-3 is approximately seven times as fast as basic MTM. However, because of its simplicity, there is little or no consideration for method, and this must be given due consideration in deciding on whether MTM-3 should be used for a particular application.

For background information see METHODS ENGINEERING; WORK MEASUREMENT in the McGraw-Hill Encyclopedia of Science and Technology.

[WILLIAM J. MATTERN]

Bibliography: R. M. Crossan and H. W. Nance, *Master Standard Data*, 1962; W. K. Hodson, Accuracy of MTM-GPD, *J. Methods Time Meas.*, pp. 9–17, September-October, 1963; W. K. Hodson and W. J. Mattern, Universal standard data, in H. B. Maynard (ed.), *Industrial Engineering Handbook*, 2d ed., 1963; J. E. Mabry, MTM-GPD general purpose data original development and future expansion, *J. Methods Time Meas.*, pp. 18–22, September-October, 1963; W. J. Mattern, R. MacDonald, and K. Knott, MTM-GPD and other second generation data, in H. B. Maynard (ed.), *Industrial Engineering Handbook*, 3rd ed., 1971; *MTM-2 Manual*, Svenska, MTM-Foreningen, Stockholm, Sweden, 1965; *MTM-3 Manual*, Svenska, MTM-Gruppen AB, Solna, Sweden, 1970.

Xenon compounds

The existence of xenon compounds has been known since 1962, when N. Bartlett prepared the first compound by reaction of xenon with platinum hexafluoride. Today the known compounds include simple and complex fluorides and oxyfluorides, oxides, and oxy and fluoro salts. Recent work in xenon chemistry has emphasized the preparation of new compounds with ligands other than oxygen or fluorine atoms and the reaction of the xenon fluorides as selective fluorinating reagents. The variety of xenon compounds is growing rapidly and the utility of xenon compounds as synthetic reagents is becoming apparent.

New xenon compounds. The available methods for the synthesis of new xenon compounds are very limited. The starting point in any preparation is with one of the xenon fluorides XeF_2, XeF_4, or XeF_6. Elemental xenon can only be oxidized with the strongest oxidizing agents, and only fluorine and certain strong oxidative fluorinating reagents, such as PtF_6, CF_3OF, and O_2F_2, have been shown to be effective. Nearly all of the xenon compounds known today have been obtained by direct reactions of the xenon fluorides or by reactions of compounds prepared initially from the three fluorides.

Until recently the known compounds of xenon contained only fluorine and oxygen atoms and OH groups as ligands bonded to xenon. Examples, including both cationic and anionic species, are shown in Table 1. The main point is that, except for $HOXeO_3^-$ and some related species which contain the OH group bonded to xenon, no compounds with more complex ligands were known until very recently.

The synthesis of new xenon compounds with more complex ligands has been accomplished by substitution reactions of the xenon fluorides. The majority of the compounds obtained have been prepared using XeF_2, but examples are also known for XeF_4 and XeF_6. The general method used involves reaction of the xenon fluoride with an appropriate oxyacid. The reactions are usually carried out stoichiometrically at low temperatures, depending on the stability of the compound formed. Another method employed successfully in one case is the reaction of the xenon fluoride with the anhydride of the oxyacid. The generalized Eqs. (1) and (2) illustrate the reactions involved with XeF_2, where ROH is an oxyacid and ROR is an anhydride. The new compounds can be thought of as xenon esters and the reactions are summarized in Table 2.

$$n\text{ROH} + \text{XeF}_2 \rightarrow$$
$$(\text{RO})_n\text{XeF}_{2-n} + n\text{HF} \quad (n=1,2) \quad (1)$$
$$n\text{ROR} + \text{XeF}_2 \rightarrow$$
$$(\text{RO})_n\text{XeF}_{2-n} + n\text{RF} \quad (n=1,2) \quad (2)$$

All the new xenon esters are solids at 22°C, except $FXeOTeF_5$, $F_2Xe(OSO_2F)_2$, and $F_4Xe(OSO_2F)_2$, which are liquids. The new esters are thermodynamically unstable with respect to decomposition but exhibit a wide range of kinetic stabilities, with $FXeOTeF_5$ and $Xe(OTeF_5)_2$ being the most stable. The decomposition reactions may involve radical intermediates of type RO, where RO is SO_3F and NO_3, for example, in the xenon fluorosulfates and xenon nitrates. A summary of the observed decomposition products is covered in Table 3, along with the temperature and qualitative rate of reaction.

The formation of a stable peroxide in the decomposition of the fluorosulfates is easily explained by the formation of an intermediate RO radical. Several of the other observed products can also be explained by invoking the formation of RO radicals, but no proof for this has been obtained.

The new xenon esters point toward the exis-

Table 2. New xenon esters

Fluoride	ROH or ROR	Esters
XeF_2	CF_3CO_2H	$FXeOC(O)CF_3$, $Xe[OC(O)CF_3]_2$
XeF_2	$HONO_2$	$FXeONO_2$, $Xe(ONO_2)_2$*
XeF_2	$HOTeF_5$	$FXeOTeF_5$, $Xe(OTeF_5)_2$
XeF_2	$HOClO_3$	$FXeOClO_3$, $Xe(OClO_3)_2$
XeF_2	$P_2O_3F_4$	$FXeOPOF_2$, $Xe(OPOF_2)_2$
XeF_2	$HOSO_2F$	$FXeOSO_2F$, $Xe(OSO_2F)_2$
XeF_4	$HOSO_2F$	$F_2Xe(OSO_2F)_2$
XeF_6	$HOSO_2F$	$F_4Xe(OSO_2F)_2$, F_5XeOSO_2F

*Tentative.

Table 1. Xenon species with F, O, and OH

Ligand	Examples
F	XeF_2, XeF_4, XeF_6, XeF^+, $Xe_2F_3^+$, XeF_5^+, XeF_7^-*, XeF_8^{2-}*
O and F	$XeOF_4$, XeO_2F_2, XeO_3F_2, XeO_3F^-, $XeOF_5^-$*, $XeOF_3^+$*
O and OH	XeO_3, XeO_4, XeO_6^{4-}, $HOXeO_3^-$

*Postulated.

tence of many other interesting xenon compounds. The variety of Xe(II) compounds can probably be extended to Xe(IV) and Xe(VI) since the latter compounds appear to be at least as stable as the corresponding Xe(II) esters. A xenon bond to elements other than oxygen or fluorine has not been conclusively demonstrated, but the synthesis of compounds containing Xe-S and Xe-N bonds, for example, may be possible.

Reactions of xenon esters. The xenon esters are all strong oxidizing agents and resemble the parent xenon fluorides in this respect. Their chemistry has been most thoroughly investigated for $FXeOTeF_5$ and $Xe(OTeF_5)_2$, and the reactions labeled (3), (4), and (5) are illustrative. Reaction (3)

$$Xe(OTeF_5)_2 + 2CF_3CO_2H \rightarrow$$
$$Xe[OC(O)CF_3]_2 + 2HOTeF_5 \quad (3)$$

$$FXeOTeF_5 + CsF \begin{array}{c} \nearrow TeF_6 + [Cs(OXeF)] \rightarrow \\ CsF + Xe + \tfrac{1}{2}O_2 \\ \searrow XeF_2 + Cs(OTeF_5) \end{array} \quad (4)$$

$$FXeOTeF_5 + AsF_5 \rightarrow (F_5TeOXe)^+(AsF_6)^- \quad (5)$$

indicates that the xenon esters may be useful in preparing other compounds if the equilibrium can be forced to the right, as is done in this case by removing the volatile $HOTeF_5$. Reaction (4) occurs by two different routes corresponding to a nucleophilic attack by fluoride ion on tellurium or xenon. Reaction (5) involves fluoride ion donation, and comparison with XeF_2 shows that $FXeOTeF_5$ is a weaker donor.

Reactions of $F_4Xe(OSO_2F)_2$ with excess fluorine and CsF are shown in Eqs. (6), (7), and (8). The for-

$$F_4Xe(OSO_2F)_2 + 2F_2 \xrightarrow[F_2]{50°} XeF_6 + 2FOSO_2F \quad (6)$$

$$F_4Xe(OSO_2F)_2 + CsF \rightarrow$$
$$CsF \cdot XeF_4(OSO_2F)_2 \quad (7)$$

$$CsF \cdot XeF_4(OSO_2F)_2 + 3CsF \rightarrow$$
$$2CsOSO_2F + CsF \cdot XeF_6 \quad (8)$$

mation of $FOSO_2F$ in the reaction with fluorine, reaction (6), is good evidence for formulating the compound as an ester. The $FOSO_2F$ results from direct reaction of fluorine with the compound and not from reactions involving decomposition prod-

Table 3. Decomposition products of xenon esters

Ester	Rate at 22°C	Products
$FXeOC(O)CF_3$	Fast	Xe, XeF_2, CO_2, C_2F_6
$Xe[OC(O)CF_3]_2$	Fast	Xe, CO_2, C_2F_6
$FXeONO_2$*	Fast (−20°C)	Xe, XeF_2, N_2O_5
$Xe(ONO_2)_2$*	Fast (−20°C)	Xe, N_2O_5
$FXeOTeF_5$	Slow (130°C)	Xe, tellurium oxyfluorides
$Xe(OTeF_5)_2$	Slow (130°C)	$Xe, O_2, F_5TeOTeF_5$
$FXeOClO_3$	Fast	Xe, XeF_2, O_2, Cl_2O_7
$Xe(OClO_3)_2$	Fast (−20°C)	Xe, O_2, Cl_2O_7
$FXeOPOF_2$	Moderate	Xe, O_2, POF_3
$Xe(OPOF_2)_2$	Moderate	$Xe, O_2, P_2O_3F_4$
$FXeOSO_2F$	Slow	$Xe, XeF_2, S_2O_6F_2$
$Xe(OSO_2F)_2$	Moderate	$Xe, S_2O_6F_2$
$F_2Xe(OSO_2F)_2$	Slow	$Xe, XeF_4, S_2O_6F_2$
$F_4Xe(OSO_2F)_2$	Slow	$XeF_4, S_2O_6F_2$
F_5XeOSO_2F	Slow	$XeF_4, XeF_6, S_2O_6F_2$

*In the presence of excess NO_2.

ucts. The xenon(IV) and xenon(VI) fluorosulfates also react with water, forming the explosive xenon trioxide, in much the same way as the parent fluorides do. Rigorous exclusion of water is thus required in all experimental work. The xenon(II) compounds also react rapidly with water but xenon and oxygen are the products, making these compounds much safer to handle. However, extreme caution is necessary in working with any xenon compound since several, including XeO_3, XeO_4, $FXeOC(O)CF_3$, and $Xe[OC(O)CF_3]_2$, are highly explosive. Any new xenon compound must be assumed to be explosive until proven otherwise.

The relatively few reactions that have been investigated for the xenon esters indicate that the compounds have an interesting chemistry, limited somewhat by their low stability. One obvious use of the new compounds would be as reagents for oxidative addition and substitution reactions. For example, the xenon fluorosulfates should be powerful fluorosulfonating reagents, and reactions such as Eqs. (9) and (10), where M is a metal or nonmetal and X is chlorine or bromine, might offer advantages over existing methods for the preparation of fluorosulfates.

$$Xe(OSO_2F)_2 + M \rightarrow M(OSO_2F)_2 + Xe \quad (9)$$
$$Xe(OSO_2F)_2 + MX_2 \rightarrow$$
$$M(OSO_2F)_2 + Xe + X_2 \quad (10)$$

Xenon fluorides and organic compounds. Another area of current interest in xenon chemistry is the use of the xenon fluorides as selective fluorinating reagents in organic chemistry. Controlled fluorination in aromatic systems is difficult by existing methods, and the xenon fluorides, particularly XeF_2, show considerable promise in this area. The reactivity of the xenon fluorides with unsaturated organic compounds increases in the order $XeF_2 < XeF_4 < XeF_6$. Xenon hexafluoride is so reactive that its practical use seems unlikely. However, XeF_2 and, to a lesser extent, XeF_4 offer promising one-step syntheses of previously difficult to obtain compounds. For example, XeF_2 and XeF_4 react with excess benzene at 25°C to give fluorobenzene and other products, as shown in Eqs. (11) and (12).

$$XeF_2 + C_6H_6 \rightarrow$$
$$C_6H_5F(68\%) + Xe + HF + \text{Other} \quad (11)$$
$$XeF_4 + C_6H_6 \rightarrow$$
$$C_6H_5F(13\%) + Xe + HF + \text{Other} \quad (12)$$

The other products consist mainly of small amounts of biphenyl, fluorobiphenyls, and unidentified tars. The reaction involving XeF_2 is strongly catalyzed by HF and is autocatalytic, once initiated by HF. Reactions with substituted benzene derivatives also occur in good yield, as indicated in Eqs. (13) and (14), with XeF_2 in carbontetrachloride solvent. The ortho/meta/para ratio is shown

$$CH_3O\text{---}C_6H_5 + XeF_2 \xrightarrow{-20°C}$$
$$CH_3O\text{---}C_6H_4F(65\%) \quad (13)$$
$$\text{o/m/p, } 12.2/1.0/25.9$$
$$CF_3\text{---}C_6H_5 + XeF_2 \xrightarrow{25°C}$$
$$CF_3\text{---}C_6H_4F(75\%) \quad (14)$$
$$\text{o/m/p, } 0/9.4/1.0$$

beneath the products. The colors observed in the reactions indicate the presence of free radicals, and radical 1+ cations have been identified by electron spin resonance measurements in the XeF_2 reactions. These reactions are, in fact, very good sources for radical cations of polyphenyls, particularly from compounds with relatively high oxidative potentials.

The product isomer distributions in the reactions of XeF_2 are those expected for electrophilic aromatic substitution reactions, and the use of XeF_2 is an attractive alternative to conventional methods for the preparation of fluoronated aromatic compounds. While xenon is expensive (~$5.00/g), the xenon can be easily recovered and converted back to the fluoride for use again. Xenon difluoride is the easiest of the xenon fluorides to prepare and handle, and increased use of XeF_2 as a fluorinating reagent in organic chemistry is likely.

For background information see XENON; XENON COMPOUNDS in the McGraw-Hill Encyclopedia of Science and Technology.

[DARRYL D. DES MARTEAU]

Bibliography: N. Bartlett et al., *Chem. Commun.*, no. 703, 1969; M. Eisenberg and D. D. DesMarteau, *J. Amer. Chem. Soc.*, 92:4759, 1970; M. Eisenberg and D. D. DesMarteau, *J. Inorg. Nucl. Chem. Lett.*, 6:29, 1970; M. J. Shaw, H. H. Hyman, and R. Filler, *J. Amer. Chem. Soc.*, 92:6498, 1970; T. C. Shieh et al., *J. Org. Chem.*, 35:4020, 1970; F. Sladky, *Monatsh. Chem.*, 101:1559, 1970.

Xylem

The study of a complex biological phenomenon such as xylem element formation can be approached from many different directions. On the physiological level one can ask how the formation of tracheary elements is regulated within the plant. That is, what are the various hormonal, nutritional, and biophysical gradients within a given plant organ and how do these gradients influence the occurrence of tracheary elements? This is very different from the molecular approach, which asks how a cell's internal activities are regulated so that, upon receiving the appropriate environmental stimuli, it may alter its biosynthetic activities to produce those macromolecules characteristic of tracheary elements. Still another approach is the structural, which seeks to determine the precise pattern and form in which these macromolecules are deposited to create the tracheary element. Most of the recent work on tracheary element formation has utilized both the structural and physiological approaches.

Hormonal influence. A number of naturally occurring chemical substances and physical conditions are known to influence the formation of tracheary elements. These include the plant hormones auxin, cytokinins, gibberellins, and abscisic acid; metabolic substrates such as sucrose and proline; mineral ions such as calcium; and physical parameters such as pressure and light quality. The now-classic work of W. P. Jacobs demonstrated that a gradient of auxin exists within the plant and that tracheary elements are formed only when the auxin concentration is maintained above a certain threshold. While relatively little is known about the affected molecular processes which make auxin essential for the formation of tracheary elements, there is little doubt that this hormone plays a very basic role in the differentiation process.

Auxin. In contrast to auxin, some of the other substances which have been shown to influence tracheary element formation probably do so quite indirectly. The physiological regulation of xylem differentiation and the role of certain hormones in this regulation are well illustrated by the recent work of Terry Shininger, who studied the factors controlling xylem fiber formation in the cocklebur (*Xanthium pensylvanicum*). He found that decapitation and removal of the leaves from the stem of this plant brought about a decline in cambial activity. Although some cambial activity continued in the stem of the decapitated plants, none of the cambial derivatives differentiated as xylem fibers. In the intact plant the majority of the cambial derivatives would have differentiated as fibers on the xylem side of the cambium.

Of course, the lateral buds began to grow after decapitation; as new leaves were formed and developed from the growing lateral buds, fiber formation from the cambial derivatives resumed. When a single leaf was allowed to develop from a single lateral bud, all subsequent leaves being destroyed early in their development, fiber formation was found to resume in the stem below the lateral bud after this single leaf reached the stage of growth in which its area was expanding at a maximum rate. As the growth rate of the leaf declined with the approach of maturity, fiber formation from the cambium of the stem ceased.

These results suggest that substances produced by rapidly growing leaves diffuse or are transported into the stem, where they regulate cambial activity and xylem formation. Since leaves are known to produce auxin, it would not be unreasonable to expect auxin to be the principle leaf substance responsible for this regulatory phenomenon. Indeed, some of Shininger's additional data support this hypothesis. He applied various growth regulators to the decapitated cocklebur plants in an effort to find a hormone, or combination of hormones, which would lead to normal cambial activity and differentiation of cambial derivatives. Some of his results are shown in the illustration. Decapitation occurred after a considerable amount of secondary tissue had been produced by the cambium; the continued division of the cambial initials after decapitation, with little or no differentiation of the cambial derivatives, produced a tissue region in the xylem which lacked tracheary elements. Several days after decapitation, auxin was applied to the stem and development was allowed to progress for several additional days before the section shown in part a of the illustration was made. It is apparent that the application of auxin restored the differentiation of xylem fibers from the cambial derivatives that were formed during the auxin treatment. Furthermore, quantitative data demonstrated that the number of cells produced by the cambium and the number of tracheary elements formed increased as the concentration of auxin in the tissue increased. Thus auxin treatment brought about two distinct responses in this system. First, it increased the number of cells which potentially could become tracheary elements by stimulating cambial cell division. Second, auxin brought about the differentiation of tracheary elements from the newly formed cambial derivatives.

Gibberellic acid. It is apparent from the behavior of this system that any factor which could further increase the rate of cambial cell division, in the presence of an optimal concentration of auxin, could thereby increase the number of tracheary elements formed in a given time period. Such a factor need not play any direct role in the initiation of tracheary elements in order to stimulate xylem element production. Gibberellic acid appears to play such a role in decapitated cocklebur plants. Applications of gibberellic acid alone stimulated the division of the cambial initials but did not lead to the formation of tracheary elements from the cambial derivatives. When decapitated plants were treated with various combinations of auxin and gibberellin, the activity of the cambium was not only increased, but there was a tendency for larger numbers of xylem fibers to be produced from the cambial derivatives over what would be expected with an auxin treatment alone.

Macromolecule synthesis. One of the assumptions that might be made about tracheary element formation is that the particular form of each of the different kinds of tracheary elements is brought about by the carefully coordinated synthesis of specific macromolecules. For example, vessel elements are characterized by the presence of a perforation plate, a region of the element where the primary cell wall has been removed between two adjacent vessel elements. Presumably the perforation plate facilitates the movement of water through the mature functioning vessels.

T. P. O'Brien showed that the removal of portions of the primary cell wall in differentiating vessel elements is brought about by a hydrolysis of the wall components which begins after secondary wall deposition is complete and when the general autolysis of the vessel element cytoplasm is initiated. Earlier he postulated that the lignification of certain regions of the primary cell wall and the secondary wall protects these structures from attack during autolysis. Only the unlignified portions of the primary wall are digested away. Of course, it is clear that such an autolysis must be brought about by enzymatic attack. Presumably cellulases, hemicellulases, and pectinases would be necessary for the complete dissolution of primary cell wall material. Recently A. R. Sheldrake demonstrated that the xylem tissue of *Acer pseudoplatanus* contains several cellulase isozymes, none of which were present in the cambium. These cellulase enzymes may represent some of the characteristic macromolecules which must be synthesized by a differentiating cell in order to produce a vessel element.

Several years ago P. B. Gahan demonstrated that differentiating xylem elements contain an abundance of lysosome-like organelles. He used the acid phosphatase localization procedure which had been shown to be diagnostic for lysosomes in animal tissues. Although Gahan did not demonstrate the localization of any additional lysosomal enzymes in these particles and although the use of acid phosphatase staining as a criterion for the

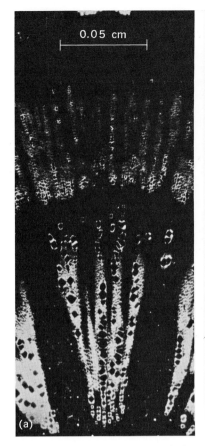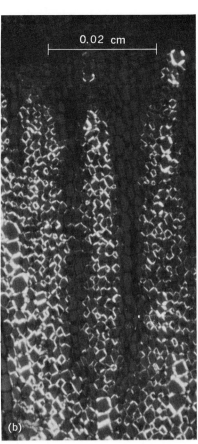

Cross sections taken from stem of auxin-treated, decapitated *Xanthium* plant. Plant was decapitated 72 hr prior to the addition of the auxin, and treatment continued for an additional 12 days. (*a*) One entire vascular bundle as seen under polarized light. Differentiated xylem elements, with their birefringent secondary walls, appear light against a dark background that represents undifferentiated parenchyma. Xylem rays are apparent on either side of the vascular bundle, the xylem of which is discontinuous due to the failure of the cambial derivatives to differentiate during the 72 hr following decapitation. (*b*) Xylem tissue under higher magnification, with differentiated xylem fibers produced during the auxin treatment.

presence of lysosomes in plants has been seriously questioned, he did demonstrate that the behavior of these particles was at least analogous to that of the lysosome. That is, the particles accumulated and remained intact during secondary wall deposition, but ruptured, releasing their contents to the cytoplasm, as autolysis began.

Recent information obtained by Sheldrake on the behavior of the xylem cellulases suggests that these enzymes might be lysosomal. He found that cellulase activity of homogenates prepared from differentiating xylem elements could be sedimented by centrifugation at 100,000g, suggesting that these enzymes were part of larger cellular organelles whose combined mass was much greater than that of the cellulase molecules alone. Furthermore, the cellulase activity could be released from these particles by detergent treatment, suggesting that these particles were membrane-bound. While considerable additional work will be necessary to clarify this point, a plausible hypothesis suggested by Sheldrake's results is that differentiating tracheary elements synthesize a number of hydrolytic, degradative enzymes which are sequestered from the remainder of the cytoplasm in lysosomes. Then, on the appropriate signal, the lysosome membrane ruptures, releasing the enzymes it contained to the cytoplasm and initiating the autolytic process. One important biochemical difference between vessel elements and tracheids may be that the lysosomes of the latter lack cellulases, while these enzymes would be present in the lysosomes of the vessel elements.

Lignin. B. J. Fergus and D. I. Goring recently demonstrated that the lignin associated with vessel secondary walls was significantly different from the lignin deposited in the secondary walls of fibers in the xylem of birch. While it is not surprising to find that there are macromolecular differences between the various kinds of tracheary elements, differences which would be manifested during the formation of the elements, the discovery that some of the different kinds of tracheary elements deposit a characteristic lignin, within a given species, was unexpected. The lignins are complex polymers, formed by the action of peroxidase on various phenylpropane derivatives. Several years ago it was shown that lignin would be formed if filter paper was incubated in a solution containing peroxidase, peroxide, and a phenylpropane derivative. The filter paper simply acted as a substrate upon which the lignin was deposited. Some recent work of P. K. Hepler showed that differentiating xylem elements excrete peroxidase, along with the carbohydrate matrix of the secondary wall. Judging from the experiment with filter paper, one might expect the lignification process simply to be an extracellular chemical reaction which occurred within the cell walls, outside the control of the living cell, once the chemical reactants had been excreted.

The apparent conflict between this picture and the existence of cell type–specific lignins may be more apparent than real. Phenylalanine acts as the principal compound from which most of the phenylpropane lignin monomers are derived. The enzyme phenylalanine ammonia lyase removes a molecule of ammonia from phenylalanine to produce cinnamic acid. Cinnamic acid is then further modified by hydroxylation and methylation to produce the principle lignin monomers. These additional reactions are also mediated by specific enzymes. Recently Philip Rubery and coworkers demonstrated that the activity of phenylalanine ammonia lyase greatly increased during the formation of xylem elements. Because of the nature of the experimental material employed for these studies, the logical explanation for this increased enzyme activity is that differentiating xylem elements synthesize or activate those enzymes necessary to produce their own phenylpropane lignin monomers. If the monomers are characteristic for the lignin of a given xylem cell type, it must be because that cell type synthesizes a unique group of phenylpropane-modifying enzymes, in addition to the basic phenylalanine ammonia lyase.

For background information see AUXIN; GIBBERELLIN; PLANT HORMONES; XYLEM in the McGraw-Hill Encyclopedia of Science and Technology.

[D. E. FOSKET]

Bibliography: B. J. Fergus and D. I. Goring., *Holzforschung*, 24:113–117, 1970; A. R. Sheldrake, *Planta*, 95:167–178, 1970; T. L. Shininger, *Amer. J. Bot.*, 57:769–781, 1970; T. L. Shininger, *Plant Physiol.*, 47:417–422, 1971; J. G. Torrey, D. E. Fosket, and P. K. Hepler, *Amer. Sci.*, 59:338–352, 1971.

McGRAW-HILL YEARBOOK OF SCIENCE AND TECHNOLOGY

List of Contributors

List of contributors

A

Alexandridis, George G. *Lockwood, Kessler & Bartlett, Syosset, N.Y.* HIGHWAY ENGINEERING.

Appelman, Dr. Evan. *Chemistry Division, Argonne National Laboratory.* HYPOFLUOROUS ACID.

Auerbach, Dr. Robert. *Department of Zoology, University of Wisconsin.* IMMUNOLOGY.

B

Bach-y-Rita, Dr. P. *Presbyterian Hospital, University of the Pacific School of Medicine, San Francisco, Calif.* PROSTHESIS (coauthored).

Bakker, Dr. Robert T. *Museum of Vertebrate Paleontology, Harvard University.* DINOSAUR (in part).

Barley, Dr. K. P. *Waite Agricultural Institute, University of Adelaide, Australia.* ROOT (BOTANY).

Bauer, Dr. E. *Physikalisches Institut der Technischen Universität Clausthal, West Germany.* ELECTRON DIFFRACTION, HIGH-ENERGY.

Baumeister, Theodore. *Consulting engineer, Yonges Island, S.C.* POWER PLANT.

Bird, Dr. John M. *Department of Geology, State University of New York at Albany.* OROGENY (coauthored).

Black, Dr. C. C., Jr. *Department of Biochemistry, University of Georgia.* PHOTORESPIRATION.

Blair, Lewis N. *The Anaconda Company, Butte, Mont.* ENVIRONMENTAL ENGINEERING (coauthored).

Bock, Dr. Walter. *Department of Biological Science, Columbia University.* STRIGIFORMES.

Borgman, Dr. Leon E. *Department of Geology, University of Wyoming.* RISK EVALUATION IN ENGINEERING.

Breed, Dr. William J. *Museum of Northern Arizona, Flagstaff, Ariz.* TRIASSIC.

Brown, Dr. Frederick, C. *Department of Physics, University of Illinois.* ABSORPTION OF ELECTROMAGNETIC RADIATION.

Burger, Dr. Max M. *Department of Biochemical Sciences, Princeton University.* CELL SURFACE (coauthored).

C

Cáceres, Dr. César A. *Department of Clinical Engineering, George Washington University.* ELECTRODIAGNOSIS.

Cairncross, Dr. Allan. *Central Research Department, E.I. du Pont de Nemours & Company, Wilmington, Del.* COPPER CHEMISTRY.

Carl, Dr. Philip. *Department of Microbiology, University of Illinois.* POLYMERASE.

Casey, Dr. H. C., Jr. *Bell Laboratories, Murray Hill, N.J.* SEMICONDUCTOR.

Chase, Dr. W. T. *Freer Gallery Laboratory, Smithsonian Institution.* SCIENCE IN ART.

Claassen, Dr. Howard. *Chemistry Division, Argonne National Laboratory.* SPECTROSCOPY.

Clark, Dr. Francis E. *Nitrogen Laboratory, U.S. Department of Agriculture, Fort Collins, Colo.* SOIL MICROORGANISMS.

Coats, Dr. Keith H. *Intercomp, Houston, Tex.* PETROLEUM RESERVOIR ENGINEERING (coauthored).

Coles, Dr. Richard W. *Tyson Research Center, Washington University.* BEAVER.

Collins, Dr. Carter. *Presbyterian Hospital, University of the Pacific School of Medicine, San Francisco, Calif.* PROSTHESIS (coauthored).

Cool, Prof. Terrill A. *Department of Thermal Engineering, Cornell University.* LASER, CHEMICAL.

Crowson, Brig. Gen. Delmar L. *Office of Safeguards, U.S. Atomic Energy Commission.* NUCLEAR MATERIALS SAFEGUARDS.

Curtin, John L. *Printing Products Division, 3M Company, St. Paul, Minn.* PRINTING PLATE (in part).

D

Dempsey, Dr. J. R. *Intercomp, Houston, Tex.* PETROLEUM RESERVOIR ENGINEERING (coauthored).

DeNoyer, Dr. John M. *Earth Observations Programs, National Aeronautics and Space Administration, Washington, D.C.* EARTH-RESOURCE SATELLITES.

DesMarteau, Dr. Darryl D. *Department of Chemistry, Kansas State University.* XENON COMPOUNDS.

Dewey, Dr. John F. *Department of Geology, State University of New York at Albany.* OROGENY (coauthored).

Downer, Hugh C. *Marine Group, Marcona Corporation, San Francisco, Calif.* SHIP, MERCHANT.

Drake, Prof. Charles L. *Department of Earth Sciences, Dartmouth College.* TECTONOPHYSICS.

Drickamer, Prof. Harry. *Department of Chemical Engineering, University of Illinois.* HIGH-PRESSURE PHYSICS.

E

El Abiad, Dr. Ahmed H. *Department of Electrical Engineering, Purdue University.* ELECTRIC POWER SYSTEM, INTERCONNECTING.

F

Fields, Dr. Paul R. *Chemistry Division, Argonne National Laboratory.* ELEMENTS (CHEMISTRY).

Finsterbusch, Gail W. *Nutrition consultant, Washington, D.C.* FOOD.

Flory, Robert E. *David Sarnoff Research Center, RCA Laboratories, Princeton, N.J.* VIDEO RECORDING AND PLAYBACK.

Foley, J. M. *Printing Products Division, 3M Company, St. Paul, Minn.* PRINTING, COLOR (in part).

Fosket, Dr. Donald E. *Department of Cell and Developmental Biology, University of California, Irvine.* XYLEM.

Freiman, Dr. David G. *Beth Israel Hospital, Harvard School of Medicine, Boston, Mass.* BERYLLIUM DISEASE.

Friend, Dr. James P. *Department of Meteorology and Oceanography, New York University.* VOLCANO.

Fripiat, Prof. J. J. *Laboratoire de Physico-Chimie Minérale, Institut des Sciences de la Terre, Université de Louvain, Belgium.* CLAY.

G

Garman, Dr. Willard H. *National Fertilizer Development Center, TVA.* SOIL (in part).
Gassmann, George J. *U.S. Air Force Cambridge Research Laboratories, Bedford, Mass.* IONOSPHERE (in part).
Gatehouse, R. N. B. *Brookes and Gatehouse, Ltd., Lymington, Hampshire, England.* NAVIGATION SYSTEMS, ELECTRONIC.
George, Harvey F. *Gravure Research Institute, Inc., Port Washington, N.Y.* PRINTING PLATE (in part).
Goodall, Dr. McChesney. *The University of Texas Medical Branch, Galveston.* PARKINSON'S DISEASE.
Goodfriend, Lewis. *Goodfriend Ostergaard Associates, Cedar Knolls, N.J.* NOISE, ACOUSTIC.
Gould, Dr. Stephen Jay. *Museum of Comparative Zoology, Harvard University.* JURASSIC.
Grant, Prof. Nicholas. *Center for Materials Science and Engineering, Massachusetts Institute of Technology.* POWDER METALLURGY.
Grégory, Dr. B. P. *Laboratoire de Physique, École Polytechnique, Paris, France.* PARTICLE ACCELERATOR (in part).
Grell, Dr. Rhoda F. *Biology Division, Oak Ridge National Laboratory.* CELL DIVISION (in part).
Griffiths, Prof. Robert B. *Department of Physics, Carnegie-Mellon University.* CRITICAL POINT.
Groom, Dr. Donald E. *University of Utah.* COSMIC RAY (in part).
Guyer, Dr. R. A. *Department of Physics and Astronomy, University of Massachusetts.* QUANTUM SOLID.

H

Hagstrum, Dr. Homer D. *Bell Laboratories, Murray Hill, N.J.* SURFACE PHYSICS.
Halpern, Dr. Jack. *Department of Chemistry, University of Chicago.* COORDINATION CHEMISTRY.
Hasse, Raymond W., Jr. *Ocean Science Department, Naval Underwater Systems Center, New London, Conn.* BUOY.
Heald, Dr. Walter R. *Northeast Watershed Research Center, U.S. Department of Agriculture, University Park, Pa.* WASTES, AGRICULTURAL.
Heindl, L. A. *U.S. National Committee for the International Hydrological Decade, National Academy of Sciences, Washington, D.C.* GREAT LAKES.
Hill, John G. *Naval architect and marine engineer, Oxford, Md.* PROPELLER, MARINE.
Hommersand, Dr. Max. *Department of Botany, University of North Carolina.* ALGAE.
Hooker, Dr. Arthur L. *Department of Plant Pathology, University of Illinois.* CORN.
Horner, Dr. Harry T., Jr. *Department of Botany and Plant Pathology Laboratory, Iowa State University.* LEAF.
Hundhausen, Dr. A. J. *High Altitude Observatory, Boulder, Colo.* SOLAR WIND.

J

Jones, Dr. Lawrence W. *Harrison M. Randall Laboratory of Physics, University of Michigan.* COSMIC RAY (in part).
Jones, Dr. Russell L. *Department of Botany, University of California, Berkeley.* SEED (BOTANY).
Juniper, Dr. Barrie. *School of Botany, University of Oxford, England.* EPIDERMIS (PLANT).

K

Kahn, Fazlur R. *Skidmore, Owings & Merrill, Chicago, Ill.* BUILDINGS.
Kastner, Dr. Jacob. *Technical Assessment Branch, Division of Radiological and Environmental Protection, U.S. Atomic Energy Commission.* COSMIC RAY (in part).
Kaufman, Dr. Larry. *ManLabs Inc., Cambridge, Mass.* EQUILIBRIUM, PHASE.
Kersh, Claude. *Rutherford Machinery Division, Sun Chemical Corporation, East Rutherford, N.J.* PRINTING, COLOR (in part).
King, John A. *Space/Aeronautics Magazine.* SPACE FLIGHT.
Kissmeyer-Nielsen, Dr. F. *Blood Bank and Blood Grouping Laboratory, Arhus Kommunehospital, Denmark.* TRANSPLANTATION BIOLOGY.
Kruger, Prof. Paul. *Department of Civil Engineering, Stanford University.* NUCLEAR EXPLOSION ENGINEERING.
Kupke, Dr. Donald W. *Department of Biochemistry, University of Virginia School of Medicine.* DENSITY.

L

Laird, Frank J., Jr. *The Anaconda Company, Butte, Mont.* ENVIRONMENTAL ENGINEERING (coauthored).
Lamb, Prof. R. C. *Physics Department, Iowa State University.* ELEMENTARY PARTICLE (in part).
Lassila, Prof. K. E. *Physics Department, Iowa State University.* ELEMENTARY PARTICLE (in part).
Levine, Dr. Barry F. *Bell Laboratories, Murray Hill, N.J.* OPTICS, NONLINEAR.
Li, Prof. Choh Hao. *Hormone Research Laboratory, University of California, San Francisco.* GROWTH HORMONE.
Lima-de-Faria, Prof. A. *Institute of Molecular Cytogenetics, University of Lund, Sweden.* CELL DIVISION (in part).
Lintz, Dr. J. *University of Nevada.* TERRAIN SENSING, REMOTE.
Lissaman, Dr. P. B. S. *Continuum Mechanics Laboratory, Northrop Corporate Laboratories, Hawthorne, Calif.* FLIGHT.

M

McAdam, Dr. Dale W. *Department of Psychology, University of Rochester.* SPEECH (coauthored).
McAshan, Dr. M. S. *High Energy Physics Laboratory, Stanford University.* PARTICLE ACCELERATOR (in part).
McCrea, Prof. W. H. *Astronomy Centre, University of Sussex, England.* ENERGY SOURCES IN GALAXIES AND QUASARS.
McCully, Dr. Kilmer S. *Department of Pathology, Massachusetts General Hospital, Boston, Mass.* ARTERIOSCLEROSIS.
McIntyre, Dr. D. S. *Commonwealth Scientific and Industrial Research Organization, Canberra, Australia.* SOIL (in part).
Martin, Stanley B. *Stanford Research Institute.* URBAN FIRES.
Mattern, William. *Maynard Research Council, Pittsburgh, Pa.* WORK MEASUREMENT.
Mayland, Dr. H. F. *Snake River Conservation Research Center, U.S. Department of Agriculture, Kimberly, Idaho.* PLANT GROWTH.
Miaggiano, Dr. Vincenzo. *Basel Institute for Immunology, Switzerland.* ANTIGEN.
Migdal, Philip N. *Micronetics Division, Teledyne Systems Company, San Diego, Calif.* MARINE NAVIGATION (in part).
Mitchell, Dr. John W. *Crops Research Division, U.S. Department of Agriculture, Beltsville, Md.* BRASSIN.

Mortenson, Dr. Leonard E. *Department of Biological Science, Purdue University.* NITROGEN FIXATION.

Mortland, Dr. M. M. *Department of Crop and Soil Science, Michigan State University.* SOIL CHEMISTRY.

Mottet, Dr. N. Karle. *Department of Pathology, University of Washington.* PATHOLOGY OF HEAVY METALS.

Mourad, Dr. A. George. *Marine Geodesy, Battelle Columbus Laboratories, Columbus, Ohio.* GEODESY.

N

Namias, Dr. Jerome. *Extended Forecast Division, National Weather Service, Washington, D.C.* MARINE INFLUENCE ON WEATHER AND CLIMATE.

Noonan, Dr. Kenneth D. *Department of Biochemical Sciences, Princeton University.* CELL SURFACE (coauthored)

Norris, Dr. H. Thomas. *Veterans Administration Hospital, Seattle, Wash.* CHOLERA.

Nowacki, Dr. Horst N. *Department of Naval Architecture and Marine Engineering, University of Michigan.* SHIP DESIGN.

O

O'Callaghan, Dr. Fred G. *Lindheimer Astronomical Research Center, Northwestern University.* ASTRONOMICAL INSTRUMENTS.

Olmsted, Leonard M. *Electrical World Magazine, McGraw-Hill Publications, New York, N.Y.* ELECTRICAL UTILITY INDUSTRY.

Ostrom, Dr. John H. *Peabody Museum of Natural History, Yale University.* DINOSAUR (in part).

P

Pallister, Dr. A. E. *Kenting Ltd., Alberta, Canada.* PROSPECTING, PETROLEUM.

Palmer, Dr. Patrick. *Department of Astronomy and Astrophysics, University of Chicago.* INTERSTELLAR MATTER.

Panish, Dr. Morton B. *Bell Laboratories, Murray Hill, N.J.* LASER, SEMICONDUCTOR INJECTION.

Patrick, Dr. W. H. *Agronomy Department, Louisiana State University.* SOIL MANAGEMENT.

Price, H. L. *U.S. Atomic Energy Commission.* REACTOR LICENSING, NUCLEAR.

Price, Dr. Harvey S. *Intercomp, Houston, Tex.* PETROLEUM RESERVOIR ENGINEERING (coauthored).

Price, Dr. Richard. *Department of Physics, University of Utah.* GRAVITATIONAL COLLAPSE.

Prothero, William A., Jr. *Institute of Geophysics and Planetary Physics, University of California, San Diego.* EARTH TIDES.

R

Rainwater, Dr. E. H. *Tenneco Oil Company, Houston, Tex.* PETROLEUM GEOLOGY.

Richards, Robert B. *De Leuw, Cather and Company, Chicago, Ill.* TRANSPORTATION ENGINEERING.

Rijke, Dr. A. M. *Department of Materials Science, University of Virginia.* FEATHER (BIRD).

Rotariu, Dr. George J. *Division of Isotope Development, U.S. Atomic Energy Commission.* PROCESS RADIATION.

Rousseau, Dr. Denis L. *Bell Laboratories, Murray Hill, N.J.* POLYWATER.

S

Sagan, Dr. Carl. *Center for Radiophysics and Space Research, Cornell University.* VENUS.

Satinoff, Dr. Evelyn. *Department of Psychology, University of Pennsylvania.* HIBERNATION.

Scraba, Dr. Douglas. *Biochemistry Department, University of Alberta, Canada.* ANIMAL VIRUS.

Seidel, Dr. Harold. *Bell Laboratories, Murray Hill, N.J.* AMPLIFIER.

Semrau, Konrad T. *Stanford Research Institute.* AIR POLLUTION.

Silfvast, Dr. William T. *Bell Laboratories, Holmdel, N.J.* LASER, METAL-VAPOR.

Silverman, Dr. Bernard A. *Stratiform Cloud Physics Branch, U.S. Air Force Cambridge Research Laboratories, Bedford, Mass.* WEATHER MODIFICATION.

Simonson, Dr. Roy W. *Soil Conservation Service, U.S. Department of Agriculture, Hyattsville, Md.* SOIL (in part).

Smith, Dr. George E. *Bell Laboratories, Murray Hill, N.J.* CHARGE-COUPLED DEVICE.

van Soest, Dr. Peter J. *Department of Animal Science, Cornell University.* SILICA, BIOLOGY OF.

Sorokin, Dr. Peter P. *Thomas J. Watson Research Center, IBM Corporation, Yorktown Heights, N.Y.* LASER, DYE.

Spencer, Dr. Arthur M., Jr. *Western Company, Richardson, Tex.* OIL AND GAS WELL COMPLETION.

Spindler, John C. *The Anaconda Company, Butte, Mont.* ENVIRONMENTAL ENGINEERING (coauthored).

Stuiver, Dr. Minze. *Departments of Geological Sciences and Zoology, Quaternary Research Center, University of Washington.* RADIOCARBON DATING.

Sturge, Dr. M. D. *Bell Laboratories, Murray Hill, N.J.* JAHN-TELLER EFFECT.

T

Tadano, Dr. Tohru. *Electronic Navigation Aids Division, The Maritime Safety Agency, Tokyo, Japan.* MARINE NAVIGATION (in part).

Tappan, Dr. Helen. *Department of Geology, University of California, Los Angeles.* FAUNAL EXTINCTION.

Taylor, Dr. J. Herbert. *Institute of Molecular Biophysics, Florida State University.* CHROMOSOME.

Temin, Dr. Howard M. *McArdle Laboratory for Cancer Research, University of Wisconsin.* DEOXYRIBONUCLEIC ACID (DNA).

Thurlbeck, Dr. William M. *Department of Pathology, McGill University Faculty of Medicine, Montreal, Canada.* EMPHYSEMA.

U

Umminger, Dr. Bruce. *Department of Biological Sciences, University of Cincinnati.* HOMEOSTASIS.

Unger, Walter H. *The Anaconda Company, Butte, Mont.* ENVIRONMENTAL ENGINEERING (coauthored).

Utlaut, William F. *Institute for Telecommunication Sciences, U.S. Department of Commerce, Boulder, Colo.* IONOSPHERE (in part).

V

Vaughan, Richard D. *Solid Waste Management Office, Environmental Protection Agency, Rockville, Md.* SOLID-WASTE MANAGEMENT.

Volek, Charles W. *Shell Oil Company, Los Angeles, Calif.* PETROLEUM SECONDARY RECOVERY.

W

Wallin, Dr. Jack R. *Iowa State University.* AGRICULTURAL METEOROLOGY.

Wang, Dr. C. G. *Massachusetts Institute of Technology.* NEUTRON STAR.

Wardrop, Prof. A. B. *Department of Botany, La Trobe University, Victoria, Australia.* COLLENCHYMA.

Wellman, Dr. Frederick L. *Department of Plant Pathology, North Carolina State University.* COFFEE.

Westbrook, Dr. Jack H. *Physical Chemistry Laboratory, General Electric Company, Schenectady, N.Y.* INTERMETALLIC COMPOUND.

Whitaker, Dr. Harry A. *Department of Psychology, University of Rochester.* SPEECH (coauthored).

Wiegand, Dr. Clyde E. *Lawrence Radiation Laboratory, University of California, Berkeley.* HADRONIC ATOM.

Woodward, Dr. J. G. *RCA Laboratories, Princeton, N.J.* STEREOPHONIC SOUND, FOUR-CHANNEL.

Wray, Dr. James. *Lindheimer Astronomical Research Center, Northwestern University.* GALAXY CLUSTERS.

Z

Ziegler, Prof. Hubert. *Technische Universität, Institut für Botanik, Munich, West Germany.* MORPHACTIN.

McGRAW-HILL YEARBOOK OF SCIENCE AND TECHNOLOGY

Index

Index

A

AA* gene 400
AA-AJ gene 400
Aagaard, Paul 39
Aarons, Leroy 13
Abell, G. O. 227
ABO blood group system 118, 399
Abscisic acid (influence on xylem tracheary elements) 422
Absorption, thin-film 99
Absorption of electromagnetic radiation 97–99
　soft x-ray absorption 97–99
　thin-film absorption 99
Abyssal Plain 303
Academia Sinica 22, 23
Accelerator: fixed-field alternating-gradient 98
　particle 308–313
　superconducting 308–310
　synthesis of new elements 203–204
Accutron (timing device) 133
Acer pseudoplatanus 423
Acetaldehyde molecule 255
Acid, hypofluorous 248–249
Aconta 109
Acoustics: measurements 132
　noise, acoustic 292–294
　stereophonic sound, four-channel 392–394
ACRS *see* Advisory Committee on Reactor Safeguards (ACRS)
Activation polarization 375
Adenosinemonophosphate (AMP) 240
3′5′-Adenosinemonophosphate 145
Adenosinetriphosphate (ATP) 121, 240, 290, 291, 324, 327
Advisory Committee on Reactor Safeguards (ACRS) 354
AEC *see* Atomic Energy Commission, U.S.
Aeration of soil 375–377
Aerial Ektachrome infrared film 398
Aerial hose 4
Aerial ladder 4
Aerobic soil layer 379
Aerodynamics (bird flight) 219–221
Aerosol (stratospheric sulfate) 412

AFCRL *see* Air Force Cambridge Research Laboratories (AFCRL)
Ag lipoprotein system 118, 119
Agglutinin: binding sites 142
　divalent 143
　monovalent 142–143
Agricultural meteorology 99–101
　computer use in forecasting 101
　rainfall 100
　relative humidity and dew 100
　temperature 100
　temperature-moisture relationship 100
　wind-flow data 100–101
Agricultural wastes 26, 413–415
Air Force Cambridge Research Laboratories (AFCRL) 417
Air pollution 101–106
　air quality management 207
　control of power plant emissions 102
　noncombustion sources emissions control 105–106
　power plant siting 334
　pretreatment of fuels 104–105
　small combustion sources 102
　sulfur dioxide removal from flue gas 102–104
　volcanic sulfur dioxide 411
Air Quality Act of 1967 206
Air quality management 207
Air-sea interactions 277
Airblast 295
Airborne techniques for geodetic coordinates 230
Airey, J. R. 268
Alden StaRRcar 403
Alder 129
Aldrich, Samuel 374
Aleurone cell 360, 361
Alfalfa (silica content) 371
Alfisol 373, 374
Algae 106–112
　cell wall formation 111
　mitotic apparatus 109–110
　neuromotor apparatus 110–111
　origin of cell organelles 111–112
　paleobotany 214–217
　photosynthetic apparatus 107–109

Algae—*cont*.
　phylogenetic relationships 112
Allosaurus 176
Alloy: high chromium 331
　new ship steels 365
　nimonic 254
Alloyed powders 330
Alnus glutinosa (L.) Gaertn. 129
Alopecia 76
Alpha-rhythm activity in EEG 391
Alpine-Himalayan orogenic belt 302–307
ALSEP *see* Apollo Lunar Science Experiment Package (ALSEP)
Altazimuth mount (telescope) 124
Aluminum gallium arsenide 273
Aluminum powder 330
Am (immunoglobulin) 118
American Public Works Association 27
American Society of Civil Engineers 29
Americas Trench 307
Amin, Mohammad 39
Aminotransferase in glycolate pathway 322
Ammonia molecule 255
Ammonite (faunal extinction) 214
Amorphous semiconductor 361
AMP *see* Adenosinemonophosphate
Amphibian, labyrinthodont 406
Amplifier 112–115
　basic features 112–113
　frequency-dependent amplifiers 114
　term feedforward 114
　wideband developments 113–114
AMTRAK 402
α-Amylase 361
Amyloplast 285
Andean orogenic belt 305
Andesite 304, 305
Andromeda Galaxy 226
Ang, Alfredo H.-S. 39
Animal anatomy (flight) 219–221
Animal assemblages 217
Animal feeds (use of food-processing wastes) 414
Animal nutrition (importance of silicon dioxide) 371

Animal pathology: beryllium disease 126–128
　cholera 144–146
　emphysema 205–206
　immunology 249–251
　Parkinson's disease 307–308
　pathology of heavy metals 68–78
Animal virus 115–118
　assembly of virion 117–118
　background 115
　particle weight 115
　protein component 115–116
　RNA component 115
　structure 116–117
Animal wastes 413–414
Animated graphics 335
Antarctica: ionosphere 257
　Triassic fossils 405
Antenna (talking beacon system) 281
Antibody synthesis: B cell 250
　T cell 250
Anticline (oil and gas fields) 313
Antiferromagnetic insulator 242
Antigen 118–121
　erythrocyte systems 119
　α_2-globulins 119
　immunoglobulins 118–119
　leukocyte systems 119–121
　β-lipoproteins 119
　recognition by surface receptors 251
　see also Immunoglobulin
Antiproton atom 239
Antiquark 198
Antiserum (HL-A system) 399
Anti-Stokes region 389
α-l-Antitrypsin 205–206
Apatosaurus 180, 181
Aphid vector 100
Apium graveolens 154
　cuticle and cell wall 209
Apollo Lunar Science Experiment Package (ALSEP) 386
Apollo Program 6, 168, 186
　Apollo 12 387
　Apollo 13 385
　Apollo 14 385–387
　Apollo 15 386, 387
Appalachian orogenic belt 302–307
Appelman, E. H. 248
Apple scab (forecasting services) 100
Approximate methods 39
Aquatic tetrapod herbivore 179

Arabica coffee 151
Archidiskodon 179
Archimedes' principle 172
Arctic ionosphere 257
Ardisia 274–276
Argille scagliose 304
Argillite 304
Argon, solid 349
Argon laser (Raman spectroscopy) 389
Argonne National Laboratory 200, 205
Aridisol 373, 374
Army Atmospheric Sciences Laboratory (ASL) 418
Aromatic ether 157
Aromatic halide 157
Arp, H. C. 226
Art, science in 10–23
Art, "scientific" 13–14
Art objects, scientific investigation of 14–20
Arteriosclerosis 121–123
   homocysteic acid and growth 122
   homocysteine metabolism 121–122
   human cell cultures 122
   prevention 123
   production 122–123
Arylcopper 158
Asarco dimethylaniline (DMA) cyclic absorption process 105
Asaro, F. 16
ASL *see* Army Atmospheric Sciences Laboratory (ASL)
ASLB *see* Atomic Safety and Licensing Board (ASLB)
Aspergillus 135
Asthenosphere 303, 397
Astronauts (exposure to neutron) 168–169
Astronomical instruments 123–125
   optical 123–124
   radio telescope 124–125
Astronomy: astronomical instruments 123–125
   energy sources in galaxies and quasars 44–55
   galaxy clusters 224–228
   gravitational collapse 231–233
   interstellar matter 254–257
   neutron star 288–290
   Venus 407–408
   x-ray 232
Astrophysics 49
Asymmetric ears (directional hearing) 394
Asymmetric gravitational collapse 232–233
Ataxia (thallium disease) 76
Atherosclerosis 77
Atlantic-Pacific Interoceanic Canal Study Commission 296
Atmospheric oxygen: soil availability in flooding 378
   theory for faunal extinction 214
Atom: antiproton 239
   hadronic 237–239

Atom—*cont.*
   Σ-hyperonic 239
   kaonic 239
Atomic, molecular, and nuclear physics: cosmic ray 162–169
   hadronic atom 237–239
   particle accelerator 308–313
Atomic Energy Commission, U.S. 191, 192, 298, 342
   licensing process 354–358
Atomic Safety and Licensing Board (ASLB) 354
Atomic shell 203–205
Atomic vibrations, surface 66
ATP *see* Adenosinetri-phosphate (ATP)
Attridge, J. 180
Auditory bulla 394
Auger effect 238
Auger processes 62–63
ϵ Aurigae 232
Australian Aborigines 119
AUTOBUOY: operational characteristics 133
   programming 133–134
AUTOKON (design system) 370
Automated fire protection 6–7
Automation in merchant ships 365
Autotape 230
Auxin 284–285
   influence on xylem tracheary elements 422
Avduevsky, V. S. 407
Awu volcano eruption 412
Azoferredoxin (AzoFd) 291
Azotobacter vinelandii 291

B

B cell 249
Bacillariophyceae 109, 110
Bacillus subtilis 147
Backus, G. 183
Bacteria 109
   decomposition of belowground plant parts 382
   leaf 274–276
   plant disease 100
   temperature-moisture relationship 100
Bacteria-plant symbiosis 274
Bacteriophage X 174 326
Baikonur Cosmodrome 388
Baker, E. A. 210
Bakker, J. W. 376
Ball, J. A. 255, 257
Baltimore, David 117, 174
Banana 324
Bang, O. 174
Barbados Oceanographic and Meteorological Experiment (BOMEX) 277
Bardeen, John 65
Barley, K. P. 359
Barley 360
Baron, J. G. 396
Barr body 138
Barrett, A. H. 255, 257

Barrett, Peter 405
Bartholomew, W. V. 382
Bartlett, N. 420
Barylambda 180
Baryon 198, 310
   formation experiments 199–200
Basalt 303, 305
BB gene 400
Beams, J. W. 171, 172
Bean: lima 324
   velvet 324
Beardmore Glacier 405
Beaulieu, A. J. 268
Beaver 125–126
   anatomy 125–126
   behavior 126
   physiology 126
Beer drinker's syndrome (cobalt toxicity) 77
Beerenberg volcano eruption 412
Begonia 284
Beidellite 149
Bellevalia romana 148
Bending magnets 98
Benioff, H. 397
Benlate 152
Bergbau-Forschung process 103
Berylliosis *see* Beryllium disease
Beryllium disease 126–128
   acute form 127
   beryllium phosphors 127
   chronic form 127–128
   effect on tissue 71
   "neighborhood" cases 128
   prevention 128
Bessho, M. 370
Beta decay 203
Beta lipoproteins 119
Beurrier, H. R. 113
Bezymjannaja volcano eruption 412
"Big-bang" cosmology 51
Binary metallic systems (phase equilibrium) 213
Binder (metal powder) 331
Bioassay (plant hormones) 128
Biochemistry: arteriosclerosis 121–123
   growth hormone 235–237
Biogenic amines (effect on hibernation) 239
Biokinetic range 323
Biology, transplantation 399–402
Biology of silica 371–372
Biomass in ecological succession 216
Biondi, M. 261
Biophysics: animal virus 115–118
   density 171–174
   electrodiagnosis 193–195
   prosthesis 348–349
Biosphere: nitrogen movement 374–377
   phosphorus movement 375
Birch (xylem) 424
Birth defect 71
Bjerknes, J. 279
Black, H. S. 112

Black, J. D. F. 376
"Black hole" 51, 231–233, 289
Blackbody radiation 288
Blasting, borehole 299–300
Blind (prosthesis) 348–349
Blueschist 304–306
Bode, H. W. 113
Bohr orbit 238
Bombyx mori 138, 139
BOMEX *see* Barbados Oceanographic and Meteorological Experiment (BOMEX)
Bone-marrow-derived cell 249–251
Bone-marrow transplantation 399
Bone tumor 128
Bonhoeffer, F. 327
Borehole blasting 299–300
Borlaug, Norman 221
Bourke, P. M. A. 100
Bovine serum albumin 143
Brachiosaurus 179–181
Brain (activity in hibernation) 241–242
Bramlage, W. H. 324
Brandt, Willy 292
Brassica napus L. 128
Brassin 128–130
   chemistry 129
   detection 128
   growth responses 129–130
   isolation 128–129
Brazil 387 (coffee strain) 152
Breccia 303
Breed, William 405
Bridgman, P. W. 242
Brillouin zone 99, 241
Bristlecone pine chronology 352
BRITSHIPS (design system) 370
Broca's area (brain) 390–392
Bromeliaceae 209
Bromotrifluoromethane 7
Brontosaur 179, 181
Brontosaurus 180
Brookhaven 30-GeV proton synchrotron 310
Brown, W. V. 134
Brueckner-Bethe-Goldstone approximation 288, 289
Brunswick Building 132
Buchsbaum, M. 391
Buhl, D. 255
Buildings 130–132
   exterior diagonal system 131
   framed tube 132
   shear wall–frame interaction 132
   systems in concrete 131–132
   systems in steel 130–131
   tube in tube 132
Bulk carrier (ship) 363
Bulla, auditory 394
Bundle sheath cell 323
Bundled tube (building) 131
Buoy 132–134
   AUTOBUOY operational characteristics 133
   geodetic positions at sea 228
Bureau of Mines, U.S. 300

Burger, M. M. 141
Burnaby, T. P. 265
Burness, A. T. H. 115
Burrill, L. C. 344
Bursell, C. G. 320

C

C-band radar 230, 231
Cable: compressed gas 188
    cryogenic 188
Cacao 324
Cadle, R. D. 411
Cadmium disease 72–73
    recent occurrence 73
    symptoms 73
Cadmium nitrate 73
Cairns, J. 326
Calcareous ooze 303
Calcium: loss in Mollisols 372–374
    loss in Ultisols 372
Calcium metabolism (role of silicon) 372
Californium-252 298
Callear, A. B. 268
Callose plug 324
Callus culture 276
Calorimetry 298
Camarasaurus 179, 181
Cambium 422, 423
Camera (in satellites) 185
Cameron, A. G. W. 232
Canada Centre for Inland Waters 234
Canadian beaver 125
Canis familiaris (heterochromatin replication) 139
Capsid, picornavirus 117
Capsomere 116
Carbon-14 167, 285
    art dating methods 21
    changes in atmosphere 352–353
    concentration in wood 352
    isotope trace 351
Carbon dioxide (photorespiration) 320–323
Carbon monosulfide molecule 255
Carbon monoxide molecule 255
Carboniferous 217
Carbonyl sulfide molecule 255
Carcinogen: cobalt 77
    metal toxicity 71–72
Cardiovirus 115
Carlisle, Edith 372
Carnosauria 176
Carotenoid 107, 108
Carrier, slurry (ship) 363
Carrier diffusion length lifetime 361
Carter, Brandon 232
de Carvalho, A. 152
Cascade, nucleonic 167
CASDOS (design system) 370
Cassegrain focus see Nasmyth focus
Cassette 338
Cassowary 180
Castor canadensis 125

Cat-Ox system 102
Catalase 322
Cataphoresis pumping 271
Cathode-ray-tube display 370
Cattle (silica content) 371
Cavalieri, L. F. 175
Cavia cobaya 139
CB see Chlorobromomethane
CCD see Charge-coupled device
Cecropia moth juvenile hormone 158
Cell antigen: ABO 118
    Diego 118
    Duffy 118
    erythrocyte systems 118
    I 118
    Kell 118
    Kidd 118
    Lewis 118
    Lutheran 118
    MNSs 118
    P 118
    Rh 118
    Xg 118
    Yt 118
Cell death (monovalent agglutinin) 143
Cell division 134–141
    chromosome replication 138–141
    meiotic and somatic chromosome pairing 134–135
Cell surface 141–143
    agglutinin binding sites 142
    cell exposure to protease 142
    molecular mechanism for site exposure 143
    monovalent agglutinin 142–143
    normal and transformed cell growth 142
    plant agglutinin role 141
Cellular response (metal toxicity) 71
Cellulase 423
Cellulose 17, 381, 415
Cenozoic 217
Centriole 110
Ceramic material: phase stability 211
    thermodynamic properties 211
Cercospora beticola 101
Cereals, surplus 222
Cerebellum 75
Cerebrum (speech) 390–392
CERN 162, 200, 202, 311, 312
Cessac, G. L. 328
Cetiosaurus 180
Chafe, S. C. 153
Chan, L. 16
Chandrasekhar, S. 288
Charge-coupled device 143–144
    applications 144
    limitations 144
    transfer of charge 143–144
Charig, A. J. 180
Charophyceae 109
Charophyta 109
Chemical clinostat 285

Chemical etching (gravure printing) 338
Chemical laser 266–268
Chemical techniques (scientific investigation of art objects) 15–18
Chemico-Basic process 102, 104
Chemisorption 64
Chemistry: coordination 154–157
    copper 157–158
    soil 377–378
Chert 305, 306
Cheung, A. C. 255
Chicken (silica content) 371
Chinchilla lanigera (heterochromatin replication) 139
Chinese bronzes (art objects study) 22–23
Chinese hamster 138
Chippewa Indians 119
Chlamydomonas 135
Chlorcholine chloride 284
Chlorfluorenol 282
Chlorobromomethane 7
p-Chloromercury benzoate (p-CMB) 327
Chloromonadophyceae 109
Chlorophyceae 109
Chlorophyll 324
Chlorophyll b 107
Chlorophyta 109
    phylogenetic relationship 112
Chloroplast 110, 112, 209
    glyoxylate reductase 322
Cholera 144–146
    action on nonintestinal tissues 146
    concepts of pathogenesis 145–146
    experimental models 145
    immunity 146
    toxin 145
Cholesterol 123, 236
Choreic athetosis (thallium disease) 76
Christensen, Tyge 106
Chromium (teratogenic effect) 71
Chromocenter 138
Chromograph 339
Chromophyta 106, 109
Chromosome 146–148
    circular 147
    classification 146–147
    damage by some metals 72
    lengths of DNA 147
    linear 147–148
    pairing in meiosis 134–136
    repetitious DNA 147
    replication 138–141
    somatic pairing 134
Chronic bronchitis 205
Chrysolaminarin 109
Chrysophyceae 109
Chrysophyta 109
Cibachrome Graphic process 337–340
Cinnamic acid 424

Civil engineering: buildings 130–132
    highway engineering 243–246
    solid-waste management 24–33
    transportation engineering 402–405
Clark, F. E. 381
Classification of fires 5
    A fire 5
    B fire 5
    C fire 5
    D fire 5
Classification of wastes 33
Claus sulfur plant 105, 106
Clay 148–150
    nature of surface atoms 149
    role of exchangeable cations 149–150
Clay crystallite 148
Clay mineral 377
Clean Air and Water Act of 1965 206
Clean Water Restoration Act of 1966 206
Climate and weather, marine influence on 277–279
Clivia (cuticle) 210
Clostridium pasteurianum 291
Clusters, galaxy 224–228
CM* gene 400
p-CMB see p-Chloromercury benzoate (p-CMB)
Coal, solvent-refined 105
Coalsack Bluff (Antarctica) 405
Coast and Geodetic Survey, U.S. 396
Coastal Bermuda grass (silica content) 371
Cobalamin 121
γ-Cobalt-60 (process radiation) 342
Cobalt chloride 77
Cobalt toxicity: beer drinker's syndrome 77
    carcinogen 77
Cochliobolus heterostrophus Drechsler 159
Cocklebur 422
Codium 112
Coelurosauria 176
Coelurus 176
Coffea arabica 150
Coffea canephora 151
Coffee 150–152
    origin and spread of rust 150–151
    rust control measures 151–152
    rust-resistant trees 152
COGO see Coordinate geometry (COGO)
"Cogwheel rigidity" (Parkinsonism) 307
Cohn, R. 391
Colbert, Edwin H. 178, 405
Colchicine 148
Cold pressing (pure metal powders) 330, 331
Colgate, Stirling 231
Collagen 206
Collapse, gravitational 231–233

Collenchyma 152–154, 209
  cell wall organization 152–153
  cytology 154
  growth and formation of cell wall 153–154
  morphogenesis 154
Colonial American pottery 17
Color duplication (gravure printing) 338
Color filter 337
Color-Key process 335, 337
Color printing 334–337
Color processes, dry photomechanical 334–335
Color scanner 335–337
  Diascan 3000 336
  Magnascan 450 336
  other types 336–337
Color transparency film 334
Colorgraphics process 334
Colymbetes 140
Coma Cluster of galaxies 226, 227
Cominco process 104, 106
Comparative physiology (homeostasis) 246–248
Composting (solid-waste management) 31–32
Compound, intermetallic 251–254
Compounds, xenon 420–422
Compressed-gas cables 188
Compsognathus 176
Compton radiation, inverse 47
Computer: analysis of EEG data 391
  control of electric power systems 189
  dead-reckoning 286–287
  design of intermetallic compounds 254
  electrodiagnosis 194
  graphics 370
  oil and gas reservoir simulation 316
  phase diagram calculation 213–214
  plant disease forecasting 101
  programs to predict nuclear explosion configuration 294–295
  ship design 366–371
  yacht navigation 287
Computer algorithm (nonlinear programming) 370
Computer-processed photographic print 12
Concanavalin A (ConA) 141
Concentration polarization 375
Concrete, reinforced 130
Concrete-polymer wall tile 343
Cone (eye) 394
Conifer 180
Conjugatae 109
Conjunctivitis 127
Conservation of art 23
Container ships 364–365
Continental drift 397
Continental glaciation (theory for faunal extinction) 214
Continental rise 303, 304
Continental shelf 303, 304

Continuous-wave chemical laser 266–267
Continuous-wave dye laser 269
Continuously tunable Raman oscillator 302
Contophora 109
Contrarotating propeller 344–345
Controllable-pitch propeller 344, 366
Cooke-Yarborough, E. H. 114
Cool, T. A. 268
Cool temperatures: negative effect on plant growth 324–325
  positive effect on plant growth 323–324
Cooling water for power plants 333–334
Cooper, P. D. 115
Coordinate geometry (COGO) 243–244
Coordination chemistry 154–157
  catalytic applications 156–157
  mechanisms 155–156
  synthetic applications 156
Copolymer produced by process radiation 343
Copper chemistry 157–158
  coupling with organic halides 158
  hydrolysis 158
  organocopper reagents 158
  oxidation 158
  preparation 157
  structure 157–158
  summary 158
  thermal decomposition 158
Copper montmorillonite 378
Copper oxychloride 152
Copper powder 330
Copper-vapor laser 272
Coproantibody (IgA) 146
Coprosma 274
Cordilleran orogenic belt 302–307
Corn 158–162, 324
  cause of disease 159–161
  comparison with other diseases 161
  disease control 161–162
  disease resistance 161
  history of blight and disease 159
  pollen hormone 128
Corona, solar 383
Corpus striatum 307, 308
Corwin, H. G. 226
Cosmic ray 162–169
  above 100 GeV 162–164
  muons 164–167
  neutrons 167–169
  particle accelerator experiment 312
  theory for faunal extinction 214
Cosmology 49, 51
Cosmos (artificial satellite) 388
  Cosmos 394–398 386
  Cosmos 400 386
  Cosmos 402 386
  Cosmos 404–405 386

Cosmos (artificial satellite)—cont.
  Cosmos 407 386
  Cosmos 409 386
  Cosmos 411–418 386
  Cosmos 421 386
  Cosmos 425–426 386
  Cosmos 429 386
  Cosmos 431 386
Cotton 324
  cross linking in cotton polyester blend by process radiation 343
Cotzias, G. 308
Coulomb crystal 350
Coulomb field 237–239
Coupled resonance (elementary particle) 200–203
Cox, A. 396
CPU-400 (fluidized-bed incinerator) 33
Crab Nebula 231
Crab Pulsar 289
Cramer, H. 39
Craspedophyceae 109
Crater lip dam 295
Creencia, R. P. 324
Cretaceous 176–181, 216–217
Cricetulus griseus 138, 139
Crick, F. H. C. 148
Crimson clover (silica content) 371
Critical point 169–171
  homogeneous or scaling functions 170–171
  indices 169–170
Crop residue 414
Crosfield Etchomatic process 338
Crosstalk 393
Cryogenic cable 188
Cryptomonad 106
Cryptophyceae 109
Crystal modulator, electrooptical 336
Crystal structure 213
  3*d* ions 264
Crystalline semiconductor 361
Crystalline solid, surface of 58
CSIRO 330
Cucumber 324
Culture, callus 276
Curie point 169–170
Cuticle 209
Cuticularization 209–211
Cutin 209–211
Cutinization 209–211
Cyanelle 112
Cyanide radical molecule 255
Cyano-acetylene molecule 255
Cyanophyceae 109
Cyanophyta 109
  phylogenetic relationship 112
Cycad 180
Cyclic AMP 145, 240
Cycloheximide 146
Cyclone (centifuge) 31
Cygnus A radio galaxy 47
Cystathionine 121
Cysteine 115
Cystine 115
Cytokinin (influence on xylem tracheary elements) 422

Cytology (collenchyma) 152–154
Cytoplasm 424
  cms-C 161
  cms-S 161
  cms-T 160
Cytosine 325–326

D

DAISY ship design program 370
Dalrymple, B. G. 396
Dam, crater lip 295
Darcy law 316
Darlington, C. D. 134
Dasberg, S. 376
Dashaveyor 403
Dating, radiocarbon 351–353
Dating studies (art objects) 20
Daughaday, W. H. 122
Davisson, C. J. 58
Dawson, C. D. 391
Day-glo colors 13
DC 300 scanner 337
De Lucia, P. 326
Dead-reckoning computer 286–287
Debye-Scherrer diagram 197
Debye temperature 350
Decarburization (metallurgy) 331
Decca (navigation system) 286
Deception Island volcanic eruption 412
Decker, J. P. 320
Decomposer reducers in soil 382
DeGeer, G. 353
Deighton, M. O. 114
Deinocheirus 176, 178
Deinonychus 176–178
Deinonychus antirrhopus 177
Deltaic sands (petroleum accumulations) 313
Dempsey, J. R. 318
Dendrochronology 20–21
Density 171–174
  applications 173–174
  methods 171–173
Deoxyribonucleic acid (DNA) 109, 140, 146–148, 174–176
  lengths in chromosomes 147
  replication 134
  replication and differentiation 139
Deoxyribonucleic acid endonuclease 175
Deoxyribonucleic acid exonuclease 175
Deoxyribonucleic acid ligase 175
Deoxyribonucleic acid polymerase, RNA-directed 174–175
  in cells 175
  in virions 175
Deoxyribonucleic acid polymerase I 325–326
Deoxyribonucleic acid polymerase II 326–327
Deoxyribonucleic acid protovirus 175

Deoxyribonucleic acid virus 142
Department of Health, Education, and Welfare 206
Dermatitis 127
  nickel 77
Deryagin, B. V. 328
Desensitized liquid nitroglycerin 300
Design, ship 366–371
Desmokontae 109
Desorption, field 60
Desorption, flash 65–66
Detector (neutron flux):
  etched track detectors 167
  liquid scintillators 167
  nuclear emulsion 167
  proportional counter 167
  thermoluminescent crystals 167
Detergents (oiled seabirds) 219
Detonation, fracture 300
Detritus feeder 216
Device, charge-coupled 143–144
Devonian 217
Devonshire pottery 17
Dew (relationship with plant disease) 100
Dew recorder 100
Dewar 181, 309
DeWitt Chestnut apartment building 131, 132
Diakinesis 134
Diarrhea 144
Diascan 3000 336
Diatom 110
Diatomaceae 109
Diazo compound 157
Dickite 149
Dictyosome (algae) 111
Die lubricants 331
Diego (immunoglobulin) 118
Dielectric relaxation time 361
Dieterici equation of state 170
Dietz, R. 396
Difolatan 152
Diffraction, electron 60–62
Diffraction electron microscopy 197–198
Digital plotter 370
Digital shift register 144
3,4-Dihydroxyphenylalanine 307
Dike 303
Dilophosaurus 407
Dimensions of galaxy clusters 224–225
bis(Dimethylglyoximato)cobalt 155
Dinitrogen fixation 290
Dinoflagellate 106
Dinophyceae 109
Dinosaur 176–181
  brontosaurs 179–181
  carnivorous 176–179
Diode, input 144
Diorite 304
Diplodocus 180, 181
Diptera 136
Direct color reproduction 334
Direction isotropy (muon) 166–167

Disease: beryllium 126–128
  cadmium 72–73
  cobalt 76–77
  emphysema 205–206
  mercury 73–75
  nickel 77–78
  thallium 76
Disease-resistant hybrid (corn) 162
Disneyland People Mover 403
Distribution function, joint 37
Disturbed solar wind 384
Dithionite 290
Divalent agglutinin 143
Division, cell 134–141
Dixon, J. S. 235
DNA *see* Deoxyribonucleic acid (DNA)
Dobrotin, N. A. 163
Dobrovolsky, Georgy 388
Doell, R. R. 396
Dome (oil and gas fields) 313
Donkey–Grévy's zebra hybrid (heterochromatin replication) 139
Dopamine (L-dopa) 307, 308
Doppler interferometry, radar delay 408
Doppler radar 230, 231
Doppler spectroscopy 408
Double-heterostructure laser 273–274
Downy mildew 100
Draco 407
Drag 220
Drew, M. C. 359
Drilling (petroleum prospecting) 346
Driography 337
  ink 340–341
  plate characteristics 340
Drive, steam 319–320
Drosophila anassae 134
Drosophila melanogaster 134, 138, 139
Dry, hot recovery systems (flue gas) 103
Dry mechanical color process:
  animated graphics 335
  signs, posters, contemporary art 335
  slide preparation and TV animation 335
Dry photomechanical color process: color transparency film 334
  direct color reproduction 334–335
Dry-powder chemical extinguisher 7
Ducted propeller 345
Duffy (immunoglobulin) 118
Dulbecco, R. 142
Dumbbells piercement 348
Dunker, A. K. 116
Duplication, color 338
Dwarf rice IR-8 221, 223
Dye laser 268–270
Dyer, A. J. 413
Dysarthria 74
Dytiscus 140

E

E region (ionosphere) 257–258
Earth magnetic field 383, 396
Earth-resource satellites 185–187
  benefits 187
  planned space flight programs 186–187
  previous experience 185–186
  remote terrain sensing 398
  science of remote sensing 185
Earth Resources Experiment Package 187
Earth Resources Technology Satellite (ERTS) 186, 398
  ERTS A 187, 398
  ERTS B 187, 398
Earth tides 181–185
  data 182–184
  sources of data inconsistencies 184
  superconducting gravimeter 181
Earthquake 306
East Central Area Reliability Group (ECAR) 190
East Pacific Rise 307
EAT *see* Experiment in Art and Technology
ECAR *see* East Central Area Reliability Group (ECAR)
ECG *see* Electrocardiogram (ECG)
Echo sounding (electronic navigation systems) 286
Eckhart, W. 142
Ecologic succession: faunal extinction 216–217
  patterns 215
Ecology: environmental engineering 206–209
  Great Lakes 233–235
  impact of nuclear power plant site 356
Ecosystem 216, 217
  stress by expanding population 223
Edelman, G. M. 119
Edema 146
EEG recording *see* Electroencephalographic (EEG) recording
Effect, Raman 302
Ehringer, H. 308
Einstein, Albert 64
Ekman transport 279
Elastin 206
  fibril 123
Electric power engineering: electric power system, interconnecting 187–190
  electrical utility industry 190–193
Electric power system, interconnecting 187–190
  computer control centers 189
  estimates for next two decades 187–188

Electric power system—*cont.*
  future transmission technology 188
  improved system reliability 190
  new control systems 189–190
  power pools 188
  regional and national coordination 188–189
  reliability and security 188
Electrical equipment (class C fires) 5
Electrical utility industry 190–193
  generation methods 191–192
  hydroelectric power 192
  money outlay 191
  power plant setting 333
  problems with environment 192–193
  transmission circuit expansion 192
  underground circuits 193
Electrocardiogram (ECG) 193
  hibernation 239–240
Electrochemistry 58
Electrocure process 342
Electrodiagnosis 193–195
  approach to computerization 194
  display 195
  storage and retrieval 195
  transmission and preprocessing 194–195
Electroencephalographic (EEG) recording 390
Electromagnetic mixing 201–202
Electron 200
Electron accelerator machines in process radiation 341
Electron diffraction 60–62, 198
Electron diffraction, high-energy 195–198
  diffraction electron microscopy 197–198
  reflection HEED 196–197
  scanning HEED 197
Electron gas, nonrelativistic 288
Electron-hole pair 144
Electron microprobe analysis 329
Electron microscopy, diffraction 197–198
Electronic behavior of solids 241
Electronic combustion-gas detector 6
Electronic engraving 339–340
Electronic navigation systems 286–288
Electronic shift-register circuits 143
Electronic structure (high-pressure studies) 242
Electronic transition 242
Electronic Video Recording (EVR) 409
Electrooptical crystal modulator 336

Element 114, heavy-ion reactions leading to 204
Elementary particle 162, 198–203
  coupled and overlapping resonances 200–203
  exotic resonances 198–200
Elementary risk model 37–38
Elements (chemistry) 203–205
  electronic structure and chemistry 204–205
  limiting half-lives 203
  new elements 203
  possible syntheses 203–204
Elephant 179
Elephas 179
Eliot, T. S. 23
Ellerman, V. 174
Elysia 112
EMC-virus 115
Emission, field 63–64
Emission, x-ray 237–238
Emphysema 205–206
  classification 205
  familial 205–206
  obstructive pulmonary 205
  pathogenesis 206
Empire State Building 130
Employee noise protection 292–293
Endeavor (spacecraft) 387
Endoplasmic reticulum 108, 154, 209
Endosome 110
Energy, solvation 378
Energy production (manure) 413
Energy sources in galaxies and quasars 44–55
  comparative properties 49
  empirical evidence 46–49
  energy sources 50–52
  redshift 49
  Seyfert nucleus 54
  theoretical working models 52–54
  theory 54–55
  total energy output 49–50
Engineering: environmental 206–209
  highway 243–246
  nuclear explosion 294–297
  petroleum reservoir 315–318
  risk evaluation in 34–43
  solid-waste management 24–33
  transportation 402–405
Engraving, electronic 339–340
Enterotoxin (cholera) 144–146
Entisol 373, 374
Environmental engineering 206–209
  air quality management 207
  governmental policy 206–207
  Great Lakes 233–235
  industrial policy 207
  land reclamation 208
  petroleum geology 313–314
  power plant siting 333
  water quality management 207–208

Environmental Protection Agency, U.S. 33, 192, 206, 207, 356
Enzyme (negative effect of cooling in plants) 325
Eocene 219
EPA *see* Environmental Protection Agency, U.S.
EPIDEM 101
Epidermis (plant) 209–211
  abnormal distributions of cuticle 209
  cuticularized layer 210–211
  epidermal cell 209–210
Equations of state 170
Equilibrium, phase 211–214
  computer techniques 213–214
  first-principle calculations 212–213
  practical approach 213
Equilibrium diagram 211
Equisetum (silica content) 371
Ericsson, John 344
Ertl, J. 391
ERTS *see* Earth Resources Technology Satellite (ERTS)
Eryngium rostratum:
  collenchyma wall 154
  leaf collenchyma 153
  lignified collenchyma 153
Esau, K. 153
Escher, Maurits C. 14
Escherichia coli 71, 175, 327
  bacteriophage T4 326
  F$^-$ cell 147
  Hfr strain 147
Ester, xenon 420
ET* gene 400
Etched-track detector 167
Etching, chemical 338
Ethyl bromide 343
Eucalyptus 210
Eucalyptus papuana 209
Eucaryota 106, 109
Euchromatin 139
Eucoxanthin 107
Euglenoid 106
Euglenophyceae 109
Euglenophyta 109
Eukaryote 146
Euplastideae 109
Eustigmatophyceae 107, 109
Eutectic systems 252
Evans, H. M. 235
Evolution: faunal extinction 214–217
  immunology 251
EVR *see* Electronic Video Recording (EVR)
Exotic resonance 198–200
Exotoxin (cholera) 145
Experiment in Art and Technology (EAT) 13
Explorer 44 (artificial satellite) 386
Explosion technology 294–295
Extinction, faunal 214–217
Extinguisher: dry-powder chemical 7
  liquid chemical 7
Extremes, statistics of 38–39

F

F$^-$ cell (*Escherichia coli*) 147
F region (ionosphere) 259, 262
Fahey, P. F. 172
Fairing tool 367
Falcon (*Apollo 15*) 387
Familial emphysema 205–206
trans-trans-Farnesol 158
Farnsworth, H. E. 61
Farrell, W. E. 183
Fault trap (oil and gas) 313
Faunal extinction 214–217
  ecologic succession 216–217
  photosynthesis influence 215–216
  theories 214–215
Feather (bird) 217–219
  rehabilitation of oiled seabirds 219
  repellency and feather structure 217–219
Federal Power Commission (FPC) 189
Federal Public Land Law Review Commission 208
Federal Radiation Council (FRC) 356
Federal Water Quality Administration 206
Federal Wilderness Act of 1964 206
Fedio, P. 391
Fedyakin, N. N. 327
Feedback 112
Feedforward 112
Feldstein, Y. I. 258–259
Fergusson, C. W. 352
Fermi energy 288
Fermi level 64, 97
Fermi solid 350, 351
Fermi surface 241
  intermetallic compounds 253
Fernandina volcano eruption 412
Ferredoxin, reduced 290
Ferric compounds 378–380
Ferroelectric materials 241
Ferromagnet 170, 171
Fertilizer: effect on productivity of new strains 221
  manure 413
Feulgen positive material 138
Feynman, Richard P. 164
FFAG accelerator *see* Fixed-field alternating-gradient (FFAG) accelerator
Fiber and food industry 413
Fick law 316
Ficus (cuticle) 210
FIDO *see* Fog incendiary dispersal operation (FIDO)
Field desorption 60
Field emission 63–64
Field emission microscope 63
Field geology (petroleum prospecting) 345–346
Field-ion microscope 64
Film: aerial Ektachrome infrared 398
  silver halide 411

Filter, color 337
Finch, J. T. 116
Fiocco, G. 413
Fire: classification 5
  cost, annual U.S. 2
  prevention and loss reduction 4–5
  research, future of 7–8
  urban 1–9
Fire-fighting methods, modern: fog or spray nozzles 4
  ventilation 4
Fire research trends 6–7
  automated fire protection 6–7
  fire retardant materials 6
Fire-suppression agents: dry-powder chemical extinguishers 7
  foams 7
  light water 7
  liquid chemical extinguishers 7
  water 7
First-principle calculations (phase equilibrium) 212–213
Fishing (talking beacon system) 281–282
Fission, spontaneous 203, 204
Fixation, nitrogen 290–291
Fixed-field alternating gradient (FFAG) accelerator 98
Fixed-temperature detector 7
FJH* gene 400
FJH-Aj gene 400
Flagellum (algae) 110–111
Flare, solar 384
Flash desorption 65–66
Flat oyster 265
Flight 219–221
  flapping flight fundamentals 219–220
  V flight formation 220–221
Flight, space 385–388
Flooded soil environment 378
Flue-gas treating systems 102–104
Fluid mechanics: air pollution 101–106
  urban fires 1–9
Fluidized-bed incinerator 33
Fluorescent dye (emission band in dye laser) 268
Fluoride, xenon 421
Fluorinated organocopper 157
m-Fluorophenylcopper 157
p-Fluorophenylcopper 157
Fly ash 31, 103
Flygare, W. H. 255
Flysch 304, 305, 306
Foam (fire-suppression agent) 7
Fog: ice fog 416
  improvement of visibility 416
  supercooled fog 416
  warm fog 416–417
Fog incendiary dispersal operation (FIDO) 417
Fog nozzle 4

Food 221–224
  aseasonal effect 221–222
  fertilizer responses 221
  second-generation problems 222–223
  stresses on Earth's ecosystem 223–224
Food and fiber industry (wastes) 413
Food chain 73
  marine 215–217
Food processing wastes 414–415
FORAN (design system) 370
Forest waste products 415
Formaldehyde molecule 255
Formamide molecule 255
Formic acid molecule 255
FORTRAN IV (computer language) 101
Fossil: Asian gliding lizard 407
  dinosaur 407
  labyrinthodont amphibian 406
  lizard 406
  thecodont 406
  tree 352
Four-channel stereophonic sound 392–394
FOUR locus (human transplantation antigen) 399
Four Quartets (poem) 23
Fourier law 316
FPC *see* Federal Power Commission (FPC)
Fracture detonation 300
Framed tube (building) 131
Frames with rigid belt trusses (building) 131
Franck-Condon principle 243
Frank, F. C. 253
Fraunhofer hologram 411
FRC *see* Federal Radiation Council (FRC)
Frere, M. H. 359
Friedman, A. N. 113
Friend, J. P. 412
Fritillaria lanceolata 138, 139
Fructose 247
Frustule 215
Fuel gases (air pollution) 106
Fundulus heteroclitus (blood freezing point) 246
Fungi: decomposition of belowground plant parts 382
  temperature-moisture relationship 100
Fusarium moniliforme 284

G

GA_3 *see* Gibberellic acid
Gabbro 304
Gabor, Dennis 292
Gahan, P. B. 423
Galaxies, energy sources in 44–55
Galaxy 166, 254
  infrared 47
  optical 46–47
  radio 47

Galaxy—*cont.*
  Seyfert 48
  x-ray 47–48
Galaxy clusters 224–228
  characteristic properties 224–225
  relationships between properties 225–228
Galin, D. 391
Gally, K. A. 119
Gametogenesis 134
Gaming techniques 39–42
Gas-liquid chromatography 129
Gas-solid interface 58
Gas-turbine plant: for electric power 191–192
  propulsion in ships 366
Gasbuggy experiment (nuclear explosion engineering) 295, 297
Gaseous sample (Raman spectroscopy) 389–390
Gasoline 75
Gastrolith 179
Gaviidae 219
Gc (immunoglobulin) 118, 119
Geisha (coffee strain) 152
Gel chromatography (brassin analysis) 129
Gelatin (use of food-processing wastes) 414
Gemini Program 186, 194
407 Gene 400
Gene amplification (heterochromatin) 140
Genetic therapy 175
Genetics: antigen 118–121
  cell division 134–141
  chromosome 146–148
  deoxyribonucleic acid (DNA) 174–176
  transplantation biology 401
Geobathymetric survey (Haakon Fiord) 346
Geochemistry: radiocarbon dating 351–353
  volcano 411–413
Geodesy 228–231
  geodetic coordinates 229–230
  geoid determination 230–231
  international symposium 231
  ocean-surface positions 229
  three-dimensional coordinates 228–229
Geodynamics Project 396
Geoid determination 230–231
Geology (surficial and historical): orogeny 302–307
Geology, petroleum 313–315
Geomagnetic field (changes affecting carbon-14 in atmosphere) 353
Geomagnetic latitude (coordinate system for ionosphere study) 257
Geomagnetic local time (coordinate system for ionosphere study) 257
Geometric probability 38–39

Geophysics: Earth tides 181–185
  geodesy 228–231
  ionosphere 257–263
  solar wind 383–385
  tectonophysics 396–397
  weather modification 416–418
GEOS-2 (artificial satellite) 231
GEOS-C (artificial satellite) 230, 231
Geothermal heat deposits 297
Geotropism (morphactin effect) 284, 285
Germanium 241
Germer, L. H. 58
Germination (negative effect of chilling) 324
Gettens, R. J. 13, 22
Giammarco-Ventrocoke process 106
Gibberellic acid (GA_3) 423
  effect on barley seed 360
Gibberellin 284
  influence on xylem tracheary elements 422
Giedt, R. R. 268
Gilbert, C. K. 302
Gilbert, F. 183
Gilding (study of art objects) 19–20
Girard, M. 115
Gland oil (bird feathers) 218
Glaucophyceae 109
Glaucophyta 109
Global solar radiation attenuation 413
α_2-Globulin systems 118
β-1,3-Glucanase 360
Glucose 145, 247
  function in supercooled fish 247
Glue (use of food-processing wastes) 414
Glycerol 247
Glycine 145, 322
Glycolate pathway (photosynthetic carbon) 321
Glycolic acid (photorespiration substrate) 320
Glycolic acid oxidase 322
Glycoprotein (as antifreeze in blood of fish) 248
Glyoxylate reductase 322
Glyoxysome 360
Gm (immunoglobulin) 118
Gold, Thomas 289
Goldstein, S. 344
Golenmeski, J. J. 112
Golgi body 209, 361
  algae 111
Goodall, McC. 307
Goodrich, R. 171
Gorgosaurus 176, 178
Goring, D. I. 424
Gottlieb, C. A. 255, 257
Gould, S. J. 265
Goulian, M. 326
Gradwell, M. W. 375
Graft: rejection 251
  skin 120
Graft versus host (GVH) reaction 399

Grams, G. 413
Granboulan, N. 115
Granite 304
Granodiorite 304
Grant, W. H. 328
Granuloma 127, 128
Grape epiphytotic 100
Graphic arts: printing, color 334–337
  printing plate 337–341
Graphics, animated 335
Graphite Peak (Antarctica) 405
Grasshopper 138
Grassland microorganisms 380
Grating, interferometrically ruled 388
Gravimeter 346
  elastic properties of Earth 181
Gravimetric analysis (scientific investigation of art objects) 15–18
Gravitational collapse 50–51, 231–233, 289
  asymmetric gravitational collapse 232–233
  detection of black holes 231–232
  supernovae and black holes 231
Gravitational energy in galaxies and quasars 50–51
Gravity field (Earth) 228, 230
Gravity loads 130
Gravity slide (orogeny) 304
Gravity survey (petroleum prospecting) 346, 347
Gravomaster 338
Gravure printing 337–340
  chemical etching 338
  color duplication 338
  electronic engraving 339–340
  etching control 338–339
  laser engraving 340
  preproofing 337–338
Gravure Research Institute (GRI) 339
Gravure resist film (printing) 338
Grazer, microbial 382
Great Lakes 233–235
  IJC report on pollution 233–234
  recent developments 234–235
"Green revolution" 222
Green's function 184
Gregg, David 268
Grell, R. F. 135
GRI *see* Gravure Research Institute (GRI)
Grignard reagent 157, 158
Grigorov, N. L. 163
Ground sloth 180
Ground squirrel 241
Growth, plant 323–325
Growth hormone 235–237
  biological properties 236
  chemistry of 235–236
  synthesis 236–237
Gryphaea 265
Gryphaea arcuata incurva 265

Guanine 325–326
Guernica (art) 20
Guinea pig 122, 140
Gulf Coast Miocene 314–315
Gullach, C. B. 101
Gunung Agung volcano eruption 412
Guseynov, O. H. 231
Guyer, R. A. 351
GVH reaction *see* Graft versus host (GVH) reaction
Gymnodinium cohnii 109

H

Haakon Fiord geobathymetric survey 346
Haber process 290
Habra-Pauly etching control 338
Hadrian (yacht computer) 286
Hadronic atom 237–239, 310
  antiproton atoms 239
  Σ-hyperonic atoms 239
  surfaces of 238–239
  x-ray emissions 237–238
Hagedorn, Rolf 164, 289
Halftone 335
Hallam, A. 265
Halon 1301 7
Ham, F. S. 263
Hammel, T. 240
Hanafusa, H. 175
Hanafusa, T. 175
Haplocanthosaurus 180, 181
Haptoglobin (locus) 119
Haptophyceae 106, 109
Hard-collision mechanism (quasars) 51
Hard magnetic materials 254
Harnden's rule 138
Harrison, J. C. 183
Harrison, Newton 13
Hartree approximation 350
Hartree-Fock-type calculation 205
Head, A. 254
Heating (fog) 417
Heavy-chain polymorphism 118–119
Heavy-ion reactions leading to element 114 204
Hecta meter-display instrument 286
HEED *see* High-energy electron diffraction (HEED)
Heirtzler, J. 396
Hekla volcano eruption 412
Helicopter (downwash mixing in fog) 417
Helio-Klischograph 339, 340
Helium: fully ionized in interplanetary plasma 384
  liquid 309
  solid 350
Helium-cadmium laser 270
Helium-selenium laser 271
Helmholtz free energy 170
Helminthosporium maydis
  Misikado and Miyake 159
Helminthosporium maydis race T 159

Helminthosporium victoriae
  Meehan and Murphy 161
Helsey, Charles 407
Hemicellulase 423
Hemicellulose 381
Hemileia vastatrix 150
Hemolymph 247
Henderson, J. H. 318
Henry, J. 255
Hepler, P. K. 424
Herbivores, aquatic tetrapod 179
Herbst corpuscle 395–396
Hercules Cluster of galaxies 227
Herzberg, Gerhard 292
Hess, H. 396
Heteroantibody 118
Heterochromatin 134
  DNA redundancy 140–141
  DNA replication 138–139
  gene amplification 140
Heterokontae 109
Heteroxanthin 107
Hexafluoropropyl-1,3-dicopper 157
Hexamethyldewarbenzene 157
Hexamethylprismane 157
HGH *see* Human growth hormone (HGH)
Hibernation 239–241
  brain activity 239–240
  hypothalamic controls 240
  premature arousal 240
  regulation during hibernation 240–241
  seasonal behavior 239
Hibrido de Timor (coffee strain) 152
Hicks, B. B. 413
High-energy electron diffraction (HEED) 60–62, 195–198
High-latitude ionosphere 257–260
High-pressure physics 241–243
  hydrostatic pressure range 241–242
  very high pressure 242–243
High-voltage direct current (HVDC) 188
Highway engineering 243–246, 402
  COGO 243–244
  esthetics 245–246
  orthophotograph 244–245
Hildemann, W. H. 119
Hill, J. G. 344
Hill, R. 291
Himalayan orogenic belt 305
Hippopotamus 179
Hiran 230
Histiocyte 127
Histocompatibility testing 400
Histone: association with DNA 109
  in chromosomes 147
Histosol 373
HL-A system (leukocyte) 120, 399
  HL-A5 gene 400
  HL-A7 gene 400
  HL-A8 gene 400

HL-A system (leukocyte)—*cont.*
  HL-A12 gene 400
  HL-A13 gene 400
Hodgins, M. G. 172
Holloway, P. J. 210
Hologram, Fraunhofer 411
Holography 271
Holoplastideae 109
Homeostasis 246–248
  antifreeze formation 247–248
  supercooling 246–247
Homeotherm 126
Homocysteic acid (growth) 122
Homocysteine (metabolism) 121–122
Homocystinuria 122
Homogeneous function 170
Homogeneous plastic (video playback system) 409
Homologous pairing 137
Homostructure laser 272
Hooke, Robert 14
Hordeum vulgare 138, 360
  heterochromatin replication 139
Horizon coordinates technique (marine geodetic measurements) 229
Hormone: dl-C_{18} cecropia moth juvenile 158
  growth 235–237
Hornykiewicz, O. 308
Horsetail 371
Horsfall, J. G. 101
Hose, aerial 4
Hot isostatic press 332
Housefly larva (growth in manure) 414
Houwink, A. L. 153
Hovertrain 404
Hubble, E. 224
Hubble system (galaxy types) 224
Huebner, R. J. 176
Human silica content 371
Human chromosomes, replicons of 139–140
Human genetics (emphysema) 206
Human growth hormone (HGH) 235
Hurricane (Offshore Oilman's Game) 40–43
Hurwitz, A. 113
HVDC *see* High-voltage direct current (HVDC)
Hybrid, disease-resistant 162
Hydrant 4
Hydrobiotite 149
Hydrodynamic resistance (ship) 370
Hydroelectric power 193, 333
Hydrogen bonding (clay minerals) 378
Hydrogen cyanide molecule 255
Hydrogen sulfide 106
  volcano 411–413
Hydrous mica 149
9-Hydroxyfluorene-(9)-carboxylic acid fluorenol 282
Hydroxyl molecule 255

Hygroscopic particle seeding (fog) 417
Σ-Hyperonic atom 239
Hypersecretion (cholera pathogenesis) 145–146
Hypofluorous acid 248–249
  properties 248–249
  synthesis 248
Hypothalamus 240
  control of hibernation 239
  neural control 239
Hypothermia 241

I

I (immunoglobulin) 118
IAA *see* β-Indoleacetic acid (IAA)
IAEA *see* International Atomic Energy Agency (IAEA)
IAG *see* International Association of Geodesy (IAG)
IAPSO *see* International Association of Physical Sciences of the Oceans (IAPSO)
IC 10 (galaxy) 225
IC 342 (galaxy) 225
IC 1613 (galaxy) 225
IC 4182 (galaxy) 53
Icarosaurus 406
Ice fog 416
ICES-COGO-I 243
ICSH *see* Interstitial cell-stimulating hormone (ICSH)
Idiopathic Parkinsonism 307
IFYGL *see* International Field Year for the Great Lakes (IFYGL)
IgA 118–119
IgD 118
IgE 118
IgG 118–119
IgM 118
Iguana 179
IJC *see* International Joint Commission (IJC)
Ilex integra 210
Illinois Agricultural Experiment Station 161
Illite 149
Imaging device 144
IMCO *see* Intergovernmental Maritime Consultative Organization (IMCO)
Immunity (cholera) 146
Immunoglobulin 146
  heavy-chain polymorphism 118–119
  light-chain polymorphism 118
  receptors 251
  see also Antigen
Immunological tolerance 250–251
Immunology 249–251
  developmental aspects of formation 249
  differences between T and B cells 250
  evolution 251

Immunology—*cont.*
  immunological tolerance 250–251
  origin of lymphocytes 249–250
  origin of variability 250
  receptors 251
  tissue culture studies 251
Inceptisol 373, 374
Incinerator 31
Incinerator, fluidized-bed 33
β-Indoleacetic acid (IAA) 284–285
Information transfer (biological systems) 174
Infrared galaxy 47
Infrared star 256
Ink: Dry Plate process 340
  tack value 341
  viscosity 341
Inorganic chemistry:
  coordination chemistry 154–157
  copper chemistry 157–158
  elements (chemistry) 203–205
  hypofluorous acid 248–249
  polywater 327–330
  spectroscopy 388–390
  xenon compounds 420–422
Input diode 144
Insect (growth in manure) 414
Insulated-gate field-effect transistor 144
Insulation (wire and cable) 343
Insulator: antiferromagnetic 242
  paramagnetic 242
Integrated circuit electronics 254
Intelsat 4F2 (artificial satellite) 386
Interatomic distance 241
Intercity travel 402
Interconnecting electric power system 187–190
Interface: gas-solid 58
  vacuum-solid 58
Interferometrically ruled grating 388
Intergalactic medium in clusters 225
Intergovernmental Maritime Consultative Organization (IMCO) 366
Intermetallic compound 251–254
  new applications 254
  new discoveries 252
  new understanding 252–254
  structure 251–252
International Association of Geodesy (IAG) 231
International Association of Physical Sciences of the Oceans (IAPSO) 231
International Atomic Energy Agency (IAEA) 294
International Field Year for the Great Lakes (IFYGL) 234
International Hydrological Decade 234

International Joint Commission (IJC) 233
International Monitoring Platform (*Explorer 43*) 386
International Rice Research Institute 221
International Satellite for Ionospheric Studies (*ISIS 2*) 386
Interplanetary plasma 384
Interrogator responsor unit (marine navigation) 280
Intersecting storage ring (ISR) 311–313
Interstellar matter 254–257
  carbon monoxide 256
  detectable molecules 254–256
  methyl alcohol 256–257
Interstitial cell-stimulating hormone (ICSH) 236
Inui, T. 370
Inv (immunoglobulin) 118
Inverse Compton radiation 47, 48
Ion neutralization 64
Ionosphere 257–263
  high-latitude ionosphere 257–260
  modification 260–263
Iron powder 330
Irrigation 415
Irwin, James B. 387
Isf (immunoglobulin) 118
Isocyanic acid molecule 255
Isostatic pressing, cold 331
Isotherm 169
ISR *see* Intersecting storage ring (ISR)
Israel, Werner 232
Iwamoto, F. 351

J

JA gene 400
Jacobs, T. A. 266
Jacobs, W. P. 422
Jahn-Teller effect 242, 263–264
  crystals with $3d$ ions 264
  simple model for Jahn-Teller distortion 263–264
  vibrational spectra of molecules 264
Japan Trench 305
Jastrow wave function 350
Jefferts, K. B. 255, 256
Jensen, James 405
Jensen, R. J. 268
Jet aircraft noise 292
John Hancock Center 131
Johnson, D. 257
Johnson, L. B. 296
JOIDES *see* Joint Oceanographic Institutions for Deep Earth Sampling (JOIDES)
Joint distribution function 37
Joint Oceanographic Institutions for Deep Earth Sampling (JOIDES) 307, 397
Junge, C. E. 412

Jupiter 388
Jurassic 180, 181, 265–266

K

Kamacite 20
Kaolinite 149
Kaon-nucleon interaction 239
Kaonic atom 238–239
Karig, D. E. 306
Karyogamy 134
Kasper, J. 253
Kasper, Jerome 266
Kawisari (coffee strain) 152
Kayenta Formation 407
Kell (immunoglobulin) 118
Kerr cell 301
Kidd (immunoglobulin) 118
Kielan-Jaworowska, Z. 178
Kikuchi diagram 197
Kilauea volcano 411
Killifish (blood freezing point) 246
King, Lester 405
Kitching, James W. 406
Klug, A. 116
Kockum's Shipyard System (design system) 370
Kodak Process Resist (KPR) 338
Kogelnik, H. 270
Kornberg, A. 325
Kort nozzle 345
Kostrama Technical Institute 327
KPR *see* Kodak Process Resist (KPR)
Krakatoa eruption 412
Kratky, O. 173
Kristensen, K. J. 375
Krypton (gas absorption) 98
Krypton laser (Raman spectroscopy) 389
Kuchemann, D. 220
Kuehneosaurus 406
Kuiper, G. 367
Kuo, J. T. 185
Kupke, D. W. 173
Kurtin, S. L. 329
Kutta-Joukowski law 220
Kuznets, Simon 292
Kwik, K. H. 367

L

LA locus (antigen) 399
Labyrinthodont amphibian 406
Ladder, aerial 4
Lake of the Clouds varve 353
Lakes, Great 233–235
Lamb, H. H. 412
Lambda 230
Laminarin 109
Lamont-Doherty Geological Observatory 397
Lamp lesion 127
Lancaster, F. W. 344
Land reclamation 208
Landau, L. D. 288
Lander, J. J. 62
Landfill, sanitary 29–31

Langmuir, Irving 58
Larodan 127 219
Lascaux Cave 11
Laser: as light source in Raman spectroscopy 388
  engraving in gravure printing 340
  lead 272
  nonlinear optics 300–302
Laser, chemical 266–268
  continuous-wave chemical laser 266–267
  metal-vapor laser 270–272
  pulsed chemical laser 268
  scientific applications 268
Laser, dye 268–270
  continuous-wave operation 269
  image amplification 269–270
  new feedback mechanism 270
Laser, metal-vapor 270–272
  continuous-wave metal-ion lasers 270–272
  pulsed metal-vapor lasers 272
Laser, semiconductor injection 272–274
  double-heterostructure laser 273–274
  homostructure laser 272
  large optical cavity laser 274
  single-heterostructure laser 273
Laser microprobe (scientific investigation of art objects) 15–18
Laser radar (studies of particles in stratosphere) 413
LASH *see* Lighter aboard ship (LASH)
Lattice structure 213
Laue diagram 197
Lead laser 272
Lead poisoning: acute form 75
  chronic form 75
Leaf 274–276
  bud relationship 274
  callus cultures 276
  identification of bacteria 275–276
  leaf relationship 274–275
  location of bacteria within plant 274
  monocotyledon 210
Leaf rust (coffee disease) 150
Leaf spot disease 100
Leather goods (use of food-processing wastes) 414
Lebedev Institute 163
Lecheosin 109
Lechtman, Heather 19
Lecolazel, P. 181
Lee-Huang, S. 175
LEED *see* Low-energy electron diffraction (LEED)
LeLevier, R. 260
Lennard-Jones potential 350
Lens, quadrupole 98
Leo I (galaxy) 225
Leo II (galaxy) 225
Leo A (galaxy) 225

LePichon, X. 396
Leucoplast 360
Leukocyte system 119–121
Lewin, Seymour Z. 18
Lewis (immunoglobulin) 118
LH see Luteinizing hormone (LH)
Li, C. H. 235
Lias, English 265
Lift 220
Ligand (coordination chemistry) 154–157
Light (correlation with photorespiration) 321
Light-chain polymorphism 118
Light-scattering (or refracting) detector 7
Light water (fire-suppression agent) 7
Lighter aboard ship (LASH) 364
Lignified collenchyma 153
Lignin 153, 415, 424
Lilley, A. E. 255, 257
LIM see Linear induction motor (LIM)
Lima bean 324
Limber, C. N. 227
Limestone-injection wet-scrubbing systems 102
Limited Test Ban Treaty 296
Linear induction motor (LIM) 404
Lines fairing 367
Ling Fang-i (art) 14
Linoscan 404 337
Liostrea irregularis 265
β-Lipoprotein systems 118
Lippincott, E. R. 328
Liquid chemical extinguisher 7
Liquid-hydrogen bubble chamber 200
Liquid metal–cooled fast-breeder reactor (LMFBR) 191
Liquid nitroglycerin, desensitized 300
Liquid scintillator 167
Liquid wastes 415
Lissaman, P. B. S. 220
Lithodesmium undulatum 110
Lithosphere 303, 306, 397
plate tectonics 305
Litter: decomposition agents 382
decomposition rate factors 381
pathways from standing vegetation 381
Litter bag, nylon net 382
Liu, W.-K. 235
LMC (galaxy) 225
LMFBR see Liquid metal–cooled fast-breeder reactor (LMFBR)
LND* gene 400
LND-AJ gene 400
Local Group (galaxies) 225, 226
Local Supercluster (galaxies) 227
LOLP see Loss of load probability (LOLP)
Long, J. A. 235

Lorac 230
Loran C 286
Lord Howe Rise 305
Los Alamos Scientific Laboratory 205
Loss-of-load probability (LOLP) 188
Loss reduction (fire prevention) 4–5
Low-energy electron diffraction (LEED) 59–62, 196
Low-sulfur fuels 104
Lowland rice 378
Loxophyceae 109
Lp (immunoglobulin) 118, 119
Lubricant, die 331
Luna 18 (artificial satellite) 386
Lunar Rover 385
Lunokhod 385
Lung: cancer 78
emphysema 205–206
Lurgi Sulfacid process 106
Luristan sword 19
Lutein 108
Luteinizing hormone (LH) 236
Lutheran (immunoglobulin) 118
Lutite 303, 305, 306
Lymphocyte function in immunological reaction 249–250
Lyons, R. M. 325
Lyons, W. R. 236
Lystrosaurus 405

M

M 1 (galaxy) 225
M 2 (galaxy) 225
M 31 (galaxy) 225, 226
M 33 (galaxy) 225
M 87 (galaxy) 47
M 147 (galaxy) 225
M 185 (galaxy) 225
M 205 (galaxy) 225
M 221 (galaxy) 225
McAdam, D. W. 391
McDonald, I. 382
Macfadyen, A. 380
McGregor Glacier (Antarctica) 406
McIntyre, D. S. 375
McMillan, B. 112
Macromolecule (synthesis in tracheary elements) 423
Maffei 1 225
Maffei 2 225
Magellanic Clouds 225
Magnascan 450 336
Magnesium: loss in Mollisols 372–374
loss in Ultisols 372
Magnesium oxide (absorbent for sulfur dioxide) 104
Magnesium sulfite 104
Magnet, bending 98
Magnetic anomaly 396
Magnetic densimetry 171–172
Magnetic field 396
Magnetic materials 241
Magnetic moment 239

Magnetic rotator (energy source in galaxies and quasars) 51
Magnetic surveys (petroleum prospecting) 345–346
Magnetic tape (storage medium for video playback system) 409
Magnetism (energy source in galaxies and quasars) 51
Magnetohydrodynamic (MHD) generation 191
Magnetometer: on Moon 386
petroleum prospecting 345
Maizel, J. V. 116
MaKi gene 400
Malone, C. R. 382
Management: soil 378–380
solid-waste 24–33
Mandl, G. 320
Manganese (mutagen) 71
Mannitol 247
Mantle 303
Manton, Irene 106
Manure 413
Marconaflo system 364
Marine influence on weather and climate 277–279
anomalous patterns 277–279
meteorological impact 279
Marine navigation 279–282
precise position measurement 279–281
talking beacon system 281–282
Marine propeller 344–345
Mariner 5 (artificial satellite) 407, 408
Mariner 8 (artificial satellite) 386, 388
Mariner 9 (artificial satellite) 386
Mariner class ship 367
Marov, M. Y. 407
Mars, missions to 387–388
Mars 1 (artificial satellite) 388
Mars 2 (artificial satellite) 386, 388
Mars 3 (artificial satellite) 386, 388
Martin, J. T. 210
Maruo, H. 370
Mason, R. B. 396
Mass distribution 201
Massachusetts Institute of Technology (MIT) 243
Master Standard Data (MSD) 418, 419
Mastigoneme 110
Mastodon 179
Matter, interstellar 254–257
Matthews, D. H. 305, 396
Max Planck Institute for Radio Astronomy 125
Mayan Fine Gray pottery 17
Mayan Fine Orange pottery 17
Mayer, Ralph 13
Mayer, S. W. 268
Maynard Research Data (MRD) 418, 419
Mayo Clinic 194
Mead, C. A. 329
Mean (probability) 37

Measurement, work 418–420
Mechanical memory (nickel-titanium compound) 254
Mechanical power engineering (power plant) 332–334
Medeiros, A. G. 151
Medical microbiology (transplantation biology) 399–402
Medical product (sterilization by process radiation) 343
Medical Systems Development Laboratory (MSDL) 193, 194, 195
Meeks, M. L. 255
Megalosaurus wetherilli 407
Meidae 219
Meiocyte 134
Meiosis 138
Meiotic chromosome pairing 134–136
Mélange (orogeny) 304
Melanoplus differentialis 138, 139
Melanosaur 180
Meltz, G. 260
Mengovirus 115
Merchant ship 363–366
Mercurialism 73–75
Mercury (planet) 388
Mercury poisoning 73–75
case histories 74
increasing prevalence of element 73–74
symptoms 74
Meristem 153
Merler, E. 119
Merrifield, R. B. 236
Mesocricetus auratus (heterochromatin replication) 139
Meson 198, 310
A_2 202–203
cascade resonances 200
K 201
Mesophyll cell 323
Mesozoic (ecologic succession) 216
Metal toxicity: carcinogens 71–72
cellular response 71
conditions 70–71
specificity 71
teratogens 71–72
Metal-vapor laser 270–272
Metallic material: phase stability 211
thermodynamic properties 211
Metallic systems, binary 213
Metallography (study of art objects) 19
Metallurgical engineering: equilibrium, phase 211–214
intermetallic compound 251–254
powder metallurgy 330–332
science in art 10–23
Metallurgy, powder 330–332
Metaphase 134
Meteor 8 (artificial satellite) 386
Meteor 9 (artificial satellite) 386

Meteorology: agricultural 99–101
 marine influence 279
Methionine 115, 121, 123
Methods, approximate 39
Methods-time measurement (MTM) 418
Methyl acetylene (propyne) molecule 255
Methyl alcohol 255–257
Methyl cyanide (acetonitrile) molecule 255
Methyl folate 121
Methyl methacrylate (polymerization by process radiation) 343
MHD generation *see* Magnetohydrodynamic (MHD) generation
Microbial biomass in soil 383
Microbial grazers in soil 382
Microbiology: nitrogen fixation 290–291
 polymerase 325–327
Microclimate (beaver) 126
Micrographia (book) 14
Micrographia (writing difficulty) 307
Micronutrient (copper) 70
Microorganisms, soil 380–383
Microprobe, laser 15–16
Microprobe analysis 20
Microscope: electron 197, 198
 field-emission 63, 64
Microtubule 110
Microtus agrestis (heterochromatin replication) 139
Microwave resonator, niobium 308
Microwave sensor (use in aircraft for remote sensing) 186
Microwave superconductor 308
Mid-Atlantic Ridge 305
Migmatite 304
Milk solids 414
Milky Way galaxy 48, 54, 225, 226, 289
Miller, C. L. 243
Miller, G. L. 114
Miller, Paul R. 100
Minamata disease 74–75
Mineral and fossil-fuel wastes 26–27
Mineral tanker 363
Minerals, clay 377
Ming dynasty 20
Minimal art 13
Miocene 314–315
Mirels, H. 266
Mississippian 217
MIT *see* Massachusetts Institute of Technology (MIT)
Mitchell, Edgar D. 385
Mitchell, J. M., Jr. 412
Mitochondrion 111, 112, 209
 negative effect of cooling in plant 324
 photorespiration 323
Mitosis: Cyanophyta 109
 Dinophyta 109

Mixed lymphocyte culture (MLC) test 401
Mixing, electromagnetic 201–202
Mizutani, Satoshi 174
MLC test *see* Mixed lymphocyte culture (MLC) test
MNSs (immunoglobulin) 118
Modeling: fire research 8
 highway engineering 245–246
 petroleum reservoir engineering 316–318
Modern fire-fighting methods 4
Modern materials in traditional art 13
Modification, weather 416–418
Modified line crossing technique (geometry and orientation of transponder arrays) 229
Moe, J. 370
Moenkopi Formation 407
MoFe protein *see* Molybdoferredoxin
Moho 303
Moisture (relationship with plant disease) 100
Molariform tooth 179
Molasse (orogeny) 304, 305, 306
Molodenski, M. S. 184
Molybdenum powder 330
Molybdoferredoxin 291
Molynia 18 (artificial satellite) 386
Monitor lizard 179
Monocotyledonous leaf 210
Monothetic group 14
Monovalent agglutinin 142–143
Monte Carlo model 39
Montmorillonite 149, 150, 378
Monzonite 304
Moore, C. B. 268
Morphactin 282–286
 influence on plant movement 284–285
 interrelations with growth hormones 284
 metabolic effects 284
 morphogenetic effects 283–284
 practical applications 286
 transport and metabolism in plants 285–286
Morrison, Philip 289
Mössbauer resonance 242
Mossop, S. C. 412
Mott, Sir Nevill 241
Mott transition 241–242
Mt. Redoubt volcano eruption 412
Mt. Spurr volcano eruption 412
Mt. Trident volcano eruption 412
Mountain belt forming 302
Mountain building (theory for faunal extinction) 214
Moving pedestrian walkway 403

MRD *see* Maynard Research Data (MRD)
MSD *see* Master Standard Data (MSD)
MSDL *see* Medical Systems Development Laboratory (MSDL)
MTM *see* Methods-time measurement (MTM)
MTM General Purpose Data (MTM-GPD) 418, 419
Mueller, E. W. 63
Mueller, W. A. 329
Muller, H. J. 135
Multidiscipline approach (petroleum prospecting) 347–348
Multispectral photography in remote terrain sensing 398
Multispectral scanner 186
Municipal wastes 26
Munk, W. 183
Muons (cosmic ray) 162–169
Mus musculus 139, 140
Music Room, The (art) 21, 22
Mycoplasma hominis 147
Myrsinaceae 274, 275

N

N 6822 (galaxy) 225
Nacrite 149
Nakamura, Takako 73
Nalidixic acid 326
Namaizawa, H. 351
NAPSIC *see* North American Power System Interconnection Committee (NAPSIC)
NASA *see* National Aeronautics and Space Administration
Nasmyth focus 124
Nastic movement 285
National Accelerator Laboratory 162
National Aeronautics and Space Administration (NASA) 6, 185, 230
National Air Pollution Control Administration 206
National Bureau of Standards 328, 352
National Electric Reliability Council (NERC) 189, 192
National Environmental Policy Act of 1969 354
National Oceanographic and Atmospheric Administration, U.S. 234
National Rail Passenger Corporation 402
National Safety Council 27
Nato 2 (artificial satellite) 386
Natural gas reservoir (stimulation) 295
Natural resources (solid-waste relationship) 28
Naughton, J. J. 411
Naval architecture: propeller, marine 344–345
 ship, merchant 363–366

Naval architecture—*cont.*
 ship design 366–371
Naval Oceanographic Office, U.S. 396
Navigable Streams Refuse Act of 1899 207
Navigation: buoy 132–134
 marine navigation 279–282
Navigation systems, electronic 286–288
 dead-reckoning computer 286–287
 echo sounding 286
 sailing computer 287–288
Negative feedback (amplifier) 112
Neodymium glass laser 301
Neogene 217
Neon laser (Raman spectroscopy) 389
Neoplasm 71
Nepenthes (cuticle) 211
Neptune 388
NERC *see* National Electric Reliability Council (NERC)
Neruda, Pablo 292
Neutralization, ion 64
Neutron, cosmic ray 167–169
Neutron activation 329
 scientific investigation of art objects 16–17
Neutron assay techniques 298
Neutron star 55, 231
 early work and recent developments 288
 physics of very dense matter 288–290
 spinars 289–290
New York City (noise control report) 293
New York City's Mayor's Task Force on Noise Control 293
Newell, R. E. 413
NGC 4151 (galaxy) 48
Nickel carbonyl 77
Nickel poisoning (industrial hazard) 77–78
Night E region (ionosphere) 258
Nilsson, S. 203
Nimonic alloy 254
Nitrogen: availability in flooded soil 379–380
 movement in biosphere 374–377
Nitrogen fixation 290–291
 control of nitrogenase 291
 mechanism 291
 nitrogenase composition 291
 requirements 290–291
Nitrogenase 290
 composition 291
 control 291
Nixon, Richard 296
NMIS *see* Nuclear Materials Information System (NMIS)
NMR *see* Nuclear magnetic resonance (NMR)
Nobel prizes 291–292
Noble, P. N. 248
Nodule, leaf 274–276

Noise, acoustic 292–294
  Federal regulation 292
  interference with creative activity 293–294
  New York City noise control report 293
  noise levels in industry 292
  protection for employed 292–293
Nonlinear optics 300–302
Nonlinear programming computer algorithm 370
Nonrecovery systems (flue gas) 103–104
Nonrelativistic electron gas 288
Noonan, K. D. 141
Noradrenaline 308
Norbornadiene 157
Norepinephrine (effect on hibernation) 240
Normal cell growth 142
Normal mouse fibroblast 142
North American Power System Interconnection Committee (NAPSIC) 188, 189
Northwestern University 205
Nosanow, L. 350
Notothenia neglecta (blood freezing point) 246
Nowacki, H. 367
Nuclear emulsion 167
Nuclear energy in galaxies and quasars 50
Nuclear engineering: nuclear explosion engineering 294–297
  nuclear materials safeguards 297–298
  process radiation 341–344
  reactor licensing, nuclear 354–358
Nuclear explosion engineering 294–297, 299
  civil construction 296
  explosion technology 294–295
  resource development 297
  safety assurance 295–296
Nuclear fuel 334
Nuclear magnetic resonance (NMR) 351
Nuclear many-body problem 288
Nuclear Materials Information System (NMIS) 298
Nuclear materials safeguards 297–298
  accounting system 298
  improving measurements 298
  nuclear reactor licensing 356
  power plant siting 334
  safeguard coverage 297–298
  systems studies 298
Nuclear power plant 354
  environmental impact 356
  inspection program 356
  radioactivity release 358
  safety 355–356
  safety criteria and standards 357–358

Nuclear structure 203–205
Nucleolus 110
Nucleon 200
Nucleonic cascade 167
Nutrient: distance traveled to each root 358
  mobility 359
  mobility from soil to water 374
  net loss in soil mantle 372
  plant 378
Nutrient element loss: Mollisols 372–374
  Ultisols 372
Nye, P. H. 359
Nyquist, H. 113

O

Oak Ridge National Laboratory 205
Oats (silica content) 371
O/B/O ship *see* Ore/bulk/oil (O/B/O) ship
O'Brien, T. P. 423
Obstructive pulmonary emphysema 205
Occipital lobe (brain electrical concomitants of language) 391
Ocean environment study 132
Ocean-surface positions (relationship to ocean bottom marks) 229
Oceanic Ridge 303
Oceanography (marine influence on weather and climate) 277–279
Office of Civil Defense 6
Office of Coal Research 105
Offshore oil and gas wells 346–347
Offshore Oilman's Game 39
Ogston, A. G. 359
Oil and gas reservoir: performance 315
  simulation 316
Oil and gas well completion 298–300
  borehole blasting 299–300
  fracture detonation 300
  system reflection dynamics 299
  trend toward exploration 346–347
Oil reservoir (stimulation) 318
Oiled seabirds, rehabilitation of 219
Olea lanceolata (cuticle) 210
Olin, Jacqueline S. 17
d'Oliveira, B. 152
Oliver, J. 397
Omega (navigation system) 286
One Shell Plaza (building) 132
One Shell Square (building) 132
O/O carrier *see* Ore/oil (O/O) carrier ship
Oocyte 136
Open dump 25
Operations research (fire research) 8–9

Ophiolite 306
Ophiolite suite 303
Oppenheimer, I. R. 288
Optical absorption 242
Optical emission spectrometry (scientific investigation of art objects) 15–18
Optical galaxy 46–47
Optical mixer 301
Optical telescope 123–124
Optical transition 243
Optics, nonlinear 300–302
  nonlinear susceptibility 301
  Raman effect 301
  ultrashort pulses 301–302
  x-ray nonlinearities 302
Optimization (ship design) 367–370
Orbiting Astronomical Observatory 2 46
Orchard residue 414
Ore/bulk/oil (O/B/O) ship 363
Ore/oil (O/O) carrier ship 363
Ore/slurry/oil (O/S/O) carrier ship 363–364
Organic compounds (interaction with clay minerals) 377
Organic liquids (class B fires) 5
Organocopper 157
Organolithium 157, 158
Orion Nebula 256
Ornithogalum 135
Ornitholestes 176
Ornstein, R. 391
Orogenic belt: Alpine-Himalayan 302–307
  Andean 305
  Appalachian 302–307
  Cordilleran 302–307
  Himalayan 305
Orogeny 302–307
  fundamental mechanism 306–307
  lithosphere plate tectonics 305
  ocean sediment accumulation 305–306
  thermal doming 304
Orthophotograph (use in highway engineering) 244–245
Os prominens 394
Oscar (artificial satellite) 386
O/S/O carrier *see* Ore/slurry/oil (O/S/O) carrier ship
Ouchterlony test 119
Overlapping resonances (elementary particle) 200–203
Owl 394–396
Oxisol 373, 374
Oxygen: atmospheric 378
  in photorespiration 320–323
Oxygen flux (effect on plant growth) 375

P

P (immunoglobulin) 118
Pachytene 134

Pacini, Franco 289
Pacinian corpuscle 395
Paint (drying by process radiation) 343
Paleobotany and paleontology: dinosaur 176–181
  faunal extinction 214–217
  Jurassic 265–266
  Triassic 405–407
Paleogene 217
Paleozoic 216, 217
Palmer, P. 255, 257
Pancreas 361
Panicum bisculatum 320
Papain (experimental induction of emphysema) 206
Paper: fiber recovery 33
  recycling 415
Papkoff, H. 235
PAPS *see* Phosphoadenosinephosphosulfate (PAPS)
Parabolic dish (telescope) 124–125
Paramagnetic insulator 242
Parametric down-conversion of x-rays 302
Parametric oscillator 301
Paramylon 108
Paranasal sinus 78
Parke, Mary 106
Parker, E. N. 383
Parkinson, James 307
Parkinson's disease 307–308
  clinical picture 307
  discoveries 307–308
  mode of action of L-dopa 308
Particle, elementary 198–203
Particle accelerator 162, 308–313
  hadronic atom 237
  intersecting storage ring 310–313
  neutron generator 298
  superconducting accelerator 308–310
  synthesis of new elements 203–204
Particle size reduction (composting) 31
Pascher, Adolph 106
Passioura, J. B. 359
Pathology of heavy metals 68–78
  cadmium 72–73
  cobalt 76–77
  lead 75–76
  mercury 73–75, 234
  nature of metal toxicity 70–72
  nickel 77–78
  thallium 76
Pathotoxin: race O 159
  race T 159
Patina (study of art objects) 18–19
Paul, E. A. 381
Pauly, Manfred 338
Pavetta 274–276
Pea (silica content) 371
Peach tree 324
Peanut (leaf spot disease) 100

Pectin 210
Pectinase 423
Penning ionization 271
Pentaceratops 180
Pentafluorophenylcopper 157
Pentose phosphate cycle 321–322
Penzias, A. A. 255, 256
Perfluoro-*t*-butylcopper 157
Perfluoroheptylcopper 157
Peridinin 107
Periodic table (predicted locations of new elements) 204
Perlman, I. 16
Permalloy 171
Permian 217, 307
Peroxisome 321–323
Peru-Chile Trench 305
Pesticide inactivated by soil clay 377
Petasites fragrans 153
Peterson, O. G. 269
Petiole 209, 210
Petroleum engineering: oil and gas well completion 298–300
   petroleum geology 313–315
   petroleum reservoir engineering 315–318
   petroleum secondary recovery 318–320
   risk evaluation in engineering 34–43
   terrain sensing, remote 397–398
Petroleum geology 313–315
   environmental control 313–314
   Gulf Coast Miocene 314–315
Petroleum reservoir engineering 315–318
   automatic history matching 317–318
   examples 318
   history matching 317
   reservoir description 316–317
   reservoir simulation 316
Petroleum secondary recovery 318–320
   steam drive 319–320
   steam soak 319
PGM_3 system *see* Phosphoglucomutase (PGM_3) system
Ph erythrocyte system 119
Phaeophyceae 109
Phaeophyta 109
Phalacrocoracidae 219
Phase diagram 211
   He^3 351
Phase equilibrium 211–214
Phenylalanine 424
Phenylalanine ammonia lyase 424
3-Phenylhexafluoropropylcopper 157
Phenylpropane-modifying enzyme 424
Philip, G. M. 265
Philip, L. R. 359
Phloem 129, 285

Phonograph record (four-channel stereo) 393
Phonon 351
Phosphoadenosinephosphosulfate (PAPS) 121
Phosphoglucomutase (PGM_3) system 401
   locus 120
3-Phosphoglyceric acid 322
Phosphorus: availability in flooded soil 380
   loss in mollisols 372–374
   loss in ultisols 372
   movement in biosphere 375
Photoelectric emission 64–65
Photoelectron spectroscopy 65
Photography, multispectral 398
Photomosaic 186
Photomultiplier 336
Photon 65, 200
Photorespiration 320–323
   apparent absence 323
   function 323
   leaf organelle 322–323
   methods of estimating 321
   physiological factors 321
   substrate 321–322
Photosynthesis: algae 107–109
   influence in faunal extinction 215–216
Phototropism (morphactin effect) 284
Phycobiliprotein 107
Phycocyanin 107
Phycoerythrin 107
Phyllosphere 380
Physical electronics: amplifier 112–115
   charge-coupled device 143–144
   electron diffraction, high-energy 195–198
   laser, chemical 266–268
   laser, semiconductor injection 272–274
   optics, nonlinear 300–302
   semiconductor 361–363
   surface physics 58–67
Physical environment in ecological succession 216
Physical techniques (scientific investigation of art objects) 18–20
Physics, high-pressure 241–243
Physics, surface 58–67
Physics of very dense matter 288–290
Physiological and experimental psychology: hibernation 239–241
   speech 390–392
Phytic acid 360
Phytoplankton productivity 215–217
Picasso, Pablo 20
Picornavirus: protein component 115–116
   RNA component 115
   structure 116–117
Pien, P. C. 370
Piezo element 410
Pigment, synthetic 13

"Pill-rolling motion" (Parkinsonism) 307
Pillow lava 304
Pimentel, G. C. 248, 266
Pionic atom 237
Pisum 284
Pisum sativum 283
Pitcher plant (cuticle) 211
Pitman, W. 396
Pittenger, E. W.
Plankton (mid-water echo source) 286
Plant, power 332–334
Plant anatomy: epidermis (plant) 209–211
   leaf 274–276
   seed (botany) 360–361
   xylem 422–424
Plant assemblages 217
Plant constituents' decay rates 381–382
Plant disease forecasting (synoptic weather chart) 100
Plant growth 323–325
   effect of oxygen flux 375
   negative effects of chilling 324–325
   positive effects of chilling 323–324
Plant hormone: brassin 128–130
   corn pollen 128
   influence on xylem tracheary elements 422
   morphactin 282–286
Plant nutrients (increased availability by flooding) 378
Plant nutrition (importance of silicon dioxide) 371
Plant pathology: agricultural meteorology 99–101
   coffee 150–152
   corn 158–162
Plant physiology: brassin 128–130
   morphactin 282–286
   photorespiration 320–323
   plant growth 323–325
Plant taxonomy (algae) 106–112
Plasma, interplanetary 384
Plasma display (art) 13
Plasmalemma 154
Plasmodesmata 360
Plasmodesmatal canal 361
Plastics (class A fires) 5
Plate, printing 337–341
Plate tectonics, lithosphere 305
Platinum hexafluoride (reaction with xenon) 420
Platinum microelectrode method (soil aeration determination) 372
Pleistocene 219
Pleuropneumonia-like organism (PPLO) 147
Plexiglas 13
Plumbism *see* Lead poisoning
Pluto 388
Pluton 304
Point, critical 169–171
Polanyi, John 267

"Polar wind" 260
Polarization: activation 375
   concentration 375
Poliovirus 117–118
Pollution, air 101–106
Polyacrylamide gel electrophoresis 327
Polyester-based camera plate 335
Polyethylene oxide (controlled degradation by process radiation) 343
Polymer: colors 13
   copolymerization by process radiation 343
   water 327–330
Polymerase 325–327
   DNA polymerase I 325–326
   DNA polymerase II 326–327
   RNA-directed DNA 174–175
Polymeric material: phase stability 211
   thermodynamic properties 211
Polysaccharide 123
Polystyrene-divinylbenzene resin 236
Polythetic group 14
Polywater 327–330
   conclusion 330
   dissent 329
   historical development 327–329
   interpretation 329–330
Pontecorvo, G. 135
Pooling power systems, benefits of 188
Porto, S. P. S. 329
Posterized x-rays 335
Potassium: loss in mollisols 372–374
   loss in ultisols 372
Potato blight 101
Pottery: colonial American 17
   Devonshire 17
   Mayan Fine Gray 17
   Mayan Fine Orange 17
Powder metallurgy 330–332
   background 330
   consolidation 331–332
   future 332
   powder production 330–331
Powell, Jon 405
Power, hydroelectric 193
Power plant 332–334
   air quality management 207
   cooling water 333–334
   emission control 102
   Rankine cycle 333
   reactor siting 354
Power pools 188
PPLO *see* Pleuropneumonia-like organism
Prandtl, L. 344
Prasinophyceae 106, 109
Preamplifier, parametric (radio telescope) 125
Precambrian 307
Precise position measurement at sea 279–281

PRELIKON (design system) 370
Primary consumers 381
Printing, color 334–337
   color scanners 335–337
   dry photomechanical processes 334–335
Printing ink (drying by process radiation) 343
Printing plate 337–341
   driography 340–341
   gravure printing 337–340
Pritchard, R. H. 135
Probability, geometric 38–39
Procaryota 106, 109
Process radiation 341–344
   growth and status 342
   recent key developments 342–344
Procolophon 406
Production experiments (meson resonances) 200
Prokaryote DNA translation 146
Proline (influence on xylem tracheary elements) 422
Pronator profundus muscle 394
Propeller, marine 344–345
   contrarotating propeller 344–345
   controllable-pitch propeller 344
   ducted propeller 345
Prophase 138
Proportional counter 167
Propulsion plants in ships 365
Prospecting, petroleum 313, 345–348
   exploration sequence 345–346
   multidiscipline approach 347–348
   petroleum reservoir engineering 315–318
   relative costs of prospecting methods 346
   remote terrain sensing 397–398
   trend to offshore exploration 346–347
Prosthesis 348–349
   recognition 348
   resolution 348–349
   stimulus 349
   working system 349
Protein metabolism (emphysema) 206
Proteoglycan structure 123
Prothero, W. A., Jr. 184
Proton 200
K^--Proton 199, 200
Proton donation *see* Protonation
Proton-proton interaction 163
Proton synchrotron 311–312
Protonation 377
Protovirus 176
Prunus laurocerasus 210
Pseudococcus obscurus 138, 139
Psychotria 274–276
Psychotria punctata 276
Public transportation in cities 402

Puerto Rico Trench 230, 231
Pulsar 288
   NP0532 289
Pulsed chemical laser 268
Pulsed metal-vapor laser 272
Pump: sputter-ion 59
   titanium sublimation 59
Pumped-storage hydro plant 193
Puntiagudo volcano eruption 412
Purkinje cell (effect of mercury poisoning) 75
Pyrex capillaries 328
Pyridoxal phosphate 121
Pyrophyllite 378
Pyrrophyta 109
Pyruvate (nitrogen fixation) 290
Pyruvic acid (nitrogen fixation) 290

Q

QSO *see* Quasar
Quadricyclene 157
Quadrupole lens 98
Quantitative chemical analysis (Raman spectroscopy) 390
Quantum solid 349–351
   macroscopic properties 350
   microscopic properties 350
   theoretical description 350–351
Quark 198
Quartz capillary 328
Quasar 48–49, 54
   3C345 289
   energy sources 44–55
Quasi-stellar object *see* Quasar
Quiet solar wind 383–384

R

R* gene 400
Rabbit 381
Race, R. R. 119
Race O pathotoxin 159
Race T pathotoxin 159
Racial senescence 215
Radar: C-band 230, 231
   Doppler 230, 231
   marine navigation 279–282
Radar altimeter (geoid determination) 230
Radar delay Doppler interferometry (Venus topography) 408
Radford, H. 255, 257
Radiant-emission detector 6–7
Radiation: blackbody 288
   inverse Compton 47
   synchrotron 47
Radio astronomy (interstellar matter) 254–257
Radio communications: talking beacon system 281
   video recording and playback 408–411
Radio galaxy 47
Radio telescope 124–125

Radio-wave absorption:
   ionosphere E region 258–259
   ionospheric modification 260
Radioactive waste treatment systems 356
Radioactivity (release from nuclear power plant) 358
Radiocarbon dating 351–353
   C^{14} changes in atmosphere 352–353
   C^{14} concentration in wood 352
   computation method 352
   dating variability 353
   tree rings 352
   varves 353
Radiometer (satellite) 185
Radius bone 394
Raff, A. D. 396
Raffinose 247
Rainfall: effect of volcanic emissions 411
   relationship with plant disease 100
Rainwater, E. H. 314
Raison, J. K. 325
Raman effect 302, 388
Raman spectroscopy 388–390
   helium-cadmium laser 271
Ramp method (sanitary landfill) 30
Rank, D. M. 255
Rankine cycle 333
Rape plant 128
Rate-of-temperature-rise detector 7
Ray, cosmic 162–169
Raydist 230
Reactive metals (class D fires) 5
Reactor licensing, nuclear 354–358
   AEC licensing process 354–356
   AEC regulatory program 354
   antitrust considerations 356–357
   civilian nuclear power program growth 354
   environmental impact 356
   power plant siting 334
   public hearings 357
   reactor inspection program 356
   safety criteria and standards 357–358
Readiness potentials (speech) 391–392
Reading machine for the blind 348
Recognition by blind of movement (prosthesis) 348
Recording and playback, video 408–411
Recovery, paper fiber 33
Recovery systems, wet scrubbing 104
Recycling: solid wastes 32
   water 208
Redlich-Kwong equation of state 170
Redshift 49

Reduced ferredoxin 290
Reducers, decomposer 382
Reed, A. 367
Reed canary grass (silica content) 371
Reeve, R. M. 153
Reflection HEED (RHEED) 196–197
Reflection seismic survey 346
Refraction seismic survey 346
Regge theory 202
Rehabilitation of oiled seabirds 219
Reichle, D. E. 382
Reinforced concrete 130
Reinluft process 103
Relativity theory 50
Remote sensing: camera 185
   radiometer 185
   terrain sensing 397–398
   visual observation 185
Repetitious DNA 147
Reptiles, Triassic 405–407
RER *see* Rough endoplasmic reticulum (RER)
Resonance: coupled 200–203
   exotic 200
   K^+p 200
   K^-p 200
   overlapping 200–203
Respiratory poison (cadmium) 73
Reuse of solid wastes 32
Reverse transcriptase 325
Rh (immunoglobulin) 118
Rhabdomyosarcoma 71
Rhaphidophyceae 109
Rhea 180
RHEED *see* Reflection HEED
Rhodamine 6G 269, 270
Rhodophyceae 109
Rhodophyta 106, 109
   phylogenetic relationships 112
Rhyolite 304
Ribonuclease 325
Ribonucleic acid (RNA) 112, 146, 174
   from bacteria in young leaves 275
   picornavirus RNA component 115
   RNA-directed DNA polymerase 174–175
   RNA-directed DNA polymerase in cells 175–176
   RNA-directed DNA polymerase in virions 175
   RNA genome 115
   RNA tumor virus 176
Ribosome 209
Rice, W. W. 268
Rice: lowland 378
   new strains 222
   silica content 371
Rich, Thomas 406
Rickman, R. W. 375
Risk estimation in structural design 39
Risk evaluation in engineering 34–43
   aptitude to guess probabilities 36

Risk evaluation in engineering — cont.
  elementary risk model 37–38
  essential concepts of probability 36–37
  gaming techniques 39–42
  risk estimation in structural design 39
  statistics of extremes 38–39
Risk model, elementary 37–38
RNA see Ribonucleic acid (RNA)
Robusta coffee 151
Rock density (drilling) 346
Rod (eye) 394
Roelofsen, P. A. 153
Roland, J. C. 154
Römisch-Germanisches Zentralmuseum 15
Roosa, Stuart A. 385
Root (botany) 358–359
  distance nutrient travels 358–359
  influence of length of root per unit volume of soil 359
  microbial activity on dead and senescing roots 382–383
  nutrient mobility 359
  root hair presence 359
  soil property and root relationship 358
Root hair 358, 359
Rosner, John 200
Rough endoplasmic reticulum (RER) 360
Rous, Peyton 174
Rous sarcoma virus 174, 175
Rousseau, D. L. 329
Rousvirus 174, 175
Rovira, A. D. 359
Ruben, John 406
Rubery, Philip 424
Rubiaceae 274, 275
Rubin, R. H. 255
Rueckert, R. R. 116
Russell, D. A. 178

S

Sabloff, J. A 16–17
Safeguards, nuclear materials 297–298
Safety criteria and standards for reactors 357–358
Sagan, Carl 407
Sailing computer 287–288
Salmon, Maurice 17
Salmon, W. D. 122
Salt dome 315
Salyut (artificial satellite) 388
  Salyut 1 386
Samson, S. 253
San Juan Trader (ship) 365
San Marco 3 (artificial satellite) 386
Sandmeyer reaction 157
Sanger, R. 119
Sanitary landfill: ramp method 30
  trench method 29–30
Sarnoff, David 408
Sasanian silver (art objects study) 22
Saslaw, W. C. 227
Satellites, Earth-resource 185–187
Saturn 388
Saurischia 176, 179
Sauropod 180, 181
Sauropoda 179
Savitskii, E. 254
Sayre, E. V. 16
Scaling function 170
Scandiatransplant 401
Scanner, color 335–337
Scanning electron microscope (study of art objects) 18
Scanning HEED 197
Schafer, E. 391
Schaller, H. 327
Schiff's base 155
Schizomycophyta 109
Schneider, G. 283
Schoenherr, K. E. 344
Schwartz, P. 257
Schwarzschild radius 51, 231, 233
Science in art 10–23
  conservation of art 23
  dating and chronological studies 20–21
  investigations of classes of objects 22–23
  modern materials in traditional art 13
  new esthetic experiences 13
  "scientific" art 13–14
  scientific investigation of art objects 14–20
"Scientific" art 13–14
Scientific investigation of art objects 14–20
  chemical techniques 15–18
  neutron activation 16–17
  physical techniques 18–20
  x-ray fluorescence spectrometry 17–18
Scott, David R. 387
SCR see Space-charge region (SCR)
Screening (composting) 32
Scurvy 122
SDS see Sodium dodecyl sulfate (SDS)
Sea-floor spreading (theory for faunal extinction) 214
Sea level fluctuation (theory for faunal extinction) 214
Seaborg, Glenn T. 205
Sears Roebuck Headquarters Building 131
Secale cereale 138, 139
Second-order galaxy cluster 225
Secondary recovery, petroleum 318–320
Sedimentary basin 313
Seed (botany) 360–361
  coat 209
  cytological changes 360
  enzymes release 361
Seed (botany) — cont.
  structural changes in wall 360–361
  structure 360
Seidel, H. 113
Seismic survey (petroleum prospecting) 346
Seismograph 346
SelectaVision system 410–411
Semiaquatic tetrapod 179
Semiconductor 65, 241, 361
  experimental verification 362–363
  reverse-bias leakage current 144
  surfaces 65
  theory 362
Semiconductor injection laser 272–274
Senescing roots (microbial activity) 382
Sensory averaged evoked responses (speech) 391
Senter, J. P. 171
Separating (composting) 31
Sephadex G-100 237
Sequential unconstrained minimization technique (SUMT) 370
Serology (HL-A system) 399
Serotonin (effect on hibernation) 240
Serpentinite 304
Serpukov 70-GeV accelerator 310
Sesamoid bone 394
Seyfert galaxy 48, 49, 54, 55
Shamoot, S. 382
Shank, C. V. 270
Sharkey, J. B. 18
Shaykewich, C. F. 376
SHEED see Scanning HEED
Sheep (silica content) 371
Sheldrake, A. R. 423
Shepard, Alan B. 385
Sheveluch volcano eruption 412
Shift register, digital 144
Shininger, Terry 422
Ship, container 364–365
Ship, merchant 363–366
  O/B/O 364
  O/S/O 363–364
  specialized container ships 364–365
  technological improvements 365–366
Ship design 366–371
  computer graphics 370
  design and production 370
  design objectives and tasks 366–367
  lines fairing and creation 367
  optimization 367–370
  systems approach 367
Shiran 230
Shollenberger, C. A. 220
Shoran 230
Shutler, N. D. 320
Side-looking radar (SLAR) 397
Silica, biology of 371–372
  animals 371–372
  plants 371
Silicic acid 372
Silicon 241
Silicon dioxide (animal and plant nutrition) 371
Silicon monoxide molecule 255
Silver halide film 411
Silver halide tape (storage medium for video playback system) 409
Simonson, Roy W. 372
Single-heterostructure laser 273
Single-screw propeller 344
Sinsheimer, R. 326
Sintering (pure metal powders) 330
Skerfving, Steffan 72
Skidmore, Owings & Merrill 132
Skin graft 120
Skylab Program 187, 230, 385, 388
Skylab A 187
SL-CM gene 400
SL-ET gene 400
SL-MaPi gene 400
SLAR see Side-looking radar (SLAR)
Slichter, L. B. 181
Sloth, ground 180
Slurry 104
  explosives 298–299
  recycling 33
Slurry carrier (ship) 363
Smelter (air pollution) 105–106
Smith, Cyril S. 19
Smith, D. 327
Smith, Gilbert 106
Smith, P. E. 235
Snavely, B. B. 269
Snodgrass, F. 183
Snyder, L. 255
Soap (use of food-processing wastes) 414
Sodium dodecyl sulfate (SDS) 116
Sodium silicate 371–372
Soft x-ray absorption spectroscopy 97
Soil 372–377
  aeration 375–377
  aerobic layer 379
  agriculture's role in pollution 374–375
  chemistry 377–378
  clay 148–150
  flooded environment 378
  food 221–224
  management 378–380
  microorganisms 380–383
  nutrient element losses 372–374
  physical properties due to flooding 378
  plant growth 323–325
  root (botany) 358–359
  silica, biology of 371–372
  soil-root relationship 358
  wastes, agricultural 413
Soil chemistry 377–378
  coordination 377–378
  hydrogen bonding 378

Soil chemistry—*cont.*
  other mechanisms 378
  protonation 377
Soil conditioner 31
Soil management 378–380
  aerobic soil layer 379
  atmospheric oxygen 378–379
  flooded soil environment 378
  plant nutrient availability 379–380
  soil physical properties 378
Soil mantle (nutrient element loss) 372–374
Soil microorganisms 380–383
  agents of litter decomposition 382
  decay of plant parts belowground 382
  decay rates of plant constituents 381–382
  decomposition of surface litter 381
  microbial biomass in soil 383
  microbial utilization of standing crop 380–381
  senescing and dead roots 382–383
  standing vegetation pathways to litter 381
Solar corona 383
Solar flare 384
Solar heating (effect on sea surface temperature) 278
Solar radiation (theory for faunal extinction) 214
Solar wind 383–385
  chemical composition 384–385
  disturbed solar wind 384
  quiet solar wind 383–384
Solid, quantum 349–351
Solid argon (quantum solid) 349
Solid helium (quantum solid) 350
Solid sample (Raman spectroscopy) 389
Solid-state physics: absorption of electromagnetic radiation 97–99
  high-pressure physics 241–243
  Jahn-Teller effect 263–264
  quantum solid 349–351
Solid waste: animal 413–414
  crop and orchard residues 414
  food-processing wastes 414
  forest waste products 415
  structure and behavior 241
Solid Waste Disposal Act of 1965 32
Solid-waste management 24–33
  fluidized-bed incinerator 33
  future 32–33
  incineration 31
  management techniques 29–32
  public health, natural resources, and economy 27–29

Solid-waste management—*cont.*
  recycling and reuse 32
  types and volume of wastes 26–27
Solomon, P. 255, 256
Solvation energy (exchangeable cations) 378
Solvent-refined coal 105
Somatic chromosome pairing 136–138
Sorbitol 247
Sorting (composting) 31
South Pole 181
Southern corn blight 158
Soybean 324
Soyuz-Salyut Program 385
  Soyuz 10 386, 388
  Soyuz 11 388
Space-charge region (SCR) 362
Space flight 385–388
  Apollo 14 385–387
  Apollo 15 387
  Cosmos and Soyuz 388
  Mars missions 387–388
Space Shuttle 388
Space technology: Earth-resource satellites 185–187
  space flight 385–388
Spark sintering 332
Spark-source mass spectrometry 329
Specificity (metal toxicity) 71
Spectrometry: incidence grazing 99
  optical emission 15
  spark-source mass 329
Spectrophotometer 377
Spectroscopy 302, 388–390
  gaseous samples 389–390
  helium cadmium laser 271
  liquid samples 389
  photoelectron 65
  soft x-ray absorption 97–99
  solid samples 389
  surface-Auger 62–63
Spectrum, flash desorption 65–66
Speech 390–392
  electrical concomitants 391
  readiness potentials and speech 391–392
  substrates of language 390–391
Spencer, D. J. 266
Sphaerocarpus 135
Spheniscidae 219
Spin resonance 264
Spinar 289–290
Spiral galaxy M 64 46
Splat cooling 252
Spline, Theilheimer 367
Spline methods 367
Spodosol 373, 374
Spontaneous fission 203, 204
Spray nozzle 4
Sputter-ion pump 59
Stachyose 247
Stack, S. M. 134
Stainless steel 331
Standing crop (utilization by microorganisms) 380

Standing vegetation (pathways to litter) 381
Stanford, George 375
Stanford Research Institute 33
Star, neutron 231, 288–290
Star clusters as energy sources 51
State University of New York at Stony Brook 202
Statistics of extremes 38–39
Statolith 285
Statsionar 1 (artificial satellite) 386
Steam drive (oil reservoir stimulation) 319–320
Steam plant 333
Steam soak (oil reservoir stimulation) 319
Steam turbine plants for ships 365
Steel powder 330
Stegosaurus 180, 181
Steinmetz, L. 181
Stephens, R. R. 268
Stereophonic sound, four-channel 392–394
  compression and sorting out 393–394
  phonograph records 393
Stereoscopic projection (highway engineering) 246
Sterilization (drugs by process radiation) 43
Stewart, W. D. P. 374
Stokes region 389
Stoma 284
Storage and retrieval systems 195
Strange eclipsing binary system ε Aurigae 232
Stratigraphic studies 313
Stratosphere 412, 413
Stratospheric sulfate aerosol 412
Stretford processes 106
Strigiformes 394–396
  os prominens 394
  osseous arch of radius 394–395
Strip mining (coal) 208
Stromberg, R. R. 328
Structural design, risk estimation in 39
Struthiomimidae 179
Struthiomimus 176, 179
Studier, M. H. 248
Studies in Perception, 1 (art) 12
Stuyvesant, Peter 3
Stylar canal 209
Stylus 410
Subclustering (galaxies) 225
Substantia nigra (degeneration in idiopathic Parkinsonism) 307
Sucrose 174, 247
  influence on xylem tracheary elements 422
Suess, Hans 353
Sulfate aerosol, stratospheric 412
Sulfation factor 122
Sulfur dioxide 102–105, 207
  in volcanic emissions 411

Sulfur oxide: air pollution 101
  air quality 207
Sulfuric acid 411
Sulfuric acid plant (air pollution) 106
SUMT *see* Sequential unconstrained minimization technique (SUMT)
Supercluster (galaxies) 225
Superconducting accelerator: current status 310
  general characteristics 308–309
  operating techniques 309–310
Superconducting gravimeter (gravity tide measurement) 181
Superconducting magnet 308
Superconducting waveguide 309
Superconductor 181
Superheavy nucleus 203
Supernova 54, 231
  black holes 231
  energy source in galaxies and quasars 52–53
Supersonic transport (neutron exposure to travelers) 167–168
Surface, crystalline solid 58
Surface atomic vibration 66
Surface Auger spectroscopy 62–63
Surface physics 58–67
  basic surface phenomena 60–66
  experiments 58–60
  goals 58
  theory 66
Surtsey volcano eruption 412
Sutherland, Earl W., Jr. 291
Swedish varve chronology 353
Swenson, G. W. 255
Sykes, L. R. 396
Sylvain, P. G. 152
Sympathetic nervous system 307
Synaptinemal complex chromosome 134
Synchrocyclotron 311
Synchrotron 98, 162
  radiation 47
Systems approach (ship design) 367

T

3T3 *see* Normal mouse fibroblast
T4 bacteriophage 71
T cell populations 249
Taal volcano eruption 412
TACV *see* Tracked air-cushion vehicle (TACV)
Taenite 20
Talc 378
TALCOR *see* Transponder array location by coplanar ranges (TALCOR)
Talking beacon system (marine navigation) 281

Tall fescue (silica content) 371
Tanker, mineral 363
Tansei 1 (artificial satellite) 386
Tappan, H. 215
Tarbosaurus 178
TD3 microwave relay 114
Techniques, chemical (art analysis) 15–18
Techniques, gaming 39–42
Technology, explosion 294–295
Tectonophysics 396–397
  dynamic picture 397
  mechanism 397
  new data 396–397
  Triassic 405
Teldec Video Disc 409–410
Teleoceras 179
Telescope: optical 123–124
  radio 124–125
Television camera 335
  prosthesis for the blind 348
Temin, H. 174, 175
Temperature: correlation with photorespiration 321
  relationship with plant disease 100
Teratogenesis (metals) 71–72
Terrain sensing, remote 346, 397–398
  multispectral photography 398
  satellite data 398
  SLAR 398
  thermal infrared segment 398
  ultraviolet signals 397–398
Tethyan Sea 305
Tetraethyllead 75
2,3,5,6-Tetrafluoro-phenylcopper 157
Tetrapod herbivore, aquatic 179
Thallium disease: occurrence 76
  symptoms 76
  teratogenic effect 71
Thallotoxicosis *see* Thallium disease
Thecodont 406
THEED *see* Transmission HEED
Theilheimer spline 367
Theoretical physics (elementary particle) 198–203
Thermal doming (orogeny) 304
Thermal infrared survey (remote terrain sensing) 398
Thermal numerical simulator model (steam drive stimulation of oil reservoir) 320
Thermo-Fax copier 334
Thermodynamics: critical point 169–171
  quantum solid 350
  Rankine cycle 333
Thermoluminescence (art objects study) 20–21
Thermoluminescent crystal 167

Thermoregulation (hibernation) 240
Theropoda 176
Thick-film resistor 254
Thin-film absorption 99
Thin-film transmission measurement 97
Tholeiite 303
Thompson, A. 260
Thomson, G. P. 58
Thorne, K. S. 231
Thornton, D. D. 255
Three Mile Island Nuclear Station 357
Thylakoid 107, 108
Thymidine 134
  chromosome studies 148
  DNA studies 148
Thymine 325–326
Thymus-derived cell 249–251
TIBA *see* 2,3,5-Triiodobenzoic acid (TIBA)
Tibia 395
Tides, Earth 181–185
Tissue culture (antibody formation) 251
Tissue-fluid antigen: Ag 118
  Am 118
  Gc 118
  α_2-globulin systems 118
  Gm 118
  immunoglobulin systems 118
  Inv 118
  Isf 118
  lipoprotein systems 118
  Lp 118
  Xm 118
Titanium sublimation pump 59
TLSP *see* Transponder location by surface positioning (TLSP)
Tolbert, N. E. 322
Tolerance, immunological 250–251
Tonalite 304
Topography, Venus surface 408
Torpedo (contrarotating propeller) 344–345
Tournesol (artificial satellite) 386
Townes, C. H. 255
Toxin: cholera 145
  corn blight race O 159
  corn blight race T 159
Tracheary element (substances influencing development) 422
Tracked air-cushion vehicle (TACV) 404
Transformed cell growth 142
Transistor, insulated-gate field-effect 144
Transition-metal cations 378
Transmission HEED (THEED) 197
Transplantation biology 399–402
  antisera problem 399–400
  clinical transplantation 401–402
  genetics 401

Transplantation biology —cont.
  HL-A typing 400
  mixed lymphocyte culture test 401
  present knowledge 400–401
Transponder array location by coplanar ranges (TALCOR) 229
Transponder location by surface positioning (TLSP) 229
Transport properties (quantum solid) 350
Transportation engineering 402–405
  highways 402
  intercity travel 402
  public transportation in cities 402
  transportation innovations 402–404
Traveling-wave tube (TWT) amplifier 112, 280
Tree ring dating 352
Trematomus borchgrevinki 247
Trench method (sanitary landfill) 29
Trends in fire research 6–7
Triassic 180, 217, 405–407
  Coalsack Bluff find 405–406
  Graphite Peak discovery 405
  McGregor Glacier discoveries 406
  other developments 406–407
Triazine 150
Tributylphosphine 158
m-(Trifluoromethyl)-phenylcopper 157, 158
o-(Trifluoromethyl)-phenylcopper 157, 158
p-(Trifluoromethyl)-phenylcopper 157
2,3,5-Triiodobenzoic acid (TIBA) 285
Trilobite (theory for faunal extinction) 214
Trimble, V. L. 231
Triphenylphosphine 158
Triton X100 (triplet quencher) 269
Tropospheric particles from volcanos 411
Trueman, A. E. 265
TT gene 400
Tuccio, S. A. 269
Tucker, V. A. 220
Tumor, bone 128
Tumor virus, RNA 176
Tungsten powder 330
"Tunnel vision" 74–75
Turbidite 303, 304
Turner, B. E. 255, 257
Turnip yellow mosaic virus (TYMV) 116
Twin-screw propeller 344
TWT amplifier *see* Traveling-wave tube (TWT) amplifier
TYMV *see* Turnip yellow mosaic virus (TYMV)
Tyrannosaurus 176, 178, 179

U

U18 gene 400
Uhuru (artificial satellite) 47
UHV transmission *see* Ultrahigh-voltage (UHV) transmission
Ullman biaryl synthesis 157
Ullman condensation 157
Ultisol 373
Ultrahigh-voltage (UHV) transmission 188
Ultrashort pulses from lasers 301
Ultraviolet light in process radiation 341
Underground transmission circuits 193
Universal Standard Data (USD) 418–419
University of Maryland 232, 328
University of Minnesota 200
University of Texas 407
University of Utah 165
Upper Mantle Project 396
Uranus 388
URBAN COGO 244
Urban fires 1–9
  classification of fires 5
  development of fire protection practices 3–4
  fire prevention and loss reduction 4–5
  future hazards 5–6
  future of fire research 7–8
  modern fire-fighting methods 4
  nature of today's fire problems 5–6
  new problems in cities 5
  trends in fire research 6–7
Uropygial gland 217
Ursolic acid (plant cuticle) 211
U.S. Government policy on air, water, and land quality standards. 206–208
USD *see* Universal Standard Data (USD)
U.S.S.R. Academy of Sciences 123

V

V formation (flight) 219–221
Vacuum-solid interface 58
Van Den Bergh, H. E. 268
Vanguard (ship) 231
Variance (probability) 37
Varo Monocab 403
Varve 353
Vasculature (cholera pathogenesis) 145
Vaucheria litorea 110
Vaucheriaxanthin 107
Vaucouleurs, G. de 224, 227
Vegetarian diet 123
Vegetation, standing 381
Velvet bean 324
Venera 4 (artificial satellite) 407
Venera 5 (artificial satellite) 407

Venera 6 (artificial satellite) 407
Venera 7 (artificial satellite) 407, 408
Veneziano, G. 289
Ventilation (urban fires) 4
Venus (art) 11
Venus (planet) 388, 407–408
  chemical composition 407–408
  temperature and pressure data 407
  topography 408
Vermiculite 149
Vertebrate anatomy: feather (bird) 217–219
  Strigiformes 394–396
Vertebrate zoology (beaver) 125–126
Vertical muon intensity 165
Vertisol 373, 374
Very dense matter, physics of 288–290
Very-long-baseline interferometry (VLBI) 229
Vibrio comma 144, 145
Vicentini-Missoni, M. 170
Vicia faba 138, 139
Victoria blight 161
Video disk 409
Video magnetic tape recorder (VTR) 408–409
Video recording and playback 408–411
  EVR 410
  SelectaVision holographic tape 410–411
  storage medium for video playback system 409
  Teldec Video Disc 409–410
  VTR 408–409
VIKING (design system) 370
Viking mission 388
Vine, F. J. 305, 396
Vinogradov, A. P. 407
Violaxanthin 108
Virgo Cluster of galaxies 227
Virus: animal 115–118
  picornavirus assembly 117–118
  temperature-moisture relationship 100
Visual prosthesis (blind users) 349
Vitamin B_{12} 155
VLBI *see* Very-long-baseline interferometry (VLBI)
Volcano 207, 411–413
  composition of volcanic emissions 411
  effects of volcanic emissions 411–412

Volcano—*cont.*
  Gunung Agung eruption 412–413
  theory for faunal extinction 214
  volcanic sulfur emissions 413
Volkoff, G. M. 288
Volkov, Vladimir 388
Vorontsov-Vel'jaminov, B. A. 226
VTR *see* Video magnetic tape recorder (VTR)
Vycor tube 341

W

van der Waals equation of state 170
Wadati, K. 397
Waggoner, P. E. 101
Wall tile, concrete-polymer 343
Walter, W. Grey 391
Wardrop, A. B. 153
Warhol, Andy 13
Warkentin, B. P. 376
Warm fog 416–418
  heating 417
  helicopter downwash mixing 417–418
  hygroscopic particle seeding 417
Wastes: agricultural 26, 413–415
  classification 33
  fossil fuel 26–27
  liquid 415
  mineral 26–27
  municipal 26
  solid 413–415
  total solid 27
Wastes, agricultural 413–415
  disposal 26
  Great Lakes 233–235
  liquid 415
  solid 413–415
Water (fire-suppression agent) 7
  light water 7
Water beetle 140
Water pollution: agriculture's role 374–377
  Great Lakes 233–235
  water quality management 207–208
Water Quality Improvement Act of 1970 354, 356
Water repellency in bird feathers 217–218
Water vapor molecule 255

Waterproofing (feather) 219
Waters, J. 257
Watson, J. D. 148
Waveguide, superconducting 309
Weather and climate, marine influence on 277–279
Weather modification 416–418
  future prospects 418
  scientific problem 416
  warm fog 416–418
Weber, Joseph 232
Wegener, Alfred 397
Wehausen, J. V. 370
Weinreb, S. 255
Weiselsberger, C. 220
Welch, W. J. 255
Well completion, oil and gas 298–300
Welles, S. P. 407
Wellman-Lord processes 106
Wernicke's area (brain) 390, 391
West, D. W. 376
Wet scrubbing recovery systems (flue gas) 104
Whalen, J. A. 257
Wheat (new strain) 221–222
Wheeler, John A. 232
Whistler, James McNeil 21, 22
Whitaker, H. A. 391
White, Richard 231
Wide-baseband amplifier 113
Wide-energy-gap semiconductor 362
Wildflysch 304
Wilkinson, G. 157
Wilson, E. B. 136
Wilson, R. W. 255, 256
Wilson, T. 396
Wimbush, M. 183
Wind, solar 383–385
de Winiwater, H. 134
Wolberg, Lewis R. 14
Wolf, E. E. 329
Wood, carbon-14 concentration in 352
Woodward-Hoffmann rules 157
Worden, Alfred M. 387
Work measurement 418–420
  MRD 419
  MSD 419
  MTM-2 419
  MTM-3 419–420
  MTM-GPD 419
  USD 418–419
World Trade Center building 131
Wrattan color filter 337

X

X chromosome 138, 139
X-ray astronomy 232
X-ray emissions 237–238
X-ray fluorescence 329
X-ray fluorescence spectrometry (scientific investigation of art objects) 17–18
X-ray galaxy 47–48
X-ray nonlinearities 302
X-ray spectrum 17
X-rays, parametric down-conversion of 302
X-rays, posterized 335
Xanthium 423
Xanthium pensylvanicum 422
Xanthophyceae 109
Xanthophyll 107, 112
Xenon compounds 420–422
  new compounds 420–421
  reactions of xenon esters 421
  xenon fluorides and organic compounds 421–422
Xg (immunoglobulin) 118
Xm (immunoglobulin) 118
Xylem 129, 422–424
  hormonal influence 422–423
  lignin 424
  macromolecule synthesis 423–424
Xylose-1-C^{14} 154

Y

Yachting (navigation systems, electronic) 286–288
Yale University 176
Yerkes system (galaxy type) 224
Yodh, G. 163
Yt (immunoglobulin) 118
Yucca (cuticle) 210

Z

Zaalberg Van Zelst, J. J. 112
Zeaxanthin 108
Zeeman spectrum 264
Zel'dovich, Y. B. 231
Zenith angle dependence (muon) 165–166
Zond 2 (artificial satellite) 388
Zuckerman, B. 255, 257
Zwicky, F. 225, 226, 227, 231
Zygotene 134